Routledge Handbook of Global Environmental Politics

This handbook brings together leading international academic experts to provide a comprehensive and authoritative survey of global environmental politics.

Fully revised, updated and expanded to 45 chapters, this book:

- Describes the history of global environmental politics as a discipline and explains the various theories and perspectives used by scholars and students to understand it.
- Examines the key actors and institutions in global environmental politics, explaining the roles of states, international organizations, regimes, international law, foreign policy institutions, domestic politics, corporations and transnational actors.
- Addresses the ideas and themes shaping the practice and study of global environmental politics, including sustainability, consumption, expertise, uncertainty, security, diplomacy, North-South relations, globalization, justice, ethics, public participation and citizenship.
- Assesses the key issues and policies within global environmental politics, including energy, climate change, ozone depletion, air pollution, acid rain, transport, persistent organic pollutants, hazardous wastes, rivers, wetlands, oceans, fisheries, marine mammals, biodiversity, migratory species, natural heritage, forests, desertification, food and agriculture.

This second edition includes new chapters on plastics, climate change, energy, earth system governance and the Anthropocene. It is an invaluable resource for students, scholars, researchers and practitioners of environmental politics, environmental studies, environmental science, geography, globalization, international relations and political science.

Paul G. Harris has authored or edited 25 books on global environmental and climate change politics, policy and justice. He is the Chair Professor of Global and Environmental Studies at the Education University of Hong Kong.

Routledge Handbook of Global Environmental Politics

SECOND EDITION

Edited by Paul G. Harris

Routledge
Taylor & Francis Group

LONDON AND NEW YORK

Cover image: © Getty Images

Second edition published 2022
by Routledge
4 Park Square, Milton Park, Abingdon, Oxon OX14 4RN

and by Routledge
605 Third Avenue, New York, NY 10158

Routledge is an imprint of the Taylor & Francis Group, an informa business

© 2022 selection and editorial matter, Paul G. Harris; individual chapters, the contributors

First edition published by Routledge 2014

British Library Cataloguing-in-Publication Data
A catalogue record for this book is available from the British Library

ISBN: 978-1-032-14580-8 (hbk)
ISBN: 978-0-367-69241-4 (pbk)
ISBN: 978-1-003-00887-3 (ebk)

DOI: 10.4324/9781003008873

Typeset in Bembo
by codeMantra

Contents

Contents

Contents

Illustrations

Figures

Tables

Contributors

Juliann Emmons Allison is Associate Professor of Gender and Sexuality Studies at the University of California-Riverside.

Derek Bell is Professor of Environmental Political Theory at Newcastle University, UK.

Katja Biedenkopf is Associate Professor of Sustainability Politics at the University of Leuven, Belgium.

Frank Biermann is a Professor of Global Sustainability Governance with the Copernicus Institute of Sustainable Development, Utrecht University, the Netherlands.

Olivia Bina is Principal Researcher at the University of Lisbon and Fellow of the World Academy of Art and Science.

Ralf Brand is Senior Sustainable Mobility Expert and leader of the People-focused Mobility Solutions team at Rupprecht Consult, Germany.

Hollie Nyseth Brehm is Associate Professor of Sociology at The Ohio State University.

Loren R. Cass is Professor of Political Science at the College of the Holy Cross.

Edward Challies is Senior Lecturer with the Waterways Centre for Freshwater Management, School of Earth and Environment, University of Canterbury, New Zealand.

Jon Marco Church is Associate Professor of Sustainability and Governance at the University of Reims, France.

Jennifer Clapp is Professor and Canada Research Chair in Global Food Security and Sustainability at the University of Waterloo, Canada.

Simon Dalby is Professor in the Balsillie School of International Affairs, Wilfrid Laurier University.

Elizabeth R. DeSombre is the Camilla Chandler Frost Professor of Environmental Studies at Wellesley College.

Contributors

Radoslav S. Dimitrov is Associate Professor at Simon Fraser University in Vancouver, Canada.

Stephen Dovers is Emeritus Professor with the Fenner School of Environment and Society, Australian National University.

Christian Downie is an Associate Professor in the School of Regulation and Global Governance (RegNet) at the Australian National University.

David Downie is Chair of the Politics Department at Fairfield University.

Hugh C. Dyer is Associate Professor of World Politics at the University of Leeds.

Helene Dyrhauge is Associate Professor of International Public Administration and Politics in the Department for Social Sciences and Business, Roskilde University.

Leandra R. Gonçalves is Assistant Professor in the Institute of Marine Science at the Federal University of São Paulo.

Peter M. Haas is a Professor in the Department of Political Science at the University of Massachusetts Amherst.

Paul G. Harris is the Chair Professor of Global and Environmental Studies at the Education University of Hong Kong.

Thomas Hickmann is Associate Senior Lecturer in the Department of Political Science, Lund University, Sweden.

David Humphreys is Professor of Environmental Policy at the Open University, UK.

David B. Hunter is Professor of Law at the American University Washington College of Law.

Karen Hussey is Associate Professor and Public Policy Fellow at the Fenner School of Environment and Society, Australian National University.

Peter J. Jacques is Professor in the School of Politics, Security and International Affairs and the National Center for Integrated Coastal Research at the University of Central Florida.

Shangrila Joshi is a Member of the Faculty of Climate Justice and Environmental Studies at The Evergreen State College.

Meri Juntti is Associate Professor in Environmental Governance in the Department of Law and Politics at Middlesex University, UK.

Andrew Karvonen is Professor of Urban Design and Planning in the Department of Architecture and the Built Environment at Lund University, Sweden.

Lada V. Kochtcheeva is Associate Professor in the School of Public and International Affairs, North Carolina State University.

Gabriela Kütting is Professor of Global Politics in the Department of Political Science and the Division of Global Affairs, Rutgers University, Newark.

Alexander K. Lautensach is an adjunct professor in the School of Education, University of Northern British Columbia, Canada.

Sabina W. Lautensach is Director of the Human Security Institute and founding editor-in-chief of the *Journal of Human Security*.

Ray Maher is a post-doctoral fellow in the Centre for Policy Futures, Faculty of Humanities and Social Sciences, the University of Queensland, Australia.

Sandra T. Marquart-Pyatt is Professor in the Department of Geography, Environment and Spatial Sciences and the Department of Political Science at Michigan State University.

Sarah J. Martin is an Associate Professor in the Department of Political Science at Memorial University, St. John's, Newfoundland and Labrador.

Volker Mauerhofer is Professor and Chair for Environmental Science with specialization in Social Science in the Department of Ecotechnology and Sustainable Building Engineering at Mid Sweden University.

Constance L. McDermott is a Jackson Senior Fellow and Associate Professor at the Environmental Change Institute, School of Geography and the Environment, University of Oxford.

Ronald B. Mitchell is Professor of Political Science and Environmental Studies at the University of Oregon.

William R. Moomaw is Emeritus Professor of International Environmental Policy at Tufts University.

Jens Newig is Full Professor and Head of the Institute of Sustainability Governance at Leuphana University Lüneburg, Germany.

Felister Nyacuru is Deputy Chief State Counsel, Office of the Attorney General, Republic of Kenya.

Kate O'Neill is Professor in the Department of Environmental Science, Policy and Management at the University of California at Berkeley.

Mihaela Papa is Adjunct Assistant Professor in Sustainable Development and Global Governance at The Fletcher School, Tufts University.

David N. Pellow is the Dehlsen Chair and Professor of Environmental Studies, and Director of the Global Environmental Justice Project, at the University of California, Santa Barbara.

Mary E. Pettenger is Professor of Political Science and the Social Sciences Division Chair at Western Oregon University.

Olivier Ragueneau is Research Professor at CNRS (the French National Center for Scientific Research).

Violet Ross is a PhD researcher in the Environmental Policy Group and the Law Group at Wageningen University in the Netherlands.

Jeannie L. Sowers is Professor of Political Science at the University of New Hampshire.

Paul F. Steinberg is Professor of Political Science and Environmental Policy at Harvey Mudd College.

Hayley Stevenson is Associate Professor of International Relations at the Universidad Torcuato Di Tella, Argentina.

Peter Stoett is the Dean of the Faculty of Social Science and Humanities, Ontario Tech University, Canada.

Peter Suechting is an interdisciplinary PhD student in Political Science and Environmental Science, Studies and Policy at the University of Oregon.

Jessica Templeton is Director of LSE100 at the London School of Economics and Political Science.

Kyla Tienhaara is Canada Research Chair in Economy and Environment at Queen's University, Canada.

Andrew Tirrell is Associate Professor of Political Science and International Relations at the University of San Diego.

Steve Vanderheiden is Professor of Political Science and Environmental Studies at the University of Colorado.

Stacy D. VanDeveer is Professor of Global Governance and Chair of the Department of Conflict Resolution, Human Security and Global Governance at the University of Massachusetts Boston.

Sofia Guedes Vaz is a researcher on philosophy and environment at the New University of Lisbon and president of the Portuguese Environmental Ethics Society.

John Vogler is Professorial Research Fellow in International Relations at Keele University.

Erika Weinthal is Professor of Environmental Policy and Public Policy, Nicholas School of the Environment, Duke University.

Jennifer Yarnold is a research fellow in the Centre for Policy Futures, Faculty of Humanities and Social Sciences, the University of Queensland, Australia.

Josephine van Zeben is Professor and Chair of the Law Group at Wageningen University in the Netherlands.

Acknowledgements

When it was published in 2014, the first edition of the *Routledge Handbook of Global Environmental Politics* was the most comprehensive book on global environmental politics yet published. This second edition, which is fully updated, revised and expanded, may well maintain this distinction. If it does so, much of the credit goes to the expert contributions of scholars from around the world. I am particularly indebted to contributors from the first edition, many of whom are now senior scholars who easily could have declined my invitation to write for the second. I trust that readers will appreciate their renewed insights and indeed those of the new contributors.

I am grateful to everyone at Routledge who has been involved in producing both the first and second editions. My thanks especially go to Hannah Rich, who prodded me to take on the second edition and helped to keep everything moving along smoothly.

As was the case with the first edition, I am grateful to Keith K.K. Chan for support at home while I worked on this one. After writing or editing more than two dozen books, I can attest with more than a little experience that such support makes the process much more bearable.

Paul G. Harris
Lantau Island
Hong Kong

Part I
Introduction

Global environmental politics

Charting the domain

Paul G. Harris

The natural environment is in very serious decline globally. With too few exceptions, environmental indicators are growing worse. For example, water and air pollution are now so poor in many developing nation-states that hundreds of millions of people are forced to drink severely tainted water and breathe toxic air (see Chapters 30 and 38). Regionally, acid rain – which has been reduced in North America and Western Europe in recent decades – is on the increase in East Asia and other developing regions, putting ecosystems and agriculture at great risk (see Chapter 30). The so-called "Asian brown cloud" of smog is so vast that it spreads across the Pacific to the Americas. Coastal seas have been overfished in most oceans, and this phenomenon has extended to regional seas in both the developed and developing worlds (see Chapter 40). Marine environments are severely degraded by polluting runoff from continents, with the world's coral reefs shrinking and ocean "dead zones" now extending along the coastlines of all continents. Wildlife around the world is under great threat, with declines and extinctions of species on the rise (see Chapter 41). Deforestation is rampant in many parts of the world, not least in parts of South America and Southeast Asia (see Chapter 42). These problems are exacerbated by climate change, which is manifested in rising global temperatures, very serious threats to agricultural productivity from droughts and floods, more severe weather events, new threats to species unable to adapt to rising temperatures and other environmental changes, declines in marine ecosystems due to warming waters and ocean acidification, and immeasurable dangers posed by sea-level rise, particularly for poor low-lying regions, states and habitats (see Chapter 32). Alas, these are only some of the environmental challenges that are increasingly prevalent around the world.

The role of politics in these challenges, whether they play out within or among states, cannot be understated. The continuing decline of the global environment can largely be put down to the failure of governments and other actors to respond effectively, if at all. When we do see successes in preventing or responding to adverse environmental changes and pollution, for example in cleaner local environments in many developed states and a handful of international successes, such as agreements among states to curb emissions of pollutants that destroy Earth's protective stratospheric ozone layer (see Chapter 31), they can often be put down to the willingness of governments and other political actors, including nongovernmental organizations (see Chapter 14) and occasionally businesses (see Chapter 13), to negotiate and

DOI: 10.4324/9781003008873-2

implement policies that prioritize environmental protection over short-term economic gain. Understanding and promoting these kinds of successes are crucially important, and in many cases vital, to the future of all societies and to natural ecosystems. This handbook is intended to be part of the process of promoting those successes: first to bolster basic understanding of environmental changes and the underlying politics that shape them, and second to provide readers with a foundation of knowledge that can help them to promote new, more environmentally sustainable relationships between humankind and the natural world.

Everyone is affected by global environmental politics, often directly through feeling the impacts of the environmental changes caused by government policies, and at least indirectly through having to watch others suffer from those changes. Many people are now affected, in positive ways, by regulations and policies that have reduced environmental pollution. The manner in which human, financial and governmental resources are used to create and hopefully reverse ecological decline, overuse of natural resources and destruction of the natural environment affects the safety of the water that people drink, the air that they breathe and the nature that they enjoy and draw from to meet their individual and community needs. Global environmental politics does and will shape the climate and even the weather of the future. Sadly, for some people, global environmental politics may be a matter of life and death. For example, the failure of governments and other global actors, such as businesses and individuals, to respond robustly to the causes and consequences of climate change means that millions of vulnerable people in the poorest parts of the world will die in the future from drought-induced famine or severe weather events, and many more will die from the diseases that will spread in a warmer and wetter world.

What this means is that global environmental politics should concern everyone. Whether one is a politician, career government official, entrepreneur, activist or student, understanding global environmental politics will help achieve policy or personal goals. Without knowledge of the global nature of environmental changes, policymakers will fail to see many of the causes of those changes, and indeed the remedies for them. Without recognition that the environment permeates other policy areas, ranging from energy supplies and national security to social justice and food production, policy responses are unlikely to succeed, least of all to be cost-effective and equitable. Without realizing that both the causes and consequences of global environmental change are highly political, being influenced by the distribution of power within and among societies, those who seek to limit pollution and destruction of natural resources will not get very far. With this in mind, this handbook brings together a large group of scholars from around the world to examine these connections and to help illuminate the causes of environmental changes and especially the ways that the world has and can respond to them. It is intended to be a comprehensive treatment of the topic. While the field of global environmental politics is much too large to be fully covered in a single volume, we have sought to survey as much of it as possible, thereby giving anyone interested in (or concerned about) it a solid foundation on which to continue with more in-depth analysis or study.

Before the contributors to this volume proceed to examine global environmental politics more widely, the remainder of this chapter briefly charts this important domain. It defines the topic and its related field of study by briefly looking at the *global*, the *environment* and the *politics* in global environmental politics. The chapter then distinguishes between the *practice* of global environmental politics and the *study* of it, in the process suggesting how the two do and should overlap. This chapter also introduces the topics that will follow in subsequent chapters, in the process showing how the field is both wide and deep, in many respects reinforcing the importance of global environmental politics for everyone.

Defining global environmental politics

What is "global environmental politics"? Global environmental politics is both an area of activity and practice, on the one hand, and a field of research and study, on the other. It is about how governments, diplomats and other actors influence the global environment, which includes local and regional environments, and how what they do is analyzed and understood by scholars, students and activists. Global environmental politics, in a plural sense, can be interpreted as the various ways in which politics are practiced in different places to alter or protect the environment. That is, there are different politics of the environment in different locations and in different issue areas. Importantly, as the term implies, global environmental politics is about the politics of the environment on a global scale.

The "global" in global environmental politics

Environmental changes and associated politics occur at all geographic and social levels. Environmental changes can occur locally and be caused by what happens locally, as when local water supplies are polluted by domestic sewage or industrial effluents. In contrast, environmental problems can be global, as in the case of global warming and associated planetary-scale manifestations of climate change. These levels of environmental change are routinely connected, sometimes intimately. For example, global climate change arising from emissions of carbon dioxide and other greenhouse gases all around the world affects local communities and individuals directly. Global problems can have local causes. Conversely, even apparently localized environmental issues can be global problems. For example, addressing local water and air pollution in poor states may require financial or technological assistance from affluent states, often those far away, or from the international community, perhaps in the form of an agency of the United Nations or an international nongovernmental organization.

These varying levels of environmental change, and the various levels of causality, impact and response, highlight the role of politics at all levels. We see different environmental politics depending on the location, scale or issue being addressed. For example, some local environmental problems can be addressed through local action, as occurs when a community implements regulatory measures to curb pollution or to protect local natural resources. Other environmental problems are regional, crossing provincial and national boundaries or entire oceans, requiring, and sometimes receiving, policy responses from a number of communities or national governments. Examples of this kind of environmental politics include responses by North American and European governments to reduce acid rain (see Chapter 30), action plans to limit pollution of the Mediterranean Sea, and management of fisheries in regional seas (see Chapter 39). Environmental problems that are more obviously global, such as stratospheric ozone depletion (see Chapter 31) and global warming (see Chapter 32), require global political responses: the governments of many states need to cooperate and collaborate to formulate and implement policy responses, and these, in turn, require action by many more subnational governmental bodies as well as non-state actors that operate globally (or nearly so), such as multinational corporations and international nongovernmental organizations (see Chapter 14).

Thus, in using the term "*global* environmental politics" we mean to encompass all levels of politics (and policy) related to the environment; a global issue is clearly global, but a local one may also be, by definition, encompassed by global politics. Put another way, global environmental politics comprises local, national, transnational, regional, international *and* geographically global environmental issues and related political activity. As such, in this volume we are interested in environmental issues at all levels and in related political activity at all levels.

The "environment" in global environmental politics

Global environmental politics is the global politics of the environment. More specifically, the environment in global environmental politics is about the human dimensions of the natural environment (or what passes for it nowadays): the human causes of environmental change, pollution and resource use, and the human approaches to solving (or trying to solve) or preventing environmental problems and resource scarcities. The "human" here often equates to government policies and the relationships between those policies and the behaviors of individuals and industries. For our purposes, the human also includes international cooperation, often resulting in environmental treaties and other agreements (see Chapters 8, 9 and 10). This connection between environment and human society, broadly defined, highlights an important point: while global environmental politics is related to the natural environment, how we define "natural" is problematic. A purist might point out that very little of genuine nature still exists; with climate change (see Chapter 32) and the spread of persistent organic pollutants (see Chapter 35), for example, nearly every part of what was once the natural world has been affected, and often utterly transformed (or destroyed), by humanity. Nevertheless, one expects that for most people "nature" can be defined as the nonhuman world, encompassing the plants, animals, minerals, air, water and ecosystems on which humanity depends for its survival and wellbeing.

Simply put, the *environment* in global environmental politics is roughly equated to "ecology" – natural systems, including humanity and all of its influences – but with the important caveat that we are interested in the human–environment relationship, often in the context of governance. This means that the environment of global environmental politics is not about the built environment per se, except insofar as this affects the natural environment. This would be the case with, for example, energy use by buildings (because most of the electricity used by buildings comes from the burning of fossil fuels, which, in turn, contributes to air pollution and climate change) and transport infrastructure (which can greatly affect air quality and local environmental habitats; see Chapter 34). In some sense, the environment in global environmental politics is about stewardship of the natural environment. Increasingly this means stewardship of the *global* environment – of the whole of planet Earth – implying that truly global cooperation is required to ensure an environmentally sustainable future for all people regardless of where they might live.

The "politics" in global environmental politics

"Politics" can be (and is) defined in a number of ways. It can refer to the struggle for and distribution of power, and thus resources, within and among national communities. This is routinely associated with the role of governments, notably their policies and actions for regulating behaviors in society, and the manner by which governments are chosen, the institutions from which they obtain their legitimacy and the way that they rule. Global environmental politics is largely about how government policies contribute to environmental problems and about specifically environmental policies (often environmental regulations) and their effects. It is about how environmental resources and pollution are distributed in society, and the role that power and influence play in that distribution. More commonly, the *politics* in global environmental politics is about international cooperation related to the environment. This might include addressing transboundary, regional and global problems through international conferences of diplomats negotiating environmental treaties (see Chapter 22), efforts by governments to manage shared resources in natural "commons" areas (such as fish in the open ocean beyond territorial waters; see Chapter 40), or attempts to formulate and implement international policies on sustainable development that benefit individual states (see Chapter 16), reduce local and global pollution, and support environmentally less harmful economic development (see Chapter 24).

Although global environmental politics routinely involves governments in some way, it is not always about governments relating to one another. It is often about non-state actors trying to influence government policies in ways that affect the environment (see Chapters 13 and 14). It may also involve struggles related to the environment by nongovernmental organizations, businesses and communities that largely ignore governments, at least directly. At the risk of upsetting purists, one must even acknowledge that the field of global environmental politics goes beyond politics strictly defined. Scholars of global environmental politics thus include those with interest and expertise in economics, sociology, other social sciences and even the humanities. Ultimately, the *politics* of global environmental politics is most often the process whereby the constellation of disparate interests – government agencies, corporations, communities and people, and some would add nonhuman species – is represented (or not) in actions that harm the natural environment or in efforts to protect it.

The practice and study of global environmental politics

The field of global environmental politics encompasses both practice (or praxis) and study (and analysis) of politics and policies related to the environment. The former interpretation tends to fit definitions of politics oriented toward activities of governments and traditional political players, although increasingly non-traditional actors, such as civil society groups, often organized via the Internet, have growing importance in environmental politics at all levels. The latter interpretation of global environmental politics is oriented toward research and teaching related to the politics of the environment, although it is important to note the overlap with practice: research about global environmental politics is routinely about, and very importantly can inform, the practice of global environmental politics, and students of global environmental politics might apply what they learn to environmental activism, work in industry or service in government.

The practice of global environmental politics

Global environmental politics is above all about activities – policies, actions, behaviors – that affect the environment, whether negatively (e.g., through pollution or harm to natural resources) or positively (e.g., by reducing or preventing pollution, or using resources sustainably). In its simplest form, the practice of global environmental politics includes those activities of governments that relate to the environment in some way. This might involve the work of environmental ministries, particularly when their work affects what happens in other states, and it would include the environment-related roles and activities of political executives (presidents, prime ministers) and legislatures, notably the environmental policies, laws and regulations they deliberate, formulate and implement. It follows that the practice of global environmental politics is also about the activities of all actors trying to influence and shape government policies related to the environment, and the responses of those and other actors to environmental regulation. Thus, the practice of global environmental politics within states includes the activities of special interests, notably corporations and, in many places, environmental advocacy groups, and the processes whereby those interests attempt to shape government policies related to the environment.

The practice of global environmental politics of course includes the actors working across national borders. For example, environmental diplomacy and the complex processes of international environmental negotiations (see Chapter 22) on all manner of issues – such as fishing (see Chapter 40), whaling (see Chapter 40), ocean pollution (see Chapter 39), trade in hazardous wastes (see Chapter 36), stratospheric ozone depletion (see Chapter 31) and climate change (see Chapter 32) – are most definitely the stuff of global environmental politics. Indeed, some scholars of global environmental politics focus almost entirely on this aspect of the topic – what might be labeled *international* environmental politics – including the roles of important or powerful national actors (such as the United States and China), foreign policy processes (including the roles of influential politicians or diplomats and their relationships with colleagues nationally and internationally; see Chapter 11), and the impact of international organizations and regimes (such as the United Nations and the constellation of international agreements and new practices associated with, say, biodiversity and especially climate change; see Chapters 8 and 9). In short, at least for some scholars, global environmental politics is primarily about what governments do at home and abroad to respond to environmental changes or to prevent them from happening (see Chapter 12).

The study of global environmental politics

As a field of analysis and learning, global environmental politics is about trying to understand and explain the practices of governments and other actors related to the environment, especially insofar as this is associated with international affairs or transboundary environmental issues. For most scholars this involves analyzing the practice of global environmental politics, finding explanations for what happens, and conveying this knowledge to others, often to the practitioners being studied. For many scholars this includes sharing their knowledge via publications of different kinds, sometimes in the form of policy papers intended to shape (and specifically to "improve") the policies of governments, international organizations and other actors, such as corporations, and to help them arrive at policies more conducive to environmental protection (see Chapter 18). Most scholars maintain a certain level of disinterestedness in their research: they attempt to find the "truth" behind environmental policies, for example, and to convey what they have learned to the scholarly and policy communities. Other scholars and researchers have more normative objectives: they want to see the environment and natural resources protected, so their research is aimed at finding ways to make that happen, possibly including advocacy work toward that end (see Chapter 27). A few (often self-styled) scholars, such as the so-called "climate skeptics" and "climate deniers," have just the opposite objective: to use their work to *prevent* governmental regulation for environmental protection.

For many scholars of global environmental politics, their work includes teaching others what they have learned about the practices of global environmental politics, notably in college and university courses (sometimes titled "global environmental politics," "international environmental politics" or something similar). These courses are often geared toward helping students who will join industries to better understand the role of environment in their future work, or to provide training for students who will join government ministries working on environmental and international affairs. Some teachers of global environmental politics no doubt hope that their students will become environmental activists. Regardless of their individual motivations, most of the contributors to this volume both conduct research on global environmental politics and teach about it.

Surveying global environmental politics

This volume brings together a diverse group of scholars from around the world. In addition to introductory and concluding sections, their contributions are organized into four parts: (1) explaining and understanding global environmental politics; (2) actors and institutions in global environmental politics; (3) ideas and themes in global environmental politics and (4) key issues and policies in global environmental politics. Together the contributors cover most topics in both the practice and study (or research) of global environmental politics, thereby giving readers, whether students, government officials, industry sustainability officers, environmentalists or ordinary concerned citizens, a scope of knowledge that is both wide and deep. Chapters describe the topic at hand in enough detail to provide a foundation for policy work and more in-depth reading and study. Most contributors also draw on their experiences to provide some assessment of real-world events. As such, on the whole, this handbook serves as a valuable primer for anyone interested in, or concerned about, humanity's relationship with the global environment.

Explaining and understanding global environmental politics

In Part II, contributors describe the theories and methods used to explain global environmental politics. In Chapter 2, Loren R. Cass provides a historical overview of global environmental politics as an academic discipline. In a wide-ranging survey of the literature, he shows how the field has advanced from one that was primarily about international environmental cooperation to one that is more inter- and multidisciplinary, encompassing a wide range of political and policy activity related to the environment while still being oriented toward international relations. In Chapters 3 and 4, John Vogler and Hayley Stevenson describe and assess all of the major theoretical approaches, and more than a few of the less common theoretical frameworks, used to analyze and understand global environmental politics. Vogler focuses on mainstream theories, notably realism and rationalism (which some might say are not always realistic or rational), that have been most commonly used by scholars, and sometimes even by practitioners, to explain the international politics of the environment. In contrast, Stevenson looks at alternative theories, including constructivism, Marxism and critical theories, which challenge the mainstream approaches. The alternative perspectives are often about showing that global environmental politics is just as much about ideas as it is about states per se.

The final chapters of Part II turn to questions of how global environmental politics is studied and taught. In Chapter 5, Juliann Emmons Allison and Thomas Hickmann draw on a wide literature to craft a framework for researching and teaching global environmental politics. They show how the theories described in the preceding chapters can be brought to bear in explaining global environmental politics to laypersons, and they propose innovative pedagogies that can be deployed to help students learn about it. In Chapter 6, Peter M. Haas, Ronald B. Mitchell and Leandra R. Gonçalves make a strong case for interdisciplinary scholarship that bridges the science–policy interface. They argue that such research is more likely to lead to publications and other outputs that will result in concrete improvements in environmental conditions. Together, the chapters in Part II serve as a theoretical foundation for the rest of this handbook and a guide for further research and study by readers of all kinds.

Actors and institutions in global environmental politics

Global environmental politics is shaped by a variety of major actors and institutions operating at all levels of human activity – from the local to the global. In Part III, contributors describe

Paul G. Harris

the most prominent actors and some of the common practices, norms and institutions they often follow in their relations with one another in the context of environmental change. In Chapter 7, Hugh C. Dyer takes a critical look at what are very likely the most important and most powerful actors, if far from the only important ones, in global environmental politics: nation-states. For some scholars and no doubt for many practitioners, especially diplomats, states are *the* chief actors, often receiving all of the attention. As Dyer points out, the international system, and the notion of state sovereignty that serves as its foundation, has the potential both to solve environmental problems and to make them much worse. What may be most interesting and most important, and is certainly germane to other chapters here, is that environmental change, while partly a consequence of the behaviors of sovereign states, is challenging the very idea of sovereignty like nothing else. It may be for this reason that states quite often find it necessary to cooperate at both regional and global levels to find common approaches to addressing environmental issues. This cooperation, and especially its manifestation in international (or, more precisely, intergovernmental) environmental organizations, is examined by Kate O'Neill in Chapter 8. She reviews both the functions and operations of regional and global international organizations, in the process examining the extent to which they are autonomous actors, independent of their member states, or more often tools used by their members to promote their own interests in global environmental politics.

One interesting aspect of global environmental politics is that states (and other actors) frequently cooperate informally. This informal cooperation can take on a life of its own. In Chapter 9, Peter Suechting and Mary E. Pettenger explore this process through an examination of international environmental regimes and some of the underlying theories that are used to explain their formation and effectiveness. While there is some disagreement among scholars about how to define international regimes, they are sometimes described as principles, norms and procedures that governments agree to follow in addressing (in this case) international environmental problems. They may have formal international organizations associated with them, and indeed the most influential regimes usually do, but this is not always the case. What is important is that states, at least the most powerful ones, sometimes recognize and accept that only through voluntarily accepting and (mostly) adhering to common approaches can they solve environmental problems. Another way that the environment-related behavior of and among states is voluntarily regulated, or at least tempered, is through international environmental law, which is described by David B. Hunter in Chapter 10. International environmental law is largely a consequence of formal agreements among states: governments voluntarily agree, through treaties, to be bound to certain behaviors, for example to stop allowing the use of certain pollutants or environmentally harmful practices within their borders that might degrade the environment beyond their borders. Having said this, international environmental law can arise in less predictable ways, whether through common practices that evolve over time or as a result of decisions taken by national and international courts. Hunter shows how these formal and informal practices have resulted in an array of commonly accepted standards in global environmental politics.

Global environmental politics is about much more than cooperation among governments at the international level. It is also about what happens within states and what happens at the domestic–international frontier where international and domestic politics and policies interface, as they do in foreign policy processes. In Chapter 11, Mihaela Papa explores the crossovers among different levels of governance by focusing on foreign policy actors. She explores two approaches to environmental foreign policy, namely one that focuses on states and the roles of government officials (such as diplomats and officials in foreign ministries) as primary actors, and another that focuses more on multilevel governance and other actors involved in global environmental politics. Moving one further step down from the purely international, in Chapter 12 Stacy D. VanDeveer, Paul F. Steinberg, Jeannie L. Sowers and Erika Weinthal describe the roles of domestic actors and institutions in global environmental politics. They do this by focusing on (and advocating) a comparative approach to analyzing global environmental politics, in the process highlighting the importance of national policies in understanding and explaining the field. It is, after all, quite often policies at this level that have the most impact on the environment.

In Chapters 13 and 14, Kyla Tienhaara and Christian Downie focus on the roles of non-state actors in global environmental politics, although it must be said that even these actors seldom act entirely independently of states. Tienhaara examines some of the actors that some scholars and observers may argue are more important than most states: corporations. She looks at how corporations wield power, influence and authority in global environmental politics, showing that sometimes businesses have inordinate ability to shape events while at other times their own conflicts leave them unable to have their way. Businesses most often work to limit environmental regulation, but occasionally they can lead in efforts to move closer to a sustainable balance between environmental and economic priorities, as is happening nowadays among some businesses that are embracing sustainability and carbon neutrality. Continuing this survey of non-state entities, in the final chapter of Part III Downie describes a variety of transnational actors in global environmental politics, including for-profit actors (like those examined by Tienhaara) and not-for-profit nongovernmental organizations, as well as other broadly civil society actors, including individuals. He shows when and how these types of actors increasingly have an impact in global environmental politics, and he helps explain why they fail to have the impact that many people might wish for. Ultimately, it is usually some amalgamation of state and non-state actors and their influences that determines environmental outcomes, whether bad or good.

Ideas and themes in global environmental politics

Like many other aspects of world affairs, people's relations with the natural environment are influenced by ideas. Even when not directly influenced by them, global environmental politics can be better understood in terms of relatively discrete ideas. For example, official and unofficial responses to environmental change have in recent decades been influenced by the notion of sustainability, or what we might define simply as the idea that there are ecological limits to economic and other human activities. Indeed, the idea of sustainability now permeates global environmental politics, although the degree to which it is implemented is debatable and certainly uneven. Similarly, a number of key themes help to characterize contemporary global environmental politics. Examples of such themes include security, which is central to other aspects of international affairs, and globalization, the powerful forces of global economic integration and opening of borders that is affecting almost every aspect of life, including as it relates to the environment. Part IV is devoted to describing these and other ideas and themes in global environmental politics.

Part IV begins in Chapter 15 with Peter Stoett and Simon Dalby's introduction to, and explication of, the "Anthropocene," a geologic epoch in which Earth's "natural" systems are increasingly affected by, and often dominated by, human activities. The human impact on the environment is so great that there is barely any part of the environment that is not affected, often extremely so in adverse ways. The notion of the Anthropocene, apart from being an accurate depiction of the current epoch, is important because it helps to highlight the importance of altering human activities if there is any real hope of preserving the environment. In Chapter 16, Jon Marco Church, Andrew Tirrell, William R. Moomaw and Olivier Ragueneau point to what doing this will require: sustainability. They define environmental sustainability and consider how ideas about it are translated into real-world action. Closely related to sustainability – arguably the most important aspect of realizing it – is the question of material consumption, which is taken up by Gabriela Kütting in Chapter 17. She recounts the history of consumption before examining the institutionalization of the idea of "sustainable consumption." Following a theme in other chapters, Kütting shows that the problems of realizing truly sustainable consumption can often be a function of politics.

Understanding sustainability, as well as the underlying ecological and human forces that are at play when environmental commons suffer declines, requires scientific knowledge. As Andrew Karvonen and Ralf Brand show in Chapter 18, scientific expertise feeds into the processes of global environmental politics and policymaking, in the process often becoming a political issue itself. This is especially the case in the United States, where a surprising number of politicians and interest groups have become "anti-science" in their efforts to deny the reality of climate change and the importance of responding to it. Closely related to questions of science is that of uncertainty. As Jennifer Yarnold, Ray Maher, Karen Hussey and Stephen Dovers point out in Chapter 19, the role of risk in political calculations and in technological responses to environmental change are influenced by the level of uncertainty. Uncertainty makes predicting the future more difficult and of course is something that science can help alleviate. It can also play a role in defining how secure people and states feel in the face of environmental change.

Conceptions of security, whether human, national or international, often underly global environmental politics. That said, whether environmental issues are considered to be threats to security is open to interpretation, as Sabina W. Lautensach and Alexander K. Lautensach reveal in Chapter 20. For example, global climate change creates enormous national and human insecurity for poor low-lying communities and coastal states that suffer its profound direct effects, not least in the form of sea-level rise (made much worse during storms), and for those that lack the ability to fully cope with these effects. For them, climate change is an immediate threat. In contrast, many developed states, while also experiencing many of the effects of climate change, are much more able to cope with its impacts and generally have more resilient societies. In other words, an environmental threat that is existential to some poor states is a relatively distant or diminished concern to some wealthy ones. At least that is what many people, including those involved in policy processes, in the latter states believe. Even such a belief has great significance in global environmental politics. Fundamentally, the question is about whether the entire Earth system can be governed effectively enough to protect human and other forms of security. Indeed, as Frank Biermann argues in Chapter 21, the new concept of "earth system governance" challenges extant approaches to global environmental politics, in terms of both policy and research.

Another very important theme in global environmental politics is, not surprisingly, that of diplomacy, which is examined by Radoslav S. Dimitrov in Chapter 22. The processes of negotiation among diplomats, whether at formal international conferences or in backroom bilateral meetings, can greatly shape outcomes. It is during such meetings that concerns about security and insecurity can be tempered or occasionally exacerbated. This is especially true in forums where diplomats from wealthy developed states confront diplomats from developing states. As Shangrila Joshi affirms in Chapter 23, diplomats' conceptions of environmental security and how to ensure it, and more generally how to respond to global environmental problems, can be quite different depending on the states they represent. For developed-state diplomats, environmental problems may be relatively simple questions of technical responses, but for diplomats from developing states they are often wrapped up with a strong sense of historical injustice as a consequence of colonialism and empire in past centuries. Closely related to these questions are those of economic globalization, addressed in Chapter 24 by Lada V. Kochtcheeva. Globalization is arguably one of the most powerful drivers of adverse environmental changes because it has enabled wealthy states to "export" their pollution by buying products from states where environmental regulations are relatively low. Related to this is the increased availability of finance, still predominantly originating in developed states, that can determine whether economic development around the world is more environmentally harmful or less so. Too often it is still the latter.

These themes – of the relative power of rich and poor states, of how states' diplomats relate to one another in environmental negotiations, and the extent to which globalization has fostered trade, often to the advantage of some over others while exacerbating environmental decline – raise very serious questions of justice, both internationally and locally. In Chapter 25, Steve Vanderheiden examines international justice in global environmental politics, in the process showing how states have both rights and obligations in the context of environmental change. Questions of environmental justice also obtain locally. As Hollie Nyseth Brehm and David N. Pellow show in Chapter 26, pollution harms some people more than others. In particular, marginalized communities and the poor are often saddled with waste and overuse of natural resources on which they may depend for their survival. But questions of what is right and wrong in the context of global environmental politics is not restricted to relations among states internationally or to interactions among individuals (and other actors) locally; they also raise questions about the roles of nonhuman species. With this in mind, in Chapter 27 Sofia Guedes Vaz and Olivia Bina describe the relationships between ethics and philosophy, on the one hand, and ecology and other species, on the other. Together, these chapters on justice and ethics show that questions of global environmental politics can often not be answered by focusing only on traditional conceptions of power and rights.

The final two chapters in Part IV look in greater detail at one set of actors that are central to global environmental politics at all levels – or should be, at least – but which sometimes get overlooked: the public. In Chapter 28, Sandra T. Marquart-Pyatt describes the role of public opinion in global environmental politics and its relationship to how and whether people participate in different forms of environmental action. She describes how public opinion related to the environment is measured and assessed, and addresses the importance of cross-national research to better understand the views of publics. Building on such themes, in Chapter 29 Derek Bell defines and analyzes environmental citizenship. He describes how environmental citizenship has been portrayed and studied in theoretical, philosophical and practical terms. Much as Marquart-Pyatt reveals the difficulties of stimulating strong public commitment to environmental causes, Bell shows that it is a challenge to foster environmental citizenship, even as some scholars question whether doing so is a good idea.

Key issues and policies in global environmental politics

Chapters in Parts II to IV lay the foundation for understanding global environmental politics and the various actors, institutions and ideas that influence it. In Part V we turn to specific issues in global environmental politics and many of the policy responses to them, in the process reinforcing and further illustrating the material in preceding parts of this handbook. In the first three chapters in Part V, contributors look at pollution that often has widespread geographic impacts. In Chapter 30, Loren R. Cass describes the causes and politics of transboundary air pollution and acid rain. While both air pollution and acid rain continue to grow worse in many world regions, such as in East Asia, in other places, for example in Europe, there have been successes in tackling both problems. Cass shows how these experiences can help scholars and practitioners understand the causes of, and solutions to, other adverse changes to the environment.

In Chapters 31 and 32, respectively, David Downie and Paul G. Harris look at the truly global environmental issues of stratospheric ozone depletion and climate change. Downie's chapter describes the negotiation of quite effective international agreements to curb ozone-destroying chemicals. Indeed, those agreements have served as the framework for addressing climate change. Alas, climate change is a far more complicated problem. Both ozone depletion and climate change are caused by pollution from all around the world. However, climate change is a far more difficult problem than stratospheric ozone depletion, both practically and politically, because the sources of greenhouse gas pollution are in the billions – everyone contributes to climate change in some way – whereas sources of ozone-destroying chemicals are relatively few – the number of factories making these pollutants is relatively limited. This may help to explain why governments have been able to agree on quite successful measures to curb ozone-destroying pollution while failing so far, at least in global terms, to reverse growing emissions of greenhouse gases. The latter is largely explained by the world's growing appetite for energy, which is examined by Hugh C. Dyer in Chapter 33. Dyer takes a critical look at the core issue of energy, including the world's reliance on fossil fuels and particularly the political and economic strategies related to energy production and consumption. One of the largest consumers of energy is transport. In Chapter 34, Helene Dyrhauge examines transport and infrastructure and considers pathways toward more environmentally sustainable mobility.

Taking on other potential sources of global pollution, in Chapter 35, David Downie and Jessica Templeton describe how persistent organic pollutants have spread throughout ecosystems, presenting very serious threats to both environmental and human health. They explore how governments, nongovernmental organizations and other actors have responded to this problem. The result is a mixed bag, with real action occurring, but not always quickly or robustly enough to keep up with increasing amounts of pollution, notably in the developing world. Katja Biedenkopf expands on this theme in Chapter 36, which is devoted to describing the global politics of hazardous waste. National policies and international cooperation have resulted in multiple avenues for governing hazardous waste, but Biedenkopf characterizes these as fragmented. As with other environmental issues, hazardous waste, particularly its movement across national borders, has been addressed through international regulation. However, this certainly does not mean that the problem has been solved, least of all in ways that have achieved environmental justice. Too often, injustices associated with this type of pollution, like so many others, have exacerbated over time.

The final chapters in Part V look at major concerns related to ecosystems and the species that live within them, and at how governments and other actors have chosen – or not chosen – to address these issues. In Chapter 37, Josephine van Zeben and Violet Ross look at the exploding global problem of plastic. In their myriad forms, plastics can be valuable resources for modern life, but they quickly become a polluting scourge that taints landscapes, waterways and seas, with profound adverse consequences for ecosystems and living things. A question that van Zeben considers is whether more plastics can be turned from wastes back into resources. In Chapter 38, Edward Challies and Jens Newig look at some of the most vital issues in global environmental politics that happen to be greatly affected by plastic waste: water, rivers and wetlands. They show how water has been managed locally and internationally through the collaboration of key actors and stakeholders. Water in lakes and rivers is often polluted, and sadly much of that pollution finds its way to the sea. This and other impacts on the ocean environment are examined in Chapter 39 by Peter J. Jacques. His chapter describes the myriad threats to the marine environment, ranging from agricultural runoff and dumping at sea to the potentially devastating effects of climate change. Extending this look at the marine environment, in Chapter 40 Elizabeth R. DeSombre describes the international and regional politics of fisheries and marine mammals. As with many other issues examined in this handbook, these have been the subjects of international agreements, sometimes at the global level. Problems persist, without a doubt, but it seems beyond doubt that things would be much worse without such agreements. To some extent the same can be said of international agreements on the protection of biodiversity, migratory species and natural heritage, which are the subjects of Chapter 41 by Volker Mauerhofer and Felister Nyacuru. As they show, environmental agreements can be successful, as demonstrated by some agreements to protect waterfowl that migrate across national borders, but these successes are greatly undermined by the relentless destruction of natural habitats.

Destruction of habitats is starkly revealed by what is happening to the world's forests, which are the subject of Chapter 42 by Constance L. McDermott and David Humphreys. McDermott and Humphreys describe how and why governments have failed to agree on a global forest treaty, in the process tying deforestation back to questions of climate change (and related international and domestic politics). One option for governments in their efforts to limit climate change is to preserve forests, which act as "sinks" for carbon dioxide, the most widespread greenhouse gas. As Humphreys shows, the question of sequestration of carbon in forests is among the most politicized environmental issues. As such, it is the stuff of global environmental politics, revealing how seemingly disconnected issues – in this case, national forest politics and the global politics of climate change – are intimately connected, becoming increasingly complex in both environmental and political terms. The final two chapters of Part V continue to make this link to climate change. In Chapter 43, Meri Juntti describes the causes of desertification around the world, in the process highlighting the politics of the problem and the roles played by key actors. International agreements have been reached to address desertification, but there is little doubt that the problem will become much worse in coming decades. In a closely related and vitally important discussion, in Chapter 44 Jennifer Clapp and Sarah J. Martin look at food and agriculture. For anyone who might still think that our reliance on the natural environment is not total, or that our connections to it are not political, Clapp and Martin's description of the politics of food should disabuse them of such thinking. Their chapter is a classic case study of how the global environment, and specifically humanity's role in undermining it while also being intimately dependent upon it, is highly politicized.

Finally, in Chapter 45 Paul G. Harris identifies many of the important themes that are revealed by the preceding chapters. He draws lessons about global environmental politics from the contributors' descriptions and analyses of events before considering the promises and prospects for both the study and practice of global environmental politics. In so doing, he also highlights some of the limits of global environmental politics – potential limits to scholars' abilities to explain it and, much more significantly, apparent limits to practitioners' abilities to *do* global environmental politics in ways that greatly lighten humanity's footprint on Earth.

Part II
Explaining and understanding global environmental politics

2

The discipline of global environmental politics

A short history

Loren R. Cass

Global environmental politics has emerged as a center of interdisciplinary work that integrates research from a range of fields including international relations, comparative politics, geography, economics, history, law, climatology and biology. This interdisciplinary approach makes it difficult to clearly define the boundaries in this rather immense and diverse field of study. This chapter will briefly review the emergence of global environmental politics as a distinct subfield within the discipline of international relations since the 1980s. Many subfields of international relations have made the environment a subject of study. As early as the mid-eighteenth century scholars were analyzing the roles of natural resources and human population dynamics in the fields of international security and political economy. By the turn of the twentieth century states had begun to address issues related to the protection of fisheries, birds and exotic animals, and to acknowledge problems related to habitat degradation and water pollution. However, environmental policy was generally viewed as a local and perhaps national problem rather than a major issue of international concern.

Global environmental politics emerged relatively recently as a distinct field of study within the larger domain of international relations. The more contemporary focus on the interactions between humans and the natural world emerged in the 1970s and it was not until the 1980s and into the 1990s that global environmental politics became established as a separate subfield with its own dedicated journals and publishers. This is a period in which scholars expanded their focus to more systematically study a range of regional and global environmental problems such as acid rain, ozone depletion, climate change, biodiversity loss, deforestation and desertification.

Global environmental problems present many unique challenges that require a variety of theoretical perspectives and analytical tools to study them. They frequently involve substantial scientific complexity and uncertainty, which has produced a wide-ranging scholarship on the relationships between science and policy. The very long timeframes of both the consequences of environmental problems as well as the efforts to address them create a number of governance challenges. While addressing environmental problems may involve decades of action, politicians and the citizens they serve in democratic systems tend to think in terms of election cycles involving two to six years. In addition, because environmental problems typically do not respect borders, they pose challenges for international cooperation, which

DOI: 10.4324/9781003008873-4

has produced a growing literature on environmental negotiation and global environmental governance. The widespread potential for massive economic, political and ecological dislocation from the consequences of global environmental problems as well as from the potential policies to address those problems have led scholars to study global environmental politics from the perspective of every paradigm within international relations as well as to draw on research from numerous other disciplines. Finally, efforts to address the consequences of environmental problems have produced controversial ethical and distributive justice questions that have generated an important philosophical and normative literature.

Global environmental politics has thus emerged as a very rich and diverse area of scholarship. The sections that follow provide a brief overview of the evolution of global environmental politics scholarship as well as offering entry points to begin exploring the great variety of topics within the field. A number of scholars have presented overviews of the emergence of global environmental politics that complement and expand upon the material presented in this chapter, including Mitchell 2013; Betsill et al. 2014; Dauvergne 2014; and Stevis 2017.

The emergence of a distinct field of global environmental politics

The scholarship on global environmental politics emerged alongside the growing international interest in environmental issues, as reflected in the 1972 United Nations Stockholm Conference on the Human Environment. Lynton Caldwell (1972), Richard Falk (1971) and Harold and Margaret Sprout (1971) represent some of the earliest scholars to publish work focused specifically on global environmental politics. Scholarly interest waned somewhat in the late 1970s and 1980s with the resurgence in the Cold War, but international relations journals continued to publish occasional articles and some books were published during this period (Westing 1986; Young 1989). However, global environmental politics achieved much greater interest from scholars as negotiations leading up to the 1992 Earth Summit came to the forefront of international politics in the early 1990s.

Perhaps the strongest indicator of the maturity of a new field of study is the creation of journals dedicated to publishing work in the area. While major international relations journals had published articles on global environmental politics during the 1970s and 1980s (and in prior decades), it was not until the 1990s that journals dedicated to international environmental political research began to emerge. Many of today's environmental politics journals trace their origins to the early 1990s. *Global Environmental Change* (first published in 1990), the *Journal of Environment and Development* (1992) and *Environmental Politics* (1992) represent some of the leading journals dedicated to global and comparative environmental issues. A second wave of journals emerged after 2000 with a specific focus on international environmental relations and international law. *Global Environmental Politics* was established in 2001 and has become a preeminent journal for environmental research within the field of international relations, and *International Environmental Agreements: Politics, Law and Economics* (2001) has established itself as an outlet for a range of research related to global environmental politics, international environmental law and policy, and comparative responses to international environmental problems. There are other more specialized journals, such as *Environmental Values* (1992), *RECIEL: Review of European Community & International Environmental Law* (1992) and the *Journal of International Wildlife Law and Policy* (1998), that were also established during this period. In addition, publishers such as Ashgate, MIT Press, Routledge, State University of New York Press and others have created environmental series to publish wide-ranging scholarship related to global environmental politics.

Another indicator of the establishment of a new field of study is the growth in courses taught on the subject in academia. The emergence of a range of textbooks devoted to global environmental politics in the 1990s signaled the growing attention that the subject was receiving. Textbooks by Porter and Brown (1991), Choucri (1993), Brenton (1994) and Conca et al. (1995) provided some of the earliest texts to address global environmental politics and they offer insights into the early focus of global environmental politics research. The textbooks typically began with an overview of the history and unique attributes of global environmental politics combined with a discussion of approaches to studying these issues. They then analyzed a set of case studies of international environmental problems and political responses. This approach has been maintained by Chasek et al. (2017) and Stevenson (2017). Reflecting the growing sophistication of the global environmental politics literature, several other textbooks, including Lipschutz (2004), O'Neill (2017) and Mitchell (2010), have approached the subject from a more theoretical perspective with less focus on case study analysis, reflecting the growing diversity of scholarship and the momentum toward a more systematic and theory-driven understanding of global environmental politics.

International relations paradigms and global environmental politics

The global environmental politics literature was inevitably shaped by the larger debates within international relations at the time of its emergence as a distinct field and, in turn, contributed to these larger debates. The 1970s were influenced by debates over the global population explosion, resource scarcity concerns and security. From the 1980s to the early 1990s was a period in which neorealist and neoliberal scholars were debating the potential ability of international institutions and international regimes to mitigate the effects of competition within the international system and to promote cooperation (see Chapter 3). Global environmental politics provided a rich field of study because there was a range of international environmental problems that offered a wealth of case studies to test hypotheses emerging from the neorealist and neoliberal debate (see Chapter 5). Much of the early scholarship in global environmental politics reflects these debates. This can be seen in works by Haas et al. (1993), Brenton (1994), Paterson (1996) and Seaver (1997). The edited volume by Haas et al. is a classic work that explores the role of international institutions in facilitating more effective international responses to emergent environmental problems (see Chapter 8). Paterson (1996) is one of the best examples from this period of attempts to apply the various international relations paradigms to the study of global environmental politics.

The neorealist paradigm of international relations focuses on the inevitable conflict that occurs among self-interested actors in an anarchic state system (a system lacking any authority above the state). Neorealist approaches to studying global environmental politics have been much less common than the more widely applied neoliberal and constructivist approaches (see Chapters 3 and 4). Scholars working within the neorealist tradition have tended to gravitate toward issues of environment and security with a focus on resource scarcity (Westing 1986; Homer-Dixon 1994; Chalecki 2017; see Deudney 1990 for a critique). Homer-Dixon's Project on Environmental Change and Acute Conflict was particularly influential in shaping the debate on security and environment. He argues that there are multiple pathways through which environmental scarcities can produce conflict and predicts that environmental conflicts will increase as a result of the growing effects of climate change. Increasingly, the security implications of global environmental problems have become a broader focus of research that bridges the divide between international security studies and global environmental politics. (See Chapter 20 for a review of issues of security in global environmental politics.)

The neorealist paradigm begins with assumptions that states act in a rational manner to secure the core interests of the state related to national security and prosperity. The most widely cited application of an interest-based approach to broader global environmental politics is Sprinz and Vaahtoranta (1994). They present a model for determining national positions in global environmental negotiations based upon a combination of the abatement costs of addressing the problem and the ecological vulnerability to the environmental threat. They argue that the higher the ecological vulnerability and the lower the abatement costs, the stronger the government's support for international action to address environmental problems will be. Conversely, the lower the ecological vulnerability and the higher the cost of abatement, the more reluctant a state will be to address a global environmental problem. There are a range of other scholars who have utilized an interest-based approach to global environmental politics (Barrett 2006; Grundig 2006; Victor 2006). While an interest-based approach remains an important element of the global environmental politics literature, most scholars have emphasized that a focus on national interests and relative power positions provides at best only a partial explanation for the observed behavior.

Beginning in the 1990s, scholars began to explore the role of international institutions and international regimes (defined as social institutions that shape actor expectations and associated behavior in a given issue area) in influencing environmental negotiations and the emergence of the dense network of international environmental agreements that emerged during this period (see Chapter 9). The "international regimes" literature emerged almost simultaneously with the widening focus on global environmental problems in the 1970s and 1980s. Global environmental politics offered case studies to test hypotheses flowing out of the work on international regimes (Young 1977 and Brown et al. 1977). This early research evolved into a broader focus on environmental governance in works such as Young (1994). One of the most influential works of this period was Haas et al. (1993). The book's analytical focus on the importance of building national capacity, improving the contractual environment and elevating governmental concern remains an important organizing focus for the study of global environmental politics.

The early research identified ways in which international organizations and regimes affect environmental politics and the behavior of actors. This then spawned a series of research projects to test the effectiveness of these regimes and the impacts of the growing web of international environmental institutions on global environmental governance.

Global environmental politics continues to be heavily influenced by research on governance and regime effectiveness (see Chapters 9 and 21). Several large-scale research projects have significantly influenced work in this area. Breitmeier et al. (2006) present findings related to regime effectiveness that emerged from the International Regimes Database project. Young et al. (2008) published the findings from the Institutional Dimensions of Global Environmental Change (IDGEC) project that studied relationships among and the effectiveness of global environmental institutions. The Global Governance Project under the direction of Frank Biermann has also produced a range of books and articles related to environmental governance and effectiveness (e.g., Biermann and Pattberg 2008 and Biermann and Siebenhüner 2009). Park et al. (2008) offer a critique of existing environmental governance structures and argue for alternative strategies based upon the principle of sustainability (see Chapter 16). Biermann (2014) and Dryzek and Pickering (2019) argue that the transition to the "Anthropocene" has fundamentally changed the context within which global governance must be understood. Busby (2017) provides a good overview of the development and current debates in the literature on environmental governance.

More recently, many global environmental politics scholars have turned to constructivist approaches to explain aspects of environmental affairs that cannot be easily explained by a focus on interests and/or international institutions. Constructivism emphasizes the role of ideas in structuring international relations with emphases on the discourses of actors as well as the identities of the actors and relationships among them (see Chapter 4). The focus is upon the social construction of reality. Constructivists have emerged as critics of the dominant theoretical paradigms that emphasize state power and international institutions as the primary variables shaping international relations. Constructivist approaches have been frequently applied to global environmental politics to try to analyze the role of science in the social construction of knowledge and the use of knowledge in making policy (Haas 2004 and Jasonoff and Martello 2004; see Chapter 18).

Scholars working within the constructivist perspective frequently split between more norm-based approaches, which emphasize social expectations regarding appropriate behavior, and discursive approaches, which focus on the use of language and its relationship to political behavior. Hajer (1995) was among the first scholars to emphasize the importance of discourse in the definition of environmental problems and solutions. Dryzek (2016) offers an updated introduction to discourse analysis. Epstein (2008) applies discourse analysis to explore the shaping of power and interests in the case of whaling. Litfin (1998) presents a variety of scholars with ties to the constructivist tradition that focus on the evolution of sovereignty and changing norms and discourses regarding how sovereignty relates to global environmental politics (see Chapter 7). Within the norms literature, Bernstein (2001) and Cass (2006) analyze the evolution of international norms and the confluence of environmental and liberal economic norms and their effects on international environmental policy. Pettenger (2007) presents the perspectives of a range of constructivist scholars, ranging from those using functionalist, international norm-based analyses to those using a discursive approach to understanding political responses to climate change.

While significantly shaped by the neoliberal paradigm, the global environmental politics literature remains a fertile field for debates among paradigms. Scholars continue to apply a range of theories to the study of global environmental politics. Dyer (2017) provides an excellent analysis of the application of international relations theories to global environmental politics with a very thorough bibliography.

Bridging the international/domestic divide

While international relations paradigms have been central to exploring the behavior of states, the global environmental politics literature is further complicated by the need to bridge the divide between the subfields of international relations and comparative politics (which examines domestic political processes; see Chapter 12). Addressing most environmental problems entails major changes to domestic regulations that cover some of the most economically important and politically controversial policy areas. For example, reducing acid rain required expensive changes in electrical power generation, transportation and manufacturing (see Chapter 33). These changes are inevitably contentious and are intimately tied to domestic political norms, processes and histories of the countries involved in the international negotiations.

Scholars of global environmental politics have frequently sought to bridge the international/domestic divide to analyze the forces shaping national positions in international environmental negotiations (see Chapter 11). Harris (2009) systematically surveys the environmental foreign policy literature and its relationship to global environmental politics. He presents a typology of theories that can explain national positions in global environmental negotiations and then offers a series of case studies that illustrate the ability of different theories operating at different levels of analysis to explain national behavior in negotiations. A number of scholars have presented case studies of national responses to global environmental problems. For example, Schreurs and Economy (1997) present a series of case studies evaluating domestic forces shaping national positions across a range of countries on climate change, ozone depletion and biodiversity loss. They argue that the internationalization of environmental protection efforts is altering domestic policymaking processes, policy outcomes and the effectiveness of policy implementation. There is thus an interactive process in which international and domestic responses to environmental problems are mutually constitutive. Rather than international politics altering domestic politics, DeSombre (2000) argues that national positions in international negotiations are significantly shaped by attempts to internationalize domestic regulations to minimize adjustment costs and improve competitiveness of domestic industry. In other words, the causal arrow points from domestic politics to international negotiations. Harris's Project on Environmental Change and Foreign Policy has produced a series of edited volumes (Harris 2007, 2009, among others) that address domestic forces shaping environmental foreign policy positions. Harrison and Sundstrom (2010) provide a series of articles addressing the comparative politics of climate policy. Balboa (2018) explores the challenges of NGOs in operating effectively across local, national and global levels of governance, which highlights the interactions among NGOs and various levels of governance.

Despite these efforts, attempts to systematically link comparative environmental politics and global environmental politics remain relatively underdeveloped. This remains a vital area of research and one that has the potential to contribute greatly to our understanding of global environmental politics.

The role of science in global environmental politics

Environmental problems are frequently characterized by scientific complexity and extensive uncertainty regarding causes and/or solutions (see Chapters 18 and 19). The integration of science into the policy process is thus a critical aspect of efforts to address global environmental politics. Unsurprisingly, scholars have produced an extensive literature to address these issues, with a number of scholars analyzing the conditions under which scientific knowledge is integrated into decision-making processes (Bocking 2004; Harrison and Bryner 2004; Dimitrov 2006; Mitchell et al. 2006).

Analysis of science and policy has a relatively long history. Haas et al. (1977) offer a very early critique of the process through which science is integrated into decision-making processes. During the 1990s the analysis of the role of science was heavily influenced by the concept of epistemic communities or groups of scientists and experts that share a common set of values and a common understanding of an environmental problem and potential solutions (Haas 1990). These groups achieve influence in situations of scientific uncertainty and have the potential to significantly shape both the framing of the environmental problem and potential responses to it. This focus on scientists as active participants in the international environmental policy process is a common theme. Boehmer-Christiansen (1995) argues that scientists must be viewed not as neutral conveyors of policy-relevant information but rather as political actors themselves who seek to shape the availability and interpretation of scientific evidence to further their interests.

International environmental negotiation

As global environmental negotiations increased in their frequency and in the range of issues being addressed, scholars turned their attention to unique attributes of global environmental problems and the difficulties that they posed for achieving effective cooperative solutions (see Chapters 11 and 22). Over 1,300 multilateral environmental agreements and more than 2,200 bilateral environmental agreements have been negotiated to date (Mitchell 2002–2021). There is thus a vast dataset of negotiations to draw upon. In analyzing these negotiations, many scholars have argued that the negotiation process itself is an important variable in determining final agreements. Susskind (1994) presented an early attempt to explore the nature of international environmental negotiations and the differences with other types of international issues. Chasek (2001) analyzes 30 years of international environmental negotiations to discern patterns in the outcomes. She argues that there are six discernible phases and five associated turning points within the process of multilateral environmental negotiation. This complex structure of negotiations affects the types of agreements that can be made and their potential for success. Chasek and Wagner (2012) analyze the changes occurring in global environmental negotiations between the 1992 Earth Summit and the 2012 UN Conference on Sustainable Development. Other scholars such as Kütting (2000), Susskind and Ali (2014), Barrett (2006) and Bodansky (2009) analyze the relative effectiveness of international environmental negotiations and offer suggestions for improving them.

While international negotiations have traditionally focused almost exclusively on the roles of states and perhaps international institutions, global environmental politics scholarship has increasingly challenged this narrow focus on the state (see Chapter 11). This is apparent in the sections above that address constructivism, science and subnational forces shaping national negotiating positions. A growing literature on non-state actors further extends the focus to actors such as nongovernmental organizations (see Chapter 14), cities and regional groups, as well as indigenous peoples (Comberti et al. 2019). Green (2014) emphasizes the importance of states delegating authority to non-state actors and the role of these actors in shaping global responses to environmental problems. This literature typically presents non-state actors as independently shaping national positions and increasingly directly influencing international negotiations. Princen and Finger (1994) provide one of the earliest analyses of the roles of environmental NGOs in global environmental politics. They argue that NGOs function as independent participants in international negotiations as well as serving as agents of social learning to shape the framing of international and subnational understandings of environmental problems and possible policy responses. Betsill and Corell (2008) offer a framework for evaluating the influence of environmental NGOs in global environmental negotiations and evaluate the degree of nongovernmental organization (NGO) influence across a variety of case studies. Levy and Newell (2005) analyze the role of business interests in global environmental politics and illustrate the ways business activity shapes and is shaped by global environmental policies. Finger and Svarin (2018) survey the literature related to transnational corporations and environmental governance and offer a thorough bibliography.

The international environmental negotiation literature is rich in case studies and nuanced analyses of the negotiating process that highlight the importance of a range of actors and variables. The literature particularly offers insights into how to improve the negotiating environment to increase the prospects of achieving successful agreements.

Methodological approaches to studying global environmental politics

Much of the early scholarship on global environmental politics was heavily influenced by individual case studies. In part this reflected the relative immaturity of the field. The lack of scholarship on global environmental politics meant that many scholars were seeking to apply a range of theories and analytical tools to the emerging field. Mitchell (2010: 7) discusses the problems that emerged during this early period of scholarship:

> Initially, deductive theories generated little follow-up in terms of operationalization and testing while inductive case studies generated useful insights that often were not framed in ways which could facilitate their application and evaluation in other environmental realms. As a result, different terminologies and taxonomies of causal factors often overlapped with, but seemed unaware of, competing or complementary ones.

The choice of case studies also reflected the sequential pattern to the emergence and definition of new environmental problems on the international agenda with scholars producing major works as each new issue achieved international prominence. Some of the earliest global environmental politics literature focused on acid rain and transboundary air pollution flows (McCormick 1985; see Chapter 30). In the 1990s and into the 2000s scholars produced major works on regional water pollution (Haas 1990; see Chapter 38), ozone depletion (Litfin 1994; Rowlands 1995; Benedick 1998; see Chapter 31) and climate change (Paterson 1996; O'Riordan and Jäger 1996; Luterbacher and Sprinz 2001 and 2018; see Chapter 32). Each work typically analyzed global environmental politics from a rather unique theoretical perspective, which provided a very diverse and intellectually stimulating range of lenses through which to study global environmental politics, but it was also difficult to determine whether the lessons derived from the individual case studies could be extrapolated to a broader set of cases.

As the scholarship in the field evolved, scholars sought to analyze multiple case studies utilizing a common theoretical approach to begin to test the broader generalizability of findings from earlier case studies. Barkin and Shambaugh (1999) looked at the nature of common pool resource issues and attempts to resolve them across a range of case studies. Haas et al. (1993) and DeSombre (2017) analyzed the role of international institutions in promoting international environmental cooperation.

With over 40 years of international environmental negotiations across a wide range of cases, scholars now have a very rich and diverse set of cases to analyze in an attempt to more systematically investigate the nature of global environmental politics. Increasingly, there have been attempts to try to apply quantitative analyses to the study of global environmental politics to empirically test the hypotheses that have emerged from the earlier case study analyses (see Chapter 5). Miles et al. (2002), Mitchell (2002), Breitmeier et al. (2006), and Young et al. (2008) represent some of the major projects to undertake quantitative analyses of various hypotheses related to global environmental politics. There is a growing diversity of methodological approaches and interdisciplinary research in global environmental politics scholarship (see Chapter 6).

Environmental ethics and justice

A vast literature on environmental ethics and justice has emerged over the last several decades (see Chapters 25–27). Global environmental problems raise a number of difficult ethical and normative challenges. What obligations do the affluent have to the less affluent in today's world? The rich consume vast quantities of energy and natural resources while the vast majority of the world's people suffer in poverty (see Chapters 17 and 23). Do the wealthy have any obligation to preserve the natural resources of our world today for use by the less affluent? Is there any obligation to use the wealth of the developed world to alleviate the environmental suffering frequently created by the exploitation of resources in developing countries?

These questions are frequently discussed in terms of "environmental justice." Schlosberg (2007) provides a particularly interesting exploration of the definition of "justice" as it relates to environmental and ecological justice. He emphasizes differences in the definitions of justice as used by American and global environmental movements and suggests ways in which environmental justice can be built into the practice of environmental policy. Parks and Roberts (2006) and Bryner (2017) present more introductory explorations of the origins and evolution of scholarship related to global environmental justice. They provide a good starting point for exploring these issues and extensive bibliographies to pursue additional research in the area. Bryner (2017) presents a useful series of frameworks for defining environmental justice, including civil rights, distributive justice, public participation, social justice and ecological sustainability.

While the authors above primarily address questions of environmental justice among the inhabitants of the world today, several scholars have sought to analyze questions of intergenerational justice. What obligation does today's generation have to future generations? Hiskes (2009) builds an argument for preserving the environment as a human right premised upon a notion of intergenerational justice. Beckerman and Pasek (2001) explore problems related to intergenerational environmental justice and highlight the need to balance the needs of the disadvantaged today with obligations to future generations.

Beyond the broader questions of environmental justice, there have been particularly intense debates surrounding the ethical foundations for addressing the problem of climate change. For example, Adger et al. (2006) address questions related to adaptation to the consequences of climate change and fairness in distributing the costs of adaptation. Page (2006) evaluates the particular problem of intergenerational justice in the case of climate change. Harris (2001) provides a critique of past American climate policy, notably its limited accounting for international equity and environmental justice considerations, and Harris (2016) examines the role of global (cosmopolitan) justice in the world's responses to climate change.

The fact that the perpetrators of environmental harm frequently do not face the full costs associated with their actions creates significant ethical problems that must be considered as a part of the international political response. However, both the effects of global environmental problems and the policy responses to those problems have the potential to have significant global redistributive consequences. They will also have important ramifications for quality of life and standards of living. These problems raise difficult equity questions as well as questions related to fairness and equity to the current generation in relationship to future generations. The literature addressing these questions is evolving, and it will be important for scholars to continue to highlight these questions in the context of international negotiations to address environmental problems. Many scholars have been highly critical of the failure to adequately incorporate ethical concerns into international decision-making processes.

Conclusion

In providing an overview of the emergence of global environmental politics as a field of study and the growing complexity of the scholarship in this area, it seems appropriate to conclude with perhaps the most important question of global environmental politics, one that relates to nearly all of the scholarship discussed above: can the existing international political economic system be sustained in the face of growing resource demands and increasing adverse impacts from the release of pollutants? (see Chapters 15 and 16). Clapp and Dauvergne (2011) and Haas (2017) provide overviews of the literature that has evolved around this question. Globalization and the associated growth in international consumption, trade, travel and migration have profound environmental implications. The environment fulfills two critical functions from a political economy perspective. It is the source of the resources that propel the global economy and create wealth, and it is a sink to absorb and process much of the waste that is generated as a side effect of global production and consumption (Cass 2012). One of the central questions of global environmental politics is whether the existing international political/economic system is sustainable in light of these increasing demands.

Supporters of the existing system and the central role of the market in allocating resources argue that the path to a sustainable world requires improving the prosperity of the world's people to create the wealth and the political will to address environmental problems (Simon 1998). Other scholars have argued that the system can be made to be compatible with sustainability. Mol (2003) has argued that a normative focus on "ecological modernization" can provide a mechanism to align the existing system with environmental protection and sustainability. He has argued that enlightened self-interest can provide the foundation for sustainability. People can come to understand that consumption patterns must be altered in ways to make the international system sustainable for future generations. Alternatively, institutionalists such as Biermann and Bauer (2005) argue that the system can be reformed and new global governance structures can be created to achieve sustainability.

While many scholars have argued that the existing system is either sustainable in its present form or can be reformed and saved, there is a radical critique within the global environmental politics scholarship that argues that the existing system is fundamentally flawed and incapable of achieving sustainability. Among the more radical critiques, Daly (1973, 1996) argues that the world must achieve "steady-state equilibrium" where the number of humans and their resource usage are reduced to environmentally sustainable levels. Such a move requires some form of population control and the rejection of the current system's emphasis on constant economic growth and associated rising consumption levels. Lipschutz (2004) argues that the existing global political/economic structures are fundamentally flawed because they create incentives for unsustainable use of resources. He argues for a radical restructuring of the system.

As we enter the 2020s a number of scholars have posited that the earth has moved into a new epoch, the "Anthropocene" (see Chapter 15). From this perspective, human activity has emerged as the dominant influence on the global environment (Lynch and Veland 2018 and Dryzek and Pickering 2019). The complexity of the emerging political, economic and environmental systems requires dramatic changes in governance systems to manage the transition to the Anthropocene. Kalantzakos (2017) argues that the transition demands greater cooperation among the great powers. Skillington (2019) argues that the dramatic environmental changes that are occurring require a fundamental redefinition of a "just society." This literature is at a relatively early stage in its maturation, but it presents a significant challenge to scholars working on global environmental governance.

The question of whether the existing international political/economic system is sustainable or can be made to be sustainable is the core foundation for much of the literature within global environmental politics. It is also one of the most important questions that will continue to motivate future scholarship in global environmental politics.

References

Adger, W.N., Paavola, J., Huq, S. and Mace, M.J. (eds) (2006) *Fairness in Adaptation to Climate Change*, Cambridge, MA: MIT Press.

Balbao, C.M. (2018) *The Paradox of Scale: Howe NGOs Build, Maintain, and Lose Authority in Environmental Governance*, Cambridge, MA: MIT Press.

Barkin, J.S. and Shambaugh, G.E. (eds) (1999) *Anarchy and the Environment: The International Relations of Common Pool Resources*, Albany: SUNY Press.

Barrett, S. (2006) *Environment and Statecraft: The Strategy of Environmental Treaty-making*, Oxford: Oxford University Press.

Beckerman, W. and Pasek, J. (2001) *Justice, Posterity and the Environment*, Oxford: Oxford University Press.

Benedick, R.E. (1998) *Ozone Diplomacy: New Directions in Safeguarding the Planet*, Cambridge, MA: Harvard University Press.

Bernstein, S. (2001) *The Compromise of Liberal Environmentalism*, New York: Columbia University Press.

Betsill, M.M. and Corell, E. (eds) (2008) *NGO Diplomacy: The Influence of Nongovernmental Organizations in International Environmental Negotiations*, Cambridge, MA: MIT Press.

Betsill, M.M., Hochstetler, K. and Stevis, D. (eds) (2014) *Palgrave Advances in International Environmental Politics*, 2nd edn, New York: Palgrave.

Biermann, F. (2014) *Earth Systems Governance: World Politics in the Anthropocene*, Cambridge, MA: MIT Press.

Biermann, F. and Bauer, S. (eds) (2005) *A World Environment Organization: Solution or Threat for Effective Environmental Governance?* Burlington, VT: Ashgate.

Biermann, F. and Pattberg, P. (2008) "Global environmental governance: taking stock, moving forward", *Annual Review of Environment and Resources*, 33: 277–294.

Biermann, F. and Siebenhüner, B. (eds) (2009) *Managers of Global Change: The Influence of International Environmental Bureaucracies*, Cambridge, MA: MIT Press.

Bocking, S. (2004) *Nature's Experts: Science, Politics and the Environment*, New Brunswick, NJ: Rutgers University Press.

Bodansky, D. (2009) *The Art and Craft of International Environmental Law*, Cambridge, MA: Harvard University Press.

Boehmer-Christiansen, S. (1995) "Britain and the International Panel on Climate Change: the impacts of scientific advice on global warming part 1: integrated policy analysis and the global dimension", *Environmental Politics*, 4: 1–18.

Breitmeier, H., Young, O. and Zürn, M. (2006) *Analyzing International Environmental Regimes: From Case Study to Database*, Cambridge, MA: MIT Press.

Brenton, T. (1994) *The Greening of Machiavelli: The Evolution of International Environmental Politics*, London: Earthscan Publications.

Brown, S., Cornell, N.W., Fabian, L.L. and Weiss, E.B. (1977) *Regimes for the Ocean, Outer Space and Weather*, Washington, DC: Brookings.

Bryner, G. (2017) "Environmental justice", *Oxford Research Encyclopedia of International Studies* Online. Available HTTP: <http://oxfordre.com/internationalstudies>

Busby, J. (2017) "International organization and environmental governance", *Oxford Research Encyclopedia of International Studies* Online. Available HTTP: <http://oxfordre.com/internationalstudies>

Caldwell, L.K. (1972) *Defense of the Earth: International Protection of the Biosphere*, Bloomington: University of Indiana Press.

Cass, L.R. (2006) *The Failures of American and European Climate Policy: International Norms, Domestic Politics and Unachievable Commitments*, Albany: SUNY Press.

Cass, L.R. (2012) "The global environment: an overview", in R. Pettman (ed.) *Handbook on International Political Economy*, London: World Scientific, 327–342.

Chalecki, E.L. (2017) "Environment and security", *Oxford Research Encyclopedia of International Studies* Online. Available HTTP: <http://oxfordre.com/internationalstudies>

Chasek, P.S. (2001) *Earth Negotiations: Analyzing Thirty Years of Environmental Diplomacy*, New York: United Nations University Press.

Chasek, P.S., Downie, D.L. and Brown, J.W. (2017) *Global Environmental Politics*, 7th edn, Boulder, CO: Westview Press.

Chasek, P.S. and Wagner, L.W. (eds.) (2012) *The Roads from Rio: Lessons Learned from Twenty Years of Multilateral Environmental Negotiations*, New York: Routledge.

Choucri, N. (ed.) (1993) *Global Accord: Environmental Challenges and International Responses*, Cambridge, MA: MIT Press.

Clapp, J. and Dauvergne, P. (2011) *Paths to a Green World: The Political Economy of the Global Environment*, 2nd edn, Cambridge, MA: MIT Press.

Comberti, C., Thornton, T.F., Korodimou, M., Shea, M. and Riamit, K.O. (2019) "Adaptation and resilience at the margins: addressing indigenous peoples' marginalization at international climate negotiations," *Environment: Science and Policy for Sustainable Development*, 61: 14-30.

Conca, K., Alberty, M. and Dabelko, G. (1995) *Green Planet Blues: Environmental Politics from Stockholm to Rio*, Boulder, CO: Westview Press.

Daly, H. (1973) *Toward a Steady-state Economy*, San Francisco, CA: W.H. Freeman.

Daly, H. (1996) *Beyond Growth: The Economics of Sustainable Development*, Boston, MA: Beacon Press.

Dauvergne, P. (ed.) (2014) *Handbook of Global Environmental Politics*, 2nd edn, Northampton, MA: Edward Elgar.

DeSombre, E.R. (2000) *Domestic Sources of International Environmental Policy: Industry, Environmentalists and U.S. Power*, Cambridge, MA: MIT Press.

DeSombre, E.R. (2017) *Global Environmental Institutions*, 2nd edn, New York: Routledge.

Deudney, D. (1990) "The case against linking environmental degradation and national security", *Millennium: Journal of International Studies*, 19: 461–76.

Dimitrov, R.S. (2006) *Science and International Environmental Policy: Regimes and Nonregimes in Global Governance*, Boulder, CO: Rowman and Littlefield.

Dryzek, J.S. (2016) *The Politics of the Earth: Environmental Discourses*, 3rd edn, Oxford: Oxford University Press.

Dryzek, J.S. and Pickering, J. (2019) *The Politics of the Anthropocene*, Oxford: Oxford University Press.

Dyer, H. (2017) "Challenges to traditional international relations theory posed by environmental change", *Oxford Research Encyclopedia of International Studies* Online. Available HTTP: <http://oxfordre.com/internationalstudies>

Epstein, C. (2008) *The Power of Words in International Relations: Birth of An Anti-Whaling Discourse*, Cambridge: Cambridge University Press.

Falk, R. (1971) *This Endangered Planet: Prospects and Proposals for Human Survival*, New York: Vintage.

Finger, M. and Svarin, D. (2018) "Transnational corporations and the global environment", *Oxford Research Encyclopedia of International Studies* Online. Available HTTP: <http://oxfordre.com/internationalstudies>

Green, J.F. (2014) *Rethinking Private Authority: Agents and Entrepreneurs in Global Environmental Governance*, Princeton, NJ: Princeton University Press.

Grundig, F. (2006) "Patterns of international cooperation and the explanatory power of relative gains: an analysis of cooperation on global climate change, ozone depletion and international trade", *International Studies Quarterly*, 50: 781–801.

Haas, E.B., Williams, M.P. and Babai, D. (1977) *Scientists and World Order: The Use of Technical Knowledge in International Organizations*, Berkeley: University of California Press.

Haas, P.M. (1990) *Saving the Mediterranean: the Politics of International Environmental Cooperation*, New York: Columbia University Press.

Haas, P.M. (2004) "When does power listen to truth? A constructivist approach to the policy process", *Journal of European Policy*, 11: 569–92.

Haas, P.M. (2017) "Environment in the global political economy", *Oxford Research Encyclopedia of International Studies* Online. Available HTTP: <http://oxfordre.com/internationalstudies>

Haas, P.M., Keohane, R.O. and Levy, M.A. (eds) (1993) *Institutions for the Earth: Sources of Effective International Environmental Protection*, Cambridge, MA: MIT Press.

Hajer, M.A. (1995) *The Politics of Environmental Discourse: Ecological Modernization and the Policy Process*, Oxford: Clarendon Press.

Harris, P.G. (2001) *International Equity and Global Environmental Politics: Power and Principles in Us Foreign Policy*, Burlington, VT: Ashgate.

Harris, P.G. (ed.) (2007) *Europe and Global Climate Change: Politics, Foreign Policy and Regional Cooperation*, Northampton, MA: Edward Elgar.

Harris, P.G. (2009) *Environmental Change and Foreign Policy: Theory and Practice*, New York: Routledge.

Harris, P.G. (2016) *Global Ethics and Climate Change*, 2nd edn, Edinburgh: Edinburgh University Press.

Harrison, K. and Sundstrom, L.M. (eds) (2010) *Global Commons, Domestic Decisions: The Comparative Politics of Climate Change*, Cambridge, MA: MIT Press.

Harrison, N.E. and Bryner, G.C. (eds) (2004) *Science and Politics in the International Environment*, Boulder, CO: Rowman and Littlefield.

Hiskes, R.P. (2009) *The Human Right to a Green Future: Environmental Rights and Intergenerational Justice*, Cambridge: Cambridge University Press.

Homer-Dixon, T. (1994) "Environmental scarcities and violent conflict: evidence from cases", *International Security*, *19*: 5–40.

Jasonoff, S. and Martello, M.L. (eds) (2004) *Earthly Politics: Local and Global in Environmental Governance*, Cambridge, MA: MIT Press.

Kalantzakos, S. (2017) *The EU, US and China Ackling Climate Change: Policies and Alliances for the Anthropocene*, New York: Rougledge.

Kütting, G. (2000) *Environment, Society and International Relations: Towards More Effective International Environmental Agreements*, London: Routledge.

Levy, D.L. and Newell, P.J. (eds) (2005) *The Business of Global Environmental Governance*, Cambridge, MA: MIT Press.

Lipschutz, R.D. (2004) *Global Environmental Politics: Power, Perspectives and Practice*, Washington, DC: CQ Press.

Litfin, Karen T. (1994) *Ozone Discourses: Science and Politics in Global Environmental Cooperation*, New York: Columbia University Press.

Litfin, Karen T. (ed.) (1998) *The Greening of Sovereignty in World Politics*, Cambridge, MA: MIT Press.

Luterbacher, U. and Sprinz, D. (eds) (2001) *International Relations and Global Climate Change*, Cambridge, MA: MIT Press.

Luterbacher, U. and Sprinz, D. (eds) (2018) *Global Climate Policy: Actors, Concepts, and Challenges*, Cambridge, MA: MIT Press.

Lynch, A.H. and Veland, S. (2018) *Urgency in the Anthropocene*, Cambridge, MA: MIT Press.

McCormick, J. (1985) *Acid Earth: The Global Threat of Acid Pollution*, London: International Institute for Environment and Development.

Miles, E.L., Underdal, A., Andresen, S., Wettestad, J., Skjaerseth, J.B. and Carlin, A.M. (eds) (2002) *Environmental Regime Effectiveness: Confronting Theory with Evidence*, Cambridge, MA: MIT Press.

Mitchell, R.B. (2002) "A quantitative approach to evaluating international environmental regimes", *Global Environmental Politics*, *2*: 58–83.

Mitchell, R.B. (2013) "International environment", in W. Carlsnaes, T. Risse-Kappen and B.A. Simmons (eds) *Handbook of International Relations*, London: Sage, 801–826.

Mitchell, R.B. (2010) *International Politics and the Environment*, Los Angeles, CA: Sage.

Mitchell, R.B. (2002–21) "International Environmental Agreements Database Project (Version 2020.1)", Online. Available HTTP: <http://iea.uoregon.edu/>

Mitchell, R.B., Clark, W.C., Cash, D.W. and Dickson, N.M. (eds) (2006) *Global Environmental Assessments: Information and Influence*, Cambridge, MA: MIT Press.

Mol, A. (2003) *Globalization and Environmental Reform: The Ecological Modernization of the Global Economy*, Cambridge, MA: MIT Press.

O'Neill, K. (2017) *The Environment and International Relations*, 2nd edn, Cambridge: Cambridge University Press.

O'Riordan, T. and Jäger, J. (eds) (1996) *Politics of Climate Change: A European Perspective*, New York: Routledge.

Page, E. (2006) *Climate Change, Justice and Future Generations*, Northampton, MA: Edward Elgar.

Park, J., Conca, K. and Finger, M. (eds) (2008) *The Crisis of Global Environmental Governance: Towards a New Political Economy of Sustainability*, New York: Routledge.

Parks, B.C. and Roberts, J.T. (2006) "Environmental and ecological justice", in M.M. Betsill, K. Hochstetler and D. Stevis (eds) *Palgrave Advances in International Environmental Politics*, New York: Palgrave, 329–60.

Paterson, M. (1996) *Global Warming and Global Politics*, New York: Routledge.

Pettenger, M.E. (ed.) (2007) *The Social Construction of Climate Change*, Burlington, VT: Ashgate.

Porter, G. and Brown, J.W. (1991) *Global Environmental Politics*, Boulder, CO: Westview Press.

Princen, T. and Finger, M. (eds) (1994) *Environmental NGOs in World Politics: Linking the Local and the Global*, London: Routledge.

Rowlands, I.H. (1995) *The Politics of Global Atmospheric Change*, Manchester: Manchester University Press.

Schlosberg, D. (2007) *Defining Environmental Justice: Theories, Movements and Nature*, Oxford: Oxford University Press.

Schreurs, M.A. and Economy, E.C. (eds) (1997) *The Internationalization of Environmental Protection*, Cambridge: Cambridge University Press.

Seaver, B. (1997) "Stratospheric ozone protection: IR theory and the Montreal Protocol on Substances That Deplete the Ozone Layer", *Environmental Politics*, *6*:31–67.

Simon, J. (1998) *The Ultimate Resource 2*, Princeton, NJ: Princeton University Press.

Skillington, T. (2019) *Climate Change and Intergenerational Justice*, New York: Routledge.

Sprinz, D. and Vaahtoranta, T. (1994) "The interest-based explanation of international environmental policy", *International Organization*, *48*: 77–105.

Sprout, H.H. and Sprout, M.T. (1971) *Toward a Politics of the Planet Earth*, New York: Van Nostrand Reinhold.

Stevenson, H. (2017) *Global Environmental Politics: Problems, Policy, and Practice*, Cambridge: Cambridge University Press.

Stevis, D. (2017) "International relations and the study of global environmental politics: past and present", *Oxford Research Encyclopedia of International Studies*. Online. Available HTTP: <http://oxfordre.com/internationalstudies>

Susskind, L.E. (1994) *Environmental Diplomacy: Negotiating More Effective Global Agreements*, Oxford: Oxford University Press.

Susskind, L.E. and Ali, S.H. (2014) *Environmental Diplomacy: Negotiating More Effective Global Agreements,* 2nd edn, Oxford: Oxford University Press.

Victor, D. (2006) "Toward effective international cooperation on climate change: numbers, interests and institutions", *Global Environmental Politics*, *6*: 90–103.

Westing, A.H. (1986) *Global Resources and International Conflict: Environmental Factors in Strategic Policy and Action*, Oxford: Oxford University Press.

Young, O. (1977) *Resource Management at the International Level: The Case of the North Pacific*, London: Pinter and Nichols.

Young, O. (1989) *International Cooperation: Building Regimes for Natural Resources and the Environment*, Ithaca, NY: Cornell University Press.

Young, O. (1994) *International Governance: Protecting the Environment in a Stateless Society*, Ithaca, NY: Cornell University Press.

Young, O., King, L. and Schroeder, H. (eds) (2008) *Institutions and Environmental Change: Principal Findings, Applications and Research Frontiers*, Cambridge, MA: MIT Press.

3

Mainstream theories

Realism, rationalism and revolutionism

John Vogler

International Relations (IR) as a distinct discipline dates from the immediate aftermath of World War I. Understandably, its preoccupation was, and remains, the problem of war and the achievement of security in what is often described as an "anarchic" system of sovereign states (see Chapter 7). Environmental issues, whether seen as transboundary disputes or the international dimension of managing common resources, were a decidedly minority interest (Stevis 2006). The natural environment provided the context, rather than the subject, of International Relations. This situation began to change from around the time of the United Nations Conference on the Human Environment, held at Stockholm in 1972. In this issue area, as in others, scholars tended to react to changes in the world of practical politics and policymaking. In developed world societies "green" politics had begun to emerge in response to various environmental disasters and public awareness of the scope of problems, such as air pollution, that were not soluble without international action (see Chapter 30). The response by students of IR was to attempt to frame such novel issues within existing theoretical traditions and to apply the same tools that had been used to analyze cooperation in managing the global economy or negotiating arms limitation in the Cold War. It is arguable that this was a mistake, and that something rather more radical would have been more appropriate – something that placed ecology or perhaps green political theory at the center of theoretical endeavor.

IR theory may be characterized as a broad, expanding and eclectic church. Its foundations can be located in the enduring traditions of conservative, liberal and radical European political thought. Mindful of this Martin Wight (1991), a leading exponent of the English School of international theory, provided a categorization of mainstream approaches in terms of the three "Rs": *realism, rationalism and revolutionism*. Realism is very well known as a predominant theoretical approach emphasizing power relations between states in an anarchic and inherently war-prone system. Rationalism denoted a reformist and liberal tradition informed by reason. Under this heading one may find liberals, internationalists and "idealists" – a characterization invented by realist antagonists that has proved remarkably resilient. The third category contains radical critics of the existing international system. Prominent among them are scholars working within a Marxist tradition who have developed a distinctive alternative to the liberal mainstream (see Chapter 4). Inevitably, when confronted with actual scholarship,

DOI: 10.4324/9781003008873-5

there are many ways in which the categories blur and overlap, but there are also some key distinctions.

The first concerns the ontological bases of theory – that which is held to exist. Whether, for example, the state or global class relations constitute the fundamental reality for theorists. In tandem with this we may also pose epistemological questions about how the various theoretical traditions claim to be able to know about reality. Here, there are important distinctions between those who follow the disciplines of social science seeking to find regularities and explain variations through the objective study of empirical evidence and those, like social constructivists or members of the English School, for whom interpretation rather than "positivistic" explanation is key (see Chapter 5). In discussions of IR theory this distinction is often presented as being between positivists or rationalists and "reflectivists." Positive IR remained the dominant approach in the United States, but even from the 1960s it has been subject to attack from various strands of" reflectivist" thought from diverse positions, including Marxist-inspired structuralism, critical theory and post-structuralism, not to mention the earlier resistance of the English School to the behavioral trend in US scholarship.

Then there are normative questions that address the purposes of theory. Usually in the study of International Relations there is such a purpose beyond a simple commitment to objective scholarship. Students of international environmental politics have frequently aimed to solve or manage problems through international cooperation. For them the ultimate test of effectiveness is whether the institutional or other arrangements devised serve to redress degradation or promote environmental quality (see Chapters 8 and 9). This was the specific intention of many research programs and of the chairman of the 1992 UN Conference on Environment and Development, who spoke of the "inescapable" need for international cooperation and, in advance of more recent discussions of the topic, called for "a world system of governance" (Maurice Strong cited in Haas et al. 1993: 6). In this way the problematic was devised beyond the academy and translated into the following, frequently referenced, formulation: "Can a fragmented and often highly conflictual political system made up of over 170 sovereign states and numerous other actors achieve the high (and historically unprecedented) levels of cooperation and policy coordination needed to manage environmental problems on a global scale?" (Hurrell and Kingsbury 1992: 1). The possibility and extent of such cooperation is something that has divided realists and liberal internationalists. For the latter the prospect of absolute gains for all parties through environmental policy cooperation is apparent, while realists have assumed that, in terms of national interests, gains will always be seen as relative (Grieco et al. 1993). However, the damage to national societies potentially inflicted by phenomena such as rapid climate change (see Chapter 32) may be such as to override such distinctions.

Realism

The realist tradition continues to animate popular and academic study of International Relations but it has had only limited impact upon the specialism of international environmental politics. One reason for this is that it tended to define the latter's subject matter out of existence. Environmental issues were for realists matters of "low politics" and the proper subject of IR was constituted by the "high politics" of statecraft, war and peace. Realists assert the primacy of the state which is assumed to pursue its national interest, famously reduced by Hans Morgenthau (1948) to the pursuit of power, but for most writers defined as the protection of its territorial integrity and the achievement of economic security and other central objectives of the state (see Chapter 7). The natural environment was, therefore, significant not in itself, but in terms of resource competition between states. Such competition is conducted within an anarchic "self-help" system where the resort to force is an ever-present possibility. Thus, the preoccupation of realist thinkers has been with the management of power balances and the achievement of some kind of order in a world of conflict. A key realignment of realist thought was inspired by the "structural" theory of Waltz (1979). Neorealism, which shares some important characteristics with neoliberal approaches (discussed below), sought to provide a parsimonious and testable theory of international power politics based upon a rational choice model of the way that any state would behave within an anarchic structure, leaving little room for factors such as discussion of the nature of national interest or prestige that had interested "classical" realists.

Realism provides one hypothesis that would be relevant to explanations of international environmental cooperation. This is the "hegemonic stability thesis," developed to account for the circumstances under which international economic cooperation could occur. It followed from realist postulates that self-interested states would only subject themselves to international rules if they were enforced by a dominant hegemon. The occupant of this role for much of the twentieth century was the United States and there was much concern from the 1970s onwards as to the future of world economic regulation, once US dominance began to erode. For students of international environmental politics this did not appear to be a plausible, still less a desirable, thesis. From the late 1980s, through the period of major construction of international environmental agreements, the United States was either absent or frequently obstructive. Major Conventions such as that for Biodiversity and even the Law of the Sea remained unratified by the US Senate. The Kyoto Protocol was opposed and in 2017 the Trump administration rejected the Paris Agreement of the United Nations Framework Convention on Climate Change (UNFCCC) that had been specifically designed to facilitate US participation. It would be more accurate to re-formulate the idea in a negative way where major states such as the United States and China exercise a veto rather than a leadership role and to observe that coalitions of the powerful typically come together to advance their own interests. This has been a recurring theme in discussions of international climate politics and notably the Copenhagen Accord of 2009 (Brenton 2013).

Many of the issues and policies covered in this volume would be considered "functional," "low politics" and thus beyond the pale of realist analysis. Yet the prominent example of climate change should alert us to the fact that a high/low politics distinction may no longer be tenable. The inextricable linkage of climate policy to energy production and issues of economic growth place it is close to the heart of national interests (see Chapter 33). Although many environmental questions may still be regarded in terms of Arnold Wolfer's (1962) "milieu" as opposed to "possession" goals, this is clearly not the case for countries threatened with the immediate existential threats of actual climate change and a widening set of related environmental problems.

There is a connection to older traditions of geopolitical analysis which centered on struggles over territorial space and resources (a distinction should be made between the established meaning of geopolitics as a form of spatial determinism and contemporary usage that employs the term as a synonym for international power politics). Geopolitics, as outlined by such scholars as Sir Halford Mackinder had clear associations with realist power political analysis. In geopolitical writing the emphasis was always on resource conflict rather than the environment *per se* although prominent political geographers Harold and Margaret Sprout (1971) managed to move on to the consideration of international environmental politics. As Stevis (2006: 20) notes, geopolitics was the predecessor of the contemporary environmental conflict and security research agenda.

The study of environmental security has produced an extensive literature over the last 30 years (Barnett 2001; Swatuk 2006; Guy et al. 2020 see also Chapter 20). It is conventionally viewed through a realist lens focused upon the relationship between environmental change and armed conflict – whether war or insurgency. Although the precise causal mechanisms are notoriously hard to pin down (Gleditsch 1998, Homer-Dixon 1999), climate change has been a preoccupation of military planners for some time and appears routinely in strategic assessments as a "threat multiplier" (Lietzmann and Vest 1999, Guy et al. 2020). This is not simply a matter of assessing potential threats associated with Intergovernmental Panel on Climate Change (IPCC) climate change scenarios projecting desertification and sea-level rise but also, in a thoroughly realist vein, of the strategic opportunities that may arise (Schwartz and Randall 2003).

It is not only the military establishment that has displayed an interest. Others may wish to raise the profile of environmental problems by "securitizing" them. (The reference here is to the Copenhagen School whose approach to the study of security highlighted the way in which political "speech acts" served to increase the salience of a particular policy by associating it with the potent idea of security (Buzan et al. 1998). For example, attempts to focus governmental attention and resources on, say, climate change would describe it as a security threat greater than that posed by terrorists (King 2004; see Chapter 28).) In 2007 and again in 2011 UN Security Council Resolutions on climate change were introduced by the United Kingdom and Germany, in order to advance collective action on greenhouse gas mitigation. They were unsuccessful because most UN members felt that this was a diversion from the proper forum for climate discussion which was the UNFCCC.

Attempts to broaden the concept of security and to shift its locus from armed conflict and relatively orthodox threats to the state have generally been resisted by realists skeptical of securitization moves that would change the referent object to individual human beings or even the planetary biosphere. However recent realist writing, while continuing to assert the primacy of the nation-state (in explicit contrast to the globalist views of liberal rationalists), has downgraded the impact of potential armed conflict arguing that the threat to Western states is "not of disappearing permanently beneath the waves" or of direct attack or resource conflict but "...of experiencing a steep decline in the legitimacy of liberal democratic orders due to a variety of factors that climate change is sure to worsen" (Lieven 2020:14).

Rationalism

The use of the term rationalism can cause confusion. It does not in this instance refer to procedural rationality of the sort that is to be found in the rational choice models employed by both realists and their opponents. Rather, the sense is that rationalists have a reasoned approach in contrast to the brutalities of power politics or the excessive zeal of those who would overturn the existing system. At the core of the rationalist tradition in IR are conceptions that can be traced back at least to Grotius, founding father of modern international law (see Chapter 10). States do not exist in a perpetual Hobbesian "war of all against all" but are capable of developing shared norms and practices that can ameliorate their condition and even develop the rights of their citizens. Classical rationalist thinkers were preoccupied by the problems of war, but the general approach has inspired the majority of studies of international environmental politics. Their progressive endeavor is to manage and resolve common problems by rationally oriented and multilateral action without the expectation of revolutionary transformation of the international system.

Liberalism and neoliberalism

Liberalism as a political and economic theory has diverse roots in the English constitutional and religious struggles of the seventeenth century and in the European enlightenment of the eighteenth. Its appeal is to the rights of the individual, the limitation of government powers and the importance not only of free association, but also of free markets. In IR it has been reflected in a progressive belief in reform of the states-system. One version is "democratic peace theory" positing that war and peace depend upon the nature of particular states, while another powerful idea, traceable to the nineteenth-century Manchester School, is that there is an equation between free economic exchange across frontiers, high levels of interdependence and a stable and pacific international system. In the interwar period liberal internationalist thinkers were in the ascendant as advocates of national self-determination and the encouragement of international law and organization as the antidote to a war-prone international system. Variants of this approach came to include "functionalism" which proposed that integration across national boundaries can be achieved by low-level socio-economic cooperation that will eventually "spill over" into the transfer of political authority beyond the nation-state. Liberals have been suspicious of the state and receptive to the idea of a more pluralist and transnational world system (see Chapter 14). This, coupled with a strong belief in the efficacy of free trade for the production of both wealth and political stability, meant that, in the aftermath of the Cold War, liberalism became the dominant ideology that both celebrated and justified the spread of economic globalization. The protection of the natural environment did not figure largely in liberal thinking. Indeed, critics will point out that liberal economics, in its encouragement of the rise of consumer capitalism, bears a major responsibility for the degradation of nature associated with economic growth. The liberal response is that free markets will provide the optimal allocation of resources in terms of efficiency and sustainability if only the environmental costs of human activity (externalities) are properly taken into account in transactions.

The fact that this does not occur and that state authorities fail to coordinate their activities in a rational way, beneficial to all in the longer term, provides a key to understanding liberal approaches to international environmental issues. Much of the intellectual inspiration for such thinking in IR derives from a preceding concern with running the international economy in the face of counterproductive "neo-mercantilist" behavior by governments. In fact proponents of liberal political economy admit that markets in themselves would not operate properly without a framework of rules. Thus governments should be encouraged to cooperate in what was assumed to be their underlying collective interest – as they had at the end of World War II with an economic settlement that put in place the Bretton Woods monetary order and the GATT/WTO free trade regime – a critical enabler of globalization. When environmental issues achieved wider salience during the 1980s liberal analysts were able to tap into existing work on the conditions required for international economic cooperation. (There were some exceptions, such as Oran Young, who had already begun to study international environmental cooperation in the preceding decade.) They adopted many of the assumptions of neoclassical economics in the study of what were defined as collective action problems. In fact it is quite difficult to distinguish between work that can be classified as IR and that which presents an essentially economic analysis.

Economists have performed extensive research not only on the viability of instruments such as emissions trading, but also into the functions of international agreements and the conditions under which they occur (Barrett 2003). Atmospheric quality was, for example, conceptualized as a global public good and climate change was described as "the greatest example of market failure we have ever seen" (Stern 2007: 1). In economic theory, public goods cannot be provided by the operation of the market and this affords a justification for cooperation between governments to ensure their supply. Key assumptions of this type of approach included the notion of rational, utility-maximizing actors who would take strategic decisions to cooperate if the incentives were right. Game theory provided a set of relevant models for such bargaining and in particular the "prisoner's dilemma" game in which actors need to overcome their mutual distrust in order to enjoy the gains available from cooperation.

Associated with this was the need to overcome the "free-rider problem" posed by actors who may profit from agreements without contributing to them. An awareness of this possibility was assumed to be a major disincentive to potential participants in an agreement (Stern 2007). The epistemological stance of these scholars of international cooperation, often confusingly referred to as "neo-liberals," was also closely aligned with mainstream economics (the neorealist confrontation with liberal critics is often referred to as the "neo-neo" controversy). Although the term neoliberal is used to denote scholars who adopted many of the assumptions of their counterparts in economics there are definitional problems. Neoliberalism has a conventional political meaning that encompasses the right-wing political philosophy of Hayek and the influential monetary analysis of Milton Freedman. Such ideas advocating a reduced role for the state and the primacy of the private sector achieved an intellectual hegemony over government thinking in the West from the 1980s that lasted through to the first decades of the twenty-first century. Some of the intellectual underpinnings of both types of neoliberalism may be similar but many of those who might be defined as neoliberal in the IR literature would certainly reject the political and economic program of neoliberalism.

Scholarship in IR sought to explain the pattern of incentives under which cooperation was possible between self-interested actors. In some ways this involved a simplification because states became the focus of analysis and other liberal preoccupations, for example with a plurality of international actors and with transnational relations (Mansbach et al. 1976), were neglected. It was sometimes observed that the difference between neorealists and liberals had been narrowed to such an extent that all that divided them was the disagreement over relative or absolute gains (Lamy 2011: 123–125).

Regimes and liberal institutionalism

It is no exaggeration to say that the mainstream position in the study of international environmental cooperation is liberal institutionalism. While sharing many of the economistic assumptions discussed above, institutionalists understand that economic activity and international cooperation necessarily occur within a framework of rules and understandings (Young 1989). This had long been the province of international law and organization but institutional theorists in IR developed the new concept of an international regime, initially in the study of the regulation of the international economy. Regimes were seen as institutions in the sociological sense of the word. They were defined as sets of norms, principles, rules and decision-making procedures around which actor expectations and behavior would converge in a given issue area (Ruggie 1975; Krasner 1983; Young 1989). International law and international organizations (often referred to as institutions in established usage) were only constituent parts of this broader concept which was designed to analyze the less formal understandings upon which cooperation was built (see Chapters 8–10). In contrast to realist analysis, regimes were seen to have an independent impact upon the calculations of governments. Also, they provided a means whereby "cooperation under anarchy" was possible without the leadership of a hegemonic power. As so often in IR theorizing, there was a real-world issue driving these concerns: the presumed loss of US hegemony following the ending of the dollar standard in 1971 and alarm at the consequent unraveling of the global monetary order. Liberal analysts argued that cooperation and stability could be achieved "after hegemony" (Keohane 1984).

Regime analysis was readily adapted to the study of international environmental cooperation (see Chapter 9); commencing with the Long Range Transboundary Air Pollution Convention 1979 and the Vienna Convention on stratospheric ozone depletion 1985 and its renowned Montreal Protocol 1987 (see Chapter 31), the production of global environmental agreements boomed. Arguably, even though the origins of liberal institutionalist scholarship on regimes lay elsewhere, many of its major developments have been located within the environmental field (Haas et al. 1993). The approach has been social scientific, searching for patterns in the empirical evidence from numerous cases of environmental cooperation (Young and Zurn 2006) and looking to explain variance and to specify independent and dependent variables. The dependent variables have been: the setting up of environmental regimes, the extent of agreement and levels of compliance and effectiveness – ultimately in the solution or amelioration of environmental problems (see Chapter 9).

At the beginning of the study of environmental regimes the question most frequently posed was the same as that posed by the economists – under what circumstances can cooperation occur (Young 1997)? From the extensive study of cases there were various answers. Perceived mutual vulnerability and a continuing interest in arrangements that safeguarded rights to use the global commons would provide one explanation. The "geometry" of agreements has been a significant theme with the proposition that small "clubs" of interested countries are likely to make most progress (Victor 2011). The continuing success of the Antarctic Treaty regime with its selective membership and the way in which the Montreal Protocol rested upon agreement among a relatively small group of chemical manufacturing companies would lend weight to this proposition. The work of Oran Young (1989 and 1994) has been preeminent in establishing the more precise dynamics of the "institutional bargaining" that underlies regime creation when consensus is required. Young presents a series of hypotheses on the conditions of success, including the absence of a specified zone of agreement and the presence of uncertainty. Other factors include the need to find solutions that are regarded as equitable as well as enforceable. External shocks increase the possibility of success and entrepreneurial leadership is a necessary condition (Young 1994: 81–116). This question of effective leadership has been extensively pursued in the literature (Andresen and Agrawala 2002; Wurzel et al. 2017). Leadership that can mobilize far-sighted international action is significant because, despite the construction of hundreds of international environmental agreements over the past decades, most of the indicators show an alarming and continuing degradation of Earth's natural systems. The underlying problem that students of international environmental cooperation have to address is, not so much the absence of international agreements, but their tendency to revert to the lowest common denominator – as formulated in Underdal's (1980) "law of the least ambitious programme." In 2012 leading scholars in the field called for "a 'constitutional moment' in the history of world politics, akin to the major transformative shift in governance after 1945" (Biermann et al. 2012: 7) Since then there have been new agreements, reviving the climate regime in Paris, establishing the UN Sustainable Development Goals, regulating mercury or amending the Biodiversity Convention (see Chapter 41), but none that could be regarded as responding to this appeal.

An important and problem-focused part of liberal scholarship investigates institutional design. This is also the province of international lawyers and covers such issues as the circumstances under which "soft law" may provide more effective solutions than a comprehensive binding agreement (see Chapter 10). Regime effectiveness has been a major focus (Victor et al. 1998) alongside related studies of appropriate policy instruments, whether "command and control" (environmental taxation) or "market based" (emissions trading). There is a recurrent conflict here between neoliberals and their opponents as evidenced by the long-running dispute over Article 6 of the Paris Agreement.

The focus upon regimes has broadened out in a number of ways (see Chapter 9) including an ambitious rebranding of institutional studies as "earth system governance" or "global environmental governance" (see Chapter 21). Sometimes in official discourse this can mean little more than a reconfiguration of existing international organizations. Witness the long-running debate on whether to raise the status of the United Nations Environment Programme (UNEP) to a specialized agency or to create a UN environment council alongside the Security and Human Rights Councils. However, in the academic world, notions of global governance denote a move away from the state-centric focus of earlier regime analysis and a recognition of the need to consider different levels of appropriate environmental governance and to include transnational actors such as NGOs, which had always received significant attention in the IR literature (Princen and Finger 1994; Keck and Sikkink 1998), and to embrace the possibility that the private sector could provide significant governance alongside or even instead of nation-states (Pattberg 2007). Frustration at the slow progress of the inter-state Climate Convention has served to accelerate this tendency (Bulkely et al. 2014). It marks a return to several key themes in liberalism that tended to be crowded out by previous attempts at parsimonious explanation through the assumption of rationally calculating unitary state actors. Furthermore, it revives a normative dimension that, in line with classical liberalism, is distrustful of the state and the possibility that it might be "greened."

Variations on the rationalist theme: cognitivism and the English School

Mirroring neoclassical economic theory, liberal institutionalist analysts tended not to delve within the state but rather to assume a set of fixed preferences. Behavioral economists have questioned this lack of interest in preference formation and in the study of international environmental politics this has been a long-standing critique. A distinct "cognitivist" approach to the understanding of regimes was evident from around 1990 and the publication of Haas's (1990) work on the Mediterranean pollution regime. Critical inquiry into the supposed linear relationship between authoritative science and policy formulation began to open up the "black box" of national policy positions, pointing out the significance of shifting discourses (Litfin 1994). Cognitive approaches to regime formation betray the influence of "reflectivist" IR and the rising interest in social constructivism (Wendt 1999). In one respect this was a challenge to liberal institutionalist orthodoxy because of its explicit rejection of the rational choice model of human behavior in favor of alternative "logics of appropriateness." Added to this is a critique of assumptions about objective natural "fact." Supposedly "positive" natural science may be seen as subject to social construction (see Chapter 18).

While such critiques must raise questions about the validity of existing liberal scholarship, attempts have been made to incorporate them into institutionalism. Regimes may themselves be viewed as social constructs with a shifting ideational and constitutive character. In some versions of constructivism there is no necessary contradiction with the epistemology of social scientific inquiry and therefore with mainstream liberal institutionalism. Alternatively, a constructivism that seeks understanding of normative evolution rather than strict explanation would seem to align both ontology and epistemology in the study of regimes (Vogler 2003). The extent to which rational choice and "reflectivist" approaches are commensurable remains one of the most disputed questions in contemporary IR theory (Smith and Owens 2008). (This is sometimes referred to as the "rationalist–reflectivist" debate. Rationalist is a shorthand for rational choice and does not refer to Wight's rationalist category used here in this chapter.)

The English School may be characterized as a "reflectivist" approach to international society. Adherents have adopted perspectives that are in some ways coincident with more recent constructivist theorists in their concern for the constitution and re-constitution of the institutions, such as sovereignty and diplomacy or, indeed, international society itself (note the distinctive definition of institutions; see Chapters 9 and 22). The English School was founded upon a rejection of the epistemological stance of US "behavioral" International Relations scholarship of the 1960s, advocating instead a more historically based interpretative approach. Its concerns with the deeper norms of an international society of states would certainly be relevant to global environmental politics, but the attention of most of its adherents was elsewhere upon the problems of war, international order and human rights. There are indications that this neglect is now being remedied for, as Falkner (2012: 509) argues, "In contrast to both realism and neo-liberal institutionalism, the English School offers a rich account of the institutional phenomena that define the durable patterns of and historically bound character of international society." This could encompass key normative issues, relating to climate justice and differentiated responsibilities that lie at the heart of international climate politics (see Chapter 26). Similarly, it provides the scope to engage with some classic concerns of statecraft, such as the pursuit of recognition and prestige, which seem to be as significant as material interests in the conduct of climate diplomacy (Vogler 2016). It is, in this respect, significant that the "pledge and review" mechanism whereby the Paris Agreement seeks to encourage greater national effort in emissions reduction relies very much on a competition for national esteem. Finally, the English School has been the site of a long running debate about the extent to which the international system has moved from a "pluralism" of antagonistic states toward "solidarism," an institutionalized concern with the reduction of mutual harm, which provides a relevant paradigm for the assessment of the evolution of global environmentalism (Falkner 2018).

Revolutionism

Alongside realism and rationalism, Martin Wight identified a revolutionist tradition in international thought. Some of those in this category, Marxist and socialist writers, had an explicitly revolutionary purpose, but others had less developed aspirations for the transformation of the inter-state system into a more congenial and pacific world system in which both individuals and communities would live in a greater degree of freedom and harmony. The unifying strand that is present in all of this work is a rejection of the status quo and with it the kind of international order that realists accepted as inevitable and rationalists sought to reform and ameliorate (see Chapter 4). Typically, sovereign states are viewed as part of the problem rather than potential promoters of a more cosmopolitan and ecologically sustainable world (see Chapter 16). To use Cox's (1981) terminology, "problem-solving" theory is the domain of rationalism, while revolutionists are "critical theorists."

The prominent critical approach has relied upon a Marxist historical-materialist tradition in which the state, far from being the center of analysis, performed as the agent of a ruling class – "the executive committee of the bourgeoisie." International politics, and in particular the imperialist struggles that characterized the contemporary epoch, were to be understood in terms of the deeper underlying contradictions of the capitalist mode of production.

As with other twentieth-century brands of IR theory, very little attention was paid to the natural environment until the final years of the decade when the relationship between capitalist accumulation, globalization and the degradation of Earth's natural systems began to crystallize (see Chapter 24). Because Marxist analysis seeks explanation through the ways in which an ever-changing system of capital accumulation determines economic activity that is fundamentally responsible for excessive resource use, loss of habitats and rising levels of pollution, it provides a powerful account of the global ecological predicament (Paterson 2001). In particular it directly challenges liberal market-based orthodoxies on solving global environmental problems and achieving justice for the dispossessed (see Chapters 26 and 28). Marxist structural analysis denies that environmental issues can be portrayed as a collective action problem between states. International regimes and schemes of global environmental governance are "epiphenomenal." They may serve a number of functions for the global capitalist system but they are a reflection of it rather than a means to ensuring that it will be less environmentally destructive. Thus, for example, the problem of climate change cannot be dealt with through the elaboration of the UN climate regime but rather through more fundamental alterations in the nature of the capitalist growth model that will provide incentives to decarbonize the global economy (Newell and Paterson 2011).

Scholarship in the Marxist tradition has often adopted a neo-Gramscian position. Gramsci has been an inspiration because his writings pay attention to the ways in which the material base and the social superstructure combine in a "hegemonic" process to manufacture consent for a prevailing order even among those whose interests would "objectively" be opposed to it (Levy and Newell 2005). There is some overlap here with a range of other literature that relocates the ecological problem beyond the structures of the existing international system and examines the implications of incorporating green political thought (Saurin 1996; Laferrière and Stoett 1999; Eckersley 2004). The supposed impossibility of reform in the face of escalating environmental crises has produced a new strain of writing that contemplates "deep adaptation" in the face of ecological and social collapse (Bendell 2018). Implications for International Relations have hardly begun to be explored, but the worry must be that they lead back to the Hobbesian world of realist imaginings.

Conclusion

In the year following the 1992 Rio Earth Summit, Steve Smith (1993) took an outsider's view of the emerging field of international environmental politics. His conclusion was that it remained "at the periphery" dominated by a liberal institutionalist orthodoxy and immune to the theoretical cross-currents so evident elsewhere in IR. The rationalist project is still prominent and strives for cumulative, evidence-based "positive" knowledge which has policy relevance to the tasks of global environmental governance. However, it is hardly alone. Critical and "revolutionist" writing has been ever present and there has been increasing interest in applications of constructivist and other theoretical approaches to global environmental problems (see Chapter 4).

Interaction between actual world politics and academic writing is an inevitable and desirable feature of IR. The darkening international situation and the apparent decline of multilateralism since the world economic crisis of 2008 have undermined confidence in rationalist projects in the face of resurgent nationalisms, renewed great power competition and seismic shifts in the structure of the global economy. How far this has been reflected in international environmental politics is difficult to discern, but the evolution of the climate regime away from the Kyoto Protocol model toward the nationally based "contributions" of the Paris

Agreement serves as an indicator. Certainly, realists have begun to take the strategic and other implications of environmental change more seriously. This may not be an entirely negative development, if as Lieven (2020) argues, it could point to ways in which national sentiment could be harnessed to the achievement of environmental security where liberal cosmopolitan sentiments have failed to deliver political action.

Recognition of the increasing political salience of the climate and related crises has meant that what might, on occasion, have been seen as technical and functional issues have been increasingly politicized or even "securitised." E.H.Carr, who penned the most celebrated realist critique of idealism, made a distinction between the purely "functional" and the "political." Not all of the business between states was "political" and he gave the example of postal services and the suppression of epidemics. From the perspective of 2020 his words are striking: "But as soon as an issue arises which involves, or is thought to involve, the power of one state over another the matter becomes political" (Carr 1939:102). As it has been with the pandemic so it has and will be with the global environment.

References

Andresen, S. and Agrawala, S. (2002) "Leaders, pushers and laggards in the making of the climate regime", *Global Environmental Change*, 12: 41–51.

Barnett, J. (2001) *The Meaning of Environmental Security: Ecological Politics and Policy in the New Security Era*, London: Zed Books.

Barrett, S. (2003) *Environment and Statecraft: The Strategy of Environmental Treaty-Making*, Oxford: Oxford University Press.

Bendell, J. (2018) *Deep Adaptation: A Map for Navigating Climate Tragedy,* University of Cumbria IFLAS Occasional Paper 2. Available online at: www.lifeworth.com/deepadaptation.pdf.

Biermann, F. et al. (2012) "Transforming governance and institutions for global sustainability: key insights from the Earth System Governance Project", *Current Opinion in Environmental Sustainability*, 4: 1–10. Available online at www.sciencedirect.com.

Brenton, T. (2013) "'Great powers' in climate politics", *Climate Policy*, 13(5): 541–546.

Bulkely, H. L., Andonova, M., Betsill, D., Compagnon, D., Hale, T., Hoffmann, M.J., Newell, P., Paterson, M., Roger, C. and Van Deever, S.D. (2014) *Transnational Climate Change Governance*, Cambridge: Cambridge University Press.

Buzan, B., de Wilde, J. and Waever, O. (1998) *Security: A New Framework for Analysis*, Boulder, CO: Lynne Rienner.

Carr, E.H. (1939) *The Twenty Years Crisis, 1919–39: An Introduction to the Study of International Relations*, London: Macmillan.

Cox, R. (1981) "Social forces, states and world orders: beyond international relations theory", *Millennium: Journal of International Studies*, 10(2): 126–155.

Eckersley, R. (2004) *The Green State: Rethinking Democracy and Sovereignty,* Cambridge MA: MIT Press.

Falkner, R. (2012) "Global environmentalism and the greening of international society", *International Affairs*, 88(3): 503–522.

Falkner, R. (2018) "International climate politics between pluralism and solidarism: an English School perspective" in O. Corry and H. Stevenson (eds) *Traditions and Trends in Global Environmental Politics; International Relations and the Earth*, London: Routledge, 26–44.

Gleditsch, N.P. (1998) "Armed conflict and the environment: a critique of the literature", *Journal of Peace Research*, 35(3): 381–400.

Grieco, J., Powell, R., and Snidal, D. (1993) "The relative-gain problem for international cooperation", *American Political Science Review*, 87(3): 727–743.

Guy, K. et al. (2020) *A Security Threat Assessment of Global Climate Change: How Likely Warming Scenarios Indicate a Catastrophic Security Future*, Washington, DC: Council on Strategic Risks.

Haas, P.M. (1990) "Obtaining environmental protection through epistemic consensus", *Millennium: Journal of International Studies*, 19(3): 347–363.

Haas, P.M., Keohane, R.O. and Levy, M.A. (eds) (1993) *Institutions for the Earth: Sources of Effective International Environmental Protection*, Cambridge, MA: MIT Press.

John Vogler

Homer-Dixon, T. (1999) *The Environment, Scarcity and Violence*, Princeton, NJ: Princeton University Press.

Hurrell, A. and Kingsbury, B. (eds) (1992) *The International Politics of the Environment: Actors, Interests and Institutions*, Oxford: Clarendon Press.

Keck, M.E and Sikkink, K. (1998) *Activists beyond Borders: Advocacy Networks in International Politics*, Ithaca, NY: Cornell University Press.

Keohane, R. (1984) *After Hegemony: Cooperation and Discord in the World Political Economy*, Princeton, NJ: Princeton University Press.

King, D.A. (2004) "Climate change science: adapt, mitigate or ignore?", *Science*, 303: 176–177.

Krasner, S.D. (ed.) (1983) *International Regimes*, Ithaca, NJ: Cornell University Press.

Kütting, G. (2004) *Globalization and Environment: Greening Global Political Economy*, Albany: SUNY Press.

Laferrière, E. and Stoett, P.J. (1999) *International Relations Theory and Ecological Thought: Towards a Synthesis*, London: Routledge.

Lamy, S. (2011) "Contemporary mainstream approaches: neo-realism and neo-liberalism", in J. Baylis, S. Smith and P. Owens (eds) *The Globalization of World Politics*, 5th edn, Oxford: Oxford University Press, 114–27.

Levy, D.L. and Newell, P.J. (2005) "A Neo-Gramscian approach to business in international environmental politics: An interdisciplimary, multilevel approach", in L.Gasser and N.Choucri (eds) *The business of global environmental governance*, Cambridge Mass.: The MIT Press, 47–69.

Lietzmann, K.M. and Vest, G. (eds) (1999) *Environment and Security in an International Context*, Brussels: NATO Committee on the Challenges of Modern Society, no. 232.

Lieven, A. (2020) "Climate change and the state: a case for environmental realism", *Survival*, 62(2): 7–26.

Litfin, K. (1994) *Ozone Discourses: Science and Politics in Global Environmental Cooperation*, New York: Columbia University Press.

Mansbach, R., Ferguson, Y. and Lampert, D. (1976) *The Web of World Politics*, Englewood Cliffs, NJ: Prentice-Hall.

Morgenthau, H.J. (1948) *Politics among Nations: The Struggle for Power and Peace*, New York: A.A. Knopf.

Newell, P. and Paterson, M. (2011) *Climate Capitalism: Global Warming and the Transformation of the Global Economy*, Cambridge: Cambridge University Press.

Paterson, M. (2001) *Understanding Global Environmental Politics: Domination, Accumulation, Resistance*, Houndmills: Palgrave.

Pattberg, P.H. (2007) *Private Institutions and Global Governance: The New Politics of Environmental Sustainability*, Cheltenham: Edward Elgar.

Princen, T. and Finger, M. (eds) (1994) *Environmental NGOs in World Politics*, London: Routledge.

Ruggie, J.G. (1975) "International responses to technology: concepts and trends", *International Organization*, 29(3): 557–583.

Saurin, J. (1996) "International relations, social ecology and the globalisation of environmental change", in J. Vogler and M. Imber (eds) *The Environment and International Relations*, London: Routledge, 77–98.

Schwartz, P. and Randall, D. (2003) *An Abrupt Climate Change Scenario and its Implications for United States National Security*, San Francisco, CA: Global Business Network for the Department of Defense.

Smith, S. (1993) "The environment on the periphery of international relations: an explanation", *Environmental Politics*, 2(4): 28–45.

Smith, S. and Owens, P. (2008) "Alternative approaches to international theory", in J. Baylis, S. Smith and P. Owens (eds) *The Globalization of World Politics: An Introduction to International Relations*, 4th edn, Oxford: Oxford University Press, 174–191.

Sprout, H. and Sprout, M. (1971) *Towards a Politics of the Planet Earth*, New York: van Nostrand Rheinhold.

Stern, N. (2007) *The Economics of Climate Change: The Stern Review*, Cambridge: Cambridge University Press.

Stevis, D. (2006) "The trajectory of the study of international environmental politics", in M.M. Betsill, K. Hochstetler and D. Stevis (eds) *Palgrave Advances in International Environmental Politics*, Houndmills: Palgrave Macmillan, 13–54.

Swatuk, L.A. (2006) "Environmental security", in M.M. Betsill, K. Hochstetler and D. Stevis (eds) *Palgrave Advances in International Environmental Politics*, Houndmills: Palgrave Macmillan, 203–236.

Underdal, A. (1980) *The Politics of International Fisheries Management: The Case of the Northeast Atlantic*, New York: Columbia University Press.

Victor, D.G. (2011) *Global Warming Gridlock: Creating More Effective Strategies for Protecting the Planet*, Cambridge: Cambridge University Press.

Victor, D.G., Raustiala, K. and Skolnikoff, E. (eds) (1998) *The Implementation and Effectiveness of Environmental Commitments*, Cambridge, MA: MIT Press.

Vogler, J. (2003) "Taking institutions seriously: how regime analysis can be relevant to multilevel environmental governance", *Global Environmental Politics*, 3(2): 25–39.

Vogler, J. (2016) *Climate Change in World Politics*, Houndsmills: Palgrave Macmillan.

Waltz, K. (1979) *Theory of International Politics*, Reading, MA: Addison-Wesley.

Wendt, A. (1999) *A Social Theory of International Politics*, Cambridge: Cambridge University Press.

Wight, M. (1991) *International Theory: The Three Traditions*, Leicester: Leicester University Press.

Wolfers, A. (1962) *Discord and Collaboration*, Baltimore, MD: Johns Hopkins University Press.

Wurzel, R.K.W., Connolly, J. and Liefferink, D. (eds) (2017) *The European Union in International Climate Change Politic: Still taking a lead?*, Abingdon: Routledge.

Young, O.R. (1989) *International Cooperation: Building Regimes for Natural Resources and the Environment*, Ithaca, NY: Cornell University Press.

Young, O.R. (1994) *International Governance: Protecting the Environment in a Stateless Society*, Ithaca, NY: Cornell University Press.

Young, O.R. (ed.) (1997) *Global Governance: Drawing Insights from the Environmental Experience*, Cambridge, MA: MIT Press.

Young, O.R. and Zurn, M. (2006) "The international regimes database: designing and using a sophisticated tool for institutional analysis", *Global Environmental Politics*, 6(3): 121–141.

4

Alternative theories

Constructivism, Marxism, critical theory and feminism

Hayley Stevenson

This chapter surveys a diverse set of theoretical approaches that scholars draw upon to study global environmental politics. What unites these scholars is their dissatisfaction with the treatment of this subject by traditional theories of international relations (IR) (see Chapter 3). Traditional IR has approached environmental problems with the same set of interests, theoretical assumptions and methodological tools that they bring to the study of any other problem in international politics (Hovden 1999). This chapter provides an overview of efforts to understand global environmental politics by venturing off the beaten track. The theories reviewed in this chapter offer different metatheoretical lenses: they "see" different things in the world (ontology) and force us to reconsider how we can understand the world (epistemology).

For some, the ontological assumptions held by traditional IR scholars are problematic and generate incomplete understandings about why environmental problems occur and how they can be overcome. "Ontology" is the theory of being; it concerns the nature of the world – its essence, boundaries and constitutive units. Traditional IR is based on a rationalist ontology, which assumes that states are unitary rational actors who interact on the basis of strict cost-benefit calculations to maximize their relative or absolute power. Power is understood in a purely material and coercive sense of one actor's ability to get another actor to do something they otherwise would not. The approaches in this chapter rest on different ontologies, which recognize the presence and significance of actors other than states, the socially constructed identities and interests of these actors, other forms of power and domination, and forms of non-instrumental rationality.

For some scholars discussed in this chapter, the epistemological commitments of traditional IR are problematic and counterproductive to the aim of ameliorating environmental degradation. Epistemology is the theory of knowledge; it concerns the potential for acquiring knowledge of the world, as well as the possible validity of this knowledge. Traditional IR is informed by a positivist epistemology, which demands that scholars approach their study of the social world as they would the natural world; namely, with the aim of generating general laws by identifying patterns of relationships across directly observable phenomena. These general laws then provide a basis for making predictions about the social world. The approaches presented in this chapter tend to analyze the power of norms, domination, discourse and patriarchy, which are contextual and not directly unobservable. These phenomena do

DOI: 10.4324/9781003008873-6

not manifest uniformly across all times and places; therefore, it is not possible to produce general laws and predictions. Yet, this does not undermine their significance for our understanding of global environmental politics.

Another assumption of positivism that is problematic for the study of environmental politics is its fact–value separation. The positivist assumption here is that scholars can and should aim for neutrality and objectivity by minimizing the potential for their own values to infiltrate and influence analyses of how the world actually is. Hovden (1999: 59) explains that this separation is problematic because "by insisting on a separation of facts from values, social scientific enquiry… implicitly becomes supportive of the *status quo*, because there is little or no room for social criticism in a positivist social scientific inquiry." The theoretical approaches presented here all engage (if only implicitly) with one of these metatheoretical critiques, while some depart from traditional global environmental politics on both ontological and epistemological grounds.

Constructivism

Constructivism is a social theory of international relations concerned with the underlying *ideas* that shape the behavior of states and other actors. We can best appreciate the distinctiveness of this approach by understanding its ontology, based principally on the mutual constitution of structure and agency. This means that structures (i.e., norms, culture and identity) constrain and enable the behavior of actors, but that structures themselves are reproduced and transformed through actors' behavior. The "ideas" that define structures are intersubjective: they are based on shared knowledge that rests on "collective intentionality" rather than individual belief (Ruggie 1998). Agency is a property that denotes an actor's capacity to act upon situations, and to formulate and implement decisions. Intersubjective meanings constitute structures, which, in turn, constitute agents.

Understanding interests and identities requires attention to the social context in which they are formed and transformed. Intersubjective meanings are of course neither universal nor static: they are specific to certain spatial, temporal and social contexts, and may be transformed. A key concern of constructivist scholars is to understand and explain processes of change in international relations, including changes within the normative structure of global governance and changes within particular states' responses to these norms. Another key concern is understanding how environmental issues are socially constructed, such as the "Anthropocene" (Schwindenhammer 2019), and how understandings of existing problems are redefined, including through "carbonization," whereby all environmental problems are connected to climate change (Foyer et al. 2018; see Chapter 32).

Norms are typically defined, following Katzenstein (1996: 5), as "collective expectations for the proper behavior of actors with a given identity." There is some agreement among both rationalists and constructivists on the existence of norms, where they differ is in their explanations for compliance with norms. The distinction is captured by March and Olson's (1998) logics of action. Rationalists explain norm conformance as driven by the "logic of consequences" whereby actors engaged in bargaining calculate the likely consequences of alternative actions and select that which best serves their exogenously given interests (March and Olson 1998: 949). Constructivists, by contrast, invoke the "logic of appropriateness" to explain norm conformance: "(h)uman actors are imagined to follow rules that associate particular identities to particular situations, approaching individual opportunities for action by assessing similarities between current identities and choice dilemmas and more general concepts of self and situations" (1998: 951).

Many scholars have analyzed global environmental politics in terms of norms that govern states' actions (see Chapter 9). Bernstein (2001) analyzed the evolution of international environmental governance during the three decades following the 1972 United Nations Conference on the Human Environment, held in Stockholm. Bernstein sought to understand why a norm-complex of "liberal environmentalism" prevailed over alternative interpretations that attributed environmental degradation to unregulated industrialization and exponential economic growth (see Chapter 24): "Liberal environmentalism accepts the liberalization of trade and finance as consistent with, and even necessary for, international environmental protection. It also promotes market and other economic mechanisms... over 'command-and-control' methods...as the preferred method of environmental management" (Bernstein 2001: 7).

Bernstein offers a "socio-evolutionary" explanation for this normative development suggesting that norm selection is a product of "*social* fitness," or the fit between new norms and the existing social structure (2001: 20–21). More recent scholarship shows that this normative understanding of "fitness," which has come to underpin constructivist scholarship, is incomplete. Normative fit is just one dimension of governance fitness: practical fit also matters. During the twenty-first century, we have witnessed a proliferation of new environmental concepts (Meadowcroft and Fiorino 2017), including many that normatively fit with the liberal order: green economy, carbon markets, ecosystem services, etc. But in a study of the institutionalization of "ecosystem services," Stevenson et al. (2021) found that the concept floundered because it failed to provide a clear and actionable program that practitioners could follow to solve the problem of biodiversity loss.

Other constructivist studies focus on how the meanings of specific environmental norms are contested (Cass 2006; Eckersley 2007; Hadden and Seybert 2016; Hoffmann 2005; Pettenger 2007). The norm of "common but differentiated responsibilities" has been widely studied and continues to attract analytical attention (Guangyu 2020; Kiessling 2019; Stalley 2018). This norm assumes that states are responsible for protecting the global environment and that all should participate in its governance, but industrialized countries bear primary responsibility on the basis of their historical pollution and/or their greater capacity to bear the costs incurred. Nevertheless, its precise prescriptions for allocating responsibility have been contested since the early 1990s. The fact that negotiators are constrained and enabled by this normative understanding only makes sense within a constructivist framework rather than a rationalist one in which states are assumed to act only on the basis of value-free calculations of costs and benefits. Betsill (2017) argues that the decision of the Trump administration to withdraw from the Paris Agreement actually revealed the strength of a climate protection norm. Rather than prompt a widespread abdication of commitment, much of the international community responded to this decision with criticism and renewed commitment.

Much constructivist work is state-centric and analyzes how state identities affect environmental negotiations, for example, how a "South" identity (Bueno 2020) or an "island vulnerability" identity (Rasheed 2019) affect climate change negotiations. However, there are no theoretical restrictions on extending the analysis to other actors in the international system. After all, constructivism is not a *substantive* theory of international relations (as in, say, realism or liberalism) but instead a metatheory. Park (2010) adopted a constructivist framework to analyze processes of socialization within the World Bank. Rather than focus exclusively on states, Park examined the ideational power of transnational environmental advocacy networks and their efforts to bring the World Bank's policies into line with environmentalist norms. Park analyzed this process of change as one of *socialization*,: "a process whereby agents internalise norms that constitute the social structure in which they exist...[this] is not a linear process but one of continuous interaction between agents and structures…[that] can lead to fundamental shifts in an organisation's identity" (Park 2010: 8). More recently, Fuentes-George (2016) combined attention to the agency of state and non-state actors in analyzing how transnational advocacy networks use justice-based arguments to help states and other actors interpret ambiguous and contradictory rules. Others, however, argue that non-state actors such as epistemic communities have more commonly promoted scientific knowledge and interest-based arguments rather than ethical discourses in environmental governance spaces (Haas 2016; Mitchell and Carpenter 2019). Green (2018) suggests this is changing with the rise of morally based "anti-fossil fuel norms," while Blondeel et al. (2019) argue that environmental norms are mostly likely to succeed when interest and moral based arguments are combined.

Epistemologically, constructivism rejects positivist assumptions. From a constructivist perspective, our understanding of the social world will always be incomplete and potentially erroneous if it is informed only by phenomena that are directly observable. Moreover, general laws are only plausible in closed systems characterized by stability and consistency. The social world is not such a system. Constructivism is instead compatible with an interpretivist epistemology, which seeks to interpret the contextual "webs of meaning" that constitute the social world (Neufeld 1993). However, constructivist scholars do not necessarily challenge the fact–value separation that is central to positivism. Constructivism itself does not entail any normative commitments concerning *how* the world should be constructed.

Constructivist scholars who wish to critique constructions of the world must import normative commitments from political theory. Stevenson (2013) employed a "green constructivist" framework to analyze interactions between state actors and underlying social structures, as well as socially constructed interests and forms of rationality. The integration of green political theory provides a foundation for critiquing existing global climate governance in terms of its institutionalization of ecological irrationality. Others, though, are not directly concerned with critique; Bernstein, for instance, states that his purpose is to "uncover how and why liberal environmentalism became institutionalized…rather than simply offering a critique of the outcome" (Bernstein 2001: 7).

Marxism

Capitalism is the "elephant in the room" among global environmental politics scholars (Newell 2011: 4). Although capitalism is deeply implicated in global environmental change and defines the parameters of permissible responses to such change, scholars generally maintain a polite silence about this system. But not all scholars have shunned capitalism in their analyses of global environmental politics. This section reveals a small literature inspired in different ways by Marx's theory and critique of capitalism. Marxism is a broad tradition encompassing a range of philosophical and political positions that share the idea that humans' experience is fundamentally shaped by the social organization of material production. This mode of organization is understood not as ahistorical and immutable. Instead, any specific mode of organization is produced, reproduced, altered or transformed by actors assuming social identities and performing inherited structural roles. This process is captured by the term "historical materialism." Marxism has traditionally critiqued capitalism as a social arrangement for material production. For some, this entails developing an alternative mode of organizing material production, generally some version of socialism. In the context of global environmental politics, this critique manifests as analyses of capitalism's contribution to environmental degradation and as analyses of corporations' influence on international environmental policy (see Chapters 13 and 24).

It has frequently been observed that Marx and Engels either had little to say about nature or viewed it in purely instrumental terms (Lipschutz 2004: 78). Contemporary Marxists have sought to refute the former and rectify the elements of truth found in the latter (Foster et al. 2010; Kovel 2007; O'Connor 1988). Kovel (2007: 9–10) argues: "[s]ince Marx emerged a century before the ecological crisis matured, we would expect its received form to be both incomplete and flawed when grappling with a society, such as ours, in advanced ecosystemic decay."

Despite the reductionism of historical materialism, Marx and Engels understood ecology as complex interacting processes and objects that could not be understood in isolation from one another (see Merchant 2008: 44). Natural objects were understood to constitute a part of human existence and consciousness, ultimately providing the material conditions for producing subsistence. Yet, the capacity for humans to master and deliberately destroy the environment through labor was also recognized: "Let us not…flatter ourselves overmuch on account of our human conquests over nature," warned Engels, "For each such conquest takes its revenge on us" (quoted in Merchant 2008: 56). Criticism was particularly directed at capitalists for their free appropriation of natural resources, which broke the unity between man (*sic*) and nature (Merchant 2008: 52–54).

The most sophisticated account of capitalism's impact on environmental degradation that Marx developed was that concerning the soil crisis. Central to this critique was Marx's concept of "social metabolism." Metabolism itself refers to "the complex biochemical process of exchange, through which an organism…draws upon materials and energy from its environment and converts these…into the building blocks of growth" (Foster et al. 2010: 402). *Social* metabolism, then, captures "the complex, dynamic interchange [of matter and energy] between human beings and nature" (Foster 2000: 158). Marx observed that capitalism generated an unavoidable "metabolic rift" in soil nutrients by rupturing the "metabolic interaction" between humans and the earth (Foster et al. 2010: 77). During earlier times when production and consumption occurred within close proximity, crops and natural wastage were returned to the land as fertilizer, thus sustaining its nutrient base and productive capacity. The accumulative imperative of capitalism was seen to concentrate land ownership, depopulate rural areas, increase the density of urban living and ultimately create an urban–rural divide that saw soil nutrients accumulate as urban waste (Foster et al. 2010: 77).

Contemporary scholars have extended Marx's theory of the "metabolic rift" to analyze the modern global economy, which is far more ecologically damaging than anything witnessed in the nineteenth century (e.g., Foster et al. 2010; Moore 2000; Weis 2010; see Chapter 24). Foster et al. (2010) observe that the "metabolic rift" has been globalized through colonialism, imperialism and market forces that all aim to maximize capital accumulation of the core states at the expense of environmental degradation in the periphery. "Rifts" have been spread throughout the system from the application of "technological fixes," including the intensive use of artificial nitrogen fertilizer to compensate for the loss of organic soil nutrients. Foster and colleagues note that the resulting airborne nitrogen compounds contribute to global warming, while soil runoff increases the concentration of nutrients in waterways causing eutrophication and marine "dead zones" (2010: 81–82). Effectively responding to this situation requires, they argue, a complete rejection of capitalism, which is inherently anti-ecological. Within the specific realm of agriculture, industrial-scale production ought to be replaced with Marx's proposal for "a society of associated producers [who] can regulate their exchange with nature in accordance with natural limits and laws, while retaining the regenerative properties of natural processes and cycles" (2010: 86). More generally, however, the ecological crisis can only be resolved through a "revolution in the constitution of human society itself…aimed at the creation of a just and sustainable society" (2010: 38, 436).

Weis argues that "the chronic biophysical contradictions of industrial capitalist agriculture are accelerating" and leading to food price volatility and "ruinous outcomes." These contradictions are generated by the organizing logic of capitalism, which prescribes ever-greater efficiency for accumulation. Yet this logic will eventually be destabilized by the system's own externalized costs: soil erosion and salinization, depletion of water supplies, biodiversity loss and greenhouse gas emissions, as well as an "intractable dependence" on finite fossilized biomass (Weis 2010: 316–317). Yet Weis sees in this instability the potential for "rebuilding biodiverse food systems and remaking and valorizing agricultural work" (2010: 315).

The second strand of capitalist-centered critique is more explicitly inspired by twentieth-century political theorist Antonio Gramsci's historical materialism. While Gramsci shared many of Marx's assumptions about capitalist processes and relations, he maintained a stronger notion of agency: social transformation would not necessarily emerge from phases of economic development, but rather could be brought about by conscious social actors. The main concept informing Gramsci's work was "hegemony," referring to "the persistence of specific social and economic structures that systematically advantage certain groups" (Levy and Newell 2002: 86). Two different understandings of power emerge from this concept: ideological and strategic. First, power is ideological because the structures that privilege a social group owe their stability to being taken for granted as "common sense." But this also creates an opening for civil society to act strategically and engage capitalists in a "war of position" that exposes the tensions and contradictions of hegemonic projects, thereby de-reifying these and presenting an alternative social order (Levy and Newell 2002).

In the 1990s and early 2000s, several scholars published influential work using a neo-Gramscian framework (Levy and Egan 2003; Levy and Newell 2002, 2005; Newell and Paterson 1998; Paterson 1996, 2000). Paterson (1996) argued that the international politics of global warming were best explained using a historical materialist framework comprising three elements. First, recognition of the structural power of capital, which emerges from what Marxists see as a primary function of states: ensuring capital accumulation. This empowers capitalists because they have the capacity to withhold or shift investment, and to construct "hegemonic ideas concerning the conditions for economic growth" (Paterson 1996: 158). A second element is Gramsci's concept of "hegemony," which "denotes the ideological struggles which occur over the projects of the dominant class designed to secure the basic conditions for accumulation. The process of securing those conditions requires that capital engages in continual ideological struggles to create a capacity to keep capitalist societies together" (Paterson 1996: 158). Blondeel (2019) shows how campaigners have sought to "delegitimise the role of fossil fuels and the industry in society" by promoting an international fossil fuel divestment norm. Paradoxically, however, to gain maximum appeal, campaigners will have to appeal to "the positive material effects of divestment, that is if it maximises profits and minimises investment risks" (Blondeel 2019), and such arguments preserve rather than transform the liberal social order.

Newell and Paterson drew on a "neo-Gramscian" framework to explain the political–economic dynamics shaping states' positions in climate negotiations (also Levy and Newell 2002, 2005). By recognizing that states are positioned within capitalist societies, in which accumulation is driven by fossil fuels, Newell and Paterson are able to appreciate the influence of fossil fuel lobbies; yet, their power is not absolute because capital itself is not a "homogenous bloc." Newell and Paterson observe how the insurance industry was able to challenge the power of the fossil fuel lobbies by establishing "tactical alliances" with environmentalists to advance other interests (1998: 680–681). This potential was limited, however, because the interests of the fossil fuel lobbies also provide most "fractions of capital" with the basic conditions for accumulation because oil and coal are so central to the global economy (1998: 692–693).

Critical theory: from the Frankfurt School to Habermas

Critical theory has had a small but increasing impact on the study of global environmental politics. This tradition took shape in the mid-twentieth century through the work of social theorists at Frankfurt's Institute for Social Research, hence the common reference to this tradition as the Frankfurt School. The tradition is typically divided into two "generations": the first generation of Theodor Adorno, Max Horkheimer and Herbert Marcuse, and the second generation led by Jürgen Habermas (Biro 2011). It is the second generation that has arguably had the most visible impact in the study of global environmental politics. First-generation Frankfurt School theorists maintained Marx's commitment to a critique of domination while recognizing that the odds of overthrowing the class system were slim. Their interpretation of domination has clear affinities with the concerns of green political theorists. Biro (2011) and Leiss (2011) argue that this interpretation can help us understand contemporary environmental crises and paradoxes. Adorno and Horkheimer criticized instrumental reason (or rationality) that was penetrating all areas of life and dominating humans and the nonhuman world alike. Reason, Horkheimer claimed, is a disease "born from man's urge to dominate nature" (Horkheimer quoted in Leiss 2011: 23). From an instrumental perspective, nature has no intrinsic value; it is merely "a stockpile of resources" to be mastered for human ends (Biro 2011: 14). Yet, in our attempts at mastering nature, "human beings distance themselves from nature" in such a way that ultimately rebounds in the domination of other humans (Leiss 2011). The contemporary significance of this has been noted in part by Saurin (1994), who argues that key elements of modernity including "distanciation," "technical–rationalism and bureaucracy," and the displacement of various local *episteme* and *techne* has led to "large-scale and systematic degradation" of ecosystems (1994: 46–49).

A second generation of critical theorists, inspired by Habermas, are more optimistic about the possibility of supplementing instrumental reasoning with substantive reasoning. Substantive reasoning entails value-infused deliberation about the goals pursued by society, not merely a value-free assessment of the means to attain pre-given goals. Some scholars have questioned whether Habermas's anthropocentric position is compatible with environmentalism, after all he argued that the only way humans can know and relate to the nonhuman world is instrumentally through labor and technology. To do otherwise is to jeopardize the survival of the human species (Eckersley 1990: 743, 753). Others are optimistic that the Habermasian idea of communicative rationality can help communities (including the international community) avoid the kind of environmental degradation that undermines human wellbeing. Through "communicative rationality," problems are addressed by seeking a reasoned consensus. Rationality is thus directed to subjecting social norms and goals to open and participatory critique rather than efficiently pursuing pre-given goals. The arguments presented in such a process ought to be generalizable beyond particular interests and have the potential to be accepted by differently positioned individuals.

The theory and practice of deliberative democracy has made a growing contribution to global environmental politics in the past decade. Much of this work has been inspired by John S. Dryzek (1987, 1994, 2006). For Dryzek, the green potential of communicative rationality requires accepting an "anthropocentric life-support approach" as a minimum basis for deliberation (1987: 35). If all human beings share an interest in sustaining the "productive, protective, and waste-assimilative value of ecosystems," then this becomes "the generalizable interest *par excellence*" (Dryzek 1987: 34, 204). There is thus strong potential for rationally legitimated social norms to be "ecologically rational." Baber and Bartlett also emphasize the importance of broadening knowledge production and decision-making beyond scientists and technocratic elites. Those opposed to an environment-related proposal will not be swayed by the availability of more information "but, rather, public involvement in the production of information through a process of discursive will-formation" (Baber and Bartlett 2005: 97). Moreover, they observe, it is impossible "to protect the environment from human degradation in the absence of a human commitment to do so"; this commitment will only come about through decentralized and democratic public discourse (2005: 98).

This theoretical work has led to numerous empirical studies of the deliberative quality of global environmental governance. Dingwerth (2007), for example, analyzed the quality of deliberation in transnational environmental networks and found trade-offs between deliberative quality and other democratic qualities including accountability, transparency and inclusiveness. Stevenson and Dryzek (2014) analyzed global climate governance in deliberative systems terms and proposed a number of ways in which democracy (understood as deliberative capacity building) can be enhanced at this scale (see Chapter 29, also Berg and Lidskog 2018). They have stressed, for example, the importance of avoiding deliberative enclaves and fostering deliberation across climate discourses, and enhancing deliberation and legitimacy in multilateral negotiations through a formula of "minilateralism plus discursive representation."

The influence of Habermas, Dryzek, and deliberative democracy is evident in many other studies, even if this remains unstated. Studies on the democratic deficit and legitimacy of various forms of environmental governance, and the quality of representation and participation have taken up an important place in this field of study (Baber and Bartlett 2015; Bäckstrand and Kuyper 2017; Dryzek and Pickering 2018; Gellers 2016; Kuyper and Bäckstrand 2016; Nasiritousi et al. 2016; Zelli et al. 2020).

Foucauldian approaches

Michel Foucault, a twentieth-century French social theorist, has inspired some scholars of global environmental politics dissatisfied with the offerings of traditional IR theory (see Chapter 3). Foucauldian-oriented studies start from the assumption that the material world, including "nature," is meaningless until it is interpreted and assigned meaning by humans. There is no deterministic relationship between the material and the meaning; to assign meaning is an act of power. Foucauldians refer to this as "productive power." This power lies in the capacity to define how humans act upon the material world. The power to assign meaning is also the power to marginalize, suppress or delegitimize other potential interpretations. In the field of global environmental politics, Foucauldian-inspired scholars have pursued two main tasks: first, discourse analysis has been employed to uncover potential meanings about the environment and the processes by which one single interpretation has been imposed and institutionalized; and, second, governmentality analyses have been carried out to expose the dominating effects of this productive power on people's lives.

One of the earliest contributions was Hajer's analysis of environmental policymaking in the United Kingdom and the Netherlands. Hajer argued that "policy-making involves much more than merely dreaming up clever ways of creating solutions. It requires first of all the redefinition of a given social phenomenon in such a way that one can also find solutions for them" (Hajer 1995: 2). He thus traces how earlier debates about the radical restructuring demanded by environmental crises were delegitimized as a new manageable way of understanding environmental degradation emerged in the late 1970s. This discourse of "ecological modernisation" established dominance with the idea that "pollution prevention pays": environmental crises do not discredit capitalist development but instead present opportunities for business to innovate and develop new markets. Unlike, say, "limits to growth," ecological modernization is supposedly compatible with existing political and economic institutional arrangements. This congruence goes a long way toward explaining its dominance as a policy discourse, at least within Europe.

But discourses do not establish dominance on the basis of rational policymakers selecting the most convenient framing of any given problem. Litfin (1994) analyzed how scientific knowledge influenced international negotiations about the problem of ozone depletion and pointed to the importance of "knowledge brokers" (see Chapter 18). These are individuals who "frame and interpret scientific knowledge" and thereby exercise significant political (and productive) power, especially under conditions of scientific uncertainty (Litfin 1994: 4). In Litfin's analysis, the deployment of "rhetorical devices," such as the metaphors of "ozone layer" and "ozone hole," helped to establish acceptance for a risk-based discourse that promoted precautionary action to phase out ozone-depleting substances (see Chapter 31).

Epstein advanced Foucauldian environmental analysis with an investigation into whaling and the power dynamics that enabled an anti-whaling discourse to supersede an earlier discourse that promoted whaling for economic, military and political ends (Epstein 2008; see Chapter 40). For Epstein, the study of discourse entails denaturalizing "what we assume to be right," thus she does not explicitly treat the anti-whaling discourse as "the truth" that prevailed over a historical "wrong." Instead, she is concerned to reveal how environmentalists "reframed perceptions and understanding by producing a new discourse on whales and whaling" that, in turn, rearticulated state identities from whaling states to anti-whaling states (Epstein 2008: 13, 94–95). The rupture in states' treatment of whales, Epstein argues, cannot be understood through a regime theoretical lens that focuses on narrow cost-benefit calculations because most states did not respond to the issue in ways that only reflected their material interests. Instead, they were "socialised" into the anti-whaling regime in ways that actually redefined their interests and identities.

Several scholars have drawn on Foucault's concept of "governmentality" to advance alternative understandings of global environmental governance (e.g., Death 2010; Fletcher 2017; Luke 2011; Methmann 2012; Oels 2005; Stripple and Bulkeley 2014). This approach has been called "analytics of government" (Dean 2010: 16). Foucault understood the term "government" to mean "conduct of conduct" whereby "conduct" is understood as a verb and a noun. Government is the calculated process of leading, directing or guiding the behavior and actions of others or of oneself (Death 2010: 18). Governmentality concerns the rationalities of government, or "how we think about governing" (Death 2010: 24). Death draws on this Foucauldian concept to analyze the "rationality of government" built into "sustainable development"; this involves "approaching sustainable development as an assemblage of practices of government which produce their own particular ways of seeing, knowing, acting and being" (Death 2010: 2). This approach allows him to focus on how "the scope, forms and identities of governmental action" were determined in part through "contests between competing rationalities of government" at the Johannesburg Summit (Death 2010: 5, 9). The prevailing rationality was an "advanced liberal rationality of government which relied upon the voluntary and responsible conduct of self-selecting partners operating at a distance from traditional centres of power" (Death 2010: 9). This builds on earlier work by Oels who argued that "climate change has been captured by advanced liberal government, which articulates climate change as an economic issue that requires market-based solutions to facilitate cost-effective technological solutions" (Oels 2005: 185). Oels identified this rationality as weak ecological modernization. By institutionalizing this discourse, climate change has been "rendered governable" by the Kyoto Protocol and the United Nations Framework Convention on Climate Change (UNFCCC) (Oels 2005: 199; see Chapter 32).

Feminism

Like all the theories reviewed in this chapter, feminism is a broad church. But what unites feminist perspectives is attention to sex-based and gender inequalities and the power structures that reinforce them. Liberal feminists are concerned about the presence and absence of women in global environmental governance. Ramstetter and Habersack (2019) questioned "whether female representatives are more likely to hold pro-environmental attitudes than their male colleagues and adjust their legislative behaviour accordingly." In a study of the European Parliament, they found that women and men "expressed similar concern for the environment, yet women were significantly more likely to support environmental legislation than men" (ibid).

Critical feminist scholars are concerned not only about the generally marginal participation of women in environmental governance, but also about the gendered assumptions embedded in such governance, as well as the implications of governance for gendered relations. Back in 1998, Bretherton observed that "attempts to 'put gender on the agenda' of global environmental politics have resulted in, not the incorporation of gender, but the addition of women" (Bretherton 1998: 85). She explains that "(g)ender analysis is not concerned per se with the incorporation of women in environmental decision-making and policy, but with…the broadly accepted, and expected, pattern of relations between men and women." Bringing women into policymaking is insufficient because "masculine values" are privileged over "feminine values" in most contexts; rectifying the sex imbalance does not rectify the enduring patriarchy (Bretherton 1998: 90). Two decades later, Bretherton's complaint remains valid. Bretherton argued that patriarchy is deeply implicated in environmental degradation. Specifically, she argues that the contemporary dominant form of "capitalist patriarchy" is based on "overlapping norms and principles of neoliberalism and Anglo-American hegemonic masculinity" that effectively "authorize conduct directly opposed to that demanded by an ethic of care for the environment" (Bretherton 2003: 103–104).

MacGregor observes an overlapping of the dominant participation of men in environmental governance and the *masculinization* of environmentalism (2009). She further explains that the masculinization of climate change is an effect of the dominant scientific and security framings of the issue, each of which "work(s) to invisibilise women and their concerns" (MacGregor 2009: 129).

Boyd's (2009) analysis of the Clean Development Mechanism reveals how "the patriarchal underpinnings of the sustainable development and climate-change policy agendas" have undermined the potential of mitigation projects. Her study of the Noel Kempff project in Bolivia found that "practical gender needs" were successfully incorporated, namely, "immediate necessities that women perceive themselves as lacking in a specific context, which would enable them to perform the activities expected of them: for example, a health post, vegetable gardens, or a water pump" (Boyd 2009: 102). However, "strategic gender needs" were neglected; this refers to "that which is necessary for women to change their status in society…: access to and ownership of land or other property, control over one's body, equal wages, or freedom from domestic violence" (Boyd 2009: 102).

More recent critical feminist scholarship uses an intersectionality lens to show how disadvantages are produced by a range of categories of differences (principally sex, gender, class, and race) (Kaijser and Kronsell 2014; Kings 2017; Perkins 2018). Much of this work remains theoretical and conceptual; however, recent work aims to make this more applicable to empirical analysis (Fletcher 2018; Walker et al. 2019).

Conclusion

For decades, the environment was treated by scholars of international politics as an issue of little or no relevance. This trend began to shift in the 1990s as the environment was increasingly recognized as a problem of international political concern and as a focus for global governance. Nevertheless, since that time the study of global environmental politics has been dominated by rationalist approaches, in particular by neoliberal institutionalism (regime) theory (see Chapter 9). This approach can undoubtedly generate important insights into institutional dynamics and inter-state cooperation over common pools and common sinks. However, this chapter has sought to uncover the fertile theoretical terrain that lies beyond this traditional theoretical foreground. For those who are attracted to the field of global environmental politics by a genuine concern for the social and ecological consequences of global environmental change, these alternative theories provide a valuable set of lenses. By surrendering a commitment to positivist social science, these approaches provide foundations for critiquing global environmental governance by exposing its blind spots and moral shortcomings. These approaches also allow scholars to try to understand the world rather than taking it as the starting point of analysis.

This points to Cox's well-known distinction between problem-solving theory and critical theory (Cox 1981). According to Cox, problem-solving theory

> takes the world as it finds it, with the prevailing social and power relationships and the institutions into which they are organised, as the given framework for action. The general aim... is to make these relationships and institutions work smoothly...Critical theory...does not take institutions and social and power relations for granted but calls them into question by concerning itself with their origins and how and whether they might be in the process of changing.
>
> *(Cox 1981: 128–129)*

Thus, the field of global environmental politics would be best served by a diversity of scholarship that is informed not only by traditional, problem-solving IR theory, but also by the critical theories examined in this chapter.

References

Baber, W.F. and Bartlett, R.V. (2005) *Deliberative Environmental Politics: Democracy and Ecological Rationality*, Cambridge, MA: MIT Press.

Baber, W.F. and Bartlett, R.V. (2015) *Consensus and Global Environmental Governance: Deliberative Democracy in Nature's Regime*, Cambridge, MA: MIT Press.

Bäckstrand, K. and Kuyper, J.W. (2017) "The democratic legitimacy of orchestration: the UNFCCC, non-state actors, and transnational climate governance," *Environmental Politics*, 26(4): 764–788.

Berg, M. and Lidskog, R. (2018) "Deliberative democracy meets democratised science: a deliberative systems approach to global environmental governance," *Environmental Politics*, 27(1): 1–20.

Bernstein, S. (2001) *The Compromise of Liberal Environmentalism*, New York: Colombia University Press.

Betsill, M.M. (2017) "Trump's Paris withdrawal and the reconfiguration of global climate change governance," *Chinese Journal of Population Resources and Environment*, DOI: 10.1080/10042857.2017.1343908

Biro, A. (ed.) (2011) *Critical Ecologies: The Frankfurt School and Contemporary Environmental Crisis*, Toronto: University of Toronto Press.

Blondeel, M. (2019) "Taking away a "social licence": Neo-Gramscian perspectives on an international fossil fuel divestment norm," *Global Transitions*, 1: 200–209.

Blondeel, M., Colgan, J. and Van de Graaf, T. (2019) "What drives norm success? Evidence from anti-fossil fuel campaigns," *Global Environmental Politics*, 19(4): 63–84.

Boyd, E. (2009) "The Noel Kempff project in Bolivia", in G. Terry (ed.) *Climate Change and Gender Justice*, Rugby: Practical Action Publishing.

Bretherton, C. (1998) "Global environmental politics: putting gender on the agenda?" *Review of International Studies*, 24: 85–100.

Bretherton, C. (2003) "Movements, networks, hierarchies: a gender perspective on global environmental governance," *Global Environmental Politics*, 3(2): 103–119.

Bueno M. (2020) "Identity-based cooperation in the multilateral negotiations on climate change," in C. Lorenzo (eds) *Latin America in Times of Global Environmental Change*. The Latin American Studies Book Series. Cham: Springer.

Cass, L. (2006) *The Failures of American and European Climate Policy: International Norms, Domestic Politics, and Unachievable Commitments*, Albany: SUNY Press.

Cox, R. (1981) "Social forces, states and world orders: beyond international relations theory", *Millennium*, 10(2): 126–155.

Dean, M. (2010) *Governmentality: Power and Rule in Modern Society*, 2nd edn, London: Sage.

Death, C. (2010) *Governing Sustainable Development: Partnerships, Protests and Power at the World Summit*, London: Routledge.

Dingwerth, K. (2007) *The New Transnationalism*, Houndmills: Palgrave Macmillan.

Dryzek, J.S. (1987) *Rational Ecology: Environment and Political Economy*, Oxford: Basil Blackwell.

Dryzek, J.S. (1994) *Discursive Democracy: Politics, Policy, and Political Science*, Cambridge: Cambridge University Press.

Dryzek, J.S. (2006) *Deliberative Global Politics*, Cambridge: Polity.

Dryzek, J.S. and Pickering, J. (2018) *The Politics of the Anthropocene*. Oxford: Oxford University Press.

Eckersley, R. (1990) "Habermas and green political thought: two roads diverging", *Theory and Society*, 19(6): 739–776.

Eckersley, R. (2007) "Ambushed: the Kyoto Protocol, the Bush administration's climate policy and the erosion of legitimacy", *International Politics*, 44: 306–324.

Epstein, C. (2008) *The Power of Words in International Relations: Birth of an Anti-Whaling Discourse*, Cambridge, MA: MIT Press.

Fletcher, A.J. (2018) "More than women and men: a framework for gender and intersectionality research on environmental crisis and conflict," in C. Fröhlich, G. Gioli, R. Cremades and H. Myrttinen (eds) *Water Security Across the Gender Divide*, Cham: Springer.

Fletcher, R. (2017) Environmentality unbound: multiple governmentalities in environmental politics, *Geoforum*, 85: 311–315.

Foster, J.B. (2000) *Marx's Ecology*, New York: Monthly Review Press.

Foster, J.B., Clark, C. and York, R. (2010) *The Ecological Rift: Capitalism's War on the Earth*, New York: Monthly Review Press.

Foyer, J., Aykut, S. and Morena, E. (2018) "Introduction", in Aykut, Foyer and Morena (eds) *Globalising the Climate. COP21 and the Climatisation of Global Debates*, London: Routledge, pp. 1–18.

Fuentes-George, K. (2016) *Between Preservation and Exploitation: Transnational Advocacy Networks and Conservation in Developing Countries*, MIT Press.

Gellers, J.C. (2016) "Crowdsourcing global governance: sustainable development goals, civil society, and the pursuit of democratic legitimacy," *Int Environ Agreements*, 16: 415–432.

Green, F. (2018) "Anti-fossil fuel norms," *Climatic Change*, 150: 103–116.

Guangyu, Q.F. (2020) "ASEAN's role expectations and the diffusion of common but differentiated responsibilities principle in the climate change context," *The Pacific Review*, DOI: 10.1080/09512748.2020.1797860

Haas, Peter M. (2016) *Epistemic Communities, Constructivism, and International Environmental Politics*, London; New York: Routledge.

Hadden, J. and Seybert, L.A. (2016) "What's in a norm? Mapping the norm definition process in the debate on sustainable development," *Global Governance*, 22(2): 249–268.

Hajer, M.A. (1995) *The Politics of Environmental Discourse: Ecological Modernization and the Policy Process*, Oxford: Oxford University Press.

Hoffmann, M.J. (2005) *Ozone Depletion and Climate Change*, Albany: SUNY Press.

Hovden, E. (1999) "As if nature doesn't matter: ecology, regime theory and international relations", *Environmental Politics*, 8(2): 50–74.

Kaijser, A. and A. Kronsell (2014) "Climate change through the lens of intersectionality", *Environmental Politics*, 23(3): 417–433.

Katzenstein, P.J. (1996) "Introduction: alternative perspectives on national security", in P.J. Katzenstein (ed.) *The Culture of National Security: Norms and Identity in World Politics*, New York: Columbia University Press.

Kiessling, C. (2019) "Internalización del principio de las responsabilidades comunes, pero diferenciadas: interpretaciones desde la sociedad civil brasileña", *Letras Verdes*, (25): 8–28.

Kings, A.E. (2017) "Intersectionality and the changing face of ecofeminism", *Ethics and the Environment*, 22(1): 63–87.

Kovel, J. (2007) *The Enemy of Nature: The End of Capitalism or the End of the World?* 2nd edn, London: Zed Books.

Kuyper, J.W. and Bäckstrand, K. (2016) "Accountability and representation: nonstate actors in UN climate diplomacy", *Global Environmental Politics*, 16(2): 61–81

Leiss, W. (2011) "Modern science, enlightenment, and the domination of nature: no exit?" in A. Biro (ed.) *Critical ecologies: The Frankfurt School and contemporary environmental crisis*, Toronto: University of Toronto Press.

Levy, D.L. and P.J. Newell (2002) "Business strategy and international environmental governance: toward a neo-gramscian synthesis." *Global Environmental Politics*, 2(4): 84–101.

Levy, D.L. and Egan, D. (2003) "A neo-Gramscian approach to corporate political strategy: conflict and accommodation in the climate change negotiations", *Journal of Management Studies*, 40(4): 803–829.

Levy, D.L. and Newell, P.J. (eds) (2005) *The Business of Global Environmental Governance*, Cambridge, MA: MIT Press.

Lipschutz, R. (2004) *Global Environmental Politics: Power, Perspectives, and Practice*, Washington, DC: CQ Press.

Litfin, K. (1994) *Ozone Discourse: Science and Politics in Global Environmental Cooperation*, New York: Columbia University Press.

Luke, T. (2011) "Environmentality", in J.S. Dryzek, R.B. Norgaard and D. Schlosberg (eds) *The Oxford Handbook of Climate Change and Society*, Oxford: Oxford University Press, pp. 97–109.

MacGregor, S. (2009) "A stranger silence still: the need for feminist social research on climate change", *Sociological Review*, 57(s2): 124–140.

March, J.G. and Olsen, J.P. (1998) "The institutional dynamics of international political orders", *International Organization*, 52(4): 943–969.

Meadowcroft, J. and Fiorino, D.J.(eds) (2017) *Conceptual Innovation in Environmental Policy*. Cambridge, MA: MIT Press.

Merchant, C. (ed.) (2008) *Ecology*, 2nd edn, Amherst, NY: Prometheus Books.

Methmann, C.P. (2012) "The sky is the limit: global warming as global governmentality", *European Journal of International Relations*, doi: 10.1177/1354066111415300.

Mitchell, R.B. and Carpenter, C. (2019) Norms for the earth: changing the climate on "Climate Change." *Journal of Global Security Studies*, 4(4): 413–429.

Moore, J.W. (2000) "Environmental crises and the metabolic rift in world-historical perspective", *Organization and Environment*, 13: 123–157.

Nasiritousi, N., Hjerpe, M. and Bäckstrand, K. (2016) Normative arguments for non-state actor participation in international policymaking processes: functionalism, neocorporatism or democratic pluralism? *European Journal of International Relations*, 22(4): 920–943.

Neufeld, M. (1993) "Interpretation and the 'science' of international relations", *Review of International Studies*, 19: 39–61.

Newell, P. (2011) "The elephant in the room: capitalism and global environmental change", *Global Environmental Change*, 21: 4–6.

Newell, P. and Paterson, M. (1998) "A climate for business: global warming, the state and capital", *Review of International Political Economy*, 5(4): 679–703.

O'Connor, J. (1988) "Capitalism, nature, socialism: a theoretical introduction", *Capitalism, Nature, Socialism*, 1(1): 11–38.

Oels, A. (2005) "Rendering climate change governable: from biopower to advanced liberal government?" *Journal of Environmental Policy and Planning*, 7(3): 185–207.

Park, S. (2010) *World Bank Group Interactions with Environmentalists: Changing International Organisation Identities*, Manchester: Manchester University Press.

Paterson, M. (1996) *Global Warming and Global Politics*, London: Routledge.

Paterson, M. (2000) *Understanding Global Environmental Politics: Domination, Accumulation, Resistance*, Houndmills: Palgrave Macmillan.

Perkins, P.E. (2018) "Climate justice, gender and intersectionality", in T. Jafrey (ed.) *Routledge Handbook of Climate Justice*. London: Routledge, pp. 349–58.

Pettenger, M.E. (ed.) (2007) *The Social Construction of Climate Change*, Ashgate: Aldershot.

Ramstetter, L. and Habersack, F. (2019) "Do women make a difference? Analysing environmental attitudes and actions of Members of the European Parliament", *Environmental Politics*, DOI: 10.1080/09644016.2019.1609156

Rasheed, A.A. (2019) "Role of small islands in UN climate negotiations: a constructivist viewpoint," *International Studies*, 56(4): 215–235.

Ruggie, J.G. (1998) *Constructing the World Polity: Essays on International Institutionalization*, London: Routledge.

Saurin, J. (1994) "Global environmental degradation, modernity and environmental knowledge", in C. Thomas (ed.) *Rio: Unravelling the Consequences*, London: Routledge.

Schwindenhammer, S. (2019) "Agricultural governance in the Anthropocene," in T. Hickmann, L. Partzsch, P. Pattberg, S. Weiland (eds) *The Anthropocene Debate and Political Science*, London: Routledge.

Stalley, P. (2018) "Norms from the periphery: tracing the rise of the common but differentiated principle in international environmental politics", *Cambridge Review of International Affairs*, 31(2): 141–161.

Stevenson, H. (2013) *Institutionalizing Unsustainability: The Paradox of Global Climate Governance*, Berkeley: University of California Press.

Stevenson, H., Auld, G., Allan, J.I., Elliott, L. and Meadowcroft, J. (2021) "The practical fit of concepts: ecosystem services and the value of nature", *Global Environmental Politics*, 21(2): 3–22.

Stevenson, H. and Dryzek, J.S. (2014) *Democratizing Global Climate Governance*. Cambridge: Cambridge University Press.

Stripple, J. and Bulkeley, H. (eds) (2014) *Governing the Climate*. Cambridge: Cambridge University Press.

Walker, H.M., Culham, A., Fletcher, A.J. and Reed, M.G. (2019) "Social dimensions of climate hazards in rural communities of the global North", *Journal of Rural Studies*, 72: 1–10.

Weis, T. (2010) "The accelerating biophysical contradictions of industrial capitalist agriculture", *Journal of Agrarian Change*, 10(3): 315–341.

Zelli, F., Bäckstrand, K., Nasiritousi, N., Skovgaard, J. and Widerberg, O. (2020) *Governing the Climate-Energy Nexus*. Cambridge: Cambridge University Press.

5

The study of global environmental politics

Strategies for research and learning

Juliann Emmons Allison and Thomas Hickmann

Contemporary research and teaching on global environmental politics (GEP) draw upon many approaches to understanding the ways in which states and societies respond to transboundary environmental problems (see Chapter 2). Approaches that emphasize the state (see Chapter 7) and international organizations (see Chapter 8) derive in a straightforward way from scholarly work in international relations (IR), with a marked focus on domestic and international institutions and on traditional determinants of power and sources of authority as key to understanding states' interests and actions. Alternatively, approaches that adopt a philosophically critical orientation toward IR are more inherently interdisciplinary, and build on theoretical frameworks developed in economics and sociology as well as in the humanities. The latter work frequently examines global environmental politics from the "bottom up," and takes seriously the notion that subnational and non-state actors, transnational networks of activist organizations as well as individuals are essential to resolving the world's pressing sustainability challenges (see Chapter 14).

This chapter describes an analytical framework informed by IR scholarship for approaching research and teaching in the GEP subfield. The framework may be adapted for use with any substantive area of interest, as well as used to develop undergraduate and graduate coursework in GEP. Elaboration of the framework itself is followed by its application to a discussion of research in GEP that both captures long-standing trends and underscores new avenues of inquiry. The section on teaching GEP that follows includes suggestions for the adoption of pedagogies – specifically fieldwork and other outdoor experiences, contemplative practices, (online) games and role-plays – that are particularly well suited to the field of GEP.

Approaches to global environmental politics

Though there are many theoretical approaches to GEP, key conceptual categories associated with other international and global issue areas are particularly useful for the study of transboundary environmental concerns. Table 5.1 provides one such conceptualization that builds on "power," "institutions" and "ideas" – generally associated with realism, liberal institutionalism and constructivism, respectively – to characterize and distinguish among more conventional and mainstream approaches to studying GEP (see Chapter 3) and more

DOI: 10.4324/9781003008873-7

Table 5.1 Key theoretical influences on generalized approaches to the study of GEP

	Conventional	Mainstream	Radical
Power and authority	State power, dominance, exclusive state authority even hegemony Potential for coercion Possibly regional	Varied determinants of power and sources of authority Power and authority as influence	Discourse paramount Persuasive power and new spheres of authority Consensus-seeking
Institutions	Largely ineffective or furthering interests of most powerful states Environmental challenges likely to cause conflict	Diffuse, while partly constraining actor behavior Law and organization constitutive of governance	Transnationalism rooted in class, cultural, racial, ethnic, gender and religious differences Importance and effectiveness of global civil society
Ideas	Self-interest/preservation Expectation of zero-sum outcomes	"Environment and development" Integrative, positive-sum cooperation anticipated	Contestation, collectivism and other alternatives to state dominance of environmental behavior

radical approaches (see Chapter 4). The various approaches illuminate the relevance of important IR concepts – i.e., power, institutions and ideas – to GEP, and together yield clearer representations of reality.

Beginning with the left column in Table 5.1, conventional approaches to studying GEP anticipate that those nations with the greatest geo-strategic, economic and/or ecological, power will use these resources to achieve their own most desired outcomes with multilateral environmental agreements (MEAs; see Chapter 22). International environmental treaty regimes and other institutional constraints on state behavior are regarded as suspect if not ineffective and will only further the interests of the most powerful states (Drezner 2008). Scientific, ecological and ethical ideas that are contrary to the policy preferences and negotiating positions of the dominant nations involved with a given environmental challenge are irrelevant. Insofar as militarized conflict is philosophically antithetical to environmental protection, traditional accounts power in world politics have not always provided compelling theoretical foundations for GEP (Barnett 2001). The West's – especially the United States' – involvement in the Middle East suggests that the world's most industrialized nations can be as prone as poorer nations to conflict over scarce resources (Klare 2009). Yet acute conflict can be blamed on lack of institutional capacity to resolve conflicts than on scarcity itself (Woertz 2014).

Mainstream orientations to the study of GEP (see Chapter 3), associated with the center column in Table 5.1, broaden the concept of power to include ecological determinants of national power, such as oceans, tropical forests and other natural habitats (see Chapters 40–42). They recognize the possibility that even developing states, if sufficiently ecologically well-endowed, can influence the negotiation of MEAs (Schreurs and Economy 1997; Steinberg 2001). Despite the proliferation of new actors and institutions with new forms of authority to address global environmental problems, such as climate change or biodiversity loss, adherents of mainstream approaches put states center stage in their analysis with a host of international environmental institutions that are deemed to provide significant limitations on state behavior (DeSombre 2017). The individual and collective behavior of states is governed by the belief that ecological sustainability will require the socio-economic and political development of the world's poorest nations (WCED 1987; see

Chapter 16). For example, the 2030 Agenda for Sustainable Development and the 17 Sustainable Development Goals (SDGs) include the principle to leave no one behind (United Nations 2015; Kanie and Biermann 2017).

The theoretical and practical success of this analytic perspective is arguably due to the wedding of the environmental and liberal economic norms to institutionalize "liberal environmentalism," which conditions sustainability on preservation of the world's liberal economy (Bernstein 2001). The economic foundation for this norm incorporates the expectation that nations become better environmental stewards over the long term as a consequence of economic growth. The Environmental Kuznets Curve (Grossman and Krueger 1995), a bell-shaped, "inverted U" curve describing the relationship between a society's economic growth and the problem of environmental degradation, suggests that "at early stages of growth, environmental degradation gets worse, but as citizens get richer, things start to get better" (Leonard 2006). In fact, the relationship between economic growth and environmental protection is tenuous, depending, at least, on the nature of pollutants, national policy prerogatives, access to control technologies and the progress of globalization (Stern 2004).

Radical GEP, represented by the right column in Table 5.1 (see Chapter 4), is distinguished by questioning technocratic solutions to environmental problems and considering redistribution of both political power and ecological resources as essential adaptations to global climate (see Chapter 32) and related changes. Radical approaches to GEP seek to deconstruct dominant thinking and practices, including our conventional understandings of development and gender (Arora-Jonsson 2013; MacGregor 2017). That is, it seeks a relatively extreme change in the norms from fairness delimited by preservation of the status quo standard of living for the world's advanced industrialized nations, toward universally applicable environmental justice (see Chapter 26) and equal opportunities for everyone, everywhere on the planet (see Chapter 25).

Theorists and social activists alike identify such global environmental justice in terms of distributional equity with respect to environmental risks and benefits, intersectional race/ethnic, class, gender and other types of social diversity among impacted communities, and widespread participation in the political processes that determine environmental policy (Schlosberg 2013). Power and authority in this context derive from the capacity to wield communication persuasively in the interest of constructing desirable norms (Gellers and Jeffords 2018).

Radical approaches to the study of GEP recognize the potential power and authority exerted by non-state actors constitutive of global civil society (see Chapter 4). Recently, "contestation" has become a catchword for a line of scholarship in IR to analyze battles about underlying norms and values (Peterson 2019). Specifically, any inequitable distribution of the ill-effects of environmental degradation and responsibility for the costs of redressing them are understood to be rooted in a pervasive lack of recognition of marginalized states, peoples, and cultures (see Chapter 23). Thus, their advocates demand "participation for those at the short end of distributional inequity, and participation by those suffering the injustice of cultural recognition" (Schlosberg 2004: 523). Such calls for massively increased participation in decision-making and knowledge production within global science institutions (Eden 2016) apply to historically less influential states as well as nongovernmental organizations (NGOs), transnational activist networks (TANs), and subnational activists and other actors.

Central areas of research

Research agendas relevant to substantive issues in GEP derive from long-standing IR foci on security, international institutions and domestic politics, subnational and non-state actors, economic growth and development, and the science–policy interface, yet bridge interdisciplinary studies to include global governances (see Chapter 21) and environmental ethics (see Chapter 27) and activism, especially in the context of climate change (Dauvergne and Clapp 2016; see Chapter 32). The rows in Table 5.1 provide a conceptualization of how scholars associated with conventional, mainstream or radical viewpoints on GEP, engage this subfield's central areas of research.

Environmental security

Environmental security may be understood in terms of national security (see Chapter 20). Insofar as environmental degradation represents an imminent threat to a nation's territorial integrity, sovereignty and security. Severe scarcity is associated not only with civil conflict abroad, but also with international war. As such, national will should be marshaled in the interest of environmental protection. Enthusiasm for this project has waned considerably since its heyday in the immediate post-Cold War era. Regardless, it never received overwhelming scholarly support, in part, simply because war is antithetical to environmental protection (Deudney 1990; Barnett 2001). Furthermore, it is increasingly clear that globalism, rather than nationalism, is needed to cope with transboundary ecological challenges ranging from traditional commons issues to pandemics (Heise 2008, see Chapter 21).

Examinations of the foundational relationship among natural resources, power and conflict have been far more fruitful than any strict identification between national and environmental security. The environmental security problematique can be understood as the threat of conflict over scarce resources (see Chapter 20). This relationship may account for the United States' incursions in the Middle East and Central Asia (Klare 2009), as well as the more frequent civil violence associated with climate change and other sources of ecological degradation and reduced access to resources, especially among women living in impoverished societies (Kaplan 1994; Barnett and Adger 2007; Nagel 2015).

While not alone among GEP scholars (see Gleditsch 2012), Homer-Dixon (2010) has been most closely associated with arguments linking to violent conflict ecological scarcity due to climate change (see Chapter 32), ozone depletion (see Chapter 31), loss of forests (see Chapter 42) and agricultural lands (see Chapters 43 and 44), depletion of fisheries (see Chapter 40), water pollution or access to fresh water (see Chapters 38 and 39), and other changes in the quantity or quality of vital resources. Klare (2012) ups the ante by charging the United States and the world's other major industrialized nations to engender reduced consumption among their citizens and drastically improve international efforts at collective security (see Chapters 20 and 21).

These concluding calls for more effective domestic political institutions and international security institutions would shift attention down a row in Table 5.1. Continuing right, instead, both mainstream and radical views on the role of power in GEP seek to expand the concept of security to incorporate something other than any traditional defense of state interests and territory (see Chapter 20). Scholars in these analytical traditions recognize that the security of contemporary states does not turn on the unilateral pursuit of generalized national interests (Kaldor 2007). Rather, security includes economic and ecological elements as well as military strategy, and expanding multilateral options for responding to contemporary security challenges (Commission on Global Governance 1995). Their radical counterparts go further, and challenge conceptualizations of security defined primarily in terms of nation-states (Matthew et al. 2009).

Human security, a conceptual approach developed by the United Nations Development Programme (UNDP), refers to the idea that global – not national – security is tied to the wellbeing of individuals, whether or not they are citizens of relatively self-sufficient advanced industrial nations (Commission on Global Governance 1995; Axworthy 2001). It is designed to reduce human suffering as well as assure security by combining humanitarian, economic, and social issues. In addition to defense against armed conflict and international intervention, human security covers protection from organized crime, criminal violence, and genocide, and provision of good government, including the guaranteed support for human rights, health services and environmental quality. Axworthy's (2001) support for this reconceptualization of security derives from recognizing that the combined effects of global economic crises and ineffective governance have made it difficult, at best, for many nations to protect their citizens. This incapacity compounds the Anthropocene dilemma: humans' extraordinary adaptability has made us the first species in the history of the planet with the power to endanger the livability of the planet (Wallace-Wells 2020; see Chapter 15). Consequently, our current political and socio-economic systems themselves arguably constitute a security threat for humanity (Fagan 2017; King and Murray 2001).

International institutions, domestic politics, and subnational and non-state actors

Moving to the second row in Table 5.1, (liberal) institutionalism identifies the most substantial GEP research agenda. Despite conventional disregard for institutions, particularly those statutes, customs, principles and precedents that are constitutive of international law (see Chapter 10), international environmental (treaty) regimes (see Chapter 9) and complexes are a foundation of institutional analyses in GEP. Regime effectiveness and the development of regime complexes, or "overlapping" or "loosely coupled" (Keohane and Victor 2011; Ward 2006) regimes in a given issues area have received particular attention among scholars. The early development of GEP consisted largely of inductive efforts to draw conclusions about the effectiveness of international environmental institutions from a range of key case studies. "Effective" refers to the degree to which the MEAs at the heart of international environmental institutions actually facilitate the domestic implementation necessary to solve identified transboundary and global sustainability problems. Haas et al.'s (1993) pioneering work suggests that increasing the capacity of developing nations to implement environmental policy, improving the contractual environment, and elevating governments' ecological concerns are key contributions of effective international environmental institutions. Assessing the effectiveness of specific international environmental treaties, related institutional arrangements and systems of governance continues to be a predominant focus of GEP research (see Victor et al. 1998; Young 1999; Young 2018). The explosion of international and transnational institutions and organizations associated with climate change and other contemporary global environmental and social problems has generated a diverse and compelling body of research on regime change and coalescence, especially in the context of energy and climate change (Keohane and Victor 2011; Colgan et al. 2012; Abbot 2014; Widerberg and Pattberg 2017; see Chapters 32 and 33).

Shifting right in Table 5.1 deepens our examination of institutions at all levels and scales from global to local to include domestic politics, subnational governments and a plethora of non-state actors, many of which recognize the class, cultural, racial, ethnic, gender and religious differences that may influence individual and group responses to environmental change. Delving into the domestic sources of nations' positions in the negotiation of MEAs both accounts for their conditions for cooperation and provides a foundation for comparative analysis of domestic environmental policy (see Chapter 12). Interactions between domestic politics and international environmental negotiations are amenable to analysis of "two-level" games (Putnam 1988; see Chapter 11). Additionally, scholars have developed more specific models for analyzing these interactions in specific regions and/or with respect to specific environmental problems or issue areas – e.g., stratospheric ozone (see Chapter 31), acid rain (see Chapter 30) climate change (see Chapter 32); and nations' regulatory structures and hazardous waste industries (see Chapter 36). The 2016 Paris Agreement alone has generated a significant and illustrative body of research in this vein (Falkner 2016)

The recent proliferation of GEP research on subnational and non-state actors, including NGOs, social movements and TANs, represented in the right column of Table 5.1, has eclipsed attention previously devoted to domestic politics. Early scholarship on these constituents of global civil society and environmental protection coincided with the unprecedented inclusion of representatives of more than 2,000 NGOs and 45,000 independent environmental and other activists – 17,000 of whom were granted consultative status to participate in the meetings – at the 1992 Earth Summit in Rio de Janeiro. A high, to date, of more than 13,000 observer organizations (International Government Organizations and UN entities as well as NGOs) participated in the 2009 Conference of Parties (COP 15) to the UN Framework Convention on Climate Change (UNFCCC) that was negotiated and signed at the Earth Summit. The 2015 COP 21 is recognized for having the largest number of representatives of UNFCCC parties and observer states (United Nations 2021). Collectively, these constituents of global civil society counterbalance government and provide a social space within which individuals and collectivities might challenge extant political and economic powers (Lipschutz 1992; Bulkeley et al. 2014; Hale 2016).

Contributors to Matthias Finger and Thomas Princen's (1994) volume on environmental NGOs and world politics provided early documentation of increased activism around ecological issues as a response to the combined effects of the end of the Cold War and rapid economic globalization on natural systems (see Chapter 24). Margaret Keck and Kathryn Sikkink's (1998) transformative work on TANs, which seek to facilitate the political and social change necessary to increase justice and improve human rights, explains the success of NGO campaigns relevant to GEP (Wapner 2002; O'Neill 2004).

Research on global civil society as the most democratic foundation for a sustainable future increasingly emphasizes environmental TANs (Betsill and Corell 2001; Wapner 2002; Pattberg 2007). It sometimes explicitly identifies regimes as a means of governance without government in this context (Biermann and Pattberg 2008). Additionally, due to the regulatory gap in the environmental policy domain, cities and other subnational governance bodies have formed coalitions to formulate best practices to promote urban sustainability (Schreurs 2008; de Oliveira 2009; Bulkeley and Betsill 2013). NGOs have created behavior standards and certify environmental-friendly practices, and corporate actors have engaged in business self-regulation (Betsill and Bulkeley 2003; Abbott et al. 2016). This development is ubiquitous, but particularly evident in the climate policy domain (Hickmann 2017, van der Ven et al. 2017; see Chapter 13).

Beyond "environment and development"

The bottom row in Table 5.1 traces the ideational bases for approaches to studying GEP discussed in this chapter, from a conventional focus on the self-interested state with sovereign authority over environmental resources within its borders, to the contemporary mainstream recognition of the identity between environmental protection and economic development, to a range of radical ideas for global governance. While mainstream ideals acknowledge calls for global environmental justice that would permit growth sufficient to afford a sustainable redistribution of resources, they are undermined by a concomitant acceptance of the world's liberal market economy. An obvious way around this conundrum would be to move from the global economy in favor of a more manageable, local alternative.

A related major debate in GEP is whether sustainable development is possible and how to reconcile the terms sustainability and development. Some scholars favor growth premised on a green economy as the solution for the overexploitation of our natural resources, and a foundation for increasing energy supplies, especially in countries of the global South, with innovative technologies (Smulders, Toman and Withagen 2014). This economic model still relies on the extraction of natural resources, albeit at a slower pace with increasingly reliance on renewable energies and recycled goods and materials. Others support the idea of limits to the prevalent growth paradigm and propose a degrowth approach. Advocates of this approach contend that economic growth is not just limited; it is the underlying cause of all sustainability problems (Meadows et al. 2004; Klein 2015). They recognize that capitalism leads to a constant increase in commodified goods – generating environmental pollution, the destruction of ecosystems, and greenhouse gas emissions (Alexander 2012). Remedies include adopting alternative worldviews and lifestyles promising minimal impact on the earth's ecosystems. For example, Buen Vivir, which originated in the cosmovisions of Andean peoples and social movements in Latin America, establishes legal standing for Earth as part of a rights-based approach to the natural environment (Walsh 2010; Acosta 2017).

The "sustainable development" oxymoron underscores the importance of smaller and lower levels of human organization (see Chapter 16). McKibben (2007) argues that we need to dissociate "growth" and "prosperity," and move intentionally toward a future where food and energy, in particular, are consumed in close proximity to where they are produced (see Chapters 17, 27 and 44). More generally, he urges more ethical consumption in the interest of responding directly and effectively to individual and community concerns about the natural environment and social justice (see Chapter 26). Attention to the impacts of our choices as consumers is central to GEP research that seeks to illuminate the environmental and social toll of Western patterns of consumption (Princen et al. 2002; Dauvergne 2010; see Chapter 17). Hess's (2009) research on "localist" movements suggests that trends toward conscious support of locally owned and operated sources of food, consumer goods, energy, transportation, and media bode well for achieving national and regional sustainability and environmental justice (see Chapters 16, 26 and 29). Global environmental governance embraces this idea of intentional localization of sustainability as a means toward democratic transformation of world politics and markets (Lipschutz 2004; Wapner 2011; see Chapter 21).

Scientific information and technological innovation

Studying the environmental impacts associated with human activity ultimately depends on the advancement of scientific knowledge and technological innovation (see Chapters 6 and 18). We are "absolutely dependent" on science and technology, according to Carl Sagan (Head 2006). That is, members of advanced industrial societies understand their experience of the natural environment through a scientific lens, characterized by the process of hypothesis testing. Winnie the Pooh's impromptu analysis of the relationship between fir cone size and the speed at which they move in water (Nordmoe 2004) highlights how we learn from an early age to make educated guesses about the causes of changes in the natural environment, and how to test these working hypotheses. Some of those who ultimately opt for careers in the natural sciences will devote their professional lives to testing relationships between human activity and degradation of the natural environment – repeatedly and with increasing stringency. As scientific certainty about human complicity in environmental degradation has increased, so has the need to better manage, if not reverse, it.

The role of scientific inquiry in the negotiation of MEAs itself represents a significant area of research within GEP. Emphasizing scientific ideas as "reflective" institutions (see Levy et al. 1995), analysis of the development of scientific consensus and its contribution to the emergence of effective international environmental regimes has proved particularly fruitful (see Chapter 9). Haas (2015), Litfin (1994), and Peterson (1992) initiated the study of epistemic communities, or networks of predominantly scientific and technical experts with policy-relevant knowledge in a specific issue area, in international environmental negotiations. These analyses examine how highly technical knowledge influences decision-making among the international and transnational actors tasked with responding to transboundary and global environmental problems (Krasner 1983; see Chapters 14 and 18).

Scientific consensus, or the position agreed upon by the vast majority of scientists and other relevant specialists concerned with a given issue, represents one of the best possible foundations for environmental policymaking (Barash 2012). Although the self-correcting process of science must remain open to alternative, yet still theoretically valid, viewpoints and the accumulation of new information, Barash explains that it is blessedly "not so open-minded as to let our brains fall out, or our planet overheat" (Barash 2012: 1). Perhaps not; unfortunately, issue-specific experts are not the only sources of influence on GEP. Broaching the contest between scientific and ideological world views, Sarewitz explains that "scientific inquiry is inherently and unavoidably subject to becoming politicized in environmental controversies" (Sarewitz 2004: 385). Facts may be manipulated to suit special interests and the ubiquity of scientific uncertainty fuels claims of scientific dissensus by skeptics and deniers (Sarewitz 2004; Jacques et al. 2008; Jacques 2012; Dunlap 2013).

Consequently, analysis of the politics of environmental skepticism, especially as it relates to climate change, represents another science-based area of research within GEP (see Chapter 32). Skepticism has generated a wealth of information on the conservative bias in skeptics' climate science, the increasing divergence between popular "belief" in global warming and scientists' consensual knowledge about the facts of this critical phenomenon, and the social reasons for these developments in GEP (Jacques et al. 2008; Jacques 2012; Engels et al. 2013). While the divergence between public perceptions and scientific evidence varies across nations – 86 percent of Indians trust climate science "a great deal" or "a lot," but less than 12 percent of Americans do (Whiting 2020) – it tends to be manifest in political partisanship. For example, while most Americans say climate change is affecting their lives, and government should do more to reduce its impact, opinion about the efficacy of existing climate policy reflects party divisions. The PEW Research Center reported (Funk and Kennedy 2020) that 71 percent of Democrats said climate policy provided net benefits for the environment, but 43 percent of Republicans said these policies make no difference. Despite the effect of this phenomenon in terms of insufficient climate action, such an intense backlash against climate science may actually underscore the success of the environmental movement (Glaeser 2014; see Klein 2015).

Teaching global environmental politics

Even the most cursory review of GEP course syllabi would reveal that in addition to facilitating students' development of a comprehensive and critical understanding of the subfield, instructors also expect them to achieve some level of competency with respect to using knowledge and skills acquired in class to make more ecologically conscious future life decisions. Hence teaching GEP demands more than simply transmitting a body of knowledge concerning how state and non-state actors interact to solve transboundary environmental problems produced as a consequence of human behavior (Mitchell 2010). In light of the dual – ecological and pedagogical – demands of the GEP classroom, this section provides suggestions for incorporating environmentally conscious practices into coursework, in addition to an overview of how the substance of the field tends to be taught on college and university campuses around the world.

While there are many ways to teach GEP, courses often provided a categorical review of the existing academic knowledge and literature. Many college undergraduates arrive on campus with little, if any, awareness of "the environment" in the context of IR or comparative politics outside of, maybe, passing knowledge of the science of global warming and its relevance for the 1.5-degree Celsius target. Consequently, in addition to covering some subset of themes and issues in the field, courses on GEP tend to incorporate a brief history of the substance of GEP, as well as include details on the politics of scientific discovery, and the international treaty-making process.

Comprehensive GEP textbooks designed for undergraduates conform to instructors' demands for this body of material and feature sections on the history or emergence of GEP (see Chapter 2), actors in GEP (see Chapters 7, 8, 13 and 14), international environmental conflict and cooperation (see Chapter 20), environmental treaty-making and/or regime formation (see Chapters 9, 10 and 22), science and scientific uncertainty (see Chapters 6 and 18), subnational actors and their roles in governance (see Chapters 21, 28 and 29), capitalism and development or sustainability (see Chapters 16, 23 and 24), and selected themes or issues (see DeSombre 2017; O'Neill 2017; Dab Elko and Conca 2019).

An alternative to such a GEP "survey" for undergraduates would be an action-oriented course, based on a text like Lipchutz's *Global Environmental Politics*, which encourages readers to act because "only politics can save the environment" (Lipschutz 2004: xi). Lipschutz's critical text defines global environmental problems in terms of our consumption choices (see Chapter 17), then deconstructs "global environment" by way of revealing the relationship between the economic globalization and environmental degradation occurring outside our own front doors (see Chapter 24). A course that focuses on a specific issue area, such as global climate change, or a locally or regionally significant problem, like air pollution in Southern California or toxic waste in China, could incorporate a selected survey of applicable theoretical and thematic work in GEP.

The use of handbooks, such as this one, and more conceptually focused texts, including Mitchell's (2010) methodologically rigorous *International Politics and the Environment,* might be particularly appropriate for graduate students. GEP handbooks consolidate the field into a volume of high quality articles that synthesize research and provide a foundation for future scholarly work. Mitchell's mainstream, institutional text provides an overview of the GEP subfield and review of theory-building in IR with a definition of international environmental problems conceptually – i.e., as commons or upstream/downstream issues. It then explains theoretical perspectives in terms of how population, affluence, and available technologies interact to impact the natural environment (the IPAT identity), before turning to the negotiation of international treaties to address these impacts, and methods for assessing their effectiveness.

Empowering student learning and action

As a sub discipline concerned with humans' interrelationships with the natural environment and nonhuman life, experiential learning opportunities are well-suited for the GEP classroom (Maniates 2002). While experiential learning in the context of higher education is frequently interpreted as "internship" or "research practicum," potential modalities include field trips and other forms of outdoor education, as well as role-play and other games that provide practice in environmental politics and policy. Given the ongoing impetus to strengthen campus relationships with surrounding communities, place-based education is particularly relevant. This pedagogical approach features "meaningful contextual experiences intended to complement and expand classroom" (Knapp 1996 1). It draws on literatures from the natural and social sciences as well as the humanities indicate that our experiences in nature influence our relationship with nonhuman life and consequent behavior (Wilson 1984; Warren 1997; Milton 2002). Incorporating place-based opportunities in a GEP course can be as easy as requiring "field" work in the community in connection with a term paper or presentation, or it could involve offering extra credit for community service related to the natural environment.

Other fruitful didactic strategies for engaging students in the substance and practice of GEP are games and role-plays that allow students to grapple with the difficulties inherent in global efforts to achieve collective action and prevent the overuse of common pool resources under conditions of anarchy and asymmetries of power and wealth (Ghilardi-Lopes et al. 2013). Games and role-plays range from full-fledged simulations of international environmental negotiations to the prisoners' dilemma and resource exploitation scenarios like "fishbanks" (Meadows 2020). These interactive tools are easily replicable and adaptable to groups of different sizes, academic backgrounds or learning levels (Brown 2018). Additionally, many options exist to adjust the framework to different learning goals. Games and role-plays enable students to apply conceptual knowledge; train and enhance their problem-solving competences; and reflect on their efforts and progress.

Responding to ecological grief, anxiety and trauma

The declining state of world can be a source of ecological grief, eco-anxiety and post-traumatic stress disorder in the wake of natural disasters (Whitmore-Williams et al. 2017; Pihkala 2020). In addition to place-based, active and outdoor pedagogies, Pihkala (2020) suggests environmental educators can facilitate the emotional growth in students necessary to process and transform these debilitating conditions by incorporating contemplative practices. Mindfulness is key to stress reduction to improve psychological and physical health and wellbeing, for instance. It has also become more familiar in the classroom, where teaching students to "quiet the mind" increases their focus, self- and collective awareness, and overall learning potential (Odahowski 2004; Zajonc 2006). The introduction of mindfulness and other contemplative practices intended "to cultivate a personal capacity for deep concentration and insight" (Zajonc 2006) is characteristic of contemplative pedagogy (Brady 2007). The simple addition of a period of silent meditation at the beginning of class supports students' efforts to become present in and attentive to lecture and class discussion. In addition, contemplative reading, mindful walking and other forms of locomotion, focused experiences in nature, yoga and a number of other contemporary physical and artistic practices are likewise contemplative.

Conclusion

This chapter's approach to research in GEP, the role of scientific inquiry in the subfield, and teaching in GEP provides a framework informed by IR for approaching studies in the field that may be adapted for use with any substantive area of interest, as well as used to develop undergraduate and graduate coursework in GEP. Although the chapter reflects the subfield's emphasis on international environmental negotiations and institutional analyses, it recognizes the ongoing shift to the left associated with novel research on the potentially enhanced role for global civil society in global environmental governance. This emerging focus in GEP research is particularly well suited to encouraging greater use of alternative pedagogies in college courses on GEP and related topics. Outdoor experiences, (online) games or role-plays, and contemplative practices are among these alternatives for consideration by those who teach global environmental politics.

References

Abbott, Kenneth. W. (2014). Strengthening the transnational regime complex for climate change. *Transnational Environmental Law* 3(1): 57–88.

Abbott, Kenneth W., Green, Jessica F. and Keohane, Robert O. (2016). Organizational Ecology and Institutional Change in Global Governance. *International Organization* 70 (2):247–277.

Acosta, Alberto. (2017). Buen Vivir, in: Rosa, Hartmut; Hennig, Christoph (eds), *The Good Life Beyond Growth: New Perspectives*. New York: Routledge.

Alexander, Samuel. (2012). Planned Economic Contraction: The Emerging Case for Degrowth. *Environmental Politics* 21(3): 349–368.

Arora-Jonsson, Seema. (2013). *Gender, Development and Environmental Governance: Theorizing Connections*. Vol. 33. New York: Routledge.

Axworthy, Lloyd. (2001). Human Security and Global Governance: Putting People First. *Global Governance* 7: 19.

Barash, David. (2012). On Scientific Consensus. *Chronicle of Higher Education. Brainstorm.* Available HTTP: <http://chronicle.com/blogs/brainstorm/on-scientific-consensus/49895>.

Barnett, Jon. (2001). *The Meaning of Environmental Security: Ecological Politics and Policy in the New Security Era*. London: Zed Books.

Barnett, Jon, and Neil Adger, W. (2007). Climate Change, Human Security and Violent Conflict. *Political Geography* 26(6): 639–55. doi:10.1016/j.polgeo.2007.03.003.

Bernstein, Steven. (2001). *The Compromise of Liberal Environmentalism*. New York: Columbia University Press.

Betsill, M. and Bulkeley, H. (2003). *Cities and Climate Change* (Vol. 4). London: Routledge.

Betsill, Michele M., and Corell, Elisabeth. (2001). NGO Influence in International Environmental Negotiations: A Framework for Analysis. *Global Environmental Politics* 1(4): 65–85. doi:10.{1162/152638001317146372}.

Biermann, Frank, and Pattberg, Philipp. (2008). Global Environmental Governance: Taking Stock, Moving Forward. *Annual Review of Environment and Resources* 33(1): 277–94. doi:10.1146/annurev.environ.33.050707.085733.

Brady, Richard. (2007). Learning to Stop, Stopping to Learn: Discovering the Contemplative Dimension in Education. *Journal of Transformative Education* 5(4): 372–94. doi:10.{1177/1541344607313250}.

Brown, Joseph M. (2018). Efficient, Adaptable Simulations: A Case Study of a Climate Negotiation Game. *Journal of Political Science Education* 14(4): 511–522.

Bulkeley, H., Andonova, L.B., Betsill, M.M., Compagnon, D., Hale, T., Hoffmann, M.J., Newell, P., Paterson, M., VanDeveer, S.D. and Roger, C. (2014). *Transnational Climate Change Governance*. New York: Cambridge University Press.

Bulkeley, H. and Betsill, M.M. (2013). Revisiting the Urban Politics of Climate Change. *Environmental Politics* 22(1): 136–154.

Colgan, J.D., Keohane, R.O. and Van de Graaf, T. (2012). Punctuated equilibrium in the energy regime complex. *The Review of International Organizations* 7(2): 117–143.

Commission on Global Governance. (1995). *Our Global Neighborhood: The Report of the Commission on Global Governance*. New York: Oxford University Press.

Dabelko, Geoffrey D., and Conca, Ken, eds. (2019). *Green Planet Blues: Critical Perspectives on Global Environmental Politics*. New York: Routledge.

Dauvergne, Peter. (2010). *The Shadows of Consumption: Consequences for the Global Environment*. Cambridge, MA: MIT Press.

Dauvergne, Peter, and Clapp, Jennifer. (2016). Researching Global Environmental Politics in the 21st Century. *Global Environmental Politics* 16(1): 1–12.

de Oliveira, J.A.P. (2009). The Implementation of Climate Change Related Policies at the Subnational Level: An Analysis of Three Countries. *Habitat international* 33(3): 53–259.

DeSombre, Elizabeth R. (2017). *Global Environmental Institutions*. New York: Taylor and Francis.

Deudney, Daniel. (1990). The Case against Linking Environmental Degradation and National Security. *Millennium: Journal of International Studies* 19(3): 461–76.

Drezner, Daniel W. (2008). *All Politics Is Global: Explaining International Regulatory Regimes*. Princeton: Princeton University Press.

Dunlap, R.E. (2013). Climate Change Skepticism and Denial: An Introduction. *American Behavioral Scientist* 57(6): 691–698.

Eden, Sally. (2016). *Environmental Publics*. London: Routledge.

Engels, A., Hüther, O., Schäfer, M. and Held, H. (2013). Public Climate-Change Skepticism, Energy Preferences and Political Participation. *Global Environmental Change* 23(5): 1018–1027.

Fagan, Madeleine. (2017). Security in the Anthropocene: Environment, Ecology, Escape. *European Journal of International Relations* 23(2): 292–314.

Falkner, R. (2016). The Paris Agreement and the New Logic of International Climate Politics. *International Affairs* 92(5): 1107–1125.

Finger, Matthias, and Princen, Thomas. (1994). *Environmental NGOs in World Politics: Linking the Local and the Global*. 1st ed. London: Routledge.

Funk, Cary and Kennedy, Brian. (2020). How Americans See Climate Change and the Environment in 7 Charts. FactTank: News in the Numbers, April 21. Available at: https://www.pewresearch.org/fact-tank/2020/04/21/how-americans-see-climate-change-and-the-environment-in-7-charts/

Gellers, Joshua C. and Jeffords, Chris. (2018). Toward Environmental Democracy? Procedural Environmental Rights and Environmental Justice. *Global Environmental Politics* 18(1): 99–121.

Ghilardi-Lopes, Natalia Pirani, et al. (2013). Environmental Education through an Online Game about Global Environmental Changes and their Effects on Coastal and Marine Ecosystems. *Proceedings of SB Games* 469–474.

Glaeser, Edward L. (2014). The Supply of Environmentalism: Psychological Interventions and Economics. *Review of Environmental Economics and Policy* 8(2): 208–229.

Gleditsch, Nils Petter. (2012). Whither the Weather? Climate Change and Conflict. *Journal of Peace Research*, 49(1): 3–9.

Grossman, G., and Krueger, A. (1995). Economic Growth and the Environment. *Quarterly Journal of Economics* 110: 353–377.

Haas, Peter M. (2015). *Epistemic Communities, Constructivism, and International Environmental Politics*. New York: Routledge.

Haas, Peter M., Keohane, Robert O. and Levy, Marc A. (1993). *Institutions for the Earth: Sources of Effective International Environmental Protection*. Cambridge, MA: MIT Press.

Hale, T. (2016). "All hands on deck": The Paris Agreement and Nonstate Climate Action. *Global Environmental Politics* 16(3): 12–22.

Head, Tom. (2006). *Conversations with Carl Sagan*. Jackson: University Press of Mississippi.

Heise, Ursula K. (2008). *Sense of Place and Sense of Planet: The Environmental Imagination of the Global*. New York: Oxford University Press.

Hess, David J. (2009). *Localist Movements in a Global Economy: Sustainability, Justice, and Urban Development in the United States*. Cambridge, MA: MIT Press.

Hickmann, Thomas. (2017). The Reconfiguration of Authority in Global Climate Governance. *International Studies Review* 19(3): 430–451.

Homer-Dixon, Thomas F. (2010). *Environment, Scarcity, and Violence*. Princeton: Princeton University Press.

Jacques, Peter J. (2012). A General Theory of Climate Denial. *Global Environmental Politics* 12(2): 9–17.

Jacques, Peter J., Dunlap, Riley E. and Freeman, Mark (2008). The Organisation of Denial: Conservative Think Tanks and Environmental Scepticism. *Environmental Politics* 17(3): 349–385. doi:10.1080/09644010802055576.

Kaldor, Mary. (2007). *Human Security*. Cambridge, UK: Polity Press.

Kaplan, Robert D. (1994). The Coming Anarchy. *The Atlantic*, February. Available HTTP: <http://www.theatlantic.com/magazine/archive/1994/02/the-coming-anarchy/304670/>.

Kanie, N., and Biermann, F. (eds.) (2017). *Governing through Goals: Sustainable Development Goals as Governance innovation*. Cambridge, MA: MIT Press.

Keck, Margaret E., and Sikkink, Kathryn. (1998). *Activists beyond Borders: Advocacy Networks in International Politics*. Ithaca, NY: Cornell University Press.

Keohane, Robert O., and Victor, David G. (2011). The Regime Complex for Climate Change. *Perspectives on Politics* 9(1): 7–23.

King, Gary, and Murray, Christopher J.L. (2001). Rethinking Human Security. *Political Science Quarterly* 116(4): 585–610. doi:10.{2307/798222}.

Klare, Michael T. (2009). *Rising Powers, Shrinking Planet: The New Geopolitics of Energy*. New York: Holt.

Klare, Michael T. (2012). *The Race for What's Left: The Global Scramble for the World's Last Resources*. New York: Metropolitan Books.

Klein, Naomi. (2015). *This Changes Everything: Capitalism vs the Climate*. New York: Simon & Schuster.

Knapp, C.E. (1996). *Just beyond the Classroom: Community Adventures for Interdisciplinary Learning*. Charleston, WV: ERIC Clearinghouse on Rural Education and Small Schools.

Krasner, Stephen D., ed. (1983). *International Regimes*. Ithaca, NY: Cornell University Press.

Levy, Marc A., Young, Oran R., and Zürn, Michael. (1995). The Study of International Regimes. *European Journal of International Relations* 1(3): 267–330. doi:10.{1177/1354066195001003001}.

Leonard, H. Jeffrey. (2006). *Pollution and the Struggle for the World Product: Multinational Corporations*, Environment, and International Comparative Advantage. Cambridge, UK: Cambridge University Press.

Lipschutz, Ronnie D. (1992). Reconstructing World Politics: The Emergence of Global Civil Society. *Millennium: Journal of International Studies* 21(3): 389–420.

Lipschutz, Ronnie D. (2004). *Global Environmental Politics: Power, Perspectives, and Practice*. Washington, DC: CQ Press.

Litfin, Karen. (1994). *Ozone Discourses: Science and Politics in Global Environmental Cooperation*. New York: Columbia University Press.

Maniates, Michael. (2002). *Encountering Global Politics: Teaching, Learning, and Empowering Knowledge*. Lanham: Roman & Littlefield.

McKibben, Bill. (2007). *Deep Economy: The Wealth of Communities and the Durable Future*. New York: Times Books.

MacGregor, Sherilyn, ed. (2017). *Routledge Handbook of Gender and Environment*. Philadelphia: Taylor & Francis.

Matthew, Richard A., Barnett, Jon, McDonald, Bryan, and O'Brien, Karen L. eds. (2009). *Global Environmental Change and Human Security*. Cambridge, MA: MIT Press.

Meadows, Dennis. (2020). Website: https://mitsloan.mit.edu/LearningEdge/simulations/fishbanks/Pages/fish-banks.aspx

Meadows, Donella H., Randers, Jorgen and Meadows, Dennis L. (2004). *Limits to Growth: The 30-Year Update*. 3rd ed. Post Mills, VT: Chelsea Green.

Milton, Kay. (2002). *Loving Nature: Towards an Ecology of Emotion*. New York: Routledge.

Mitchell, Ronald B. (2010). *International Politics and the Environment*. Thousand Oaks, CA: Sage.

Nagel, Joane. (2015). *Gender and Climate Change: Impacts, Science, Policy*. New York: Routledge.

Nordmoe, Eric D. (2004). Of Poohsticks and p-Values: Hypothesis Testing in the Hundred Acre Wood. *Teaching Statistics* 26(2): 56–8. doi:10.1111/j.1467–9639.2004.00163.x.

Odahowski, Marga. (2004). Mindfulness in Higher Education. http://faculty.virginia.edu/odahowski/moddocs/MindfulnessinHigherEd.pdf.

O'Neill, Kate. (2004). Transnational Protest: States, Circuses, and Conflict at the Frontline of Global Politics 1. *International Studies Review* 6(2): 233–52. doi:10.1111/j.1521–9488.2004.00397.x.

O'Neill, Kate. (2017). *The Environment and International Relations*. New York: Cambridge University Press.

Pattberg, Philipp. (2007). *Private Institutions and Global Governance. The New Politics of Environmental Sustainability*. Cheltenham: Edward Elgar.

Peterson, Mildred J. (1992). "Whalers, Cetologists, Environmentalists, and the International Management of Whaling." *International Organization* 46(1): 147–86.

Peterson, Mildred J. (2019). *Contesting Global Environmental Knowledge, Norms and Governance*. London: Routledge.

Pihkala, P. (2020). Eco-Anxiety and Environmental Education. *Sustainability* 12(23), p.10149.

Princen, Thomas, Maniates, Michael and Conca, Ken eds. (2002). *Confronting Consumption*. Cambridge, MA: MIT Press.

Putnam, Robert D. (1988). Diplomacy and Domestic Politics: The Logic of Two-Level Games. *International Organization* 42(3): 427–60.

Sarewitz, Daniel. (2004). How Science Makes Environmental Controversies Worse. *Environmental Science & Policy* 7: 385–403.

Schlosberg, David. (2004). Reconceiving Environmental Justice: Global Movements and Political Theories. *Environmental Politics* 13(3): 517–40.

Schlosberg, D. (2013). Theorising Environmental Justice: The Expanding Sphere of a Discourse. *Environmental Politics* 22(1): 37–55.

Schreurs, M.A. (2008). From the Bottom Up: Local and Subnational Climate Change Politics. *The Journal of Environment & Development* 17(4): 343–355.

Schreurs, Miranda A., and Elizabeth Economy, eds. (1997). *The Internationalization of Environmental Protection.* Cambridge: Cambridge University Press.

Smulders, Sjak, Toman, Michael and Withagen, Cees. (2014). Growth Theory and 'Green Growth'. *Oxford Review of Economic Policy* 30(3): 424–446.

Steinberg, Paul F. (2001). *Environmental Leadership in Developing Countries: Transnational Relations and Biodiversity Policy in Costa Rica and Bolivia.* Cambridge, MA: MIT Press.

Stern, David. (2004). The Rise and Fall of the Environmental Kuznets Curve. *World Development* 32(8): 1419–1439.

United Nations. (2015). Transforming Our World: The 2030 Agenda for Sustainable Development. A/RES/70/1. New York: United Nations General Assembly.

United Nations. (2021). Statistics on Participation and in-session Engagement. https://unfccc.int/process-and-meetings/parties-non-party-stakeholders/non-party-stakeholders/statistics-on-non-party-stakeholders/statistics-on-participation-and-in-session-engagement

van der Ven, Hamish, Steven Bernstein, und Matthew Hoffmann. (2017). Valuing the Contributions of Nonstate and Subnational Actors to Climate Governance. *Global Environmental Politics* 17(1):1–20.

Victor, David, Kal Raustiala, and Eugene B. Skolnikoff, eds. (1998). *The Implementation and Effectiveness of International Environmental Commitments: Theory and Practice.* Cambridge, MA: MIT Press.

Wallace-Wells. (2020). *The Uninhabitable Earth: Life after Warming.* New York: Tom Duggan Books.

Walsh, Catherine. (2010). Development as Buen Vivir: Institutional Arrangements and (De)Colonial Entanglements. *Development* 53(1): 15–21.

Wapner, Paul. (2002). Horizontal Politics: Transnational Environmental Activism and Global Cultural Change. *Global Environmental Politics* 2(2): 37–62. doi:10.{1162/15263800260047826}.

Wapner, Paul. (2011). Civil Society and the Emergent Green Economy. *Review of Policy Research* 28(5): 525–30. doi:10.1111/j.1541–1338.2011.00520.x.

Ward, Hugh. (2006). International Linkages and Environmental Sustainability: The Effectiveness of the Regime Network. *Journal of Peace Research* 43(2): 149–66. doi:10.{1177/0022343306061545}.

Warren, Karen J., ed. (1997). *Ecofeminism: Women, Culture, Nature.* Bloomington: Indiana University Press.

WCED (World Commission on Environment and Development) (1987). *Our Common Future.* Oxford: Oxford University Press.

Whiting, Kate. (2020). 3 charts that show how attitudes to climate change vary around the world. World Economic Forum, January 22. Available at: https://www.weforum.org/agenda/2020/01/climate-science-global-warming-most-sceptics-country/.

Whitmore-Williams, Susan Clayton, Christie Manning, Kirra Krygsman and Meighen Speiser. (2017). Mental Health and our Changing Climate: Impacts, Implications and Guidance. ecoAmerica. Available at: https://www.apa.org/news/press/releases/2017/03/mental-health-climate.pdf

Widerberg, O. and Pattberg, P. (2017). Accountability Challenges in the Transnational Regime Complex for Climate Change. *Review of Policy Research* 34(1): 68–87.

Wilson, Edward O. (1984). *Biophilia.* Cambridge, MA: Harvard University Press.

Woertz, Eckart. (2014). Environment, Food Security and Conflict Narratives in the Middle East. *Global Environment* 7 (2): 490–516.

Young, Oran R., ed. (1999). *The Effectiveness of International Environmental Regimes: Causal Connections and Behavioral Mechanisms.* Cambridge, MA: MIT Press.

Young, Oran R. (2018). Research Strategies to Assess the Effectiveness of International Environmental Regimes. *Nature Sustainability* 1(9): 461–465.

Zajonc, Arthur. (2006). Contemplative and Transformative Pedagogy. *Kosmos Journal* 5(1): 1–6.

6

Advanced scholarship
Interdisciplinary research at the science–policy interface

Peter M. Haas, Ronald B. Mitchell and Leandra R. Gonçalves

How can we more effectively organize and publish meaningful research to help us better understand and respond to the global environment problems we face? This chapter provides suggestions for successful interdisciplinary research on international environmental politics, based on a review of published and unpublished works in the field. Usable science and knowledge are essential for devising effective environmental policies to address major global environmental threats, including climate change (see Chapter 32 and others chapters in Part V of this volume). Most policy analysts believe that better public discourse and elite deliberations require reliable knowledge that is accurate and socially legitimate (Haas 2004a; Mitchell et al. 2006). Accurate knowledge in the environmental domain must be interdisciplinary in order to capture the complex array of interactions between social and physical drivers that give rise to global environmental threats. Legitimate knowledge must enjoy a social pedigree, which in practice is often the peer-review process. For example, the Intergovernmental Panel on Climate Change (IPCC), the United Nations Environment Program's Global Environmental Outlook and the Intergovernmental Science-Policy Platform on Biodiversity and Ecosystem Services require that all information that it presents be published or accepted in peer-reviewed journals and books.

While this requirement leads to a lag in the dissemination of scientific knowledge to policy making, it does enforce the legitimacy of the knowledge that is being presented. Consequently, despite efforts by the well-known "climate denialists" (see Chapter 32) and ultimately to science denialists to delegitimize science over the last several years in any country, the integrity of the science was upheld by the courts and high-level oversight panels in each country.

Many scientists are frustrated that their work is not readily recognized in the policy community (Hulme 2009; Schneider 2009; Bradley 2011). One recent approach to science communication focuses on the rhetorical presentation of science and the psychological factors that influence its reception (Boykoff and Boykoff 2004; Leiserowitz et al. 2006; Boykoff 2011). Others look at the political constraints operating on governments that impede the reception of new information, which may require costly new measures (Hulme 2009), or from entrenched domestic interests in the United States (Oreskes 2007; Schneider 2009; Oreskes and Conway 2010; Bradley 2011). In this chapter we focus on the instrumental means by which

DOI: 10.4324/9781003008873-8

usable knowledge is generated and circulated (see also Chapter 18). Elsewhere, Haas has argued that credible science is provided by epistemic communities (Haas 2001, 2004a, 2004b, 2007, 2016). We focus on the published medium by which epistemic communities may better make their voices heard in the public discourse. We draw largely on experiences from 21 published manuscripts from the MIT Press series on Science, Politics and the Environment.

Although the causes and effects of global environmental problems tend to be multidisciplinary and interdisciplinary, modern scholars too often are disciplinary. The complexity of environmental issues – in terms of the number of interactions among variables, the length of causal chains, and the extent of interactions across time, space and scale – requires insights from multiple disciplines to capture accurately the extensive and multiple understandings of their causes, causal mechanisms and effects (Jacobson and Price 1990; Wiman 1991; Consortium for International Earth Science Information Network (CIESIN) 1992; Price 1992; National Research Council 1999a; Brewer and Stern 2005; Biermann 2007). The international community is starting to recognize that a complex global policy environment requires more sophisticated interdisciplinary insights. The 2030 Agenda and the 17 Sustainable Development Goals (SDGs) recently articulated under the auspices of the United Nations highlight the interdependent challenges for global society (Kanie and Bierman 2017; Sachs et al. 2019; Van wees et al. 2019).

Despite this, most scholars are trained – and often continue to think – in ways that are strongly disciplinary (Snow 1962). As Gary Brewer cleverly quipped, "the world has problems, but universities have departments" (Brewer 1999: 328). Addressing this disconnect between the problems we face and the solutions we offer is akin to reconciling different "epistemic cultures," i.e., the habits and beliefs associated with different academic disciplines (Knorr-Cetina 1999). Given this, how can we better organize and publish meaningful research to help us better understand and respond to the global environmental problems we face? (For more on research strategies, see Chapter 5.)

Interdisciplinarity and Sustainability Science

Since environmental problems emerged on the scholarly agenda in the 1970s, academics have debated the proper way to analyze their causes and effects. Views about the proper training of environmental scholars have changed significantly over time, with corresponding changes in terminology from "generalists" to "multidisciplinary," "interdisciplinary," "transdisciplinary" and "sustainability" scientists.

Alvin Weinberg called for "transdisciplinary" work that went beyond single discipline studies of environmental issues (Weinberg 1972). Others promoted the virtues of multidisciplinary work that drew on various disciplines. Tribe and colleagues noted that variation in analyses of a given environmental problem was likely to reflect, in large measure, the disciplinary values and perspectives of the analysts rather than real variation in the problem unless an interdisciplinary approach was used to help those from different disciplines converge on common values and methods (Tribe et al. 1976). Integrated assessment modelers, particularly in Europe in the 1990s, frustrated by their lack of influence on policymakers, argued for interdisciplinary work that included policymakers and stakeholders at the outset. Indeed, some have argued that environmental complexity exceeds the limits of traditional policy analysis and can only be meaningfully addressed through dialogs among such diverse groups (Ravetz 1986; Funtowicz and Ravetz 1991, 2001; Kasemir et al. 2003).

Training generalists was difficult in a disciplinary-based world. Universities lacked tenure track jobs for such individuals, either failing to hire them or placing them in programs (rather than departments) in which they trained few if any graduate students who could reproduce, develop and refine their ideas. It soon became clear that few individuals could master the array of tools and scope of knowledge to conduct environmental research.

By the 1980s, multidisciplinary had become the professional mantra, largely in response to the institutional incentive and individual capacity problems mentioned above. This approach saw the answer as building teams of scholars from diverse social science disciplines who individually could receive tenure and promotions within existing university structures but who collectively could shed better light on the complex environmental problems in question (Keohane and Ostrom 1995; Young 1997, 1999; Miles et al. 2002; Young et al. 2008; Young 2017). It was hoped that teams composed of individuals well versed in their own disciplines but interested in working with those from other disciplines could generate better insights by creating analytic synergies and identifying and removing disciplinary blind spots.

During the 1990s, this multidisciplinary perspective transitioned into an interdisciplinary one that sought to bridge the disciplinarian chasm that traditionally divides the social sciences from the natural sciences and engineering (Social Learning Group 2001a, 2001b; Miller and Edwards 2001; Schellnhuber et al. 2003; Jasanoff and Martello 2004). This shift urged greater collaboration across this chasm in an effort to progressively remedy the problem that social scientists often got the natural science wrong and natural scientists and engineers often got the social science wrong, with either error posing the risk that the science would be wrong and/or irrelevant to policymakers.

Policymakers have increasingly expressed their desire for "usable" science that was not only ecologically sound but was also politically, economically and sociologically informed, while scholars demonstrated an increasing desire to contribute to policy debates and a frustration that their work so rarely did so. Increasing attention was paid to those who were calling for transdisciplinary work. Such work sought to generate new theoretical frameworks for understanding social–ecological relationships rather than, as earlier work was accused of doing, simply trying to better understand the causes and effects of particular social–ecological problems (Jasanoff 2003, 2004; Kasemir et al. 2003; Brewer and Stern 2005). Such an approach aspires to forging a new theoretical framework for understanding environmental complexity that is drawn from a hands-on dialog between practitioners, civil society advocates and active scientists across the full spectrum of natural and social sciences and humanities. It also cautions against the hubris of a physics-based nomothetic approach to knowledge cumulation, rather focusing on deeper understandings of specific important problems through participatory learning.

More recently, scholars have called for interdisciplinary, international research teams that encompass not only academic researchers but also policymakers under an umbrella of Sustainability Science (Kates et al. 2001; Gallopin 2006; Clark and Harley 2020; see Chapter 16). It offers an interdisciplinary focus on the interactions between natural and social systems, and on how those interactions affect the challenge of sustainability (Kates et al. 2001). It is also a problem-driven field, that seeks to contribute with practical solutions that span from global to local scales, and it includes different perspectives from the global south and north. A review article of the evolution of Sustainability Science summarizes its purpose as "A science of sustainability necessarily requires collaboration between perspectives in developed and developing human societies, among theoretical and applied scientific disciplines, and must bridge the gap between theory, practice and policy" (Bettencourt & Kaure 2011: 19540). Sustainability Science does not just study the interactions between natural and social systems, but also aspires to govern them sustainably. It focuses on the salient spatial and temporal scales of the interplay, as well as imbuing decision-makers with the skills to govern such features by putting sustainability scientists in positions of authority.

Sustainability Science has been refined and promoted by the Harvard Kennedy School of Government, and the US National Academy of Sciences, with opportunities for publication in *PNAS*. It has been actively adopted by Future Earth and the International Council of Scientific Unions (ICSU), and the Earth System Governance project (https://www.earthsystemgovernance.org) (see Chapter 21). The Belmont Forum (https://www.belmontforum.org) has become a funding source for such activities. The 2019 Global Sustainable Development Report devoted a chapter to science for sustainable development, trumpeting the virtues of Sustainability Science for governing global issues, as well as the UN Sustainable Development Goals in particular (United Nations 2019). Sustainability Science also undergirds some major international policy documents, including the World Conservation Strategy, the Brundtland Report and Agenda 21. Still, it remains far more popular in the global north than south, as fewer scholars have been trained or taken jobs in the developing world.

For interdisciplinary research to be successful, it must involve individuals from a range of disciplines, each of whom is well trained in their own discipline; has some familiarity with the core concepts of other relevant disciplines; and is skilled in making the core concepts of their discipline accessible to other scholars, policymakers and stakeholders. Assembling teams of such scholars is thought to promote progressive research that generates new knowledge and new frameworks of understanding that could not, or would be unlikely to, emerge from a single discipline's perspective.

The US National Academy of Sciences proposed a division of labor for social–ecological research. In the National Academy's rubric, the social sciences can help explain the causes (or driving forces) of human behaviors that lead to global environmental change. The social sciences can also help explain the processes by which societies and decision-makers respond to identified threats and thus help better understand the likelihood, means and conditions that foster or inhibit alternative collective responses. The natural sciences can help explain how problems unfold and identify goals for sustainable responses. In turn, different disciplines can contribute in ways that relate to their core concepts: power and institutions from political science, markets and price signals from economics, public opinion and social attitudes from sociology and political science, local knowledge and organization from anthropology, issues of law and enforcement from legal scholars, and the like. Similarly, distinct fields of natural science can contribute insights into the behavior of different types of ecosystems (Rayner and Malone 1998; National Research Council 1999b; Biermann 2007). Including Indigenous and local community knowledge appears key for the better governance of biodiversity, deforestation and desertification (Xavier et al. 2018; Turnhout, Tuinstra, and Halffman 2019).

Such calls for interdisciplinarity and transdisciplinarity, of whatever sort, complement rather than replace more traditional disciplinary efforts. A full understanding of social–ecological systems will always require the deep disciplinary research that stays within more traditional disciplinary boundaries. For instance, in political science, Institutions for the Earth (Haas et al. 1993), a team-based project undertaken by political scientists, looked at the question of how international institutional design can improve the management of shared ecosystems, as well as international public goods (see Chapters 8 and 9). It found that institutions that enhance cooperation, concern and capacity were more likely to yield beneficial results than those without. Other groups of political scientists have confirmed that regimes with organized scientific involvement (epistemic communities) yield more comprehensive regulatory commitments and also better environmental outcomes than those without (Andresen et al. 2000; Miles et al. 2002; Haas 2007; Biermann and Pattberg 2012; see Chapter 18).

Conducting effective environmental policy research

How can effective research on global environmental issues be conducted? A key conclusion from this review of the philosophy of science for social–ecological research suggests at the very least that meaningful work is best performed by teams of scholars. Several recent books have also tried to develop heuristics for effective environmental policy research (Benda et al. 2002; Bergmann et al. 2005; Morin et al. 2020). Our judgments are based on our experiences as authors, as participants in interdisciplinary research projects, as editors of journals and book series, and as peer-reviewers for journals, publishers and foundations. For present purposes, we consider research as effective when it provides new insights into the causes or consequences of global environmental problems in ways that foster, in the short or long term, human society's ability to mitigate or adapt to those problems. Achievements in this realm can be observed (if not measured) by reference to the degree that research is published in peer-reviewed journals or with university presses, trains new scholars, and leads policymakers and stakeholders to accept new understandings of a problem and respond in more effective ways to mitigate or adapt to those problems.

The results of most past collective research projects in the global environmental politics arena, usually published as edited volumes, have tended to involve multiple chapters written by different, often multiple, scholars from various disciplines and countries. Such volumes often include authors at different career stages, from graduate students to senior professors. Building on our distinctions above, we distinguish two classes of research: interdisciplinary projects involving scholars from distinctly different disciplines including both social and natural scientists; and multidisciplinary projects involving scholars from a single discipline or a narrow range of cognate disciplines within the social (or natural sciences), such as political science, sociology, law and economics (Choucri 1993; Winter 2006).

To date, most published work has been multidisciplinary. Interdisciplinary work is more difficult to achieve, as discussed below, because of the difficulties in spanning disciplinary cultures and vocabularies. In general, while these efforts highlight insights from individual disciplines about a problem, they fail more generally to integrate them into a more coherent picture or even clearly to articulate the compatibility or tensions between different approaches (Cebon et al. 1998; Social Learning Group 2001a, 2001b). In short, truly interdisciplinary work remains in its infancy with considerable room for improvement.

To foster progress in that venture, the following section reflects our thoughts for improving, and publishing, both multidisciplinary and interdisciplinary work on global environmental problems. While successful multidisciplinary and interdisciplinary work may generate new integrated wisdom, it may also reveal uncertainties and fundamental differences in understanding between actors and disciplines.

Applications of interdisciplinarity

Here we provide three exemplars of interdisciplinary books whose findings exceed the conventional views of single disciplines. *Changing the Atmosphere* (Miller and Edwards 2001) has ten chapters written by nine authors, ranging from PhD candidates to full professors. The authors come from information sciences, philosophy, social studies of science, biology and climate science. The research was well supported by a variety of grants. This collection was one of the earlier social science investigations of the production and use of climate science for policy. Thus, it had a comprehensive introduction, providing an overview of the critical social studies of science literature, but lacked a concluding chapter. The empirical chapters demonstrate the greater role of interpretation and uncertainty associated with scientific advice and the IPCC than was generally recognized by hard scientists and policy analysts (see Chapters 18 and 19). It developed the finding that science and science–policy does not directly mirror the natural world, but rather that it interprets the world for policy and political consumers in ways that are socially and politically shaped. The effective provision of scientific information requires political and social inquiry about the frames and context within which policymakers solicit and understand scientific advice. Policy studies need to better understand the degree of distortion involved in the knowledge being delivered, and to focus on the political processes by which choices about knowledge claims are made and the knowledge is itself interpreted by less technically trained policymakers.

The Reflexive Governance for Global Public Goods (Brousseau et al. 2012) provides an interdisciplinary investigation of global public goods; an analytic category that includes climate change. *Reflexive Governance* has 15 chapters as well as an introduction and conclusion, written by 21 international contributors, drawn from research fellows, assistant professors to full professors, and one government official. Substantively, they include economics, ecological economics, philosophy, politics, and interdisciplinary training in environment change. The interdisciplinary approach to global public goods complements conventional studies of international public goods that seek to internalize the costs of environmental degradation through hierarchical controls, market arrangements to internalize costs, or institutional arrangements to concentrate the environmental consequences. By studying a number of public goods occurring at different scales and with different participants, the authors find that the provision of organized scientific knowledge is capable of educating political actors to change their behavior and take account of environmental externalities, which remain economically low cost. In this regard the volume is "reflexive" in documenting knowledge about how knowledge may be usefully integrated by national-level decision-makers to learn about climate change, and to embark on new policies that are more sustainable. Such collective reflection requires democratic participation, scientific information and a lengthy social process of deliberation (Dedeurwaerdere et al. 2012: 316–317).

Governance and Environmental Planning: adaptation and public policies in the Macrometrópole Paulista (Torres et al. 2020) is a collective effort to discuss climate change adaptation from an interdisciplinary point of view. The authors include economists, engineers, biologists, social scientists, lawyers, urban planners and oceanographers and all chapters aiming to discuss how a climate change adaptation policy should be carried to promote environmental governance within different perspectives. The book presented various conceptualization approaches, methods and critical thinking in an attempt to integrate epistemologies.

Improving interdisciplinary and multidisciplinary research

What are the factors that support (or not) publishing interdisciplinary research in peer-reviewed journals? In our view, conducting and publishing effective research requires that the scholars design the research in ways that meet the three criteria delineated. This conclusion is confirmed by an analysis that additionally mentioned six factors that would ideally foster an interdisciplinary publication, such as "(1) a strong, interdisciplinary coordinator, (2) a clear shared vision of integration and a common framework, (3) flexibility in terms of money and time, (4) a certain sense of timing regarding when and how to exchange results and knowledge, (5) subject editors who are familiar with the specific project and its interdisciplinary merits, and (6) reviewers who are open minded about interdisciplinary efforts" (Pohl et al. 2015).

Yet, where peer-reviewed articles may raise challenges to publication, books, introduce a great opportunity to assemble a team of experts to approach a subject within an interdisciplinary perspective.

Selecting participants

The first step in developing successful interdisciplinary research is the selection of the research team. Individuals should be chosen on the basis of their depth of disciplinary expertise and their ability to communicate clearly about their discipline with those from other disciplines. Individuals also should be chosen to create an "expert team" rather than a "team of experts." An expert team consists of a set of scholars who have individual skills but also, collectively, represent the range of disciplines necessary to accurately evaluate and analyze the environmental problem in question and who also have the interpersonal skills that help a team run well. These include the ability and willingness to provide honest yet constructive feedback to others, to listen and respond quickly and well to such feedback from others, and to contribute to the project's overall goals, especially when that means altering individual research approaches and processes to foster those goals.

In addition, several benefits arise from having multiple ranks represented within a team. Junior scholars benefit from the explicit and implicit training and mentoring from more senior scholars with more extensive and varied experience who can demonstrate various solutions to the inevitable problems that arise in collective research. Senior scholars benefit from the intense exposure to and interaction with those trained in the most current research and methodological developments and by being challenged to respond to, rather than merely read about, alternative perspectives on various issues. Such interactions may help overcome the theoretical myopia that can develop in senior researchers who have worked within their own traditionally defined boundaries for most of their careers.

Additionally, beyond generational balance within a team, it is also valuable to consider gender balance as well as a regional balance among the north and south countries. This will support a wider view of the problem, and it will bring a range of experiences to contribute to the team.

There are several obstacles to building such a team. One is that most networks of scholars are built within rather than across disciplines. Most scholars' networks include those who went to graduate school together and those who meet by going to the annual conventions of their own discipline. Institutional incentives reinforce the need to write papers that will be published in one's own discipline's journals and to "build a reputation" in that discipline and discourage the time "wasted" going to conferences, engaging in collaborations, and networking with those from other disciplines. The challenge is to identify and recruit people who either have found ways to achieve traditional measures of disciplinary success while retaining both the time and inclination to engage in interdisciplinary work or have found less traditional research trajectories in places such as the Santa Fe Institute. Few graduate or undergraduate programs yet provide meaningful training.

We believe that policymakers and stakeholders can make significant contributions to interdisciplinary research teams. One useful model involves having policymakers and stakeholders involved in initial research project meetings to ensure that the research questions are framed in ways that promote salient research results that stand some chance of contributing to upcoming policy decisions in ways that are sensitive to existing political, financial, and social constraints and perspectives (Mitchell et al. 2006). Briefing these policymakers and stakeholders at regular intervals during the research process also allows for "course corrections" that can improve the "uptake" of the ultimate conclusions without making them susceptible to the influence of these groups. An obstacle that may need to be overcome exists in the relatively brief job tenure and demanding time schedule of individual policymakers and civil society members. Thus, involving individuals in such an enterprise runs the risk of discontinuities as members drop off and replacements bring in new agendas. Having briefing sessions with a broader community at the beginning and end of the research process, rather than relying directly on a cadre of individuals, offers an alternative solution (see Chapter 14).

Finally, we believe there is a "Goldilocks" problem in terms of team size. Interdisciplinary teams, to be successful, must contain sufficient expertise to address the array of perspectives and disciplines that can contribute to analyzing the problem in truly interdisciplinary ways. At the same time, teams that exceed 10 to 15 individuals can present a range of cost and logistical problems that can prove challenging for the organizers and can undermine team members' sense that their contributions are crucial to the team goals.

Building a team

Once participants have been selected, the next step in effective interdisciplinary research is building a team. Perhaps most important to doing so is the need to develop effective communication among team members, taking time to understand both the terminology and perspectives of the other scholars involved. Different disciplines can use the same word or phrase to mean completely different things and, at times, can use different words or phrases to mean the same thing (consider the difference in what a "climate regime" means to an atmospheric scientist and a political scientist). Equally important, but often harder to get at, are the more subterranean assumptions, methodologies and "ways of thinking" that are deeply embedded in each discipline. Without intending to stereotype, economists may be more comfortable monetizing certain human values, physicists may see the world in more mechanistic terms, anthropologists may be less comfortable generalizing across different cultures, etc. Mutual understanding of and, equally important, respect for, these "cultural differences" requires an ongoing process that tends to require considerable in-person interaction and may take a year or more. Open and explicit discussions of disciplinary semantics and methodologies can help identify often broad and deep divergences in outlooks and approaches. Such efforts are crucial to development of a common but integrated understanding of the environmental problem that the scholars seek to understand.

The success of "team-building" also requires explicitly and directly addressing the task of designing an internally consistent framework that accurately and usefully integrates the different disciplines and perspectives of the scholars involved. When such efforts are undertaken and succeed, truly interdisciplinary work can emerge that creates synergies from the contributing scholars. When such efforts fail, edited volumes whose chapters nominally address the same problem may prove quite non-cumulative, with insights from many chapters being ignored, misunderstood, or not taken advantage of with the result that meaningful communication across disciplines fails to emerge.

Overcoming these problems often benefits from strong editorial leadership that develops support for, and if necessary, imposes, a common framework for analyzing the problem, either with all contributing scholars applying the same framework or each scholar accurately using their own disciplinary tools to contribute to the overall framework. Procedurally, this often requires frequent face-to-face meetings throughout the course of the research project – and often more meetings than seem necessary – to develop a coherent common framework, to ensure collective understanding of that framework, to foster consistent application of that framework within individual chapters, and to develop careful cross-chapter insights as the project moves toward conclusion.

Sources of knowledge

One great challenge to promote interdisciplinarity is to build knowledge including different sources, from academic peers as well as diverse actors, such as Indigenous and local communities with knowledge from practice. A more recent example of this effort is The Intergovernmental Science–Policy Platform on Biodiversity and Ecosystem Services. Although they have accumulated some lessons learned, still there are challenges to promoting inclusion and to working with Indigenous, local and scientific knowledge (Pascual et al. 2017; Díaz-Reviriego, Turnhout, and Beck 2019; Hill et al. 2020).

Yet improving participation and inclusiveness has been highlighted as important to the co-production of knowledge and influence the decision-making process (Fischhoff 2019), which involves the collaborative creation of knowledge that may be recognized as usable knowledge. Still, there are challenges and barriers to involve and incorporate different sources of knowledge, such as collating and validating local knowledge, and ensuring the appropriate and fair participation of various stakeholder groups (Sutherland et al. 2014).

Developing coherent and collective findings

To ensure a project generates strong interdisciplinary insights and presents them in a coherent manuscript requires iterative interactions among those contributors analyzing the individual cases and the editors developing the collective conclusions. Reinforcing the need for "strong leadership" noted above, the need for a strong leader or team of leaders becomes particularly important as a project moves to completion. These individuals must, from the outset, clarify both the standards and deadlines they will use for including or excluding chapters in any final published manuscript. Projects are too often delayed by one or two scholars who deliver their manuscripts late or provide manuscripts of demonstrably lower quality than others planned for inclusion. Although telling a team member that their contribution will not be included is unlikely to be pleasant for either party, they are easier when the criteria for such a decision have been delineated and understood at the outset.

Beyond these logistical points, the editors of collective volumes owe an obligation to their contributors to engage in the careful cross-case comparisons that are necessary to identifying common patterns and themes and to deriving both backward-looking conclusions and forward-looking conjectures. Editors should plan on blocking out the requisite three to six months of time needed to carefully read the contributed analyses, identify and write up interesting patterns, analyze the comparisons carefully, have their findings reviewed by all contributing authors, and revise the conclusions and introduction so that they simultaneously meet the goals of abstracting from the individual cases without doing injustices to the empirical evidence from those cases.

Training scholars

Beyond their intellectual benefits, interdisciplinary research projects that contain both senior and junior scholars provide excellent opportunities for mentoring. In-person interactions as well as those by phone or email, provide excellent opportunities for senior scholars to advise junior scholars on "threading the needle" of conducting research that is publishable in disciplinary journals and fosters professional advancement, that contributes to interdisciplinary understanding of important environmental problems, and that helps stakeholders and policymakers improve human responses to the environmental problems being studied.

Equally important, relationships that develop over the two- to ten-year timelines common to such projects provide the basis for respected senior scholars to write compelling letters of recommendation for interdisciplinary junior scholars seeking jobs or promotion in a world that remains, unfortunately, highly disciplinary.

These training and mentoring benefits can be fostered, especially for junior scholars, by developing a common team identity. This can be promoted by having a central institutional home for the research team, with a critical mass of PhD candidates, post-docs, and faculty that can interact regularly over the course of two or three years. Where such intensive inter-actions are not possible, ensuring that dedicated research team meetings are combined with more ad hoc meetings involving those team members that happen to be at annual conven-tions, particularly when team findings are presented at those meetings, can help considerably. Annual "retreats" at relatively isolated locations can also improve team esprit de corps and promote possibilities for following up themes more carefully than can occur in briefer more structured settings and can also facilitate more serendipitous interactions with benefits in terms of concept formation, analytic insights and development of future collaborations.

Crossing the academic–policy divide

A crucial aspiration of many scholars involved in studying social–ecological systems is to have their scholarship contribute to the mitigation and resolution of specific environmental problems and, more generally, to the improvement of the relationship humans have with the natural world. Yet understanding the conditions under which and processes by which good scholarship becomes usable and used knowledge remains a poorly understood element of social–ecological work (Mitchell et al. 2006). Indeed, the current popularity of Sustainability Science reflects, at least in part, an effort to improve the ways social–ecological scholarship is produced and presented to make it more usable and thereby overcome existing political disinterest and resistance that fail to lead to usable knowledge actually being used.

In the short term and at an initial level, scholars can increase the contribution they make to policy by self-consciously attempting to understand, and conduct their research in ways that reflect and respond to, the political and policy opportunities and constraints that often are the cause of scholarly irrelevance. Research often fails to be "salient," in the sense of being relevant to current policy decisions – it comes in before the policy recommendations being offered have any chance of success or after the policy "window of opportunity" has closed (Kingdon 2003; Mitchell et al. 2006). Equally important, scholars often confuse what "should be" the constraints and opportunities with what are those constraints and opportu-nities. In this vein of "small changes," it certainly also makes sense for scholars to carefully develop "summaries for policymakers," to provide policy briefings to those working on the issue, and to entertain the wide range of other opportunities to communicate with and provide inputs to policymakers and decision-makers. Dual conclusions, aimed at academic researchers and policymakers, are another an imaginative technique (Miles et al. 2002).

Conclusion

The ability for scholars to have a larger and more long-lasting influence with policymakers and stakeholders requires a deeper change in how research is conducted and how scholars are being trained. Notions of "co-production" of knowledge and of "adaptive management" involve ongoing interactions among scholars (both natural and social scientists), policymak-ers, diverse stakeholders (e.g., Indigenous and local communities) and resource managers

(Jasanoff 2004). In this model, the sequestered generation of knowledge by scholars that is published and handed off to policymakers and others in policy briefings is replaced by efforts to build social institutions that involve relatively frequent interactions over several years in which trust and understanding can develop in ways that are designed to avoid political pressures influencing scientific findings while, at the same time, ensuring that political constraints are recognized as creating important boundaries within which policy recommendations must fall (even if, over the longer term, those boundaries themselves may be subject to change). Such co-production institutions and bridging organizations allow policymakers and stakeholders to realize the value of, and better understand natural and social science insights; provide managers with better insights into novel techniques for addressing their day-to-day problems; and help scholars have a better sense of existing policy constraints and opportunities and why they exist.

These approaches are likely to be more challenging, more time-consuming and slower to "bear fruit" than more traditional strategies of publishing scholarship and hoping it has influence. But they offer the promise of allowing scholars to have significantly more influence than they would otherwise. Such strategies also require scholars to think carefully about how they maintain their scientific impartiality and credibility while improving their policy-relevance, what Stephen Schneider has called the "double ethical bind" of being politically effective while being scientifically accurate and honest (Russill 2010).

Acknowledgments

Peter Haas recognizes the Wissenschaftczentrum Berlin for support during early work on this chapter. Leandra R. Gonçalves acknowledges the Fundação de Amparo à Pesquisa do Estado de São Paulo (Fapesp) for financial support (FAPESP LRG: 2018/00462-8). Our thanks to Clay Morgan for assisting with our publications database.

References

Andresen, S., T. Skodvin, A. Underdal, and J. Wettestad (2000). *Science and Politics in International Environmental Regimes*. Manchester: Manchester University Press.

Benda, L. E., N. L. Poff, C. Tague, M. A. Palmer, J. Pizzuto, S. D. Cooper et al. (2002). How to Avoid Train Wrecks When Using Science in Environmental Problem Solving. *BioScience* 52(12): 1127–1136.

Bergmann, M., B. Brohmann, E. Hoffmann, M. C. Loibl, R. Rehaag, E. Schramm et al., Eds. (2005). *Quality Criteria of Transdisciplinary Research*. Frankfurt am Main, Germany: Institute for Social–Ecological Research (ISOE).

Bettencourt, L. M., and J. Kaure (2011). Evolution and Structure of Sustainability Science. *PNAS* 108(49): 19540–19545.

Biermann, F. (2007). 'Earth System Governance' as a Cross-Cutting Theme of Global Change Research. *Global Environmental Change* 17: 326–337.

Biermann, F. and P. Pattberg, Eds. (2012). *Global Environmental Governance Reconsidered*. Cambridge, MA: MIT Press.

Boykoff, M. T. (2011). *Who Speaks for the Climate?* Cambridge: Cambridge University Press.

Boykoff, M. T. and J. M. Boykoff (2004). Balance as Bias: Global Warming and the US Prestige Press. *Global Environmental Change* 14(2): 125–136.

Bradley, R. S. (2011). *Global Warming and Political Intimidation*. Amherst: University of Massachusetts Press.

Brewer, G. and P. C. Stern, Eds. (2005). *Decision Making for the Environment*. Washington, DC: National Academies Press.

Brewer, G. D. (1999). The Challenges of Interdisciplinarity. *Policy Sciences* 32: 327–337.

Brousseau, E., T. Dedeurwaerdere, and B. Siebenhüner, Eds. (2012). *Reflexive Governance for Global Public Goods*. Cambridge, MA: MIT Press.

Cebon, P., U. Dahinden, H. Davies, D. Imboden, and C. C. Jaeger, Eds. (1998). *Views from the Alps: Regional Perspectives on Climate Change*. Cambridge, MA: MIT Press.

Choucri, N. (1993). *Global Accord*. Cambridge, MA: MIT Press.

Clark, W.C., and A.G. Harley (2020). Sustainability Science: Towards a Synthesis, Sustainability Science Program. Consortium for International Earth Science Information Network (CIESIN) (1992). *Pathways of Understanding*. University Center, MI: CIESIN.

Dedeurwaerdere, T., E. Brousseau, and B. Siebenhüner (2012). Conclusion. In E. Brousseau, T. Dedeurwaerdere, and B. Siebenhüner, Eds., *Reflexive Governance and Global Public Goods*. Cambridge, MA: MIT Press.

Díaz-Reviriego, Isabel, E. Turnhout, and S. Beck (2019). Participation and Inclusiveness in the Intergovernmental Science–Policy Platform on Biodiversity and Ecosystem Services. *Nature Sustainability* 2 (6): 457–464. https://doi.org/10.1038/s41893-019-0290-6.

Fischhoff, Baruch. (2019). Evaluating Science Communication. *Proceedings of the National Academy of Sciences* 116 (16): 7670–7675. https://doi.org/10.1073/pnas.1805863115.

Funtowicz, S. O. and J. R. Ravetz (1991). A New Scientific Methodology for Global Environmental Issues. In R. Costanza, Ed., *Ecological Economics*. New York: Columbia University Press, 137–152.

Funtowicz, S. O. and J. R. Ravetz (2001). Global Risk, Uncertainty, and Ignorance. In J. X. Kasperson and R. Kasperson, Eds., *Global Environmental Risk*. London: Earthscan.

Gallopin, G. C. (2006). Linkages between Vulnerability, Resilience and Adaptive Capacity. *Global Environmental Change* 16(3): 293–303.

Haas, P. M. (2001). Epistemic Communities and Policy Knowledge. In N. J. Smelser and P. B. Baltes, Eds., *International Encyclopedia of Social and Behavioral Sciences*. Oxford: Elsevier Science, 11578–11586.

Haas, P. M. (2004a). Science Policy for Multilateral Environmental Governance. In N. Kanie and P. Haas, Eds., *Emerging Forces in Environmental Governance*. Tokyo: UNU Press, 115–136.

Haas, P. M. (2004b). When Does Power Listen to Truth? A Constructivist Approach to the Policy Process. *Journal of European Public Policy* 11(4): 569–592.

Haas, P. M. (2007). Epistemic Communities. In D. Bodansky, J. Brunnee, and E. Hey, Eds., *Oxford Handbook of International Environmental Law*. New York: Oxford University Press, 791–806.

Haas, P. M. (2016). *Epistemic Communities, Constructivism, and International Environmental Politics*. New York, NY: Routledge.

Haas, P. M., R. O. Keohane, and M. A. Levy, Eds. (1993). *Institutions for the Earth*. Cambridge, MA: MIT Press.

Hill, R., Ç. Adem, W. V. Alangui, Z. Molnár, Y. Aumeeruddy-Thomas, P. Bridgewater, M. Tengö, R. Thaman, C.Y. Adou Yao, F. Berkes, J. Carino, M. Carneiro da Cunha, M.C. Diaw, S. Díaz, V.E. Figueroa, J. Fisher, P. Hardison, K. Ichikawa, P. Kariuki, M. Karki, P.O. Lyver, P. Malmer, O. Masardule, A.A. Oteng Yeboah, D. Pacheco, T. Pataridze, E. Perez, M.M. Roué, H. Roba, J. Rubis, O. Saito, and D. Xue (2020). Working with Indigenous, Local and Scientific Knowledge in Assessments of Nature And Nature's Linkages with People. *Current Opinion in Environmental Sustainability* 43: 8–20. https://doi.org/10.1016/j.cosust.2019.12.006

Hulme, M. (2009). *Why We Disagree about Climate Change*. Cambridge: Cambridge University Press.

Jacobson, H. K. and M. F. Price (1990). *A Framework for Research on the Human Dimensions of Global Environmental Change*. Paris: International Social Science Council.

Jasanoff, S. (2003). Technologies of Humility: Citizen Participation in Governing Science. *Minerva* 41: 223–244.

Jasanoff, S. Ed. (2004). *States of Knowledge: The Co-Production of Science and Social Order*. New York: Routledge.

Jasanoff, S. and M. B. Martello, Eds. (2004). *Localizing and Globalizing: Knowledge Cultures of Environment and Development*. Cambridge, MA: MIT Press.

Kanie, N. and F. Biermann (2017). *Governing through Goals Sustainable Development Goals as Governance Innovation*, 1st ed. London: MIT Press Series.

Kasemir, B. J., J. Jäger, C.C. Jaeger, and M.T. Gardner, Eds. (2003). *Public Participation in Sustainability Science*. Cambridge: Cambridge University Press.

Kates, R.W., W.C. Clark, R. Corell, J.M. Hall, C.C. Jaeger, I. Lowe, J.J. Mccarthy, H.J. Schellnhuber, B. Bolin, N.M. Dickson, S. Faucheux, G.C. Gallopin, A. Grubler, B. Huntley, J. Jager, N.S. Jodha,

R.E. Kasperson, A. Mabogunje, P. Matson, H. Mooney, B.M. Ill, T.O. Riordan, and U. Svedin (2001). *Sustainability Science* 292: 641–642.

Keohane, R. O. and E. Ostrom Eds. (1995). *Local Commons and Global Interdependence*. Thousand Oaks, CA: Sage Publications.

Kingdon, J. W. (2003). *Agendas, Alternatives, and Public Policies*. New York: Longman.

Knorr-Cetina, K. (1999). *Epistemic Cultures: How the Sciences Make Knowledge*. Cambridge, MA: Harvard University Press.

Leiserowitz, A. A., R. W. Kates and T. Parris (2006). Sustainability Values, Attitutudes, and Behaviors: A Review of Multinational and Global Trends. *Annual Review of Environmental Resources* 31: 413–444.

Miles, E. L., S. Andresen, E. M. Carlin, J. B. Skjærseth, A. Underdal and J. Wettestad, Eds. (2002). *Environmental Regime Effectiveness: Confronting Theory with Evidence*. Cambridge, MA: MIT Press.

Miller, C. A. and P. N. Edwards (2001). *Changing the Atmosphere*. Cambridge, MA: MIT Press.

Mitchell, R. B., W. C. Clark, D. W. Cash and N. M. Dickson, Eds. (2006). *Global Environmental Assessments: Information and Influence*. Cambridge, MA: MIT Press.

Morin, Jean-Frédéric, Amandine Orsini and Sikina Jinnah (2020). *Global Environmental Politics: Understanding the Governance of the Earth*. United Kingdom: Oxford University Press.

National Research Council (1999a). *Our Common Journey: A Transition toward Sustainability*. Washington, DC: National Academy Press.

National Research Council (1999b). *Human Dimensions of Global Environmental Change. By Committee on the Human Dimensions of Global Change and Committee on Global Change Research, National Research Council*. Washington, DC: National Academy Press.

Oreskes, N. (2007). The Scientific Consensus on Climate Change. In J. F. C. DiMention and P. Doughman, Eds., *Climate Change*. Cambridge, MA: MIT Press.

Oreskes, N. and E. M. Conway (2010). *Merchants of Doubt*. New York: Bloomsbury Press.

Pascual, U., P. Balvanera, S. Díaz, G. Pataki, E. Roth, M. Stenseke and N. Yagi (2017). Valuing Nature's Contributions to People: The IPBES Approach. *Current Opinion in Environmental Sustainability* 26: 7–16.

Pohl, C., G. Wuelser, P. Bebi, H. Bugmann, A. Buttler, C. Elkin, A. Grêt-Regamey, C. Hirschi, Q. B. Le, A. Peringer, A. Rigling, R. Seidl and R. Huber (2015). How to Successfully Publish Interdisciplinary Research: Learning from an Ecology and Society Special Feature. *Ecology and Society* 20(2): 23. http://dx.doi.org/10.5751/ES-07448-200223

Price, M. F. (1992). The Evolution of Global Environmental Change. *Impact of Science on Society* 166: 171–182.

Ravetz, J. R. (1986). Usable knowledge, Usable Ignorance: Incomplete Science with Policy Implications. In W. C. Clark and R. E. Munn. Eds., *Sustainable Development of the Biosphere*. New York: Cambridge University Press, 415–432.

Rayner, S. and E. L. Malone, Eds. (1998). *Human Choice and Climate Change*. Colombus, OH: Batelle Press.

Russill, C. (2010). Stephen Schneider and the 'Double Ethical Bind' of Climate Change Communication. *Bulletin of Science, Technology and Society* 30: 60–69.

Sachs, J.D., G. Schmidt-Traub, M. Mazzucato, D. Messner, N. Nakicenovic and J. Rockström (2019). Six Transformations to Achieve the Sustainable Development Goals. *Nature Sustainability* 2: 805–814. https://doi.org/10.1038/s41893-019-0352-9

Schellnhuber, H.J., P.J. Crutzen, W.C. Clark, M. Claussen and H. Held, Eds. (2003). *Earth System Analysis for Sustainability*. Cambridge, MA: MIT Press.

Schneider, S.H. (2009). *Science as a Contact Sport*. Washington, DC: National Geographic.

Snow, C.P. (1962). *The Two Cultures and the Scientific Revolution*. New York: Cambridge Univeristy Press.

Social Learning Group, Ed. (2001a). *Learning to Manage Global Environmental Risks, Volume 1: A Comparative History of Social Responses to Climate Change, Ozone Depletion and Acid Rain*. Cambridge, MA: MIT Press.

Social Learning Group, Ed. (2001b). *Learning to Manage Global Environmental Risks, Volume 2: A Functional Analysis of Social Responses to Climate Change, Ozone Depletion and Acid Rain*. Cambridge, MA: MIT Press.

Sutherland, W. J., Gardner, T. A., Haider, L. J., & Dicks, L. V. (2014). How can local and traditional knowledge be effectively incorporated into international assessments. *Oryx* 48(1): 1–2.

Torres, P.H.C., P.R. Jacobi, F. Barbi, and L.R. Gonçalves (2020). *Adaptation and Public Policies in the São Paulo Macrometropolis: A Science-Policy Approach*, 1st ed.; São Paulo. ISBN 978–86923-59-3.

Tribe, L.T., C. Schelling and J. Voss, Eds. (1976). *When Values Conflict.* Cambridge, MA: Ballinger.

Turnhout, Esther, Willemijn Tuinstra and Willem Halffman. 2019. *Environmental Expertise: Connecting Science, Policy, and Society.* Cambridge; New York: Cambridge University Press.

United Nations. (2019). *Global Sustainable Development Report.* New York: United Nations.

Van Wees, S.L., H. Målqvist and R. Irwin (2019). Achieving the SDGs through Interdisciplinary Research in Global Health. *Scandinavian Journal of Public Health* 47: 793–795. https://doi.org/10.1177/1403494818812637

Weinberg, A. (1972). Science and Trans-Science. *Minerva* 10: 2009–2222.

Wiman, B.L.B. (1991). Implications of Environmental Complexity for Science and Policy. *Global Environmental Change* 1: 235–247.

Winter, G., Ed. (2006). *Multilevel Governance of Global Environmental Change.* Cambridge: Cambridge University Press.

Xavier, L.Y., P.R. Jacobi, and A. Turra (2018). On the Advantages of Working Together: Social Learning and Knowledge Integration in the Management of Marine Areas. *Marine Policy* 88: 139–150.

Young, O.R. (1999). *The Effectiveness of International Environmental Regimes: Causal Connections and Behavioral Mechanisms.* Cambridge, MA: MIT Press.

Young, O.R., Ed. (1997). *Global Governance: Drawing Insights from the Environmental Experience.* Cambridge, MA: MIT Press.

Young, O.R. (2017). *Governing Complex Systems. Social Capital for the Anthropocene.* Cambridge, MA: MIT Press.

Young, O.R., L.A. King and H. Schroeder., Eds. (2008). *Institutions and Environmental Change: Principal Findings, Applications, and Research Frontiers.* Cambridge, MA: MIT Press.

Part III

Actors and institutions in global environmental politics

7

States

Nations, sovereignty and the international system

Hugh C. Dyer

The idea of a "nation" suggests a degree of unity among a defined group of people. Like other "ideal types," the term is used for simplicity or clarity, but the idea of nationhood disguises the complexity of social history. There are few groups of people so homogeneous in their origins and identities as to be accurately summarized as belonging to one nation. Indeed, nationalism has an unfortunate history because of its use by powerful political actors to impose a unitary identity, which includes some people while excluding others, in order to achieve a political goal. This has sometimes been in aid of developing peaceful solidarity among diverse peoples, but it has also been used to justify and support international political violence, as in the case of national socialism and the German Reich 1933–1945.

The idea of the "state" is likewise a somewhat arbitrary term to describe a political–legal entity, and though often traced back to ancient city-states as models of civic order, its international significance is usually associated with seventeenth-century European political settlements designed to attach ultimate political authority ("sovereignty") to particular rulers within defined territories. Such authority over a territory is easier to justify if those people being ruled are seen (and see themselves) as a unitary group somehow naturally belonging to the state. Thus, a hybrid notion of a sovereign territorial nation-state emerged, and it was in due course exported globally, either by agreement or more often by conquest and imposition – particularly in areas of European colonization – as the standard model of political association. Since relations between states, or international relations, are based on mutual recognition of their sovereign territorial status (however they may be governed internally), this gives rise to a particular kind of international system of equally sovereign entities without an overarching higher authority – at least in principle.

Because the central concepts of "nation," "sovereignty" and "international system" are abstract ideal types and historically contingent, they do not accurately reflect or describe reality. To begin with, both nation and state are concepts imposed on people, even if accepted or adopted by them. There are few if any examples of an ethnically, culturally, linguistically, religiously homogeneous "nation" that matches the territorial boundaries of a "state," and indeed few if any genuinely unitary states in terms of political authority. The state itself is a legal abstraction, and takes many forms internally and only stands up to claims of sovereign equality as a convenient fiction. For example, Palestine might be considered a state for some

DOI: 10.4324/9781003008873-10

purposes, by some actors, but not for others. Political authority expressed as sovereignty is in fact fragmented or absent – various internal actors may exercise political influence, while the fiction of equality in relation to external actors belies the huge variations in the resources and capabilities of states.

In relation to global environmental issues, the idea that states could exercise sovereign command authority over the eco-sphere is absurd, but in a globalized world there are ample opportunities for political cooperation and even the prospect of global governance. At the heart of the matter are the limitations of the state as a mechanism for delivering environmental policy, which at the international level may either be resolved with a liberal perspective on a constructive role for states (Beardsworth 2020) or be reflected in new forms of traditional challenges for inter-state relations (Dyer 2017). In *Environment and the Nation State*, Liefferrink (1996) argues that increased ecological interdependence challenges states' ability to control "not actually the borders but rather the quality or what may be called the 'ecological sustenance base' or 'eco-capacity' of their territories" (Liefferink 1996: 26). It is this ecological perspective that raises difficulties for the economistic and anthropocentric habits of nations, states and the international system, and for their political and economic practices. In short, states may have to extend their concern for the "other" beyond other nations of humans to include nonhuman nature (Wapner 2002; see Chapter 27). The impact of human activities and the manner in which they are organized in the international system is a major environmental factor, perhaps the determining factor in the "Anthropocene" period of human-dominated global ecology (see Chapter 15).

The global environment and the limits of the nation-state

The division of the planet into sovereign states does not reflect the interdependencies of ecosystems crossing state borders. This is important in respect of transnational relations between multiple actors. The state, as a legal entity, is responsible for its own jurisdiction, but it can also be held responsible for pollution beyond its borders (see Chapter 10). State policies adjust to pressure from lobbying groups, but they are also subject to other domestic and international pressures (see Chapter 12). It is widely expected that environmental problems will be managed by governments and "that states are willing and able to assume this managerial role" (Lipschutz and Conca 1993: 19). Is this a reasonable or realistic expectation? Given the "very prevalent suspicion of the state on the part of many ecologists" (Hurrell 2006: 166), is it even desirable? It may be difficult for states to reconcile the different aspects of their responsibilities, creating an unmanageable situation which drives global environmental politics, even so far as precipitating a "constitutional moment" in terms of where legitimate decision-making authority resides, and thus provoking fundamental revisions in governance (UNEP 2012: 6).

There is a growing web of economic, cultural, social and political relations among states and between states and other actors. While this web creates a changing political context in which other actors become more significant, the nation-state is not likely to disappear. (Indeed, as a consequence of demands for independence, many new states emerged in the second half of the twentieth century as a result of decolonization.) As long as environmental agreements are negotiated by states, the national interest will play a significant part in decisions that are made. Because environmental issues have challenged many of the social and political assumptions of national life, they are arguably important aspects of the national interest.

Barry and Eckersley (2005: 261–263) point to the tension between the accumulation and legitimation functions of the state in relation to ecological modernization, that is, updating policy and practice in relation to ecological goals. The accumulation function is supported by weak "win–win" versions of ecological modernization – policies aimed at both economic and ecological objectives – that support the globally competitive position of the state and supply-side concern about efficient production. At the same this increases pressure on the legitimation function arising from expectations of higher environmental standards (and demand-side concern with consumption). This points to a stronger version of ecological modernization focused on ecological objectives, implying a need for more clearly transnational political and economic practices. There may also be more fundamental doubts about whether and which interests are served by the technological optimism (the hope that technologies will emerge to deal with ecological problems), reformism (the hope that modest reforms will suffice) and "statism" (the prioritizing of state interests over ecological interests) of ecological modernization policies (Fisher and Freudenburg 2001; Mol 2001; York and Rose 2003). The state as an authoritative and typically legitimate decision-maker (to the extent they can aggregate resources and promote collective action) will remain vitally important for environmental protection, but its context and position relative to other actors (such as international organizations and civil society nongovernmental organizations operating in the realm of "global governance" [see Chapters 8 and 14]) may change.

The traditional political goals of society that the state purports to serve, such as health, wealth and security, are likely to be viewed differently in an environmental light, requiring development of sustainability policies, albeit within the constraints of existing social and political systems in the first instance (see Chapter 16). It may be that participatory representative government can rise to the challenge, whether by means of "environmental democracy" (reforming rather than radically transforming current institutions) or a more radical "ecological democracy" (critical of existing liberal democratic institutions) (Pickering, Bäckstrand and Schlosberg 2020; see Chapter 28). However, achieving success in environmental policy is likely to require substantial change in habitual political practices of decision-making and agenda-setting, and the means of wealth creation and protection of national interests – for example, the familiar bargaining between groups with a vested interest in current economic and political practices. Equally, or in parallel, there are challenges to existing social practices, such as uneven distribution of resources, the character of the capitalist economic system in respect of profit motive and pressures to increase productivity, and the corollary of economic growth manifested in increased consumption (see Chapter 17). This may amount to a "glass ceiling" that inhibits the necessary political and economic transformations; that is, invisible structural barriers that prevent the merely *environmental* state from developing into a *green* or *eco*-state (Blühdorn 2020; Hausknost 2020). In many respects, therefore, both the state and society may be wrestling with a set of conflicting goals, such as the desire for progress through material wealth creation set against aspirations for environmental sustainability, though this might yet be addressed if there remains scope for reframing identities and interests in more ecological terms.

To exacerbate this situation on conflicting political–economic and ecological objectives, there are various constraints on social and political change, at individual, institutional and international levels. Constraints exist in the embedded assumptions and habits of individuals and in the attitudes of the general public to environmental issues (for example, that change is difficult and expensive, and that the environment is a secondary issue not linked to our everyday practices). These may evolve, but perhaps not very quickly. The existing version of the capitalist system serves and supplies individual "needs," and so reinforces and is reinforced by individual and public attitudes about the appropriateness of economic growth, individualism, competition and self-interest. At the institutional level (established formal or informal national practices) there are constraints in that the attitudes of individuals are rooted in political, social and economic institutions that are not designed or developed to implement sustainable goals (thus such ecological goals are seen as a challenge to individual interests). At the international level, the constraints relate to the authority of the state in decision-making processes, economic competition between states, and the relative weakness of international regimes that establish shared expectations (see Chapter 9).

Eckersley (2004) points to three core challenges for the prospect of a "green" state, by which she means the state as an ecological steward and facilitator of transboundary democracy: anarchy in the international system of state, promotion of capitalist accumulation and democratic deficits in the liberal state. She argues that the key to transformation toward "green" statehood is increased accountability of states to both global civil society (e.g., citizens and other actors; see Chapter 14) and international society (e.g., international state-based organizations and institutions; see Chapter 8). The logic of the existing political structure of states and inter-state relations is challenged by emerging environmental multilateralism, sustainable development strategies and environmental advocacy, though crucially the success of such a challenge is dependent on a distinctly green conception of state governance (Eckersley 2004: 14–15). We can see some evidence of this tension between accountability and the pursuit of capital accumulation in recent attempts to establish more ambitious carbon reduction targets coinciding with liberalization of transatlantic air travel, with increased competition likely leading to increase in the number of flights. Kostic et al. (2012: 41) argue that conventional "liberal state- and nation-building" does not bring societal integration nor take account of post-conflict environmental problems. In the context of climate change (see Chapter 32), time is not on our side and the importance of longer-term transformation may be displaced by the urgency of shorter-term action – such is the difficulty of escaping the immediate logic of established political and economic practices in order to adopt a more ecological perspective.

The existing structure of both the international legal and political systems rests heavily on independence and autonomy of states. Collective environmental management poses politically sensitive challenges involving the creation of rules and institutions that reflect the rather different idea of shared responsibilities. There may also be a range of apparently reasonable grounds for resistance by states to an environmental supranational authority: the state remains a source of human identity, and it is a significant means of political expression, which gives claims to national sovereignty their moral credibility. The significance of environmental challenges, though important, may not be sufficient reason to abolish sovereignty when it is anyway not clear that supranational authority would lead to efficient environmental management.

Litfin has indicated that experience of environmental regimes "warrants a healthy scepticism about whether the nation-state system can smoothly adapt to ecological interdependence via traditional forms of multilateral, state-centric institutions" (Litfin 1993: 111). There may be other reasons to abandon strict versions of sovereignty (the absolute authority of the nation-state over its territory and people), including the absence of any choice about the matter. Under the heading "Sovereignty and the inadequate state," Elliott (2004: 109) argues that it is "not simply that the unilateral state cannot meet the challenges of global environmental change through self-help when the causes of that change lie outside its borders. It is that the state itself – its autonomy, capacity and legitimacy – is being eroded, or at least challenged, by the very nature of environmental problems which do not respect territorial borders" (Elliott 2004: 109).

The international system per se is only one factor in the management of the global environment, and there are signs of change in states in terms of policies on pollution and waste management (e.g., via taxation or regulation) and increased environmental awareness among citizens, at least in the limited terms of managing environmental risk (immediate and visible environmental damage or change that directly harms human interests). Individuals have undergone changes to attitudes and practices, with consumer activism in the global "North" and producer activism in the global "South" (for example, through "fair trade" or "sustainable" product labeling), and individuals are increasing their political leverage through public demand for increased transparency and involvement in environmental policy formulation (Princen 2010; see Chapter 14). Even those aspects of the international system that are able to escape a purely state-centered perspective may influence the behavior of states and other actors; for example the role of multi-actor environmental regimes in creating and disseminating norms. International engagement can both help to promote domestic policy goals and underwrite international law (see Chapters 10 and 12), as exemplified by nationally determined contributions to combatting climate change under the Paris Agreement.

If the changing position of the state can be attributed to public demand rather than governmental initiative, then we should perhaps consider the importance of public opinion in the creation and formulation of state policies for the environment, and equally how public pressure plays a similar role at the global level (see Chapter 28). Of course, public opinion can be difficult to assess, and there is a question as to what constitutes a relevant "public" (if we cannot assume that this is already constituted by the state in terms of citizenship and electoral registers, or indeed by the idea of a "nation"). It would be problematic to assume that "the public" is constituted by unelected elites or unrepresentative activists, whatever their environmental credentials.

Nevertheless, there is plenty of evidence for public opinion carrying weight on environmental issues even if it is often difficult to distinguish from the influence of organized non-state actors. A good example is the change led by influential nongovernmental organizations mobilizing world public opinion behind the rights of whales (see Chapters 14 and 40). Other familiar examples include seal culling, disposal of the Brent Spar oil storage container, French nuclear testing in the South Pacific and so forth. In such cases, governments are required to balance "national interests" with the need for public support. Such cases of public opinion driven by environmental activism have normative significance and reflect "universalistic moral concern and a conception of collective human interest" (Vogler 1995: 201). Vogler suggests that it is easy to be cynical about moral positions in politics, but public support for such positions cannot be discounted. We could even consider extending the moral community, and indeed some form of representation, to the nonhuman realm in an ecocentric approach, notwithstanding the challenges of integrating such positions into current practice (Eckersley 1992).

There is further evidence of this trend in non-state politics in the emergence and relative significance of environmental social movements (see Chapter 14). Dryzek et al.'s (2003) *Green States and Social Movements* suggests that social movements are influenced by the kind of state they relate to, and conversely that states may be transformed by incorporation of, or resistance to, social movements. This has implications for the choice of political strategies for environmental movements (Dryzek et al. 2003). If the environmental situation is a cause of political behavior, and if it is not improving, then the drive toward environmentalism is one that states may not be able to resist. Equally, as more of the world's population now lives in cities, such subnational actors may increasingly be the locus or political space for environmental politics (Murthy 2019).

National identities and the environment

There are good ethical and political reasons for privileging people over the state or any other form of political authority. This could begin with individuals, on the simple premise that each "has an overriding obligation to be morally autonomous" such that a "legitimate state is a logical impossibility" (Wolff 1998: vii). Individuals may well hold the secret to dealing with the environmental crisis, either as challengers of technocratic society (Roszak 1979) or as a source of ethical and political meaning (Peterson 2001). However, if individuals are fundamental in ethical and political terms, they typically are so in a wider national context. It may be that the conventional political location of individuals within the defining purview of the state is inadequate for the purposes of environmental politics, and as Beitz (1979: 180), for example, notes, "the critique of the idea of state autonomy clear[s] the way for the formulation of a more satisfactory normative international political theory." However, he elsewhere notes that political theory should guide rather than replace practical judgment (Beitz 1989: 227), and we may feel that people are practical in ways that political institutions are often not.

Nations, even in the context of global environmental politics, are a significant political fact. What is more, nations have a convincing claim to be both source and content of value, and so it is only by finding a place for individuals and their nations in global environmental politics that we can determine the source and content of the relevant political values. While a more holistic ecocentric perspective would certainly challenge fragmented individualism and nationalism, ecocentric values will need to be held politically. Individuals and nations may still be the political home for such values. Out of this emerges perhaps a story about global environmental politics in which national identity is the source or locus of political values. It remains a weakly anthropocentric story because politics is an anthropocentric exercise in which such story telling is important. Peterson identifies humans (in the context of a socially created ecocrisis) as "storytelling culture dwellers," in contrast to rational self-interested agents (Peterson 2001: 8).

A focus on the individual actor and national identity has implications for theorizing global environmental politics and reorients our understanding of global politics more generally. The political implications of this perspective may be rather more trans-social than inter-state, more global–local than international, while allowing that the individual retains moral and political standing in ecological politics. This still leaves us with some room for considering the social influences of environmental politics, and hence its impact on national identities and the international system. Kütting (2000) connects environmental degradation to social origins, and she shows that the failure to recognize the centrality and complexity of this connection has resulted in its externalization from our central concerns, through concentration on the study of international institutional developments. To the extent that the literature tends to reflect scientific and rational analysis, which largely ignores underlying social issues, the implications for the study of international relations are quite broad, and in particular point less to the "international" and more to the "social."

If we cannot evade the political, we can nevertheless see that the institutional and the social are implicated in our understanding of global politics. Elliott's (2004) survey of *The Global Politics of the Environment* begins by noting two "simple aphorisms" – that "global environmental problems … require global solutions," and that there are "no simple solutions" – and she puzzles about "how we should understand the 'global' as an organising principle" (Elliott 2004: 3–4). We are perhaps most concerned about the direction of political causation for good practical and ethical reasons. In this lies both potential for change and potential for danger. The case of "global governance" is a useful test, describing something short of government in its state-centric sense and indicating formal and informal structures and processes, all in aid of (potentially global) political order, but it remains a form of "politics from above" (Maiguashca 2003: 5). In its global manifestation this may well amount to hegemony or imperialism if global environmental governance is understood as a device for protecting existing power structures rather than changing them – a "globalisation from above" (Elliott 2004: 111–112). Such global governance initiatives have encountered some resistance from individual activists, social movements, non-state actors and weak-state actors, representing a form of "politics from below" (Maiguashca 2003: 5). This hierarchical "above" or "below" seems typically political, but governance issues may be better understood as social. In any case the environmental crisis adds a particular additional consideration to ethical, social, political and economic (anthropocentric) tests that we might apply to any scheme – specifically that of ecological integrity.

Schemes to improve the human condition do not always work out according to plan, even if it is possible to point to progress in some respects. There is no reason to think that we will cope much better with environmental problems than we have with problems of inequality, not least because they are linked. However, there may yet be some progress over the long term. In years to come, there may be a call for expressions of regret for contemporary environmental practice in much the same way as there have been some belated apologies for historical injustices. The comparison is useful, not because expressions of regret will right a wrong (or that there is moral equivalence between cases), but because it illustrates how behaviors that were once widely accepted can become unthinkable. It also illustrates changes in authoritative values and practices. The nature of struggles to change practices and values are seldom linear, uniform or complete. It further illustrates the tensions between economic forces and the proper exercise of political (or moral) authority, even if there is consensus on the issues.

Thus, environment politics may follow a pattern of social change, in the context of its own times, such that what may be viewed today as unrealistically burdensome constraints on behavior are in the future seen to be clear moral requirements and become both commonplace and common sense. What is more, facing the environmental challenge need not be seen in negative terms of constraint; it can be readily understood in positive terms of opportunity. Princen has convincingly shown that sufficiency, rather than "efficiency," is an entirely practical goal that results not in merely surviving but in actually thriving (Princen 2005: 3; see Chapter 16). The challenge for people and politics is to underwrite such opportunities as being legitimate. The nature of illegitimacy (or political distance) is illustrated by Perkins (2000): if the political scale of decision-making (at the level of the nation-state) is at odds with the ecological scale of environmental impacts, the result is a democratic deficit. The consequence of the environmental crisis for the state and international politics is that it "calls into question both the practical viability and the moral adequacy of this pluralist conception of a state-based global order," and this has already elicited a partial response in that it has "pushed states towards new forms of international law and global governance" (Hurrell 2006: 167), even if it has not yet brought about fundamental change in the world's social structure (see Chapter 10).

The sovereign state and the environment

The state remains at the center of debate. Dauvergne (2005) discusses the possibility of a "secure world of states, institutions and regimes" with circumspection, noting arguments that global institutions and regimes cannot constrain the self-interested behavior of states (which damage common resources), and other arguments supporting global governance on the basis of complex drivers and constraints on states, and the rational choice of states to cooperate through management regimes and institutions (Dauvergne 2005: 13–16). Consequently, the critique of the discipline of International Relations is its conventional preoccupation with the state as a constitutive central actor in the practice of international relations. In particular, the traditional attribute of the state – sovereignty – is seen as a constitutive concept of international relations. Karen Litfin's (1998) edited volume, *The Greening of Sovereignty in World Politics*, examines the less than obvious relationships between sovereignty and ecology, countering the commonplace assumption that the two are irreconcilable by pointing to ways in which sovereignty is "revisioned," reoriented and problematized in respect of relevant socio-political boundaries and concepts. Specifically, she argues that "conceiving of sovereignty in terms of autonomy, control, and authority usefully decenters the state," and that sovereignty "can be an attribute of various political entities, not just the state" (Litfin 1998: 9).

In his study of the relationship between the extremes of inter-state regimes and global civil society, Conca (2005) sets out two challenging observations on the role of the state while acknowledging that the state does have some role to play (see Chapter 9). The first challenge is a poor track record of centralization and of industrialization at the expense of the environment. The second challenge is that globalization has cost the state some potential ability to respond to environmental issues, even if states have been complicit in the deregulation and liberalization of a transnationalized world economy (Conca 2005: 181–182; see Chapter 24). Globalization is uneven and hierarchical, and the competitive aspects of international relations remain, supported by notions of relative gain in a zero-sum situation of scarcity (see Chapter 3). However, in the context of global environmental change absolute gains are more likely in the long term even if immediate costs imply relative gains in the short term. In this respect, the stakes are high for those state actors considering political and economic integration, as they must take short-term risks for long-term gains, and convince their populations of the merits. For those inclined to protect their borders and economies from the effects of globalization (and environmental change), this approach to protecting sovereignty may come at a high price as the avoidance of short-term risks may lead to long-term losses.

If environmental governance is tethered to broader processes of globalization and associated forms of global governance, and if these can draw attention away from state-centric concerns toward the global–local, then the question is whether globalization can be good for the environment. If it could be, could it also be good for people? This may still be an open question; but even if the anti-globalization movement provides an obvious case of resistance, this is focused on exploitation and inequality rather than constructive global cooperation. Clearly, if there is any emerging consensus on a new global environmental order, then one or more alternative existing orders would be displaced by it. Perhaps the existing orders are "development" (which means different things to different peoples) and "sovereignty" (which has always been elusive), and there are those who would be happy to see the end of some versions of either or both of these.

Keuhls (1996) argues that it is conceptually questionable to create a sovereign object out of land, to create "sovereign territory" with all its political implications. For example, "environment and development" has become an established (if unclear) agenda of international relations for the twenty-first century. That this agenda reflects earlier agendas of colonization, decolonization and uneven development, all long-standing "North–South" issues, perhaps reduces its novelty, but it adds environmental concern even if also disguising the political interests of the North (see Chapter 23). This agenda points to the complexity of environmental issues and their interdependence with other functional issue-areas, such as agriculture and education and health care, all set against the economic and political challenges of countries pursuing industrial growth. It also raises again the question of what (and if) global institutional arrangements might ameliorate the situation. Young argues for a common research agenda and a unified theory of environmental governance for localized common-property systems in small-scale settings (Young 2005: 176), and for "conditions under which environmental regimes will produce outcomes that fulfill various criteria of sustainability, efficiency or equity" (Young 2005: 178). This suggests that subtle, complex and systemic solutions may be needed to square environment and development. This also suggests a case for humility in the face of such political challenges.

With any global political scheme comes the danger of substituting one overarching discourse for another. In the current political imagination, global environmental governance is constrained by, or aligned with, the desire for deregulation and liberalization such that win–win solutions (which may have merit if winning is ecological) are promoted under the banner of economic efficiency. Perhaps eco-capitalism is the appropriate charge against current global environmental governance, rather than eco-imperialism. Even so, assuming the hegemonic aspects of "primitive accumulation" (expropriation of the direct producers, for example through colonialism) remains problematic in a world where social transformation is so varied (Shilliam 2004). Lipschutz argues that "it is the relationships between ruler and ruled, and the mechanisms of rule, that are important," and he cites two models of (American) empire: "neo-liberal institutionalism" (power and bargaining in collective inter-state decision-making through international institutions) and "new sovereignty" (pressure on states to comply with treaty obligations; a kind of imperial sovereignty, or power) (Lipschutz 2004: 21). Here again we may ask if global environmental policy represents a new form of imperial governance to be resisted or a gradual transformation in the mechanisms of authority and legitimacy of states – that is, something that might be welcomed.

Conca argues that there remains some exercise of state authority even as states are displaced to some degree by transnational civil society (Conca 2005: 183; see Chapter 14). However, this occurs in a rather different context of institutionalized politics in which the state is not irrelevant but its authority is contested by other non-state actors (Conca 2005: 194–196). He concludes that governance is increasingly transnational, institutions of governance are more complex, and exercise of authority is more fluid (Conca 2005: 202–203). Weiss and Jacobson (1998) point out that compliance with environmental agreements alters relationships between individual and collective actors in the international system which should tell us something about the way international relations is changing – and how this might change our thinking about it. Perhaps, not surprisingly, implementation of environmental agreements turns out to be the greatest challenge. In order to judge the success of environmental diplomacy it may be necessary to know how and why implementation and compliance with such agreements varies. Interestingly, the international political environment remains a dominant factor even in the practical exercise of state capacities in implementing and complying.

While states are obviously central to implementation and compliance with environmental agreements, none can or will act in a vacuum. The term "engagement" goes some way to capturing the political dynamic and prescription for achieving compliance (Weiss and Jacobson 1998). Compliance is typically both a legal and technical issue, but ultimately behavior modification is what counts, and in this respect nongovernmental actors and communities of technical experts make considerable contributions to what may be seen as both transnational (nongovernmental) and transgovernmental (inter-state) activity (Vogler 2000).

Stiles take a pluralist (multiple independent actor) view of the relationship between civil society actors and states (and their intergovernmental institutions). He argues that "the interests and identities of major players tend not to change over time, only their strategies and tactics depending on the general distribution of power and resources" (Stiles 1998).

This pluralist perspective on the system of multiple independent state actors, and its relation to other non-state actors, may be both accurate and politically appealing because "it is important to note the normative claims made for this kind of [classical] pluralism" (Hurrell 2006: 166). Yet the pluralist view, and its practice in terms of independent actors, may be fundamentally (if unintentionally) reactionary: "Indeed, green arguments that economies should be brought back under firm national control and that 'excessive' immigration should be resisted attest to the continued power of the pluralist impulse" (Hurrell 2006: 167). This seems to make sense in terms of contemporary practices of sovereign independence (in the individual national interest), but Smith (1993) warned against an unchallenged pluralist consensus and absence of more critical engagement as being a cause of the environment's marginal place in the academic discipline of International Relations – which may say more about the academic discipline than about environmental politics (see Chapters 4 and 5).

Brenton (1994), in considering the role of the state, suggests some caution in regard to supranational rather than subnational, local, individual approaches to environmental problems. He suggests that the collective and integrative perspective of environmentalism has made it too easy to accept regulations and grand schemes, sometimes at the expense of liberty. Some goals can only be achieved by international cooperation given the global environmental and political–economic context, but the modes and consequences of such cooperation remain an issue. Reflecting widely held concern about the nature of political authority and accountability in the supranational context, Brenton goes on to note that "replacement of the judgement of the individual by that of the state raises problems of its own," and he notes that this is compounded by transfer of authority to the international level, "placing it still further from the people it is intended to serve" (Brenton 1994: 268). A deeper critique attacks claims of state sovereignty over the natural world, as this resource-based orientation is the source of many environmental problems (Smith 2011). This growing challenge to the notion of sovereignty supports "post-sovereign" forms of global environmental governance (Karkkainen 2004; see Chapter 4).

Conclusion

The "state" and the "international" represent a set of political structures with roots in political self-determination and independence rather than in environmental concerns. If concern with the role of the state and international relations becomes less relevant because of the global nature of environmental challenges, concern with human practices and the role of political economy and transnational civil society becomes more so. It may be felt that escaping state-centric structures is merely creating a new difficulty: if we did not have such administrative structures we would have to invent something similar to manage our collective decision-making. Nevertheless, this may allow us to acknowledge the emergence of "an epoch defined primarily by globalization and de-territorialization" (Lawson 2006: 415–416), in which collective decision-making would not be best organized around sovereign territorial jurisdictions of nation-states.

The complexity and ubiquity of environmental issues provide environmental concern with a potential new avenue of expression and application, and with environmental values running through modern political discourse there is the possibility for constructively subversive developments in global environmental politics (Dyer 1996). One illustration is Tim Hayward's *Political Theory and Ecological Values* (1998), which argues that environmental values can be supported by enlightened human interests, that this link must exist if ecological goods are to be promoted, and that there are profound implications of fully integrating environmental issues into our disciplinary concerns (Hayward 1998). If political analysis is concerned with the transformation of political community (Linklater 1998) rather than the preservation of the environment, once environmental issues are introduced the fundamental problematic becomes the transformation of the human relationship to the environment. States may not be able to reflect enlightened human interests, as opposed to merely national interests defined in terms of sovereignty and state power, and thus increasing concern with the environment may lead to decreasing concern with the state.

As Biermann and Dingwerth (2004) argue, understanding global environmental governance requires reconsidering key concepts such as sovereignty. A consequence of engaging with the environment is that international politics is less about conflict and cooperation being means for promoting states' interests and more about coping with competing values and the practical means of dealing with them (Dyer 2000). In this sense we are already concerned more with "global politics" than with "international politics," and if this amounts to a transformation, emerging patterns of global governance suggest a somewhat different model of world politics driven in part by environmental concerns (Sonnenfeld 2008). The prospects for global governance in issue areas like climate change may illustrate the boundaries and possibilities of our inherited political designs (Haas 2008; see Chapters 32 and 33). Plans for climate governance collapsed dramatically in a diplomatic failure at the Copenhagen climate conference in 2009, with subsequent meetings suggesting little improvement. More generally, states have not had a very good record of achieving sustainability over the decades since the 1992 Earth Summit in Rio de Janeiro. However, the legacy of Rio goes beyond formal agreements because it encouraged a wider range of political actors and spaces that have challenged conventional notions of national sovereignty (Andonova and Hoffman 2012).

As Hobson and Hobden put it, we are obliged "to rethink the origins of international systems, states and international institutions as well as to denaturalize such historical forms, and to consider the potential and actual processes which are reconstituting, if not transforming, the present into possible and desirable futures" (Hobson and Hobden 2002: 283). Bigo and Walker observe that a sociological approach to politics and international relations offers the benefit of "emphasis on the study of practices," including discourses, rather than lapsing into engagement "with systems, states, sovereignties and so on as more or less disembodied structures, even abstractions" (Bigo and Walker 2007: 5). Environmental change exacerbates the situation of states by creating different contexts which are not state-centric, but for which states are responsible. Demands for state action on the environment create tensions between established institutional practices and environmental responsibility (Falkner 2012). Nevertheless, with the increasing significance of non-state contexts and civil society actors, the environment may be the determining factor in the end.

Nations, states, sovereignty and the international system should be viewed differently in the ecological context of global environmental politics. With processes of global environmental governance cutting across political, social and economic boundaries at different levels and scales of politics, the resilience of nation-states as a political form will be thoroughly tested.

References

Andonova, L.B. and Hoffman, M.J. (2012) "From Rio to Rio and beyond: innovation in global environmental governance", *Journal of Environment Development* 21: 57–61. Online. Available HTTP: <http://jed.sagepub.com/content/21/1/57.full.pdf+html>

Barry, J. and Eckersley, R. (2005) "W(h)ither the green state?", in J. Barry and R. Eckersley (eds) *The State and the Global Ecological Crisis*, Cambridge, MA: MIT Press: 255–273.

Beardsworth, R. (2020) " Climate science, the politics of climate change and futures of IR", *International Relations* 34(3): 374–390. Online. Available HTTP <https://doi.org/10.1177/0047117820946365>

Beitz, C. (1979) *Political Theory and International Relations*, Princeton, NJ: Princeton University Press.

Beitz, C. (1989) *Political Equality: An Essay in Democratic Theory*, Princeton, NJ: Princeton University Press.

Biermann, F. and Dingwerth, K. (2004) "Global environmental change and the nation state", *Global Environmental Politics* 4: 1–22.

Bigo, D. and Walker, R.B.J. (2007) "Editorial", *International Political Sociology* 1: 5.

Blühdorn, I (2020) "The legitimation crisis of democracy: emancipatory politics, the environmental state and the glass ceiling to socio-ecological transformation", *Environmental Politics* 29(1): 38–57. DOI: 10.1080/09644016.2019.1681867

Brenton, T. (1994) *The Greening of Machiavelli*, London: Earthscan.

Conca, K. (2005) "Old states in new bottles? The hybridization of authority in global environmental governance", in J. Barry and R. Eckersley (eds) *The State and the Global Ecological Crisis*, Cambridge, MA: MIT Press: 181–206.

Dauvergne, P. (ed.) (2005) *Handbook of Global Environmental Politics*, Cheltenham: Edward Elgar.

Dryzek, J., Downes, D., Hunold, C., Schlosberg, D. and Hernes, H.K. (2003) *Green States and Social Movements*, Oxford: Oxford University Press.

Dyer, H.C. (1996) "Environmental security as a universal value: implications for international theory", in J. Vogler and M.F. Imber (eds) *The Environment and International Relations*, London: Routledge: 22–39.

Dyer, H.C. (2000) "Coping and conformity in International Relations: environmental values in the post-Cold War world", *Journal of International Relations and Development* 3(1): 6–23.

Dyer, H.C. (2017) "Security politics and climate change: the new security dilemma", in O. Corry and H. Stevenson (eds) *Traditions and Trends in Global Environmental Politics*, London: Routledge: 154–170.

Eckersley, R. (1992) *Environmentalism and Political Theory: Towards an Ecocentric Approach*, London: UCL Press.

Eckersley, R. (2004) *The Green State: Rethinking Democracy and Sovereignty*, Cambridge, MA: MIT Press.

Elliott, L.M. (2004) *The Global Politics of the Environment*, Houndmills: Palgrave.

Falkner, R. (2012) "Global environmentalism and the greening of international society", *International Affairs* 88: 503–522.

Fisher, D.R. and Freudenburg, W.R. (2001) "Ecological modernization and its critics: assessing the past and looking toward the future", *Society and Natural Resources* 14: 701–709.

Haas, P.M. (2008) "Climate change governance after Bali", *Global Environmental Politics* 8(3): 1–7.

Hausknost, D. (2020) "The environmental state and the glass ceiling of transformation", *Environmental Politics* 29(1): 17–37, DOI: 10.1080/09644016.2019.1680062

Hayward, T. (1998) *Political Theory and Ecological Values*, Cambridge: Polity Press.

Hobson, J.M. and Hobden, S. (2002) "On the road towards an historicised world sociology", in S. Hobden and J.M. Hobson (eds) *Historical Sociology of International Relations*, Cambridge: Cambridge University Press: 265–285.

Hurrell, A. (2006) "The state", in A. Dobson and R. Eckersley (eds) *Political Theory and the Ecological Challenge*, Cambridge: Cambridge University Press: 165–182.

Karkkainen, B.C. (2004) "Post-sovereign environmental governance", *Global Environmental Politics* 4: 72–96.

Keuhls, T. (1996) *Beyond Sovereign Territory: The Space of Ecopolitics*, Minneapolis: University of Minnesota Press.

Kostic, R., Krampe, F. and Swain, A. (2012) "Liberal state-building and environmental security: the international community between trade-off and carelessness", in R. Amer, A. Swain and J. Öjendal (eds) *The Security–Development Nexus: Peace, Conflict and Development*, London: Anthem Press: 41–64.

Kütting, G. (2000) *Environment, Society and International Relations: Towards More Effective International Environmental Agreements*, London: Routledge.

Lawson, G. (2006) "The promise of historical sociology in International Relations", *International Studies Review* 8(3): 397–423.

Liefferink, D. (1996) *Environment and the Nation State*, Manchester: Manchester University Press.

Linklater, A. (1998) *The Transformation of Political Community*, Cambridge: Polity Press.

Lipschutz, R.D. (2004) "Imitations of empire", *Global Environmental Politics* 4(2): 20–23.

Lipschutz, R.D. and Conca, K. (1993) "Act I: the state and global ecological interdependence", in R.D. Lipschutz and K. Conca (eds) *The State and Social Power in Global Environmental Politics*, New York: Columbia University Press: 19–23.

Litfin, K. (1993) "Ecoregimes: playing tug of war with the nation-state", in R.D. Lipschutz and K. Conca (eds) *The State and Social Power in Global Environmental Politics*, New York: Columbia University Press: 94–117.

Litfin, K. (1998) "The greening of sovereignty: an introduction", in K. Litfin (ed.) *The Greening of Sovereignty in World Politics*, Cambridge, MA: MIT Press: 1–27.

Maiguashca, B. (2003) "Governance and resistance in world politics: introduction", *Review of International Studies* 29(1): 3–28.

Mol, A.P.J. (2001) *Globalization and Environmental Reform: The Ecological Modernization of the Global Economy*, Cambridge, MA: MIT Press.

Murthy, S.L. (2019) "States and cities as 'Norm Sustainers': a role for subnational actors in the Paris Agreement on climate change", *Virginia Environmental Law Journal, 2019*. Online. Available at <http://dx.doi.org/10.2139/ssrn.3308613>

Perkins, E. (2000) "Equity, economic scale, and the role of exchange in a sustainable economy", in R.M. M'Gonigle and F.P. Gale (eds) *Nature, Production, Power*, Cheltenham: Edward Elgar: 183–94.

Peterson, A.L. (2001) *Being Human: Ethics, Environment and Our Place in the World*, Berkeley: University of California Press.

Pickering, J., Bäckstrand, K. and Schlosberg, D, (2020) "Between environmental and ecological democracy: theory and practice at the democracy-environment nexus", *Journal of Environmental Policy & Planning* 22(1): 1–15. DOI: 10.1080/1523908X.2020.1703276

Princen, T. (2005) *The Logic of Sufficiency*, Cambridge, MA: MIT Press.

Princen, T. (2010) "Consumer sovereignty, heroic sacrifice: two insidious concepts in an endlessly expansionist economy", in M. Maniates and J.M. Meyer (eds) *The Environmental Politics of Sacrifice*, Boston, MA: MIT Press: 145–164.

Roszak, T. (1979) *Person/Planet: The Creative Disintegration of Industrial Society*, London: Gollancz.

Shilliam, R. (2004) "Hegemony and the unfashionable problematic of 'primitive accumulation'", *Millennium: Journal of International Studies* 33(1): 59–88.

Smith, M. (2011) *Against Ecological Sovereignty: Ethics, Biopolitics and Saving the Natural Environment*, Minneapolis: University of Minnesota Press.

Smith, S. (1993) "The environment on the periphery of International Relations: an explanation", *Environmental Politics* 2: 28–45.

Sonnenfeld, D.A. (ed.) (2008) Special issue on "Globalisation and environmental governance", *Global Environmental Change: Human and Policy Dimensions* 18(3): 341–538.

Stiles, K. (1998) "A rational choice model of grassroots empowerment", International Studies Association paper. Online. Available HTTP: <http://www.ciaonet.org/conf/stk01/>

UNEP (2012) *21 Issues for the 21st Century: Result of the UNEP Foresight Process on Emerging Environmental Issues*, Nairobi: United Nations Environment Programme: Online. Available HTTP: <http://www.unep.org/publications/ebooks/foresightreport/>

Vogler, J. (1995) *The Global Commons: A Regime Analysis*, Chichester: John Wiley.

Vogler, J. (2000) *The Global Commons: Environmental and Technological Governance*, Chichester: John Wiley.

Wapner, P. (2002) "The sovereignty of nature? Environmental protection in a postmodern age", *International Studies Quarterly* 46: 167–187.

Weiss, E.B. and Jacobson, H.K. (eds) (1998) *Engaging Countries: Strengthening Compliance with International Environmental Accords*, Cambridge, MA: MIT Press.

Wolff, R. (1998) *In Defence of Anarchy*, Berkeley: University of California Press.

York, R. and Rose, E.A. (2003) "Key challenges to ecological modernization theory", *Organization and Environment* 16(3): 273–288.

Young, O.R. (2005) "Why is there no unified theory of environmental governance?", in P. Dauvergne (ed.) *Handbook of Global Environmental Politics*, Cheltenham: Edward Elgar: 170–184.

8

International organizations

Global and regional environmental cooperation

Kate O'Neill

Global environmental governance does not occur in a vacuum. It is initiated, encouraged, coordinated, strengthened and monitored by a series of intergovernmental organizations (IGOs), with exclusive or partial environmental mandates. The environment as an issue for global governance organizations appeared on the international agenda in the late 1960s, and the intersection between environment and development has gone on to become a central focus of global governance ever since. IGOs have played a critical role in this process.

This chapter reviews the functions and operation of existing global and regional IGOs as they pertain to global environmental governance. It examines lines of inquiry into their role in the international political system, both individually and as a collective whole, and how they have been applied in research on environmental IGOs to further our understanding of this set of actors. These include the extent to which IGOs are autonomous actors, how to assess their performance and impacts, how to manage links and overlap between them, and the emergence of regional IGOs, adding a new dimension to this field. Finally, this chapter outlines possible future trajectories and reforms for this complex institutional terrain.

International organizations and global governance

Intergovernmental organizations have long played a critical role in creating, coordinating and steering inter-state cooperation and global governance (O'Neill 2019). They also play a role in creating and enforcing international law and principles (see Chapter 10). In other words, nation-states delegate the business of managing and implementing global political processes to IGOs. Most simply, IGOs are "organizations that include at least three states among their membership, that have activities in several states, and that are created through a formal intergovernmental agreement such as a treaty, charter or statute" (Karns and Mingst 2010: 5). The terms "international organization" and "international institution" are not synonymous. Institutions more broadly are the "rules of the game that serve to define social practices, assign roles and guide interactions among the occupants of those roles" at the global level (Young 1994: 15), and may or may not take on formal shape. This chapter focuses on "concrete" environmental IGOs whose role and actions in the international system

DOI: 10.4324/9781003008873-11

are shaped, at least in part, by these social institutions, as well as by their member states (see Chapter 7) – and vice versa.

Intergovernmental organizations have existed in recognizable form since the nineteenth century, but most current IGOs came into existence after World War II. In 1909, there were thought to be 39 IGOs in existence, in 2017, this number increased to 7653 (Park 2018). They may be global in membership, in the sense that any country may join, such as the United Nations (UN) or the World Trade Organization (WTO). They may also be limited in membership or in mandate. The Organization for Economic Cooperation and Development (OECD) requires applicant countries to meet certain conditions before being allowed to join. Smaller IGOs are tied to a specific issue area – the International Whaling Commission (established in 1946) is one such example, the Secretariat of the Ramsar Wetlands Convention (1971) is another. I use these examples as they are autonomous of the more modern environmental governance regimes, whose secretariats tend to be nested in broader UN organizations, most importantly, the main "anchor organization" for global environmental governance, the United Nations Environment Programme (UNEP).

Intergovernmental organizations may be regional, covering a (usually) contiguous group of states, such as the European Union (EU), the North American Free Trade Agreement (NAFTA) or the Association of Southeast Asian Nations (ASEAN). They are engaged in many different spheres of global governance, from peace and security to trade, world health, development financing and the environment (see Chapters 20 and 22).

Many IGOs share structural similarities. They are managed by (often tiny) secretariats, often with the assistance of standing or ad hoc committees. They have a particular mandate, or purpose, as delineated in their founding charter or constitution. They have a budget, funded by governments, either directly or through the UN, and mechanisms whereby member state representatives meet and make policy decisions or organizational changes. This may be through permanent representation, or via regular conferences of the parties that bring national representatives together, or some combination. In most IGOs, day-to-day affairs are run by a permanent secretariat staffed by full-time employees, while major decisions are (officially) made by state representatives. Voting rules differ across organizations: some are "one member, one vote" (e.g., the UN General Assembly), others, particularly financial institutions, are weighted according to states' financial contributions (e.g., the World Bank, where donor states collectively hold the bulk of decision-making authority).

Theoretical perspectives on intergovernmental organizations

The proliferation of IGOs in general since the end of World War II and of environmental IGOs since the early 1970s has given rise to a variety of theoretical perspectives and debates about their identity, functions and impacts (see, for example, Barnett and Finnemore 1999, 2004; Diehl 2005; Karns and Mingst 2010; Hurd 2011; and for an earlier overview, Kratochwil and Ruggie 1986; see also *International Organization* and *Global Governance* journals. For an overview, see Park 2018, Chapter 2). More recently, studies of environmental issue and regime linkages across IGOs, and the management of those linkages, helped to revive and refresh a bureaucratic politics literature on IGOs (Biermann and Siebenhüner 2009; Jinnah 2010, 2012).

The earliest post-World War II scholarship saw IGOs through a (neo)-functionalist lens: created by states in a process of integration, in order to fulfill particular tasks that they could not accomplish on their own (Haas 1964; Schmitter 1969; see Chapter 3). This perspective flowed relatively easily into the neoliberal institutionalist perspective on international relations, which put states back at the center of analysis but still saw important functions for IGOs in easing inter-state cooperation and managing interdependence (see, for example, Keohane 1984; Abbott and Snidal 1998; see Chapter 7). According to neoliberal institutionalists, IGOs are created by states in order to reduce the transaction costs of international cooperation, by coordinating meetings, collecting information, running day-to-day operations, and creating mechanisms to ensure transparency and accountability (see Chapter 3). These activities help counteract the possibility that states will cheat or free-ride on the efforts of other states, which reduces the likelihood of effective or lasting cooperation.

Institutionalist theories, unlike their realist counterparts, accord IGOs some "life of their own," in that they outlast the constellation of national interests that created them (Keohane 1984), but often ascribe that longevity to inertia (Krasner 1988) or to state actors continuing to value their functions (see Chapters 3 and 4). In terms of environmental IGOs, the neoliberal institutionalist approach has a great deal of applicability. Addressing global environmental problems, certainly at the outset, seemed more a question of coordination among nation-states, and creating rules and norms that made them take into account global and transboundary environmental degradation. This degree of interdependence clearly (in a semi-functionalist sense) demanded some form of international organization in charge.

Constructivist scholarship, to the extent it examines the co-constitution of organizations, issues and identities in the international system, examines how IGOs are more than forums for collective decision-making by nation-states, acknowledging that IGOs act as autonomous agents, often exceeding their existing mandate (Barnett and Finnemore 1999, 2004). As with other sorts of bureaucracies, including at the domestic level, it became important to address the way that IGOs develop their own autonomy, their own goals – their own continued existence being one – and begin, indeed, to take on agency and power in a political realm thought to be dominated by nation-states (see Chapter 7). Some have also examined the conditions under which the actions of IGOs become dysfunctional – or pathological, as when, for example, they fail in their mission. Such failures can be explained, for instance, by the development of a stagnant or perhaps too insular bureaucratic culture (Barnett and Finnemore 2004).

The IGO-as-Actor approach is clearly reflected and extended in the study of environmental IGOs, most particularly in works that examine overlap, or interplay, management across international regime boundaries, where critical personnel within the organizations often take on an entrepreneurial or leadership role in governing this process (Biermann and Siebenhüner 2009; Jinnah 2010, 2011). In this sphere, IGO power and authority clearly and at least in part derives from the expertise provided by secretariats and associated regime bodies (outlined below). Such expertise – scientific and otherwise – is critical for effective governance of environmental problems (see Chapter 18), but it is hard to acquire and build on without an IGO willing to coordinate transnational scientific efforts, for example. Further, in a world populated by a large number of small agencies often sharing space in the same city, collaboration across units is both effective and likely. Finally, we have seen instances of creative and effective leadership by individuals, which have helped bolster global environmental governance, as well as instances where less effective leadership has contributed to less effective governance. Ivanova (2010) discusses this with respect to UNEP.

Others study the factors influencing IGO performance and impacts, how their performance is assessed, and how they evolve over time (Barnett and Finnemore 2004; Ivanova 2010). In particular, many are interested in how, or whether, IGOs can learn over time, from assessments, or from each other, and under what conditions (Haas 1990; Greene 1998; Siebenhüner 2006), or can act quickly in response to global crises. Again, given the emphasis in the global environmental politics literature on the effectiveness and impacts of environmental regimes (O'Neill, 2017; see Chapter 9), work in this field has yielded important insights into issues of environmental IGO performance and learning that have broader applicability to IGOs in general. Finally, researchers examine IGOs as a collective whole, asking the perhaps inevitable question: are environmental IGOs greater or less than the sum of their parts? Is the system too fragmented, with duplicative or conflicting mandates and activities across IGOs? Are actors within the system working to forge linkages across IGOs, as in, for example, regime complexes (Keohane and Victor 2011)? Or does the system require some sort of reform, perhaps centralization into a World Environment Organization (Biermann 2001; Najam 2003; Biermann and Bauer 2005)?

These perspectives are complicated by the emergence of regional organizations as nodes of environmental governance, through, for example, the EU's environmental governance structures, or through regional organizations and agreements connected with international environmental regimes (see Chapter 9). Interest in multilevel and/or cross-scale governance is growing (Balsiger and VanDeveer 2012), especially given the perceived failure of, or deadlock within, global environmental governance processes (Conca 2012). Therefore, the conditions under which regional governance (or the devolution of governance capabilities across levels) is appropriate and effective for addressing transboundary or global environmental problems has become a new focal point of research for scholars of IGOs.

Mapping environmental intergovernmental organizations: functions, nesting and linkages

Many international organizations and agencies have full or partial mandates to address global environmental problems. This creates a complex terrain and sets of interactions for the researcher to depict, despite the fact that most are nested in some way within the UN system, or work closely with it. This section maps the major international environmental organizations, and their functions, from UNEP to individual secretariats and other regime bodies, to IGOs with or that have developed environmental governance functions. It also identifies some of the regional IGOs that have an environmental mandate within and across their member states. It demonstrates the linkages – horizontal and vertical – and interrelationships across environmental IGOs, as well as some of the insights and perspectives on their work, goals and influence analyzed by the leading researchers in this field.

The United Nations Environment Program

The UN Environment Program is the anchor organization for global environmental governance (Ivanova 2007). It was created in 1972 at the United Nations Conference for Humans and the Environment, held in Stockholm, Sweden. The UN convened this conference to bring states together to discuss and implement a coordinated legal framework to address global and transboundary environmental problems, the extent of which had only recently become clear. Despite some calls at the time for a form of "International Environment Organization" (Kennan 1970) that would be more centralized and have more enforcement powers, UNEP was established as a UN program under the auspices of the UN General Assembly and Economic and Social Council (Biermann 2001: 46–47), reflecting a certain amount of pragmatism on the part of its architects. As a program, UNEP is designed to be nimble and responsive (Ivanova 2010), but it lacked the authority and autonomy to make binding decisions on its members (Bauer 2009), as a specialized UN agency such as the World Health

Organization can. It cannot make funding allocations, compared with the World Bank, nor enforce treaty provisions when violated or settle disputes, compared with the UN Security Council and the WTO.

UNEP is the first UN agency to be based in a developing country, with headquarters in Nairobi, Kenya, although many of its offices are based in Geneva and other European cities, and its associated secretariats and offices are distributed worldwide. Its functions are to serve as a focal point for and coordinator of international environmental organizations, to engage in monitoring, assessment and early warning, to foster compliance with international agreements, and engage in long-term, capacity-building efforts. UNEP is also tasked with fostering linkages across the UN system, and is largely funded through voluntary, not assessed, contributions by member states (Ivanova 2010: 33–34).

In 2012, at the Rio+20 Summit, UNEP received a much-needed upgraded to "assembly' status. Previously its membership was restricted to a governing council of 58 UN member states. Now it is open to all 193 member states, with the creation of the UN Environment Assembly, with the promise of enhanced funding and structural capacity to fulfill its mandate (O'Neill 2019, p. 43).

Many have questioned UNEP's performance, given its financial and political constraints, and assessments are decidedly mixed (as pointed out in Bauer 2009; Downie and Levy 2000; Ivanova 2010: 36–37; and Najam 2003). In her assessment of UNEP's performance across the different areas and goals of its mandate, Ivanova also finds its performance to be mixed (Ivanova 2010; see also Ivanova 2007). Although in many ways its activities reflect the pragmatic nature of its design, it has not been able to push much beyond its original mandate, and it has been excluded from some critical global governance processes. While it has failed to become *the* main single international environmental organization, it has been more successful in monitoring and assessing the state of the global environment and in establishing and managing many different international environmental regimes and negotiating processes (Ivanova 2010: 46). Najam (2003) finds that, despite some deserved criticism, UNEP has functioned well in the light of its budgetary constraints and overwhelming mandate, and by comparison with similar international agencies.

A number of factors help explain this performance record. For example, leadership has been important for UNEP. Its key achievements correlate with the terms of office of particular executive directors – such as Maurice Strong or Mostafa Tolba – who were often lauded (or criticized) for pushing the global environmental agenda (Benedick 2007). Not all directors have been considered as effective. Ivanova (2010) builds on the Barnett and Finnemore (2004) framework outlined above to identify features that determine UNEP's performance. She examines elements of UNEP's design and operation (as established by member states), its internal leadership and organizational culture – and adds its distant location in order to understand the challenges it faces, and why it has not been able to go above and beyond its mandate.

Treaty secretariats and other regime bodies

Other important international environmental organizations are nested within international environmental regimes (see Chapter 9). Each treaty-based regime, from ozone depletion (Chapter 31) to biodiversity (Chapter 41) to toxic chemicals (Chapter 35) to climate (Chapter 32), is managed by its own secretariat, which, in turn, reports to the regime's Conference of the Parties (COP), and has its own permanent staff. Many of these are nested within UNEP. The UN manages some, such as the secretariat of the UN Framework Convention on Climate Change (UNFCCC), while others, such as the Ramsar Convention Secretariat, exist entirely outside the UN system (Jinnah 2012). Long dismissed as merely functional agencies, which coordinate treaty-related paperwork and run Conferences of the Parties, it has become clear that in many cases, they wield considerable power behind the scenes, but that power, or, more accurately, influence, varies across environmental regimes (Biermann and Siebenhüner 2009; Munoz et al. 2009; Jinnah 2010, 2012). For example, they have been able in many cases to exercise considerable (but not explicit) leeway in terms of steering their member states toward particular outcomes (for example, by supplying draft text).

Both Jinnah (2010, 2011) and the authors of the essays in Biermann and Siebenhüner (2009) are particularly interested in how secretariats are directly engaged in overlap or interplay management – where the sphere of action of one regime cuts across another. This may be across environmental regimes (in the same issue area or different), or between environmental and other international governance arenas, in particular the trade regime (see Chapter 22). Secretariats are especially important because international law and politics have little provision for what happens when regime processes overlap – despite the potential for conflict – or for mutual advantage in such cases. Thus they have been able to exert agency in shaping a whole new area of global governance activity, albeit often through informal means. Jinnah (2012) examines, for example, how the secretariat of the 1992 Convention on Biological Diversity has worked with secretariats of other biodiversity-related regimes to strengthen joint activity on overlapping goals (see Chapter 41). She also points out how the secretariats of the various chemicals treaties have cooperated to create an ad hoc joint working group to look for ways to make them work more closely together (see also Selin 2019; and Chapter 35).

Many treaty-based environmental regimes also contain subsidiary bodies, often for scientific and technical advice (Kohler et al. 2012; see Chapter 18). Many of these bodies are permanent, some are ad hoc. The UN Framework Convention on Climate Change and the Kyoto Protocol have a particularly complex combined administrative structure (UNFCCC 2012). Permanent and ad hoc committees serve each agreement under the overall authority of the UNFCCC secretariat, reporting to the Conference of the Parties (UNFCCC) and the Meeting of the Parties (Kyoto). Two subsidiary bodies – for Scientific and Technical Advice, and for Implementation – are permanent bodies, along with other committees on compliance, funding mechanisms and so on. However, ad hoc working groups have played a very important role, too, in steering the progress of the UNFCCC and the Kyoto Protocol. The most prominent international scientific body associated with the climate regime, the Intergovernmental Panel on Climate Change (IPCC) operates outside the UNFCCC. Co-established by UNEP and the World Meteorological Organization (WMO) in 1988, it collects, summarizes and assesses global scientific research on climate change (Hulme and Mahony 2010).

Other regimes have similar, if perhaps not quite so complex, structures. The Convention on Biological Diversity has, for example, its Subsidiary Body on Scientific, Technical and Technological Advice (SBSTTA), and the Article 8j Working Group, whose mandate is to integrate local knowledge and knowledge holders into the regime. Even in regimes that are not anchored by a multilateral agreement, more informal international bodies – such as the UN Forum on Forests – provide venues and assistance for multilateral dialog and advice. Established in 2013, the Intergovernmental Panel on Biodiversity and Ecosystem Services (IPBES) now plays a similar role to the IPCC in the biodiversity arena (Borie and Hulme 2015).

Understanding these subsidiary bodies is not only important for drawing an accurate map of this institutional landscape. They provide important information and advice to treaty secretariats, UNEP (and other IGOs), and Conferences of the Parties that is used as the basis for new measures within regimes. And they show how, and under what conditions, IGOs can act autonomously of their member-states, and to what effect (Jinnah 2012).

Finally, a small number of IGOs work across environmental regimes. The Commission on Sustainable Development (CSD) connected the world of global environmental governance with that of global development and development goals (Park 2018, Chapter 5). The Global Environment Faculty (GEF) coordinates funding and capacity-building projects across several regimes and issue areas: climate change (Chapter 32), ozone depletion (Chapter 31), biodiversity (Chapter 41), oceans (Chapter 39), persistent organic pollutants (Chapter 35) and desertification or land degradation (Chapter 43). GEF is administered by UNEP and UNDP, with funding coordinated by the World Bank, although it has its own council and decision-making body. As a capacity-building organization, it has also been assessed (and found wanting) on its performance – but has been lauded as a moderately successful experiment in terms of cooperation between three different agencies (Lattanzio 2010). On a far smaller scale, the Green Customs Initiative works to train customs officials in developing countries to be able to identify and prevent smuggling of various goods and substances prohibited across different environmental regimes, from ozone-depleting substances to hazardous wastes to wildlife and genetically modified organisms. Managed through partnering of a variety of international agencies, including regime secretariats, Interpol, the World Customs Organization and others, it is a small but potentially innovative agency within this landscape, albeit under-studied.

Other intergovernmental organizations with environmental links and functions

Issues of global environmental protection, politics and sustainable development have spilled over into the mandates and activities of many other IGOs. Some have a long association with global environmental issues, and indeed have been instrumental in helping get environmental problems on to the international policy agenda. The World Meteorological Organization, for example, worked with UNEP on early meetings around developing a regime to combat ozone layer depletion (see Chapter 31), and with the UN to establish the IPCC. The United Nations Economic and Social Council (UNESCO) oversees the 1972 World Heritage Convention, an early conservation agreement that protects sites of natural and cultural importance worldwide. The UN Food and Agriculture Organization (FAO) monitors the world's forests. It is increasingly engaged in sustainability debates around world agricultural production (see Chapter 44). Finally, the International Maritime Organization (IMO), among other important functions to do with maritime security and safety, oversees the International Convention for the Prevention of Pollution by Ships (MARPOL 1973/78). The Organization for Economic Cooperation and Development (OECD), while not a law-making organization, issues guidelines and data on environmental practices and performance, primarily but not wholly for and on its member states.

Other IGOs have taken on environmental responsibilities far more reluctantly, often in the wake of extensive criticism of the environmental and social impacts of their previous work (O'Neill 2017). Perhaps most famously, the World Bank was forced to address, starting in the late 1980s and continuing through the 1990s, the environmental degradation and social dislocation that had followed many of the large-scale infrastructure projects it had funded. Nongovernmental organizations (NGOs), both local to the affected communities and transnational, were able to apply pressure to politicians in donor countries to get the Bank to start integrating environmental assessment into its funding process (Fox and Brown 1998). While it has made progress in these tasks, it still faces criticism on a variety of fronts, including its technocratic approach to environmental management (Goldman 2005) and its continued funding of "brown" development projects (see Clapp and Dauvergne 2011 for an overview).

The WTO and its predecessor, the General Agreement on Tariffs and Trade, faced particular criticism in the 1990s for high-profile rulings against US regulatory actions to restrict imports of tuna (from Mexico) and shrimp (from South East Asia) on environmental grounds (see Chapters 24). These cases generated fears that any environment-related trade restriction, including those under multilateral environmental agreements, might be struck down in the interests of fostering global trade liberalization (see Chapter 24). In fact, these rulings were either never enforced or overturned on appeal (O'Neill and Burns 2005). Furthermore, the secretariats of both UNEP and the WTO have started working together in recent years to minimize conflicts and manage overlap between their respective jurisdictions (Jinnah 2010; Gehring 2011). The WTO was a significant presence both at the 2002 World Summit on Sustainable Development and the 2012 "Rio+20" summit. From an organizational perspective, these activities both demonstrate active, and relatively autonomous, work by regime secretariats over and above their individual mandates. The results of the cases reflect how the GATT/WTO dispute settlement process can be contingent as well as how it has changed over time, as in the creation of appellate panels under the 1995 WTO agreements.

Therefore, in the cases of the World Bank and the GATT/WTO, we see some progress in incorporating sustainable development goals into their initial mandates – economic development for the former, and trade liberalization for the latter – albeit in ways that fit with rather than depart radically from their overall economic ideologies (see Chapter 24). These developments stand, for instance, in stark contrast to the other main international financial organization, the International Monetary Fund (IMF), which has remained relatively resistant to societal pressures. However, developments in this arena speak to two of the theoretical debates brought up at the start of this chapter. First, they reflect the way norms have diffused across international organizations and policy arenas – in this case with respect to sustainable development (Bernstein 2000, 2002; see Chapter 16). Second, they speak to overall debates about fragmentation of global environmental governance, as international economic organizations become new governance sites (O'Neill 2017), as well as to efforts to overcome such fragmentation and potential conflict.

Regional intergovernmental organizations and global environmental governance

Regional environmental governance – governance arrangements across several (usually contiguous) states, terrestrial ecosystems (such as mountain ranges) or shared bodies of water – and the role of regional governmental organizations in environmental governance have of late garnered more attention from analysts and policymakers (Balsiger and VanDeveer 2012). This is perhaps an unsurprising development, given the ways global governance processes have stalled in recent years and the perception that global and transboundary environmental problems might be addressed more effectively by smaller groups of actors who share common characteristics, thus ameliorating collective action problems (Conca 2012). Proponents of multilevel governance, and of integrating the work of local, regional and global organizations and actors are also interested in the role of regional governance organizations in reaching across scales. This renewed interest is both generating and bringing together a rich set of research studies, which can be only briefly surveyed here.

While regional environmental governance arrangements have a long history (e.g., Rhine River management arrangements in the nineteenth century), regional organizations have recently started to take on more, and more diverse, environmental governance functions. Most generally, regional agreements are "those bilateral or multilateral agreements which are signed by at least two countries that share territorial or maritime borders, or that govern a contiguous, transnational region" (Balsiger and VanDeveer 2012: 5, citing Balsiger et al. 2012). Their organizational components vary widely in size and capacity, from a huge institutional apparatus with strong enforcement powers (e.g., the EU) to very tiny units with few employees and resources. Based on an Internet search by the author, some regional fisheries management organizations (RFMOs), for example, are lucky to have even three permanent employees.

As Balsiger and VanDeveer (2012) point out, some of these initiatives are part of autonomous organizations, such as environmental policies within the EU, the Arctic Council, ASEAN, or environmental bodies within free trade associations such as NAFTA. Others are part of multilevel governance arrangements, such as regional centers established under chemicals treaties (see Selin 2012) or regional treaties formed under the umbrella of a broader global regime. Examples of the latter include various sub-regimes associated with the 1975 Convention on Migratory Species (see Chapter 41), such as Eurobats (which monitors the European bat population and engages in educational activities and came into force in 1994) or ACCOBAMS (the Agreements for the Conservation of Cetaceans of the Black Sea, Mediterranean Sea and Contiguous Atlantic Areas, created in 1996). Others still are autonomous regional governance arrangements, although they may be networked with similar groups, such as RFMOs. Another example of regional governance arrangements around specific environmental issue areas is the 1979 Convention on Long Range Transboundary Air Pollution, whose activities are largely based in Europe, but with another North American sub-regional organization (Levy 1993; see Chapter 30).

In many cases of regional environmental governance, issue-area-based regional organizations often cover (or are extended to cover) a cluster of concerns, including environmental, but also those related to sustainable development, border control, regional security and others (Elliott 2012; see Chapter 20). Examples include transboundary mountain regions, such as the Himalayas or the Swiss Alps (Balsiger 2012; Matthew 2012) and regional seas (Chapter 39), such as the Barents Sea and the Mediterranean (P. Haas 1990; Stokke et al. 1999).

In terms of analyzing regional environmental IGOs, many of the same themes that occur in the broader literature apply to them, but perhaps play out in different ways. Some research examines how regional organizations reflect and/or shape identities across borders or within certain eco-regions (Balsiger 2012, on the European Alps). Others examine how well existing organizations are able to build environmental concerns into their existing governance activities and structures (see Aggarwal and Chow 2010 on the ASEAN Agreement on Transboundary Haze Pollution, which entered into force in 2003, and Tobing 2017). Yet others analyze the impacts and effectiveness of regional agreements and organizations. Specifically, there is quite a lot of literature on RFMOs in this context (e.g., Cullis-Suzuki and Pauly 2010). A 2006 study published by the World Wildlife Fund and the wildlife NGO TRAFFIC (Wilcock and Lack 2006) recommended in particular the development of some coordinated management of RFMOs, for example in terms of monitoring and compliance, or providing scientific input, and noted a need for high level discussions about creating some sort of global fisheries agency (see Chapter 40). Although this has not happened, nor does it seem likely, in this case of regional governance effectiveness appears low, to at least some extent because of its regional level.

The possibility of adding a regional dimension to the environmental IGO landscape raises some more questions specific to this issue. First, where, and under what conditions,

Kate O'Neill

is regional governance more effective or more appropriate than global governance? What sorts of relationships exist between global and regional organizations? Across regional organizations? Are they more vertical (hierarchical), or horizontal (networked)? Can regional organizations in similar areas learn from each other and adapt more easily than global IGOs can? Stokke et al. (1999), in their study of the bilateral regime managing the Barents Sea, suggest that while learning is possible, careful attention needs to be paid to the contextual characteristics of specific regional arrangements, which may make policy or institutional diffusion much harder. Either way, these questions about regional IGOs deserve further exploration.

Conclusion

In practice and in theory, environmental IGOs have recently been experiencing a moment in the sun – even if that sunlight has sometimes shown up their flaws. This chapter has described the complex landscape, at global and regional levels, of IGOs with environmental governance responsibilities. It has also described some of the linkages, formal and informal, that exist between them, for example the nested nature of the major UN-related agencies, from UNEP to the secretariats and other regime bodies, or their growing interrelationship across regime lines with the WTO. The emergence of multilevel governance as a very visible phenomenon has also raised questions of linkages across scales of global governance – between global and regional.

Factors and themes that have been important to understanding individual IGOs have to do with their performance and impacts, and their role in reducing the transaction costs of inter-state cooperation. In terms of performance, the evidence, as discussed above, is mixed, though several studies find that performance is better than expected, given the various constraints IGOs face, and have made progress in identifying what factors are important in determining performance (Finnemore and Sikkink 1998). Recent research also addresses the (often growing) extent to which IGOs are able to exert agency, over and above the functions delegated to them by member states. In the environmental arena, while research shows that UNEP is more constrained, a more recent set of studies are showing how bureaucracies at a lower level – such as treaty secretariats – are able to push environmental governance agendas in particular directions. Quite often this is through informal means – such as the development of shared norms and understandings in the management of overlap between regimes.

Collectively, environmental IGOs face a number of challenges. The main one is quite simply the complexity of this landscape, which makes for potential overlaps and conflicts between organizations and across jurisdictions, as well as unnecessary fragmentation of governance activities. Adding regional IGOs to this mix raises the possibility for conflicts across scales too. Debates have turned to the possible future trajectories for this institutional system or complex. This demonstrates the extent to which environmental IGOs are now entrenched in the global political system. One of the possible trajectories for this system is toward deliberate centralization, building, for example, an overarching authority to coordinate global environmental governance such as a World Environment Organization (Biermann 2001). Others point out problems with this model (e.g., Najam 2003), and there is some doubt that such deliberate reform, even if the political will existed, would lead to the desired results (O'Neill 2012). Conversely, some (e.g., Conca 2012) have identified forces for decentralization given the perceived failings of traditional state-led global environmental governance, in this case, pursuing governance solutions at a regional level. One possible shift has to date been under-represented: the incorporation of NGOs and other non-state actors directly into

114

IGO decision-making processes (see Chapter 14). While some studies – with respect to partnerships (Joyner 2005), multi-stakeholder commissions such as the World Commission on Dams (Ottaway 2001), and specific global institutions, such as the World Bank (Park 2010, 2018) – address the possibilities of a more hybrid form of international organization (international "governance" organizations, perhaps), this possibility needs more exploration.

In sum the role of IGOs in global and regional environmental cooperation has been shown to be significant, and, largely, positive. Their impacts and activities go well beyond realist or neoliberal institutionalist formulations of their role and activities (see Chapter 3). Many challenges to – and critiques of – their activities do exist. Practical constraints are, of course, important, and, ultimately, without member states on board, they cannot fulfill their mission. Finally, if we are to assume that they will take on more autonomy as time goes on, we will have, sooner or later, to address their legitimacy, or lack thereof, in this role as global policymakers, and whose interests they are going to represent (Jinnah 2010).

References

Abbott, Kenneth W., and Duncan Snidal. (1998) "Why States Act through Formal International Organizations." *Journal of Conflict Resolution 42.1*: 3–32.

Aggarwal, Vinod K., and Jonathan T. Chow. (2010) "The Perils of Consensus: How ASEAN's Meta-Regime Undermines Economic and Environmental Cooperation." *Review of International Political Economy 17.2*: 262–290.

Balsiger, Jörg. (2012) "New Environmental Regionalism and Sustainable Development in the European Alps." *Global Environmental Politics 12.3*: 58–78.

Balsiger, Jörg, Miriam Prys, and Niko Steinhoff. (2012) "The Nature and Role of Regional Agreements in International Environmental Politics." *GIGA Working Paper No. 208*, Hamburg: German Institute of Global and Area Studies.

Balsiger, Jörg, and Stacy VanDeveer. (2012) "Navigating Regional Environmental Governance." *Global Environmental Politics 12.3*: 1–17.

Barnett, Michael N., and Martha Finnemore. (1999) "The Politics, Power, and Pathologies of International Organizations." *International Organization 53.4*: 699–732.

Barnett, Michael N., and Martha Finnemore. (2004) *Rules for the World: International Organizations in World Politics.* Ithaca, NY: Cornell University Press.

Bauer, Steffen. (2009) "The Secretariat of the United Nations Environment Programme: Tangled up in Blue." *Managers of Global Change: The Influence of International Environmental Bureaucracies.* Eds. Biermann, Frank and Bernd Siebenhüner. Cambridge, MA: MIT Press.

Benedick, Richard E. (2007) "Science, Diplomacy and the Montreal Protocol." *Encyclopedia of Earth.* Ed. Cleveland, Cutler J. Washington, DC: Environmental Information Coalition, National Council for Science and the Environment.

Bernstein, Steven. (2000) "Ideas, Social Structure and the Compromise of Liberal Environmentalism." *European Journal of International Relations 6.4*: 464–512.

Bernstein, Steven. (2002) *The Compromise of Liberal Environmentalism.* New York: Columbia University Press.

Biermann, Frank. (2001) "The Emerging Debate on the Need for a World Environment Organization: A Commentary." *Global Environmental Politics 1.1*: 45–55.

Biermann, Frank, and Steffen Bauer, eds. (2005) *A World Environment Organization: Solution or Threat for Effective International Environmental Governance?* London: Ashgate Publishing.

Biermann, Frank, and Bernd Siebenhüner, eds. (2009) *Managers of Global Change: The Influence of International Environmental Bureaucracies.* Cambridge, MA: MIT Press.

Borie, Maud, and Mike Hulme. (2015) "Framing Global Biodiversity: Ipbes between Mother Earth and Ecosystem Services." *Environmental Science & Policy 54*: 487–496.

Clapp, Jennifer, and Peter Dauvergne. (2011) *Paths to a Green World: The Political Economy of the Global Environment.* Cambridge, MA: MIT Press.

Conca, Ken. (2012) "The Rise of the Region in Global Environmental Governance." *Global Environmental Politics 12.3*: 127–133.

Cullis-Suzuki, Sairka, and Daniel Pauly. (2010) "Failing the High Seas: A Global Evaluation of Regional Fisheries Management Organizations." *Marine Policy 34*: 1036–1042.

Diehl, Paul, ed. (2005) *The Politics of Global Governance: International Organizations in an Interdependent World*, 3rd ed. Boulder, CO: Lynne Rienner Press.

Downie, David, and Marc A. Levy. (2000) "The UN Environment Programme at a Turning Point: Options for Change." *The Global Environment in the Twenty-First Century: Prospects for International Cooperation*. Ed. Chasek, Pamela S. Tokyo: UN University Press.

Elliott, Lorraine. (2012) "ASEAN and Environmental Governance: Strategies of Regionalism in Southeast Asia." *Global Environmental Politics 12.3*: 38–57.

Finnemore, Martha, and Kathryn Sikkink. (1998) "International Norm Dynamics and Political Change." *International Organization 52.4*: 887–917.

Fox, Jonathan A., and L. David Brown, eds. (1998) *The Struggle for Accountability: The World Bank, NGOs and Grassroots Movements*. Cambridge, MA: MIT Press.

Gehring, Thomas. (2011) "The Institutional Complex of Trade and Environment: Toward an Interlocking Governance Structure and a Division of Labor." *Managing Institutional Complexity: Regime Interplay and Global Environmental Change*. Eds. Oberthür, Sebastian and Olav Schram Stokke. Cambridge, MA: MIT Press.

Goldman, Michael. (2005) *Imperial Nature: The World Bank and Struggles for Social Justice in the Age of Globalization*. New Haven, CT: Yale University Press.

Greene, Owen. (1998) "The System for Implementation Review in the Ozone Regime." *The Implementation and Effectiveness of International Environmental Commitments: Theory and Practice*. Eds. Victor, David G., Kal Raustiala, and Eugene B. Skolnikoff. Cambridge, MA: MIT Press.

Haas, Ernst B. (1964) *Beyond the Nation State: Functionalism and International Organizations*. Stanford, CA: Stanford University Press.

Haas, Ernst B. (1990) *When Knowledge Is Power: Three Models of Change in International Organizations*. Berkeley: University of California Press.

Haas, Peter M. (1990) *Saving the Mediterranean: The Politics of International Environmental Cooperation*. New York: Columbia University Press.

Hulme, Mike, and Martin Mahony. (2010) "Climate Change: What Do We Know about the IPCC?" *Progress in Physical Geography 34.5*: 705–718.

Hurd, Ian. (2011) *International Organizations: Politics, Law, Practice*. Cambridge: Cambridge University Press.

Ivanova, Maria. (2007) "Designing the United Nations Environment Programme: A Story of Compromise and Confrontation." *International Environmental Agreements 7*: 337–361.

Ivanova, Maria. (2010) "UNEP in Global Environmental Governance: Design, Leadership, Location." *Global Environmental Politics 10.1*: 30–59.

Jinnah, Sikina. (2010) "Overlap Management in the World Trade Organization: Secretariat Influence on Trade-Environment Politics." *Global Environmental Politics 10.2*: 54–79.

Jinnah, Sikina. (2011) "Marketing Linkages: Secretariat Governance of the Climate–Biodiversity Interface." *Global Environmental Politics 11.3*: 23–43.

Jinnah, Sikina. (2012) "Singing the Unsung: Secretariats in Global Environmental Politics." *The Roads from Rio: Lessons Learned from Twenty Years of Multilateral Environmental Negotiations*. Eds. Chasek, Pamela S. and Lynn M. Wagner. New York: Routledge.

Joyner, Christopher C. (2005) "Rethinking International Environmental Regimes: What Role for Partnership Coalitions?" *Journal of International Law & International Relations 1.1–2*: 89–119.

Karns, Margaret P., and Karen A. Mingst. (2010) *International Organizations: The Politics and Processes of Global Governance*, 2nd ed. Boulder, CO: Lynne Rienner.

Kennan, George F. (1970) "To Prevent a World Wasteland: A Proposal." *Foreign Affairs 48*: 401–413.

Keohane, Robert O. (1984) *After Hegemony: Cooperation and Discord in the World Economy*. Princeton, NJ: Princeton University Press.

Keohane, Robert O., and David G. Victor. (2011) "The Regime Complex for Climate Change." *Perspectives on Politics 9.1*: 7–23.

Kohler, Pia M., Alexandra Conliffe, Stefan Jungcurt, Maria Gutierrez, and Yulia Yamineva. (2012) "Informing Policy: Science and Knowledge in Global Environmental Agreements." *The Roads from Rio: Lessons Learned from Twenty Years of Multilateral Environmental Negotiations*. Eds. Chasek, Pamela S. and Lynn M. Wagner. New York: Routledge.

Krasner, Stephen D. (1988) "Sovereignty: An Institutional Perspective." *Comparative Political Studies* *21.1*: 66–94.

Kratochwil, Friedrich, and John G. Ruggie. (1986) "International Organization: A State of the Art on an Art of the State." *International Organization 40.4*: 753–775.

Lattanzio, Richard K. (2010) *Global Environment Facility (GEF): An Overview.* Washington, DC: Congressional Research Service.

Levy, Marc A. (1993) "European Acid Rain: The Power of Tote-Board Diplomacy." *Institutions for the Earth: Sources of Effective International Environmental Protection.* Eds. Haas, Peter M., Robert O. Keohane and Marc A. Levy. Cambridge, MA: MIT Press.

Matthew, Richard A. (2012) "Environmental Change, Human Security, and Regional Governance: The Case of the Hindu Kush/Himalaya Region." *Global Environmental Politics 12.3*: 100–118.

Munoz, Miquel, Rachel Thrasher, and Adil Najam. (2009) "Measuring the Negotiation Burden of Multilateral Environmental Agreements." *Global Environmental Politics 9.4*: 1–13.

Najam, Adil. (2003) "The Case against a New International Environmental Organization." *Global Governance 9*: 367–384.

O'Neill, Kate. (April 2012) "From Stockholm to Copenhagen: The Evolving Meta-Regime of Global Environmental Governance." Paper presented at the Annual Meeting of the International Studies Association, San Diego.

O'Neill, Kate. (2017) *The Environment and International Relations, Second Edition.* Cambridge: Cambridge University Press.

O'Neill, Kate. (2019) "Architects, Agitators, and Entrepreneurs: International and Nongovernmental Organizations in Global Environmental Politics." *The Global Environment: Institutions, Law, and Policy, Fifth Edition.* Eds. Axelrod, Regina S. and Stacy D. VanDeveer. Los Angeles: Sage/CQ Press.

O'Neill, Kate, and William C. G. Burns. (2005) "Trade Liberalization and Global Environmental Governance: The Potential for Conflict." *Handbook of Global Environmental Governance.* Ed. Dauvergne, Peter. Cheltenham: Edward Elgar.

Ottaway, Marina. (2001) "Corporatism Goes Global: International Organizations, Nongovernmental Organization Networks, and Transnational Business." *Global Governance 7*: 265–292.

Park, Susan. (2010) *World Bank Group Interactions with Environmentalists: Changing International Organization Identities.* Manchester: Manchester University Press.

Park, Susan. (2018) *International Organizations and Global Problems: Theories and Explanations.* Cambridge: Cambridge University Press.

Schmitter, Philippe C. (1969) "Three Neo-Functional Hypotheses about International Integration." *International Organization 23*: 161–166.

Selin, Henrik. (2012) "Global Environmental Governance and Regional Centers." *Global Environmental Politics 12.3*: 18–37.

Selin, Henrik. (2019) "Global Politics and Policy of Hazardous Chemicals." *The Global Environment: Institutions, Law and Policy, Fifth Edition.* Eds. Axelrod, Regina S. and Stacy D. VanDeveer. Los Angeles: Sage/CQ Press.

Siebenhüner, Bernd. (2006) "Can Assessments Learn, and If So, How? A Study of the IPCC." *Assessments of Regional and Global Environmental Risks: Designing Processes for the Effective Use of Science in Decision-making.* Eds. Farrell, Alexander E. and Jill Jäger. Cambridge, MA: MIT Press.

Stokke, Olav Schram, Lee G. Anderson, and Natalia Mirovitskaya. (1999) "The Barents Sea Fisheries." *The Effectiveness of International Environmental Regimes: Causal Connections and Behavioral Mechanisms.* Ed. Young, Oran. Cambridge, MA: MIT Press.

Tobing, Dio Herdiawan. (2017) "Indonesia drags its feet on ASEAN Haze Treaty" *The Conversation,* September 8, 2017 at https://theconversation.com/indonesia-drags-its-feet-on-asean-haze-treaty-81779.

Wilcock, A., and M. Lack. (2006) *Follow the Leader: Learning from Experience and Best Practice in Regional Fisheries Management Organizations.* Washington, DC: WWF International and TRAFFIC International.

Young, Oran R. (1994) *International Governance: Protecting the Environment in a Stateless Society.* Ithaca, NY: Cornell University Press.

9

International environmental regimes

Formation, effectiveness, trends and challenges

Peter Suechting and Mary E. Pettenger

Human activity over the last century has significantly damaged Earth's atmosphere, water, land and biodiversity. Environmental problems addressed in other chapters of this handbook, such as stratospheric ozone depletion (see Chapter 14), air pollution and acid rain (see Chapter 30), climate change (see Chapter 32), collapse of fish stocks (see Chapter 40), and water and ocean pollution (see Chapters 38 and 39), represent significant challenges requiring action. A central focus of these and other chapters in this handbook is explaining and understanding human behavior that has led to environmental problems, and the theories, methods and tools to alter this behavior. This chapter focuses on international environmental regimes as influential actors in international environmental politics and as a research topic.

International environmental regimes have been a very active topic of study, spurred on by rapid growth in international institutions, over the past four decades (Mitchell et al. 2020). This trend appeared in decline at the beginning of the twenty-first century with the formation rate of new international institutions dropping off significantly. Keohane (2020), reflecting on the hard times facing international cooperation, concluded that the "arc of history" no longer bends toward "constitutional democracy, increased globalization, and multilateralism" (Keohane 2020: 11). However, in a post-2015 Paris Agreement (see Chapter 32) world in which a 2° Celsius increase in mean global temperatures, or much worse, is likely, regime formation and design are key loci of research on global environmental politics by scholars seeking the means to respond (Allan 2019; Lenton et al. 2019).

International regime theory (IRT) emerged in the international relations field during a period of theoretical debate (see Chapters 3 and 4). It was influenced by liberalism's rejuvenation (e.g., Keohane and Nye's 1977 "complex interdependence"), and significant international challenges, such as declining US hegemony and rising awareness of transboundary pollution. IRT has followed in the steps of neoliberal institutionalism, neofunctionalism, social constructivism and global governance (see below and chapters in Part II of this handbook). Over time, IRT has shifted slightly from its primary focus on state-based behavior to the study of numerous issues, *inter alia*, the environment, human rights, refugees, trade, monetary policy, nuclear non-proliferation, food security, space, telecommunications and intellectual property rights.

DOI: 10.4324/9781003008873-12

International regimes are no longer perceived as a passing "fad" (Strange 1982), with four decades of articles and books, and the presence of international regimes as a theme within numerous textbooks on international relations, international organizations and environmental politics. In congruence, the bellwether event of the 1972 United Nations Conference on the Human Environment was arguably the stimulus of many subsequent multi- and bilateral environmental agreements, and of numerous international environmental regimes. However, recent trends seem to indicate a decline in the creation of environmental international regimes and an increase in competition with alternative international institutional forms, as will be discussed below.

International environmental regime theory

This chapter surveys international environmental regime theory's (IERT) key scholars, foundations, successes, criticisms, current applications and future challenges. IERT has been applied to numerous environmental issues, including deforestation (see Chapter 42), stratospheric ozone depletion (see Chapter 31), the Arctic, whales, marine pollution (see Chapter 40), climate change (see Chapter 32), air pollution (see Chapter 30) and space. We focus on the application of IERT rather than on individual environmental issues, which are covered in detail in other chapters of this handbook (see especially Part V).

Much of the early to mid-phase IERT literature served to answer important questions. For example, are regimes effective (can they change state behavior)? What makes them more or less effective (what improves compliance and legitimacy, how are they formed, what is the influence of different states participating)? Associated with this is the investigation of how regimes challenge and overcome state sovereignty; many environmental problems extend over state borders and may require a global response.

A central theme of the IERT literature that connects this chapter with others in this handbook is whether regimes matter – are they effective in increasing cooperation and improving the environment? However, environmental degradation has continued since the first edition of this handbook. Is this a failure of international regimes, or are international regime scholars coming closer to identifying how regimes can and will solve environmental problems? Phase One of the chapter expounds on the beginnings of IERT scholarship. Phase Two discusses a reorientation, beginning in the 1990s, to regime effectiveness and expansion of research foci. Phase Three elucidates the focus on acquisition of data and creation of large datasets. Phase Four describes the consolidation of core themes and concepts, and examines scholars' reaction to the emergence of alternative theoretical frameworks, in particular, global governance. The final phase, titled Today, presents current trends in response to contemporary environmental problems, and describes a shift beyond the state to analyzing multilevel actors, networks and regime complexes.

Phase One: definitions, theory and focus

The binding characteristic of the first phase of IRT and IERT is disagreement. Early efforts focus on three areas: (a) definition/conceptualization of international regimes, (b) theoretical orientation and (c) regime function, formation and persistence. The definitional divide is emblematic of one of the fundamental controversies in the IRT field. Different researchers, while conceptualizing international regimes, adapt countervailing theories and proposed definitions that include or exclude significant terms (Hasenclever 1997: 8–22). In short, without agreement on the concept, it was difficult to design rigorous studies that could identify if, how and why environmental regimes could be effective.

Discussion of international regimes began in earnest in the 1970s (Haas 1975; Ruggie 1975). However, a 1982 special issue of *International Organization* journal, subsequently published as a book, *International Regimes* (Keohane 1983), propelled the research topic to prominence. The journal/book introduced Stephen Krasner's oft-quoted definition: "Regimes can be defined as sets of implicit or explicit principles, norms, rules, and decisionmaking procedures around which actors' expectations converge in a given area of international relations" (Keohane 1983: 2). In contrast, Oran Young's early definition focuses more directly on the social and ideational aspect of regimes: "Regimes are social institutions governing the actions of those interested in specifiable activities (or accepted sets of activities)" (Young 1982a: 93). Even more broadly, Kratochwil and Ruggie define regimes as "governing arrangements constructed by states to coordinate their expectations and organize aspects of international behavior in various issue-areas" (Kratochwil and Ruggie 1986: 759). Thus, for some scholars, regimes are socially constructed institutions that promote greater cooperation, or collective action, between states on issues of shared concern. For others, they are rule-based structures, formed by states in the anarchical international system, which alter state interests (Keohane 1989).

International regimes are human constructs and are not corporeal phenomena such as a table or a tree. Therefore, in order to "observe" a regime, one has to look for signs of its impact or influence. International regimes are a form of institution because routinized behavior and social practices, based on mutual expectations of state behavior (i.e., rules, norms and principles), can be observed. However, they exist between the formal (physical) institutions of international organizations, such as the North Atlantic Treaty Organization, which has a staff and buildings, and the informal (nonphysical) institutions of conventions, such as driving on the left (or the right) side of the road. Kratochwil and Ruggie (1986) may be right when they say that "regimes are conceptual creations not concrete entities" and as such will remain a "contestable concept" (Kratochwil and Ruggie 1986: 763–764).

Much of the early literature on regimes can be found in the journal *International Organization* as well as in legal journals, such as the *American Journal of International Law* and *Yale Law Journal*. For the purposes of this chapter, many environmental regulatory regimes have formed out of multilateral legal agreements, for example the Montreal Protocol and the stratospheric ozone regime (see Chapter 31), or the Law of the Sea Convention and the fisheries regime (see Chapters 39 and 40). Thus, there was a significant early focus on the legal aspects of these agreements, for example parties to the agreement and their interests, and soft/hard law components, such as compliance, enforcement and regulations.

Those planning to study international regimes should acquaint themselves with the ontological and epistemological differences within the field (Kratochwil and Ruggie 1986; Haggard and Simmons 1987; Hasenclever 1997; Zürn 1998; Mitchell 2002b; Keohane 2020). The scope of this chapter prevents explicit discussion, but there are a number of important early (and potentially current) debates to note. First, disagreement is found concerning the function of regimes as intervening variables between "causal variables" and "behaviors and outcomes" (Keohane 1983: 1–9). That is, are regimes independent actors or "autonomous variables" that can overcome state interests (Haggard and Simmons 1987: 392; Keohane 1983: vii and 355–368)? Or are they dependent on state-based power structures (Puchala and Hopkins 1983) and thus never independent (Rittberger et al. 2012: 5)? In other words, can they overcome state sovereignty and, if so, under what conditions can they be an effective means of solving environmental problems?

Second, in this debate is a deeper question of the theoretical foundation for examining regimes, such as interest-based theories (neoliberalism), power-based theories (neorealism) and knowledge-based (cognitive) theories (Rittberger 1995; Hasenclever 1997; see Chapters 3 and 4). In short, the epistemological and ontological differences between these theoretical approaches may leave the field unable to move toward a common framework and understanding of the effectiveness of regimes. Zürn (1998) encourages side-stepping this issue and moving to a more significant focus on effectiveness. After all, do we not know a regime when we see one?

A third focus was on the function and purpose of regimes, regime formation, persistence, implementation and compliance (List and Rittberger 1992; Rittberger 1995). Such studies established the means by which regimes "collectively manage conflicts" over environmental issues, for example vis-à-vis principles, norms, rules, expectations, and "guided behavior" (Young 1989: 89). Regimes have been examined for their regulatory functions, and their ability to promote norms against certain behaviors. Numerous functions of regimes have been analyzed including reducing transaction costs and providing information (Axelrod and Keohane 1986: 250), reducing uncertainty and risk (Keohane 1983: 161–162), and nurturing and strengthening the continuance of a regime, such as "providing high-quality information to policy makers" (Keohane 1983: 165).

A fourth focus concerns the assumption that international regimes will create greater cooperation between states and mitigate environmental problems, what some call "fairy tales" (Paterson 1999: 793). Keeley notes in his call to "develop a nonliberal alternative" that "Liberal approaches assume, rather than establish, regimes as benevolent, voluntary, cooperative and legitimate" (Keeley 1990: 90). There appears to be an even deeper assumption that regimes can be created to confront environmental degradation. Young notes that the "naïve hopes concerning the efficacy of social engineering in the realm of international regimes constitute a common and serious failing among policy makers and students of international relations alike" (Young 1982a: 281). In addition, during the growing pains of this phase several useful studies apply IRT to environmental issues as exemplified by single issue-specific studies.

Oran Young has been extremely prolific in documenting and theorizing about international environmental regimes. His early work, *Resource Regimes* (1982b), was one of the first large-scale studies to position regimes as important variables between state sovereignty and natural resources. His research continued (Young 1989) with an exploration of regimes "in theory" and "in practice." Prognosticating today's studies of global governance, his study continued with an analysis of international governance that examines the climate regime and resource regimes in the Arctic (Young 1998).

In conclusion, Phase One identifies many of the tensions and theoretical differences among approaches to international environmental regimes. However, by the end of the 1980s, calls were made for more rigorous applications of the approach. For example, Haggard and Simmons call for "Large-n studies" and conclude that "current theories of international regimes have ignored domestic political processes" (Haggard and Simmons 1987: 513–515). Such calls were heard, as discussed below.

Phase Two: effectiveness, expansion and empiricism

External factors play a significant influence on the focus and relevance of IERT, such as the end of the Cold War when environmental crises were elevated in importance, perhaps even becoming "high" political issues. As the available scope of topics ballooned, so too did the identification of potential benefits of regimes, multilevel actors and cooperation beyond the sovereign state. Articles examining environmental regimes began to appear outside *International*

Organization and within issue-specific journals, such as the *Journal of Environmental Management, International Security, International Studies Quarterly, International Environmental Affairs* and later *Global Environmental Politics*. Additionally, concentrated efforts were made to unite European–US IRT research approaches and agendas (Rittberger 1995; Miles et al. 2002).

The primary focus of this second phase of international environmental regime research is a reorientation away from the definitional and theoretical debates, and toward regime effectiveness and research that identifies and documents push/pull factors with rigorous case studies and empirical analysis (Mitchell 2002b). Additionally, during this phase, IERT became closely linked with policy studies and solutions. Zürn (1998) contrasts the second generation of environmental politics research, which "broadened both the scope of the issues and the empirical observations of the field," with the first generation, which "identified the preeminence of the environment for the analysis of international relations" (Zürn 1998: 618).

Germane to this phase is Zürn's identification of a second-generation focus on regime consequences and effectiveness that "has the most potential for producing an enduring research program" (Zürn 1998: 620). The question that took on greater prominence in the 1990s persists today: "Do regimes matter?" (Zürn 1998: 632). If regimes are designed to solve environmental problems, do they? How are regime consequences different from regime effectiveness? Zürn defines regime effectiveness as "those intended and issue-area-specific outcomes of the regime," implying a narrower action with intent. Regime consequences "refer mainly to the more general impacts of the regime, whether intended or unintended, issue-area specific or general" (Zürn 1998: 632). Setting a high standard for international environmental regimes, he states that "Institutional effectiveness occurs when the quality of the environment is improved because of the institution" (Zürn 1998: 637). Rittberger and Zürn (1991) went further to stipulate that "the institutionalization of cooperative conflict management needs to develop *durability* and a high degree of *rule-effectiveness*" (Rittberger and Zürn 1991: 166, italics added). Likewise, numerous variables have been examined in the efforts by scholars to operationalize environmental regime effectiveness (Mitchell 2002a: 507–512; Wettestad 2006).

Wettestad (2001) notes that IERT needs to move beyond "do regimes matter" to "more specifically to what extent and how regimes possibly matter. In other words, are we able to say anything more specific about types of regimes and their specific regime features that are likely to make more impact and to contribute to higher effectiveness than others" (Wettestad 2001: 317). For many scholars, regime effectiveness revolves around the regime's ability to "problem solve," to create international cooperation or collective action to overcome an environmental problem. In order to define "effectiveness" one also has to identify both ends of the causal equation, the independent variable, such as the characteristics of an environmental problem that needs to be solved (Hisschemoller and Gupta 1999) or sources of influence, for example science and knowledge, and the dependent variable, for example the measurable signs of effectiveness, such as compliance or providing information (Mitchell 2002a). Some even dispute using effectiveness as a dependent variable (see discussion of the Oslo–Potsdam debate below).

The operationalization of "effectiveness" has been an exhaustive and controversial issue for IERT. Sprinz (2001) discusses capturing effectiveness by including decision-making, reporting and compliance, funding mechanisms and development components. Luterbacher and Sprinz (2001) propose including science, international actors and bargaining power, equity, institutional setup, side payments and regime linkages (interplay) (Luterbacher and Sprinz 2001: 300), such as in their discussion of the United States' refusal to join the Kyoto Protocol (Luterbacher and Sprinz 2001: 298).

In conclusion, Phase Two displays an increased emphasis on the extent to which and how regimes matter. As noted, effectiveness emerged as a contested concept, and greater emphasis is placed on developing more rigorous qualitative and quantitative studies of environmental regimes.

Phase Three: databases and operationalizing effectiveness

The third phase of IERT contributes significantly to the maturation of IERT. The most important contributions of Phase Three are the operationalization of regime effectiveness and the advent of large empirical studies. Several teams have assembled extensive databases to identify the factors contributing to regime effectiveness, these include the Institutional Dimensions of Global Environmental Change (IDGEC) project (Young 2008), the Oslo–Seattle Project Database (Miles et al. 2002) and the International Regimes Database (IRD) (Breitmeier 2006).

As IERT advanced, the need to operationalize regime effectiveness became more urgent (Young 2011). For example, why did the ozone regime succeed (see Chapter 31), while the climate change regime has failed (see Chapter 32)? Answering these questions is important for researchers and policymakers. Yet, the assumption persists that regimes are the tools with which to advance positive change, and IERT scholars seem to prescribe future actions.

In Phase Three, heated but productive debate emerges on the proper means to empirically examine regime effectiveness. One example is the "dialogue, or what is called 'good trouble,'" between Young (2003) and Hovi et al. (2003) regarding the "Oslo–Potsdam solution to measuring regime effectiveness." Hovi et al. (2003) propose a "formula" premised on game theory to measure regime effectiveness based on the variables of "a no-regime counterfactual," a "measure of actual performance" and a determination of what they call a "collective optimum" (Hovi et al. 2003: 75). Young praises the steps forward to devising measures that can be easily quantified and "allow comparison," but critiques their variables and the focus on regime effectiveness as the dependent variable with causal implications (Young 2003).

A second focus of Phase Three (and into Phase Four) IERT effectiveness research is on the interplay (or linkages) between environmental regimes and other types of regimes, such as the relationship between the World Trade Organization and marine mammal regimes, and the interplay or linkages among international environmental regimes themselves (Stokke 2001; Andersen 2002; Ward 2006). These studies highlight the types of interplay, such as embedded, nested, clustered and overlapping regimes (Young et al. 1996), utilitarian, normative and ideational factors (Stokke 2001: 10–11) and "time dimensions" (Andersen 2002). The purpose is to examine "whether such interplay will be supportive or obstructive to problem-solving efforts under international regimes" (Stokke 2001: 23).

Oberthür and Stokke's (2011) edited volume provides another clear example of the directions the field takes regarding interplay. Their book connects empirical data from the IDGEC project, numerous environmental case studies (e.g., climate change, Arctic resources and biodiversity) and analysis of the interplay management of regimes (e.g., trade and environmental regimes) (Oberthür and Stokke 2011). Raustiala and Victor (2004) examine the influence of regime "density" in the issue area of plant genetic resources. They present the concept of "a *regime complex*: an array of partially overlapping and nonhierarchical institutions governing a particular issue area" (Raustiala and Victor 2004: 279) that led to "legal inconsistencies" (2004: 306) and lack of effective action. Likewise, a 2011 *Global Environmental Politics* special issue is devoted to linkages between numerous environmental regimes and the climate change regime (e.g., biodiversity, fisheries and desertification). The journal examines "how regime overlap is managed by political actors through the creation of strategic linkages between international regimes" (Jinnah 2011: 4).

A third focus is on the questions of why regimes were formed in some issue areas, why they were not in others (non-regimes) and why some regimes have failed. Several researchers in this area employ counterfactuals to assess the impact of the appearance or absence of a regime, that is, they ask what would happen if the regime did not exist (Helm and Sprinz 2000: 633–635). Others seek to identify the factors that led to the lack of a regime (Dimitrov et al. 2007; Wilkening 2011) or regime failure (Harris 2007, 2021).

International environmental regime theory is also challenged by critical theories and constructivism (see Chapter 4) that foreshadow Phase Four discussed below. For example, Paterson (2009) raises an interesting criticism of "institutionalist conclusions" directed specifically at environmental regime literature. While the number of international environmental institutions grew from the 1980s to 2000s, so too did the level of environmental damage. Does this signify a weakness of IERT to capture the means by which to bring positive change? Or, rather, is the world changing so that new patterns of behavior, such as global governance, and more importantly "global environmental governance," are changing environmental politics, and regimes themselves (Paterson 2009: 264–266)?

Eckersley (2004: 28–52) discusses the theoretical limits of regime theory in relation to "critical constructivism." Because regimes by definition incorporate shared rules, norms and principles, and affect behavior, the process by which shared (intersubjective) meaning is formed fits clearly within the cognitive/social learning IERT approach, as well as that of social constructivism. Paterson (1999) critiques regime theory's inability or unwillingness to include additional levels, such as nongovernmental organizations, in its analysis. These criticisms of regime theory questioned its ability in Phase Three to explain and alter human behavior because it focused solely on cooperation between states. This myopia means IERT might miss shifting "power structures" that are beyond state sovereignty, for example resistance to capitalism and its detrimental effects on the environment (Paterson 1999: 798–800). Nevertheless, much of the IERT literature adopts a positivist approach, namely, neo-institutionalism or neofunctionalism, and for some critical theorists the theoretical differences "are incommensurable (making empirical comparisons unhelpful, if not impossible)" (Walsh 2004: 12). In addition, it may be counterproductive for IERT to transform its epistemological and ontological foundations to absorb a post-structuralist or post-modernist perspective. Such an effort could return the field back to the morass of Phase One.

In conclusion, regime effectiveness remained an important focus of IERT in Phase Three. Yet dialogue (or controversy) over the conceptualization of effectiveness continues to dominate the field, and may do so into the foreseeable future.

Phase Four: consolidation and diversification

In Phase Four, IERT consolidated concepts while utilizing new methodologies, new databases and new variables to answer critical questions. The interlinked concepts of regime design (including epistemological foundations), state participation and effectiveness are central foci. Furthermore, the core elements of IERT are increasingly organized around the theme of climate change with a sustained focus on the United Nations Framework Convention on Climate Change (UNFCCC) (see Chapter 32) and related Multilateral Environmental Agreements (MEA). Though historically a strongly positivist and empiricist research tradition, IERT has adopted a more prescriptive orientation, sizing up the complex system of international climate governance and assessing how to move it in the direction of decarbonization. In sum, IERT in Phase Four has focused simultaneously on expanding upon the theory itself while confronting the existential threat of climate change.

The first concept, regime design, has been a perennial concern of IERT. For this reason, scholars new to IERT may find themselves adrift among the many conflicting epistemological frameworks with which to understand regime design, and might find it beneficial to consult Voeten's (2019) summary of the "efforts to think theoretically about institutional design" (Voeten 2019: 148) for potential new avenues of research. Broadly speaking, international institutions,' such as international regimes,' design theories can be grouped into four families:

- *Rational functionalism*: regime design is viewed as an equilibrium outcome, and as the consequence of the structural characteristics of both a problem and the institutions seeking to resolve it.
- *Distributive rationalism*: regime design is viewed as an equilibrium outcome, and as the product of purposive agents seeking to structure future interactions in their favor.
- *Structural processualism*: regime design is viewed as a continually evolving process, and as the consequence of both a problem and the institutions seeking to resolve it.
- *Historical institutionalism*: regime design is viewed as a continually evolving process, and as the product of purposive agents seeking to structure future interactions in their favor (Voeten 2019).

Historically speaking, the scholars most closely associated with the development, testing and elaboration of IERT typically address regime design from a rational functionalist perspective. Often, these studies focus on aspects of cooperation problems (like distribution, enforcement and uncertainty), which are used to explain institutional design outcomes (captured in dependent variables like centralization, membership and flexibility). The distributive (functionalist) view contrasts with the rational (functionalist) view because it accounts for the often-contentious politics of who gets what, when and how, an aspect of regime design often elided in functionalist accounts (Voeten 2019).

Despite multiple epistemological frameworks, most IERT scholars rely on distributive rationalist and rational functionalist frameworks to assess the impacts of regime design on states' participation in international environmental regimes. In Phase Four, addressing the second concept – participation – scholars link specific treaty provisions like funding mechanisms, technical and financial assistance, and dispute resolution procedures to an increased probability of a state ratifying a treaty (Mohrenberg et al. 2018); however, clearly stated targets, monitoring, and enforcement may spark concerns over the degree of surrendered sovereignty required for ratification and thus can decrease the probability of ratification (Bernauer et al. 2013). Scholars also link a number of domestic-level characteristics of states, for example, democratic regime type, below-average income levels, and a legislative supermajority requirement for treaty ratification to a decreased probability of a state ratifying a treaty (Bohmelt and Butkute 2018; Mohrenberg et al. 2018).

Regarding the third concept, the debates over regime effectiveness that characterized Phase Two and Phase Three have resulted in progress. In large part, research tends to adopt different definitions of effectiveness to suit particular research designs and theoretical scaffolding. Rendering "effectiveness" a contextual concept – dictated by the epistemological and ontological presuppositions embedded in particular research questions and designs – can transcend much of the debate over its exact definition. In other words, convergence on a shared idea (rather than definition) of "effectiveness" enables it to be operationalized to suit each different study's parameters while preserving the core conceptual coherence of the term (see Bernstein and Cashore 2012).

For example, Köppel and Sprinz (2019) use a classical definition: effectiveness in the sense of environmental impact, interpreted as positive or negative changes in environmental quality, that is, "the biophysical environment itself" (Köppel and Sprinz 2019: 3), which is a definition that fits neatly into the one developed by Zürn (1998). Distinguishing between Legally Binding Agreements (LBAs) and Legally Non-binding Agreements (LNBAs), Köppel and Sprinz (2019) show LBAs are more effective than LNBAs at improving indicators of water quality. In functional terms, Köppel and Sprinz (2019) employ the Oslo-Potsdam solution (see Phase Three) to measuring regime effectiveness, using a specified measure of the collective optimum, a measure of a no-regime counterfactual, and a measure of actual progress (Hovi et al. 2003; Köppel and Sprinz 2019). Brandi et al. (2019), by contrast, employ a definition of effectiveness rooted as well in earlier phases (see Phase Two), that implementation of a treaty's provisions is usually necessary for treaty effectiveness. They ask whether ratification of MEAs leads to domestic legislation ensconcing treaty provisions into law at the state-level to determine "whether, when, and by which countries" are implemented, and find that MEAs are positively associated with passage of domestic environmental legislation, but the effect is only present mostly in developing countries, more pronounced after a treaty formally goes into effect, and shows significant variation across issue areas (Brandi et al. 2019: 14).

Current trends in IERT point as well to a connection of regime theory with the study of global governance (Vogler 2003; Biermann 2006; Young et al. 2008; R. Kim 2019). Betsill and Bulkeley (2006) have critiqued IERT for its limited ability "to adequately engage the concept of governance, especially the increasingly complex interactions between supranational and subnational state and nonstate actors" (Betsill and Bulkeley 2006: 142). Likewise, Young (2008) and the IDGEC research team seems to push for IERT to expand itself to see "governance systems" of regimes at multiple levels, not just those that are state-based "governmental or intergovernmental in nature" (Young 2008).

To date, the movement to redefine international regimes within the realm of governance has been problematic. It is unclear if this movement is a recognition that IERT is deficient because it focuses too narrowly on states that are declining in influence, or if particular regimes themselves are ineffective and need to be redesigned to include actors beyond states in their formation and implementation (e.g., Okereke et al. 2009). Regimes seem to have become another variable or level of global governance for some. Have they become part of the network or web that will bind state behavior and subjugate sovereignty? If so, how far will this process go? Global governance has moved beyond the nation-state. Conversely, IERT remains mostly focused on the roles of states and their behavior, though it does show signs of movement toward less state-centric and more systemic, relational modes of thinking.

In sum, do international environmental regimes matter? In Phase Four, the answer to this question is often yes, but almost always must be accompanied by: how and to what extent do regimes matter? Ultimately, the answers to these questions contribute to the shared ideas underpinning each strand of contemporary research on regime design, participation, and effectiveness.

Today: trends, roadblocks and opportunities

IERT faces important internal and external challenges. As noted, the creation of international environmental regimes seems to be in decline (Mitchell et al. 2020), alternative and sometimes conflictual organizational forms (e.g., preferential trade agreements) are increasing

(Morin et al. 2018), and, meanwhile, planetary tipping points are approaching (Lenton et al. 2019) and the climate change regime appears dangerously insufficient to slow this trajectory (Allan 2019; Harris 2021).

IERT, however, has shown promising signs in its ability to incorporate elements from "beyond the regime" into its conceptual framework. Recent scholarly work expands the traditional set of variables employed in previous phases of research on regime design, formation and effectiveness to include multilevel variables – such as economic variables capturing aspects of market structure (e.g., the degree of concentration of industrial production, and the preferences of firms) (Vormedal 2010; Ovodenko 2017; Vormedal et al. 2020), and domestic variables "unboxing" the state to capture, among other things, the political economy of its rule-making processes (Ovodenko 2016). Furthermore, the regime complex framework (and regime interaction, more generally) has probably become the predominant paradigm by which scholars make sense of regime design, formation and effectiveness in the twenty-first century (Young 2010; Orsini et al. 2013; Oh 2020). Importantly, the incorporation of new (to the discipline) methodologies such as network analysis illustrates that this increased attention on regime complexes is actually driven by scholars' increased reliance on systemic, relational and complexity-based modes of thinking – that is, it is not merely terminological change masking substantive stagnation (Orsini et al. 2013; R. Kim 2019).

In 2010, Oran Young wrote that "the next phase of research" (2010: 185) would be "the study of institutional dynamics" (Young 2010: 192). While not abandoning regime effectiveness and empirical studies, Young examines the patterns of institutional change. He argues that regimes are "complex and dynamic systems" that are always changing, and he seeks to identify the "determinants" of these changes (Young 2010: 192). While Young's prediction strictly in terms of substance has not been realized, the spirit of the prediction – that regimes should be viewed as ever-evolving and in relation to other institutions and systems – has fared better. Regimes have clearly moved a great distance from their early stages when researchers fought over definitions of the concept, to a stage where regimes are actors that influence other actors. In response to an increasingly densely layered and ever-shifting international institutional landscape, the center of gravity in the field of IERT has moved away from static portraits of single regimes, and toward the study of regime complexes and interactions.

First developed during Phase Three, the study of regime complexes has only grown more relevant to contemporary IERT. A special issue of *Global Governance* published in 2013 provides the most precise definition to date: "a *regime complex* [is] a network of three or more international regimes that relate to a common subject matter; exhibit overlapping membership; and generate substantive, normative, or operative interactions recognized as potentially problematic whether or not they are managed effectively" (Orsini et al. 2013: 29). This definition leads to an important distinction in the literature exploring regime interaction, "one studying a dyadic relation, and [the] other studying a system level relation" (Oh 2020: 562).

Dyadic regime relations have been referred to differently in the recent literature, from "interplay" (Andersen 2002) to "linkage" (Jinnah 2011), "engagement" (Hall 2015) and "interaction" (Oh 2020). Regarding systemic level interactions, the focus and terminology of the literature has shifted quite a bit in recent years, as Oh (2020) explains:

> ...from the institutional design of a singular institution to the design of institutional interaction between or among multiple institutions.... Institutional interaction has two structural characteristics: institutional integration and institutional fragmentation. Integration implies the existence of a core institution that is hierarchically linked with

non-core institutions, whereas fragmentation indicates multiple institutions interacting in neither a hierarchically nor an institutionally linked manner. Any interaction is situated somewhere on the continuum between integration and fragmentation.

(Oh 2020: 562)

The study of regime interaction may be terminologically divergent, but it also exhibits a strong pattern of methodological convergence. Early on, Orsini et al. (2013) noted an opportunity for network analysis techniques to be applied to regime complexes: "Given that regime complexes are a network structure, network analysis – a tool already used in various disciplines ranging from physics to sociology – offers an interesting resource for the analysis of regime complexes" (Orsini et al. 2013: 6). The network approach distinguishes between individual nodes (in which the unit of analysis is a single treaty, regime, IGO/NGO or international bureaucracy), links between those nodes (unit of analysis: an institutional interlinkage, interaction or other form of interactivity / entanglement), clusters (unit of analysis: a regime complex, treaty clusters or lineage), and the overall network(s) (unit of analysis: structure and its dynamics, evolutions, adaptations) (R. Kim 2019: 8).

Despite the popularity of the network approach, it is not without limitations. Network approaches account only for the "structural backbone of a complex system" (R. Kim 2019: 17). Indeed, they are most accurately characterized as more of a qualitative approach rather than a quantitative approach. Nevertheless, the structure of a system is an important component for understanding the relationships between actors, institutions and their broader systemic context, and how those webs of relationships condition those actors' and institutions' interactions with each other.

Conclusion

As in the first edition of this handbook, we conclude by asking important questions for the future: Should we continue to study international environmental regimes? Have we proven that regimes matter, and now can we focus on how and why? Will global governance transcend international environmental regimes? And most importantly, will environmental regimes effectively protect our environment? The IERT field appears to have matured to a point where scholars are focusing on the intricacies of regime complexes, formation, design, participation and the ever-present concept of effectiveness. Moreover, at the same time IERT has mostly abandoned the pretense of prescriptive regime design – that is, the pursuit of a blueprint for an effective regime – preferring to deliver *better* answers to the same questions rather than fruitlessly chasing the *perfect* answer. Regimes, though they may have been initially conceptualized as a *deus ex machina* saving an anarchic international system from environmental degradation, may no longer hold this type of panacean promise for scholars of IERT.

The world has become more complicated and complex during the lifetime of IERT which may make the task of resolving international problems through international regimes, either a cause or consequence of the theory, more confounding. The field has demonstrated longevity, and theoretical rigor, endurance and flexibility. With the decline in the creation of international institutions and the existential threat of environmental problems such as climate change growing, the future is opaque. The need for cooperation, with more agreements and institutions, seems essential, progress may remain slow, but nonetheless will persist. IERT, if it follows the first four plus decades of its existence, will continue to adapt and mature, confronting with genuine sincerity the need for positive change.

References

Allan, J. (2019) "Dangerous Incrementalism of the Paris Agreement", *Global Environmental Politics*, 19: 4–11.

Andersen, R. (2002) "The Time Dimension in International Regime Interplay", *Global Environmental Politics*, 2: 98–117.

Axelrod, R. and Keohane, R. (1986) "Achieving Cooperation Under Anarchy: Strategies and Institutions", in K. Oye (ed), *Cooperation Under Anarchy*, Princeton, N.J.: Princeton University Press.

Bernauer, T., et al. (2013) "Is there a 'Depth versus Participation' Dilemma in International Cooperation?", *The Review of International Organizations*, 8: 477–497.

Bernstein, S. and Cashore, B. (2012) "Complex Global Governance and Domestic Policies: Four Pathways of Influence", *International Affairs*, 88: 585–604.

Betsill, M. and Bulkeley, H. (2006) "Cities and the Multilevel Governance of Global Climate Change", *Global Governance*, 12: 141–159.

Biermann, F. (2006) "Global Governance and the Environment", in M. Betsill, K. Hochstetler, and D. Stevis (eds) *International Environmental Politics*, New York: Palgrave Macmillan.

Bohmelt, T. and Butkute, E. (2018) "The Self-Selection of Democracies into Treaty Design: Insights from International Environmental Agreements", *International Environmental Agreements-Politics Law and Economics*, 18: 351–367.

Brandi, C., Blümer, D., and Morin, J. (2019) "When Do International Treaties Matter for Domestic Environmental Legislation?", *Global Environmental Politics*, 19: 14–44.

Breitmeier, H. (2006) *Analyzing International Environmental Regimes : From Case Study to Database*, Cambridge, Mass.: MIT Press.

Dimitrov, R., et al. (2007) "International Nonregimes: A Research Agenda", *International Studies Review*, 9: 230–258.

Eckersley, R. (2004) *The Green State : Rethinking Democracy and Sovereignty*, Cambridge, Mass.: MIT Press.

Haas, E. (1975) "Is there a Hole in the Whole? Knowledge, Technology, Interdependence, and the Construction of International Regimes", *International Organization*, 29: 827–876.

Haggard, S. and Simmons, B. (1987), "Theories of International Regimes", *International Organization*, 41: 491–517.

Hall, N. (2015) "Money or Mandate? Why International Organizations Engage with the Climate Change Regime", *Global Environmental Politics*, 15: 79–97.

Harris, P. G. (2007) "Collective Action on Climate Change: The Logic of Regime Failure", *Natural Resources Journal*, 47: 195–224.

Harris, P. G. (2021) *Pathologies of Climate Governance: International Relations, National Politics and Human Nature*, Cambridge: Cambridge University Press.

Hasenclever, A. (1997) *Theories of International Regimes*, Cambridge; New York: Cambridge University Press.

Helm, C. and Sprinz, D. (2000) "Measuring the Effectiveness of International Environmental Regimes", *The Journal of Conflict Resolution*, 44: 630–652.

Hisschemoller, M. and Gupta, J. (1999) "Problem-Solving through International Environmental Agreements: The Issue of Regime Effectiveness", *International Political Science Review*, 20: 151–173.

Hovi, J., Sprinz, D., and Underdal, A. (2003) "The Oslo-Potsdam Solution to Measuring Regime Effectiveness: Critique, Response, and the Road Ahead", *Global Environmental Politics*, 3: 74–96.

Jinnah, S. (2011) "Climate Change Bandwagoning: The Impacts of Strategic Linkages on Regime Design, Maintenance, and Death Introduction", *Global Environmental Politics*, 11: 1–9.

Keeley, J. (1990) "Toward a Foucauldian Analysis of International Regimes", *International Organization*, 44: 83–105.

Keohane, R. (1983) "The Demand for International Regimes", in S. Krasner (ed) *International Regimes*, Ithaca, New York: Cornell University Press.

Keohane, R. (1989) *International Institutions and State Power : Essays in International Relations Theory*, Boulder, Colorado: Westview Press.

Keohane, R. (2020) "Understanding Multilateral Institutions in Easy and Hard Times", in M. Levi and N. Rosenblum (eds), *Annual Review of Political Science*, 23: 1–18.

Kim, R. (2019) "Is Global Governance Fragmented, Polycentric, or Complex? The State of the Art of the Network Approach", *International Studies Review*, 4: 903–931.

Köppel, M. and Sprinz, D. (2019) "Do Binding Beat Nonbinding Agreements? Regulating International Water Quality", *Journal of Conflict Resolution*, 63: 1860–1888.

Kratochwil, F. and Ruggie, J. G. (1986) "International Organization: A State of the Art on an Art of the State", *International Organization*, 40: 753–775.

Lenton, T., et al. (2019) "Climate Tipping Points – Too Risky to Bet Against", *Nature*, Available https://www.nature.com/articles/d41586-019-03595-0

List, M. and Rittberger, V. (1992) "Regime Theory and International Environmental Management", in A. Hurrell and B. Kingsbury (eds), *The International Politics of the Environment : Actors, Interests, and Institutions*, Oxford : New York: Clarendon Press.

Luterbacher, U. and Sprinz, D. (eds) (2001) *International Relations and Global Climate Change*, Cambridge, MA: MIT Press.

Miles, E., et al. (2002) *Environmental Regime Effectiveness: Confronting Theory with Evidence*, Cambridge, MA: MIT Press.

Mitchell, R. (2002a) "A Quantitative Approach to Evaluating International Environmental Regimes", *Global Environmental Politics*, 2: 58–83.

Mitchell, R. (2002b), "International Environment", in W. Carlsnaes, T. Risse, and B. Simmonds (eds), *Handbook of International Relations*, London: SAGE Publications.

Mitchell, R., et al. (2020) "What We Know (and Could Know) About International Environmental Agreements", *Global Environmental Politics*, 20: 103–121.

Mohrenberg, S., Koubi, V., and Bernauer, T. (2018) "Effects of Funding Mechanisms on Participation in Multilateral Environmental Agreements", *International Environmental Agreements: Politics, Law and Economics*, 19: 1–18.

Morin, J., Dur, A., and Lechner, L. (2018) "Mapping the Trade and Environment Nexus: Insights from a New Data Set", *Global Environmental Politics*, 18: 122–139.

Oberthür, S. and Stokke, O. (2011) *Managing Institutional Complexity : Regime Interplay and Global Environmental Change*, Cambridge, MA: MIT Press.

Oh, C. (2020) "Contestations over the Financial Linkages between the UNFCCC's Technology and Financial Mechanism: Using the Lens of Institutional Interaction", *International Environmental Agreements-Politics Law and Economics*, 20: 559–575.

Okereke, C., Bulkeley, H., and Schroeder, H. (2009) "Conceptualizing Climate Governance Beyond the International Regime", *Global Environmental Politics*, 9: 58–78.

Orsini, A., Morin, J., and Young, O. (2013) "Regime Complexes: A Buzz, a Boom, or a Boost for Global Governance?", *Global Governance*, 19: 27–39.

Ovodenko, A. (2016) "Governing Oligopolies: Global Regimes and Market Structure", *Global Environmental Politics*, 16: 106–126.

Ovodenko, A. (2017) *Regulating the Polluters : Markets and Strategies for Protecting the Global Environment*, New York, NY: Oxford University Press.

Paterson, M. (1999) "Interpreting Trends in Global Environmental Governance", *International Affairs*, 75: 793–802.

Paterson, M. (2009) "Green Politics", in S. Burchill, A. Linklater, and R. Devetak (eds) *Theories of International Relations*, 4th edn, New York: Palgrave Macmillan.

Puchala, D. and Hopkins, R. (1983) "International Regimes: Lessons from Inductive Analysis", in S. Krasner (ed) *International Regimes* Ithaca, New York, Cornell University Press.

Raustiala, K. and Victor, D. (2004) "The Regime Complex for Plant Genetic Resources", *International Organization*, 58: 277–309.

Rittberger, V. (1995) *Regime Theory and International Relations*, Oxford : Oxford University Press.

Rittberger, V., Zangl, B., and Kruck, A. (2012) *International Organization*, 2nd edn, New York: Palgrave Macmillan.

Rittberger, V. and Zürn, M. (1991) "Regime Theory: Findings from the Study of 'East-West' Regimes", *Cooperation and Conflict*, 26: 165–183.

Ruggie, J. (1975), 'International Responses to Technology: Concepts and Trends', *International Organization*, 29 (3), 557–583.

Sprinz, D. (2001) "Comparing the Global Climate Regime With Other Global Environmental Accords", in U. Luterbacher and D. Sprinz (eds) *International Relations and Global Climate Change*, Cambridge, MA: MIT Press.

Stokke, O. (2001) "The Interplay of International Regimes: Putting Effectiveness Theory to Work", *The Fridtjof Nansen Institute (FNI Report)*, 14: 1–29.

Strange, S. (1982) "Cave! Hic Dragones: A Critique of Regime Analysis", in S. Krasner (ed) *International Regimes*, Ithaca, NY: Cornell University Press.

Voeten, E. (2019) "Making Sense of the Design of International Institutions", *Annual Review of Political Science*, 22: 147–163.

Vogler, J. (2003) "Taking Institutions Seriously: How Regime Analysis can be Relevant to Multilevel Environmental Governance", *Global Environmental Politics*, 3: 25–39.

Vormedal, I. (2010) "States and Markets in Global Environmental Governance: The Role of Tipping Points in International Regime Formation", *European Journal of International Relations*, 18: 251–275.

Vormedal, I., Gulbrandsen, L., and Skjærseth, J. (2020) "Big Oil and Climate Regulation: Business as Usual or a Changing Business?", *Global Environmental Politics*, 20: 1–23.

Walsh, V. (2004) *Global Institutions and Social Knowledge: Generating Research at the Scripps Institution and the Inter-American Tropical Tuna Commission 1900s–1990s*, Cambridge, MA: MIT Press.

Ward, H. (2006) "International Linkages and Environmental Sustainability: The Effectiveness of the Regime Network", *Journal of Peace Research*, 43: 149–166.

Wettestad, J. (2001) "Designing Effective Environmental Regimes: The Conditional Keys", *Global Governance*, 7: 317–341.

Wettestad, J. (2006) "The Effectiveness of Environmental Politics", in M. Betsill, K. Hochstetler, and D. Stevis (eds), *International Environmental Politics*, New York: Palgrave Macmillan.

Wilkening, K. (2011) "Science and International Environmental Nonregimes: The Case of Arctic Haze", *The Review of Policy Research*, 28: 125–148.

Young, O. (1982a) "Regime Dynamics: The Rise and Fall of International Regimes", *International Organization*, 36 (2), 277–297.

Young, O. (1982b) *Resource Regimes: Natural Resources and Social Institutions*, Berkeley, CA: University of California Press.

Young, O. (1989) *International Cooperation : Building Regimes for Natural Resources and the Environment*, Ithaca, NY: Cornell University Press.

Young, O. (1998) *Creating Regimes : Arctic Accords and International Governance*, Ithaca, NY: Cornell University Press.

Young, O. (2003) "Determining Regime Effectiveness: A Commentary on the Oslo-Potsdam Solution", *Global Environmental Politics*, 3: 97–104.

Young, O. (2008) "Institutions and Environmental Change: The Scientific Legacy of a Decade of IDGEC Research", in O.R. Young, L.A. King, and H. Schroeder (eds), *Institutions and Environmental Change : Principal Findings, Applications, and Research Frontiers*, Cambridge, MA: MIT Press.

Young, O. (2010) *Institutional Dynamics: Emergent Patterns in International Environmental Governance*, Cambridge, MA: MIT Press.

Young, O. (2011) "Effectiveness of International Environmental Regimes: Existing Knowledge, Cutting-Edge Themes, and Research Strategies", *Proceedings of the National Academy of Science, USA*, 108: 19853–19860.

Young, O., Demko, G., and Ramakrishna, K. (eds) (1996) *Global Environmental Change and International Governance*, Hanover, NH: University Press of New England.

Young, O., King, L., and Schroeder, H. (eds) (2008) *Institutions and Environmental Change: Principal Findings, Applications, and Research Frontiers*, Cambridge, MA: MIT Press.

Zürn, M. (1998) "The Rise of International Environmental Politics: A Review of Current Research", *World Politics*, 50: 617–649.

10

International environmental law
Sources, principles and innovations

David B. Hunter

International environmental law is generally considered to be comprised of (1) a large number of bilateral, regional and global treaties negotiated and ratified by the states that choose to be bound by them; (2) certain customary or general principles of environmental law and (3) a growing number of international judicial or arbitral decisions resolving inter-State disputes. More controversially, international environmental law can also be seen as including an array of norms, standards and associated compliance mechanisms that are adopted through multi-stakeholder processes and hold State and non-State actors accountable in specific contexts. In this chapter, each of these sources of international environmental law is discussed, in turn, after an introduction to international law.

The sources of public international law

International environmental law is a subset of the broader field of public international law – the law that governs the relationship between nation-states. International law is bound by centuries-old traditions and relies fundamentally on the consent of the sovereign states. The international law-making system is far less developed than the more familiar national systems. Each State is independent and sovereign. No centralized legislative or law-making body exists, except arguably the United Nations Security Council, which rarely addresses environment-related issues. The subjects of international law are generally limited to states, rather than private enterprises or individuals. States are thus both the international lawmakers and the subjects of the law they make, and they must consent to limits on their sovereignty. While consent can sometimes be inferred, states that do not explicitly agree to be bound generally are not. And even when they initially agree to be bound, they can in most cases withdraw their consent later if their governments choose to do so. The United States' entry into, departure from, and re-entry into the Paris Agreement on climate change is a case in point (see Chapter 32).

The sources of international law are typically considered to be those sources accepted by the International Court of Justice (ICJ) – the primary judicial organ of the United Nations system. The ICJ acts as both a legal advisory body and a court for the settlement of disputes between states. Its 15 judges are chosen to represent geographic regions and types of legal

DOI: 10.4324/9781003008873-13

system. Article 38(1) of the charter establishing the ICJ identifies four sources of international law that the Court will consider in resolving international disputes:

The Court, whose function is to decide in accordance with international law such disputes as are submitted to it, shall apply:

a   international conventions, whether general or particular, establishing rules expressly recognized by the contesting states;
b   international custom, as evidence of a general practice accepted as law;
c   the general principles of law recognized by civilized nations; and
d   …judicial decisions and the teachings of the most highly qualified publicists of the various nations, as subsidiary means for the determination of rules of law (International Court of Justice 1945: Art. 38(1)).

The first three sources – treaty, custom and general principles of international law – create binding legal obligations for states. Judicial decisions and the writings of publicists are subsidiary means for understanding what the law is, but do not create generally binding obligations on states. This chapter next reviews the contribution of each of these sources of international law to the field of international environmental law.

Treaties

Treaties create specific legal obligations between those states that have consented to become treaty parties. The Vienna Convention on the Law of Treaties is the primary source of rules governing the negotiation, interpretation, amendment, termination and related aspects of treaties. It defines a treaty as "an international agreement concluded between states in written form and governed by international law, whether embodied in a single instrument or in two or more related instruments and whatever its particular designation" (United Nations 1969: Art. 2.1(a)). The instrument need not be called a treaty; it can be called an agreement, convention, pact, covenant or virtually any other name. Four basic steps are inherent in the development of any international treaty: (1) identification of needs and goals; (2) negotiation; (3) adoption and signature and (4) ratification. Even after these steps are completed, treaties must be implemented through national law, monitored for compliance, and enforced.

A State is bound by the terms of a treaty only if it takes affirmative steps to demonstrate its consent to be bound. For multilateral agreements, consent is typically demonstrated by ratification, which is usually done by depositing an "instrument of ratification" with the United Nations or another designated depositary organization. In many states, a treaty must be approved through domestic political processes before the treaty is ratified. In the United States, for example, Senate ratification requires a two-thirds vote. As a result, the United States has failed to ratify several major environmental treaties, including the Kyoto Protocol, the Convention on Biological Diversity, and the UN Convention on the Law of the Sea. Thus, a State's signature on a treaty is only part of the battle. Until the treaty is ratified and has entered into force, the State is not obligated to comply with it.

Treaties are the principal method for creating binding rules of international law in the environmental field. By most estimates more than 500 treaties relate to environmental protection. Although most environmental treaties are bilateral or regional, more than a dozen significant multilateral environmental agreements (MEAs) have been negotiated in the past few decades, most of which enjoy nearly universal acceptance by countries around the

Table 10.1 Widely adopted global environmental agreements (as of 2019)

Treaty	Number of State Parties	Opened for Signature	Entered into Force
Convention on Biological Diversity	196	1992	1993
Cartagena Protocol on Biosafety	171	2000	2003
Convention on International Trade in Endangered Species	183	1973	1987
Convention on Migratory Species	128	1979	1983
Basel Convention	187	1989	1992
Montreal Protocol	197	1985	1988
UN Framework Convention on Climate Change	197	1992	1994
Kyoto Protocol	192	1997	2005
Convention to Combat Desertification	197	1994	1996
Ramsar Convention	170	1971	1975
UNESCO Heritage Convention	193	1972	1975
Law of the Sea Convention	168	1982	1994
Stockholm Convention on POPs	184	2001	2004
Rotterdam Convention on PIC	164	1998	2004
Minamata Convention on Mercury	105	2013	2017

Sources: The information in this table was drawn from the websites of each of the conventions listed above, which were accessed in December 2020.

world (see Table 10.1). As reflected in Table 10.1, environmental treaties address a wide range of issues, including for example: ozone depletion (see Chapter 31); climate change (see Chapter 42); the protection of the oceans and its resources (see Chapters 39 and 40); management of transboundary rivers or other shared natural resources (see Chapter 38); conservation of migratory wildlife or biodiversity (see Chapter 41), and the control of hazardous substances (see Chapters 35–37).

Many contemporary environmental treaty regimes reflect the relatively dynamic framework/protocol approach to treaty-making. Used to develop the treaty regimes that address ozone depletion, climate change, the loss of biological diversity and persistent organic pollutants, the framework/protocol approach allows states first to adopt a general framework agreement that generally commits the parties to cooperate on a specific issue, without imposing significant substantive obligations. The framework agreement typically explains the purpose for the treaty regime; identifies the principles governing the regime; establishes the institutional architecture for responding over time to the environmental problem; and sets out reporting and data collection requirements for meeting anticipated information needs. Participation in the framework conventions also builds international support and capacity for taking stronger actions over time. These actions are taken in the form of protocols, amendments or revisions adopted under the general framework agreement. The Vienna Ozone Convention, Montreal Protocol and subsequent revisions and amendments that comprise the ozone treaty regime are collectively considered to be the first example to adopt the framework/protocol approach (see Chapter 31). The UN Framework Convention on Climate Change with the subsequent Kyoto Protocol and Paris Agreement is another example (see Chapter 32).

Although states are the predominant actors in the treaty-making process, international governmental organizations (IGOs), civil society organizations, and other non-State actors are playing an increasingly significant role, particularly in identifying environmental issues appropriate for international cooperation, building the political will for countries to negotiate treaties, and monitoring implementation and compliance.

Customary principles of law

In addition to this substantial array of international treaties, many principles of environmental law are emerging either as customary or general principles of international law (discussed in the next section). International law can be formed by the customary practice of states where such practice is done under the belief that it is required by law. To prove that a customary norm exists, a court must establish general acceptance of the rule: first, by demonstrating that State practice is consistent with the rule; and, second, by demonstrating that states act in accordance with the rule from a sense of legal obligation to do so. This sense of legal obligation is known as *opinio juris*. Both State practice and *opinio juris* are required to prove the existence of a customary rule of international law.

While there is no precise definition of what constitutes State practice, the ICJ has required that practice be both extensive and virtually uniform and include those states that are particularly affected by the proposed norm. It is not necessary that State practice continue over a long period of time. Nor must State practice rigorously and consistently conform to the rule at issue. However, it must be clear that State conduct that is inconsistent with the customary practice has generally been treated as a breach of the rule.

For the norm to be recognized as a rule of customary international law, it must further be shown that the State practice follows from a sense of legal obligation rather than from a sense of moral obligation or political expediency. The existence of such *opinio juris* is a factual matter that can be determined by consideration of a wide range of evidence, including diplomatic correspondence, government policy statements and press releases, opinions of official legal advisors, national legislation or judicial decisions, legal briefs endorsed by the State, a pattern of treaties, and resolutions and declarations by the United Nations.

Once the customary norm is established it becomes binding on all states, regardless of whether those states contributed to the formation of the custom. However, a State may exclude itself from the obligations of a particular customary rule by persistent conduct exhibiting an unwillingness to be bound by the rule or a refusal to recognize it as law (American Law Institute 1987: § 102, cmt. B).

When a principle has reached the status of customary law can be ambiguous and open to debate. The environmental field has a relatively large number of principles, often repeated in various international treaties, declarations, and resolutions – but only a few appear to reflect the dual characteristics needed to form custom, namely, widespread state practice and opinion juris. Although their frequent reiteration in international instruments of every kind can provide evidence of possible *opinio juris*, practice in the environment may be insufficient or variable to satisfy the consistent State practice requirement.

Nevertheless, several environmental principles have been widely recognized as legally binding, including in judicial and arbitral opinions. Foremost among these are the obligation of a State not to cause significant environmental harm outside its territory and the obligation to conduct an environmental impact assessment for projects that could have significant transboundary harm. The ICJ has helpfully determined that these two principles are part of international environmental law (although the analyses are unclear whether the Court considered them as customary law or general principles, as described immediately below). The status and substance of these and other environmental law principles are discussed further below.

General principles of law

General principles of law "recognized by civilized nations" are another source of international law recognized by the ICJ (ICJ 1945: Art. 38(1)(c)). Ian Brownlie, a leading international law scholar, states that general principles may refer to "rules accepted in the domestic law of all civilized states," or alternatively, to the general principles of private law used within all or most states (Brownlie 2008: 16). General principles are primarily used to fill gaps in international law that have not already been filled by treaty or custom.

In the environmental field, the category of general principles seems to be expanding to include substantively important principles that do not meet the criteria for custom. Environmental principles that are frequently cited as being general principles of international law include the obligation to pursue sustainable development, defined as integrating environmental concerns into development decisions; the precautionary principle; and the obligation to have effective environmental legislation. Some environmental scholars argue that general principles in the environmental field can emerge from multilateral consensus when reaffirmed in different multilateral forums or international instruments, rather than relying solely on a principle's acceptance in national law. Charney, for example, argues that this approach leads to "general international law" or "universal international law" that can bind non-parties even without their consent (Charney 1993: 543). Bodansky similarly refers to this in the environmental context as "declarative law" (Bodansky 2010: 200).

Judicial decisions and the writings of eminent publicists

The final sources of international law are "judicial decisions and the writings of eminent publicists," which are subsidiary means for determining international law (ICJ 1945: Art.38(1) (d)). The writings of publicists and jurists can codify the law to help states and courts discern what the law is, but they have no independent force to create law. The ICJ opinions, for example, are binding only on the states before the Court in the specific dispute being decided. For other states, such decisions only provide some evidence of what the law is. Having said this, ICJ opinions are cited as authority so frequently that the distinction between simply identifying the law and actually making it has been blurred.

Although judicial decisions have been relatively uncommon in the environmental field, they have been important for the development of international environmental law and for shaping the responsibilities of states that share transboundary resources. In addition, to the ICJ, which has general jurisdiction over international disputes (when both parties agree to jurisdiction), other more specialized tribunals, such as the International Law of the Sea Tribunal, the World Trade Organization Appellate Body, or the European Court of Justice, also play an increasingly important role in furthering the development of international environmental law. The following are brief examples of two of the best-known international environmental law decisions.

The Trail Smelter Arbitration (1941). The Trail Smelter Arbitration, the most famous international environmental adjudication, involved transboundary sulfur dioxide emissions from a smelter located in Trail, British Columbia, just a few miles north of the US–Canada border. During the 1930s, the Trail Smelter emitted approximately 250,000 tons of sulfur dioxide per year into the air. This plume traveled across the border and damaged the property of apple growers in Washington State. For a variety of jurisdictional reasons, Washington State residents could not bring a lawsuit either in Washington State or in British Colombia, so they asked the US government to intervene on their behalf in 1927. Ultimately, the Tribunal would side with the United States, ruling that "under the principles of international law... no State has the right to use or permit the use of its territory in such a manner as to cause injury by fumes in or to the territory of another or the properties or persons therein, when the case is of serious consequence and the injury is established by clear and convincing evidence" (Trail Smelter Case 1941: 1965). The Tribunal accordingly held Canada responsible under international law for damages caused by the air pollution. Canada was forced to pay compensation and to take measures to reduce the pollution. The obligation not to cause environmental harm to a neighboring State would later be recognized as part of customary law by the ICJ.

The Pulp Mill Case (Argentina v. Uruguay) (2010). In 2010, the ICJ ruled on a dispute involving the anticipated pollution from two pulp mills proposed in Uruguay into the River Uruguay, which forms the boundary between Uruguay and Argentina. Argentina alleged a series of violations of a 1975 treaty between the two countries that set forth the regime for the shared use of the river. The 1975 treaty established a bilateral river commission, the Comision Administradora del Rio Uruguay (CARU), which provided the institutional framework for mutually achieving the rational use and development of the river. In a comprehensive decision, the Court ruled that Uruguay had breached procedural, but not substantive, obligations owed to Argentina in planning and constructing the pulp mills. The Court found that the obligation to inform CARU of planned developments was the first step in the whole procedural mechanism established by the treaty and could not be replaced by some alternative form of notification. The ICJ ultimately found that Uruguay had violated its procedural obligations to notify and consult with Argentina under the 1975 treaty. The Court affirmed that the obligations not to cause environmental harm and to conduct an environmental assessment of transboundary impacts were binding principles of international law, but that Argentina had not demonstrated that Uruguay had violated these obligations. It further ruled that Argentina should receive no remedy beyond the finding of a violation. The pulp mills could still operate without any restitution (ICJ 2010).

Notwithstanding these two examples, international environmental disputes are typically not brought to courts or other formal tribunals. Several factors account for this. First, the jurisdictional and enforcement authority of formal tribunals may be inadequate to ensure a meaningful remedy in complex environmental matters. Formal dispute mechanisms can be slow and costly, and simply may be inappropriate for reaching effective and practical solutions to the technical and difficult issues frequently posed by environmental issues. It may be more efficient to resolve such disputes through informal negotiations between the parties, through the good offices of regional institutions, or at periodic conferences of treaty parties. The overriding reason, though, likely resides in a simple truism of international law – states are generally unwilling to cede their sovereignty by submitting to the jurisdiction of third-party arbitration or judicial settlement. Moreover, the substantive rules of international environmental law are not yet totally clear, so predicting which State will prevail is difficult and may deter some states from bringing judicial cases. For these reasons, recent treaties have focused as much on mechanisms to *avoid* disputes (e.g., through facilitating compliance) as on the procedures for litigating them.

Key principles of international environmental law

As already noted, international environmental law includes a large number of principles and concepts that are reiterated in various treaties, declarations, and resolutions, and that are applied in many different contexts. A few of them have been identified as binding principles of law in specific international cases – either as custom or as general principles of national law.

Others are still "emerging" as legal principles and are frequently identified as non-binding or "soft" law.

Principles and concepts do not have to be binding, however, to have a significant impact. Arguing that a principle is customary law or a general principle of national law is absolutely essential for resolving transboundary environmental disputes before the ICJ and other tribunals, but in other contexts whether a principle has met the ICJ's formal criteria for being "hard law" may be less relevant. For example, these principles shape negotiations and implementation of environmental treaties, contribute to the development and convergence of national and subnational environmental laws, and the integration of international environmental law with other fields, such as international trade or human rights (Hunter et al. 2015). Following are some of the most prominent international environmental legal principles.

State sovereignty

State sovereignty in the legal sense signifies independence – that is, the right to exercise, within a portion of the globe and to the exclusion of other states, the functions of a State, such as the exercise of jurisdiction and enforcement of laws over persons therein (see Chapter 7). A bedrock principle of international environmental law is that countries have the sovereign right to exploit their natural resources pursuant to their own environmental and developmental policies. Territorial sovereignty extends to the geographic borders of the country and to the underlying subsoil as well as the airspace overhead. States have sovereignty over inland waters, including groundwater, wholly within their boundaries, and they have substantial sovereign rights with respect to shared watercourses. Sovereignty over resources also extends outward through the Exclusive Economic Zone (EEZ), the area within 200 nautical miles of the State's coast. International environmental law reflects the fundamental tension between a State's interest in protecting its independence (i.e., its sovereignty) and the recognition that certain problems, in this case regional and global environmental problems, require international cooperation. In this respect, most international environmental treaties by their very nature constrain a State's sovereignty.

Common heritage of humankind

State sovereignty and the principles and rights that derive from it have historically been applied to the natural resources within a State. Yet, over half of the world's surface area lies outside the national borders of any one State. Those areas beyond the limits of national jurisdiction – the high seas, the seabed, Antarctica, outer space, and sometimes the outer atmosphere, including the ozone layer – are frequently referred to as the "global commons." Resources in the global commons are outside the territorial reach of states, and the concept of sovereignty does not readily apply. For many global commons resources, most notably the high seas fisheries, the general rule has been the right of capture – that is, whoever captures a fish or other resource has the right to it. Concerned that this right of capture penalizes developing and landlocked states, participants in the Law of the Sea Convention and other negotiations perceived a need for a new conceptual framework to address resources in the global commons (see Chapter 39). This framework became known as the "common heritage of mankind" (or, more accurately, humankind). Areas governed by the principle of common heritage cannot be appropriated by any State and must be used only for peaceful purposes and for the shared benefit of all states. The application of this principle today is limited primarily to Antarctica, outer space and the moon, certain cultural landmarks, and possibly certain plant genetic resources.

Common concern

The global environment is increasingly viewed as a "common concern" of humankind, which reflects the growing consensus that the planet is ecologically interdependent, and therefore humanity may have a collective interest in certain activities that take place within, or resources that are located within, State boundaries. Thus, for example, the recognition that nations have a common concern in the global environment has provided a critical conceptual framework for treaties addressing climate change and biological diversity (see Chapters 32 and 41). Common concern has limited legal content for resolving disputes or clarifying State obligations, but for some global environmental challenges it does provide the primary conceptual justification for why states should forgo their sovereignty in favor of international cooperation.

Duty not to cause environmental harm

A central principle in international environmental law is the obligation of states to ensure that activities within their jurisdiction or control do not cause damage to the environment of other countries or of areas beyond the limits of national jurisdiction. This obligation not to cause environmental harm has been elaborated in the *Trail Smelter* decision (as noted above), in Article 21 of the 1972 *Stockholm Declaration* and Article 2 of the 1992 *Rio Declaration*, and in several ICJ opinions. The principle is now considered a part of customary international law and is thus binding on all countries. Unfortunately, many questions remain that are critical to applying the principle in specific cases – for example, is the obligation only triggered by significant damage? Is the obligation satisfied if a country has taken all reasonable steps to prevent the damage? What does "all reasonable steps" entail? Answers to these and related questions will determine the legal rights and responsibilities in most transboundary environmental disputes.

State responsibility

Under the principle of State responsibility, states are generally responsible for breaches of their obligations under international law. Under the International Law Commission's Draft Articles, states responsible for an internationally wrongful act are under an obligation to make restitution (i.e., to re-establish the situation which existed before the wrongful act was committed), to compensate for any damage caused, and to give satisfaction (for example, to acknowledge the breach, express regret, or formally apologize). Thus, State responsibility comes into play as a complementary rule that explains the remedies one State has against another that has violated its international legal obligations.

Common but differentiated responsibilities

According to the principle of common but differentiated responsibilities, all states have common responsibilities to protect the environment and promote sustainable development, but because of different social, economic and environmental situations, countries should shoulder different responsibilities. The principle reflects core elements of equity, placing more responsibility on wealthier countries and those that are more responsible for causing specific global environmental problems (see Chapter 23). Differentiated responsibility also allows for ecological differences in countries for example, the particular vulnerability of small-island states to the sea-level rise resulting from global warming. Common but differentiated responsibilities are not a principle for resolving specific disputes. Instead it presents a conceptual framework for compromise and cooperation in negotiations to meet complex environmental challenges. It allows countries that are in different positions with respect to specific environmental issues to be treated differently.

The polluter-pays principle

As reflected in the *Rio Declaration*, national authorities should promote the internalization of environmental costs by taking those actions necessary to ensure that polluters and users of natural resources bear the full environmental and social costs of their activities. The principle integrates environmental protection and economic activities, by ensuring that the full environmental and social "external" costs (including costs associated with pollution, resource degradation, and environmental harm) are reflected in the ultimate market price for a good or service. Environmentally harmful or unsustainable goods will tend to cost more, and consumers will switch to less-polluting substitutes. In addition, if all states require their industries to pay for pollution and other impacts, then no State will have a significant competitive trade advantage by allowing their companies to pollute freely. This principle is thus not a formula for resolving environmental disputes between two states, but rather serves to integrate the goals of trade liberalization and environmental protection (see Chapter 24).

Intergenerational equity

The principle of intergenerational equity requires that we take into consideration the impact of our activities on future generations, giving them a "seat at the table" when making current decisions. At a minimum, implementing this principle requires using natural resources sustainably and avoiding irreversible environmental damage (see Chapter 27). It may also lead to expanding our concepts of judicial standing to future generations. Although primarily a principle of fairness, several national courts have relied on the principle to uphold legal standing on behalf of future generations in environmental cases.

Environmental impact assessment

The ICJ and other tribunals have recognized that states are under an obligation to conduct an environmental impact assessment (EIA), at least where there are potentially significant impacts on transboundary resources (ICJ 2010). Although many details of how an EIA is conducted remain questions of national law, at a minimum the EIA process should ensure that before granting approval for a project that could significantly harm the environment, the appropriate government authorities have fully considered the environmental effects of proposed activities under their jurisdiction and control, and affected citizens have an opportunity to understand the proposed project or policy and to express their views to decision-makers.

The precautionary principle

The precautionary principle states that "where there are threats of serious or irreversible damage, lack of full scientific certainty shall not be used as a reason for postponing cost-effective measures to prevent environmental degradation" (United Nations 1992: Principle 15). The precautionary principle addresses how environmental decisions should be made in the face of scientific uncertainty. It provides a framework for governments to set preventative policies where existing science is incomplete or where no consensus exists regarding a particular threat (see Chapter 19). The principle is often viewed as conflicting with science-based decision-making, but the principle operates where there is a lack of scientific knowledge or where there is significant uncertainty; decision-makers are not excused from considering what science does exist. In most instances, the precautionary principle has been used to *allow* or *authorize*, but not to *require*, policy measures. Outside of the United States, the precautionary principle is widely viewed as a binding rule of law.

Public participation, access to information and access to justice

In Principle 10 of the *Rio Declaration*, governments began to recognize three pillars of environmental democracy, also known as the access rights: the right to access environmental information, the right to participate in environmental decisions that affect their lives and the right to access justice in environmental cases. Most of Europe endorsed these rights in the 1998 Aarhus Convention (UNECE 1998) and much of Central and South America endorsed them in the 2018 Escazú Agreement (UNECLAC 2018). Rooted in both concepts of human rights and sustainable development, the rights have been upheld in domestic legislation and judicial decisions in many other countries (see Chapter 28).

Notification and consultation

States should provide prior and timely notification to, and consult with, potentially affected States regarding activities that may have a significant adverse transboundary environmental effect. The principle of prior notification obliges States planning a potentially damaging activity to transmit to affected States all necessary information sufficiently in advance so that the latter can prevent damage to their respective territory. The principle of consultation requires States to allow potentially affected states an opportunity to review and discuss a planned activity that may have potentially damaging effects. The acting State is not necessarily obliged to conform to the interests of affected states, but it should take them into account (see Chapter 25).

Sustainable development

Sustainable development is viewed as the general goal of international environmental policy, guiding the integration of environment and development at the international and national levels (see Chapter 16). In recent years, the concept has also taken on a legal nature, requiring the integration of environmental concerns into treaties that were negotiated prior to the emergence of environmental consciousness. Thus, for example, Belgium was allowed to develop a transboundary railroad across the Netherlands under a decades-old treaty, but only if it considered and mitigated the environmental impacts of the project (Permanent Court of Arbitration 2005). Environmental issues, which were not explicitly included in the treaty, were implicitly required due to the application of sustainable development as a legal principle.

The limitations of international law have left it open to criticism by those who aspire for a more effective and nimble response to environmental challenges. The State-centered focus of international law limits its effectiveness, particularly given the critical role of the private sector and civil society organizations in the pursuit of environmental protection (see Chapters 13 and 14). The cumbersome processes of international law have led non-State actors to adopt norms and standards through multi-stakeholder processes that are often more flexible, inclusive and tailored to the specific context. There is controversy whether the norms adopted through these processes should be considered a part of "international environmental law" – and by formal definitions they clearly are not – but we include them here because in certain contexts these norms constrain behavior, create accountability and protect the environment more than traditional sources of international law. They are part of a broader new governance model, in which multi-stakeholder processes allow new forms of norm creation and broader conceptions of law, compliance and accountability.

New governance and innovations in international environmental law

To many observers, the formalistic, non-participatory, consensus-based nature of the international law system has hindered efforts to formulate an effective international response to our global environmental crisis (Speth 2005). International law ascribes "hardness" to treaties and custom that meet certain forms, but it leaves little room for normative development outside those strict categories. Yet treaties and custom are not nimble and require a State's consent, so they are ill-suited for holding states accountable in many environmental contexts.

Moreover, the primary behavioral changes needed to address global environmental challenges are frequently those of corporations, consumers, and other private actors – not necessarily governments (see Chapters 13 and 14). Private actors are only indirectly the subject of traditional international environmental law and thus escape direct accountability.

The inherent limitations of international law for addressing global environmental challenges have led advocates to seek more flexible "new governance" models of norm creation. These new approaches are inclusive, frequently relying on multi-stakeholder processes that may include governments, international organizations, private sector companies, civil society organizations and community groups (Dupuy 1991; Mattli and Woods 2009).

Environmental standards now come in many forms, targeting specific projects, corporations, industry sectors or general behaviors. Some of these international standards may be wholly voluntary, require public reporting, or be part of elaborate certification systems that include third-party monitoring. Others may be issued as standards or rules and associated with some means of ensuring compliance. Although not in a form cognizable as binding law by the ICJ, these standards/compliance-mechanism frameworks have the compulsory pull we expect from "law" – sometimes even more than treaties. For this reason, we can include those frameworks in an expanded view of international law (Weiss 2010; Hunter 2020). The following are several examples of new forms of international environmental law that constrain behavior and provide contextual accountability in a fast-changing world (Weiss 2010; Hunter 2020).

Environmental and social standards in international finance

Over the past three decades, the requirement to have environmental and social standards and an independent mechanism for monitoring compliance has become a norm for any institution active in development finance. The World Bank, for example, adopted a set of environmental and social safeguard policies beginning in the late 1980s, and in 1993 created the Inspection Panel, an independent unit for investigating allegations of non-compliance brought by persons affected by the Bank's project (IBRD/IDA 1993). With establishment of the Panel, the World Bank became the first international organization to allow non-state actors, particularly affected people, to raise objections to the organization without first having to go through their government first (Hunter 2020).

Today, dozens of financial institutions have established similar frameworks comprised of environmental and social standards monitored by an accountability framework. The leading model is framework of the International Finance Corporation (IFC), the private sector lending arm of the World Bank Group. Because many of its projects have potentially negative development impacts, the IFC was forced to adopt a set of environmental and social performance standards for its borrowers. These standards, which were adopted in 2006 and revised in 2012, arguably became the most important set of standards for international project finance. The IFC's approach includes a "Policy on Social and Environmental Sustainability" and eight "Performance Standards" covering environment, labor, resettlement, indigenous peoples, and community health issues (IFC 2012). The policy applies to the IFC's review and due diligence of the project, and the performance standards apply to the borrowers. The heart of the performance standards is a requirement that all borrowers must have an environmental and social management system, including in most cases a project-specific action plan that is negotiated with each borrower to ensure that the project meets the other performance standards over time. The environmental and social action plans and annual monitoring reports must be released to the public.

In 1999, the IFC created the Compliance Advisor and Ombudsman (CAO), which allows people affected by IFC projects to raise their concerns, including whether the IFC has complied with its environmental and social policy (CAO 2010). The CAO will mediate the claim if both the affected people who raised the complaint and the private sector borrower agree. If one of the parties do not consent to dispute resolution, the CAO assesses the complaint to determine whether IFC Performance Standards have been met. In the case of non-compliance, IFC is expected to develop a response to bring the project into compliance.

Virtually every international source of development finance now has a similar set of environmental and social standards they expect their borrowers the CAO. This includes the World Bank, all major regional development banks, a growing number of bilateral export agencies, several UN agencies like the United Nations Development Programme, the Global Environment Facility and the Green Climate Fund, and the World Wildlife Fund and several other large NGOs. Even the world's largest private financial institutions have a common set of policies (known as the Equator Principles) to determine, assess and manage environmental and social risks in project finance. For projects with significant environmental and social risks, borrowers from Equator Banks must, in consultation with the public, prepare an Environmental Management Plan (EMP) that includes the management, mitigation and monitoring of environmental and social impacts (Equator Principles Association 2020). The Equator Banks have not yet agreed to an independent accountability mechanism with the authority to monitor compliance.

The OECD guidelines for multinational enterprises

The Organization of Economic Cooperation and Development (OECD) Council of Ministers first adopted a set of voluntary rules of conduct for multinational corporations that included environmental concerns in 1991. Known as the OECD Guidelines, these standards are aimed at ensuring that multinational corporations operate in a way that is compatible with the expectations of host countries by establishing a baseline of standards (OECD 2011). An environmental chapter was added in 1991 with further amendment in 2000 and 2010. Implementation of the Guidelines is voluntary, but a series of National Contact Points

(NCP), typically housed within government agencies, receive complaints from interested parties who believe the Guidelines have been violated. Each NCP has its own rules for responding to such requests, but in general they provide a forum for the parties to negotiate or discuss the application of the principles. If the discussions at this level do not resolve the issue between the parties, it can be passed to the OECD's Investment Committee, which is ultimately responsible for adjudication and development of the Guidelines. In this way, the OECD's Investment Committee can clarify the scope and meaning of the Guidelines in specific instances. The Committee's judgments do not "enforce the Guidelines" against either of the parties. Instead it uses its findings in specific cases to clarify the meaning of how a provision in the Guidelines should be applied in the future.

Supply chains, eco-labels and certifications

In recognition of the limits to the state-centric approach of international law and given the increasing role of corporations in an era of globalization, environmental and human rights advocates increasingly focused their attention on the private sector. They targeted companies with important consumer brands by highlighting the destructive impact of their company's operation and the operations of other companies in their supply chain. These pressures manifested themselves in industries seeking to define environmental and social standards that would meet public concerns and a mechanism to ensure these standards are met.

One of the leading examples of these initiatives is the certification of timber products by that meet the Forest Stewardship Council (FSC) standards for sustainable forest management. FSC is a non-profit association established in 1993 to promote environmentally responsible, socially beneficial and economically viable management of the world's forests by establishing a worldwide standard of recognized and respected Principles of Forest Stewardship. The FSC accredits independent organizations to certify timber operations and products as meeting FSC standards. Some major retailers and users of timber products will only purchase FSC-certified timber. Millions of hectares in over 100 countries have been certified according to the FSC standards, and products carrying the FSC trademark had an estimated value of more than $20 billion (Forest Stewardship Council 2012).

Producers of many other commodities and goods, including, for example, soy, palm oil, cocoa, diamonds and gold, have followed the path of the FSC and developed their own standards often through extensive stakeholder dialogs. Reporting and certification standards now cover companies exceeding $4 trillion in market capitalization (Forest Trends 2015). In a shift from earlier certification efforts like the FSC, the drive for certification now comes more from retail companies directly from consumers. Where retailers require certified products, they can enforce these standards through explicit agreements in supply contracts. Consumers and the public also "enforce" adherence to these certification schemes by threatening a company's reputation and the value of its brand.

Conclusion

Although the past three decades have seen significant political commitment to international environmental law, as evidenced by a variety of international environmental treaties and other instruments, the period has also witnessed a marked decline in virtually every important global environmental indicator (UNEP 2019). This has called into question the effectiveness of international environmental law to meet the demands of a changing world with significant global environmental challenges (Speth 2005). Moving beyond a focus on formal law-making provides new opportunities for creativity, innovation and flexibility in crafting a more effective response to increasingly complex international and global environmental challenges.

New forms of international environmental standards that are both prescriptive and enforceable have developed in many different contexts. Whether an environmental standard is found in a treaty, for example, may be relevant to determine whether it is binding in a dispute between two states – but not relevant to a dispute brought to the Compliance Advisor/Ombudsman involving the International Finance Corporation, its private sector borrower, and people forcibly evicted by the IFC-financed borrower. In that context, the IFC environmental and performance standards are binding norms, enforceable against the borrower as conditions in the loan contract and enforceable by the communities against the IFC through the CAO. Similarly, retail stores may contractually require that all forest products in their supply chains be certified as meeting FSC's standards, regardless in which country they were harvested. Viewed this way, we see the scope of international environmental law as expanding to encompass international contexts beyond the traditional state-state disputes, where environmental norms are written with clarity and there is a process that can compel compliance.

References

American Law Institute (ALI) (1987) *Restatement (Third) of Foreign Relations Law of the United States, § 102, cmt, b*.
Bodansky, D. (2010) *The Art and Craft of International Environmental Law*, London: Harvard Press.
Brownlie, I. (2008) *Principles of Public International Law*, 7th edn, New York: Oxford University Press.
Charney, J. (1993) "Universal International Law" *American Journal of International Law* 87: 529.
Compliance Advisor Ombudsman (CAO) (2010) *The CAO at 10: Annual Report FY 2010 and Review FY 2000–10*.
Dupuy, P. M. (1991) "Soft Law and the International Law of the Environment" *Michigan Journal of International Law* 12: 420.
Equator Principles Association (2020) "Equator Principles: EP4".
Forest Stewardship Council (FSC) (2012) "Global FSC Certificates: Type and Distribution".
Forest Trends, Supply Change: Commitments that Count (March 2015).
Hunter, D., Salzman, J., and Zaelke, D. (2015) *International Environmental Law and Policy*, 5th edn, New York: Foundation Press.
International Bank for Reconstruction and Development & International Development Agency (IBRD/IDA 1993) "The World Bank Inspection Panel" *Resolution No. IBRD 93-10, Resolution No. IDA 93–6* (Sept. 22, 1993).
International Court of Justice (ICJ) (1945) *Statute of the International Court of Justice (1945)*.
Hunter, D. (2020) "Contextual Accountability, the World Bank Inspection Panel, and the Transformation of International Law in Edith Brown Weiss's Kaleidoscopic World" *Georgetown Environmental Law Review* 32: 439.
ICJ (2010) *The Pulp Mill Case (Argentina v. Uruguay)* (2010).
International Finance Corporation (IFC) (2012) *Performance Standards on Environmental and Social Sustainability*.
Mattli, M. and Woods, N. (2009) *The Politics of Global Regulation,* Princeton, NJ: Princeton University Press.
Organization of Economic Cooperation and Development (OECD) (2011) *Guidelines for Multinational Enterprises*.
Permanent Court of Arbitration (PCA) (2005) *Belgium v. Netherlands*, Arbitral Award.
Speth, J. G. (2005) *Red Sky at Morning: America and the Crisis of the Global Environment*, 2nd edn, New Haven, CT: Yale University Press.
Trail Smelter Case (U.S. v. Can.) (1941) *Reports of International Arbitral Awards*.
United Nations (1969) *Vienna Convention on the Law of Treaties*.
United Nations (1992) *UN Declaration on Environment and Development* (the "Rio Declaration").

UNECE (1998) *Convention on Access to Information, Public Participation in Decisionmaking and Access to Justice in Environmental Matters,* Doc. ECEBCEPB43 (June 25, 1998).

UNECLAC (2018) *Regional Agreement on Access to Information, Public Participation and Justice in Environmental Matters.*

United Nations Environment Programme (UNEP) (2019) *Global Environmental Outlook 6.*

Weiss, E. B. (2010) "On Being Accountable in a Kaleidoscopic World" *American Society of International Law Proceedings* 104: 477.

11

Environmental foreign policy

Crossovers among levels of governance

Mihaela Papa

Diplomacy, as a common saying goes, is the art of sending people to hell in such a way that they look forward to the trip. Yet in few subfields of diplomacy are negotiation skills as needed, the trip as long and the hell as close as in environmental diplomacy. Countries cannot directly negotiate with nonhuman species and ecosystems or ask the planet to fight for its own future. Environmental and human wellbeing depends on our policies and effective mobilization to prevent or address environmental harm. Thus, at the very heart of environmental foreign policy (EFP) is the challenge of operating both at the nexus between local and global governance, and between human systems (e.g., economic, social) and natural systems.

In a seminal article, Robert Putnam (1988) used the metaphor of "two-level games" to describe how government officials conduct foreign policy by simultaneously engaging in negotiations at two levels: domestic and international. This idea of foreign policy decision-makers as "brokers" along the domestic–foreign (or international) frontier has since gained significant ground in the academic literature. Studies looking more closely at both the domestic and the international level subsequently acknowledged that these levels are themselves made up of different levels involving various administrative units (Cottier and Hertig 2003; Piattoni 2010; Woolcock 2011). The increasing complexity of world politics – the growth of actors, issues and cooperative arrangements – enables EFP officials to operate at the crossovers among multiple negotiation processes. This not only helps expand their reach and bargaining space, but also raises questions about their roles, choices and impact.

This chapter examines the theoretical approaches to EFP in the context of multiple levels of governance and investigates how policymaking operates at the crossovers. This is a particularly timely topic due to the growing pressures of environmental degradation, and the role of foreign policy as an important mechanism of political agency in global environmental politics. The chapter begins by introducing EFP and the two main approaches used to describe, analyze and explain its operation across multiple levels. The first approach is a state-centered analysis. It considers the state as the primary arena of political power, and explains how EFP officials seek to take advantage of the different levels of cooperation. The second approach, focused on multilevel governance, takes shared competencies as a point of departure, and EFP officials are one group of actors, but not necessarily the leading ones, that assume policy responsibilities in this realm.

DOI: 10.4324/9781003008873-14

Following the analysis of these two approaches, the chapter discusses the changing politics of foreign policy in light of greater complexity of the system of global governance. It points out several foreign policy challenges at the crossovers between the domestic and the international, such as defining environmental problems, strategizing about institutional choice, dealing with transscalar civil society, and responding to equity concerns arising from the system's growth. The chapter concludes by discussing the promising areas for future EFP research.

Theoretical debates: environmental foreign policy and multiple levels of governance

Environmental foreign policy refers to state policy directed to matters beyond state borders as it aims to protect, preserve and improve the environment (Papa 2009). From an environmental perspective, the very concept of the state is artificial because the environment and environmental problems disregard political borders. During much of the twentieth century, the notion of the state reflected the centralization of authority, and foreign policy officials were the sole brokers between the national and the international realm. In the twenty-first-century politics, state actors engaging in EFP come from various ministries (Morin et al. 2020, 104–106), and various functional departments of the government directly engage with their counterparts abroad (Slaughter 1997). Moreover, multi-stakeholderism has become common as two or more classes of actors – states, international organizations, firms or civil society actors – collaborate on common governance enterprises concerning public policy issues (Raymond and DeNardis 2015). Contemporary governance thus refers to new ways of achieving social objectives in which states participate and may, but do not necessarily have to, play a leading role (Rosenau and Czempiel 1992; Rhodes 1996; Stoker 1998). When these objectives relate to addressing environmental challenges or resolving environmental conflicts by creating, changing or reaffirming institutional arrangements, scholars talk about environmental governance (Davidson and Frickel 2004; Paavola 2007).

This section first uses a *state-centric perspective* to examine how foreign policy officials engage at multiple levels of cooperation. Second, it introduces the concept of multilevel governance, which uses *shared competences* between foreign policy officials and other actors as an entry point for analyzing EFP.

Policy engagement at multiple levels: states as points of departure

Government officials conducting EFP simultaneously engage with a large number of environmental issues at multiple levels of governance. A state's foreign policy can be conceived of as a portfolio of policies implemented at different levels and designed to achieve foreign policy outcomes that the state wants (Palmer and Morgan 2006). While states' interests, environmental vulnerabilities, capacities and ambitions differ, their foreign policy approaches are similar. Common types of diplomatic engagement include unilateralism, bilateralism, regionalism, multilateralism, and minilateralism.

- *Unilateralism.* Unilateral actions are one-sided or undertaken by a single state. They are used as an expression of a commitment toward a policy. A case in point is Russia's effort to prevent the extinction of fur seals in 1893: Russia issued a decree prohibiting the taking of fur seals just outside its territorial waters in reaction to British and North American fishing in that area. Unilateral measures can also be used as a way to spearhead policy change in multilateral forums and to act when effective multilateral cooperation is impossible (Bodansky 2000). For example, parties to the Convention on International Trade in Endangered Species are allowed to unilaterally impose restrictions on wildlife trade that may be considerably stricter than those imposed by the Convention as a result of their own concerns about contributing to the decline of species that are consumed within their territory. When a critical mass of countries adopts unilateral measures, a multilateral response becomes more likely.

- *Bilateralism*. Bilateral actions are two-sided or undertaken by two governments. EFP officials often take advantage of this form of cooperation to manage shared resources together with their neighbors; for example, they negotiate treaties on joint river development (e.g., Dinar et al. 2011; see Chapter 38). Sometimes neighboring countries that do not share common interests have a common aversion to environmental harms, so environmental cooperation may be a way to improve their relations (Ali 2007). For example, in 1999 Botswana and South Africa signed a historic bilateral treaty to form southern Africa's first peace park, the Kgalagadi Transfrontier Park. The countries decided to manage their adjacent national parks as a single ecological unit: there are no physical barriers between the two parks and animals can move freely. Bilateral environmental cooperation also occurs between non-neighboring states: environmental assistance arrangements or joint development of clean energy technologies are cases in point.
- *Regionalism*. EFP officials engage in regional cooperation by entering issue-specific regional agreements or by developing environmental cooperation within broader regional integration processes. For example, the transboundary nature of acid rain in Europe required regional cooperation to ensure that both polluters and the most vulnerable countries jointly addressed the problem (see Chapter 30). The resulting 1979 Convention on Long-Range Transboundary Pollution used the United Nations Economic Commission for Europe as its secretariat. Environment can also be a significant aspect of regional integration processes: countries joining the European Union (EU) have been required to harmonize their environmental policies with EU standards, resulting in regional norm diffusion and greater regulatory standardization. While regional environmental cooperation is a particularly active field of practice, academics have sought to pay more attention to its conceptualization and systematically measure its patterns over time (Balsiger and Prys 2016; Church 2020).
- *Multilateralism*. Foreign policy officials take advantage of multilateral cooperation when the involvement of a larger number of countries is necessary. Environmental problems like climate change are global in scope: greenhouse gas emissions from anywhere in the world contribute to the rising global mean temperature, which then entails differential but substantive risks for all countries (see Chapter 32). Multilateral cooperation helps better understand the problem, facilitates the development of comprehensive policy responses, and it can establish new norms. Large-scale multilateralism involving over 100 governments has been a common feature of environmental diplomacy. It has been used to create agreements ranging from addressing transboundary movement of chemicals to defining the rights and responsibilities of nations in their use of the world's oceans. Many multilateral environmental processes have been driven by the United Nations Environment Programme (UNEP), a global anchor institution for the environment over the past five decades (Ivanova 2021). UN mega-conferences dedicated to the environment and development are some of the largest events in world politics: the 1992 UN Conference on Environment and Development in Rio de Janeiro and the 2012 "Rio+20" UN Conference on Sustainable Development each gathered more than 170

governments and numerous other actors to strategize the implementation of the global transition to sustainability (see Chapter 22).

– *Minilateralism (e.g., clubs and groups)* is a targeted approach to international cooperation, where the idea is to "bring to the table the smallest possible number of countries needed to have the largest possible impact on solving a particular problem" (Naim 2009, see also Spies 2019, 73). This approach runs against the preoccupation of environmental governance with legitimacy, accountability and inclusiveness. Nonetheless, it has become a popular way to accelerate action when cumbersome multilateral talks fail, experiment with new initiatives and launch strategic partnerships. Minilateral groups comprise a range of self-standing G-clubs such as G8 and G20; transregional groupings including BRICS and IBSA; and, bounded multilateral meetings inside and outside larger multilateral forums such as BASIC. While they often cover many cooperation areas, minilateral entities are becoming more prominent in environmental cooperation, particularly on climate change (Papa and Gleason 2012; Hjerpe and Nasiritousi 2015; Tienhaara and Downie 2019).

How to explain states' engagement in environmental cooperation at multiple levels? International relations scholarship draws on power, interest and knowledge-based theories of international cooperation (Hasenclever et al. 1997; Barrett 2003). From the most simplified perspective, power-based theories argue that power differences shape the level of cooperation, its rules and payoffs. Power-centered approaches often emphasize the importance of relative gains and security concerns in environmental cooperation, for example in cases of powerful riparian states sustaining and defending their privileged shares of transboundary waters (see Chapter 3). Interest-based theories treat EFP officials as rational utility maximizers acting to overcome collective action problems and engaging in cooperation to avoid suboptimal outcomes. Well-designed institutions (e.g., for the protection of the ozone layer; see Chapter 31) can produce mutual gains for countries and change their incentives to exploit the environment. Choosing between different levels of cooperation is a function of the trade-off between each instrument's relative flaw: multilateralism can be a solution to high transaction costs, and bilateralism or regionalism can help avoid free-riding or exclusion in the case of a public good (Thompson and Verdier 2014). Finally, knowledge-based theories focus on the way in which knowledge shapes EFP-makers' behavior and identities. This was the case in whaling, where powerful anti-whaling discourses influenced policymakers' positions on the need to save whales and helped recast whales into an issue of global concern (Epstein 2008).

The conduct of EFP depends on the effectiveness of environmental leadership and problem management. EFP officials can assume various roles in negotiations, as leaders/pushers or draggers/laggards if they are more invested, or swing states or bystanders otherwise (Sprinz and Vaahtoranta 1994; Chasek et al. 2010; Wurzel et al. 2019). Policymaking is a dynamic process that requires (re)negotiations and coalition building both at the national and international level. This is particularly the case when problems are malignant (e.g., due to power asymmetries, science complexities and time lags, or other concerns) and present significant management challenges, so governments lack the intention or capacity to engage (Brown Weiss and Jacobson 1998). The goal is to create adaptive institutions that can facilitate participation and accommodate changes in political and environmental circumstances (see also Moomaw et al. 2017). When institutions are in place, states need to comply with their provisions to ensure that domestic implementation affects environmental outcomes. Tracking EFP outcomes has become more systematic, particularly with the International Environmental Agreements Database, which includes over 1,300 multilateral, 2,200 bilateral and 250 other environmental agreements (Mitchell 2002–2021), and the Environmental Performance Index that ranks 180 countries on 32 performance indicators across 11 issue categories covering ecosystem health and ecosystem vitality (Wendling et al. 2020). While the existing literature acknowledges that environmental agreements have coevolved with the increasing complexity of environmental challenges (e.g., Kim 2013), it would be useful to systematically analyze EFP officials as brokers and measure how they link actors at various levels of governance.

The concept of multilevel governance: shared competencies as points of departure

A second analytical approach to EFP is that of multilevel governance. In this approach, the point of departure is the existence of overlapping competencies among multiple levels of government (see Marks et al. 1996: 41). EFP decision-makers are one of the actors sharing policy responsibilities. Multilevel governance describes decision-making processes that involve the simultaneous mobilization of public authorities at different jurisdictional levels as well as that of nongovernmental organizations and social movements (Piattoni 2010). While it is a contested concept, its broad appeal reflects a shared concern with increased complexity, proliferating jurisdictions, the rise of non-state actors, and the related challenges to state power (Bache and Flinders 2004: 4–5). Multilevel governance thinking has spread to many subfields of political science, including EU studies, international relations, federalism and public policy, comparative politics, political economy and normative political theory (Hooghe and Marks 2003; Zürn et al. 2010).

The operation of multilevel governance can be illustrated in Baltic Sea fisheries, a case of commons governance that involves multiple regulatory processes operating at different scales (Burns and Stöhr 2011). This governance system is under the EU Common Fisheries Policy framework, and decisions about regulations for the Baltic Sea are negotiated among the ministers of EU members. The European Commission (i.e., the Directorate-General for Maritime Affairs and Fisheries) prepares and proposes regulations for the EU Council of Ministers, which is the highest decision-making body determining broad policy measures that are to be implemented by the member-state fishing ministries. EU member states play an important role in the allocation of the annual total allowable catches, but they are faced with the substantial power of the Commission in setting up and managing institutional arrangements, and the powers of a multilevel system of member states (including non-coastal states), as well as powers at the grassroots level.

Similarly, other issues that work across multiple spatial scales in their ecological dimensions, like water and climate, have been analyzed from a multilevel perspective (Moss and Newig 2010). For example, Schreurs (2010) argued that national, regional and local governments have both distinct and complementary roles in developing climate mitigation and adaptation strategies and explained how cities and provinces in China, Japan and South Korea initiate their own climate action plans and join local, national and international networks for climate change. Finger and his colleagues (2006) described the politics of transnational water resource management through case studies of the Aral Sea basin and the Danube, Euphrates and Mekong river basins as a multi-governance effort to collectively solve public problems by involving a variety of relevant actors, from the local to the global level, including institutions, states, civil society and businesses.

Multilevel governance approaches have also been used to analyze regulatory processes at different levels to obtain a comprehensive picture of human responses to environmental problems and analyze their effectiveness. Studies using such approaches have given rise to a number of analytical debates about the "right" scale for addressing a problem, the centralization of decision-making across levels, the interplay among levels, and the rise of institutional complexity. These debates are briefly outlined below:

– *Scales and subsidiarity*. One of the central questions of multilevel governance is what tasks should be "scaled" to which level of jurisdiction. The EU has experimented for decades with the principle of subsidiarity, which says that tasks should rest at the lowest effective level of governance unless relocating them to a higher level would ensure greater comparative effectiveness (Jordan 2000). In the international arena, there is no one-size-fits-all approach to matching tasks and levels, and this matching can happen by design or by default (automatically due to path dependence and daily operations of the communities of practice). Ensuring that political–administrative characteristics of institutions for collective action match the characteristics of the biogeophysical systems with which they interact is crucial (Young 2002). While an optimal level for addressing a problem might exist from a functional perspective, environmental issues are continuously created, regulated and contested between, across and among scales (Delaney and Leitner 1997; Bulkeley 2005: 876; Gupta 2008). The resulting governance arrangement is the outcome of negotiations about the right definition: Is regulating the number of children people have a household issue or a global environmental concern? Is the Amazon a global resource – the lungs of the world – or a local resource? Is whaling a regional issue for whaling countries or an issue of global concern? Is the funding for adaptation to climate change a global responsibility of the key historical emitters or a local responsibility of affected communities?
– *Center(s) of decision-making*. A monocentric system is one controlled by a central predominant authority (e.g., a comprehensive governmental authority), while a decentralized, polycentric system comprises multiple governing authorities at different scales in nonhierarchical relationships (Ostrom 2010; Morrison et al. 2019). The existence of a central government at the domestic level makes the application of a top-down, command-based approach possible. Internationally, there is no world government, so monocentricity refers to top-level rulemaking through global regimes, where international negotiators find global solutions through centralized multilateral negotiations (e.g., how to protect the ozone layer) and the lower levels of government carry out mandates. Polycentric governance at the international level connotes many centers of decision-making that are formally independent of each other, compete, cooperate, self-organize and engage in mutual adjustment. It is frequently discussed in areas where multilateral rulemaking is either non-existent or inadequate, so the focus is on other levels (e.g., forest governance), or where it is deadlocked, so participants tend to seek other terrains for political intervention (e.g., climate change). The debates in this field assess the positive and negative aspects of monocentric versus polycentric governance and the ways to integrate top-down and bottom-up approaches to produce policy (Tal and Cohen 2007; Howlett et al. 2011).
– *Institutional interplay and the rise of regime complexity*. Scholars have acknowledged the importance of understanding how the interplay among international governance arrangements affects policymaking. Young (2002) distinguished between horizontal and vertical interplay in order to denote the horizontal linkages among distinct institutional arrangements at the same level of social organization and the vertical linkages across levels. With the proliferation of international institutions, the number of such linkages has proliferated. Greater attention to institutional interactions has led some scholars to

change focus from individual institutions to sets of institutions that simultaneously in-fluence an issue area. Raustiala and Victor's (2004) "regime complex" describes arrays of partially overlapping and nonhierarchical institutions governing a particular issue area. Oberthür and Stokke (2011) discuss "institutional complexes," Biermann and his colleagues (2009) use global governance "architectures," and Kahler (ed. 2009) re-energized the debate on networked politics as a feature of contemporary interna-tional relations scholarship. The "rise of international regime complexity" over the past 18 years in particular has made it common to analyze environmental problems by mapping the complex, all of the relevant agreements and institutions and how they in-teract (Alter and Raustiala 2018, 331). International interactions, as argued, can result in conflicts or synergies between or among institutions: for example, the interaction between the ozone and climate regimes is conflictual, and the one between the bio-diversity and climate regime complementary (see also Kanie and Haas 2004; Morin et al. 2020, 245). The consequences of increasing complexity for policymaking remain contested: while some argue that complexity creates greater leeway for opportunistic behavior by states and undermines institutions (Alter and Meunier 2009), others ar-gue that institutional divisions of labor are more stable than expected as international organizations tend to orchestrate institutional interactions (Oberthür and Stokke 2011; Abbott et al. 2015).

These debates highlight both structural and procedural features of multilevel governance and the importance of examining its dynamics and mechanisms (Behnke et al. 2019). Multilevel governance features an ongoing interplay of continuity and change, as actors use various change mechanisms including authority migration and institutional layering (Gerber and Kollman 2004; Van der Heijden 2011; Benz and Broschek 2013).

Overall, when shared competencies are the point of departure for investigating EFP, the analytical emphasis moves from the pursuit of state goals across levels to responding to an environmental problem through multilevel governance. EFP is thus embedded into broader problem-solving efforts, which can extend its reach when policymakers operate effectively across scales. However, some levels may not interact regularly and some processes (e.g., busi-ness engagement) are occurring "away from the negotiating table" (Lax and Sebenius 2006), which can shape situations in ways that distract from the desired foreign policy goals.

The politics of foreign policymaking at the crossovers

EFP decision-makers face competing claims and demands at the crossovers among multiple levels of governance. They need to choose where and how they may act, what difference their action makes, to whom they are responsible and to what degree (Hill 2003: 284). They need to decide how to position issues, make institutional choices, manage non-state actors and preserve the legitimacy and equity of the governance system.

Positioning issues: (re)scaling as a political pursuit

The ability of foreign policy decision-makers to derive gains from international cooperation depends on their ability to reach consensus with others on the nature of the problem they are addressing. Even before the start of negotiations, they may have an issue-framing prefer-ence that reflects the amount of control and responsibility they are ready to assume. While agenda promotion and problem framing have traditionally been a characteristic of a robust foreign policy, the increasing complexity of governance means that issues can be taken to multiple venues by multiple actors and (re)defined through political contestation. Foreign policy officials can behave strategically and engage in upscaling and downscaling of issues to shape policymaking in ways that meet their needs, but they also need to be alert to similar efforts of other actors to block them if needed (Gupta 2008). Rescaling occurs at any stage of the international cooperation process, both as a result of self-interested opportunism and as a pragmatic search for more effective solutions to collective action problems (Spector and Zartman 2003).

Making "right" institutional choices

EFPs have traditionally been oriented toward institutions that would act as focal points for states in the international system, and states would coordinate their bargaining and their expectations around them (Schelling 1960). Yet the increasing institutional density means that international institutions seldom stand alone, and new arrangements are not negotiated on a clear institutional table (Orsini et al. 2013, 27–28). As a result, deciding where and how to act raises questions of institutional choice from among multiple institutional alternatives and judgment about institutional relevance. EFP has largely involved interactions with environmental institutions, but scholars have found that these institutions are not necessarily the institutions most important in shaping human behavior that drives environmental change (Underdal 2008). Given the common formula for the aggregate impact of human activities on the environment (i.e., impact = population × affluence × technology), the institutional arrangements most important to the environment are likely to be those that influence major economic activities, technological change and collective systems of beliefs, values and practices, such as the World Trade Organization (WTO), the EU, and transnational religious and cultural communities (Ehrlich and Holdren 1971; Underdal 2008). Yet such institutions might not perceive themselves as appropriate venues or might not be sufficiently equipped to address environmental issues: tensions between trade and environment regimes and efforts to pursue environmental agendas in the WTO's dispute settlement body illustrate such challenges.

Dealing with transscalar civil society

As the societal regulation now spans and interlinks multiple spaces – global, regional, national, provincial, local – civil society has more opportunities for influence and can use them to drive state agendas (Scholte 2010; see Chapter 14). This contributes to the lack of predictability in policymaking: the boundary between domestic and foreign policy can blur rapidly and create inconsistencies in governmental responses as different parts of the government pursue different agendas. For example, Alcañiz and Gutiérrez (2009) illustrated how civil society can push a small conflict over the planned construction of two pulp plants on the Uruguay River, shared by Uruguay and Argentina, to grow and spread across multiple regional and global forums. Pralle (2006) demonstrated how, in the conflict over preserving old-growth forests in Clayoquot Sound in Canada, various political actors – local and national civil society, timber companies and different levels of government – created and reconfigured alliances and drew in different governmental institutions to pursue their goals. Civil society activism across levels can make EFP more accountable as officials seek to defend their decisions and appear supportive of environmental values in global information space (Chong 2007: 197). It can also help address policy inconsistencies, for example, when governments advocate for a certain principle in one forum and resist it in another one (e.g., the United States advocating for the precautionary principle in the whaling regime and downplaying it in the climate regime during the Bush administration).

Moving toward more power politics and less democracy or rethinking sustainability of global governance?

Questions of equity and justice, particularly regarding the common but differentiated responsibility norm, have been prominent in environmental diplomacy. The increasing complexity of environmental governance highlights two additional equity concerns. One is that the very nature of multilevel governance conflicts with existing norms of democratic legitimacy because, although policy bargaining in this context encompasses multiple levels, it lacks the control of elected politicians and can lack transparency and democratic legitimacy (Jordan 2000). Furthermore, as Roger (2020, 8) argues, the current prevalence of informality in the system of global governance – where informal organizations now constitute as much as a third of all the currently active international organizations – is historically unique and suggests that the legal foundations of global governance are themselves shifting. The other concern is that the proliferation of international agreements at all levels may negatively affect the rule of law in the long term because it leads to normative fragmentation, which weakens legal obligations and challenges the integrity of international law (ILC 2006). Some argue that countries with greater resources have an advantage in this context due to their capacity to navigate across various arenas and manage spillovers among them (e.g., Pistorius 1995 illustrates this point on negotiations over plant genetic resources in three international arenas). Others assert that politics within a regime complex may also benefit less-powerful actors engage in strategic behavior (Alter and Raustiala 2018, 340). Finally, while sustainability scholars are right to be concerned about how global governance processes affect sustainability, it is also important to apply sustainability principles when we consider reforming the system of global governance: why not reduce, reuse and recycle international institutions (Papa 2015)?

The practice of EFP-making is concerned with both making the right policy choices and improving these choices over time. When decision-makers operate at the crossovers among multiple negotiation processes, the menu for choice enlarges and highlights the question of the relevance of various cooperation channels. At the same time, the exposure of foreign policy decision-makers to transscalar civil society can increase, raising the bar for civil society to optimize its influence across multiple forums. Greater awareness of equity concerns illustrates the importance of examining whether EFP has a role in ensuring that the system of global governance itself develops in a more sustainable way.

Conclusion

In order to reveal the various ways in which EFP operates at the crossovers among multiple levels of governance, as well as to offer greater conceptual clarity in this context, this chapter has looked at EFP from a state-centric perspective and from the perspective of multilevel governance. Drawing on these two approaches highlights the challenges that foreign policy decision-makers face as brokers at the domestic–foreign frontier. The politics of operating at the crossovers among various levels of governance requires that officials consider how they position and scale their policy agendas; make institutional choices; manage non-state actors; and preserve legitimacy and equity of governance.

As the literature on EFP continues to develop, three directions seem particularly relevant both for advancing the current state of knowledge and thinking about the evolution of this field. First, at a time when one of the major universal diplomatic endeavors – the effort to implement Sustainable Development Goals (SDGs) – faces implementation delays and the world is affected by Covid-19, it is important to strengthen the conceptualization of "sustainable development foreign policy" and its links to EFP. Previous research has established the growing relevance of the environment in foreign policymaking (Harris 2009) and documented the global spread of environmental ministries (Aklin and Urpelainen 2014). Promising new research examines the bodies in charge of SDG implementation, illustrating the relevance of new entities and foreign ministries (Breuer et al. 2019). The imperative for future research is to empirically address the "who question" of EFP and assess if, how, and under what conditions its participants conceive themselves as a community (see, e.g., Hooghe et al. 2019).

Second, China is becoming an increasingly important policy innovator domestically and internationally, and its policies are likely to have effects beyond its borders. Since 2014, a variety of "social credit systems" have been implemented in China in order to steer the behavior of Chinese individuals, businesses, social organizations and government agencies (Kostka 2019). While some suggest that reputation tracking jeopardizes privacy, others welcome improved government efficiency, enforcement and intragovernmental agency control (ibid). Internationally, China and its partners, especially other rising powers, are rethinking EFPs in the context of new infrastructure investments through recently established development banks and as they promote sustainable industrialization (see Chapter 34). Investing in research on Chinese EFP-making at the crossovers would contribute new non-Western perspectives on EFP and multilevel governance.

Finally, the artificial intelligence (AI) revolution is under way, presenting potentially transformative opportunities to address Earth's environmental challenges including climate change, food and water security, and biodiversity loss across various levels of governance (WEF 2018). AI also highlights the question of robot rights as the next frontier of nonhuman rights in light of the lessons learned from the political struggles over animal rights and the rights of nature (Gellers 2021). If steered well, AI can improve EFP and accelerate SDG implementation; in the future, we might even need a new Universal Declaration of Entity Rights as diplomacy moves toward managing coupled human–environment–AI systems.

The promise of EFP in the context of multilevel governance lies in states' ability to seize the new opportunities for action and improve existing policy responses. Greater attention to this endeavor can help states and processes of international cooperation become viable instruments for the political innovation that is required to cope with global environmental change.

References

Abbott, Kenneth W., Philipp, Genschel, Duncan, Snidal and Bernhard, Zangl (eds) (2015) *International Organizations as Orchestrators*, Cambridge, UK: Cambridge University Press.

Aklin Michaël and Urpelainen, Johannes (2014) "The Global Spread of Environmental Ministries: Domestic—International Interactions," *International Studies Quarterly* 58(4): 764–780.

Alcañiz, Isabella and Gutiérrez, Ricardo A. (2009) "From Local Protests to the International Court of Justice: Forging Environmental Foreign Policy in Argentina," in Paul G. Harris (ed.) *Environmental Change and Foreign Policy: Theory and Practice*, London: Routledge, pp. 109–120.

Ali, Saleem H. (ed.) (2007) *Peace Parks: Conservation and Conflict Resolution*, Cambridge, MA: MIT Press.

Alter, Karen J. and Meunier, Sophie (2009) "The Politics of International Regime Complexity," *Perspectives on Politics* 7(1): 13–14.

Alter, Karen J. and Raustiala, Kal (2018) "The Rise of International Regime Complexity," *Annual Review of Law and Social Science* 14(1): 329–349.

Bache, Ian and Flinders, Matthew (2004) "Themes and Issues in Multi-Level Governance," in Ian Bache and Matthew Flinders (eds) *Multi-Level Governance*, Oxford: Oxford University Press, pp. 1–11.

Balsiger, Jörg and Prys, Miriam (2016) "Regional Agreements in International Environmental Politics," *International Environmental Agreements* 16: 239–260.

Barrett, Scott (2003) *Environment and Statecraft: The Strategy of Environmental Treaty-Making*, Oxford: Oxford University Press.

Behnke, Nathalie, Broschek, Jörg and Sonnicksen, Jared (eds.) (2019) *Configurations, Dynamics and Mechanisms of Multilevel Governance*, Cham, Switzerland: Palgrave-Macmillan.

Benz, Arthur and Broschek, Jörg (2013) *Federal Dynamics: Continuity, Change and the Varieties of Federalism*, Oxford University Press.

Biermann, Frank, Pattberg, Philipp and van Asselt, Harro (2009) "The Fragmentation of Global Governance Architectures: A Framework for Analysis," *Global Environmental Politics* 9(4): 14–40.

Bodansky, Daniel (2000) "What's So Bad about Unilateral Action to Protect the Environment?" *European Journal of International Law* 11(2): 339–347.

Breuer, Anita, Leininger, Julia and Tosun, Jale (2019) *Integrated Policymaking: Choosing an Institutional Design for Implementing the Sustainable Development Goals (SDGs)*, Discussion Paper, Bonn, Germany: German Development Institute.

Brown Weiss, Edith, and Jacobson, Harold K. (eds) (1998) *Engaging Countries: Strengthening Compliance with International Environmental Accords*, Cambridge, MA: MIT Press.

Bulkeley, Harriet (2005) "Reconfiguring Environmental Governance: Towards a Politics of Scales and Networks," *Political Geography* 24: 875–902.

Burns, Tom R. and Stöhr, Christian (2011) "Power, Knowledge, and Conflict in the Shaping of Commons Governance. The Case of EU Baltic Fisheries," *International Journal of the Commons* 5(2).

Chasek, Pamela S., Downie, David Leonard and Brown, Janet Welsh (2010) *Global Environmental Politics*, Boulder, CO: Westview Press.

Chong, Alan (2007) *Foreign Policy in Global Information Space: Actualizing Soft Power*, New York: Palgrave.

Church, Jon Marco (2020) *Ecoregionalism: Analyzing Regional Environmental Agreements and Processes*, United Kingdom: Taylor & Francis.

Cottier, Thomas and Hertig, Maya (2003) "The Prospect of 21st Century Constitutionalism," *Max Planck Yearbook of United Nations Law* 7: 261–328.

Davidson, Debra, J. and Frickel, Scott (2004) "Understanding Environmental Governance: A Critical Review," *Organization & Environment* 17(4): 471–492.

Delaney, David and Leitner, Helga (1997) "The Political Construction of Scale," *Political Geography* 16(2): 93–97.

Dinar, Shlomi, Dinar, Ariel and Kurukulasuriya, Pradeep (2011) "Scarcity and Cooperation along International Rivers: An Empirical Assessment of Bilateral Treaties," *International Studies Quarterly* 55(3): 809–833.

Ehrlich, Paul R. and Holdren, John P. (1971) "Impact of Population Growth," *Science* 171(3977): 1212–1217.

Epstein, Charlotte (2008) *The Power of Words in International Relations: Birth of an Anti-Whaling Discourse*, Cambridge, MA: MIT Press.

Finger, Matthias, Tamiotti, Ludivine and Allouche, Jeremy (eds) (2006) *The Multi-Governance of Water: Four Case Studies*, New York: SUNY Press.

Gellers, Joshua C. (2021) *Rights for Robots: Artificial Intelligence, Animal and Environmental Law*, New York: Routledge.

Gerber, Elisabeth R. and Kollman, Ken (2004) "Introduction" (to the Symposium: Authority Migration: Defining an Emerging Research Agenda), *PS: Political Science and Politics* 37(3): 397–401.

Gupta, Joyeeta (2008) "Global Change: Analyzing Scale and Scaling in Environmental Governance," in Oran Young, Leslie A. King and Heike Schroeder (eds) *Institutions and Environmental Change: Principal Findings, Applications, and Research Frontiers*, Cambridge, MA: MIT Press, pp. 225–258.

Harris, Paul G. (ed.) (2009) *Environmental Change and Foreign Policy: Theory and Practice*, Abingdon: Routledge.

Hasenclever, Andreas, Mayer, Peter and Rittberger, Volker (1997) *Theories of International Regimes*, Cambridge: Cambridge University Press.

Hill, Christopher (2003) *The Changing Politics of Foreign Policy*, New York: Palgrave Macmillan.

Hjerpe, Mattias and Nasiritousi, Naghmeh (2015) "Views on Alternative Forums for Effectively Tackling Climate Change. *Nature Climate Change* 5, 864–867.

Hooghe, Liesbet and Marks, Gary (2003) "Unraveling the Central State, but How? Types of Multi-Level Governance," *American Political Science Review* 97(2): 233–243.

Hooghe, Liesbet, Lenz, Tobias and Marks, Gary (2019) *A Theory of International Organization: A Postfunctionalist Theory of Governance, Vol. IV.* Oxford: Oxford University.

Howlett, Michael, Rayner, Jeremy, Goehler, Daniela, Heidbreder Eva, Perron-Welch, Frederic, Rukundo, Olivier, Verkooijen, Patrick and Wildburger, Christoph (2011) "Overcoming the Challenges to Integration: Embracing Complexity in Forest Policy Design through Multi-Level Governance," in Jeremy Rayner, Alexander Buck and Pia Katila (eds) *Embracing Complexity: Meeting the International Forest Governance Challenge*, Vienna: IUFRO World Series, vol. 27, pp. 93–110.

ILC (International Law Commission) (2006) "Fragmentation of International Law: Difficulties Arising from the Diversification and Expansion of International Law," Report of the ILC Study Group chaired by Martti Koskenniemi. Fifty-eighth session, Geneva, 1 May–9 June and 3 July–11 August.

Ivanova, Maria (2021) *The Untold Story of the World's Leading Environmental Institution: UNEP at 50*, Cambridge, MA: The MIT Press.

Jordan, Andrew (2000) "The Politics of Multilevel Environmental Governance: Subsidiarity and Environmental Policy in the European Union," *Environment and Planning A* 32(7): 1307–1324.

Kahler, Miles (ed.) (2009) *Networked Politics: Agency, Power, and Governance*, Ithaca, NY: Cornell University Press.

Kanie, Norichika and Haas, Peter M. (eds) (2004) *Emerging Forces in Environmental Governance*, Tokyo: United Nations University Press.

Kim, Rakhyun E. (2013) "The Emergent Network Structure of the Multilateral Environmental Agreement System," *Global Environmental Change* 23(5): 980–991.

Kostka, Genia (2019) "China's Social Credit Systems and Public Opinion: Explaining High Levels of Approval," *New Media & Society* 21(7): 1565–1593.

Lax, David A. and Sebenius, James K. (2006) *3-D Negotiation: Powerful Tools to Change the Game in Your Most Important Deals*, Cambridge, MA: HBS Press.

Marks, Gary, Nielsen, F., Ray, L. and Salk, J. (1996) "Competencies, Cracks and Conflicts: Regional Mobilization in the European Union," in Gary Marks, Fritz W. Scharpf, Philippe H. Schmitter and Wolfgang Streck (eds) *Governance in the European Union*, London: Sage, pp. 40–63.

Mitchell, Ronald B. (2002–2021) *International Environmental Agreements Database Project (Version 2020.1)*. Available at: http://iea.uoregon.edu/

Moomaw, William R., Bhandary, Rishikesh Ram, Kuhl, Laura and Verkooijen, Patrick (2017) "Sustainable Development Diplomacy: Diagnostics for the Negotiation and Implementation of Sustainable Development," *Global Policy* 8(1): 73–81.

Morin, Jean-Frédéric, Orsini, Amandine and Jinnah, Sikina (2020) *Global Environmental Politics*, Oxford, UK: Oxford University Press.

Morrison, Tiffany et al. (2019) "The Black Box of Power in Polycentric Environmental Governance," *Global Environmental Change* 57(101934): 1–8.

Moss, Timothy and Newig, Jens (2010) "Multi-Level Water Governance and Problems of Scale: Setting the Stage for a Broader Debate," *Environmental Management* 46: 1–6.

Naim, Moises (2009) "Minilateralism: The Magic Number to Get Real International Action," *Foreign Policy* 173: 135–136.

Oberthür, Sebastian and Stokke, Olav Schram (eds) (2011) *Managing Institutional Complexity: Regime Interplay and Global Environmental Change*, Cambridge, MA: MIT Press.

Orsini, Amandine, Morin Jean- Frédéric and Young, Oran (2013) "Regime Complexes: A Buzz, a Boom, or a Boost for Global Governance?" *Global Governance* 19: 27–39.

Ostrom, Elinor (2010) "Polycentric Systems for Coping with Collective Action and Global Environmental Change," *Global Environmental Change* 20(4): 550–557.

Paavola, Jouni (2007) "Institutions and Environmental Governance: A Reconceptualization," *Ecological Economics* 63: 93–103.

Palmer, Glenn and Morgan, T. Clifton (2006) *A Theory of Foreign Policy*, Princeton, NJ: Princeton University Press.

Papa, Mihaela (2009) "Environmental Foreign Policy: Towards a Conceptual Framework," in Paul G. Harris (ed.) *Environmental Change and Foreign Policy: Theory and Practice*, London: Routledge, pp. 202–220.

Papa, Mihaela (2015) "Sustainable Global Governance? Reduce, Reuse, and Recycle Institutions," *Global Environmental Politics* 15(4): 1–20.

Papa, Mihaela and Gleason, Nancy W. (2012) "Major Emerging Powers in Sustainable Development Diplomacy: Assessing Their Leadership Potential," *Global Environmental Change* 22(4): 915–924.

Piattoni, Simona (2010) *The Theory of Multi-Level Governance. Conceptual, Empirical, and Normative Challenges*, Oxford: Oxford University Press.

Pistorius, Robin (1995) "Forum Shopping: Issue Linkages in the Genetic Resources Issue," in Robert V. Bartlett, Priya A. Kurian and Madhu Malik (eds) *International Organizations and Environmental Policy*, Westport, CT: Greenwood Press, pp. 209–222.

Pralle, Sarah B. (2006) *Branching Out, Digging In: Environmental Advocacy and Agenda Setting*, Washington, DC: Georgetown University Press.

Putnam, Robert (1988) "Diplomacy and Domestic Politics: The Logic of Two-Level Games," *International Organization* 42: 427–460.

Raustiala, Kal and Victor, David G. (2004) "The Regime Complex for Plant Genetic Resources," *International Organization* 58(2): 277–309.

Raymond, Mark and DeNardis, Laura (2015) "Multistakeholderism: Anatomy of an Inchoate Global Institution," *International Theory* 7(3): 572–616.

Roger, Charles B. (2020) *The Origins of Informality: Why the Legal Foundations of Global Governance are Shifting, and Why It Matters*, Oxford, UK: Oxford University Press.

Rhodes, R.A.W. (1996) "The New Governance: Governing without Government," *Political Studies* 44: 652–667.

Rosenau, James N. and Czempiel, Ernst-Otto (eds) (1992) *Governance without Government: Order and Change in World Politics*, Cambridge: Cambridge University Press.

Schelling, Thomas C. (1960) *The Strategy of Conflict*, Cambridge, MA: Harvard University Press.

Scholte, Jan Aart (2010) "Civil Society in Multi-Level Governance," in Henrik Enderlein, Michael Zürn and Sonja Wälti (eds) *Handbook on Multi-Level Governance*, Cheltenham: Elgar, pp. 383–396.

Schreurs, Miranda (2010) "Multi-Level Governance and Global Climate Change in East Asia," *Asian Economic Policy Review* 5(1): 88–105.

Slaughter, Anne Marie (1997) "The Real New World Order," *Foreign Affairs*, September/October.

Spector, Bertram I. and Zartman, I. William (eds) (2003) *Getting it Done: Post-agreement Negotiation and International Regimes*, Washington, DC: United States Institute of Peace Press.

Spies, Yolanda Kemp (2019) *Global South Perspectives on Diplomacy*, Cham, Switzerland: Palgrave Macmillan.

Sprinz, Detlef F. and Vaahtoranta, Tapani (1994) "The Interest-Based Explanation of International Environmental Policy," *International Organization* 48(1): 77–105.

Stoker, Gerry (1998) "Governance as Theory: Five Propositions," *International Social Science Journal* 50(155): 17–28.

Tal, Alon and Cohen, Jessica A. (2007) "Bringing 'Top-Down' to 'Bottom-Up': A New Role for Environmental Legislation in Combating Desertification," *Harvard Environmental Law Review* 31: 163–217.

Thompson, Alexander and Verdier, Daniel (2014) "Multilateralism, Bilateralism, and Regime Design," *International Studies Quarterly* 58(1): 15–28.

Tienhaara, Kyla and Downie, Christian (2019) "Green Theory and the G20," in Steven Slaughter (ed.) *The G20 and International Relations Theory: Perspectives on Global Summitry*, Northampton, MA: Edward Elgar Publishing Inc., pp. 183–207.

Underdal, Arild (2008) "Determining the Causal Significance of Institutions: Accomplishments and Challenges," in Oran R. Young, Heike Schroeder and Leslie A. King (eds) *Institutions and Environmental Change: Principal Findings, Applications, and Research Frontiers*, Cambridge, MA: MIT Press, pp. 49–78.

Van der Heijden, Jeroen (2011) "Institutional Layering: A Review of the Use of the Concept," *Politics* 31(1): 9–18.

Wendling, Zachary A., Emerson, John W., de Sherbinin, Alex, Esty, Daniel C. et al. (2020) *2020 Environmental Performance Index*, New Haven, CT: Yale Center for Environmental Law & Policy.

Woolcock, Stephen (2011) "Multi-Level Economic Diplomacy: The Case of Investment," in Nicholas Bayne and Stephen Woolcock (eds) *The New Economic Diplomacy. Decision-Making and Negotiation in International Economic Relations*, 3rd edn, Aldershot: Ashgate, pp. 131–151.

World Economic Forum (2018) "Harnessing Artificial Intelligence for the Earth, Fourth Industrial Revolution for the Earth Series," In Collaboration with PwC and Stanford Woods Institute for the Environment.

Wurzel, Rüdiger K.W., Duncan Liefferink and Diarmuid Torney (2019) "Pioneers, Leaders and Followers in Multilevel and Polycentric Climate Governance," *Environmental Politics* 28(1): 1–21.

Young, Oran R. (2002) *The Institutional Dimensions of Environmental Change: Fit, Interplay, Scale*, Cambridge, MA: MIT Press.

Zürn, Michael, Wälti, Sonja and Enderlein, Henrik (2010) "Introduction," in Henrik Enderlein, Sonja Wälti and Michael Zürn (eds) *Handbook on Multi-Level Governance*, Cheltenham: Elgar, pp. 1–17.

12

Comparative environmental politics

Contributions from an emerging field

Stacy D. VanDeveer, Paul F. Steinberg,
Jeannie L. Sowers and Erika Weinthal

Comparative environmental politics (CEP) is the systematic study and comparison of environmental politics in different countries around the globe; these manifest through differences in political processes, institutions, governance systems and actors of various types. The importance of a comparative approach to global environmental politics (GEP) stems from the fact that the political processes that promote or impede trends like deforestation (see Chapter 42), biodiversity loss (see Chapter 41), desertification (see Chapter 43) and climate change (see Chapter 32) play out every day in places as diverse as the American state of Maine, middle-income countries such as Chile and China, poor and unstable political systems like those of Haiti and Guinea, and wealthy industrialized countries such as Germany and Japan. GEP is largely shaped by domestic politics around the world, while local and national politics are, in turn, influenced by global politics.

To understand GEP requires analyzing the common ways in which states and societies have approached the challenges of domestic environmental governance as well as attention to the specificities of policymaking in particular places and at specific historical junctures. This requires using theoretical tools and conceptual frameworks employed in the field of comparative politics to help us make sense of this complexity and diversity. This chapter makes three arguments. First, comparative inquiry brings a great deal to the study and practice of GEP and should occupy a more prominent position in the field (see Chapter 2). Second, to realize this potential and cohere into a cumulative literature, comparative environmental research should give greater attention to the theories and methods that undergird the study of comparative politics. Third, an emphasis on CEP is critical for understanding how diverse political, economic, legal, and social influences shape countries' environmental policy choices. Combined, a robust CEP subfield is emerging, which includes excellent, empirically grounded and theoretically rich work from around the world, covering a broad array of issue areas.

By connecting environmental research to the broader scholarly tradition of comparative politics, we gain important insights into practical questions of environmental governance while addressing enduring concerns of social science. How do government institutions affect the substance and style of environmental policymaking (Scruggs 2001; Vanhala 2013) What role do business actors and civil society movements play in shaping domestic and regional

DOI: 10.4324/9781003008873-15

environmental regulation and governance (Andonova et al. 2019; Dryzek et al. 2003; Hochstetler and Keck 2007)? What domestic processes influence the diffusion of policy ideas across borders (Busch and Jörgens 2005)? How do profits from the extraction of natural resources like oil and timber shape domestic institutions and politics (Dauvergne 1997; Jones Luong and Weinthal 2010)? When does decentralizing decision-making produce better environmental outcomes (Herrera 2017; Ribot 1999)? How do movements for environmental and social justice influence environmental outcomes (Marion Suiseeya forthcoming, Fuentes George 2016)?

Comprehensive introductions and surveys of this important field are provided elsewhere (Steinberg and VanDeveer 2012; Sowers, VanDeveer and Weinthal, forthcoming). Our goal here is to demonstrate why theories of comparative politics are indispensable for understanding human dimensions of global environmental problems and collective efforts to address them, and how we might bring more comparative analysis and research methods into global environmental politics research (see O'Neill et al. 2013, see Chapter 5).

Seeing global environmental politics through comparative lenses

As this handbook demonstrates, the most prolific area of political science research on the environment has been in the field of international relations. Research grounded in international relations theory expanded our understanding of the prospects for international cooperation (see Chapters 3 and 4), the interplay of state and non-state actors (see Chapters 7, 13 and 14), the effectiveness of international regimes (see Chapter 9), and the distributive results among participants (see Dauvergne 2005; Mitchell 2002; K. O'Neill 2017). As GEP research expands its disciplinary and methodological interests, CEP research sheds light on the governance of environmental issues at both local and transboundary levels (Fuentes-George 2016; Hochstetler 2021; Neville 2021; Sowers, VanDeveer and Weinthal, forthcoming). Indeed, comparativists have advocated such an agenda for a long time (Kamieniecki and Sanasarian 1990; Lundqvist 1978; McBeath and Rosenberg 2006; Sowers 2012b; Steinberg 2001; Steinberg and VanDeveer 2012).

Comparative political inquiry is a rich intellectual tradition encompassing diverse areas of research that share two distinctive features. First, comparative politics research takes seriously the role of domestic politics. Second, comparative politics draws on and contributes to an understanding of political phenomena in more than one country. Together these two commitments allow the field to pay careful attention to national and subnational contexts while promoting a broader understanding of politics that transcends national boundaries. To say that comparative research focuses on domestic politics does not mean that it downplays international phenomena. Indeed, comparativists are keenly interested in how transnational processes – such as activist movements, trade flows, scientific expertise, war, colonialism, and regional organizations – influence and are shaped by domestic politics.

What does comparative politics offer the field of GEP? Scholars of GEP have a strong interest in understanding the domestic sources of state preferences, divergent national implementation of international accords, and the rapidly evolving role of non-state actors, include private sector actors and social movements (Allan 2020; Green 2013). Cross-national comparison demonstrates the importance of studying these issues in light of their political contexts – and this context varies substantially across borders and localities (Goodin and Tilly 2006). Nations, cities and regions display distinctive combinations of interests, ideas and institutions that can be analyzed comparatively using common themes and concepts. In every locale, public socialization, interpretive frameworks, traditions, and expectations based on historical experience shape what citizens and policymakers see as feasible, normal and right. Thus, even seemingly similar economies and societies like those of Europe and North America approach issues like climate change (see Chapter 32), hazardous waste (see Chapter 36), and genetically engineered food (see Chapter 44) differently, despite having access to the same scientific and technical information (Jasanoff 2005; Schreurs 2002; see Chapter 18). As countries continue to address global environmental challenges, their domestic political economy, history, and public opinion will influence policy choices, for example, in the energy sector. Differences in legal traditions, administrative structures, and political interests may produce divergent outcomes, even in instances when international agreements, multinational corporations and transnational actor networks push for political and regulatory convergence across borders.

In short, there are a host of good reasons to compare political systems and outcomes. We focus on four areas that stand to benefit from greater interaction between comparative politics research and the field of GEP: (1) gaining greater insight into causal processes linking international and domestic politics (see Chapter 11); (2) appreciating the enduring importance of domestic politics, including the state, the political regime and ties to society (see Chapter 7); (3) placing non-state actors in their broader social and historical contexts (see Chapter 14) and (4) expanding the political imagination by studying social change and institutional innovation in geographically and historically diverse settings (see Chapter 5).

Understanding causal processes

To be effective, international environmental initiatives – be they treaties, conservation projects, citizen campaigns, or transboundary governance arrangements – require an understanding of the behaviors and social relations driving environmental outcomes (Young 1999). Effectiveness requires an appreciation for what it takes to bring about change in a given political system. These are precisely the sorts of insights found in the field of comparative politics, which boasts an intellectual pedigree stretching back a century and includes analyses of state structures, policymaking styles, modes of social mobilization, state–society relations, institutional change, and the origins of public preferences (for an overview, see Kopstein and Lichbach 2009). If we wish to understand water governance in Egypt, for example, we need to know something about institutional change and multilevel governance in authoritarian systems (Sowers 2012b). If we want to know why the United States and Europe have responded differently to climate change, we need to appreciate differences in the structure of their federal systems of government (Selin and VanDeveer 2012, 2021).

Comparative research into causal processes can shed light on a central question in GEP – namely, why do countries support or shun international environmental cooperation (Bättig and Bernauer 2009; Raustiala 1997; see Chapter 8)? If a country's support for a climate change treaty is a function of national interests, as Sprinz and Vaahtoranta (1994) argue, we are still left with questions about where national interests come from. Material conditions alone do not provide a satisfactory answer. Only through the comparative study of political parties, think tanks, legislatures, electoral systems, bureaucracies, and social movement influence that we can begin to come to terms with the domestic origins of national interests (Aklin and Mildenberger 2020; Moravcsik 1997). This is all the more important in the era of the 2015 Paris Agreement on climate change, in which participating countries voluntarily submit national goals and pledges of action (Selin and VanDeveer 2022).

The enduring relevance of domestic politics

The growth of transnational activity by non-state actors is among the most important developments in GEP in recent decades. In the rush to understand this new phenomenon, it would be a mistake to overlook the enduring relevance of government institutions, domestic political economy and associated political processes. National governments and economic conditions have a profound impact on the size, tactics and impacts of nongovernmental organizations (NGOs) and social movements (Schreurs 2002; Sowers 2018; Steinberg 2005; see Chapter 14). The field of comparative politics has further produced a vast literature devoted to modern states and their evolving policy priorities (Esping-Andersen 1990; Herbst 1990; Krasner 1984). Drawing on this tradition, comparative scholars explore the prospects for the "greening" of states and societies as environmental governance becomes part and parcel of what it means to be a modern state (Duit 2016; Meadowcroft 2012). Comparative political inquiry can infuse the broader field of GEP with a greater understanding of how a variety of states, populist movements and political coalitions advocate ideas like the green new deal, cap and trade schemes, and environmental justice.

From non-state actors to social histories

Too often NGOs are described in GEP in ways that offer little context or understanding of the societies in which they operate. Yet environmental movements are shaped by connections with other domestic social movements, by access to domestic political parties and state institutions, and by strategic choices about how to mobilize constituencies and connect their concerns with established national discourses (Dalton 1994; Dryzek et al. 2003; Weinthal and Watters 2010). Rather than study NGOs only when they appear on the international stage, we gain a deeper understanding of the role of non-state actors in GEP through comparative historical analysis, by combining in-depth histories of social actors in particular places with theories that facilitate comparisons across borders (Lipschutz 2001). Lee and So (1999) use such an approach to help us understand the cultural repertoires that environmental activists throughout Asia draw on when pressing for change. Comparative social history enhances our appreciation for how transnational exchanges of resources and ideas take place in the context of domestic environmental movements and opportunity structures shaped by trends of democratization or lack thereof, economic development, urbanization, decentralization and state-building. Such approaches can help us understand emergent climate movements among young people around the globe or the transnational connections between indigenous communities and their demands for social, economic and environmental justice (Allan 2020).

Expanding our political imagination

Given the well-documented challenges facing efforts at global environmental cooperation – from least-common denominator agreements to contentious debates pitting national sovereignty against the provision of global public goods – it is not uncommon for students of GEP to become discouraged about the prospects for positive change. Yet comparative research on domestic policy innovation reveals more complex and encouraging stories. While states from the North and South are often at loggerheads in multilateral diplomatic venues, far removed from the halls of the United Nations one finds countless collaborations between actors from developing and industrialized countries, ranging from the creation of innovative institutions for financing conservation to transnational partnerships among air quality regulators.

One also finds environmental and energy leadership in all corners of the globe and CEP research has sought to explicate the reasons for climate policy variation among countries (Tobin 2017). The contrast between international stalemate and domestic action can be seen with the US position toward climate change (see Chapter 32). At the international level, the United States vacillates between climate change policy leader and obstructionist pariah (Selin and VanDeveer 2021), with an often dismal record of ratifying and implementing the environmental treaties it has signed (Schreurs et al. 2009). Domestically, however, many US cities, universities, and states engaged in substantial efforts to reduce greenhouse gas emissions over the past two decades. Many adopted reduction targets more stringent than those set by global agreements (Rabe 2004; Selin and VanDeveer 2009, 2020).

At a broader level, the comparative study of institutional reform expands the political imagination, helping researchers and practitioners to see how outcomes often assumed to be technically or politically impossible are achieved in different settings. Innovative environmental governance initiatives involving private, public and civil society actors take place in a diverse array of countries, including in many developing and post-communist countries that have made significant strides despite the challenges of building effective institutions in these settings (Carmin and VanDeveer 2005; Kauffman 2017; Steinberg 2012). Renewable energy in Portugal, industrial policy in China, energy transitions in Germany, pollution reduction in Indonesia, local water governance in Brazil, biodiversity conservation in Costa Rica, corporate environmental leadership in Mexico, and right-to-know legislation in the United States all offer models worthy of careful study (e.g., Grant 1997; Nahm 2021; Pulver 2007; Steinberg 2001). If some countries have well-managed and popular national park systems that incorporate local economic development and social justice needs, what can these places teach those who do not (Fuentes-George 2016)? When a green party has electoral success and legislative influence, what can other parties learn from their accomplishments? (M. O'Neill 2012) How do ideas and activism around critical environmental justice connect to Black Lives Matter or prison and judicial system reform (Pellow 2018)?

Better connecting the global to comparative environmental politics

Reviews of the CEP literature find a large and rapidly growing number of publications on domestic environmental politics around the globe (Sowers, VanDeveer and Weinthal forthcoming; Steinberg and VanDeveer 2012). Much of this research is interdisciplinary and has been published outside of conventional political science, international relations and comparative politics journals. Certainly the field of comparative politics has played its part in this disjuncture; despite claims to straddle the worlds of theory and social practice, leading research venues in comparative politics have often paid less attention to environmental issues than to such classic topics as labor relations. Nevertheless, efforts to establish a distinctive tradition in CEP are bearing fruit, with topical journals such as *Environmental Politics* offering ample coverage of CEP and scholars highlighting the need for greater coverage of environmental issues in broad political science journals (e.g., Javeline 2014). Building on Sowers, VanDeveer and Weinthal (forthcoming), we categorize this work in terms of the five overarching themes below.

States and environmental governance in comparative perspective

Born of unique social histories and reflecting distinctive constellations of political demands, domestic political institutions channel domestic interests, respond to international pressures, and shape social and environmental outcomes. In Western Europe, green parties have articulated the demands of environmental movements and have sometimes reshaped traditional party politics (Kitschelt 1989; O'Neill 2012). Comparativists have examined how voting rules affect the development of new parties, how the type and numbers of parties affect implementation of environmental regulations, the salience of environmental ideas within traditional party politics, and how Green parties handle the competing demands of accountability to grassroots constituencies and the parliamentary requirements of bargaining, professionalization, and electoral success (M. O'Neill 2012; Su et al. 2020).

Beyond political parties, analysts have compared environmental regulatory styles and policy processes across various types of political systems (e.g., corporatist, pluralist, parliamentary, presidential, federalist, and authoritarian) to better understand how change occurs in diverse settings and whether some political systems are more amenable to achieving environmental goals than others (Aguilar 1993; Hochstetler 2012, 2017; Scruggs 2001; Sowers 2012a; Steinberg 2012). Studies of the European Union show considerable disparities in national regulatory styles and environmental outcomes despite strong incentives for regulatory convergence (Andonova and VanDeveer 2012), reflecting differences in size, public opinion, and types of welfare state regimes (Andersen 2019; Carter, Little and Torney 2019; Farstad 2019). Extensive research in countries around the world have sought to specify what types of institutions and forms of state capacity help achieve environmental goals (e.g., see Jänicke and Weidner 1997). Other researchers have assessed the merits of specific environmental policy approaches, comparing national experiences with instruments such as tradable permits, environmental taxes, the polluter-pays principle, right-to-know legislation, product certification, voluntary industry agreements, the precautionary principle and participatory scientific assessments (Blackman 2008; de Bruijn and Norberg-Bohm 2005; Cashore et al. 2004; Eckley and Selin 2004; Jordan, Wurzel and Zito 2003; Harrington et al. 2004).

A growing literature also explores how authoritarian or illiberal regimes erect significant obstacles to environmental activism and adequate environmental enforcement (Henry 2010; Li and Shapiro 2020; Sowers 2018). Where political elites and leaders have eroded democratic norms and institutions, as in Hungary and Russia, or in illiberal regimes, such as Myanmar, China and Egypt, political elites often allow state-sanctioned forms of environmentalism while silencing environmental activists, muzzling independent media and undermining the rule of law. These repressive measures undermine social trust, allow well-connected firms to escape compliance, and enable leaders to pursue grandiose "prestige" projects with significant environmental impacts (van der Camp 2020).

Ideas, ethics, ideologies and law

Environmental problems are shaped by predominant cultural understandings, modes of knowledge production and transmission, and the values and interpretive frameworks of specific actors and institutions (Harris 2011; see Chapter 18). Thus, it is hardly surprising that environmental advocates often seek to instill in publics and elites new ways of thinking. Comparative research has documented how public and elite attitudes in different national settings change over time, in the process debunking widespread misconceptions, such as the notion that people in industrialized societies care more about environmental protection than do citizens of developing countries (Dunlap and York 2012; Steinberg 2001; see Chapter 29). Many scholars have moved beyond basic comparisons of survey responses to craft theories that identify clusters of attitudes, values and interpretive frameworks, their relation to actual behaviors, and how these viewpoints differ across social groups and national borders (Brechin 1999; Dunlap et al. 2000; Oreg and Katz-Gerro 2006). Indeed, many of the canonical theories of policy change emerged from studies of the connections between ideas and institutions in environmental policymaking (Baumgartner and Jones 2009; Weible et al. 2009; see by Stevenson). Comparative research also explores how norms, regulatory styles, and legitimizing discourses shape environmental movement tactics and state policy (O'Neill 2000). Research on third-party certification and private regulation within global food supply chains has explicated the role of information to consumers as well as the importance of trust and social capital of these third-party certifiers' credibility (Starobin and Weinthal 2010).

Environmental governance is often associated with vast quantities of technical information, raising questions about how knowledge is generated and interpreted in different national settings and by diverse social groups ranging from indigenous peoples to land-use planners (Jasanoff 2005; Wilkening 2004; see Chapter 18). Comparative research reveals how domestic institutions (such as those mediating the relationship between technical experts and law-makers) shape the manner in which policymakers, interest groups, and the public at large interpret and respond to scientific information. Comparative inquiry also sheds light on how ideas move across borders, including dynamics of policy convergence and the cross-national diffusion of social movement demands (Busch and Jörgens 2005; McAdam and Rucht 1993).

Movements, activism and non-state actors

Environmental movements have attracted the attention of comparative social movement scholars in political science and sociology, who offer perspectives on the goals, tactics and impacts of environmental movements around the world (K. O'Neill 2012; see Chapter 14). Scholars have steadily expanded their analysis of environmental social movements and public opinion from an initial focus on affluent countries to poorer countries, where environmental mobilization is driven in part by widespread exposure to water, air and soil pollution as many people live in proximity to harmful environmental activities (Eisenstadt and West 2017; Sowers 2018). Comparativists have explored the relationship between environmental movements and domestic social mobilization around gender, organized labor, nationalism, social justice, race, indigenous peoples, and democratization (Bennett et al. 2005; Carmin and Bast 2009; Carruthers 1996; Dawson 1996; Kim 2000; Obach 2002; Taylor 2000; see Chapter 14 and 26). Much of the observed cross-national variation in movement activity can be understood as strategic responses to these other movements – making common cause with mass movements for democracy, for example – and as attempts to cope with barriers and opportunities created by states (Dryzek et al. 2003; Schreurs 2002). Environmental movements can transform social and institutional contexts, causing states and social actors to take greater account of environmental quality.

A primary concern for comparativists is whether and how international governance arrangements affect a country's interests, ideas and institutions, and how domestic political forces mediate these influences (Cass 2006; DeSombre 2006; O'Neill 2000; O'Neill et al. 2004; Schreurs 2002; see Chapters 8 and 9). Particularly in Europe, with its increasingly integrated states and societies and extensive participation in multilateral institutions, researchers have focused on the mechanisms underlying "Europeanization" (Andonova and VanDeveer 2012). Even where states seek harmonized policy outcomes, comparative research reveals diverse patterns of convergence, divergence, and hybridization over time, rather than a simple process of homogenization (Knill and Lenschow 1998). Moreover, domestic interests frequently shape the content and structure of trans-jurisdictional institutions (DeSombre 2000; Schreurs and Tiberghien 2007).

Environmental justice, equity and rights

Central to the study of CEP are questions pertaining to equity, rights and justice (see Chapters 25–27). The study of comparative environmental justice draws not only on political science, but also on traditions of engaged critical studies in political ecology, anthropology and sociology, particularly in the subfields of rural and "peasant" studies (see, e.g., Pellow 2016). This body of work underscores the complex relationships and historical institutional legacies that yield to environmental and health disparities alongside unequal access to natural resources. Justice scholarship has long shed light on issues of environmental racism and marginalization, owing to the deliberate siting of polluting industries in poor communities, communities of color, Indigenous lands and informal urban areas (Auerbach et al. 2018; Bullard 2000). Likewise, scholars have documented forms of injustice tied to societies' dependence on fossil fuels; for example, Michael Watts (2004) and others have highlighted the negative effects of oil and gas pollution on the Ogoni people in the Niger Delta, while the rise in shale gas exploration and production in rural America has exposed poor communities to pollution from unconventional gas wells (e.g., Ogneva-Himmelberger and Huang 2015).

Another body of burgeoning research has focused on decolonizing institutions governing land use and water governance. Common-pool resource regimes, in which local communities devise internal norms to allocate access to resources such as water and pasture, are shaped by larger-scale political institutions and processes (Agrawal 2012). Comparative social scientists have also explored the impact of state structures on customary and local forms of resource ownership and use (Baver and Lynch 2006; Gadgil and Guha 1985). Increasingly, scholars have explored how recognizing Indigenous and local land rights not only serves to address injustices experienced by indigenous peoples and marginalized groups but also improves environmental management and prospects for sustainability (Jackson 2018). For example, scholars have documented how recognizing local rights to forests, land, and water can foster partnerships with local communities and deepen substantive processes of democratization (Kashwan 2017).

Rights-based scholarship has provided opportunities to analyze why some countries and not others seek to implement a human right to water or a human right to a healthy environment. Increasingly, scholars have explored the extension of rights-based legal frameworks to nonhuman entities, such as specific rivers, watersheds, and sites held sacred by Indigenous peoples. Such "rights to nature" are expanding at municipal, provincial and national levels, and comparative scholars have explored under what conditions such new legal frameworks contribute to enhanced environmental protection (Kauffman and Martin 2017). Finally, a large and dynamic literature has arisen around climate justice and injustice across scale (Barrett 2014; Shi et al. 2016).

Markets, firms and political economy

Natural resources have long shaped the political economies of countries, and scholars have found that different resources matter in different ways. Literature on oil, coal, gold and other extractive industries has proliferated over the last few decades, allowing scholars to examine why some states that depend largely on these "resource rents" experience negative political and economic outcomes, such as unbalanced economic growth, authoritarian durability, patronage networks, and corruption, while other states manage to avoid these pitfalls (Jones Luong and Weinthal 2010; Luciani 1994; Mahdavi 2019; Mazaheri 2016; Wantchekon and Jensen 2004). Notably, as extractive industries and large projects expand into seemingly remote areas such as the Amazon and the Arctic, some Indigenous peoples have engaged in new forms of mobilization (Henry 2018; Jaskoski 2020).

Comparative political economy has also long explored how institutional differences in business–state relations shape governance and policy outcomes. For example, scholars have studied how the uptake of renewable technologies such as wind and solar varies depending on how states structure markets in such diverse settings as China, South Africa and Brazil (Hochstetler and Kostka 2015; Hochstetler 2021). Comparative scholars have also examined the proliferation of business and market initiatives purported to promote more effective or efficient environmental policies. These include studies focusing on transparency schemes (Auld et al. 2018) and certification efforts so as to better track supply chains, mitigate corruption, and provide for more stakeholder participation in the economies of resource-rich countries.

Conclusion

International relations scholars have long recognized the need for research on domestic politics and institutions and their interactions with international policy processes (see, e.g., Moravcsik 1997). Within the field of GEP, the literature on domestic–international linkages has grown steadily over the past two decades (see, e.g., Harris 2003; Harrison and Sundstrom 2010; O'Neill 2017; Schreurs and Economy 1997; Weinthal 2002), but has yet to affect many of the basic premises of the field. For too long, analyses of GEP were confined to international negotiations and the challenge of implementing international agreements, paying lip service to the role of domestic politics and institutions without engaging in theoretically grounded empirical research on these topics. Yet a distinctive tradition in CEP has emerged. This new field will likely constitute a leading edge of the next generation of research on GEP even as it develops into its own research area. At a time when the discipline of political science is witnessing increasing integration between international relations and comparative politics, we do not propose the creation of a new fiefdom. Rather, we hope that the insights from CEP enrich and engage with the major theoretical debates and issues areas within GEP.

References

Agrawal, Arun. (2012). Local Institutions and the Governance of Forest Commons. In *Comparative Environmental Politics: Theory, Practice, and Prospects*, edited by Paul F. Steinberg and Stacy D. Vandeveer, 313–340. Cambridge, MA: MIT Press.

Aguilar, Susana. (1993). Corporatist and Statist Designs in Environmental Policy: The Contrasting Roles of Germany and Spain in the European Community Scenario. *Environmental Politics* 2(2): 223–247

Aklin, Michaël and Matto Mildenberger. (2020). Prisoners of the Wrong Dilemma: Why Distributive Conflict, Not Collective Action, Characterizes the Politics of Climate Change. *Global Environmental Politics* 20: 4, 4–27

Allan, Jen Iris. (2020). *The New Climate Activism: NGO Authority and Participation in Climate Change Governance*. Toronto: University of Toronto Press.

Andersen, Mikael Skou. (2019). The Politics of Carbon Taxation: How Varieties of Policy Style Matter. *Environmental Politics* 28(6): 1084–1104.

Andonova, Liliana B., Hale, T. N., and Roger, C. B. (Eds.). (2019). *The Comparative Politics of Transnational Climate Governance*. London: Routledge.

Andonova, Liliana B. and Stacy D. VanDeveer. (2012). EU Expansion and the Internationalization of Environmental Politics in Central and Eastern Europe. In *Comparative Environmental Politics: Theory, Practice, and Prospects*, edited by Paul F. Steinberg and Stacy D. Vandeveer, 287–312. Cambridge, MA: MIT Press.

Auld, Graeme, Michele Betsill and Stacy D. VanDeveer. (2018). Transnational Governance for Mining and the Mineral Lifecycle. *Annual Review of Environment and Resources*, 43: 425–453.

Auerbach, A. M., LeBas, A., Post, A. E., and Weitz-Shapiro, R. (2018). State, Society, and Informality in Cities of the Global South. *Studies in Comparative International Development*, 53(3): 261–280.

Barrett, S. (2014). Subnational Climate Justice? Adaptation Finance Distribution and Climate Vulnerability. *World Development* 58: 130–142.

Bättig, Michèle B. and Thomas Bernauer. (2009). National Institutions and Global Public Goods: Are Democracies More Cooperative in Climate Change Policy? *International Organization* 63(2): 281–308.

Baumgartner, Frank and Bryan D. Jones. (2009). *Agendas and Instability in American Politics*, 2nd Edition. Chicago, IL: University of Chicago Press.

Baver, Sherrie L. and Barbara Deutsch Lynch, eds. (2006). *Beyond Sun and Sand: Caribbean Environmentalisms*. Piscataway, NJ: Rutgers University Press.

Bennett, Vivienne, Sonia Davila-Poblete, and Maria Nieves Rico. (2005). *Opposing Currents: The Politics of Water and Gender in Latin America*. Pittsburgh, PA: University of Pittsburgh Press.

Blackman, Allen. (2008). Can Voluntary Environmental Regulation Work in Developing Countries? Lessons from Case Studies. *Policy Studies Journal* 46(4): 119–141.

Brechin, Steven R. (1999). Objective Problems, Subjective Values, and Global Environmentalism: Evaluating the Postmaterialist Argument and Challenging a New Explanation. *Social Science Quarterly* 80(4): 793–809.

Bullard, Robert D. (2000). *Dumping in Dixie: Race, Class, and Environmental Quality*. 3rd Edition. Westview Press.

Busch, Per-Olof and Helge Jörgens. (2005). The International Sources of Policy Convergence: Explaining the Spread of Environmental Policy Innovations. *Journal of European Public Policy* 12(5): 860–884.

Carmin, JoAnn and Elizabeth Bast. (2009). Cross-movement Activism: A Cognitive Perspective on the Global Justice Activities of US environmental NGOs. *Environmental Politics* 18(3): 351–370.

Carmin, JoAnn and Stacy D. VanDeveer. (2005). *EU Enlargement and the Environment: Institutional Change and Environmental Policy in Central and Eastern Europe*. London: Routledge.

Carruthers, David. (1996). Indigenous Ecology and the Politics of Linkage in Mexican Social Movements. *Third World Quarterly* 17(5): 1007–1028.

Carter, Neil, Conor Little, and Diarmuid Torney. (2019). Climate Politics in Small European States. *Environmental Politics* 28(6): 981–996.

Cashore, Benjamin, Graeme Auld, and Deanna Newsom. (2004). *Governing Through Markets*. New Haven, CT: Yale University Press.

Cass, Loren R. (2006). *The Failures of American and European Climate Policy: International Norms, Domestic Politics, and Unachievable Commitments*. Albany: SUNY Press.

Dalton, Russell J. (1994). *The Green Rainbow: Environmental Groups in Western Europe*. New Haven, CT: Yale University Press.

Dauvergne, Peter. (1997). *Shadows in the Forest: Japan and the Politics of Timber in Southeast Asia*. Cambridge, MA: MIT Press.

Dauvergne, Peter. (2005). Research in Global Environmental Politics: History and Trends. In *Handbook of Global Environment Politics*, edited by Peter Dauvergne, 8–32. Northampton: Edward Elgar.

Dawson, Jane I. (1996). *Eco-Nationalism: Anti-Nuclear Activism and National Identity in Russia, Lithuania and Ukraine*. Durham, NC: Duke University Press.

De Bruijn, Theo and Vicky Norberg-Bohm, eds. (2005). *Industrial Transformation: Environmental Policy Innovation in the United States and Europe*. Cambridge, MA: MIT Press.

DeSombre, Elizabeth R. (2000). *Domestic Sources of International Environmental Policy: Industry, Environmentalists, and U.S. Power*. Cambridge, MA: MIT Press.

DeSombre, Elizabeth R. (2006). *Flagging Standards: Globalization and Environmental, Safety and Labor Regulations at Sea*. Cambridge, MA: MIT Press.

Dryzek, John S., David Downes, Christian Hunold, and David Schlosberg. (2003). *Green States and Social Movements: Environmentalism in the United States, United Kingdom, Germany, and Norway*. New York: Oxford University Press.

Duit, A. (2016). The Four Faces of the Environmental State: Environmental Governance Regimes in 28 Countries. *Environmental Politics* 25(1): 69–91.

Dunlap, Riley E., Kent D. Van Liere, Angela G. Mertig, and Robert Emmet Jones. (2000). New Trends in Measuring Environmental Attitudes: Measuring Endorsement of the New Ecological Paradigm: A Revised NEP Scale. *Journal of Social Issues* 56(3): 425–442.

Dunlap, Riley E. and Richard York. (2012). The Globalization of Environmental Concern. In *Comparative Environmental Politics: Theory, Practice, and Prospects*, edited by Paul F. Steinberg and Stacy D. Vandeveer, 89–112. Cambridge, MA: MIT Press.

Eckley, Noelle, and Henrik Selin. (2004). All Talk, Little Action: Precaution and European Chemicals Regulation. *Journal of European Public Policy* 11(1): 78–105.

Eisenstadt, Todd A., and Karleen Jones West. (2017). Public Opinion, Vulnerability, and Living with Extraction on Ecuador's Oil Frontier: Where the Debate between Development and Environmentalism Gets Personal. *Comparative Politics* 49(2): 231–251.

Esping-Andersen, Gøsta. (1990). *The Three Worlds of Welfare Capitalism*. Princeton, NJ: Princeton University Press.

Farstad, Fay Madeleine. (2019). Does Size Matter? Comparing the Party Politics of Climate Change in Australia and Norway. *Environmental Politics* 28(6): 997–1016.

Fuentes-George, Kemi. (2016). *Between Preservation and Exploitation: Transnational Advocacy Networks and Conservation in Developing Countries*. Cambridge, MA: MIT Press.

Gadgil, Madhav and Ramachandra Guha. (1985). *Ecology and Equity: The Use and Abuse of Nature in Contemporary India*. New York: Routledge.

Green, Jessica. (2013). *Rethinking Private Authority: Agents and Entrepreneurs in Global Environmental Governance*. Princeton, NJ: Princeton University Press.

Goodin, Robert E., and Charles Tilly, eds. (2006). *The Oxford Handbook of Contextual Political Analysis*. New York: Oxford University Press.

Grant, Don S. (1997). Allowing Citizen Participation in Environmental Regulation: An Empirical Analysis of the Effects of Right-to-Sue and Right-to-Know Provisions on Industry's Toxic Emissions: Research on the Environment. *Social Science Quarterly* 78(4): 859–873

Harrington, Winston, Richard D. Morgenstern, and Thomas Sterner, eds. (2004). *Choosing Environmental Policy: Comparing Instruments and Outcomes in the United States and Europe*. London: RFF Press/Earthscan.

Harris, Paul G., ed. (2003). *Global Warming and East Asia: The Domestic and International Politics of Climate Change*, Routledge Research in Environmental Politics. New York: Routledge.

Harris, Paul G., ed. (2011). *China's Responsibility for Climate Change: Ethics, Fairness and Environmental Policy*. Portland, OR: Policy Press.

Harrison, Kathryn, and Lisa McIntosh Sundstrom. (2010). *Global Commons, Domestic Decisions: The Comparative Politics of Climate Change*. Cambridge, MA: MIT Press.

Herbst, Jeffrey. (1990). War and the State in' Africa. *International Security* 14: 117–139.

Henry, Laura. (2010). *Red to Green: Environmental Activism in Post-Soviet Russia*. Ithaca, NY: Cornell University Press.

Henry, Laura. (2018). Oil Extraction and Benefit Sharing in an Illiberal Context: The Nenets and Komi-Izhemtsi Indigenous Peoples in the Russian Arctic, (with Maria Tysiachniouk, Machiel Lamers, and Jan P.M. van Tatenhove), *Society and Natural Resources* 31(5): 556–579.

Herrera, Veronica. (2017). *Water and Politics: Clientelism and Reform in Urban Mexico*. Ann Arbor, MI: University of Michigan Press.

Hochstetler, Kathryn. (2017). Tracking Presidents and Policies: Environmental Politics from Lula to Dilma. *Policy Studies* 38(3): 262–276.

Hochstetler, Kathryn. (2021). *Political Economies of Energy Transition: Wind and Solar Power in Brazil and South Africa*. Cambridge: Cambridge University Press.

Hochstetler, Kathryn. (2012). Democracy and the Environment in Latin America and Eastern Europe. In *Comparative Environmental Politics: Theory, Practice, and Prospects*, edited by Paul F. Steinberg and Stacy D. Vandeveer, 199–230. Cambridge, MA: MIT Press.

Hochstetler, Kathryn and Margaret E. Keck. (2007). *Greening Brazil: Environmental Activism in State and Society*. Durham, NC: Duke University Press.

Hochstetler, Kathryn and Genia Kostka. (2015). Wind and Solar Power in Brazil and China: Interests, State-Business Relations and Policy Outcomes. *Global Environmental Politics* 15(3): 74–94.

Jackson, Sue. (2018). Indigenous Peoples and Water Justice in a Globalizing World. In Ken Conca and Erika Weinthal, eds. *Oxford Handbook of Water Politics and Policy*. Oxford, UK: Oxford University Press.

Jänicke, Martin and Helmut Weidner, eds. (1997). *National Environmental Policies: A Comparative Study of Capacity-Building*. Berlin: Springer.

Jasanoff, Sheila. (2005). *Designs on Nature: Science and Democracy in Europe and the United States*. Princeton, NJ: Princeton University Press.

Jaskoski, Maiah. (2020). Participatory Institutions as a Focal Point for Mobilizing: Prior Consultation and Indigenous Conflict in Colombia's Extractive Industries. *Comparative Politics* 52(4): 537–556.

Javeline, Debra. (2014). The Most Important Topic Political Scientists are not Studying: Adapting to Climate Change. *Perspectives on Politics* 12(2): 420–434.

Jones Luong, Pauline and Erika Weinthal. (2010). *Oil is Not a Curse: Ownership Structure and Institutions in Soviet Successor States*. Cambridge, UK: Cambridge University Press.

Jordan, Andrew, Rüdiger K. W. Wurzel, and Anthony R. Zito. (2003). "New" Instruments of Environmental Governance: Patterns and Pathways of Change. *Environmental Politics* 12(1): 1–24.

Kamieniecki, Sheldon and Eliz Sanasarian. (1990). Conducting Comparative Research on Environmental Policy. *Natural Resources Journal* 30(2): 321–339.

Kashwan, Prakash. (2017). *Democracy in the Woods: Environmental Conservation and Social Justice in India, Tanzania and Mexico*. New York: Oxford University Press.

Kauffman, Craig. (2017). *Grassroots Global Governance: Local Watershed Management Experiments and the Evolution of Sustainable Development*. Oxford: Oxford University Press.

Kauffman, Craig M. and Pamela L. Martin. (2017). Can Rights of Nature Make Development More Sustainable? WhySome Ecuadorian Lawsuits Succeed and Others Fail. *World Development*, 92, 130–142

Kim, Sunhyuk. (2000). Democratization and Environmentalism: South Korea and Taiwan in Comparative Perspective. *Journal of Asian and African Studies* 35(3): 287–302.

Kitschelt, Herbert P. (1989). *The Logics of Party Formation: Ecological Politics in Belgium and West Germany*. Ithaca, NY: Cornell University Press.

Knill, Christoph and Andrea Lenschow. (1998). Coping with Europe: The Impact of British and German Administrations on the Implementation of EU Environmental Policy. *Journal of European Public Policy* 5(4): 595–614.

Kopstein, Jeffrey and Mark Lichbach, eds. (2009). *Comparative Politics: Interests, Identities, and Institutions in a Changing Global Order*. Cambridge, New York: Cambridge University Press.

Krasner, Stephen D. (1984). Approaches to the State: Alternative Conceptions and Historical Dynamics. *Comparative Politics* 16(2): 223–246.

Lee Yok-Shiu F. and Alvin Y. So, eds. (1999). *Asia's Environmental Movements: Comparative Perspectives*. Armonk, NY: M.E. Sharpe.

Li, Yufei and Judith Shapiro. (2020). *China Goes Green: Coercive Environmentalism for a Troubled Planet*. Cambridge, UK: Polity Press.

Lipschutz, Ronnie D. (2001). Environmental History, Political Economy and Change: Frameworks and Tools for Research and Analysis. *Global Environmental Politics* 1(3): 72–91.

Luciani, Giacomo. (1994). The Oil Rent, the Fiscal Crisis of the State and Democratization. In *Democracy without Democrats?*, edited by Ghassan Salamé, 130–155. London: I.B. Tauris.

Lundqvist, Lennart J. (1978). The Comparative Study of Environmental Politics: From Garbage to Gold? *International Journal of Environmental Studies* 12(2): 89–97.

Mahdavi, Paasha. (2019). Institutions and the Resource Curse: Evidence from Cases of Oil-Related Bribery. *Comparative Political Studies* 53(1): 3–39.

Marion Suiseeya, Kimberly R. Forthcoming. Towards a Comparative Politics of Environmental Justice: Critical Perspectives on Representation, Equity, and Rights. In Jeannie Sowers, Stacy VanDeveer, Erika Weinthal, eds. *Oxford Handbook of Comparative Environmental Politics*.

Mazaheri, Nimah. (2016). *Oil Booms and Business Busts: Why Resource Wealth Hurts Entreprenuers in the Developing World*. Oxford: Oxford University Press.

McAdam, Doug and Dieter Rucht. (1993). The Cross-National Diffusion of Movement Ideas. *Annals of the American Academy of Political and Social Sciences* 528: 56–74.

McBeath, Jerry and Jonathan Rosenberg. (2006). *Comparative Environmental Politics*. New York: Springer.

Meadowcroft, James. (2012). Greening the State? In *Comparative Environmental Politics: Theory, Practice, and Prospects*, edited by Paul F. Steinberg and Stacy D. Vandeveer, 63–87. Cambridge, MA: MIT Press.

Mitchell, Ronald. (2002). International Environment. In *Handbook of International Relations*, edited by Thomas Risse, Beth Simmons, and Walter Carlsnaes, 500–516. Thousand Oaks, CA: Sage Publications.

Moravcsik, Andrew. (1997). Taking Preferences Seriously: A Liberal Theory of International Politics. *International Organization* 51(4): 513–553.

Nahm, Jonas (2021). *Collaborative Advantage: Forging Green Industries in the New Global Economy.* New York: Oxford University Press.

Neville, Kate J. (2021). *Fueling Resistance: The Contentious Political Economy of BioFuels and Fracking.* Oxford: Oxford University Press.

Obach, Brian K. (2002). Labor–Environmental Relations: An Analysis of the Relationship between Labor Unions and Environmentalists. *Social Science Quarterly* 83(1): 82–100.

O'Neill, Kate. (2000). *Waste Trading among Rich Nations: Building a New Theory of Environmental Regulation.* Cambridge, MA: MIT Press.

O'Neill, Kate. (2012). The Comparative Study of Environmental Movements. In *Comparative Environmental Politics: Theory, Practice, and Prospects*, edited by Paul F. Steinberg and Stacy D. Vandeveer, 115–142. Cambridge, MA: MIT Press.

O'Neill, Kate. (2017). *The Environment and International Relations*, Second Edition. Cambridge, UK: Cambridge University Press.

O'Neill, Kate, Jörg Balsiger, and Stacy D. VanDeveer. (2004). Actors, Norms, and Impact: Recent International Cooperation Theory and the Influence of the Agent-Structure Debate. *Annual Review of Political Science* 7: 149–175.

O'Neill, Kate, E. Weinthal, K. Marion Suiseeya, S. Bernstein, A. Cohn, M. Stone, and B. Cashore. (2013). Methods and Global Environmental Governance. *Annual Review of Environment and Resources* 38: 441–471.

O'Neill, Michael. (2012). Political Parties and the "Meaning of Greening" in European Politics. In *Comparative Environmental Politics: Theory, Practice, and Prospects*, edited by Paul F. Steinberg and Stacy D. Vandeveer, 171–195. Cambridge, MA: MIT Press.

Ogneva-Himmelberger, Yelena and Liyao Huang. (2015). Spatial Distribution of Unconventional Gas Wells and Populations in the Marcellus Shale in the United States. *Applied Geography* 60: 165–174

Oreg, Shaul and Tally Katz-Gerro. (2006). Predicting Proenvironmental Behavior Cross-Nationally: Values, the Theory of Planned Behavior, and Value-Belief-Norm Theory. *Environment and Behavior* 38(4): 462–483.

Pellow, David N. (2016). Environmental Justice and Rural Studies: A Critical Conversation and Invitation to Collaboration. *Journal of Rural Studies* 47: 381–386.

Pellow, David N. (2018). *What Is Critical Environmental Justice?* Cambridge, UK: Polity Press.

Pulver, Simone. (2007). Importing Environmentalism: Explaining Petroleos Mexicanos' Cooperative Climate Policy. *Studies in Comparative International Development* 42(3–4): 233–255.

Rabe, Barry. (2004). *Statehouse and Greenhouse: The Emerging Politics of American Climate Change Policy.* Washington, DC: Brookings Institution Press.

Raustiala, Kal. (1997). Domestic Institutions and International Regulatory Cooperation. *World Politics* 49(4): 482–509.

Ribot, Jesse C. (1999). Decentralisation, Participation, and Accountability in Sahelian Forestry: Legal Instruments of Political-Administrative Control. *Africa* 69(1): 23–65.

Schreurs, Miranda. (2002). *Environmental Politics in Japan, Germany, and the United States.* New York: Cambridge University Press.

Schreurs, Miranda and Elizabeth Economy, eds. (1997). *The Internalization of Environmental Protection.* New York: Cambridge University Press.

Schreurs, Miranda, Henrik Selin, and Stacy D. VanDeveer. (2009). *Transatlantic Environmental and Energy Politics.* Burlington, VT: Ashgate Press.

Schreurs, Miranda and Yves Tiberghien. (2007). Multi-Level Reinforcement: Explaining European Union Leadership in Climate Change Mitigation. *Global Environmental Politics* 7(4): 19–46.

Scruggs, Lyle. (2001). Is There Really a Link Between Neo-Corporatism and Environmental Performance? Updated Evidence… *British Journal of Political Science* 31(4): 686–692.

Selin, Henrik and Stacy D. VanDeveer. (2009). *Changing Climates in North American Politics: Institutions, Policymaking and Multilevel Governance.* Cambridge, MA: MIT Press.

Selin, Henrik and Stacy D. VanDeveer. (2012). Federalism, Multilevel Governance, and Climate Change Politics across the Atlantic. In *Comparative Environmental Politics: Theory, Practice, and Prospects*, edited by Paul F. Steinberg and Stacy D. Vandeveer, 341–368. Cambridge, MA: MIT Press.

Selin, Henrik and Stacy D. VanDeveer. (2021).. Climate Change Politics and Policy in the United States: Forward, Reverse and through the looking glass. In *Climate Governance Across the Globe*, edited by Rudiger K.W. Wurze, Mikael, Skou Andresen and Paul Tobin, 123–141. London: Routledge.

Selin, Henrik and Stacy D. VanDeveer, (2022).. Global Climate Change Governance Can the Promise of Paris be Realized? In *Environmental Policy: new Directions for the Twenty-First Century*, edited by Norman J. Vig, Michael E. Kraft and Barry G. Rabe, 275–299. Los Angeles: Sage.

Shi, Linda, Chu, Eric, Anguelovski, Isabelle, et al. (2016). Roadmap towards Justice in Urban Climate Adaptation Research. *Nature Climate Change* 6: 131–137.

Sowers, Jeannie. (2012a). *Environmental Politics in Egypt: Activists, Experts, and the State*, 2013. New York: Routledge.

Sowers, Jeannie. (2012b). Institutional Change in Authoritarian Regimes: Water and the State in Egypt. In *Comparative Environmental Politics: Theory, Practice, and Prospects*, edited by Paul F. Steinberg and Stacy D. Vandeveer, 231–254. Cambridge, MA: MIT Press.

Sowers, Jeannie. (2018). Environmental Activism in the Middle East and North Africa. In *Environmental Politics in the Middle East,* edited by Harry Verhoeven, 27–52. Oxford: Oxford University Press/ Hurst Publishers.

Sowers, Jeannie, Erika Weinthal and Stacy D VanDeveer. Forthcoming. *Oxford Handbook of Comparative Environmental Politics*. Oxford, UK: Oxford University Press.

Sprinz, Detlef and Tapani Vaahtoranta. (1994). The Interest-Based Explanation of International Environmental Policy. *International Organization* 48(1): 77–105.

Starobin, Shana and Erika Weinthal. (2010). The Search for Credible Information in Social and Environmental Global Governance: The Kosher Label. *Business and Politics* 12(3): 1–35.

Steinberg, Paul F. (2001). *Environmental Leadership in Developing Countries: Transnational Relations and Biodiversity Policy in Costa Rica and Bolivia*. Cambridge, MA: MIT Press.

Steinberg, Paul F. (2005). From Public Concern to Policy Effectiveness: Civic Conservation in Developing Countries. *Journal of International Wildlife Law and Policy* 8: 341–365.

Steinberg, Paul F. (2012). Welcome to the Jungle: Political Theory and Political Instability. In *Comparative Environmental Politics: Theory, Practice, and Prospects*, edited by Paul F. Steinberg and Stacy D. Vandeveer, 255–284. Cambridge, MA: MIT Press.

Steinberg, Paul F. and Stacy D. VanDeveer, eds. (2012). *Comparative Environmental Politics: Theory, Practice, and Prospects*. Cambridge, MA: MIT Press.

Su, Zheng, et al. (2020). Electoral Competition, Party System Fragmentation, and Air Quality in Mexican Municipalities. *Environmental Politics* 29: 1–21.

Taylor, Dorceta E. (2000). The Rise of the Environmental Justice Paradigm: Injustice Framing and the Social Construction of Environmental Discourses. *American Behavioral Scientist* 43(4): 508–580.

Tobin, Paul. (2017). Leaders and Laggards: Climate Policy Ambition in Developed States. *Global Environmental Politics* 17(4): 28–47.

van der Camp, Denise. (2020). Can Police Patrols Prevent Pollution? The Limits of Authoritarian Governance in China. *Comparative Politics* 53(3): 403–433.

Vanhala, Lisa. (2013). The Comparative Politics of Courts and Climate Change. *Environmental Politics* 22(3): 447–474.

Wantchekon, Leonard and Nathan Jensen. (2004). Resource Wealth and Political Regimes in Africa. *Comparative Political Studies* 37: 816. DOI: 10.1177/0010414004266867

Watts, Michael. (2004). Resource Curse? Governmentality, Oil and Power in the Niger Delta, Nigeria. *Geopolitics* 9 (1): 50–80.

Weible, Christopher M., Paul A. Sabatier, and Kelly McQueen. (2009). Themes and Variations: Taking Stock of the Advocacy Coalition Framework. *Policy Studies Journal* 37(1): 121–140.

Weinthal, Erika. (2002). *State Making and Environmental Cooperation: Linking Domestic and International Politics in Central Asia*. Cambridge, MA: MIT Press.

Weinthal, Erika and Kate Watters. (2010). The Transformation of Environmental Activism in Central Asia: From Dependent to Interdependent Activism. *Environmental Politics* 19(5): 782–807.

Wilkening, Kenneth E. (2004). *Acid Rain Science and Politics in Japan*. Cambridge, MA: MIT Press.

Young, Oran R., ed. (1999). *The Effectiveness of International Environmental Regimes: Causal Connections and Behavioral Mechanisms*. Cambridge, MA: MIT Press.

13

Corporations
Business and industrial influence

Kyla Tienhaara

On 11 January 2021, HRH The Prince of Wales of the United Kingdom unveiled a "Terra Carta," or Earth Charter, at the One Planet Summit hosted by France. In his forward to the Terra Carta, he wrote: "Today must be the decisive moment that we make sustainability the growth story of our time, while positioning Nature as the engine of our economy… I am calling on CEOs [chief executive officers of corporations] from around the world to engage and play their part in leading the global transition." Supporting CEOs listed at the launch included those that led Bank of America and HSBC bank, institutional investor Blackrock, and consumer goods giant Unilever (Sustainable Markets Initiative 2021). As has occurred with many similar initiatives, the reaction to the Terra Carta was mixed. While some applauded the effort as a step in the right direction, others denounced it as greenwashing. The inclusion of the CEO of oil-and-gas giant BP on the list of signatories, in particular, sparked immediate backlash from environmental activists. Aside from its position as a major emitter of greenhouse gases, at the time of the launch BP was embroiled in controversy over its stonewalling of an investigation into an oil spill in Mauritius. This was even noted in *Forbes* (a business-focused magazine), where one journalist argued: "The intent for *Terra Carta* is the right one, but the way it is currently structured allows too much risk that this effort will be sabotaged by industry insiders. Several major UN and international organizations have already been infiltrated by industry bodies seeking to greenwash their corporate commitments" (Degnarain 2021).

Indeed, the Terra Carta is but one in a long line of voluntary codes of conduct that corporations have been encouraged to sign up to over the years. Notable examples include the OECD Guidelines for Multinational Enterprises, the UN Norms on the Responsibilities of Transnational Corporations and other Business Enterprises with regard to Human Rights, and the Global Compact (for an overview of these initiatives, see Clapp 2005a; Tienhaara 2009: ch. 2). Sector-specific "roundtables" have also flourished in areas such as palm oil, meat and soy (Schouten and Glasbergen 2011, Schouten, Leroy and Glasbergen 2012, Buckley et al. 2019). Despite the proliferation of pledges and soft rules, corporate actors continue to find themselves the targets of activist campaigns on issues ranging from climate change (see Chapter 32) to plastic pollution (see Chapter 37). They are singled out for their role not only in creating environmental problems, but also for actively working – at the local, national and international level – to prevent or delay serious policy responses to these problems.

DOI: 10.4324/9781003008873-16

This chapter explores these dynamics with a focus on global corporations. Newell (2020: 3) points out that "95% of private-sector companies in most industrialized economies are small and medium-sized businesses with fewer than 250 employees" and that collectively these businesses "play a crucial, but understated and under-researched, role in sustainability transitions." Nevertheless, global corporations are particularly significant because of their size and their ability to dominate markets. As Clapp (2018: 25) notes, the scale of some firms "means that their business decisions have an enormous impact on national economies. In this context, policymakers cannot help but pay attention to their preferences." In particular, a subset of global corporations has a disproportionate influence over environmental outcomes, making them "keystone actors of the Anthropocene" (Österblom et al. 2015: 1; see Chapter 15). While Österblom et al. (2015: 1) observed this in the fisheries sector (see Chapter 40), where "thirteen corporations control 11–16% of the global marine catch (9–13 million tons) and 19–40% of the largest and most valuable stocks," keystone actors are also visible in other areas. For example, Heede (2014) identified 90 "carbon majors" that were responsible for 63 percent of cumulative worldwide emissions of industrial CO_2 and methane between 1751 and 2010. In December 2020, an update of Carbon Majors Database demonstrated that just 20 companies have collectively contributed to 35 percent of all fossil fuel and cement emissions worldwide since 1965 (Climate Accountability Institute 2020). Clapp (2018) points to even greater concentration of power in the agricultural input sector, where six firms controlled 75 percent of the market in 2009 and this had shrunk (through "mega-mergers") to four by 2018. In terms of plastic waste, Greenpeace (2017) has highlighted that one company – Coca Cola – is responsible for a sixth of all plastic drink bottles sold around the world (see Chapter 11). In addition to these examples of corporations that *directly* impact the environment, it can also be argued that large players in the finance sector – which can play a major role in facilitating or blocking environmentally destructive projects (see, e.g., RAN et al. 2020) – are also keystone actors in global environmental politics.

While a focus is given in this chapter to keystone actors that work to delay and dilute environmental policy outcomes, it is important to acknowledge at the outset that these corporations are not representative of the business community as a whole. Even within a given sector, corporations can have different preferences and adopt different approaches to environmental issues (Meckling 2015; Downie 2017). Corporate positions and strategies are shaped by factors such as: the country or region in which the corporation is based; the sector in which it operates; its position on a particular supply chain; access to markets and technologies; its exposure to environmental risks; and the structure and style of the corporation's leadership (Clapp 2005a; Newell and Levy 2006; Jones and Levy 2007; Downie 2017). Recent research has also highlighted that "business interests are not static but subject to change as a result of shifting technological, market, and policy conditions" (Vormedal et al. 2020: 161).

The sections that follow will first address the different political strategies that corporations can adopt in global environmental politics: opposition, hedging, support, and non-participation. I then move to the question of the types of power, influence, and authority that corporations wield when they opt to engage in the political process.

Corporate political strategies

Environmental regulation is typically assumed to imply costs for industry and is, therefore, likely to be opposed. However, in reality, as Witte (2020: 179) argues, the effects of environmental regulation "differ across industries and companies, leading to an uneven distribution of benefits and the creation of winners and losers." Opposition to environmental policy by firms is certainly something that can be empirically observed on a regular basis. However, even when this strategy is successful it may have negative consequences for a firm, for example, in terms of reputational damage.

In some cases, corporations will view opposition as either fruitless or too risky and will instead choose to proactively engage in a policy process in order to minimize compliance costs ("hedging") (Meckling 2015). Firms may also engage in this strategy if they are, or expect to be, regulated through domestic legislation and want to ensure a level playing field with their global competitors. In terms of what types of policies corporations are likely to push for with this strategy, scholars have noted a clear preference for voluntary initiatives and self-regulation over binding rules, and market-based policies over "command and control" regulations (Clapp 2005b; Newell 2008).

Some firms will benefit from environmental regulation, such as first movers in alternative technologies (e.g., renewable energy, substitutes for CFCs); these firms will "support" environmental policy if the benefits of participating in the political process outweigh the costs (Meckling 2015). Corporations involved in "dirty" industries may also support particular environmental policy frameworks if they see an opportunity for strategic market expansion. For example, while much of the literature suggests that global oil majors have supported carbon pricing in an effort to stave off more radical or costly regulation ("hedging"), Vormedal et al. (2020) argue that these corporations are actually seeking market expansion through increased demand for gas (as a "bridge fuel") at the expense of coal.

Finally, Meckling (2015) notes that due to the costs associated with participation in policy making, particularly at the international level, some firms that stand to benefit from regulatory action will choose not to participate. Unfortunately, costs are more likely to deter emerging industry players that would be more likely to favor stricter environmental policies than older established incumbent industries.

Dimensions of corporate power

In addition to a variety of strategies, there are a number of different forms of power that corporations can draw on when they engage in the political process: structural, instrumental, discursive, technological and institutional (Levy and Egan 1998; Fuchs 2007; Falkner 2008; Clapp and Fuchs 2009; Tienhaara, Orsini and Falkner 2012).

The structural dimension of corporate power stems from the central role that business plays in the economy. In capitalist countries corporations are the main sources of economic growth, employment and innovation. These are issues of paramount importance to governments. Regulators endeavor to avoid hurting business or creating competitive disadvantages for companies that might encourage them to move offshore (Falkner 2008). The extent to which companies actively exercise their structural power – punishing and rewarding countries by moving investments in response to regulation – is contested. In the environmental sphere, there was an intense debate in the 1990s over the occurrence of regulation-induced capital flight to "pollution havens" (for an overview of the literature, see Clapp 2002). Although structural power can be effective even if it is not actively employed by a corporation, the mere threat of capital flight, particularly in a globalized world characterized by high capital mobility, can be sufficient to shape public policy. However, although the structural power that corporations possess arguably gives them a privileged position in relation to other non-state actors such as NGOs, the fact that environmental regulations exist at all suggests that it is not all-encompassing.

Instrumental power (also referred to as "relational power" by Falkner 2008) concerns the power of one individual (or group) over another (Fuchs 2007; Falkner 2008). Lobbying is a classic example of the exercise of instrumental power (see Chapter 12). Although various kinds of interest groups engage in lobbying, corporations (especially large ones) have several advantages over their non-profit counterparts, including: access to substantial material resources; possession of specialized expertise; and access to the actors with the power to make decisions (policymakers) through relational networks and close contacts (which are enhanced by the "revolving door" between government and industry) (Newell 2000; Newell and Levy 2006; Falkner 2008; Sell 2009). The issue of access is particularly important; as noted by Geels (2014: 27) "frequent contacts may lead policymakers to internalize the ideas and interests of industries, which is a more subtle mechanism of influence."

Corporations have long engaged in lobbying domestic government delegations that attend international meetings, but increasingly they lobby directly at the international level as well (Clapp 2005b). Vormedal (2008: 43) outlines that one particularly important form of lobbying that corporations engage in is "information-based" lobbying, which "may involve providing governments with expert advice, technical reports and position papers, and assisting decision-makers directly with policy formulation and the writing of legal texts." In addition to making their preferences known to decision-makers, lobbying of this sort can also remind government officials of the potential costs of regulation for the economy.

The organizational strength of corporations enhances their instrumental power (Falkner 2008). In this respect, it is important to highlight the rise in coalition-building at the international level. There are business coalitions, often referred to as Business and Industry Non-Governmental Organizations (BINGOs), business–state coalitions, and business–NGO coalitions (Meckling 2011: 31). BINGOs provide corporations with a number of benefits: they allow for cost sharing and information pooling; they can establish and maintain access channels to forums and negotiations that individual companies would not be invited to; and they can monitor negotiation processes to ensure that industry perspectives are provided at every opportunity (Vormedal 2008). Examples of BINGOs that have been actively involved in global environmental politics include the International Chamber of Commerce (ICC) and the World Business Council for Sustainable Development (WBCSD).

BINGOs can be sector-specific, but they are increasingly issue-specific. For example, the Global Climate Coalition (GCC), which operated from 1990 to 2002, represented major producers and users of fossil fuels such as mining, oil and gas companies, car manufacturers, and the chemicals industry (Levy and Egan 2003). The GCC engaged in lobbying against action on climate change and commissioned a number of economic studies that gave high estimates for the costs associated with reducing emissions (Levy and Egan 2003; Ihlen 2009). In addition to organizing with one another, individual corporations and BINGOs also enhance their instrumental power by mobilizing state allies (Meckling 2011). For example, the fossil fuel lobby has cooperated closely with the Organization of Petroleum-Exporting Countries (OPEC), and biotech companies often work with the "Miami group" of states that export genetically modified organisms (Newell and Levy 2006; see Chapter 41).

A third key dimension of corporate power relates to the "potency of the frames that actors use to couch their preferences," which has been referred to as discursive power (Sell 2009: 188). Geels (2014: 29) identifies three particular forms that this takes: "a) diagnostic framing, which identifies and defines problems; b) prognostic framing, which advances solutions to problems; and c) motivational framing, which provides a rationale for action and serves as a 'call to arms'." While corporations do not have a monopoly on public discourse, they do have substantial financial resources to devote to advertising, to funding studies that support their view of a particular environmental issue, or even to funding specific individuals (e.g., climate skeptics) who will promote their views (Newell 2000; Newell and Levy 2006; Falkner 2008). These last two strategies are particularly important in light of the link between discursive power and political legitimacy (Clapp and Fuchs 2009: 10–11). If a corporation makes a discursive claim, that claim may be easily dismissed by a skeptical and cynical public. However, if the same claim is made by an "expert," it can be more influential (see, e.g., Oreskes and Conway 2011). Falkner (2008) points out that corporations also derive discursive power from their central position in the technological innovation process; in effect, they can define the discourse on what is possible in terms of technological solutions to environmental problems (prognostic framing).

A clear example of corporations attempting to exert discursive power is the long-running campaign of climate denial by oil majors such as Exxon (see exxonknew.org). However, in recent years, just as many firms have shifted from "opposition" to "hedging" political strategies, they have also shifted from discourses of denial (a diagnostic framing that suggests either that there is no problem or that it is not caused by humans) to discourses of delay (a prognostic framing that focuses on corporate-favored "solutions"). As Lamb et al. (2020: 1) explain, discourses of delay "accept the existence of climate change, but justify inaction or inadequate efforts." An example is the narrative of "clean coal" that, for a time, was used to push the idea that carbon capture and storage (CCS) technology was a viable solution to climate change. Marshall (2016: 229) puts it plainly: "CCS primarily serves as a defensive fantasy preserving the current political and social order."

Although climate denial and delay (see Chapter 32) are perhaps the most obvious examples of corporations using discursive power to undermine environmental action, they are by no means the only ones. Another prominent example is the way that the plastics industry has helped to shape the public understanding of recycling as being an effective environmental solution (Lerner 2019). Public belief that the problem is solved leads to an increase in consumption of plastic goods and corresponding increase in the production of waste (Ma et al. 2019). This discursive power, combined with effective lobbying, has likely undermined efforts to deal with the plastic waste crisis at the global level (see Chapter 37).

Finally, "institutional power" derives from the ability of global corporations to shift environmental issues from "regulatory institutions," like multilateral environmental agreements, to "enabling institutions" that are designed to facilitate economic processes such as trade and investment (Levy and Egan 1998; Levy and Prakesh 2003). International investment agreements are especially important in this regard because they provide global corporations direct recourse to international dispute settlement – a process that has been used extensively to challenge environmental regulation (Tienhaara 2009). The North American Free Trade Agreement (NAFTA), which was host to a large number of these disputes, was replaced in 2019 by the US-Mexico-Canada Agreement (USMCA) which eliminated the possibility of such disputes between investors from the United States operating in Canada and vice versa. However, new rules on "regulatory cooperation" included in USMCA may increase the lobbying power of corporations seeking to influence environmental policy development (Tienhaara 2019). Furthermore, thousands of other investment agreements remain in place. Recent research has highlighted the potential for fossil fuel companies to utilize these treaties to pressure states to provide compensation when their assets are stranded by climate policies such as coal-power phase-outs (Tienhaara and Cotula 2020).

Evidence of corporate influence

Corporate power, in its various forms, is considerable and is frequently wielded in local, national and international political forums. However, it does not always result in actual influence over the outcome of political processes. Business actors often have to contend with "countervailing forces," particularly from the non-profit sector (Levy and Egan 2003). Although NGOs are generally not as well-resourced as large global corporations, they have developed sophisticated strategies to cajole or pressure corporations into more environmentally friendly behavior, giving rise to a new form of "world civic politics" (Wapner 1996) or "civil regulation" of business (Newell 2001) (see Chapter 14).

Additionally, Falkner (2005, 2008) has stressed that conflict within the business community can limit corporate influence (see also Orsini 2012; Downie 2017). Discord is most likely to be sown between: international and national firms; market leaders and laggards; and firms at different points in the production or supply chain (Falkner 2009, 2010). In the case of the biodiversity regime, it has largely been different industry sectors that have clashed over issues such as access to genetic resources and benefit sharing (see Chapters 41 and 44). There are not only two BINGOs – the Global Industry Coalition and the International Grain Trade Coalition – that are competing to be the voice of business in the negotiations, but also divisions within the former body between pharmaceutical and agricultural biotech companies (Tienhaara et al. 2012). While corporations have certainly had some success in shaping the biodiversity regime, their influence has arguably been limited by a lack of business unity on key issues.

Measuring business influence in global environmental politics is methodologically challenging. Arguably, one can search for causal links between the exercise of instrumental power (lobbying and campaign finance) and policy outcomes (Falkner 2008). However, scholars need to be careful not to assume causal effects – that is, if the outcome of a policy process is favored by business, it does not necessarily indicate that corporate influence was a decisive factor in that outcome. As Falkner (2008: 31) points out, weak environmental regimes do not reflect solely the preferences of corporations but also "the inherent difficulties of multilateral policy-making…as well as the complex political trade-offs that societies have to make between environmental protection, technological innovation, economic development and poverty reduction."

Furthermore, tracing the links between non-relational forms of power, such as structural power and discursive power, and influence is even more complicated. In the view of Fuchs (2007: 140), discursive power is potentially the dimension of greatest significance in terms of corporate influence on policy. This is because it is the power not only to pursue interests but also to create them (Clapp and Fuchs 2009: 10). However, as Meckling (2011: 38) points out, public discourse emerges from a "cacophony of voices," not single actors. This is arguably the case with the rise of discourses on ecological modernization, corporate social responsibility and sustainability, which may very well "legitimize and consolidate the power of large corporations" (Banerjee 2008: 51) but have been promoted not only by corporations but also by other actors, including academics.

One of the earliest cases of very clear corporate influence in global environmental politics was the negotiation of the Montreal Protocol in 1987. The Protocol was a key step in the development of the regime to combat stratospheric ozone-layer depletion, which is widely considered the most successful international environmental regime (see Chapter 31). Scholars have documented how companies shifted from an initial stance of fierce opposition to any regulation of ozone-depleting substances like chlorofluorocarbons to the support of their phase-out, arguing that this transformation was critical to the ultimate success of the regime (Litfin 1994; Oye and Maxwell 1995; Levy 1997).

In the area of climate change, it has been argued that the absence of binding targets in the United Nations Framework Convention on Climate Change (UNFCCC) was a victory for business, although an incomplete one as they would have preferred that no treaty was adopted at all (Newell 2000; Falkner 2010; see Chapter 32). Corporate influence has also been linked with the refusal of the United States to sign the Kyoto Protocol (Falkner 2008). On the other hand, the vocal support for the Paris Agreement by a number of carbon majors was not enough to keep the Trump administration from withdrawing from it (Vormedal et al. 2020). At a broader level, corporations have arguably played a key role in shaping the discussions at the international level, which emphasize technological solutions and market-based climate policy (Newell 2008; Meckling 2011).

Although many scholars of global environmental politics focus their research exclusively on "environmental" regimes and conferences, there are other international forums with implications for environmental governance where corporations have influence. Rules on trade, intellectual property and foreign investment "serve to narrow the menu of regulatory choices open to governments" (Newell 2007: 75), and, unlike many multilateral environmental agreements, they are highly enforceable through mechanisms of binding dispute settlement (Tienhaara 2009). Many of the successes that corporations can claim from their lobbying efforts in the environmental realm relate to their ability to shift issues between forums and maintain a hierarchy in which norms on free trade and the protection of investments and property rights trump environmental ones (Sell 2009; see Chapter 24).

Finally, in addition to these instances of influence over the shape of regimes, it is important to look at those areas where global cooperation has not been achieved. The absence of any global convention on corporate accountability is particularly notable in this regard. There was a concerted effort in the 1970s and 1980s by the United Nations Commission on Transnational Corporations (UNCTC) to codify the duties of global corporations, including their environmental responsibilities (Hansen 2002; Correa and Kumar 2003). However, due to opposition from developed countries, an economic recession, and the debt crisis, the drive to adopt the UNCTC Code faded in the 1980s and the body was officially dismantled in 1992. Thereafter, the main venues for discussions on corporate accountability have been UN "Earth Summits." Corporations had a notable presence at the UN Conference on Environment and Development in Rio de Janeiro in 1992 and the World Summit on Sustainable Development (WSSD) in Johannesburg in 2002 and again in Rio in 2012 at the United Nations Conference on Sustainable Development (UNCSD or "Rio +20"). The Business Council for Sustainable Development (BCSD) reportedly drafted most of the language on business roles and responsibilities in the original Rio documents (Pattberg 2007: 89). Provisions on environmental regulations for business that had been developed by the UNCTC were not incorporated into the conference declaration, "Agenda 21"; instead, there was an exclusive focus on voluntary initiatives (Clapp 2005a; Newell and Levy 2006).

Efforts to create a convention on corporate accountability failed again at the WSSD (Newell and Levy 2006). In this case, it was Business Action for Sustainable Development (BASD) – formed by the ICC and the WBCSD and representing over 150 global corporations – that led the lobbying efforts against the convention (Clapp 2005a). In the lead-up to Rio+20, it was clear that the role that business plays in UN summits had, if anything, grown since the 2002 Summit in Johannesburg. An entire day of summit was designated as a "business day," and the "official United Nations coordinator of business and industry" for the conference was the BASD (BASD 2012).

While all of these examples suggest a strong role for business in shaping global environmental politics, much progress has been made in some areas despite the resistance of corporations. For example, the Cartagena Protocol on Biosafety entered into force even though there was strident opposition from biotech firms (Falkner 2009; see Chapter 41). Corporations have at times also overestimated the extent of their discursive power. Attempts to shape public discourse through public relations and advertising not only can fail but also can prove counterproductive. Newell (2007) notes that the efforts of biotech companies to make inflated claims about the ability of biotechnology to tackle poverty and food insecurity resulted in a backlash from NGOs and were generally viewed skeptically by the public (see Chapter 44). The same could be said of the attempts of mining and utilities companies to rebrand coal as "clean"; this spurred the launch of a large "Beyond Coal Campaign" by NGOs in the United States, as well as a number of fake advertisements that mocked the terminology "clean coal."

Private authority

In addition to influencing the development of rules by state actors, corporations also make and enforce rules of their own (Pattberg 2007). As Witte (2020: 169) notes "TNCs are at the end of increasingly complex and global supply chains, with higher sway over suppliers' environmental practices than governments through requiring compliance with private standards in contracts." Dauvergne and Lister (2013: 83) provide the example of Walmart which, in 2010, announced to its 100,000 major suppliers that it would be cutting 20 million metric tonnes of carbon emissions from its supply chain by 2015.

There are also a number of private and public–private regulatory efforts at the global level in the form of: reporting schemes (e.g., the Global Reporting Initiative); certification and labeling schemes (e.g., the Forest Stewardship Council, the Marine Stewardship Council); investor-governance networks (e.g., the Carbon Disclosure Project/CDP, Coalition for Environmentally Responsible Economies); sets of voluntary principles and codes of conduct (e.g., ICC Business Charter for Sustainable Development); partnerships (e.g., WSSD "type 2" outcomes) and standards regimes (e.g., International Association of Oil and Gas Producers Standards Solution, which drafts standards relevant to the industry for the International Organization for Standardization) (see Chapters 40 and 42; Cashore et al. 2004; Gulbrandsen 2004; Pattberg 2007; Levy et al. 2010; MacLeod and Park 2011; Pattberg et al. 2012).

Scholars have argued that these initiatives are a manifestation of "private authority" (Cutler et al. 1999; Hall and Biersteker 2002; Cashore et al. 2004). What distinguishes authority from mere decision-making power is that the exercise of power is considered legitimate (Cutler et al. 1999: 5). The rise of private authority has presented a challenge to traditional theories of International Relations (see Chapter 3) because legitimate power has traditionally been considered the sole purview of sovereign states (see Chapter 7). It has fed into broader debates about neoliberalism and the "retreat of the state," but it has also sparked scholarly interest in the accountability structures (or lack thereof) of these new forms of governance (Chan and Pattberg 2008). However, in terms of concrete environmental gains, the evidence thus far is decidedly mixed. For Dauvergne and Lister (2013: 135), the bottom line is that "Eco-business is not turning brands into sustainable companies. Nor will it solve the world's problems."

One sector where further research on private authority is merited is finance (Bowman 2015). Major banks, institutional investors and insurance companies are de facto banning certain activities (e.g., building new coal plants and mines) and projects in certain locations (e.g., oil drilling in the Arctic) by ruling out providing finance for them (Matthews and McCaffrey 2020). Action by these institutions is undoubtedly undertaken, in part, in response to activist campaigns (e.g., RAN et al. 2020). However, it also stems from a "business case" which suggests that is prudent for financial institutions to reduce their exposure to assets that will be "stranded" in a climate-constrained world.

Conclusion

Global corporations and, in particular, a small number of very large firms, have considerable power to influence the development of environmental policies at the local, national, and international level. If they perceive that they will be a loser if an environmental policy is adopted, they may decide to oppose it. Alternatively, if opposition is futile or counterproductive, they may adopt a hedging strategy to ensure that the most favorable form of policy will be adopted. If they perceive that they will be a winner if an environmental policy is adopted, they may actively support it. However, as engagement in the political process is costly, some firms may choose non-participation regardless of the costs or benefits of the policy. Corporations that do actively participate in political processes will often engage in direct lobbying ("instrumental" or "relational" power), but their capacity to influence governments extends far beyond this. The ability of corporations to frame the parameters of debate ("discursive power") is particularly important.

Efforts to improve the environmental performance of global corporations – from voluntary codes of conduct to certification and labeling schemes – have thus far failed to make substantial headway. As a consequence some scholars have called for a more radical re-think of the capitalist corporate model, which is centered on profit and growth (Banerjee 2008; Ihlen 2009). Newell (2020: 8) argues that there are "alternative models of ownership from social enterprises to mutuals, community owned enterprises, charities, B-Corps, and cooperatives that might warrant further attention from researchers and businesses to better understand the strategies by which environmentally motivated enterprises might scale up their positive impacts."

Some important work in this vein has already been undertaken (Phelan et al. 2012). For example, MacArthur (2016) explores cooperatives in the renewable energy sector in Canada. However, most research focuses on the impacts of such shifts for local or national environmental politics. The implications that a proliferation of business models would have on global environmental politics are not clear. It is quite possible that small B-Corps and community-owned enterprises would choose the "non-participation" strategy with respect to global negotiations, in light of the costs associated with engagement. On the other hand, they may form regional or transnational alliances in order to bolster their instrumental power. More research on such issues would be welcome.

It is also worth highlighting that the study of global environmental politics is increasingly moving away from a narrow focus on regimes. The broad question of whether global capitalism (as we know it) is compatible with environmental sustainability is being addressed through engagement with issues such as over-consumption (see Chapter 17; Dauvergne 2008) and a resurgence of concern about the limits to growth (Jackson 2009; Hickel 2020). Further research that connects these issues with the role of business in society would greatly enrich the debate.

Acknowledgment

This research was undertaken, in part, thanks to funding from the Canada Research Chairs Program.

References

Banerjee, B. (2008) "Corporate Social Responsibility: The Good, the Bad and the Ugly", *Critical Sociology*, 34(1): 51–79.

BASD (Business Action for Sustainable Development) (2012) "Key Role for Private Sector Acknowledged: Governments and Business Representatives Discuss Role of Private Sector During UN Intersessional Meeting in New York". Online. Available HTTP: <http://basd2012.org>

Bowman, M. (2015) *Banking on Climate Change*. Alphen aan den Rijn: Kluwer Law International.

Buckley, K. Newton, P., Gibbs, H., McConnel, I. and Ehrmann, J. (2019) "Pursuing Sustainability through Multi-Stakeholder Collaboration: A Description of the Governance, Actions, and Perceived Impacts of the Roundtables for Sustainable beef", *World Development*, 121: 203–217.

Cashore, B., Auld, G., and Newson, D. (2004) *Governing through Markets: Forest Certification and the Emergence of Non-State Authority*, New Haven, CT: Yale University Press.

Chan, S. and Pattberg, P. (2008) "Private Rule-Making and the Politics of Accountability: Analysing Global Forest Governance", *Global Environmental Politics*, 8(3): 103–121.

Clapp, J. (2002) "What the Pollution Havens Debate Overlooks", *Global Environmental Politics*, 2(2): 11–19.

Clapp, J. (2005a) "Global Environmental Governance for Corporate Responsibility and Accountability", *Global Environmental Politics*, 5(3): 23–34.

Clapp, J. (2005b) "Transnational Corporations and Global Environmental Governance", in P. Dauvergne (ed.) *Handbook of Global Environmental Politics*, Cheltenham: Edward Elgar.

Clapp, J. (2018) "Mega-Mergers on the Menu: Corporate Concentration and the Politics of Sustainability in the Global Food System", *Global Environmental Politics*, 18(2): 12–33.

Clapp, J. and Fuchs, D. (2009) "Agrifood Corporations, Global Governance, and Sustainability: A Framework for Analysis", in J. Clapp and D. Fuchs (eds) *Corporate Power in Global Agrifood Governance*, Cambridge, MA: MIT Press.

Climate Accountability Institute (2020) "Update of Carbon Majors 1965–2018", Press Release, 9 December 2020, https://climateaccountability.org/pdf/CAI%20PressRelease%20Dec20.pdf

Correa, C. and Kumar, N. (2003) *Protecting Foreign Investment: Implications of a WTO Regime and Policy Options*, London: Zed Books.

Cutler, A.C., Haufler, V., and Porter, T. (1999) "Private Authority and International Affairs", in A.C. Cutler, V. Haufler, and T. Porter (eds) *Private Authority in International Affairs*, Albany: SUNY Press.

Dauvergne, P. (2008) *The Shadows of Consumption: Consequences for the Global Environment*, Cambridge, MA: MIT Press.

Dauvergne, P., and Lister, J. (2013) *Eco-Business: A Big-Brand Takeover of Sustainability*, Cambridge, MA: MIT Press.

Degnarain, N. (2021) "Has Prince Charles' Nature Pledge Become A Platform For BP To Greenwash Oil Spills?", *Forbes*, 17 Jan.

Downie, C. (2017) "Fighting for King Coal's Crown: Business Actors in the US Coal and Utility Industries", *Global Environmental Politics*, 17(1): 21–39.

Falkner, R. (2005) "The Business of Ozone Layer Protection: Corporate Power in Regime Evolution", in D.L. Levy and P. Newell (eds) *The Business of Global Environmental Governance*, Cambridge, MA: MIT Press.

Falkner, R. (2008) *Business Power and Conflict in International Environmental Politics*, Basingstoke: Palgrave Macmillan.

Falkner, R. (2009) "The Troubled Birth of the 'Biotech Century': Global Corporate Power and Its Limits", in J. Clapp and D. Fuchs (eds) *Corporate Power in Global Agrifood Governance*, Cambridge, MA: MIT Press.

Falkner, R. (2010). "Business and Global Climate Governance: A Neo-Pluralist Perspective", in M. Ougaard and A. Leander (eds) *Business and Global Governance*, London: Routledge.

Fuchs, D. (2007) *Business Power in Global Governance*, London: Lynne Rienner.

Geels, F. (2014) "Regime Resistance against Low-Carbon Transitions: Introducing Politics and Power into the Multi-Level Perspective", *Theory, Culture & Society*, 31(5): 21–40

Greenpeace (2017) "Plastic Pollution – Why Coca-Cola Bears Responsibility", https://www.greenpeace.org/canada/en/story/492/plastic-pollution-why-coca-cola-bears-responsibility/

Gulbrandsen, L. (2004) "Overlapping Public and Private Governance: Can Forest Certification Fill the Gaps in the Global Forest Regime?", *Global Environmental Politics*, 4(2): 75–99.

Hall, R.B. and Biersteker, T. J. (2002) "The Emergence of Private Authority in the International System", in R.B. Hall and T.J. Biersteker (eds) *The Emergence of Private Authority in Global Governance*, Cambridge: Cambridge University Press.

Hansen, M. (2002) "Environmental Regulation of Transnational Corporations: Needs and Prospects", in P. Utting (ed.) *The Greening of Business in Developing Countries: Rhetoric, Reality and Prospects*, London: Zed Books.

Heede, R. (2014) "Tracing Anthropogenic Carbon Dioxide and Methane Emissions to Fossil Fuel and Cement Producers, 1854–2010", *Climatic Change*, 122: 229–241.

Hickel, J. (2020) *Less is More*. London: Penguin Random House.

Ihlen, O. (2009). "Business and Climate Change: The Climate Response of the World's 30 Largest Corporations", *Environmental Communication: A Journal of Nature and Culture*, 3(2): 244–62.

Jackson, T. (2009) *Prosperity without Growth: Economics for a Finite Planet*, London: Earthscan.

Jones, C. and Levy, D.L. (2007) "North American Business Strategies towards Climate Change", *European Management Journal*, 25(6): 428–40.

Lamb, W. et al. (2020) "Discourses of Climate Delay", *Global Sustainability*, 3: e17, 1–5. https://doi.org/10.1017/sus.2020.13

Lerner, S. (2019) "Waste Only: How the Plastics Industry Is Fighting to Keep Polluting the World", *The Intercept*, 20 July, https://theintercept.com/2019/07/20/plastics-industry-plastic-recycling/

Levy, D.L. (1997). "Business and International Environmental Treaties: Ozone Depletion and Climate Change", *California Management Review*, 39(3): 54–71.

Levy, D.L., Brown, H.S., and de Jong, M. (2010). "The Contested Politics of Corporate Governance: The Case of the Global Reporting Initiative", *Business & Society*, 49(1): 88–115.

Levy, D.L. and Egan, D. (1998) "Capital Contests: National and Transnational Channels of Corporate Influence on the Climate Change Negotiations", *Politics & Society*, 26(3): 337–61.

Levy, D.L. and Egan, D. (2003) "A Neo-Gramscian Approach to Corporate Political Strategy: Conflict and Accommodation in the Climate Change Negotiations", *Journal of Management Studies*, 40(4): 803–29.

Levy, D.L. and Prakesh, A. (2003) 'Bargains Old and New: Multinational Corporations in Global Governance', *Business and Politics*, 5(2): 131–50.

Litfin, K. (1994) *Ozone Discourses: Science and Politics in Global Environmental Cooperation*, New York: Columbia University Press.

Ma, B., Li, X., Jiang, Z. and Jiang, J. (2019) "Recycle More, Waste More? When Recycling Efforts Increase Resource Consumption", *Journal of Cleaner Production*, 206: 870–877.

MacArthur, J. (2016) *Empowering Electricity: Co-operatives, Sustainability, and Power Sector*, Vancouver: UBC Press.

MacLeod, M. and Park, J. (2011) "Financial Activism and Global Climate Change: The Rise of Investor-Driven Governance Networks", *Global Environmental Politics*, 11(2): 54–74.

Marshall, J. (2016) "Disordering Fantasies of Coal and Technology: Carbon Capture and Storage in Australia", *Energy Policy*, 99: 288–298.

Matthews, C. and McCaffrey, O. (2020) "Banks' Arctic Financing Retreat Rattles Oil Industry", *The Wall Street Journal*, 8 Oct. https://www.wsj.com/articles/banks-arctic-financing-retreat-rattles-oil-industry-11602157853

Meckling, J. (2011) *Carbon Coalitions: Business, Climate Politics and the Rise of Emissions Trading*, Cambridge, MA: MIT Press.

Meckling, J. (2015) "Oppose, Support, or Hedge? Distributional Effects, Regulatory Pressure, and Business Strategy in Environmental Politics", *Global Environmental Politics*, 15(2): 19–37.

Newell, P. (2000) *Climate for Change: Non-state Actors and the Global Politics of the Greenhouse*, Cambridge: Cambridge University Press.

Newell, P. (2001) "Managing Multinationals: The Governance of Investment for the Environment", *Journal of International Development*, 13: 907–919.

Newell, P. (2007) "Corporate Power and 'bounded autonomy' in the Global Politics of Biotechnology", in R. Falkner (ed.) *The International Politics of Genetically Modified Food: Diplomacy, Trade and Law*, Basingstoke: Palgrave Macmillan.

Newell, P. (2008) "The markeTization of Global Environmental Governance: Manifestations and Implications", in J. Park, K. Conca, and M. Finger (eds) *The Crisis of Global Environmental Governance: Towards a New Political Economy of Sustainability*, London: Routledge.

Newell, P. (2020) "The Business of Rapid Transition", *Wires Climate Change*, 2020: e670. https://doi.org/10.1002/wcc.670

Newell, P. and Levy, D.L. (2006) "The Political Economy of the Firm in Global Environmental Governance", in C. May (ed.) *Global Corporate Power*, London: Lynne Rienner.

Oreskes, N. and Conway, E. (2011) *Merchants of Doubt: How a Handful of Scientists Obscured the Truth on Issues from Tobacco Smoke to Climate Change*, New York: Bloomsbury.

Orsini, A. (2012) "Business as a Regulatory Leader for Risk Governance? The Compact Initiative for Liability and Redress under the Cartagena Protocol on Biosafety", *Environmental Politics*, 21(6): 960–979.

Österblom, H., Jouffray, J.-B., Folke, C., Crona, B., Troel, M., Merrie, A., et al. (2015) "Transnational Corporations as 'Keystone Actors' in Marine Ecosystems", *PLoS ONE*, 10(5): e0127533. https://doi.org/10.1371/journal.pone.0127533

Oye, K.A. and Maxwell, J.H. (1995) "Self-Interest and Environmental Management", in R.O. Keohane and E. Ostrom (eds) *Local Commons and Global Interdependence: Heterogeneity and Cooperation in Two Domains*, Newbury Park, CA: Sage.

Pattberg, P. (2007). *Private Institutions and Global Governance: The New Politics of Environmental Sustainability*, Cheltenham: Edward Elgar.

Pattberg, P., Biermann, F., Chan, S., and Mert, A. (eds) (2012) *Public–Private Partnerships for Sustainable Development. Emergence, Influence and Legitimacy*, Cheltenham: Edward Elgar.

Phelan, L., McGee, J., and Gordon, R. (2012) "Cooperative Governance: One Pathway to a Stable-State Economy", *Environmental Politics*, 21(3): 412–31.

Rainforest Action Network (RAN), BankTrack, Indigenous Environmental Network (IEN), Oil Change International, Reclaim Finance, and the Sierra Club (2020) *Banking on Climate Change: Fossil Fuel Finance Report 2020*, https://www.ran.org/wp-content/uploads/2020/03/Banking_on_Climate_Change__2020_vF.pdf

Schouten, G., and Glasbergen, P. (2011) "Creating Legitimacy in Global Private Governance: The Case of the Roundtable on Sustainable Palm Oil", *Ecological Economics*, 70(11): 1891–1899.

Schouten, G., Leroy, P., and Glasbergen, P. (2012) "On the deliberative capacity of private multi-stakeholder governance: The Roundtables on Responsible Soy and Sustainable Palm Oil", *Ecological Economics*, 83: 42–50.

Sell, S.K. (2009) "Corporations, Seeds, and Intellectual Property Rights Governance", in J. Clapp and D. Fuchs (eds) *Corporate Power in Global Agrifood Governance*, Cambridge, MA: MIT Press.

Sustainable Markets Initiative (2021) *Terra Carta*, https://www.sustainable-markets.org/terra-carta

Tienhaara, Kyla. (2009) *The Expropriation of Environmental Governance: Protecting Foreign Investors at the Expense of Public Policy*, Cambridge: Cambridge University Press.

Tienhaara, Kyla (2019) "NAFTA 2.0: Good or Bad News for the Environment?" *Earth System Governance*, 1: 1–4.

Tienhaara, K. and Cotula, L. (2020) "Raising the Cost of Climate Action? Investor-State Dispute Settlement and Compensation for Stranded Fossil Fuel Assets", International Institute for Environment and Development (IIED), London.

Tienhaara, K., Orsini, A., and Falkner, R. (2012) "Global Corporations", in F. Biermann and P. Pattberg (eds) *Global Environmental Governance Reconsidered*, Cambridge, MA: MIT Press.

Vormedal, I. (2008) "The Influence of Business and Industry NGOs in the Negotiation of the Kyoto Mechanisms: the Case of Carbon Capture and Storage in the CDM", *Global Environmental Politics*, 8(4): 36–65.

Vormedal, I., Gulbrandsen, L. and Skjærseth, J. (2020) "Big Oil and Climate Regulation: Business as Usual or a Changing Business?" *Global Environmental Politics*, 20(4): 143–166.

Wapner, P. (1996) *Environmental Activism and World Civic Politics*, Albany: SUNY Press.

Witte, D. (2020) "Business for Climate: A Qualitative Comparative Analysis of Policy Support from Transnational Companies", *Global Environmental Politics*, 20(4): 167–191.

14

Transnational actors

Nongovernmental organizations, civil society and individuals

Christian Downie

The modern era of global environmental politics coincided with contemporary scholarship on transnational actors. The 1972 United Nations Conference on the Human Environment took place in Stockholm one year after a special issue of *International Organization* journal was released entitled "Transnational Relations and World Politics" (Nye and Keohane, 1971). Since then, there has been a dramatic growth both in the involvement of transnational actors in environmental politics and research on their activities. The growing presence of transnational actors has been evident at the principal global environmental conferences. At the Stockholm Conference in 1972 some 170 nongovernmental organizations were present, in 1992 around 1,400 were registered at the Rio Earth Summit, 8,000 were at the World Summit on Sustainable Development in Johannesburg in 2002, and 9,856 were at the Rio+20 Summit held in 2012. It is estimated that transnational organizations generally grew from 2,795 in 1970 to 48,220 by 2010 (Andonova, 2011). By the time of the international climate negotiations in Paris in 2015, transnational actors were being incorporated as a core part of the negotiating process in a bid to boost the climate action of non-state actors (Hale, 2016).

The ubiquitous presence of transnational actors reflects the increasingly cross-border nature of environmental problems. It is widely recognized that across environmental issues, from climate change to biodiversity loss, that humans are now operating outside what scientists refer to as a "safe operating space" (Steffen et al., 2015). Indeed, such has been the level of human impact on the environment that scholars now conclude that we have entered a new geological epoch known as the Anthropocene (see Chapter 15; Biermann et al., 2012; Steffen et al., 2015). To address the scale of the environmental challenge global cooperation is needed. It is no surprise then that a wide range of transnational actors with varying motivations and pursuing different strategies have been a constant presence in the world of environmental politics. The aim of this chapter is to survey the role of these actors. It seeks to consider the types of transnational actors, their strategies and their influence across the field of environment politics. In doing so, this chapter moves beyond traditional debates about whether the rise of transnational actors requires that we replace a state-centered view of the world with a society-dominated view. Rather the discussion supports the view of many scholars in the field that global environmental problems cannot be solved without governments, and hence, networks of state and non-state actors are required.

DOI: 10.4324/9781003008873-17

The chapter proceeds as follows. The next section charts the evolving body of literature on transnational actors. It then considers three types of transnational actors – for-profit, non-profit and individual actors – and the role they have played in environmental politics. The focus is on civil society actors given the attention paid to corporate actors in the previous chapter. Drawing on this discussion, the chapter turns to reflect on the principal question most scholars seek to answer, that is, under what conditions do transnational actors influence policy outcomes? The chapter concludes with some reflections on how we are to understand the role of transnational actors in environmental politics and world politics more generally.

What are transnational actors? An evolving literature

In the international sphere, especially in the world of environmental politics, the growth in the number of transnational actors since the end of the Cold War has led to a burgeoning literature on their role and impact. Generally used to refer to nongovernment actors that organize in network forms across state borders, the term transnational actors includes for-profit actors such as multinational corporations (MNCs) and business associations (see Chapter 13) and non-profit actors such as environmental NGOs and advocacy networks (Bexell et al., 2010; Jönsson and Tallberg, 2010). Scholars in this tradition argue that "transnational relations matter in world politics" and that state behavior in international relations cannot be understood without taking account of the cross-boundary activities of sub-units of government and non-state actors (Risse-Kappen, 1995b: 280). In this view an intergovernmental approach to world politics is too narrow because it implies limited access to the international system, which "no longer holds true in many issue areas" (Keck and Sikkink, 1998: 4). In other words, we must look inside and outside state borders.

The concept of transnational actors, which came to prominence with the work of Robert Keohane and Joseph Nye in the 1970s, presented a direct challenge to the conventional view of realists and neorealists, among others, that the state is a primary actor in a system characterized by anarchy (Keohane and Nye, 1972) (see Chapter 3). This literature is based on a more substantive critique of intergovernmental approaches in arguing that states have lost control over non-state actors who can organize and move across national borders, be they individuals, multinational corporations or advocacy networks (Lake, 2008). However, the research agenda proffered by Keohane and Nye did not prosper in the short term, especially in the field of environmental politics, with much of the literature concentrating on the role of multinational corporations and economic issues (Keohane and Nye, 1972). In the 1980s, the dominance of neorealist approaches under the influence of Waltz (1979) and the intensification of the Cold War, meant that much scholarly work returned to focus on nation-states and security issues (Jönsson, 2010).

It was not until the 1990s, with Risse-Kappen's (1995a) volume, *Bringing Transnational Relations Back In*, that a renewed interest was taken in transnational actors. He defined transnational relations as "regular interactions across national boundaries when at least one actor is a non-state agent or does not operate on behalf of a national government or an intergovernmental organization" (Risse-Kappen, 1995a: 3). The 1990s also marked a turning point in the literature, with earlier disputes about whether transnational actors influence outcomes replaced by a focus on the conditions under which their influence is felt. Indeed, Risse-Kappen's volume set out the broad terrain for transnational relations research in asking: under what conditions do transnational coalitions and actors succeed or fail in changing the policy outcomes of states in a specific issue-area? In the study, the success of transnational actors was dependent on the domestic structure of the state and the role of international regimes.

In the years that have followed, the growing presence of transnational actors alongside a recognition that global environmental politics is highly fragmented, in part because there is no single world environment organization, has led scholars to interrogate the forms of transnational governance (Roger and Davergne, 2016). For example, scholars have employed a range of theoretical concepts to capture and analyze the role of actors that operate across national boundaries, such as fragmentation, regime complexity and polycentric governance, to name a few (Biermann et al., 2009; Keohane and Victor, 2011; Pattberg and Widerberg, 2016; Dorsch and Flachsland, 2017). One particular focus of this work has been to examine the relationships between actors that interact across national boundaries. As the so-called "governance triangle" highlights, state and non-state actors can pool their resources to perform governance functions across borders, achieving more than they could if they acted alone (Abbott and Snidal, 2009).

The nature of the types of relationships can vary considerably, from hierarchical, to competitive, or cooperative, for example. They can also be direct or indirect, and they can be formal, informal or ad hoc. Some studies have used the concept of enrolment to capture the breadth of these relationships (Downie, 2020). Others have focused on particular forms of relationships, such as transnational delegation in environmental politics, that is, the delegation of authority by states acting collectively to non-state actors, to examine when states choose non-state actors and for what purpose (Green, 2018). One popular conceptualization is orchestration, which seeks to capture a particular type of relationship, namely indirect relationships with soft forms of control (Abbott et al., 2015). For example, Hale and Roger (2014) have used the concept to unpack the different types of relationships that emerge when international organizations help transnational actors to resolve collective action problems.

Irrespective of the ultimate approach taken, almost all of this work is consistent with the contemporary view that rather than transnational actors replacing the state, new forms of transnational governance can reorganize relationships between state and non-state actors and with it the location of power, authority and accountability. In this context, there has also been a contemporary debate about the democratic legitimacy of international institutions, and whether the involvement of transnational actors, specifically civil society actors, offers a source of democracy (see Chapter 28). For instance, former UN Secretary Generals Boutros Boutros-Ghali and Kofi Annan have both argued that the participation of non-state actors in international institutions can help to reinvigorate such institutions and enhance their democratic legitimacy (Tallberg and Jönsson, 2010: 7–8). While not all scholars accept that transnational actors offer the solution, debates about the democratizing potential of these actors has been a common theme, especially among those concerned with transnational governance as opposed to transnational actors themselves (Bexell et al., 2010; Tallberg and Jönsson, 2010; Bäckstrand and Kuyper, 2017). For example, there has been considerable concern that as more decision-making authority moves from the national to the international and on to the transnational level, citizens become further detached from decision-makers and the democratic accountability deficit grows (Schleifer et al., 2019).

Three types of transnational actors

While there is no definitive typology within the literature on transnational actors, a distinction is typically made between for-profit actors and non-profit actors. In essence, this is a distinction based on motivations. For-profit actors, such as MNCs and various business associations, are primarily motivated by instrumental goals, normally the pursuit of profit for their owners or shareholders (see Chapter 13). Non-profit actors, on the other hand, such as epistemic communities (see Chapter 18), environmental NGOs and advocacy networks, often referred to as civil society, lay claim to a common good. In environmental politics this is commonly a precautionary approach to environmental protection (Risse, 2002; Jönsson, 2010; Oberthür et al., 2002). Of course, such distinctions are never perfect, and some scholars have challenged this distinction based on instrumental motivations (Sell and Prakash, 2004). For example, for-profit actors can also lay claim to a common good. In addition, some scholars categorize transnational actors according to their structure rather than their motivation (Oberthür et al., 2002). Nevertheless, for the purpose of this chapter actors will be distinguished based on their motivations in line with the majority of scholars.

For-profit actors

While most of the research on transnational actors in environmental politics is concerned with civil society actors, it is important to consider the role of transnational for-profit actors briefly here, not only because studies of MNCs shaped much of the early transnational relations literature, but also because MNCs tend to invest in environmentally sensitive areas, such as the energy sector. MNCs and business associations have been prominent players in international environmental discussions. Despite the limited voice of business at the Stockholm Conference in 1972, the influence of individual MNCs and business associations has been more than evident at the Rio Earth Summit and every major forum since. For example, the World Business Council for Sustainable Development (WBCSD), which first came to the fore in 1995 (after an earlier merger) and includes some of the largest and most powerful companies in the world, such as General Motors, DuPont, Deutsche Bank, Coca-Cola, BP and Wal-Mart, has been an active player in discussions on everything from climate change to biosafety (Clapp, 2005) (see Chapters 33 and 41).

While some for-profit actors outwardly support sustainable development and have worked cooperatively on environmental initiatives, in many critical cases business groups have either hedged their positions, or outright opposed initiatives, for example, to address climate change (Downie, 2014a; Downie, 2019; Brulle, 2018; Brulle, 2020; Cory et al., 2020). It is no surprise then that scholars have sought to explain under what conditions, transnational business actors have succeeded in limiting the ambition of global environment governance and eschewing the need for business regulation. The literature commonly converges on power and strategies (see Chapter 13). First is the power of business and the dimensions of its power (structural, instrumental and discursive). For example, in the tradition of the critical theories, some scholars focus on the dominant position of these actors in the global economy (see Chapters 4 and 24). When a group with the membership of the WBCSD stakes out its position, as it did, for example, in opposition to a global corporate accountability agreement, which was raised at multiple sustainable development forums, governments take notice (Clapp, 2005).

Second, almost all studies of for-profit actors point to business strategies, which are informed by their power. In the literature on transnational actors, two strategies stand out. One is lobbying, which reflects the instrumental power of these actors (Renckens, 2019) (see Chapter 12). In the case of climate change, for example, US-based groups such as the Global Climate Coalition (GCC) and the Climate Council, which largely represented the interests of fossil fuel companies, such as Exxon and Shell, had a well-documented strategy of combining domestic and international lobbying to thwart agreement in the lead up to the United Nations negotiations in Kyoto in 1997. As well as domestic lobbying in the United States, these groups organized across borders to form alliances with the Organization of the Petroleum Exporting Countries (OPEC), principally Saudi Arabia, who had a similar interest in seeing the negotiations stall (Leggett, 1999; Newell, 2000; Downie, 2014a). Another is attempts to promote green business ideologies and voluntary sustainability codes and guidelines. For example, the WBCSD, the International Chamber of Commerce, the International Business Leadership Forum and the World Economic Forum have all promoted such initiatives (Andonova and Mitchell, 2010). And transnational voluntary codes are often developed in conjunction with other state and non-state actors, as is the case with the Forest Stewardship Council (Lambin and Thorlakson, 2018).

Non-profit actors

Since the 1990s and the renewed interest in transnational actors, the vast majority of work has been concerned with civil society actors. One of the most influential attempts to analyze the effectiveness of these actors was Peter Haas's pioneering work on epistemic communities, which he defined as a "network of professionals with recognised expertise and competences in a particular domain and an authoritative claim to policy-relevant knowledge within that domain or issue-area" (Haas, 1992: 3). In his early work, Haas used the case of the Mediterranean Action Plan, a regime for marine pollution control in the Mediterranean Sea, to argue that countries that were most supportive of the plan were those in which the epistemic community had been strongest (Haas, 1989). Haas argued that the language of science is becoming a worldview that penetrates politics everywhere and hence could affect how states' interests are defined (see Chapters 7 and 18). This would be especially so in issue-areas with high complexity and uncertainty, though he recognized that there must be demand for such knowledge from policymakers. The empirical inquiries of Haas and others showed that the involvement of epistemic communities can promote organizational learning by helping to create shared understandings in their specialized field and hence to improve state cooperation (Braithwaite and Drahos, 2000; Raustiala and Bridgeman, 2007).

Perhaps the most influential epistemic community of the last 20 years in environmental politics has been the United Nations Intergovernmental Panel on Climate Change (IPCC). Established in November 1988 around a small core of scientific experts, its scientific assessments are considered to have been the catalyst for much of the diplomatic activity on climate change and, at least in the 1990s, for shifting the consensus among key policy elites (Boehmer-Christiansen, 1994; Agrawala and Andresen, 1999; Andresen and Agrawala, 2002; see Chapter 32). For example, a large study of the role of the United States and the European Union (EU) in the international negotiations found that the IPCC helped to establish a consensus among government leaders and policy elites that human influence was the cause of climate change, which did not exist in the 1980s. This was one of the reasons that the administration of President Clinton agreed to accept binding greenhouse gas emission targets in 1996 (Downie, 2014a), and a key reason why, 20 years later, states agreed to the Paris Agreement on climate change that, despite opposition from some, included an explicit intention to limit the global temperature increase to 1.5°C above pre-industrial levels (Ourbak and Tubiana, 2017).

Other scholars have focused on "transnational advocacy networks," which are formed around shared principled ideas, instead of scientific knowledge and expertise (True and Mintrom, 2001; Smith and Wiest, 2005; Tarrow, 2005; Schroeder, 2008). According to Keck and Sikkink (1998: 2), a transnational advocacy network "includes those relevant actors working internationally on an issue, who are bound together by shared values, a common discourse, and dense exchanges of information and services." They argue that these networks are created, for example, when domestic actors find their influence over a nation-state is blocked. Because transnational advocacy networks are not powerful in a traditional sense, they rely on persuasion or socialization. This entails more than reasoning with opponents, but also bringing pressure, arm-twisting, encouraging sanctions and shaming. Keck and Sikkink claim that these networks' influence derives from strategies of persuasion through the quick movement of information and the framing of particular problems, staging symbolic events, calling on powerful actors for leverage and holding states to account for international commitments (Keck and Sikkink, 1998). Importantly, the research on transnational advocacy networks directly addressed the question posed by Risse-Kappen (1995a), that is, under what conditions do advocacy networks have influence? Keck and Sikkink (1998) identify five stages of network influence: issue creation and agenda setting; influence on the discursive positions of states and international organizations, on institutional procedures, on policy change of target actors and on state behavior.

Much of the literature that has followed has sought to investigate the success transnational advocacy networks have had under such conditions when adopting different strategies. For some time now the consensus has been that these networks are most influential during the agenda-setting phase of the policy cycle (Finnemore and Sikkink, 1998). In the field of environmental politics, and across other issue-areas such as trade and human rights, there is considerable evidence to indicate that the use of strategic framing is a particularly successful strategy under such conditions (Joachim, 2003; Sell and Prakash, 2004). For example, in their study of international environmental negotiations, Corell and Betsill (2008) have highlighted the importance of issue framing during the course of the negotiation process (see Chapter 22). Others have pointed to the success that environmental NGOs have had using such strategies during the international climate change negotiations (Newell, 2000; Downie, 2014b).

However, within the tradition of the transnational relations literature, some scholars have argued that the research on epistemic communities and transnational advocacy networks remains wedded to the state-centric view of the world because these actors are only relevant in so far as they impact state policies. One of the leading proponents of this view, Wapner (1995), argues that the best way to think about transnational activists is through the concept of "world civic politics" where activists work to change conditions without directly pressuring states. For example, he uses the anti-whaling campaigns led by Greenpeace and the Sea Shepherds Conservation Society to argue that these actors disseminate an ecological sensibility not restricted to governments, but circulated throughout all areas of collective life (see Chapter 40). In highlighting the ways civil society groups seek to influence other non-state actors, such as MNCs and individuals, this work was a precursor to much of the scholarships that followed in the 2000s on transnational governance more generally, described above.

Nevertheless, most of the literature on transnational actors remains focused on the relations between state and non-state actors. This is further evident in recent studies that draw attention to the growth in transnational partnerships, which have proliferated in the areas of climate change and biodiversity, among other issue-areas. In essence, such partnerships represent "soft agreements between state and non-state actors on specific governance objectives and on means to advance them across borders" (Andonova, 2011: 2; see Pattberg and Widerberg, 2016). For example, the agreement between the World Bank and WWF to establish the Amazon Regional Protected Areas is one such partnership. While there remain serious reservations about the environmental effectiveness of some of these partnerships, the range of actors that now engage in global environmental politics, vertically across geographical and jurisdictional space, and horizontally across networks of state and non-state actors, has led some to suggest that we are witnessing a "rescaling of global environmental politics" (Andonova and Mitchell, 2010: 256). It also highlights how far the literature on transnational actors has moved beyond earlier debates about whether we need to replace a state-centered view of the world with a society-dominated view.

Individuals

Much less theorized in the transnational relations literature is the role of individuals. While individuals may not possess the institutional power they had in feudal and early modern times, as Braithwaite and Drahos (2000: 495) point out, "we must still be wary of an institutional analysis of TNCs, states, NGOs and business organizations that treat them as institutional actors, writing their enrolment by individuals out of the script." That said, to the extent that individuals are considered in the transnational literature, it is generally as non-profit actors motivated by a common good. Much of the work in environmental politics focuses on the relationship between individuals and civil society groups. For example, Tarrow (2005: 28) argues in his work on transnational activists that some of these individuals are "seeking the development of a global civil society or a world polity; but many others are people who are simply following their domestically formed claims into international society when these claims can no longer be addressed domestically." The stories of Chico Mendes, Wangari Maathai or Ken Saro are all instances of activists seeking to increase the awareness of local environmental problems by exploiting transnational networks. In doing so, the efforts of such individuals also have the potential to influence the ideas and norms of global environmental politics (Andonova and Mitchell, 2010: 623–4).

In the last decade, perhaps the best illustration has been the development of the fossil fuel divestment movement (Ayling and Gunningham, 2015). The creation of the Internet and the proliferation of social media, such as Facebook, YouTube and Twitter, has provided individuals with new opportunities to have an influence across borders via pathways that did not exist a decade ago. In 2012, environmental activist Bill McKibben authored a critique of the fossil fuel industry's contribution to climate change in the magazine *Rolling Stone*, which subsequently went viral (McKibben, 2012). In it he called for a movement to divest from fossil fuels and shortly after helped to establish 350.org, an environmental NGO seeking to do just that. Other environmental activists, such as Greta Thunburg, the Swedish teenager who launched the student strikes for climate change, which has now gone global, rely heavily on social media to link individuals, share information, and build a global grassroots movement for action on climate change.

Of course, individuals at the helm of international institutions have also proved important transnational actors. The leadership of Mostafa Tolba, former executive director of the United Nations Environment Programme (UNEP) during the negotiations on ozone depletion, is often pointed to as an example of the role individuals can play to further international efforts to address global environment problems (Braithwaite and Drahos, 2000: ch. 12). Likewise the literature on the Kyoto Protocol negotiations is almost unanimous in its praise for the role Ambassador Raúl Estrada Oyuela of Argentina played in bringing the negotiations to a successful conclusion. In a section of their book on the negotiations entitled "The Estrada Factor," Oberthur and Ott (1999: 54) claim that "the outcome of the Kyoto process cannot be fully understood without paying tribute to Chairman Estrada." Former American and European negotiators have also described how his use of the gavel at critical junctures was "brilliant," with one stating that he "stitched together a deal all by himself, it was unbelievable" (Downie, 2014a: 106).

Nevertheless, for the most part, scholars have and continue to focus on transnational networks, with the role of individuals therein generally consigned to that of anecdotes. This is not to say that individuals are unimportant as transnational actors, but it does mean that we know much less about under what conditions their actions have proved successful. It is to this general question that we now return.

Under what conditions do transnational actors influence environmental outcomes?

Broadly speaking, three sets of conditions that affect transnational actor influence can be identified. First is the domestic structure of the "target state." To affect state policies transnational actors have to access the political system of the target state and they must be able to contribute to the creation of "winning coalitions" within that polity (see Chapters 11 and 12). According to this logic, it follows that the more open the domestic policymaking process and the more pluralist the society, the easier it should be for transnational actors to access the decision-makers and build coalitions (Risse-Kappen, 1995a). For example, in the relatively open political systems of the United States and the EU, environmental NGOs and business groups have been particularly successful at infiltrating orthodox policy networks to affect state policies (Downie, 2014a; Boström et al., 2015) (see Chapters 8, 13 and 33). Further, recent studies have emphasized how state policies at the national level create incentives for nonstate actors in engage in complementary forms of transnational governance. For instance, the participation of non-state actors in transnational activities appears to be highest in states with strong national climate policies (Andonova et al., 2017).

However, there are limits to how much domestic conditions can explain. For one thing, as Keck and Sikkink have argued, "they cannot tell us why some transnational networks operating in the same context succeed and others do not" (Keck and Sikkink, 1998: 202). While many scholars have attempted to address this problem, particularly from a constructivist perspective (see Chapter 4), by looking at norms and ideas, at the very least it is clear that domestic conditions are not all that matter for assessing the impact of transnational actors (Risse, 2002).

Accordingly, a second set of international conditions are commonly identified in the literature. Research has shown that international regimes and institutions can facilitate the efforts of transnational actors by facilitating the formation of coalitions and legitimating their attempts to influence policy outcomes (see Chapter 9). For example, Risse-Kappen has argued that the more an issue-area is regulated by international norms of cooperation, the more permeable state boundaries become for transnational activities. He states that "highly regulated and cooperative structures of international governance tend to legitimize transnational activities and to increase their access to the national polities as well as their ability to form 'winning coalitions' for policy change" (Risse-Kappen, 1995a: 6–7). Others have gone as far as to suggest that the access international regimes and agreements grant to networks may be as important as the content of the agreement itself (Hafner-Burton et al., 2009: 573).

However, facilitating access does not equate to influence, nor is it certain that when access becomes more difficult the influence of transnational actors declines. As Risse (2002: 268) points out, "we probably need to differentiate among various phases in the international policy cycle." On this front, there is a consensus, as discussed, that transnational actors are most influential during the agenda-setting phase of the international policy cycle because of their capacity to affect ideas and norms. Studies in the field of environmental politics have also highlighted other conditions that warrant further research. For example, Betsill and

Corell (2008) suggest that environmental NGOs could be more influential when the political stakes of an international negotiation are relatively low or, for instance, that environmental NGOs may have greater difficulty exerting influence when there is a high level of contention over entrenched economic interests. In the course of prolonged international environmental negotiations, these ideas have been taken further to suggest that there are strategic opportunities for highly networked actors to influence state behavior depending on the elements of long negotiations (Downie, 2012; Downie, 2014b). Much of this work also recognizes that non-state actors should exploit the potential of the "two-level game" (Putnam, 1988), while at the same time building transnational coalitions, be it with states or other non-state actors (see Chapter 11).

Third, putting aside the structural conditions of the domestic and international sphere, the characteristics of transnational actors themselves will also mediate their influence. The knowledge and expertise of transnational actors, as we have seen with epistemic communities such as the IPCC, can be critical to creating shared understandings among policy elites about the nature of a problem (see Chapter 18). However, the influence of knowledge will also be dependent on the demand for it from other actors. In environmental politics, in particular, where many issues are characterized by high complexity and uncertainty, knowledge is likely to be a more powerful resource. In addition, as Keck and Sikkink (1998: 28) argue, the networks that transnational actors participate in will "operate best when they are dense, with many strong actors, strong connections among groups in the network, and reliable information flows." While this may be so, other studies have shown that coordination among environmental NGOs does not necessarily increase their influence (Corell and Betsill, 2008). Similarly, studies of for-profit actors have highlighted how factors internal to firms can propel or constrain their participation in transnational climate governance. For example, global businesses with a sustainability champion at the managerial or executive level are more likely to participate in voluntary climate action and disclosure (Hsueh, 2017).

Conclusion

Transnational actors are now central players in world politics. Over the last 50 years, for-profit actors, various civil society groups and individuals have all helped shape the modern era of global environmental politics (see Table 14.1). The literature on transnational actors no longer debates whether these actors matter, nor does it debate whether a state-centered view of the world should be replaced by a society-centered view. Instead scholarship focuses more squarely on the interactions between state and non-state actors, whether they be formally or informally organized, and their influence on governance outcomes. In particular, this literature is characterized by a focus on networks – networks of scientific experts, advocacy networks of environmental NGOs, or collaborative partnerships between state and non-state actors, to name a few examples. In this sense research on transnational actors is very much part of the discussion about "governance without government" (Börzel and Risse, 2010).

Accordingly, much of the recent scholarship is concerned with the conditions under which transnational actors influence policy outcomes. A question of continued importance, given comparisons of transnational activity across policy domains, highlights the prevalence of this form of activity in global environmental politics compared to other domains (Reinsberg and Westerwinter, 2019). As this survey shows, three sets of conditions can be broadly identified: domestic conditions, such as the political structure of the target state; international conditions, including the role of international institutions; and the characteristics of transnational actors themselves, such as the density of their networks.

Table 14.1 Transnational actors and influence

Type of actor	Conditions under which they have influence
For-profit (e.g., World Business Council for Sustainable Development, Global Climate Coalition, ExxonMobil) *Non-profit* (e.g., Intergovernmental Panel on Climate Change, Climate Action Network, WWF, Greenpeace) *Individuals* (e.g., Bill McKibben, Greta Thunburg, Chico Mendes, Wangari Maathai, Mostafa Tolba)	• Domestic conditions (e.g., political structure of the target state) • International conditions (e.g., role of international institutions) • Characteristics of transnational actors (e.g., density of transnational networks)

In pursuing these lines of inquiry, scholars continue to debate what role these actors might play in "democratizing global governance" (see Chapters 28 and 29). Even a cursory glance of journals in the fields of international relations, global governance and environmental politics highlights this as an important area of scholarship (Cerny, 2009; Bäckstrand and Kuyper, 2017). This chapter has not focused on this broader issue because the primary concern has been with transnational actors in environmental politics rather than transnational relations more generally. Nevertheless, it is clear that the outcomes of these debates will be critical for transnational actors given the underlying assumption of many scholars that they can contribute to democracy and that better global governance is a cornerstone of their legitimacy.

Further, in the era of the Anthropocene, where scientists have shown that human society is at increasing risk of crossing planetary boundaries that will destabilize the earth system on which we depend to exist, the power, authority and accountability of all actors could be quickly upended (Steffen et al., 2015). The effect could be to limit the space for transnational activity as states exercise their authority with renewed vigor, or it could be to increase the importance of transnational activity should some states lose control of their borders, for example, due to new migration flows forced by climate change. Either way, charting the presence of transnational activity and the conditions under which these different actors have influence will be critical. To what extent this occurs is the province of future research, but the study of transnational actors in such a setting is likely to offer important theoretical insights for scholars that extends far beyond the world of environmental politics.

References

Abbott, K. & Snidal, D. (2009). The Governance Triangle: Regulatory Standards Institutions and the Shadow of the State. *In:* Mattli, W. & Woods, N. (eds.) *The Politics of Global Regulation.* Princeton: Princeton University Press.

Abbott, K. W., Genschel, P., Snidal, D. & Zangl, B. (2015). Orchestration: Global Governance through Intermediaries. *In:* Abbott, K. W., Genschel, P., Snidal, D. & Zangl, B. (eds.) *International Organizations as Orchestrators.* Cambridge: Cambridge University Press.

Agrawala, S. & Andresen, S. (1999). Indispensability and Indefensibility? The United States in the Climate Treaty Negotiations. *Global Governance,* 5, 457–482.

Andonova, L., Hale, T. & Roger, C. (2017). National Policy and Transnational Governance of Climate Change: Substitutes or Complements? *International Studies Quarterly,* 61, 253–268.

Andonova, L. & Mitchell, R. (2010). The Rescaling of Global Environmental Politics. *Annual Review Environment and Resources,* 35, 255–282.

Andonova, L. B. (2011). Boomerangs to Partnerships? Explaining State Participation in Transnational Partnerships for Sustainability. *Conference on Research Frontiers in Comparative and International Environmental Politics.* Princeton University.

Andresen, S. & Agrawala, S. (2002). Leaders, Pushers and Laggards in the Making of the Climate Regime. *Global Environmental Change,* 12, 41–51.

Ayling, J. & Gunningham, N. (2015). Non-State Governance and Climate Policy: The Fossil Fuel Divestment Movement. *Climate Policy,* 17, 131–149.

Bäckstrand, K. & Kuyper, J. W. (2017). The Democratic Legitimacy of Orchestration: The UNFCCC, Non-State Actors, and Transnational Climate Governance. *Environmental Politics,* 26, 764–788.

Betsill, M. & Corell, E. (2008). Introduction to NGO Diplomacy. *In:* Betsill, M. & Corell, E. (eds.) *NGO Diplomacy: The influence of Nongovernmental Organizations in International Environmental Negotiations.* Cambridge: MIT Press.

Bexell, M., Tallberg, J. & Uhlin, A. (2010). Democracy in Global Governance: The Promises and Pitfalls of Transnational Actors. *Global Governance,* 16, 81–101.

Biermann, F., Abbott, K., Andresen, S., Backstrand, K., Bernstein, S. & AL., E. (2012). Navigating the Anthropocene: Improving Earth System Governance. *Science,* 335, 1306–7.

Biermann, F., Pattberg, P., Van Asselt, H. & Zelli, F. (2009). The Fragmentation of Global Governance Architectures: A Framework for Analysis. *Global Environmental Politics,* 9, 14–40.

Boehmer-Christiansen, S. (1994). Scientific Uncertainty and Power Politics: The Framework Convention on Climate Change and the Role of Scientific Advice. *In:* Spector, B., Sjostedt, G. & Zartman, I. W. (eds.) *Negotiating International Regimes: Lessons Learned from the United Nations Conference on Environment and Development (UNCED).* London: Graham & Trotman/Martinus Nijhoff.

Börzel, T. A. & Risse, T. (2010). Governance Without a State: Can It Work? *Regulation & Governance,* 4, 113–134.

Boström, M., Rabe, L. & Rodela, R. (2015). Environmental Non-Governmental Organizations and Transnational Collaboration: The Baltic Sea and Adriatic-Ionian Sea Regions. *Environmental Politics,* 24, 762–787.

Braithwaite, J. & Drahos, P. (2000). *Global Business Regulation.* Cambridge, Cambridge University Press.

Brulle, R. J. (2018). The Climate Lobby: A Sectoral Analysis of Lobbying Spending on Climate Change in the USA, 2000 to 2016. *Climatic Change.*

Brulle, R. J. (2020). Networks of Opposition: A Structural Analysis of U.S. Climate Change Countermovement Coalitions (1989–2015). *Sociological Inquiry,* 0.

Cerny, P. G. (2009). Some Pitfalls of Democratisation in a Globalising World: Thoughts from the 2008 Millennium Conference. *Millennium-Journal of International Studies,* 37, 767–790.

Clapp, J. (2005). Transnational Corporations and Global Environmental Governance. *In:* Dauvergne, P. (ed.) *Handbook of Global Environmental Politics.* Cheltenham: Edward Elgar.

Corell, E. & Betsill, M. (2008). Analytical Framework: Assessing the Influence of NGO Diplomats. *In:* Corell, E. & Betsill, M. (eds.) *NGO Diplomacy: The influence of Nongovernmental Organizations in International Environmental Negotiations.* Cambridge: The MIT Press.

Cory, J., Lerner, M. & Osgood, I. (2020).Supply Chain Linkages and the Extended Carbon Coalition. *American Journal of Political Science,* early version published online.

Dorsch, M. J. & Flachsland, C. (2017). A Polycentric Approach to Global Climate Governance. *Global Environmental Politics,* 17, 45–64.

Downie, C. (2012). Toward an Understanding of State Behavior in Prolonged International Negotiations. *International Negotiation,* 17, 295–320.

Downie, C. (2014a). *The Politics of Climate Change Negotiations: Strategies and Variables in Prolonged International Negotiations,* Cheltenham: Edward Elgar.

Downie, C. (2014b). Transnational Actors in Environmental Politics: Strategies and Influence in Long Negotiations. *Environmental Politics,* 23, 376–394.

Downie, C. (2019). *Business Battles in the US Energy Sector: Lessons for a clean Energy Transition.* London: Routledge.

Downie, C. (2020). Steering Global Energy Governance: Who Governs and What Do They Do? *Regulation & Governance,* n/a.

Finnemore, M. & Sikkink, K. (1998). International Norm Dynamics and Political Change. *International Organization,* 52, 887–917.

Green, J. F. (2018). Transnational Delegation in Global Environmental Governance: When Do Non-State Actors Govern? *Regulation & Governance,* 12, 263–276.

Haas, P. (1989). Do Regimes Matter? Epistemic Communities and Mediterranean Pollution Control. *International Organization,* 43, 377–403.

Haas, P. (1992). Introduction: Epistemic Communities and International Policy Coordination. *International Organization,* 46, 1–35.

Hafner-Burton, E., Kahler, M. & Montgomery, A. (2009). Network Analysis for International Relations. *International Organization,* 63, 559–592.

Hale, T. (2016). "All Hands on Deck": The Paris Agreement and Nonstate Climate Action. *Global Environmental Politics,* 16, 12–22.

Hale, T. & Roger, C. (2014). Orchestration and Transnational Climate Governance. *The Review of International Organizations,* 9, 59–82.

Hsueh, L. (2017). Transnational Climate Governance and the Global 500: Examining Private Actor Participation by Firm- Level Factors and Dynamics. *International Interactions,* 43, 48–75.

Joachim, J. (2003). Framing Issues and Seizing Opportunities: The UN, NGOs, and Women's Rights. *International Studies Quarterly,* 47, 247–274.

Jönsson, C. (2010). Capturing the Transantional: A Conceptual History. In: Jönsson, C. & Tallberg, J. (eds.) *Transnational Actors in Global Governance: Patterns, Explanations, and Implications.* London: Palgrave Macmillan.

Jönsson, C. & Tallberg, J. (eds.) (2010). *Transnational Actors in Global Governance: Patterns, Explanations, and Implications,* London: Palgrave Macmillan.

Keck, M. & Sikkink, K. (1998). *Activists Beyond Borders: Advocacy Networks in International Politics,* Ithaca: Cornell University Press.

Keohane, R. & Nye, J. (eds.) (1972). *Transnational Relations and World Politics,* Cambridge: Harvard University Press.

Keohane, R. & Victor, D. (2011). The Regime Complex for Climate Change. *Perspectives on Politics,* 9, 7–23.

Lake, D. (2008). The State and International Relations. In: Reus-Smit, C. & Snidal, D. (eds.) *Oxford Handbook of International Relations.* Oxford: Oxford University Press.

Lambin, E. F. & Thorlakson, T. (2018). Sustainability Standards: Interactions Between Private Actors, Civil Society, and Governments. *Annual Review Environment and Resources,* 43, 369–93.

Leggett, J. (1999). *The Carbon War: Dispatches from the End of the Oil Century.* London, Allen Lane.

Mckibben, B. (2012). Global Warming's Terrifying New Math. *Rolling Stone.*

Newell, P. (2000). *Climate for Change: Non-state Actors and the Global Politics of the Greenhouse.* Cambridge, Cambridge University Press.

Nye, J. S. & Keohane, R. O. (1971). Transnational Relations and World Politics: An Introduction. *International Organization,* 25, 329–349.

Oberthür, S., Buck, M., Müller, S., Pfahl, S., Tarasofsky, R. G. & Werksman, J. (2002). Participation of Non-Governmental Organisations in International Environmental Governance: Legal Basis and Practical Experience. Berlin: Institute for International and European Environmental Policy.

Oberthur, S. & Ott, H. (1999). *The Kyoto Protocol: International Climate Policy for the 21st Century.* Berlin: Springer.

Ourbak, T. & Tubiana, L. (2017). Changing the Game: the Paris Agreement and the Role of Scientific Communities. *Climate Policy,* 17, 819–824.

Pattberg, P. & Widerberg, O. (2016). Transnational Multistakeholder Partnerships for Sustainable Development: Conditions for Success. *Ambio,* 45, 42–51.

Putnam, R. (1988). Diplomacy and Domestic Politics: The Logic of Two-Level Games. *International Organization,* 42, 427–460.

Raustiala, K. & Bridgeman, N. (2007). Nonstate Actors in the Global Climate Regime. *Public Law and Legal Theory Paper Research Series,* Research Paper No. 07–29.

Reinsberg, B. & Westerwinter, O. (2019). The Global Governance of International Development: Documenting the Rise of Multi-Stakeholder Partnerships and Identifying Underlying Theoretical Explanations. *The Review of International Organizations.*

Renckens, S. (2019). The Instrumental Power of Transnational Private Governance: Interest Representation and Lobbying by Private Rule-Makers. *Governance,* 33, 657–674.

Risse, T. (2002). Transnational Actors and World Politics. In: Carlsnaes, W., Risse, T. & Simmons, B. (eds.) *Handbook of International Relations.* London: Sage.

Risse-Kappen, T. (1995a). Bringing Transnational Relations Back In: introduction. In: Risse-Kappen, T. (ed.) *Bringing Transnational Relations Back In: Non-State Actors, Domestic Structures and International Institutions.* Cambridge: Cambridge University Press.

Risse-Kappen, T. (1995b). Structures of Governance and Transnational Relations: What Have We Learned? *In:* Risse-Kappen, T. (ed.) *Bringing Transnational Relations Back In: Non-State Actors, Domestic Structures and International Institutions.* Cambridge: Cambridge University Press.

Roger, C. & Davergne, P. (2016). The Rise of Transnational Governance as a Field of Study. *International Studies Review,* 18, 415–437.

Schleifer, P., Fiorini, M. & Auld, G. (2019). Transparency in Transnational Governance: The Determinants of Information Disclosure of Voluntary Sustainability Programs. *Regulation & Governance,* 13, 488–506.

Schroeder, M. (2008). Transnational NGO Cooperation for China's Climate Politics. *Cambridge Review of International Affairs,* 21, 505–525.

Sell, S. & Prakash, A. (2004). Using Ideas Strategically: The Contest Between Business and NGO Networks in Intellectual Property Rights. *International Studies Quarterly,* 48, 143–175.

Smith, J. & Wiest, D. (2005). The Uneven Geography of Global Civil Society: National and Global Influences on Transnational Association. *Social Forces,* 84, 621–639.

Steffen, W., Richardson, K., Rockström, J., Cornell, S., Fetzer, I. & AL., E. (2015). Planetary Boundaries: Guiding Human Development on a Changing Planet. *Science,* 347, 1259855.

Tallberg, J. & Jönsson, C. (2010). Transnational Actor Participation in International Institutions: Where, Why, and with What Consequences? *In:* Jönsson, C. & Tallberg, J. (eds.) *Transnational Actors in Global Governance: Patterns, Explanations, and Implications.* London: Palgrave Macmillan.

Tarrow, S. (2005). *The New Transnational Activism.* New York, Cambridge University Press.

True, J. & Mintrom, M. (2001). Transnational Networks and Policy Diffusion: The Case of Gender Mainstreaming. *International Studies Quarterly,* 45, 27–57.

Waltz, K. (1979). *Theory of International Politics.* New York, Random House.

Wapner, P. (1995). Politics Beyond the State: Environmental Activism and World Civic Politics. *World Politics,* 47, 311–340.

Part IV

Ideas and themes in global environmental politics

15

The Anthropocene

Rethinking humanity's role in the earth system

Peter Stoett and Simon Dalby

Few concepts have risen in popular usage with the rapidity of the Anthropocene; and while it remains contentious and polarizing for some observers of current developments, there is widespread recognition of its significance. It could be argued that it is simply a new term coined to reflect a popular proposition: that humankind has significantly altered the biosphere, to the extent that earth systems are no longer evolving independent of that influence. Our indelible mark on the geology, biology, topography, hydrology and meteorology of the earth (to list just a few of the more obvious spheres) is increasingly evident and undeniable. The more penetrating issue, for scholars of global environmental politics, is whether a collective awareness of this condition, bundled into a conceptual package such as the Anthropocene, will actually change the way we approach nature, each other, and the planet as a whole.

In general, the Anthropocene, as a guiding concept, has come from a rather bad place: it stands as formal recognition not just of the tremendous impact of human ingenuity, but of ecological, economic and cultural imperialism, the folly of a utilitarian approach to nature, and other forms of both empirical and epistemological violence visited upon the earth (Todd 2015; Grove 2019); and it is characterized by the distinct threat of an unprecedented mass extinction that could easily include our own species (Kolbert 2014). It is as much an explicit recognition of the mess that humanity has made, as well as our obligation to clean it up, as it is an objective scientific designation. It mirrors or even repositions conceptions of stewardship derived from major religions in this limited manner. It is certainly an immediately political, and not just geological, term, inseparably linked with the thematic anxiety of a *crisis* peculiarly forged by modernity (Hamilton, Bonneuil, and Gemenne 2015).

While there are many research agendas flowing from the rise of the concept (see Hoffman and Jennings 2015), this chapter will examine the conceptual evolution of its employment within the context of global environmental politics; and develop the notion that the Anthropocene is emerging as a key guiding concept in related scholarly analysis. It is certainly a central and animating assumption of the Earth Systems Governance literature (see Chapter 21), for example. Indeed, journals are emerging with the word in their title. But are scholars interrogating the term? Harrington (2017) concluded several years ago that the concept has been "largely absent" from the study of international relations (IR). Is it a largely heuristic

DOI: 10.4324/9781003008873-19

device at this point, or is it seeping into not only mainstream literature but also related policy guidance? As importantly: is it also being used (quietly or with great enthusiasm from some) to render inevitable the need for geoengineering our way out of the climate crisis? And is one of the most important normative dimensions of the field of global environmental politics, the quest for justice in various forms (intra- and intergenerational; see Chapters 25 and 26) enhanced or diminished by the growth in popularity of thinking about the Anthropocene as a guiding contextual element?

The Anthropocene in global environmental politics: scholarly usage

While the Anthropocene has become a widely used term in popular culture and in some parts of the academy (Clark and Szerszynski 2021), especially in scholarly efforts explicitly focused on the Earth Systems Governance project (see Chapter 21), it has not been as prominent in the wider international relations literature. A recent survey found that the major journals in the field have been largely silent on the topic: "Apart from *Millennium,* no prominent IR journal, including *International Organization, International Security, World Politics* or the *Journal of Conflict Resolution,* has published an article [specifically] about the Anthropocene as of May 2019. The absence of the Anthropocene in the discussions of high-ranking IR journals suggests that the topic occupies a peripheral position in the discipline" (Simangan 2020, 215). This has not changed much since; *Millennium* remains a notable exception, having published papers by Dipesh Chakrabarty, Bruno Latour, Cameron Harrington and Scott Hamilton that addressed the theme and the related Planet Politics Manifesto (see below) (Burke et al. 2016).

Nonetheless, as Simangan (2020) documents, the term is spreading in the international relations field in other venues, including the *European Journal of International Affairs, Global Environmental Politics, Global Society* and elsewhere. Also noteworthy, though probably not surprising to those who have studied global environmental politics for any length of time, is the fact that much of the discussion comes from authors in developed states, and that the Anglosphere is over-represented; the appearance of Anthropocene-related papers mirrors the larger geographic pattern of funded research and other contingent factors. But clearly the discipline has been a laggard in taking up this theme, a point made forcefully in the 2016 "Planet Politics Manifesto" (Burke et al. 2016), which generated an acerbic critique (Chandler, Cudworth and Hobden 2018) and subsequent rejoinder (Fishel et al. 2018). The literature engaging the topic is growing, but as of early 2021 it was still fair to claim that mainstream IR had not undertaken many substantial investigations of matters that, given the disciplinary focus on international security, regime formation, and related concerns, might be expected. Though managing environmental collective action problems has been a popular theme, the more novel context of a disrupted earth system has not penetrated core disciplinary concerns.

In parallel, there is a growing sense that the term *environment* has outlived its usefulness as a term for policy formulation; *sustainability* is the increasingly popular term (Biermann 2020). Even the "environmental policy" paradigm is now fundamentally challenged by the Anthropocene discussion, not least because earth system science, in particular among contemporary academic research fields, has made notions of an external environment somehow separate from humanity increasingly untenable. Many of the discussions of environmental matters, biodiversity loss (see Chapter 41), and climate change (see Chapter 32) are within the ambit of the discipline although these studies are frequently not using the exact term Anthropocene, nor engaging with earth system science. Simangan (2020) notes how few of the abstracts at International Studies Association annual meetings have included the term. Simultaneously, however, the Anthropocene discussion has spun off new journals, such as *Anthropocene, Elementa, The Anthropocene Review* and *Earth's Future*, and generated novel academic research organizations. Most notably, the Future Earth and the Earth System Governance Project have attracted scholars in many disciplines interested in themes encompassed or even inspired by the Anthropocene term. Green thinking and environmental politics, frequently without an explicit IR dimension, have flourished (Nicholson and Jinnah 2016; Pattberg and Zelli 2016; Arias-Maldonado and Trachtenberg 2019) and especially so under the auspices of Earth System Governance (Biermann and Lovbrand 2019; Linner and Wibeck 2019; see Chapter 21).

If the Anthropocene as a concept is in essence just the continuation of pleas for recognizing the fundamental interconnection between humans and their environment, we are not moving much beyond the calls for more ecological thinking that have resonated in some corners of mainstream IR since the 1960s (see Laferrière and Stoett 1999 for a discussion preceding the rise of the term); but perhaps the magnitude of the concept effectively diminishes the more simplistic "back to nature" visions that emerged from both green authoritarianism and deep ecology. If, however, the Anthropocene signifies a scientific (or even paradigmatic) turn, one which renders past managerialist approaches inherently problematic, then we should see this reflected in global environmental political discourse and even diplomatic activity. But as yet this trend, if it is to happen, has been slow to emerge.

The Anthropocene in global political discourse

The term Anthropocene has had a very uneven appearance in matters of global environmental politics. One might argue that the term had already come of age when it appeared as the cover story of *The Economist* magazine in May of 2011, with a detailed description of the profound implications of the concept for rethinking many things in international politics. But the uptake was limited, and the Paris Climate Agreement of 2015 (see Chapter 32) is effectively silent on the term, where one might expect it to be a prominent overarching conceptualization of the new conditions of the planet that require the comprehensive approach to climate change that the Paris Agreement promised. The contemporaneous Sustainable Development Goals do not explicitly incorporate the Anthropocene formulation either.

In contrast, a few years later German Chancellor Angela Merkel highlighted the term Anthropocene in her high-profile speech to the 2019 Munich Security conference. In Merkel's terms, "This means that we are living in an age in which humankind's traces penetrate so deeply into the Earth that future generations will regard it as an entire age created by humans. These are traces of nuclear tests, population growth, climate change, exploitation of raw materials, and of microplastics in the oceans" (Merkel 2019). She linked the formulation of the Anthropocene back to Alexander von Humboldt's motto of "everything is interaction" and then used it as a framework to emphasize the interconnected contexts that presented the North Atlantic Treaty Organization with its security problems. Given the Chancellor's background as a scientist perhaps this is not surprising. Nonetheless, its formal use in such forums is relatively rare.

The United Nations Secretary General's major speech in December 2020 on "The State of the Planet," which outlined numerous serious environmental crises and the need for comprehensive efforts to counter climate change, biodiversity loss, and other threats to environmental and human security, did not explicitly use the term Anthropocene. Neither did the

January 2021 One Planet Summit focusing on biodiversity, although given the focus on such themes as the "Great Green Wall" and modes of global green (and blue) economic development, the summit's overarching view of biodiversity and global change certainly reflected themes encapsulated in prevalent discussions of the meaning and implications of the Anthropocene (Government of France 2021).

The silence on the theme of the Anthropocene in these high-profile political speeches is especially noteworthy because they occurred simultaneously with the publication of the United Nations *Human Development Report* of 2020 (United Nations Development Programme 2020), which adopts the concept of the Anthropocene as its overarching framework. This is not the first UN document to address the Anthropocene; the *UNESCO Courier* adopted the slogan from the front cover of the *Economist* seven years earlier – "Welcome to the Anthropocene" – in devoting its April–June issue in 2018 to the theme. But the 2020 *Human Development Report* was an altogether more ambitious treatment, endorsed by governments contributing to the United Nations Development Programme, and its more than 400 pages explicate why this formulation is key to understanding the contemporary context in which traditional notions of development have to be reconsidered.

The *Human Development Report* does not mince words about the failures of traditional approaches to the issues that have generally been classified as environmental problems, and it suggests that the conventional narratives around "solutions" to discrete "problems" have failed to grapple with the scope of development challenges. This is the case not least because in separating out discrete problems and suggesting that they are somehow external to humans and the technosphere of human industry and innovation, traditional approaches have not provided the appropriate contextualization:

> Once solutions are discovered, the storyline goes, we need only implement them as panaceas everywhere. Technology and innovation matter – and matter a lot, as the Report argues – but the picture is much more complex, much more non-linear, much more dynamic than simple plug-and-play metaphors. There can be dangerous unintended consequences from any single seemingly promising solution. We must reorient our approach from solving discrete siloed problems to navigating multidimensional, interconnected and increasingly universal predicaments.
>
> *(United Nations Development Programme 2020: 5)*

Understood in terms of the Anthropocene, and the "pressures" that the earth system is under, the report goes on to argue that the Covid-19 pandemic is only a warning about what is to come if human development is not oriented to deal with both inequities and those "pressures" that fossil-fueled economic "development" has unleashed. The encroachment of human activities into relatively "wild" spaces, habitat destruction and the use of wild animals as a food source are part of the expansion of the global economy and simultaneously the cause of increased dangers of future pandemics (IPBES 2020; see Chapter 24).

The 2020 *Human Development Report* emphasizes the point that environmental pressures are having the most deleterious consequences for poor and marginal peoples. These trends are now magnified by the setbacks and dislocations, especially for women, caused by both the pandemic and attempts to slow its spread. Many of the gains in terms of human development are being lost by the simultaneous impacts of disease and disasters. This, the report forcefully argues, requires a fundamental rethink, and the overarching formulation of the Anthropocene offers a way to grapple with the scale of the human transformation of the planetary system and its implications for humanity in future decades.

Earth system science and existential risk

In Earth System Science terms, we know that large-scale organized arrangements of humans can flourish in the relatively stable ecological situation of the Holocene Epoch. In comparison, prior glacial periods were interspersed by brief warm periods, and then the return of glacial conditions. While humans survived and spread across the globe in these earlier circumstances, it is not clear that such volatility is conducive to large-scale human civilization. While urban arrangements in tropical forests may predate the spread of grain and domestic animal-based agriculture in the Holocene, and suggest a longer trajectory of small-scale anthropogenic environmental changes than was recognized until recently (Taylor et al. 2020), the major expansion of the human population occurred within Holocene parameters. This is the climatic and ecological "sweet spot" that we know we can flourish within (Steffen et al. 2015); the radical alterations to earth systems that is currently underway as a consequence of modern ecological imperialism and the global extension of industrial combustion have already pushed the earth system beyond the parameters of the Holocene and into a new set of geological circumstances beyond the historical conditions of glaciation that have dominated the planet while humans have existed. We are in a new, rapidly changing set of circumstances (Lewis and Maslin 2018).

Whether geochronological timekeepers decide that humanity continues to live in the Meghalayan Age of the Holocene Epoch (which stabilized in temperature roughly 10,000 years ago) or a new construction called the Anthropocene, is not what is most important. Regardless of the accepted geochronological designation the important points are that since the last global retreat of glacial ice about 12,000 years ago, the global climate has been remarkably stable in comparison to the previous hundreds of thousands of years when glaciers advanced and retreated repeatedly. But now this stable period is being replaced by a much more dynamic and less predictable earth system (Zalasiewicz et al. 2019). Debates on the starting date of the Anthropocene – the beginning of agriculture (and thus land use change), the advent of the industrial revolution (especially, the start of extensive coal burning), the so-called Great Acceleration (the post-World War II era with its massive expansion of the global economy, automobiles and petroleum use; see Crutzen 2002; Ruddiman 2003; Zalasiewicz et al. 2015) – have largely been overtaken by the acceptance of the term as an apt description of the current state of affairs. When the Anthropocene will end, however, remains anyone's guess. Are we at the start of a deep-time era, or will the Anthropocene prove a mere blip in geological time? The rapid reduction of biodiversity, the emergence of new sedimentary structures, such as the "plastiglomerates" where plastic waste is being consolidated into what will be new rocks in the long-term future (Corcoran, Moore and Jazvac 2014; Gerhardt 2018), novel biological entities formed around plastics (Amarel-Zettler, Zettler, and Moncer 2020) and the long-term spike in carbon dioxide, methane and other greenhouse gases in the atmosphere, suggest a very long-lasting impact on Earth's history.

While individual civilizations have flourished and fallen (in some cases essentially committing environmental auto-destruction [see Diamond 2005]), *homo sapiens* has generally thrived, proving to be a highly adaptive invasive species of the first order. In the past substantial human populations have been eliminated by disease outbreaks, and by the repeated violence of conquest, but the unprecedented expansion of humanity over the last few generations suggests extraordinary, by historical standards, if markedly unequal, recent success for humans. However, the current global biodiversity crisis (see Chapter 41), climate change (see Chapter 32) and the persistent dangers of a major nuclear war present novel hazards to humanity, threatening it with catastrophe. These are potentially existential risks in so far as human extinction is a possibility (Ord 2020). The success of the human species (as the apex biological invader) has come at the price of destabilizing many of the ecological conditions that permitted the great successes of human civilization, and current conditions present potentially catastrophic risks and are related quite directly to the construction of political arrangements of world order. Human use of combustion in its varied forms, to both change materials and human circumstances, is a new geophysical element in the earth system that needs to be understood as an earth-changing novelty (Dalby 2018). Existential risk prevention is, in international relations terms, about global security in the Anthropocene (Bostrom 2013); global environmental governance is now central to international relations even if the traditional study of the topic has been slow to understand this new contextualization (Simangan 2020).

The diversity of empirical problems presented in the Anthropocene should not be lost: although biodiversity and climate change tend to receive most of the attention in global environmental discussions, major ecological threats to human existence include various perturbations to the earth system, such as possible future ocean anoxic episodes where much of oceanic life dies due to widespread oxygen depletion or the spread of smaller dead zones resulting from the profligate use of potash fertilizer in terrestrial agricultural systems (Baum and Handoh 2014). Humanity has managed to cope with the stratospheric ozone depletion crisis (see Chapter 31) caused by the widespread use of chlorofluorocarbons, a very notable and entirely necessary success in global environmental governance, but one that appears hard to emulate. Hence the importance of thinking in terms of the larger canvass of the Anthropocene period, not just threats from climate change, even if it is the most high-profile risk (Wallace Wells 2019). International policy studies that are focused on more specific transnational environmental problems may not contextualize themselves within the Anthropocene problematique, but they can contribute to the larger discussion even if formulated in different terms.

Ecomodernism and international relations

So, too, can age-old questions that have animated IR theory, such as the causal role of human beings – as both individuals and collectivities. As Christopher Bonneuil reminds us, human agency is at the heart of the concept of the Anthropocene. Indeed, the term is "an anomaly in the stratigraphic nomenclature: until now, geological divisions were named after flora and fauna composition, not after any causal agent" (Bonneuil 2015:19). This novel appreciation of the human context requires much more all-encompassing thinking than traditional environmental discussions have usually entailed; it is about much more than preserving a given external nature from human, or more specifically industrial, depredations. Relatedly, however,

> Both ecomodernists and would-be engineers describe the Anthropocene as a new evolutionary moment … in which human beings are at last in the driving seat both of human and natural history. [This] fosters hot humility but arrogant hubris of the kind that recalls the cosmological assumptions of the Baconian vision of science as redemption. [But] if rising sea levels inundate cities and ports, and droughts destroy much presently viable cropland, the Anthropocene will turn out to be an era in which human power over nature is greatly reduced … nature will have wrested back control…
>
> *(Northcott 2015:104)*

Recent earth system research strongly suggests that the eco-modernist formulation of solutions as a matter of technical fixes, market mechanisms, pollution controls, and the rhetoric of sustainable development, is simply not adequate to the task of shaping the increasingly artificial world that the global economy is making in ways that will allow a relatively stable configuration of the earth system.

The term *technosphere* is increasingly used by Earth System Science to highlight these new elements in the earth system made by the massive use of technology in the global economy (Donges et al. 2017). How this will be shaped is a key question for global environmental politics.

Human survival is going to require the flourishing of numerous other life forms (Erdelen 2020), beyond humans and those used for our food supply and pets. Slipping into the Anthropocene as signifier of our victory over nature not only replicates a false dualism but also reifies the arrogant tendencies that led to such a disagreeable state in the first place. We should not exaggerate the deep time period of human dominance; as Peter Brannon writes, "the most enduring geological legacy [of the Anthropocene] will be the extinctions we cause" (2019; see also Mitchell 2016) – including, perhaps, our own.

How this novel recognition of the scale of the contemporary difficulties and the need for a major transformation of economic activity might be formulated in terms of governance is now the key question. Clearly many traditional notions of environmental governance, useful in specific cases, are not anything close to adequate as a means of collectively shaping the future so that the earth system loosely approximates Holocene conditions. In Peter Dauvergne's (2016) terms, a selective "environmentalism of the rich" which constrains the worst excesses of immediately harmful pollution, but allows or promotes the overall growth of industrial disruption and ignores the destruction of numerous landscapes and their inhabitants, is not fit for purpose in these novel circumstances of Anthropocene transformations (see also Todd 2015).

If many environmental governance institutions are effectively "empty" or, worse, "decoy" entities that fail to accomplish their purposes (Dimitrov 2020) then a very substantial rethink is needed, one that goes beyond traditional notions of protecting particular ecosystems from industrial disruptions or imposing constraints on the emissions from particular industries. While there are now vast amounts of data available on environmental change, it is far from clear that this works to facilitate ecological governance rather than just feeding into processes of further exploitation (Bakker and Ritts 2018). Metrics and measurements do not necessarily translate into useful policies or political actions that are capable of tackling the questions of how to effectively shape the future trajectory of the technosphere.

Geoengineering the future

As suggested above, hovering over all these considerations is the looming issue of geoengineering, the intentional anthropogenic interference in earth systems, which will present unique collective action problems in international relations. In the face of rapidly rising climate change and growing weather-caused disruptions, the case for artificially modifying climate seems at least superficially compelling, and has attracted well-funded research (Khanna and Ferrari 2020), but it is currently subject to little more than "governance-by-default" (Talberg, Christoff, Thomas and Karoly 2018). Such initiatives raise numerous ethical and technical problems in addition to governance puzzles (Blackstock and Low 2019). Normally solar radiation management involving the artificial modification of the planet's albedo by (for example) injecting sulfuric droplets into the stratosphere, is distinguished from plans to reduce carbon dioxide levels in the atmosphere by reforestation and regenerative agriculture. The latter involve land-use management and farming conducted in novel ways to soak up carbon dioxide rather than emit it, and are less likely to be controversial in terms of governance. Solar radiation management – literally adjusting the amount of solar radiation reaching the surface of the earth – is much more likely to generate international conflict and is a matter for increasing consideration in terms of governance (Reynolds 2019).

There are strong doubts that solar radiation management could deliver what its advocates suggest (Pierrehumbert 2019); and once started, programs for earth cooling would have to be maintained indefinitely to prevent sudden disastrous climate disruptions should they be ended – the so-called termination problem. In terms of governance, there are also serious concerns that, should climate change continue to accelerate, and states or corporations attempt geoengineering unilaterally, the whole matter could become increasingly militarized and a source of international conflict (Surprise 2020). In terms of the Anthropocene formulation, the question this raises is whether the earth system discussion actually facilitates suggestions that geoengineering can offer technical solutions to environmental problems, or if a more comprehensive understanding of the Anthropocene, wherein the complexity of the dynamic interactions of the components of the earth system is foregrounded, will actually lead to caution in advocating strategies of artificial climate modification. The latter would seem to be necessary if science is taken seriously.

Similarly, other issue-areas, from plastic pollution (see Chapter 37) to biodiversity conservation (see Chapter 41) to melting permafrost to overfishing (see Chapter 40), elude quick, reliable, and easily implementable engineered fixes. If we cannot find universal technical solutions to the transnational problems that characterize the Anthropocene, or if the solutions we find are politically problematic, the pursuit of global environmental governance will remain a fundamental aspect of the survival of humankind. What this might mean for world order is not clear, but in rough outline it requires a quite literal focus on how the technosphere (and thus the planet) is shaped in coming decades (Haff 2014).

All this has been made more complicated by the destabilizations that human actions have already introduced into the earth system. It has been clear to at least hydrologists for some time that they can no longer use records of past meteorological conditions as a reliable guide for planning for the future. While weather fluctuates frequently, until recently, in most places, the range of past conditions was a reliable guide as to what could be expected in future years. This condition of "stationarity" where rainfall and temperature fluctuated within predictable ranges has been replaced by conditions of "non-stationarity" which is making planning practical things like how to construct dams and bridges much more difficult (Milly et al. 2008). A stable planetary context for world order can no longer be taken for granted. This is a profound shift, urgently demanding governance thinking about how to facilitate trajectories toward a stabilized earth system (Steffen et al 2018).

In Avin et al.'s (2018) terms, the conditions for grappling with global catastrophic risks involve thinking about critical systems whose boundaries might be breached, the mechanisms by which a threat to these boundaries might spread, and the manners in which human arrangements might fail to prevent or mitigate the threats. Given the interconnected ecological factors that humanity relies on, there are numerous linkages between systems that make it difficult to designate which of them are most critical. The planetary boundaries framework is designed to minimize the risks by preventing human activity that is approaching the thresholds for major dangerous system change. But where exactly the thresholds are, and how then to specify the appropriate boundary, remains a major research task within Earth System Science (Biermann and Kim 2020). That said, it is clear that dangerous thresholds for climate change and biodiversity loss are not in the far distance but are proximate if current trends are not reversed soon (IPCC 2018). Natural scientists have been repeatedly raising the alarm about the urgency and difficulties of dealing with numerous global environmental matters, but the conventional policy response, even in the aspirational statements in the Sustainable Development Goals, has been far short of what is needed (Dalby et al. 2019). All of which suggests the need for rethinking environmental governance, and extending the discussion very substantially. Using the Anthropocene as a lens through which to do this presents considerable potential.

Justice in the Anthropocene?

However, one of the limitations that Biermann (2020) and others have noted about the conventional environmental policy framework is its failure to grapple with issues of justice and equality. Global justice and human security (including the vital significance of indigenous peoples) cannot be overlooked if the Anthropocene is to mean anything beyond a reason for reifying great-power status and directing tax revenue toward geoengineering contracts. After all, we are not all equally responsible for the state of the global environment today. Indeed, one concern that is raised in an international context is a concern that the conceptual shift to a universalized era of the Anthropocene will replicate earlier "one earth" imagery that negates the historical climate and environmental injustice that has led to the current state of affairs:

> Global statistics, so central to the Anthropocene thesis, create the image of a global humanity united by carbon dioxide, thereby erasing the incommensurability of responsibilities. Indeed, a quick glance at carbon emissions data reveals that, up to 1980, the *anthropos* of the Anthropocene seems to have a very strong English accent ... Historically speaking, the Anthropocene could well have been called the Anglocene.
>
> *(Fressoz: 2015:70–71)*

Reinventing this Northern, or, perhaps better put, imperial view of the world – of the governance of the peripheries as processes driven by metropolitan decisions and technological design and local (often coerced) implementation – in order to define our understanding of the Anthropocene, will only compound the physical and epistemological violence that ushered in this era. This vision will be understandably despised by those caught in harm's way by the rising hazards of an increasingly disrupted world (Chaturvedi and Doyle 2015). Justice, both in terms of the recognition of the historical contribution to climate change by capitalist economies where industrialization conjoined a mode of production with fossil fuel combustion, and the current discrepancies in vulnerability to risk, has been a key theme in international climate negotiations, even if they frequently get downplayed by media and politicians in the North (see Chapter 25).

Generally, it is probably safe to assume that communities that have been subjected to the dispossession and destruction of habitats, which has been part and parcel of colonization (and, subsequently globalization [see Chapter 24]), will not be moved by expressions of existential alarm by the rich and powerful (Grove 2019; Stoett 2019). As a field of research distinct from mainstream IR, global environmental politics has always had a strong justice orientation in terms of gender, race, intergenerational legacy, North–South relations, and many other themes (see Chapters 25–27), and it will be interesting to see how the wider usage of the terminology of the Anthropocene will affect the continued conceptualization of these concepts. It is only fitting that the UN's *Human Development Report* referred to earlier includes a key section on "empowering people for equity, innovation, and stewardship of nature" and calls for nature-based development and "a just transformation that expands human freedoms while easing planetary pressures" (United Nations Development Program 2020: 10). The contemporary field of global environmental politics is animated primarily by this agenda, and its agents of change – scholars, activists, progressive politicians, international bureaucrats – now toil in the complex context of the Anthropocene.

Conclusion

While many artists and performers have embraced the concept of the Anthropocene as a lightning rod for creative inspiration for new modes of artistic expression, and scholars have embraced the term to explore interdisciplinarity (see, for example Kaya and Keane 2019), it has yet to have a significant impact on mainstream IR or on global environmental discourse (though the 2020 UN Human Development Report is a major exception). Indeed, the term has already been employed with widespread regularity in the social and natural sciences (a formal division that is increasingly weathered itself – see Mitchell 2015) to emphasize humanity's novel circumstances, yet the Anthropocene as a concept has done little more in IR than replicate earlier calls to reconceptualize the Westphalian state system to better confront collective action problems. We suspect the term will be employed with growing frequency in the study of global environmental politics, of course; whether this will spill over into the mainstream with any durability is less predictable.

The world political system is apparently stuck in a dysfunctional path dependency (Dryzek and Pickering 2019); territorial boundaries are genuine conceptual fences if, for example, climate adaptation is seen as a solely national matter (Benzie and Persson 2019). The living plethora of international arrangements, treaties, agreements, protocols and technical specifications related to industrial standards that shape the production and trading systems that drive the multitudinous commodities that constitute modern human life in consumption cultures suggests a semi-autonomous realm of global governance exists (Zurn 2018; see chapter by Kutting). Yet these arrangements are inadequate for either constraining the scale of fossil-fuel combustion or the widespread ecological changes that mark the emergence of the technosphere, involving at least 30 trillion tons of artificial materials which have polluted all parts of the atmosphere, hydrosphere, and much of the biosphere and lithosphere.

Formulating all of this in terms of the Anthropocene makes it clear that environmental policy can no longer be a matter of protecting parts of the world from industrial depredation, as much of traditional environmental governance has been about (Dalby 2020). It is a matter of much bigger considerations of climate, biodiversity and how to facilitate ecological flourishing in a world already being quickly transformed (Kareiva and Fuller 2016). Neither is discussing this agenda in terms of resilience, as in the sense of adaptations allowing for the return to a status quo ante after a disaster, the appropriate formulation for sensible policy in a world undergoing rapid change (Chandler, Grove and Wakefield 2020).

In one sense there is a useful precedent for consideration in this context: the mutual restraint around the use of nuclear weapons due to the condition of common endangerment (Deudney 2007). Political decision-makers have strong incentives for self-restraint in the field of nuclear weapons; the condition of mutually assured destruction makes the likely consequences of using nuclear weapons so catastrophic as to preclude their deliberate use as a mode of rational political action. Recent calls for a fossil-fuel non-proliferation treaty and for a ban on coal (Burke and Fishel 2020) are following this path, bringing arms-control formulations into the discussion of earth governance. These initiatives are suggesting that the climate emergency is being envisaged in terms that parallel the dangers of major nuclear war, and which hence need similar strategies to constrain dangerous trajectories.

The key challenge for environmental governance is to entice policy makers to act in ways that are appropriate for planetary boundaries. If the liberal international order has to a considerable degree been based on fossil-fueled globalization (see Chapter 24), what comes after it if climate change is taken seriously as a threat to humanity (Duncombe and Dunne 2018)? Put another way: can the institutions of the liberal order function to shape a relatively benign trajectory in the Anthropocene (Simangan 2021)? Linking the material circumstances of human life and community with institutional and political order seems to be key to finding a safe operating space, but the assumption of "continuationism" – that fossil-fueled economic growth can continue to be the premise for international arrangements – needs to be abandoned (Albert 2020). It must be replaced by a vision that embraces human diversity, progressive ingenuity, and an idea much older than that of the Anthropocene: reasonable and equitable limits to growth (see Chapter 16).

References

Albert, M. (2020). "Beyond Continuationism: Climate Change, Economic Growth and the Future of World (dis)order." *Cambridge Review of International Affairs* Latest articles.

Amarel-Zettler, L., Zettler, E. and Mincer, T. (2020). "Ecology of the Plastisphere." *Nature Reviews: Microbiology*, 18: 139–151.

Arias-Maldonado, M and Trachtenberg, Z. (eds) (2019). *Rethinking the Environment for the Anthropocene: Political Theory and Socionatural Relations in the New Geological Epoch.* New York: Routledge.

Avin, S., Wintle, B.C., Weitzdorfer, J., O'hEigeartaigh, S.S., Suthrenland, W.J. and Rees, M.J. (2018). "Classifying Global Catastrophic Risks." *Futures*, 102: 20–26.

Bakker, K and Ritts, M. (2018). "Smart Earth: A Meta-Review and Implications for Environmental Governance." *Global Environmental Change*, 52: 201–211.

Baum, S. D. and Handoh, I. C. (2014). "Integrating the Planetary Boundaries and Global Catastrophic Risk Paradigms." *Ecological Economics*, 107: 13–21.

Benzie, M. and Persson, A. (2019). "Governing Borderless Climate Risks: Moving beyond Territorial Framing of Adaptation." *International Environmental Agreements: Politics, Law and Economics*, 19: 369–393.

Biermann, F. (2020). "The Future of 'environmental' Policy in the Anthropocene: Time for a Paradigm Shift." *Environmental Politics*, DOI: 10.1080/09644016.2020.1846958

Biermann, F. and Kim, R. (2020). "The Boundaries of the Planetary Boundary Framework: A Critical Appraisal of Approaches to Define a "Safe Operating Space" for Humanity." *Annual Review of Environment and Resources*, 45: 497–521.

Biermann, F. and Lovbrand, E., eds. (2019). *Anthropocene Encounters: New Directions in Green Political Thinking.* Cambridge: Cambridge University Press.

Blackstock, J.J. and Low, S., eds. (2019). *Geoengineering our Climate? Ethics, Politics, and Governance.* London: Routledge.

Bonneuil, C. (2015). "The Geological Turn: Narratives of the Anthropocene." In Hamilton, Bonneuil, and Gemenne (2015). 17–31.

Bostrom, N. (2013). "Existential Risk Prevention as Global Priority." *Global Policy*, 4(1): 15–31.

Brannon, P. (2019). "The Anthropocene is a joke: on geological timescales, human civilization is an event, not an epoch." *The Atlantic*, August 13: https://www.theatlantic.com/science/archive/2019/08/arrogance-anthropocene/595795/

Burke, A. and Fishel, S. (2020). "A Coal Elimination Treaty 2030: Fast Tracking Climate Change Mitigation, Global Health and Security." *Earth System Governance*, 3. 100046.

Burke, A., Fishel, S., Mitchell, A., Dalby, S. and Levine, D. (2016). "Planet Politics: A Manifesto from the End of IR." *Millennium*, 44(3): 499–523.

Chandler D., Cudworth, E. and Hobden, S. (2018). "Anthropocene, Capitalocene and Liberal Cosmopolitan IR: A Response to Burke et al.'s 'planet politics'." *Millennium: Journal of International Studies*, 46(2): 190–208.

Chandler, D., Grove, K. and Wakefield, S. eds. (2020). *Resilience in the Anthropocene: Governance, and Politics at the End of the World.* New York: Routledge.

Chaturvedi, S. and Doyle, T. (2015). *Climate Terror: A Critical Geopolitics of Climate Change.* New York: Palgrave Macmillan.

Clark, N. and Szerszynski, B. (2021). *Planetary Social Thought: The Anthropocene Challenge to the Social Sciences.* Cambridge: Polity.

Corcoran, P., Moore, C. and Jazvak, K. (2014). "An Anthropocentric Marker Horizon in the Future Rock Record." *Geological Society of America Today*, 26(4): 4–8.

Crutzen, P. (2002). The Geology of Mankind." *Nature*, 415: 23

Dalby, S. (2018). "Firepower: Geopolitical Cultures in the Anthropocene." *Geopolitics*, 23(3): 718–742.

Dalby, S. (2020). *Anthropocene Geopolitics: Globalization, Security, Sustainability.* Ottawa: University of Ottawa Press.

Dalby, S., Horton, S., Mahon, R. and Thomaz, D. eds. (2019). *Achieving the Sustainable Development Goals: Global Governance Challenges.* London: Routledge.

Dauvergne, P. (2016). *Environmentalism of the Rich.* Cambridge MA, MIT Press.

Deudney, D. (2007). *Bounding Power: Republican Security Theory from the Polis to the Global Village.* Princeton: Princeton University Press.

Diamond, J. (2005). *Collapse: How Societies Choose to Fail or Succeed.* New York: Viking.

Dimitrov, R. (2020). "Empty Institutions in Global Environmental Politics." *International Studies Review*, 22(3): 626–650.

Donges, J.F., Lucht, W., Muller-Hansen, F. and Steffen, W. (2017). "The Technosphere in Earth System Analysis: A Coevolutionary Perspective" *The Anthropocene Review*, 4(1): 23–33.

Dryzek, J.S. and Pickering J. (2019). *The Politics of the Anthropocene.* Oxford: Oxford University Press.

Duncombe, C. and Dunne, T. (2018). "After the Liberal Order" *International Affairs*, 94(1): 25–42.

Erdelen, W.R. (2020). Shaping the Fate of Life on Earth: The Post-2020 Global Biodiversity Framework *Global Policy*, doi: 10.1111/1758-5899.12773

Fishel, S., Burke, A., Mitchell, A., Dalby, S. and Levine, D. (2018). "Defending Planet Politics." *Millennium*, 46(2). 209–219.

France, Government of. (2021). One Planet website: https://www.oneplanetsummit.fr/en/our-approach-125

Fressoz, J-B. (2015). "Losing the Earth Knowingly: Six Environmental Grammars Around 1800." In Hamilton, Bonneuil, Gemene (2015), 70–83.

Gerhardt, C. (2018). "Plastic and the Pacific: Midway Atoll, Plastiglomerate and Love of Place." *Mosaic*, 51(3): 123–140.

Grove, J. (2019). *Savage Ecology: War and Geopolitics at the End of the World.* Durham, NC: Duke University Press.

Haff, P.K. (2014). "Humans and Technology in the Anthropocene: Six Rules" *The Anthropocene Review*, 1(2): 126–136.

Hamilton, C., Bonneuil, C. and Gemenne, F. eds. (2015). *The Anthropocene and the Global Environmental Crisis: Rethinking Modernity in a New Epoch.* Routledge Environmental Humanities Series.

Harrington, C. (2017). "The Ends of the World: International Relations and the Anthropocene." *Millennium: Journal of International Studies*, 44(3): 478–498.

Hoffman, A. and Jennings, P.D. (2015). "Institutional Theory and the Natural Environment: Research In (And On) the Anthropocene." *Organization and Environment*, 28(1): 8–31.

IPBES. (2020). Workshop Report on Biodiversity and Pandemics of the Intergovernmental Platform on Biodiversity and Ecosystem Services. Daszak, P., das Neves, C., Amuasi, J., Hayman, D., Kuiken, T., Roche, B., Zambrana-Torrelio, C., Buss, P., Dundarova, H., Feferholtz, Y., Foldvari, G., Igbinosa, E., Junglen, S., Liu, Q., Suzan, G., Uhart, M., Wannous, C., Woolaston, K., Mosig Reidl, P., O'Brien, K., Pascual, U., Stoett, P., Li, H., Ngo, H. T., IPBES secretariat, Bonn, Germany, DOI:10.5281/zenodo.4147317

IPCC. (2018). *Global Warming of 1.5°C* https://www.ipcc.ch/sr15/.

Kareiva, P. and Fuller, E. (2016). "Beyond Resilience: How to Better Prepare for the Profound Disruption of the Anthropocene." *Global Policy*, 7(S1): 107–118.

Kaya, B. and Keane, J. (2019). *Creative Measures of the Anthropocene: Art, Mobilities, and Participatory Geographies*. London: Macmillan.

Khanna, P. and Ferrari, M. (2020). "Geoengineering Is the Only Solution to Our Climate Calamities." *Wired* December. https://www.wired.com/story/geoengineering-is-the-only-solution-to-our-climate-calamities/?utm_source=sendinblue&utm_campaign=Motherload__PKcom_Geoengineering_Is_the_Only_Solution_to_Our_Climate_Calamities&utm_medium=email

Kolbert, E. (2014). *The Sixth Extinction: An Unnatural History.* New York: Henry Holt.

Laferrière, E. and Stoett, P. (1999). *International Relations Theory and Ecological Thought: Towards Synthesis.* London: Routledge.

Lewis, S. L. and Maslin, M. A. (2018). *The Human Planet: How We Created the Anthropocene.* London: Penguin Random House.

Linner, B-O. and Wibeck, V. (2019). *Sustainability Transformations: Agents and Drivers across Societies.* Cambridge: Cambridge University Press.

Merkel, A. (2019). Speech to the Munich Security Conference https://www.bundesregierung.de/breg-en/chancellor/speech-by-federal-chancellor-dr-angela-merkel-on-16-february-2019-at-the-55th-munich-security-conference-1582318

Milly, P. C. D., Betancourt, J., Falkenmark, M., Hirsch, R. M., Kundzewicz, Z. W., Lettenmaier, D. P. and Stouffer, R. J. (2008). "Stationarity Is Dead: Whither Water Management?" *Science*, 319(5863): 573–574.

Mitchell, A. (2015). "Thinking Without the Circle: Marine Plastic and Global Ethics." *Political Geography*, 47: 77–85.

Mitchell, A. (2016). "Beyond Biodiversity and Species: Problematizing Extinction." *Theory, Culture, and Society.* 33(5): 23–42.

Nicholson, S. and Jinnah, S. eds. (2016). *New Earth Politics: Essays from the Anthropocene.* Cambridge, MA: MIT Press.

Northcott, M. (2015). "Eschatology in the Anthropocene: From the Chronos of Deep Time to the Kairos of the Age of Humans." In Hamilton, Bonneuil and Gemenne, eds., (2015). 100–111.

Ord, T. (2020). *The Precipice: Existential Risk and the Future of Humanity.* New York: Hachette.

Pattberg, P. and Zelli, F., eds (2016). *Environmental Politics and Governance in the Anthropocene: Institutions and Legitimacy in a Complex World.* New York: Routledge.

Pierrehumbert, R. (2019). "There is No Plan B for Dealing with the Climate Crisis." *Bulletin of the Atomic Scientists*, 75(5): 215–221.

Reynolds, J.L. (2019). *The Governance of Solar Geoengineering: Managing Climate Change in the Anthropocene.* Cambridge: Cambridge University Press.

Ruddiman, W. (2003). "The Anthropogenic Greenhouse Era Began Thousands of Years Ago." *Climatic Change*, 61: 261–293.

Simangan, D. (2020). "Where Is the Anthropocene? IR in a New Geological Epoch." *International Affairs*, 96(1): 211–224.

Simangan, D. (2021). "Can the Liberal International Order Survive the Anthropocene? Three Propositions for Converging Peace and Survival." *The Anthropocene Review* (Early view)

Steffen, W. et al. (2015). "Planetary Boundaries: Guiding Human Development on a Changing Planet." *Science*, 347(6223): 10.1126/science.1259855.

Steffen, W. et al. (2018). "Trajectories of the Earth System in the Anthropocene." *Proceedings of the National Academy of Sciences*, 115(33): 8252–8259.

Stoett, P. (2019). *Global Ecopolitics: Crisis, Justice, and Governance.* Second Edition. Toronto: University of Toronto Press.

Surprise, K. (2020). "Geopolitical Ecology of Solar Geoengineering: From a 'Logic of Multilateralism' to Logics of Militarization." *Journal of Political Ecology*, 27: 213–235.

Talberg, A., Christoff, P., Thomas, S. and Karoly, D. (2018). "Geoengineering Governance-by-Default: an Earth System Governance Perspective." *International Environmental Agreements: Politics, Law and Economics*, 18: 229–253.

Taylor, P., O'Brien, G. and O'Keefe, P. (2020). *Cities Demanding the Earth: A New Understanding of the Climate Emergency.* Bristol: Bristol University Press.

Todd, Z. (2015). "Indigenizing the Anthropocene." In H. Davies and E. Turpin, eds., *Art in the Anthropocene: Encounters Among Aesthetics, Politics, Environment and Epistemologies* London: Open Humanities Press, 241–254.

United Nations Development Programme. (2020). *The Next Frontier: Human Development and the Anthropocene.* New York: United Nations.

United Nations Secretary General. (2020). The State of the Planet. https://www.un.org/sg/en/content/sg/statement/2020-12-02/secretary-generals-address-columbia-university-the-state-of-the-planet-scroll-down-for-language-versions

Wallace-Wells, D. (2019). *The Uninhabitable Earth: Life After Warming.* New York: Tim Duggan Books.

Zalasiewicz, J., et al. (2015). "When Did the Anthropocene Begin? A Mid-Twentieth Century Boundary Level Is Stratigraphically Optimal." *Quaternary International,* 383: 196–203.

Zalasiewicz, J., Waters, C.N., Williams, M. and Summerhayes, C.P. (eds) (2019). *The Anthropocene as a Geological Time Unit A Guide to the Scientific Evidence and Current Debate.* Cambridge: Cambridge University Press.

Zurn, M. (2018). *A Theory of Global Governance: Authority, Legitimacy and Contestation.* Oxford: Oxford University Press.

16

Sustainability

From ideas to action in international relations

Jon Marco Church, Andrew Tirrell,
William R. Moomaw and Olivier Ragueneau

This chapter considers the origins of the concept of sustainability and its evolution toward becoming one of the defining concepts of our times. It then presents sustainable development as a means to develop society and the economy while protecting the environment and other life support systems in the long term. Finally, the chapter explores the global and regional practice of sustainable development and describes the scientific knowledge needed to pursue it, particularly sustainability science. Sustainability, however, does not exist in a vacuum. Other paradigms, such as human rights and resilience, often complement it, and sometimes compete against it, as movers of collective action in international relations.

In recent decades, a number of institutions and policy instruments have been crafted to promote sustainability, including multilateral environmental agreements, environmental assessments, global funds, the High-Level Political Forum to the Sustainable Development Goals (SDGs), and more (Haas et al. 1993; Young 2002). These endeavors to keep humanity within planetary boundaries and preventing ecosystem collapse have met with mixed results. Fulfilling the SDGs requires effective diplomacy around a number of "wicked" – that is, extremely difficult to solve – socio-ecological challenges.

Sustainable Development Diplomacy is a collection of approaches toward the negotiation and implementation of measures to achieve sustainability. Drawing on various disciplines – including international negotiations, global environmental governance, and socio-ecological systems – this chapter identifies core principles to assess ongoing sustainable development diplomacy (SDD) efforts toward achieving sustainability. To be successful, such approaches must incorporate diverse stakeholders and flexible solutions that match the growing complexity and scale of current sustainability challenges, while avoiding reliance on panaceas that are inadequate to the task.

The emergence of sustainability concerns

Human societies have always faced sustainability challenges. For example, settlements depend on a sustainable supply of water and wood, and successful management of livestock relied on avoiding overgrazing. Some traditional societies collapsed due to the unsustainable use of natural resources. Prominent cases include Easter Island in the Pacific due to deforestation

DOI: 10.4324/9781003008873-20

(see Chapter 42) and invasive species (Diamond 2011, but see DiNapoli et al. 2020 for a critique of this claim), and the Ancient Near East due to land degradation from overexpansion of farming and pastoral activities (Cordova 2020; see Chapter 44). Other traditional societies prospered thanks to the sustainable use of natural resources. For example, Venice carefully managed the lagoon and forests upon which it depended. However, it is with the industrial revolution in the eighteenth century that the modern concept of sustainability emerged in England and Germany. In fact, some saw that the development of industry depended on the availability of timber: sustainable extraction could not exceed the regeneration capacity of forests (Caradonna 2014). This led to the concept of sustainable yield, which was later applied to agriculture (see Chapter 44), fisheries (see Chapter 40) and other domains. Similar considerations appeared with respect to the dependency of population growth on food supply. In the late eighteenth century, this led Thomas Malthus to conclude that, given that food production increased more slowly than population growth, population would eventually shrink due to lack of food supply (Malthus 1798). The concept of limits to growth emerged from these perspectives (Meadows et al. 1972).

The westward expansion of the US frontier brutally devastated indigenous societies and rapidly degraded the lands that were seized from them. In the late nineteenth century, national parks were created in the western United States to protect remaining natural areas from destruction. In Europe, where the population was already dense and wilderness rare, national parks were often created from hunting reserves. A similar approach was used in colonial settings in Africa and other parts of the world. In fact, timber extraction was not the only threat to forests. Large-scale hunting and fishing led to the collapse of animal populations, such as the American bison, and sometimes to the extinction or near extinction of certain species, such as the Asian elephant. In the late nineteenth century, the emergence of nature conservation – the movement and science aimed at sustaining wildlife and later ecology – linked the protection of species and the protection of the ecosystems on which they depended (Matson et al. 2016). The acceleration of the industrial revolution in the late nineteenth and early twentieth centuries turned humanity into the major driver of ecological degradation on Earth (see Chapter 15). A series of natural disasters, such as the Dust Bowl that affected American prairies in the 1930s due to soil mismanagement, created societal awareness of human impacts on the environment. Concepts such as sustainable yield and nature conservation inspired action to reduce the negative impacts of human activity.

The later twentieth century witnessed a progressive increase of awareness about the international dimensions of environmental protection on issues such as transboundary waters (see Chapter 38), sea exploration (see Chapter 39), pollution (see Chapters 35–37), migratory birds (see Chapter 41), fisheries (see Chapter 40), the poles, international trade in endangered species (see Chapter 41) and wetlands (see Chapter 38). Action by such prominent organizations as the International Union for the Conservation of Nature (IUCN), the World Wildlife Fund (WWF) and the UNESCO Man and Biosphere (MAB) program contributed to the internationalization of research and the circulation of norms and ideas across borders, permeating even the Iron Curtain during the Cold War. These efforts culminated in the 1972 United Nations Conference on the Human Environment in Stockholm (see Chapter 22). A defining tension emerged in Stockholm, growing out of Indian Prime Minister Indira Gandhi's question: "Are not poverty and need the greatest polluters?" The Stockholm Conference exposed conflicts between those who defined the problem of sustainability as too much economic development across human societies, mostly in richer countries, and those who believed that the root cause of many sustainability issues was too little development, mostly in poorer nations. In subsequent years, this clash materialized as competition between the

United Nations Environment Programme (UNEP), which was created after Stockholm, and the United Nations Development Programme (UNDP), which had been created in 1965 and which remains to this day much larger than UNEP.

The invention and practice of sustainable development and its science

While publications such as the *Limits to Growth* report (Meadows et al. 1972) rekindled the Malthusian debate about the limits to population and economic growth in a context of limited resources, the economic turmoil that followed the oil crises of 1973 and 1979 undermined environmentalist momentum around the world, as priorities shifted to geopolitics, inflation and unemployment. Toward the end of the Cold War, however, the idea emerged that the environment and development were not antithetical to each other. If current development degrades the natural resources and other support systems that life depends on, it also must undermine future development. This led to the emergence of the concept of sustainable development in the 1980s. Through this concept, the United Nations saw an opportunity to reconcile the environment/development divide. In 1983, it established the World Commission on Environment and Development to encourage countries to embrace the concept of sustainable development. The resulting *Our Common Future* report from 1987 stated that:

> Environment is where we live; and development is what we all do in attempting to improve our lot within that abode. The two are inseparable. ... Humanity has the ability to make development sustainable: to ensure that it meets the needs of the present without compromising the ability of future generations to meet their own needs.
>
> *(World Commission on Environment and Development 1987: 14)*

This definition, which added an intergenerational dimension, is considered the starting point of mainstream sustainable development. However, many issues remained unclear, such as how far these considerations should reach into the future and who should make decisions for humanity. Moreover, this definition did not clarify whether needs should be limited to basic needs or if they should extend further (and if so, how much further). It also did not explain what should be sustained and what should be developed. By 2005, the Millennium Ecosystem Assessment, another report commissioned by the United Nations, tried to clarify some of these open questions. It identified human wellbeing and poverty reduction as what should be developed and ecosystem services (including their provisioning, regulating, supporting and cultural functions) as what should to be sustained. On this basis, Bill Clark provided the following working definition: "Sustainable development improves inclusive human wellbeing, while conserving the Earth's life support systems over the long run" (Clark 2010, 85).

Sustainable development must not, therefore, be confused with sustained economic growth. Sustainable development is a more holistic concept, including human development and wellbeing. Moreover, development must be inclusive of all, leaving no one behind (United Nations 2015). This understanding of sustainable development follows the emergence of the concept of socio-ecological systems and the study of ecosystems at all levels, from the local landscapes to the planet itself, which is increasingly marked by anthropogenic pressure. In the age of the Anthropocene (see Chapter 15), sustainable development considers all ecosystems to be socio-ecological systems. Conservation and traditional approaches to sustainability, on the other hand, focus on ecosystems and do not engage much with their social dimensions.

Sustainable development, therefore, must simultaneously focus on enhancing nature conservation and improving human wellbeing. (This perspective has been criticized as "soft" sustainability, compared to "hard" sustainability, which considers ecosystem services as not substitutable through technical solutions. The debate between hard and soft sustainability remains open.) This is no easy task, as socio-ecological systems are complex adaptive systems, which means that they face path dependency (i.e., the range of potential outcomes is constrained by previous events), behave in a non-linear manner (i.e., patterns may suddenly change) and possess different potential outcomes (i.e., the same input can produce different outputs). Given these challenges, scholars and practitioners have developed new methods to study complex adaptive systems and address wicked problems such as sustainable development (Levin et al. 2012). Much as agronomy emerged in the nineteenth century to help agriculture tackle the challenge of a growing population and economy, sustainability science emerged at the beginning of the new millennium with a goal of better understanding socio-ecological systems to guide sustainable development (Matson et al. 2016). Building on all branches of natural and social sciences (interdisciplinarity) and bringing together scientific research and knowledge of practice (transdisciplinarity), sustainability science is highly relevant for global environmental politics because it addresses planetary dynamics, such as climate change (see Chapter 32) and biodiversity loss (see Chapter 41). It also considers regional environmental processes through the study of large socio-ecological systems or ecoregions, such as transboundary rivers (see Chapter 38), mountain ranges and regional seas (see Chapter 39). Finally, it studies how socio-ecological systems are connected with each other from a distance (telecoupling), which is highly relevant to international trade (see Chapter 24).

Competing but complementary paradigms for collective action: human rights and resilience

By the 1992 Earth Summit in Rio de Janeiro, and affirmed at the 2002 World Summit on Sustainable Development in Johannesburg, sustainable development had become an organizing concept for the international community. Agenda 21, which was adopted at the Earth Summit, lists measures aimed at pursuing sustainable development. In fact, this new concept helped to ease the tensions between richer and poorer countries that had emerged in Stockholm and promote cooperation between developed and developing countries. Agenda 21 became such a cornerstone for United Nations activities in the new millennium that following up the Millennium Development Goals (MDGs) (United Nations 2000) – the blueprint for economic development from 2000–2015 – the UN created the Sustainable Development Goals (SDGs) (United Nations 2015), a list of 18 goals set at the Rio+20 conference in 2012 to guide international development through the year 2030.

Sustainable development has not been the only organizing concept for the international community in recent decades, however. After the end of World War II, human rights, as set forth in the 1948 Universal Declaration of Human Rights, emerged as a global organizing concept, and provided both justifications and a model for global action on sustainable development. For instance, after the 1992 Rio Conference, the United Nations set up a Commission on Sustainable Development in the mold of the United Nations Commission on Human Rights (they were replaced, respectively, by the High Level Political Forum on Sustainable Development in 2012, and the Human Rights Council in 2006). However, competition between sustainable development and human rights remains. In principle, the goals of sustainable development and human rights should not be mutually exclusive. In fact, since the achievement of human rights will be undermined in the long term if development is not sustainable, they should mutually reinforce each other. This is especially true for communities that are most vulnerable to the impacts of climate change, biodiversity loss and other sustainability challenges. At the same time, if human rights are not upheld, development cannot be sustainable because serious violations of human rights lead to backlash and societal disruption. In practice, however, human and financial resources are limited, sometimes creating competition between sustainable development and human rights actors. For instance, while in the last two decades funding to pursue sustainable development has steadily increased, the same cannot be said for human rights.

Another complementary paradigm is *resilience* (Berkes et al. 2003), which should not be confused with mere resistance to stresses and shocks. Like sustainability, it stems from the scientific literature on socio-ecological systems. In the words of Carl Folke, "Resilience reflects the ability of people, communities, societies, and cultures to live and develop with change, with ever-changing environments. It is about cultivating the capacity to sustain development in the face of change, incremental and abrupt, expected and surprising" (Folke 2016, 44).

Resilience and sustainability share many concepts, and consequently some organizations use the terms interchangeably. For example, while the whole of the United Nations, including UNDP, has used the SDGs as their framework for development assistance in recent years, one of UNDP's latest mottos was "empowered lives, resilient nations," and "build resilience to shocks and crises" was one of its three broad development settings from 2018–2021. Moreover, the United Nations Development Assistance Frameworks (UNDAF) were converted into the United Nations Sustainable Development Assistance Frameworks (UNSDAF) (ECOSOC 2014). Similarly, the Stockholm Resilience Center, one of the main hubs of resilience thinking, presents itself as "a reference point for research on global sustainability" (Stockholm Resilience Center 2018). Perhaps most tellingly, the words "sustainability" and "resilience" are often used in the same breath, as if they are synonymous (or at least share very close meanings). For instance, a chapter of the Special Report of the Intergovernmental Panel on Climate Change (IPCC) on *Managing the Risks of Extreme Events and Disasters to Advance Climate Change Adaptation* is entitled "Toward a Sustainable and Resilient Future," and the two concepts are used interchangeably throughout the document (IPCC 2012).

One final concept that is gaining currency alongside sustainability is *antifragility* (Taleb 2012). This term has emerged out of the many economic, political and health crises that the world has faced in recent decades. Nassim Nicholas Taleb defines antifragility this way:

> "Fragility" can be defined as an accelerating sensitivity to a harmful stressor: this response plots as a concave curve and mathematically culminates in more harm than benefit from random events. "Antifragility" is the opposite, producing a convex response that leads to more benefit than harm.
>
> *(Taleb 2013, 430)*

Antifragility invites us to shift our focus to the capacity of shocks and crises to make systems stronger, not weaker. For instance, while wildfires can be very destructive, they can also help to regenerate some forests, increasing their biodiversity. While antifragility may seem suitable for analyzing socio-economic dynamics, it goes against the grain of most of the literature from the fields of global environmental politics and sustainability science, which start from the assumption that shocks and crises threaten systems. It is not yet clear if the concept of antifragility will become a widely adopted paradigm guiding collective action on environmental issues.

Negotiating and implementing sustainability agreements

We live within a social and economic system that is proving to be less sustainable with each passing moment. Polar ice caps melt and raise sea levels; droughts, fires and intense storms threaten lives daily; local and global temperatures continue to break records; seas are becoming warmer and more acidic: and regional epidemics caused inter alia by human encroachment on nature morph into global pandemics. Yet a global economic system supported by government subsidies continues to reward firms that make and distribute the products and energy responsible for the environmental changes that we are observing (see Chapter 24). This system is destroying the natural world upon which all life, including that of the ever-growing human population, depends.

Despite a definition of sustainable development as "development that meets the needs of the present without compromising the ability of future generations to meet their own needs" (World Commission on Environment and Development 1987: 14), most humans pursue their own desires (regardless of need), degrading the capacity of the planet to support us all – a highly unsustainable outcome. The realization of this conundrum in the 1980s resulted in a process driven by nation-states for addressing unsustainable development practices in a collaborative manner (Keohane and Ostrom 1995).

Traditional diplomacy is often designed to avoid conflict. It frequently does so by defending the sovereignty and territoriality of individual nations. To reduce conflict among nations, treaties and other sources of international law establish rules of conduct to restrict aggression or interference within the boundaries of another nation. To promote economic development through increased growth, some treaties provide a set of provisions to promote free trade and encourage increased production and consumption with little consideration of environmental or social costs (see Chapters 17 and 24).

Since the 1970s, a wave of international agreements aimed at supporting more sustainable outcomes that increase human wellbeing, rather than promoting only economic growth, have come to fruition. The aforementioned Universal Declaration of Human Rights was one of the first instruments of this "New Diplomacy" (Najam, Christopoulou and Moomaw 2004), holding that the international community has an interest in transcending sovereign power to promote human welfare. Global environmental issues, including air and water quality (see Chapters 30 and 38), toxic substances (see Chapter 36), stratospheric ozone (see Chapter 31), climate protection (see Chapter 32), biological diversity (see Chapter 41) and oceans (see Chapter 39) also transcend territorial boundaries and agreements that regulate them are often at odds with national sovereignty.

One of the most successful legally binding sustainability regimes is the stratospheric ozone protection mandated by the 1987 Montreal Protocol and its multiple strengthening amendments (see Chapter 31). The Montreal Protocol was the first treaty of any type to have nearly universal membership. It resulted in the replacement not only of ozone-depleting substances but also, through the Kigali Amendments, substitutes that contribute to global warming. Banning (in a legally binding way) the trade of ozone-destroying substances was a significant factor in the protocol's success, as was the provision for an international fund to assist developing nations to shift to replacements used for refrigeration and air-conditioning.

Successful outcomes have also been achieved at times through non-binding agreements (Kanie and Biermann 2017). For example, the MDGs targeted reductions in poverty and hunger and increases in education, in order to improve health, gender equality, environmental sustainability and development. Progress was made toward several of these goals, including educational improvements, advances in vaccinations, bringing several global epidemics

Table 16.1 The Sustainable Development Goals categorized by their primary sustainable development dimensions

Economy	Environment	Society
SDG 1 No poverty	SDG 6 Clean water and sanitation	SDG 2 Zero hunger
SDG 8 Decent work and economic growth	SDG 7 Affordable and clean energy	SDG 3 Good health and wellbeing
SDG 9 Industry, innovation and infrastructure	SDG 11 Sustainable cities and communities	SDG 4 Education
SDG 12 Responsible consumption and production	SDG 13 Climate	SDG 5 Gender equality
	SDG 14 Life below water (oceans)	SDG 10 Reduced inequalities
	SDG 15 Life on land	SDG 16 Peace justice and strong institutions
		SDG 17 Partnerships for the goals

under control, and reducing poverty in many least-developed nations. The more comprehensive SDGs span all three dimensions of sustainable development: economy, environment and society. Four of them relate to the economy, six to environment and seven to society (see Table 16.1).

Developing a set of procedures to implement the Sustainable Development Goals depends greatly on establishing a successful diplomatic process to usher them in as global priorities. In short, if the SDGs are to be achieved, they will require advances in sustainable development diplomacy.

Solutions and challenges

Sustainable development diplomacy

One possible solution for the inadequacy of international action on global environmental problems is the emergence of sustainable development diplomacy (SDD). SDD is an approach that aims to create a response to socio-ecological challenges that transcends merely treating them as yet another subject of negotiation within international diplomacy. It recognizes not only that socio-ecological challenges have the potential to pose an existential threat to humanity, but also that they are intractably interwoven with nearly every sphere of human activity. SDD also recognizes that comprehensive solutions to environmental issues will invariably touch on profound questions of how we meet basic needs (food, water, health and safety), how we define what a "good life" is (wealth, happiness and ethics), and how we share not only the resources of the world, but also (and perhaps more importantly) the capacity of our planet to process the waste that humans are creating in ever-increasing quantities.

Sustainable development diplomacy differs from traditional diplomacy in three important ways. First, it embraces the aforementioned definition of sustainable development endorsed by the Brundtland Report. In other words, it is a development process that can endure into the indefinite future to meet societal needs, maintain an effective economic system that manages the exchange of goods and services and an environment that can continue to supply essential resources and other ecosystem services.

Second, SDD seeks to address underlying causes rather than treating symptoms. Understanding the social, economic and ecological aspects of sustainability alluded to above is a key element of treating root causes. Too many past sustainability efforts have failed to meet their full potential because they do not meet social or economic needs, and so many of the ecological challenges that we face are driven by economic and social policies that have discounted their true costs by ignoring the externalities they produce (often in the form of waste products). SDD calls for a full accounting of the costs of a given action, while still acknowledging that policies must meet human needs or be limited in their success.

Third, SDD is involved in all phases of the negotiation, implementation and enforcement processes of sustainability measures. That is, it engages in the complete life cycle of sustainability agreements and/or policies, and often takes an iterative approach. It moves from initial design, to monitoring and evaluation, and, not uncommonly, reassessment of what works, and how policy design can be improved in the continuing or next implementation. In this way, it is able to remain flexible and dynamic in order to meet the frequent shifts in the nature of the socio-ecological challenges we are facing, and incorporate improvements in our understanding of those challenges.

Fourth, SDD aims to be inclusive of stakeholders. Rather than focusing exclusively on state actors, SDD incorporates actors from the private sector, civil society, and subnational political units. Moreover, it also strives to achieve outcomes that confer benefits to this range of diverse stakeholders. This latter point is essential to creating diplomatic agreements that have broad support, making them easier to implement and enforce. Moreover, such holistic agreements are better equipped to work at different levels of scale and to flourish in the wide-ranging contexts where sustainability measures are implemented.

A justification for the inclusive approaches of the SDD approach can be found in Raworth's Doughnut Model (see Figure 16.1), based on the planetary boundaries laid out in "A Safe Operating Space for Humanity" (Rockström et al. 2009; see Figure 16.2). The model positions sustainability, described by Raworth as "the safe and just space for humanity," between a social foundation and an ecological ceiling. The social foundation sets a floor to ensure that human needs such as food, water, energy, education, health and equity are met, while the ecological ceiling is in place to ensure that humans do not blunder into disasters such as climate change, ocean acidification, biodiversity loss, and nitrogen and phosphorous overloads. Raworth describes the balance found between the social foundation and the ecological ceiling as a "regenerative and distributive economy" (Raworth 2012). By accounting for the economic, social, and ecological pillars of sustainability, and by including a broad range of stakeholders with interests that lie across these three domains, the SDD approach fosters conditions that help to keep humans within the safe and just space Raworth describes.

Sustainability challenges

One inherent challenge of sustainability is that, even when we take the best approaches, we often do not know if an action is truly sustainable until we see how it turns out. This highlights the importance of monitoring and evaluation, especially in an iterative approach that allows for redesign ranging from modest tweaks to complete overhauls. Moreover, even an initially successful sustainability measure can lose efficiency or effectiveness over time as circumstances change. Adaptability is absolutely essential in order to keep pace with changing conditions. However, in some cases, adapting may not be enough. It may not be sufficient

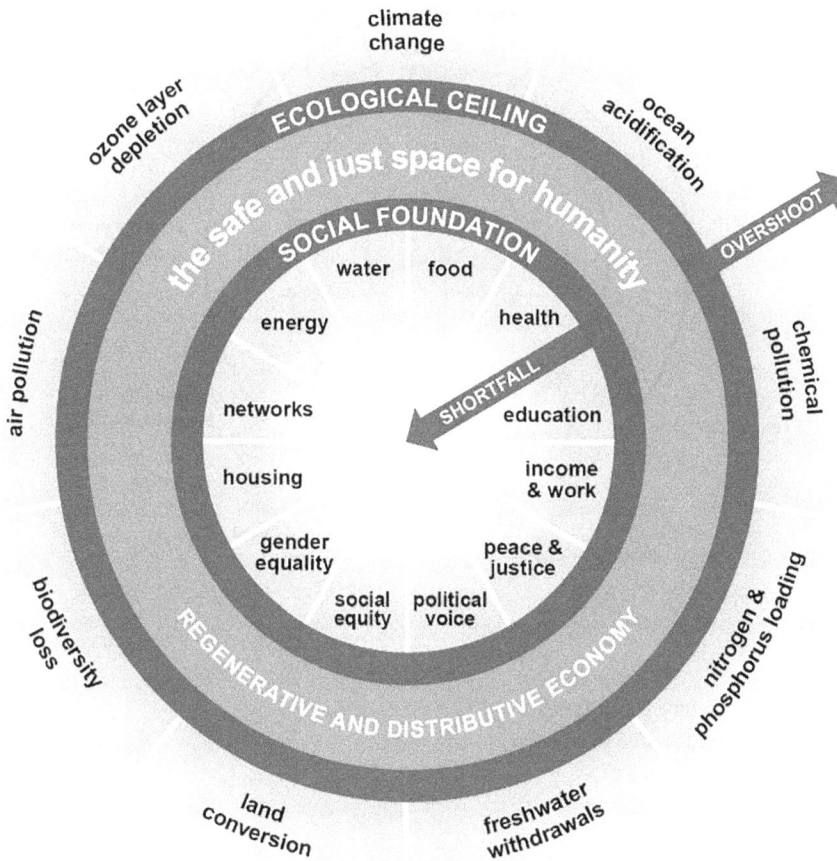

Figure 16.1 Raworth's Doughnut Model

Source: Raworth (2012) at https://commons.wikimedia.org/wiki/File:Doughnut_(economic_ model).jpg.

to embrace yet another socio-technical transition (i.e., a shift to a new technology), like our current transition from internal combustion engines to electric cars, so that we do not have to change our way of life. In some cases, we may need to embrace more radical solutions, so-called socio-ecological transformations, that change our relationship with the environment. In these cases, SDD might need to go well beyond business-as-usual and envision more profound change.

Less obvious, but equally important, are the connections between various levels of scale and diverse actors. It is often easier to spot policy weaknesses, or to realize gains in effectiveness, from the diverse vantage points enjoyed when a problem is tackled by multiple stakeholders, in numerous ways, and at different levels of society. Moreover, this varied approach builds in redundancies in problem-solving that can be invaluable when one or more efforts lose effectiveness. By creating systems that are adaptive and redundant we are borrowing some of the very best tools for resiliency found in nature.

Biosphere integrity
Climate change
Genetic diversity
Novel entities
Functional diversity
High risk
Increasing risk
Unsure
Safe
?
?
Land-system change
Stratospheric ozone depletion
?
Freshwater use
Atmospheric aerosol loading
Phosphorus
Nitrogen
Ocean acidification
Biochemical flows

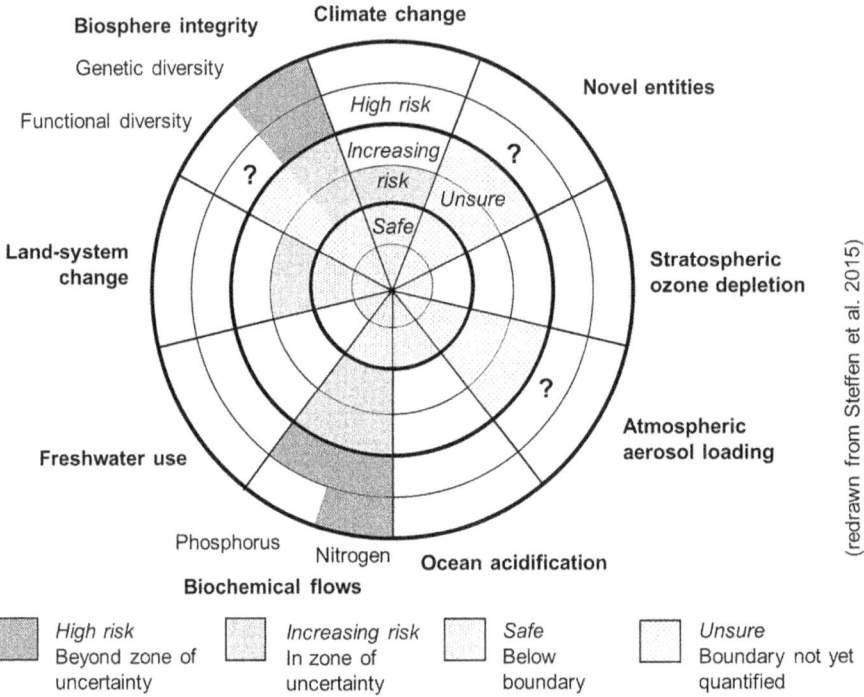

(redrawn from Steffen et al. 2015)

High risk
Beyond zone of uncertainty

Increasing risk
In zone of uncertainty

Safe
Below boundary

Unsure
Boundary not yet quantified

Figure 16.2 Planetary boundaries
Source: Redrawn from Steffen et al. (2015).

Conclusion

The ozone regime and its Montreal Protocol proved to be successful to address a problem where the science was clear, a technological solution was available and no country benefited from the status quo (Jasanoff and Martello 2004; see Chapter 31). Regimes such as climate, biodiversity, fisheries and forestry, both at the global and regional levels, tried to replicate this success, establishing a pernicious path dependency along a trajectory that cannot possibly work for every socio-ecological challenge. Unfortunately, we must accept that panaceas, silver bullets and one-size-fits-all solutions are not the answer, and instead we must craft bespoke solutions for the diverse and complex challenges that confront us (Ostrom et al. 2007; Webster 2017). Most sustainable development challenges are wicked problems, growing out of production and consumption patterns that connect places, people and socio-ecological systems (see Chapter 17). This means that action for sustainable development takes place in a context where knowledge is imperfect, evolving and asymmetrical (see Chapter 18). However, there is no need for perfect information to act. Research in sustainability science, and translating that knowledge into action (Kerkhoff and Lebel 2006) must go hand in hand with sustainable development diplomacy to overcome the many sustainability challenges that the world faces.

References

Berkes, F., Colding, J., and Folke, C. (Eds.) (2003). *Navigating Social-Ecological Systems: Building Resilience for Complexity and Change*. Cambridge: Cambridge University Press.
Caradonna, J.L. (2014). *Sustainability: A History*. New York: Oxford University Press.

Clark, W. C. (2010). Sustainable Development and Sustainability Science. In Levin, S. A., and Clark, W. C., (Eds.) *Toward a Science of Sustainability*. Center for International Development Working Papers 196. Cambridge, MA: Harvard University, John F. Kennedy School of Government, 82–104.

Cordova, C. E. (2020). The Degradation of the Ancient Near Eastern Environment. In Snell, D. C. (Ed.) *A Companion to the Ancient Near East* (2nd ed., pp. 65–84). Hoboken, NJ: Wiley Blackwell.

Diamond, J. (2011). *Collapse: How Societies Choose to Fail or Succeed*. New York: Penguin.

DiNapoli, R. J., Rieth, T. M., Lipo, C. P., and Hunt, T. L. (2020). A Model-Based Approach to the Tempo of "Collapse": The Case of Rapa Nui (Easter Island). *Journal of Archaeological Science*, 116: 105094.

ECOSOC. (2014). The Future We Want. The UN System We Need. *ECOSOC Dialogue on the Longer-Term Positioning of the United Nations Development System*. Papers by the Independent Team of Advisors to the ECOSOC Bureau. New York: United Nations.

Folke, C. 2016. Resilience (Republished). *Ecology and Society* 21(4):44.

Haas, P.M., Keohane, R.O., and Levy, M.A. (Eds.). (1993). *Institutions for the Earth: Sources of Effective International Environmental Protection*. Cambridge, MA: MIT Press.

IPCC. (2012). Managing the Risks of Extreme Events and Disasters to Advance Climate Change Adaptation. *A Special Report of Working Groups I and II of the Intergovernmental Panel on Climate Change*. Cambridge: Cambridge University Press.

Jasanoff, S., and Martello, M. L. (Eds.). (2004). *Earthly Politics: Local and Global in Environmental Governance*. Cambridge, MA: MIT Press.

Kanie, N., and Biermann, F. (Eds.). (2017). *Governance through Goals: New Strategies for Global Sustainability*. Cambridge, MA: MIT Press.

Keohane, R. O., and Ostrom, E. (Eds.). (1995). *Local Commons and Global Interdependence: Heterogeneity and Cooperation in Two Domains*. London: Sage.

Kerkhoff, L.v., and Lebel, L. (2006). Linking Knowledge and Action for Sustainable Development. *Annual Review of Environment and Resources* 31(1): 445–477.

Levin, K., Cashore, B., Bernstein, S., and Auld, G. (2012). Overcoming the Tragedy of Super Wicked Problems: Constraining Our Future Selves To Ameliorate Global Climate Change. *Policy Sciences* 45(2): 123–152.

Malthus, T. R. (1798). *An Essay on the Principle of Population*. London: Joseph Johnson.

Matson, P.A., Clark, W.C., and Andersson, K. (2016). *Pursuing Sustainability: A Guide to the Science and Practice*. Princeton, NJ: Princeton University Press.

Meadows, D.H., and Club of Rome. (1972). *The Limits to Growth: A Report for the Club of Rome's Project on the Predicament of Mankind*. New York: Universe Books.

Najam, A., Christopoulou, I., and Moomaw, W. R. (2004). The Emergent "System" of Global Environmental Governance. *Global Environmental Politics* 4(4): 23–35.

Ostrom, E., Janssen, M.A., and Anderies, J.M. (2007). Going beyond Panaceas. *Proceedings of the National Academy of Sciences* 104(39): 15176–15178.

Raworth, K. (2012). A Safe and Just Space for Humanity: Can We Live within the Doughnut. *Oxfam Policy and Practice: Climate Change and Resilience* 8(1): 1–26.

Rockström, J., et al. (2009). A Safe Operating Space for Humanity. *Nature* 461(7263): 472–475.

Steffen, W., Richardson, K., Rockström, J., Cornell, S. E., Fetzer, I., Bennett, E. M., Biggs, R., Carpenter, S. R., De Vries, W., De Wit, C. A., Folke, C., Gerten, D., Heinke, J., Mace, G. M., Persson, L. M., Ramanathan, V., Reyers, B. and Sörlin, S. (2015). Planetary Boundaries: Guiding Human Development on a Changing Planet. *Science* 347(6223): 1259855.

Stockholm Resilience Centre. (2018). A Reference Point for Research on Global Sustainability. www.stockholmresilience.org.

Taleb, N.N. (2012). *Antifragile: Things that Gain from Disorder*. New York: Random House.

Taleb, N.N. (2013). 'Antifragility' as a Mathematical Idea. *Nature* 494: 430

United Nations. (2000). *Millennium Declaration*. A/RES/55/2.

United Nations. (2015). *The Future We Want*. A/RES/66/288.

Webster, D.G. (2017). Scape Goats, Silver Bullets, and Other Pitfalls in the Path to Sustainability. *Elementa Science of the Anthropocene* 5: 7.

World Commission on Environment and Development. (1987). *Our Common Future*. Oxford: Oxford University Press.

Young, O. R. (2002). *The Institutional Dimensions of Environmental Change: Fit, Interplay, and Scale*. Cambridge, MA: MIT Press.

17

Consumption

Institutions and actors

Gabriela Kütting

Consumption is an economic and social activity that can be analyzed as being at the end of the production chain but also as a political and economic institution. It is a generally accepted fact that the world consumes at a rate exceeding the planet's capacity to regenerate itself. This is why consumption is such an important topic and at the heart of the global environmental politics concern. Academics who study consumption address the type of economic system we have, the distribution of resources and the planet's ability to deal with climate change and environmental problems in general. They also analyze the behavioral habits of the "consuming classes." This chapter will provide an overview of consumption in global environmental politics. It is organized in the following way: First, consumption will be outlined in a historical context. Then the political institutions of consumption in a global framework will be analyzed. Finally, the chapter will critically analyze consumption as a social activity and explore the role of the individual, in the process exploring the notion of "sustainable consumption."

The history of consumption

Depending on how consumption is defined, it is of course an activity that is integral to human survival. If we did not consume water and food, we would not be able to sustain ourselves. However, with increasing technological progress, people have started to extract more resources from the planet than it has the capacity to replenish. This results in sterile soils, depleted seas, polluted air and water and ultimately climate change. It can also mean that in certain localities people who are dependent on their immediate environment for sourcing cannot fulfill their needs any longer.

Traditionally, consumption has been concentrated in the global North with the global economy organized to service these sites of high consumption. In the past 20 years consumption pattern have become much more varied. Both China and India have become important consumption centers in the world (even though consumption within these countries is obviously extremely uneven). The number of people with high consumption habits has risen exponentially. For example, global meat consumption has doubled between 1990 and 2020 and trebled in Asia alone (FAO 2020). Likewise, world energy consumption has doubled

DOI: 10.4324/9781003008873-21

between 1990 and 2020 (EEA 2020), although China's lead here is obviously related to it being the production center of the world.

Historically, different ideological approaches to consumption date the point differently where consumption becomes unsustainable from an environmental point of view. Some approaches see it as a scientific point where carrying capacity of the planet is reached; other approaches see it as a systemic point where the economic activity of a society went beyond production for subsistence (production for accumulation) or where the production system per se was based on an unsustainable model (such as the modern capitalist system based on mass production and a fossil fuel economy). All of these approaches, and incidentally many other approaches to global environmental politics that do not put consumption to the forefront of their concerns, agree that earth is out of balance in terms of the effects of human activity. They agree that the damaged relationship between society and nature, or between human beings and the environment, needs to be improved.

From a historical perspective, one approach is to talk about the rise of consumerism as opposed to consumption itself, implying that consumerism is qualitatively different from the act of consumption. Campbell (1987) argues that the rise of a romantic and a Protestant ethic coincides with the evolution of the spirit of modern consumerism. McKendrick, Brewer and Plumb (1982) discuss the consumer revolution and the commercialization of economics, particularly in the field of fashion. They argue that, in addition to economic changes that came with the industrial revolution, industrialized society also underwent a cultural change that made people more consumption-oriented or consumerist. Furthermore, consumption has obviously been integrated into economic analysis through the inclusion of the "demand curve."

Literature suggests that changes in consumerism can be equated with the rise of modern capitalism and that the evolution of a consumerist ethic contributed to the success of the industrial revolution and modern forms of economic organization. In the post-war period, a definite change in the ethics of consumption can be discerned in the 1960s and 1970s with the spread of post-Fordism (see below), which, in turn, coincided with what is conceived of as the rise of globalization (see Chapter 24). Both Campbell (1987) and McKendrick et al. (1982) trace changes in consumption patterns in the late eighteenth century back to a shift in the nascent middle classes, which aspired to emulate the spending behavior of the upper classes. The early industrial revolution produced consumer rather than capital goods, and by the eighteenth century most people in Britain, the home of the industrial revolution, had disposable incomes that they tended to spend on consumer goods (Campbell 1987: 19–25). Although the phenomena of consumer spending and of emulating higher classes are not new in history, the changing income structure of early modern capitalism led to yet more disposable income in the middle classes.

McKendrick et al. (1982) also note the increasing velocity of changes in fashion as a key contribution to increased consumerism. Up to about 1750 ladies' fashions took decades or longer to change, while between 1753 and 1757 fashion changed dramatically in the course of only four years, and between 1776 and 1777 the change took only one year (1982: 56). The enthusiasm to be fashionable permeated all spheres of society and was carefully manipulated by the fashion industry. This was a Western phenomenon; fashions in other parts of the world, such as Japan, China, North Africa or the Muslim countries, remained virtually the same for hundreds of years (1982: 36). Another increase in the velocity of fashion can be identified with the rise of globalization and post-Fordism in the 1970s when fashion/clothing gradually became cheaper and consumption rates of fashion items increased exponentially, further accelerating fashion changes in the 1980s and 1990s.

This connection between socio-cultural factors in consumption and production is particularly obvious in the case of commodity chain analysis. Loosely based on world systems theory, this approach takes a linear view of the production process with consumption as the final stage of production. Commodity chains are a good analytical tool for tracing the global nature of the production process, without which the nature of consumption cannot be understood. This type of analysis takes a commodity as the starting point for analyzing the political, economic, social and, to a much lesser extent, ecological linkages between the different production and consumption stages. Therefore commodity chain analysis is not exclusively concerned with the different stages of production in the life of a commodity per se; it also places commodities in a social context.

Hopkins and Wallerstein describe commodity chains as networks "of labor and production processes whose end result is a finished commodity" (1986: 159 quoted in Gereffi and Korzeniewicz 1994). Thus, a global commodity chain comprises not only the different production processes from raw material to finished product but also links households, firms, states and social actors across spatial and temporal boundaries and analyzes their relationship with each other. There are producer-driven and consumer-driven commodity chains. Commodity chain analysis comprises a strong historical component in that it sees variances in the production process over time and it is generally a world systems approach and can also be seen as a general historical materialist approach. As such, it obviously places its emphasis on production rather than consumption and sees consumption primarily as a spatial issue in the context of unequal social relations.

As the primary venue of consumption, households are more or less integrated into commodity chain analysis as the final destination of the product but do not link back into the chain as a factor influencing production processes. Thus commodity chain analysis takes a linear approach rather than letting the ethic of consumption feed back into the production process through attitude changes, taste, social consciousness and so forth. Consumption by consumers/households is seen as the last link in the chain rather than as a structural force or agent influencing production or other social processes.

The commodity chain approach traditionally operates without locating itself in its environmental context. This means it regards resources as inputs, disregards waste output and sidelines the finite nature of resources and sinks. In short, it does not take account of the fact that the social construct of an economy is physically located within the global ecosystem and is dependent on drawing on its resources as well as on putting its waste into this system. The finite nature of the resources used for production and the fragile nature of the ecosystem as a recipient of waste products (in the form of pollution) are complicating factors that need to be incorporated into commodity chain analysis because they are also part of the chain. It could be argued that part of this task is actually fulfilled in a life cycle analysis approach. Life cycle analysis "measures the environmental impacts of products over their entire life cycle from cradle to grave" (Berkhout 1997). However, life cycle analysis is part of the environmental management school of thought rather than part of commodity chain analysis.

Consumption or consumerism is often reduced to the spending power of an individual, which is dependent on her or his position within the division of labor. Therefore the increasingly global division of labor spatially distanciates the locations of production and consumption, meaning that there is no (or only a distant) relationship between the production and the consumption process. In the twentieth century, this division of labor was characterized by Taylorism, Fordism and post-Fordism (Lipiètz 1997: 2). All three models are based on making consumer goods available to a wider user circle through a revised wage structure and through mass production, thus making consumer goods cheaper.

While Taylorism was mainly about the streamlining of the production process, Fordism had a more definite consumerist argument to it. The idea behind Fordism was not only the automation of the production process but also the making available of mass-produced consumer goods to a wider base. The rationale was simple: there was a limited market for capital goods such as cars because these products have a relatively long life span and there were only a limited number of consumers that could afford these products. Therefore new markets needed to be created and the logical solution was to make luxury consumer goods available to workers by reducing the prices through mass production and increasing workers' wages. Thus cars became available for much larger segments of society. In the 1950s and 1960s Fordism was also characterized by stable jobs for life, wage settlements that meant steadily increasing wages every year, a general rise in the standard of living, redistributive state policies and institutionalized collective bargaining. These conditions were meant to secure an outlet for production and led to drastically increased consumption behavior as households spent their income, as well as leading to vastly increased expectations in terms of standards of living.

However, this increase in consumption and standard of living was limited to the developed world, and mostly countries of the Organization for Economic Cooperation and Development (OECD). Although production was becoming internationalized at this stage, the consuming classes were still almost exclusively situated in the North/West. At the end of the 1960s, markets in industrialized countries were slowly becoming saturated. Consumer spending was down and profitability of companies went down, too (Mittelman 1997). Logically, the perks of job security and ever-increasing wages in real terms could not be sustained in this period. Rather than expanding markets in the search for new consumers, other changes happened. As Hoogvelt summarizes:

> By the late 1960s that distinctive period of mass production and Fordist accumulation had come to an end. The rigidities of the Fordist regime showed up with irrepressible frequency. There were many instances of rigidity at all levels but the most important was undoubtedly the deepening global inequalities. These put a limit on the further expansion of that particular system of mass production. There was a global demand crisis and thus capitalism had to reconstitute itself on an entirely new basis. In a world economy where 20 per cent of the population has 150 times the spending capacity of the poorest 20 per cent, clearly a new production system was needed that could fully exploit consumer demand from the 'have-lots' in an ever fiercer climate of global competition.
>
> *(Hoogvelt 1997: 93)*

Enter the era of post-Fordism, which is characterized by economies of flexibility. The economic instability of the 1970s made workers in industrialized countries lose many of the perks they had quickly become accustomed to, such as more-or-less permanent jobs, generous wage settlements, social benefits. As Cox puts it: "The new strategies emphasized a weakening of trade union power, cutting of state budgets (especially for social policy), deregulation, privatization and priority to international competitiveness" (1996: 22). The new economic organization was much leaner and based on maximizing profits. Conca sees the post-Fordist mode of production to be based on flexible capital, vertical disintegration and select markets, that is, "flexible specialization" (2001: 61). This was done not so much by broadening the consumer base but by making products cheaper and more easily available for those who had the spending power.

The post-Fordist flexible mode of production has not only led to a separation of the activities of production and consumption, with the consequence that consumers are not aware of the ethical and environmental conditions under which the product they consume was made. It has also led to a further globalizing of production, which is not matched by a globalized consumption pattern (Kütting 2004). While there is a very definite rise in consumption in many developing countries and consuming middle classes have established themselves, overall consumption and production trajectories still show that the vast bulk of consumption takes place in high GNP countries (although this is changing). As is often said, the workers who make the shirts or electronic gadgets in the factories in many Asian countries could never afford to wear or use them.

Thus the rise of consumerism is embedded into the economic system. It has increased exponentially with the rise of modern capitalism and the industrial revolution. It reached new heights with the beginning of the age of globalization in the 1970s, which is closely linked to the principles of a post-Fordist way of production. This process goes hand in hand with the institutionalization of consumption at the global level.

Institutionalization and the concept of sustainable consumption

The focus on production as the precursor to consumption is obviously a natural method of analysis but some authors argue that it is not a useful starting point because it neglects certain angles of the problem of overconsumption. Of course there are fewer producers than individual consumers, and from that point of view it is easier to regulate production processes to make them more sustainable. It is also politically more palatable to propose policies regulating production than to talk about curbing consumption and questioning lifestyles. By regulating production and not framing the problem in terms of unsustainable consumption, it is also possible to avoid a discussion on the central question, namely whether a system based on infinite economic growth is sustainable. It also does not address the question of global equity, where extremely unequal patterns of consumption mean that an individual in the United States consumes 80 times as many resources as an individual in one of the world's poorest countries. These issues are particularly pertinent in the negotiations on the Kyoto Protocol and per capita greenhouse gas emissions in rising powers such as India and China, but also when we measure greenhouse gas emissions by the final recipient of the goods produced rather than by the producer (see Chapters 25 and 32).

Consumption as a political institution of importance has been recognized by international organizations dealing with sustainable development. But, like sustainable development in general, it suffers from definitional problems: what is sustainable can be defined in vastly different ways depending on the author's environmental ideology (see Chapter 16). Doris Fuchs points out that in 1994 the Oslo Roundtable defined sustainable consumption as "the use of services and related products which respond to basic needs and bring a better quality of life while minimizing the use of natural resources and toxic materials as well as the emissions of waste and pollutants over the life cycle of the service or product so as not to jeopardize the needs of further generations" (Ministry of the Environment Norway 1994 quoted in Fuchs and Boll 2018). Sustainable consumption as a political concept was integrated into the global governance agenda with the United Nations Conference on Environment and Development's *Agenda 21* (1992), which dedicated a chapter to the topic. As a result, several international organizations and nongovernmental organizations (NGOs) have espoused the topic of consumption, but limited progress in institutionalizing the concept has been made although sustainable consumption is part of the Sustainable Development Goals (SDGs), goal 12.

The UN's Commission on Sustainable Development is one of the policy drivers in the field of sustainable consumption. In 1995 it published a report, the *International Work Program on Changing Consumption and Production Patterns*, and it commissioned several research projects on consumption trends and policies. In particular, it advocated the establishment of sustainable consumption indicators as well as consumer protection guidelines. However, critics argue that none of these efforts made their way into official reports and documents (Fuchs and Boll 2010). The United Nations Environment Program (UNEP) also has a Sustainable Consumption Program, which is located in the Production and Consumption Unit of the Division of Technology, Industry, and Economics (DTIE). It aims to study and analyze the forces behind global consumption patterns in order to help businesses and other stakeholders find suitable strategies to improve their activities. Its work is focused on helping businesses innovate and achieve efficiency gains in production that filter through to consumption activities. UNEP published a report on "Consumption Opportunities" in 2001, raising the issue of consumption quantity as opposed to consumption quality, but it did not take this initiative any further and also did not specifically address the issue of highly consuming countries. With its *Global Status Report* (UNEP 2002), UNEP highlighted six core areas in which future work or research on sustainable consumption ought to be concentrated. These deal with definitional issues, development of indicators, and the interplay between local and global policies. However, UNEP has so far not indicated a desire to be a leading actor in the consumption issue.

Nevertheless, the central outcome of the last global summit on sustainable development, which took place in New York in 2015, were the Sustainable Development Goals. Of these, Goal 12 calls on governments to stress and strengthen their commitments regarding their policies of sustainable consumption and production and to develop initiatives to accelerate the shift toward sustainable consumption and production.

Another international organization that has taken on the issue of consumption is the OECD. This is a particularly pertinent organization since only about 20 percent of the world's population lives within the borders of the OECD region, yet they consume about 80 percent of the world's resources. The OECD integrated Agenda 21's aims into its work by setting up a work program in 1995 called *Environmental Impacts of Production and Consumption*. Again, the focus was on efficient use of resources and encouraging suitable technological change. In other words, this work program was about making economic growth more sustainable rather than debating the nature of growth per se. The OECD referred to this as "sustainable consumption." The OECD focused its efforts on development of policy instruments for sustainable consumption in the fields of tourism, food, energy and water, as well as the volume of waste generated. In 2008, the OECD widened its scope and included equity aspects in its policy research. If followed through consistently, such an angle could lead to a new life and different focus in its efforts to develop a sustainable consumption framework.

The European Union (EU) arguably has the world's most rigorous environmental policy as well as having enshrined the precautionary principle and the polluter-pays principle in its framework. Not surprisingly it also adopted a *Sustainable Development Strategy*, which included dimensions on sustainable production and consumption. Nevertheless, it was not until 2007 that a monitoring report was first published by Eurostat. This report paid homage to the idea of sustainable growth through greater technological efficiency. It made the point that effective change can also be achieved through changing production and consumption patterns. Arguably, the EU has the most promising framework of all international governance institutions to tackle the problem of consumption. However, as Fuchs (Fuchs and Boll 2018) argues, while the EU has taken a number of initiatives and in many ways is the leading actor promoting sustainable consumption, it has not consistently followed through with this intention, and it is still linked to the belief of sustainable economic growth as the way forward.

Gabriela Kütting

The definition, analysis and institutionalization of the concept of consumption are not confined to the role of international organizations and attempts to produce a global governance framework. National governments can also be instrumental in this field. The Norwegian and Danish governments tried to push the global consumption agenda by hosting workshops and facilitating the efforts of global governance institutions. However, these efforts failed to make much headway. The UK government established a Commission on Sustainable Development. As part of its remit, the Commission published a report in 2009 called "Prosperity without Growth?," which "analyses the complex relationships between growth, environmental crises and social recession" (Sustainable Development Commission 2009). The Commission's contribution to the field of sustainable consumption was seminal in its own right, but its work was cut short with the change of government in the United Kingdom in 2010. However, the report constitutes one of the most important documents in providing guidance toward a transition to an economic system not dependent on infinite economic growth.

Many initiatives from outside the OECD countries have also developed interesting and important frameworks addressing consumption, consumerism and the growth economy. Bhutan developed its Gross National Happiness Index as early as 1972 because it was not satisfied with the exclusive material focus of the standard GNP measurement. The index defines happiness as inclusive of harmonious social relations and harmony between nature and society. While this is not a blueprint for sustainable development, it has become an oft-cited model for alternatives to the sole focus on material wellbeing engrained in conventional measures. Another more recent model is the concept of *buen vivir* that has been enshrined in at least two national constitutions in Latin America. Early forms of the *buen vivir* concept arose in resistance to classical development policies applied in Latin America, which were perceived not to be working economically as well as having a negative impact on society and environment. As a result, a strand of social thinkers felt it was time to abandon the classical development model embedded through the Washington Consensus. *Buen vivir* finds its origins in contributions from indigenous knowledge, which culturally lack concepts of development and progress. Indigenous concepts in Ecuador and Bolivia stress the fullness of life in a community setting, in harmony with society and nature (Gudynas 2011). Both countries incorporated the concept of *buen vivir* into their constitutions in 2008 and 2009 respectively. In Bolivia, it is included as part of the "moral and ethical principles describing the values, ends and objectives of the State" (Gudynas 2011: 442). The constitution of Ecuador uses a framework of rights to integrate *buen vivir*. As Gudynas puts it, the concept "is not an ethical principle for the State as in Bolivia, but a complex set of several rights, most of them found in the Western tradition, although fitted in a different framework. These are in the same hierarchy level with another set of rights, that include, among others, those of freedom, participation, communities, protection, and also the rights of Nature" (2011: 443).

Contributions of the national governments of the United Kingdom (under the Labour government), Bhutan, Boliva and Ecuador are seminal. They are probably the most promising starting points for an institutionalization of the concept of sustainable consumption.

Nongovernmental organizations have also tried to make their contributions to the consumption problematique. They are at liberty to ask fundamental questions about lifestyles and the underlying values of the consumer society. Movements such as *voluntary simplicity* or *need not want*, local currency organizations, fair trade networks, the freecycle movement and others have drawn substantial support, but there is no collaboration between these movements to organize larger action. While many of these movements and NGOs have a large following and even a global resonance, this does not translate into the means to contribute to a global governance of consumption (see Chapters 14 and 21).

The business sector has also addressed the issue of consumption, for example through a consultation exercise with consumers and report by the World Business Council for Sustainable Development (WBCSD 2002). Not surprisingly, the business community has seen itself as responsible for increasing ecological efficiency, but it has firmly placed responsibility for levels of consumption with the consumer. So, one could argue that consumption is a difficult sell for many actors because it threatens many interests. From that point of view, it is dependent on consumers making rational and ethical choices for the benefit of future generations and less fortunate contemporaries. However, social psychology research in other fields has shown that only a minority of the population will actively make these choices. They need an institutional framework as an incentive, yet an institutional framework will only be constructed when there is sufficient political pressure to do so (see Chapters 14, 28 and 29).

Likewise, while sustainable consumption is a Sustainable Development Goal, by 2020 most of the indicators for sustainable consumption under this development goal are yet to be developed, demonstrating the reluctance of business actors to making a full commitment (Gasper, Shah and Thanka 2019). The strongest push for sustainable consumption as a SDG came from developing countries that wanted the inequality of production/consumption institutionally reflected.

Sustainable consumption: the problems

Consumption as a subject of political economy and as a subject of global environmental politics has received substantial attention from a variety of perspectives. First of all, consumption as a political activity has been highlighted, whether in shareholder activism, sustainable consumption, the sociology of consumption, or the relationship between consumption and production. The role of unequal consumption has been raised as an equity issue. Likewise, the ethics of consumption has seen a wealth of writings. While some of these are normative (Schor and Holt 2000; Jackson 2006), some are empirical and deal with increasing consumer choices and ethical consumerism (Dauvergne 2008; Kütting 2010). The literature can be condensed into two questions: How to consume? and How much to consume? The lion's share of the literature is on the former question, while the latter question is generally avoided because it is so contested. A path-breaking work on the topic of consumption is Princen, Maniates and Conca's *Confronting Consumption* (2002). The book addresses the concept and issue of consumption from a variety of angles, but is based on the premise that some resources are finite and that the capacity of sinks (places to put waste and pollution) is also finite.

There is an ever-increasing world population. While the issue of population growth is contested, with some arguing that the population increase will put dramatic pressure on resources, others argue that it is an issue of distribution and that there would be plenty left over for everybody if unequal consumption levels could be tackled. While it is uncontested that the world's richest people consume far more than the poorest, it is also true that the ranks of the high consumers are swelling. Therefore the problem of population is real, even if not quite in the terms presented by those that point to population growth in an undifferentiated manner.

Clearly, for a sustainable future, the pressure on resources and sinks needs to ease. Modern technology and ecological modernization may provide at least some of the tools to achieve this, but they do not supply the distribution mechanisms that will provide more equal access to resources and sinks, nor do they address the question of how such a redistribution could be handled given the entrenched political and economic interests behind a consumerist economic system. Recent technological developments have shown that the availability of more sustainable technology alone is not enough to achieve more equity and sustainability, or to eradicate poverty, because the cost and access to such technologies makes them unattainable to the majority of people who need them most. Some would go much further, as, for example, David Harvey, who argues that the current international or global system can only be described as "accumulation by dispossession" (Harvey 2003: 137). Thus, the existence of solutions to a particular problem does not necessarily solve it. And this is exactly where the notion of sacrifice becomes important because it highlights the chasm between technological capability and political reality.

Michael Maniates and John Meyer (2010) have identified the concept of sacrifice as an important issue that society ought to engage with, yet they have not yet addressed the question of a redistribution of resources. Thomas Princen's *Logic of Sufficiency* (2006) is an important text on consumption issues, in part because it does not suggest sacrifice or radical lifestyle change. Rather, it is based on questioning the logic that efficiency, as defined by economies of scale and instantly maximized profits without regard for the future, is the best organizing principle for economy and society. Consumer psychology writers have conducted studies that show that instant gratification and indiscriminate material consumption actually lead to less rather than more happiness. A burgeoning literature on the ethics of consumption has questioned neoliberal lifestyles. Of course there are various civil society movements doing the same. Princen's work contributes to and defines this literature in a new way by using the logic of sufficiency (see Chapter 16). He illustrates with case studies that the concept of sufficiency can indeed by applied to mainstream economic, social and political scenarios, using diverse examples of citizens of a Toronto island, Maine fishers and a West coast logging company. However, his studies are all of a particularly local nature and deal with the local part of society's interactions. It is not clear whether the concept of sufficiency can be applied to the global level or whether it would generate the kind of political consensus needed for a global framework.

A relatively new school of thought, also touching on consumption, is found in the literature on "degrowth." Academics write about degrowth from a variety of perspectives, basing their ideas around three pillars. The first pillar is culturalist. Like the *buen vivir* concept, it is based on the understanding that it may not be the most appropriate path for developing countries to follow in the footsteps of the most industrialized countries. This goes back to earlier critiques of modernization theory. The second pillar of degrowth takes its source from a more democratic and pluralist understanding of how economic and social systems ought to function. The third pillar is derived from ecology and the need for environment–society relations that reflect the finite nature of many ecosystems and the human dependence on them. Degrowth is the logical pathway to take in order to ensure the sustainability of resources and the capacity of sinks. The concept has long been integrated into the fundamentals of ecological economics and suggested by writers such as Herman Daly (1973, 1996) and Joan Martinez-Alier (2002). More recently, Fuchs and Lorek (2013) have argued for strong sustainable consumption governance as a path toward degrowth. However, the discussion on how degrowth of the economy is needed as a first step toward social and ecological transformations is new. It is fundamentally different from the market-based idea of a green economy, which was one of the key projects at the Rio+20 conference in June 2012.

Conclusion

The study of consumption is very much in flux, and it is also very much an inter- and trans-disciplinary field of study. Researchers are still trying to create a bigger picture of consumer behavior, consumer motivations, consumer rationality and of course the social relations between consumer and "the system," or rather between the various political, economic, social and ecological systems in which consumption is embedded. As Fuchs puts it,

> the structural contexts of the consumer environment strongly influence the characteristics of the available options for decisions regarding consumption. In order to not overestimate the responsibility and ability for change of the individual consumer, sustainable consumption research has to take an integrated perspective and link consumer decisions to their societal environment as well as develop a joint production–consumption strategy.
>
> *(Fuchs and Boll 2018: 84)*

It is clear that in order to tackle global environmental problems that form the most formidable challenges of the twenty-first century, we cannot avoid looking at consumption, the distribution of consumer power, and the sociology and culture of consumption.

References

Berkhout, F. (1997) *Life cycle assessment and industrial innovation*. ESRC Global Environmental Change Programme Briefing No. 14. Swindon: Economic and Social Research Council.

Campbell, C. (1987) *The romantic ethic and the spirit of modern consumerism*, London: Blackwell Publishers.

Conca, K. (2001) "Consumption and environment in a global economy", *Global Environmental Politics* 1(3), 53–71.

Cox, R. (1996) *Approaches to world order*, Cambridge: Cambridge University Press.

Daly, H. (1973) *Toward a steady state economy*, San Francisco, CA: Freeman.

Daly, H. (1996) *Beyond growth – the economics of sustainable development*, Boston, MA: Beacon Press.

Dauvergne, P. (2008) *The shadows of consumption: consequences for the global environment*, Cambridge, MA: MIT Press.

European Environment Agency (eea.europa.eu)

Food and Agriculture Organization of the United Nations (fao.org)

Fuchs, D. and Boll, F. (2018) "Sustainable consumption," in G. Kütting (ed.) *Global environmental politics – concepts, theories, case studies*, 2nd ed., London: Routledge.

Fuchs, D. and Lorek, S. (2013), "Strong sustainable governance – a precondition for degrowth?", *Journal of Cleaner Production* 38, 36–43.

Gasper, D., Shah, A. and Thanka, S. (2019), "The framing of sustainable production and consumption in SDG 12", *Global Policy* 10(1), 83–95.

Gereffi, G. and Korzeniewicz, M. (eds) (1994) *Commodity chains and global capitalism*, London: Praeger.

Gudynas, E. (2011) "Buen vivir: today's tomorrow", *Development* 54(4), 441–447.

Harvey, D. (2003) *The new imperialism*, Oxford: Oxford University Press.

Hoogvelt, A. (1997) *Globalisation and the postcolonial world*, Basingstoke: Macmillan.

Jackson, T. (2006) *The Earthscan reader on sustainable consumption*, London: Earthscan/James & James.

Kütting, G. (2004) *Globalization and environment, greening global political economy*, Albany, NY: SUNY Press.

Kütting, G. (2010) *The global political economy of the environment and tourism*, Basingstoke: Macmillan Palgrave.

Lipiètz, A. (1997) "The post-Fordist world: labour relations, international hierarchy and global ecology", *Review of International Political Economy* 4(1), 1–41.

McKendrick, N., Brewer, J. and Plum, J.H. (1982) *The birth of a consumer society*, London: Europa.

Maniates, M. and Meyer, J. (2010) *The environmental politics of sacrifice*, Boston, MA: MIT Press.

Martinez-Alier, J. (2002) *The environmentalism of the poor*, Cheltenham: Edward Elgar.

Mittelman, J.H. (ed.) (1997) *Globalisation, critical reflections*, London: Lynne Rienner.

Princen, T. (2006) *The logic of sufficiency*, Cambridge, MA: MIT Press.

Princen, T., Maniates, M., and Conca, K. (eds) (2002) *Confronting consumption*, Cambridge, MA: MIT Press.

Schor, J. and Holt, D. (eds) (2000) *The consumer society reader*, New York: New Press.

Sustainable Development Commission (2009) "Prosperity without growth? – The transition to a sustainable economy". Available HTTP: <http://www.sd-commission.org.uk/publications.php?id=914>

UNEP (2002) *A global status report*, Paris: United Nations Environmental Programme.

United Nations Department of Economic and Social Affairs Division for Sustainable Development (1992) *Agenda 21.* Available HTTP: <http://www.un.org/esa/dsd/agenda21/res_agenda21_00.shtml>

United Nations Division for Sustainable Development (UNDSD) (1995) The CSD Work Programme on Indicators of Sustainable Development. Available HTTP: <http://www.unece.org/fileadmin/DAM/stats/documents/2001/10/env/wp.27.e.pdf>

WBCSD (2002) *Sustainable production and consumption: a business perspective*, Geneva: World Business Council for Sustainable Development.

18

Expertise

Specialized knowledge in environmental politics and sustainability

Andrew Karvonen and Ralf Brand

The pervasive role of technology in contemporary societies requires the public to rely on individuals with specialized knowledge to invent, design, manufacture, and maintain complex artifacts and networks. As Stilgoe et al. (2006: 16) note, "Our everyday lives are played out through a series of technological and expert relationships." This is particularly evident in current issues related to the climate crisis, the digitalization of society, collective responses to pandemics, growing social inequalities, and other "wicked problems" of the twenty-first century (see Chapters 15 and 32). At the same time, there has been a gradual erosion of trust between the public and technical experts, particularly since the late 1960s, as contemporary environmental, social and economic problems have revealed the limitations and unintended consequences of scientific and technological development. This growing distrust of experts has been amplified in the last two decades with the rise of the World Wide Web and the emergence of "fake news," "alternative facts," anti-intellectualism and conspiracy theories (Fischer 2019; Mede and Schäfer 2020; Merkley 2020). Thus, the role of technical experts in contemporary society is in flux and by extension, the legitimacy of those political decisions that are based on expert advice.

In this chapter, we draw upon theoretical and empirical contributions from environmental politics, sustainability studies, and Science and Technology Studies (STS) to examine how the specialized knowledge of technical experts and the informal knowledge of non-experts influence debates about sustainability and the environment. Of particular interest is how experts from different disciplines interact with one another and the public they are ostensibly chartered to serve. The work reflects our interest in technology and expertise as it relates to current debates on urban sustainable transitions and transformations, socio-ecological systems and resilience, and urban infrastructure studies (e.g., Coutard and Rutherford 2015; Hölscher and Frantszeskaki 2020; Metzger and Lindblad 2021). However, the ideas in this chapter are also relevant to broader debates about the contested ways that knowledge is produced and applied.

We begin the chapter with an overview of the ascendancy of the technical expert in contemporary society and a summary of some of the most prominent critiques of expertise. We then provide a brief discussion of sustainability, with particular emphasis on how it differs from previous conceptualizations of environmental problems, and question how traditional

DOI: 10.4324/9781003008873-22

models of expertise engage in knowledge production practices to address this paradigm. Finally, we present three models of expertise that have been applied since the 1980s to create more sustainable modes of human life and conclude with a discussion of the implications that sustainability has for technical experts in the future.

The rise of the technical expert

The beginnings of technocracy – or perhaps more accurately termed "expertocracy" – can be traced to the Enlightenment when individuals began to acquire or were granted the power to shape and direct societies through scientific and technological development. Their efforts produced large complex systems including gas, electricity, water, sewage, and transit networks, making technical experts particularly influential in public policy and city-building activities (Seely 1996; Graham and Marvin 2001). Experts served as the "human face" of these technological networks, symbolizing the founding tenets of modernity including efficiency, stability, functionality, objectivity, and perhaps most importantly, progress (Hickman 1992). The rise of the technical expert in modern societies resulted in a privileged status for those with specialized knowledge while stabilizing social relations in fundamental ways.

Today, the most conspicuous technical experts in developed countries include natural scientists and engineers whose specialized knowledge is based on the formal study of a scientific or technical discipline. They are joined by many other experts including medical professionals, lawyers, policymakers, financial analysts and so on. In all cases, the social power of the expert is derived from a combination of professional status (e.g., engineers and architects), adherence to the scientific method (natural scientists), or simply the mastery of a specialized field of knowledge through formal training (urban planners). The technical expert is differentiated from non-experts by the possession of a "core set" of specialized knowledge as well as an elevated position in society, with non-experts deferring to the expert's superior judgment. As Whatmore (2009: 588) argues, "Expert knowledge claims, and the technologies through which these become hardwired into the working practices of industry and government, manifest themselves in the products and policies we live with and the sociomaterial environments we inhabit." In other words, expertise is closely tied to practices of knowledge production and is "inextricably tied to its social utility" (Selinger and Crease 2006: 230).

The pursuit of expertise has the social effect of elevating the individual to semi-god status, but it inevitably comes at the expense of a narrowed perception through specialization. Experts are celebrated for their microscopic, specialized analysis of problems rather than emphasizing a macroscopic, holistic perspective. A well-known joke states that, "An expert is one who knows more and more about less and less until she knows absolutely everything about nothing." Meanwhile, there are long-standing appeals for knowledge production practices to go beyond mono-disciplinarity to engage in multi-, inter- and trans-disciplinarity practices (Evans and Marvin 2006; Petts et al. 2008; Ramaswami et al. 2018). This has important implications for all experts, not only academic researchers but also those who generate and apply specialized knowledge in the public and private sectors and civil society.

Critiques of expertise

The sacrifice of breadth for depth seems the logical price to pay for the acquisition of expert knowledge. It creates a division of labor among the various (sub-)disciplines and a pragmatic approach to managing the increasingly complex technical artifacts and systems that support contemporary societies. However, the specialized worldview of the technical expert has been critiqued for several reasons. At the most basic level, the limited perspective of the expert is problematic because of the inability to "see the forest for the trees." As Lane and McDonald (2005: 724) argue, "technical knowledge simultaneously sharpens our focus and obscures our vision." And specialized knowledge has deeper problems beyond its atomistic worldview, four of which we discuss briefly in the following paragraphs.

First, ontological and epistemological critiques of expertise challenge commonly held assumptions about knowledge generation practices. The ontological assumption of traditional forms of expertise is that of a knowable and unequivocally re-presentable world "out there" that is independent of context, the basis of the positivist perspective (see Chapter 3). The positivist approach to problem-solving involves the application of universal knowledge. In contrast, post-positivist and post-structural scholars argue that knowledge is plural rather than unitary (Gibbons et al. 1994; Sandercock 1998; Nowotny et al. 1999; Schlosberg 1999; Miller et al. 2008). As Barry et al. (2008: 23) argue, "Knowledge production has always occurred in a variety of institutional sites and geographically dispersed assemblages, not just in the apparently enclosed space of the humanist's study or the scientific laboratory."

A second critique of expertise is the assumption that scientific and technical knowledge is value-free and neutral. Technical experts tend to be portrayed as objective actors in policymaking activities, transcending partisan interests and "speaking truth to power" (Fischer 2006). However, the existence of multiple forms of formal knowledge, and the inherent political character of this knowledge, is readily apparent in environmental conflicts where a high degree of uncertainty and the presence of conflicting values and interpretations are both common and unavoidable. This has resulted in the emergence of counter-experts who dispute technical experts on their own terms (Yearley 2000; Bäckstrand 2004). One of the most famous and influential environmental counter-experts is Rachel Carson who assembled a network of researchers and scientific evidence in the 1950s and 1960s to make the case against the indiscriminate use of pesticides in the United States (Lytle 2007, see Chapter 35). Her approach of "fighting science with science" helped to spawn the emergence of counter-expertise that is now practiced by environmental nongovernmental organizations as well as private corporations who introduce competing interpretations of a particular environmental issue. Counter-experts challenge the authority of experts by using equivalent methods and language (see Chapters 13 and 14) while creating uncertainty for the receivers of specialized knowledge. As Fischer (2000: 61) notes, "After having long trusted experts generally, citizens are confronted with the task of choosing which experts to believe and trust."

A third critique of expertise concerns the existence and undeniable relevance of experiential, local and tacit knowledge arising from personal experience and exploration outside the confines of established knowledge institutions and without strict adherence to the scientific method. Multiple forms of formal knowledge are joined by multiple forms of informal knowledge. Scott (2020: 320) differentiates between formal and informal knowledge using the classical notions of techne and metis where the former involves "impersonal, often quantitative precision and a concern with explanation and verification," while the latter refers to indigenous knowledge, meaning, experience, and practical results. Stakeholders who lack formal knowledge are often portrayed as being "incapable of grasping the technical nuance and methodological complexity of science" (Kleinman 2000: 139). In this regard, Turner (2001: 123) observes that, "expertise is treated as a kind of possession which privileges its possessors with powers that the people cannot successfully control, and cannot acquire or share in." In contrast, others argue that knowledge is not an entity that can be obtained and applied but rather is the outcome of multiple social and material interactions (Irwin 1995; Latour 1999; Rydin 2007).

The recognition of different forms of knowledge highlights the tensions between democratic forms of governance and technical expertise. Holders of experiential, local and tacit knowledge are generally not granted a seat at the decision-making table to discuss scientific and technical problems due to an institutional bias toward formal and seemingly objective knowledge. The centrality of the technical expert in political systems is commonly referred to as "technocracy," where technical experts rule by virtue of their specialized knowledge and position in the dominant political and economic institutions (Habermas 2015; Raco and Savini 2019). Here, expert knowledge is applied to the task of governance and promotes technical solutions to political problems, with the technical expert assumed to be positioned above partisan politics and an ignorant general public (Fischer 1990).

Fourth and finally, there are important practical issues that cannot be resolved through the application of technical expertise (Petts et al. 2008). For example, Ulrich Beck (1992) argues that the question of whether we should use nuclear energy can never be answered with an objective "yes" or "no" because issues of risk and risk perception require "soft" and culturally specific responses (see Chapter 19). This recognizes that values and politics are an inherent part of knowledge production processes (Karvonen 2020). This is clearly the case in contemporary scientific disputes related to climate change, genetically modified organisms, human cloning, nanotechnology, artificial intelligence and the like. A technocratic response to these conflicts is to portray critics of scientific and technical solutions as irrational and the mission of technical experts is to educate objectors to the "facts" of a particular problem. Meanwhile, alternative configurations of society and technology as well as the values that underpin them are dismissed or ignored.

The aforementioned critiques of technical expertise reveal multiple challenges involving epistemology and ontology, objectivity, political power and practical matters. This becomes even more pronounced when we consider how knowledge production relates to sustainability, as described in the next section.

Sustainability as a challenge to the technical expert

Sustainability has multiple meanings and interpretations, although most advocates would probably agree that it involves a holistic approach to solving complex, interrelated and multidimensional problems (see Chapter 16). Dryzek (2013: 150) argues that the main accomplishment of sustainability has been "to combine systematically a number of issues that have often been treated in isolation, or at least as competitors." Thus, the principal advantage of sustainability is a pluralistic and inclusive approach to problem-solving, as opposed to conventional problem-solving that limits its focus to particular elements while overlooking unintended consequences as well as the proverbial "big picture." The conceptual comprehensiveness of the sustainability agenda is an attempt to address the interrelated quality of seemingly independent problems.

This is reflected in the Three E model that describes sustainability as the triad of Economic viability, Environmental protection, and social Equity. The model is intended to highlight the challenge of simultaneously accommodating a multiplicity of competing demands. In other words, the openness of the sustainability concept to various claims and concerns comes at the expense of compromise. Campbell (1996) highlights a crucial implication of this model by identifying the inherent conflicts between each pair of "Es" and the pressing need for strategies to resolve these tensions. From this perspective, sustainability issues involve the management of conflict through a "restless, dialectical process" of open discussion and negotiation (Healey 2004: 95). Recognizing the importance of negotiation between competing interests reveals sustainability as a highly political endeavor that is inherently context specific (Prugh et al. 2000; Moore 2007; Moore and Karvonen 2008). Thus, the conventional model of technical expertise that purports to be objective, apolitical, and universal is not an ideal fit for the indelibly political and situated character of sustainability.

Despite the inherent politics of sustainability, the Western world has generally addressed this challenge by relying heavily on technical expertise (see Tate et al. 1998). Technical experts have been tapped to develop more efficient and effective technologies to avoid having to make compromises, to resolve conflicts between stakeholders, and to avoid unintended consequences. A prime example of the reliance on experts to "solve" sustainability challenges is the development of renewable energy strategies to replace fossil fuels (see Chapter 33). This is the underlying message of ecological modernization advocates who argue that industrial society's harmful aspects can be expunged through the application of improved technologies (Hajer 1995; Brand and Fischer 2013). The attractiveness of ecological modernization stems from its implicit assumption that environmental and social problems can be overcome without leaving the path of modernization (Hannigan 2014) or – in colloquial language – that we can "have our cake and eat it too." Sustainability is framed as a technocratic endeavor that reinforces the dominance of the political, economic and knowledge elites while strengthening the societal role of technical experts (Dryzek 2013).

Bottom-up notions of ecological democracy provide an alternative to the ecological modernization approach to sustainable development by championing public deliberation, communication and participation by civil society (see Chapter 14). Proponents of these bottom-up approaches to sustainability do not call for the wholesale abandonment of technical expertise but call for technological development to be directed by society as a whole rather than being imposed from above by powerful elites (Smith and Seyfang 2013; Smith et al. 2014). This requires the creation of political communities to deliberate on the inherent conflicts of sustainability and to transform them via equitable and lasting solutions. In other words, there is a strong emphasis on the social equity aspects of sustainability (Agyeman 2013, see Chapters 4 and 26).

Finding common ground between technical experts and sustainability

Conventional notions of expertise are not an optimal fit for the deliberative, bottom-up approaches to sustainability as described above. However, rather than abandoning specialized knowledge outright, we argue that there are possibilities to renovate technical expertise to align with the goals of sustainability. In the following sections, we describe three types of expertise that have emerged since the 1970s to reorient individuals with specialized knowledge toward more sustainable goals. We label these approaches the Communicative Expert, the Collegial Expert, and the Collaborative Expert, and argue that each makes unique and helpful contributions to the renovation of conventional forms of expertise and the pursuit of a more sustainable world.

The communicative expert

One response to the eroding credibility and legitimacy of technical experts has been a call for a more scientifically literate public. This movement, first taken up in the 1980s, focuses on issues of risk and uncertainty in science and technology, and is frequently referred to as "the public understanding of science" (Snow and Dibner 2016). The intent has been to improve the communication of scientific and technical knowledge to affected citizenry and, in turn, to educate the public about the importance of this knowledge (Wynne 1996; Turner 2001; see Chapters 28 and 29). This is an appealing model because it promotes the development of a more educated and informed public.

One way that scientists and technical experts have imparted their knowledge to the public has been through Science Shops that proliferated in the United Kingdom, the Netherlands, and other Northern European countries in the 1980s and 1990s (Irwin 1995; Leydesdorff and Ward 2005). The Communicative Expert makes contact and fosters relations with those who are implicated in knowledge production processes to support, educate and increase awareness. The approach frames scientific and technical organizations (and universities in particular) as repositories of wisdom, reaching out to those who are implicated in the application of specialized knowledge. Today, Communicative Experts engage in "outreach and engagement" activities involving policy briefs, toolkits, science festivals, researcher nights, science parliaments, videos, blogposts, podcasts, and social media to disseminate specialized knowledge to policymakers and the general public (Haywood and Besley 2014). An example of this is the European Science Engagement Association (EUSEA), an organization that aspires to become the institutional home of "all experts involved in the design, organisation and implementation of public engagement activities across Europe" (EUSEA 2021).

The dissemination of specialized knowledge by the Communicative Expert can be useful to resolve some of the tensions between experts and non-experts. It has the potential to level the playing field of knowledge generation to some degree by opening up debates over shared issues and by rebuilding trust between those with specialized knowledge and the general public. However, this model has significant shortcomings. First, it does little to address existing power differentials between experts and non-experts, and instead falls back on the conventional "sage on the stage" model of modern scientific and technological development. The expert continues to know "what is best" for the public and has a tendency to reinforce paternalistic, positivist notions of knowledge production. Furthermore, it implies that the public, through its ignorance of science and technology, is largely to blame for scientific and technical failures, further exacerbating the lack of trust between experts and non-experts. This approach can be seen as a tokenistic reform of technical expertise because its sole emphasis is to bring the public "up to speed" while failing to question the social position and power of experts.

The collegial expert

A second option for accommodating and aligning expertise with the discursive and political character of sustainability is to increase the permeability between existing disciplinary boundaries and overcome "disciplinary silos." The Collegial Expert fosters partnerships involving two or more disciplines to address a specific sustainability issue, resulting in a new core set of knowledge that is related to but independent of the core sets of each individual. This is commonly described as interdisciplinary knowledge production that "brings into play two or more established disciplines so that they interact dynamically to allow the complexity of a given object of study to be described, analyzed and understood" (Darbellay 2015: 165; see Chapter 6). The aim here is not to abandon specialized knowledge but rather to work in the borderlands between existing disciplines (Ramadier 2004; Castán Broto et al. 2009) to develop new insights on the world. The Collegial Expert reinforces and extends the legitimacy and power of expert knowledge through an alliance between two or more domains of expertise. There are, of course, many formidable barriers to overcome in the pursuit of interdisciplinary knowledge production including but not limited to jargon, epistemological assumptions, funding protocols and the attribution of reputational credit arising from joint projects (Evans and Marvin 2006; Petts et al. 2008).

The Collegial Expert is an intermediary, broker or boundary-spanner that juggles the sundries of multiple specialized knowledges to address the multivalent character of sustainability issues (Evans and Marvin 2006; Guy et al. 2011; Acuto et al. 2019; Kivimaa et al. 2019). The Collegial Expert is tasked with identifying potential linkages and facilitating their co-discovery by mediating the knowledge claims of multiple technical experts. Unlike the traditional expert who retains a core set of specialized knowledge, the Collegial Expert coordinates many core sets to devise a meta-set of knowledge. One could imagine public policy experts, sociologists, anthropologists and geographers as particularly well-suited for such intermediary roles (Jepson 2019).

Similar to the above model of the Communicative Expert, the Collegial Expert has merit but again fails to challenge the idea of a core set of specialized knowledge being retained by technical experts. The sage on the stage now involves multiple sages that interact on a new "meta-stage" before sharing their collective wisdom with an unenlightened public. Problem-solving remains in the elitist province of the alma mater and challenges the boundaries between disciplines but not the boundaries between experts and non-experts. There is a continued promotion of formal knowledge and no recognition of experiential, local and tacit forms of knowledge.

The collaborative expert

The previous two models of expertise have advantages over traditional models of expertise because they improve the non-expert understanding of scientific and technical knowledge (the Communicative Expert) or increase collaboration and understanding between experts (the Collegial Expert). However, neither of these models challenges the privileged status of the expert in society or acknowledges that other forms of knowledge are also valid and useful. In other words, they do not question the expertocratic mode of decision-making and do not call for experts to engage with non-experts. The Collaborative Expert does exactly this by recognizing the value of experiential knowledge and by facilitating two-way engagement between experts and non-experts. This does not entail the abandonment of formal expertise but rather an appreciation of a wider *range* of expert opinion through transdisciplinarity rather than interdisciplinarity. Darbellay (2015: 166) describes transdisciplinarity as "a method of research that brings political, social and economic actors, as well as ordinary citizens, into the research process itself, in a 'problem-solving' perspective." It champions participation to highlight the social contingency of knowledge production processes and recognizes the need for transparency and accountability (Sclove 1992; Bäckstrand 2004). From this perspective, the top-down authority of the expert personified by technocratic forms of politics is replaced by democratic politics where experts and non-experts function as collaborators or partners in problem-solving. This arrangement does not guarantee an equitable distribution of power between stakeholders but, at the very least, allows for the possibility of non-expert voices to be heard.

A number of promising techniques were developed in the 1980s and 1990s to advance the notion of the Collaborative Expert, including constructive technology assessment, strategic niche management, citizen panels, and responsible research and innovation (Schot and Rip 1997; Brown 2009; Owen et al. 2012; Stilgoe et al. 2013). Today, similar approaches can be found in practices of participatory action research, and public engagement (Polk 2015; Foley et al. 2020) as well as interventions labeled as urban experiments, living laboratories, collaboratories, maker spaces and so on (Karvonen and van Heur 2014; Evans et al. 2016; Marvin et al. 2018). Of particular note here is the rise of citizen science and the related "quadruple helix innovation model" (Carayannis and Campbell 2010; Hecker et al. 2018). Examples of this approach include the European Network of Living Labs (ENOLL 2021) and the European Union's Green Deal Funding program to "enable[e] citizens to act on climate change, for sustainable development and environmental protection through education, citizen science, observation initiatives, and civic engagement" (European Union Green Deal 2021). These approaches are aimed at extending policymaking practices beyond formal experts to include citizen voices in scientific and technological knowledge production and application (see Rip et al. 1995). It rejects the conventional "end-of-pipe" model where the public reacts to the consequences of scientific and technological development and instead, reveals the wide range of assumptions, values and visions that underpin knowledge production (Wilsdon and Willis 2004). This approach has also been described as "interactional expertise" where formal and informal experts come together to co-produce knowledge (Collins and Evans 2002; Carolan 2006; Ribeiro and Lima 2016).

Proponents of transdisciplinarity argue that this mode of knowledge production is not antithetical to science and technological development but instead embraces the foundational scientific tenets of skepticism, exploration and inquiry. Collaborative Experts raise new questions about scientific and technological development to develop more useful and socially relevant scientific and technological solutions where experts should be "on tap, not on top" (Stilgoe et al. 2006). Forester (1999: 143) argues that transdisciplinary experts "work to encourage practical public deliberation – public listening, learning and beginning to act on innovative agreements too – as they move project and policy proposals forward to viable implementation or decisive rejection." This involves "second-order learning" and critical reflection on the assumptions that underpin the pursuit of factual and technical first-order learning (Autio et al. 2008; Robinson 2008). The involvement of citizens in decision-making broadens expertise by not only asking questions of "how" but also of "why."

Collaborative Experts include a wide range of actors and perspectives where "different experiences, knowledges, and politics are all included in an integrated, holistic, approach to a complex problem or set of problems" (Evans and Marvin 2006: 1012). Knowledge is no longer the sole domain of experts but involves the generation of socially distributed expertise (Nowotny et al. 2001; Bäckstrand 2004; Jasanoff 2004; Rydin 2007). There is an emphasis on processes of co-production, co-design, and co-creation to improve decision-making via "the intelligence of democracy" (Liu et al. 2013; Simon et al. 2018). It is intended to improve accountability of knowledge generation and application while reducing public resistance and criticism. It can also foster innovation by opening up future scenarios to more experiences and insights while strengthening community organization and citizen empowerment (Barry et al. 2008; Albrechts 2012; Popa et al. 2015).

The Collaborative Expert and transdisciplinary knowledge generation face significant barriers related to entrenched power relations as well as a lack of familiarity and experience with deliberative practices among all involved parties, experts and non-experts alike. Likewise, there is a significant epistemological difference because knowledge emerges from deliberation rather than being imparted by the technical expert to non-experts (Klein 1996; Wynne 1996; Castán Broto et al. 2009). These more democratic forms of knowledge generation have emerged in political cultures such as Denmark and the Netherlands where citizen participation in political decision-making processes is encouraged but continues to be an exception to the rule even in these countries.

Toward an ecosystem of expertise

Table 18.1 provides a summary of the three models of sustainable expertise described above. The Communicative Expert translates scientific and technical knowledge to the public, the Collegial Expert works to identify synergies between disciplines to address sustainability challenges, and the Collaborative Expert opens up knowledge production to formal and informal experts. Each model has specific epistemological and disciplinary assumptions as well as attitudes toward other experts and the public, and preferred approaches to knowledge production and application. As a whole, this comprises an "ecosystem of expertise" where different niches are filled by different interpretations of what it means to be an expert in contemporary society (Brand and Karvonen 2007).

There are clearly merits to each approach and a general conclusion we forward is that it is not important to determine which model is most effective. In other words, all of the models should be welcomed because they challenge the traditional expert in different ways. They encourage holders of specialized knowledge to consider their multiple roles as experts, citizens, and participants in democratic politics, to reflect on their individual strengths and weaknesses, and to orient their work and allegiance toward one or more of these models. For example, those of us who are more adept at identifying synergies between knowledge domains might choose the Collegial Expert model while those of us who are better at communicating with the public might choose the Communicative Expert model or the Collaborative Expert model.

We recognize that such a pluralist attitude toward expertise is an idealized perspective whose implementation faces numerous hurdles in terms of institutional incentives, allocation of research money, social status, public visibility, vested interests, power gradients and so on. We describe these modes of expertise in the hope of arousing debate among all of those individuals and organization who are engaged and implicated in knowledge production processes, and as an invitation to strategize on methods to overcome these barriers.

Table 18.1 Three models of expertise to address sustainability

	The communicative expert	The collegial expert	The collaborative expert
Cliché role	"The educator"	"The broker"	"The democrat"
Mode of knowledge production	Monodisciplinary	Interdisciplinary	Transdisciplinary
Epistemological assumptions	Core set of scientific principles	Synergism of core sets	Emergent from discourse between experts and non-experts
Attitude toward other experts	Competitors	Necessary partners	One of many sources of knowledge
Attitude toward the public	Receivers of expert wisdom	Not considered	Co-designers in generating solutions
Knowledge flow	Top-down	Lateral and discursive	Multidirectional and discursive
Role of power	Competition between disciplines for the exclusive claim to truth	Emergent from collaboration between disciplines	Shared and contested between experts and non-experts

Conclusion

In this chapter, we explored the rise of expertise and how specialized knowledge informs sustainability issues and decision-making. We presented a framework with three modes of sustainable expertise that are based on monodisciplinary, interdisciplinary, and transdisciplinary modes of knowledge production. Each of these approaches to specialized knowledge provides different opportunities to address sustainability issues while enhancing the relationship between experts and non-experts.

An important question that lurks in the background of this framework is the motivation for technical experts to change their attitudes toward other disciplines and the public. Why should experts sacrifice their relatively privileged position in society? Three points come to mind that may make a pluralist approach to expertise more appealing. First, the models of expertise presented above can potentially help to reverse the erosion of trust between experts and the general public. Sustainability problem-solving can be seen as a way to bridge the gulf between those with specialized knowledge and those who are affected by the application of this knowledge (Karvonen 2011). Second, the quest for more sustainable solutions can appeal to the problem-solving disposition shared by most, if not all, experts. The promise of more socially acceptable and, in essence, more effective solutions is worth the work required to renovate existing approaches to knowledge generation (Nowotny 2003). Finally, some experts are professionals with an explicit ethical responsibility to serve society and our proposed models of expertise offer a way to fulfill this social contract. For experts without professional standing, an ethical argument can appeal to the citizen within the technical expert (see Chapter 27). We leave these normative dimensions of expertise for future study but recognize that this is perhaps the most formidable barrier to adopting these models.

In conclusion, we propose two challenges to all experts. First, it is important to maintain a bird's-eye view of the whole "ecosociotechnical" system (Moore et al. 2019) and resist the temptation to simplify complex processes. There are advantages to all of these models of expertise, and the goal of experts should be to acknowledge and appreciate these advantages and then seek strategic alliances to develop new modes of practice. Second, a daunting barrier to the further development of these modes of expertise involves the institutional barriers that inhibit new modes of knowledge production (Petts et al. 2008). There is a need to lobby for the dissolution or at least lowering of these barriers if sustainable approaches to scientific and technological development are to become widespread. The former is an individual challenge, the latter a political one.

References

Acuto, M., Steenmans, K., Iwaszuk, E., and Ortega-Garza, L. (2019) "Informing Urban Governance? Boundary-Spanning Organisations and the Ecosystem of Urban Data," *Area 51*: 94–103.

Agyeman, J. (2013) *Introducing Just Sustainabilities: Policy, Planning, and Practice.* London: Zed Books.

Albrechts, L. (2012) "Reframing Strategic Spatial Planning by Using a Coproduction Perspective," *Planning Theory 12*: 46–63.

Autio, E., Kanninen, S., and Gustafsson, R. (2008) "First- and Second-Order Additionality and Learning Outcomes in Collaborative R&D Programs," *Research Policy 37*: 59–76.

Bäckstrand, K. (2004) "Scientisation vs. Civic Expertise in Environmental Governance: Eco-feminist, Eco-modern and Post-modern Responses," *Environmental Politics 13*: 695–714.

Barry, A., Born, G., and Weszkalnys, G. (2008) "Logics of Interdisciplinarity," *Economy and Society 37*: 20–49.

Beck, U. (1992) *Risk Society: Towards a New Modernity*, Newbury Park, CA: Sage Publications.

Brand, R. and Fischer, J. (2013) "Overcoming the Technophilia/Technophobia Split in Environmental Discourse," *Environmental Politics 22*: 235–254.

Brand, R. and Karvonen, A. (2007) "The Ecosystem of Expertise: Complementary Knowledges for Sustainable Development," *Sustainability: Science, Practice, and Policy 3*: 21–31.

Brown, M. B. (2009) *Science in Democracy: Expertise, Institutions, and Representation*, London: The MIT Press.

Campbell, S. (1996) "Green Cities, Growing Cities, Just Cities? Urban Planning and the Contradictions of Sustainable Development," *Journal of the American Planning Association 62*: 296–312.

Carayannis, E. G. and Campbell, D. F. (2010) "Triple Helix, Quadruple Helix and Quintuple Helix and How Do Knowledge, Innovation and the Environment Relate to Each Other?: A Proposed Framework for a Trans-Disciplinary Analysis of Sustainable Development and Social Ecology," *International Journal of Social Ecology and Sustainable Development 1*: 41–69.

Carolan, M. S. (2006) "Sustainable Agriculture, Science and the Co-production of 'Expert' Knowledge: The Value of Interactional Expertise," *Local Environment: The International Journal of Justice and Sustainability 11*: 421–431.

Castán Broto, V., Gislason, M. and Ehlers, M. H. (2009) "Practising Interdisciplinarity in the Interplay between Disciplines: Experiences of Established Researchers," *Environmental Science & Policy 12*: 922–933.

Collins, H. M. and Evans, R. (2002) "The Third Wave of Science Studies: Studies of Expertise and Experience," *Social Studies of Science 32*: 235–296.

Coutard, O. and Rutherford, J. (eds.) (2015) *Beyond the Networked City: Infrastructure Reconfigurations and Urban Change in the North and South*. London: Routledge.

Darbellay, F. (2015) "Rethinking Inter- and Transdisciplinarity: Undisciplined Knowledge and the Emergence of a New Thought Style," *Futures 65*: 163–174.

Dryzek, J. S. (2013) *The Politics of the Earth: Environmental Discourses*, 3rd Ed., New York: Oxford University Press.

ENOLL (European Network of Living Labs) (2021) European Network of Living Labs website, https://enoll.org.

European Union Green Deal (2021) European Union Green Deal website, https://ec.europa.eu/info/strategy/priorities-2019-2024/european-green-deal_en.

EUSEA (European Science Engagement Platform). (2021) European Science Engagement Platform website, https://eusea.info.

Evans, R. and Marvin, S. (2006) "Researching the Sustainable City: Three Modes of Interdisciplinarity," *Environment and Planning A 38*: 1009–1028.

Evans, J., Karvonen, A., and Raven, R. (eds.) (2016) *The Experimental City*. London: Routledge.

Fischer, F. (1990) *Technocracy and the Politics of Expertise*, Newbury Park, CA: Sage Publications.

Fischer, F. (2000) *Citizens, Experts, and the Environment: The Politics of Local Knowledge*, Durham, NC: Duke University Press.

Fischer, F. (2006) "Environmental Expertise and Civic Ecology: Linking the University and its Metropolitan Community," in A. C. Nelson, B. L. Allen, and D. L. Trauger (eds.) *Toward a Resilient Metropolis: The Role of State and Land Grant Universities in the 21st Century*, Alexandria, VA: Metropolitan Institute at Virginia Tech.

Fischer, F. (2019) "Knowledge Politics and Post-Truth in Climate Denial: On the Social Construction of Alternative Facts," *Critical Policy Studies 13*: 133–152.

Foley, R., Rushforth, R., Kalinowski, T., and Bennett, I. (2020) "From Public Engagement to Research Intervention: Analyzing Processes and Exploring Outcomes in Urban Techno-Politics," *Science as Culture 29*: 1–26.

Forester, J. (1999) *The Deliberative Practitioner: Encouraging Participatory Planning Processes*, Cambridge, MA: MIT Press.

Gibbons, M., Limoges, C., Nowotny, H., Schwartzman, S., Scott, P., and Trow, M. (1994) *The New Production of Knowledge: The Dynamics of Science and Research in Contemporary Societies*, London: Sage Publications.

Graham, S. and Marvin, S. (2001) *Splintering Urbanism: Networked Infrastructures, Technological Mobilities and the Urban Condition*, London: Routledge.

Guy, S., Marvin, S., and Medd, W. (eds.) (2011) *Shaping Urban Infrastructures: Intermediaries and the Governance of Socio-Technical Networks*, London: Routledge.

Habermas, J. (2015) *The Lure of Technocracy*, London: John Wiley & Sons.

Hajer, M. A. (1995) *The Politics of Environmental Discourse: Ecological Modernization and the Policy Process*, Oxford: Clarendon Press.

Hannigan, J. (2014) *Environmental Sociology: A Social Constructionist Perspective*, 3rd ed., New York: Routledge.

Haywood, B. K. and Besley, J. C. (2014) "Education, Outreach, and Inclusive Engagement: Towards Integrated Indicators of Successful Program Outcomes in Participatory Science," *Public Understanding of Science 23*: 92–106.

Healey, P. (2004) "Creativity and Urban Governance," *Policy Studies 25*: 87–102.

Hecker, S., Haklay, M., Bowser, A., Makuch, Z., and Vogel, J. (eds) (2018) *Citizen Science: Innovation in Open Science, Society and Policy*, London: UCL Press.

Hickman, L. A. (1992) "Populism and the Cult of the Expert," in L. Winner (ed.) *Democracy in a Technological Society*, Dordrecht, the Netherlands: Kluwer Academic Publishers.

Hölscher, K. and Frantzeskaki, N. (eds) (2020) *Transformative Climate Governance*, London: Palgrave Macmillan.

Irwin, A. (1995) *Citizen Science: A Study of People, Expertise and Sustainable Development*, New York: Routledge.

Jasanoff, S. (2004) *States of Knowledge: The Co-Production of Science and Social Order*, London: Routledge.

Jepson, E. J., Jr. (2019) "Sustainability Science and Planning: A Crucial Collaboration," *Planning Theory & Practice 20*: 53–69.

Karvonen, A. (2011) *Politics of Urban Runoff: Nature, Technology, and the Sustainable City*, London: The MIT Press.

Karvonen, A. (2020) "Urban Techno-Politics: Knowing, Governing, and Imagining the City," *Science as Culture 29*: 417–424.

Karvonen, A. and van Heur, B. (2014) "Urban Laboratories: Experiments in Reworking Cities," *International Journal of Urban and Regional Research 38*: 379–392.

Kivimaa, P., Boon, W., Hyysalo, S. and Klerkx, L. (2019) "Towards a Typology of Intermediaries in Sustainability Transitions: A Systematic Review and a Research Agenda," *Research Policy 48*: 1062–1075.

Klein, J. T. (1996) *Crossing Boundaries. Knowledge, Disciplinarities, and Interdisciplinarities*, Charlottesville: University Press of Virginia.

Kleinman, D. L. (2000) "Democratizations of Science and Technology," in D. L. Kleinman (ed.) *Science, Technology, and Democracy*, Albany: SUNY Press.

Lane, M. B. and McDonald, G. (2005) "Community-Based Environmental Planning: Operational Dilemmas, Planning Principles and Possible Remedies," *Journal of Environmental Planning and Management 48*: 709–731.

Latour, B. (1999) *Pandora's Hope: Essays on the Reality of Science Studies*, London: Harvard University Press.

Leydesdorff, L. and Ward, J. (2005) "Science Shops: A Kaleidoscope of Science–Society Collaborations in Europe," *Public Understanding of Science 14*: 353–372.

Liu, L. Y. J., Lindblom, C. E., and Woodhouse, E. J. (2013) *The Policy-Making Process*, Englewood Cliffs, NJ: Prentice-Hall.

Lytle, M. H. (2007) *The Gentle Subversive: Rachel Carson, Silent Spring, and the Rise of the Environmental Movement*, New York: Oxford University Press.

Marvin, S., Bulkeley, H., Mai, L., McCormick, K., and Palgan, Y. V. (eds.) (2018) *Urban Living Labs: Experimenting with City Futures*, London: Routledge.

Mede, N. G. and Schäfer, M. S. (2020) "Science-Related Populism: Conceptualizing Populist Demands Toward Science," *Public Understanding of Science 29*: 473–491.

Merkley, E. (2020) "Anti-intellectualism, Populism, and Motivated Resistance to Expert Consensus," *Public Opinion Quarterly 84*: 24–48.

Metzger, J. and Lindblad, J. (eds) (2020) *Dilemmas of Sustainable Urban Development: A View from Practice*, London: Routledge.

Miller, T. R., Baird, T. D., Littlefield, C. M., Kofinas, G., Chapin III, F. S. and Redman, C. L. (2008) "Epistemological Pluralism: Reorganizing Interdisciplinary Research," *Ecology and Society 13*: 46.

Moore, S. A. (2007) *Alternative Routes to the Sustainable City: Austin, Curitiba, and Frankfurt*, Lanham, MD: Lexington Books.

Moore, S. A. and Karvonen, A. (2008) "Sustainable Architecture in Context," *Science Studies 21*: 29–46.

Moore, S. A., Torrado, M., and Joslin, N. (2019) "Knowledge Production for Interdependent Critical Infrastructures: Constructing Context-Rich Relationships Across Ecosociotechnical Boundaries," *Environmental Science & Policy 99*: 97–104.

Nowotny, H. (2003) "Democratising Expertise and Socially Robust Knowledge," *Science and Public Policy 30*: 151–156.

Nowotny, H., Scott, P., and Gibbons, M. (2001) *Re-thinking Science: Knowledge and the Public in an Age of Uncertainty*, Malden, MA: Blackwell.

Owen, R., Macnaghten, P. and Stilgoe, J. (2012) "Responsible Research and Innovation: From Science in Society to Science for Society, with Society," *Science and Public Policy 39*: 751–760.

Petts, J., Owens, S., and Bulkeley, H. (2008) "Crossing Boundaries: Interdisciplinarity in the Context of Urban Environments," *Geoforum 39*: 593–601.

Polk, M. (2015) "Transdisciplinary Co-production: Designing and Testing a Transdisciplinary Research Framework for Societal Problem Solving," *Futures 65*: 110–122.

Popa, F., Guillermin, M., and Dedeurwaerdere, T. (2015) "A Pragmatist Approach to Transdisciplinarity in Sustainability Research: From Complex Systems Theory to Reflexive Science," *Futures 65*: 45–56.

Prugh, T., Costanza, R., and Daly, H. (2000) *The Local Politics of Global Sustainability*, Washington, DC: Island Press.

Raco, M. and Savini, F. (eds.) (2019) *Planning and Knowledge: How New Forms of Technocracy are Shaping Contemporary Cities*, Chicago: Policy Press.

Ramadier, T. (2004) "Transdisciplinarity and its Challenges: The Case of Urban Studies," *Futures 36*: 423–439.

Ramaswami, A., Bettencourt, L., Clarens, A., Das, S., Fitzgerald, G., Irwin, E., Pataki, D., Pincetl, S., Seto, K., and Waddel, P. (2018) *Sustainable Urban Systems: Articulating a Long-Term Convergence Research Agenda: A Report by the Advisory Committee for Environmental Research and Education.* Washington, D.C.: National Science Foundation.

Ribeiro, R. and Lima, F. P. (2016) "The Value of Practice: A Critique of Interactional Expertise," *Social Studies of Science 46*: 282–311.

Rip, A., Misa, T. J., and Schot, J. (eds.) (1995) *Managing Technology in Society: The Approach of Constructive Technology Assessment*, New York: St. Martin's Press.

Robinson, J. (2008) "Being Undisciplined: Transgressions and Intersections in Academia and Beyond," *Futures 40*: 70–86.

Rydin, Y. (2007) "Re-examining the Role of Knowledge within Planning Theory," *Planning Theory 6*: 52–68.

Sandercock, L. (1998) *Towards Cosmopolis,* London: Wiley.

Schlosberg, D. (1999) *Environmental Justice and the New Pluralism*, New York: Oxford University Press.

Schot, J. and Rip, A. (1997) "The Past and Future of Constructive Technology Assessment," *Technological Forecasting and Social Change 54*: 251–268.

Sclove, R. (1992) "The Nuts and Bolts of Democracy: Democratic Theory and Technological Design," in L. Winner (ed.) *Democracy in a Technological Society*, Dordrecht, the Netherlands: Kluwer Academic Publishers.

Scott, J. C. (2020) *Seeing Like a State: How Certain Schemes to Improve the Human Condition Have Failed*, New Haven, CT: Yale University Press.

Seely, B. E. (1996) "State Engineers as Policymakers: Apolitical Experts in a Federalist System," in J. R. Rogers, D. Kennon, R. T. Jaske, and F. E. Griggs, Jr. (eds.) *Civil Engineering History: Engineers Make History*, New York: ASCE.

Selinger, E. and Crease, R. P. (2006) "Dreyfus on Expertise: The Limits of Phe-nomenological Analysis," in E. Selinger and R. P. Crease (eds.) *The Philosophy of Expertise*, New York: Columbia University Press.

Simon, D., Palmer, H., Riise, J., Smit, W., and Valencia, S. (2018) "The Challenges of Transdisciplinary Knowledge Production: From Unilocal to Comparative Research," *Environment and Urbanization 30*: 481–500.

Smith, A., Fressoli, M. and Thomas, H. (2014) "Grassroots Innovation Movements: Challenges and Contributions," *Journal of Cleaner Production 63*: 114–124.

Smith, A. and Seyfang, G. (2013) "Constructing Grassroots Innovations for Sustainability," *Global Environmental Change 23*: 827–829.

Snow, C. E. and Dibner, K. A. (2016) *Science Literacy: Concepts, Contexts, and Consequences.* Washington, DC: National Academies Press.

Stilgoe, J., Irwin, A., and Jones, K. (2006) *The Received Wisdom: Opening Up Expert Advice*, London: Demos.

Stilgoe, J., Owen, R., and Macnaghten, P. (2013) "Developing a Framework for Responsible Innovation," *Research Policy 42*: 1568–1580.

Tate, J., Mulugetta, Y., Sharland, R., and Hills, P. (1998) "Sustainability: The Technocratic Challenge," *Town Planning Review 69*: 65–86.

Turner, S. (2001) "What is the Problem with Experts?" *Social Studies of Science 31*: 123–149.

Whatmore, S. J. (2009) "Mapping Knowledge Controversies: Science, Democracy and the Redistribution of Expertise," *Progress in Human Geography 33*: 587–598.

Wilsdon, J., and Willis, R. (2004) *See-through Science: Why Public Engagement Needs to Move Upstream*, London: Demos.

Wynne, B. (1996) "May the Sheep Safely Graze? A Reflexive View of the Expert–Lay Knowledge Divide," in S. Lash, B. Szerszynski, and B. Wynne (eds.) *Risk, Environment and Modernity: Towards a New Ecology*, Thousand Oaks, CA: Sage Publications.

Yearley, S. (2000) "Making Systematic Sense of Public Discontents with Expert Knowledge: Two Analytical Approaches and a Case Study," *Public Understanding of Science 9*: 105–122.

19

Uncertainty

Risk, technology and the future

Jennifer Yarnold, Ray Maher, Karen Hussey and Stephen Dovers

We are witnessing a "New Renaissance" of science and innovation as technologies converge to create ever more powerful platforms, products and processes (Roco and Bainbridge 2013). Rapid transformations arising from scientific discovery are not new: in modern history, the agricultural revolution, industrial revolution and the information revolution altered the fabric of society through socio-economic, political and environmental impacts (Boyden 1987; Hussey et al. 2019). Human ingenuity – driven by desperation, commercial drive, or both – spawned remarkable achievements in medicine, food, infrastructure and quality of life.

Yet, modern technologies have brought unintended and even catastrophic outcomes. The consequences of our ingenuity on the environment and human life have made names such as "Chernobyl," "Bhopal," "Fukushima" and "Exxon Valdez" synonymous with disasters. Then there are more chronic and diffuse impacts: deforestation (see Chapter 42) to biodiversity (see Chapter 41), plastics to marine systems (see Chapters 37 and 39), invasive species to ecosystem health, greenhouse gases to climate (see Chapter 32), various pollutants and wastes (see Chapters 35 and 36) to ozone depletion (see Chapter 31) and contamination of soil, air (see Chapter 30) and water (see Chapter 38). We have learned of the risks that modern innovations can bring, particularly in terms of scale and irreversibility. These events spurred developments in how we assess and manage risk and uncertainty; other events in future will too. As Rose (1991) observed, reform is "so often contingent upon an exogenous crisis."

Adverse effects aside, many technologies exist or are being developed that could address environmental problems (Hennen 1999), while others have enabled assessment techniques that reduce uncertainty through better risk analysis (Mitter and Hussey 2019) (see Table 19.1). Population growth, economic growth and consumption growth (see Chapter 17) are eroding our collective future and threaten social–ecological collapse. While addressing these underlying drivers is essential to a sustainable future, *yet more innovative technology* will also feature. The increasing demands on natural resources and vital earth systems are more complex, interconnected and urgent than ever before (see Chapters 15–17), requiring multiple channels through which solutions are sought (see the chapters in Part III of this volume). Increasingly, policymakers and regulators are in unchartered waters where they must act *promptly* but with *caution*, weighing potential benefits and harms.

DOI: 10.4324/9781003008873-23

Jennifer Yarnold et al.

Uncertainty with regard to environmental and human health has become more pervasive since global environmental politics emerged as a field of research in the 1980s (see Chapter 2). Some scholars describe the twenty-first century as the "era of VUCA": volatility, uncertainty, complexity and ambiguity (Hart 2020). The foci has shifted from discrete issues, to more holistic themes around human interactions with earth systems, such as the Anthropocene (see Chapter 15) and the "planetary boundaries" which define the "safe space" in which humanity must operate (Rockström et al. 2009). Throughout these developments, climate change (see Chapter 32) and the response of governments has dominated the discourse in global environmental politics, often framed around threats to food, water, energy security and the stability of global political systems (see Chapter 20 and Barnett et al. 2008). These developments have inspired new, more integrated ways of thinking toward sustainability, emphasizing the need for earth system governance (see Chapter 21), interdisciplinary and cross-sector collaboration (Brandt et al. 2013; Clark 2007; Miller 2013), systems approaches (Meadows 1999) and design thinking (Birkeland 2012; Maher et al. 2018).

The remainder of this chapter explores emerging technologies, including uncertainties around the risk and benefits they present. It looks at how regulators and policymakers manage these uncertainties and the problems they pose for existing and future governance arrangements. We begin with a sketch of significant "cutting-edge" technology platforms, their potential application to some emerging global problems and the factors that challenge governments. We examine the current treatment of uncertainty in policy and regulatory design and risk management approaches, and gaps in existing governance arrangements, and conclude with possible solutions.

Technology innovation and global challenges

Governments and industries worldwide are investing heavily in new areas of science and technology, including artificial intelligence (AI) and information technology, renewable energy and storage, nanotechnology, biotechnology, and synthetic biology (SynBio). During the first half of 2020, equity investment in SynBio increased 57 percent from the year before amidst the economic fallout of Covid-19 (Cumbers 2020), while AI's contributions to the global economy are estimated to exceed US$15.6 trillion by 2030 – more than the total of China's and India's contributions combined (Kohli 2019). From these platforms, numerous applications are emerging, spanning multiple sectors, and addressing many global challenges (see Table 19.1).

Table 19.1 Global trends, technology innovations and technology/science platforms

Global trends

• Global climate change	• Rising populism, nationalism, geopolitical tensions
• Growing consumption of resources & waste, associated resource scarcity	• Circular economy & bioeconomy
• Declining health of ecosystems & ecosystem services	• Biodiversity loss, 6th Mass Extinction
• Covid-19 and emerging diseases	• Water scarcity, unequal access
• Energy demand and transition	• Sustainable business, policy & lifestyles
• Improving food production, nutrition	• Corporate global citizenship
• Shifting centres of economic activity	• Social life in a technological world
• Increasing wastes, pollution impacts	• Demographic change
• Prolific information sharing	• Disinformation

Technology innovations

• Vaccines	• Smart drugs	• Telehealth	• Advanced diagnostics
• Chatbots & virtual assistants	• Driverless cars & drones	• Facial and language recognition	• Immersive communications
• Bottom-up manufacturing	• Substitute materials	• Smart materials	• High-conductivity materials
• Strong, lightweight materials	• Battery storage	• Biofuels	• Eco-industrial parks
• Point-of-use energy generation	• Renewable energy	• Safer nuclear power	• Smart grids
• Intensive food production	• Resilient crops	• Crop & soil monitoring	• Targeted pesticides
• Climate control	• Better food preservation	• High value crops	• Smart irrigation
• Water desalination	• Thermal insulators	• Efficient resource extraction	• Carbon sequestration
• Automated traffic management	• Automated services & targeted marketing	• Advanced prosthetics	• At-source water purification & water separation
• Social media	• Cryptocurrencies	• Sustainable production processes	• Sorting & resource recovery technologies

Technology/science platforms

• Nanotechnology	• Synthetic biology	• Information technology	• Bio-interfaces
• Geo-engineering	• Artificial intelligence & Robotics	• Biotechnology	• Web 2.0
• Cognitive technology	• Computational chemistry & biology	• Blockchain / distributed ledger systems	• Data interfaces

Source: Adapted from Maynard and Harper (2011).

Novel solutions are emerging to address intractable environmental problems. Examples include synthetic enhancement of carbon metabolism in crops to enable net long-term carbon storage in soil (Wurtzel et al. 2019), using nanomaterials for high-power rechargeable battery systems (Xin et al. 2017) and AI-optimization of distributed energy systems (Mou 2019). More radical solutions being explored are climate geoengineering approaches such as marine cloud brightening (Ahlm et al. 2017) or solar radiation management to reduce the amount of sunlight absorbed by the atmosphere (Jinnah and Nicholson 2019), and using gene drives to eliminate invasive species by propagating lethal genes through populations (Esvelt et al. 2014). Such technologies are controversial, with justifiable concern around the ethics and implications of "messing with nature." Incomplete understanding of complex, evolving natural systems (Weis 2008) means that we cannot foresee the impacts of these technologies confidently.

Challenges of emerging technologies to traditional regulation

From a governance perspective, there are distinct features of emerging technologies and how they interface with today's globally connected society that present both old and new challenges for policymakers. Hussey et al. (2019) define these features as (1) technology splitting, (2) technology convergence, (3) new economic value and (4) regulatory divergence. Technology splitting occurs when a technology developed for one application or sector migrates to other applications and/or sectors. For example, drones originally developed for defense and surveillance are now widely used in agriculture for crop monitoring; gene technologies first applied to improve food crops are now used for a range of medicinal and industrial applications and are being advanced to create new synthetic life forms. Most pervasive are digital technologies, including the Internet and Internet-connected devices, which are now embedded in nearly every aspect of our lives.

Then there is technology convergence, the phenomenon where two or more technologies combine to create a novel application that could otherwise not occur. This differs from mere "piggybacking" in which one technology is used to improve another. For example, integrating AI systems into driverless cars ultimately produces a vehicle with the same purpose (to get from one place to another), despite its sophistication (piggybacking). In contrast, nano-encapsulated bioactive medicines and non-toxic pesticides have only been made possible through the convergence of nanotechnology and biotechnology (Mitter et al. 2017). Some believe that transformational breakthroughs will emerge in these spaces "between" platforms, creating new synergistic possibilities (see Figure 19.1).

This combination of technology convergence and splitting has led to an exponential increase in applications spanning multiple sectors and industries, where regulation and governance have struggled to keep pace. This so-called "pacing problem" has often led to decision-makers falling back on outdated policy frameworks and institutionalized ways of thinking and/or incremental changes in law and regulation that are not fit-for-purpose (Downes 2009; Dunlop 2010).

While splitting and convergence relate to a technology itself, the next two challenges relate to the world in which they are applied. The combination of low-cost inputs and new infrastructure underpin the ubiquitous nature of disruptive technologies. As demand for a new technology increases, so does the demand for the key inputs which the technology relies upon, creating *new economic value* for such commodities, sometimes where no value existed before. The most palpable example, of course, is the demand for fossil fuels following the invention of the internal combustion engine, subsequent dependence upon it, and the consequences for global warming. More recently, a growing "circular economy" movement – driven by diminishing virgin resources and increasing wastes – creates new economic value for biological and technical waste streams once destined for landfill (Stahel 2016). The advent of smart phones and laptops has seen a doubling of demand for tantalum in recent decades, and with it an increased risk of mineral conflicts in countries such as the Congo and Rwanda (Nassar 2017). Such inputs need not be tangible: data that drives AI systems, and genetic sequences used in biotechnology, have become highly sought-after commodities. Decision-makers must not only consider the impacts of the processes and applications of emerging technologies, but also of the resources underpinning them.

Finally, economic globalization (see Chapter 24) and digitalization have created highly complex supply chains and greatly diminished the role of physical borders in jurisdictional containment, enabling the rapid transfer of technologies and their diffusion to other parts of the world (the so-called speed of transfer). Consequently, regulatory divergence between

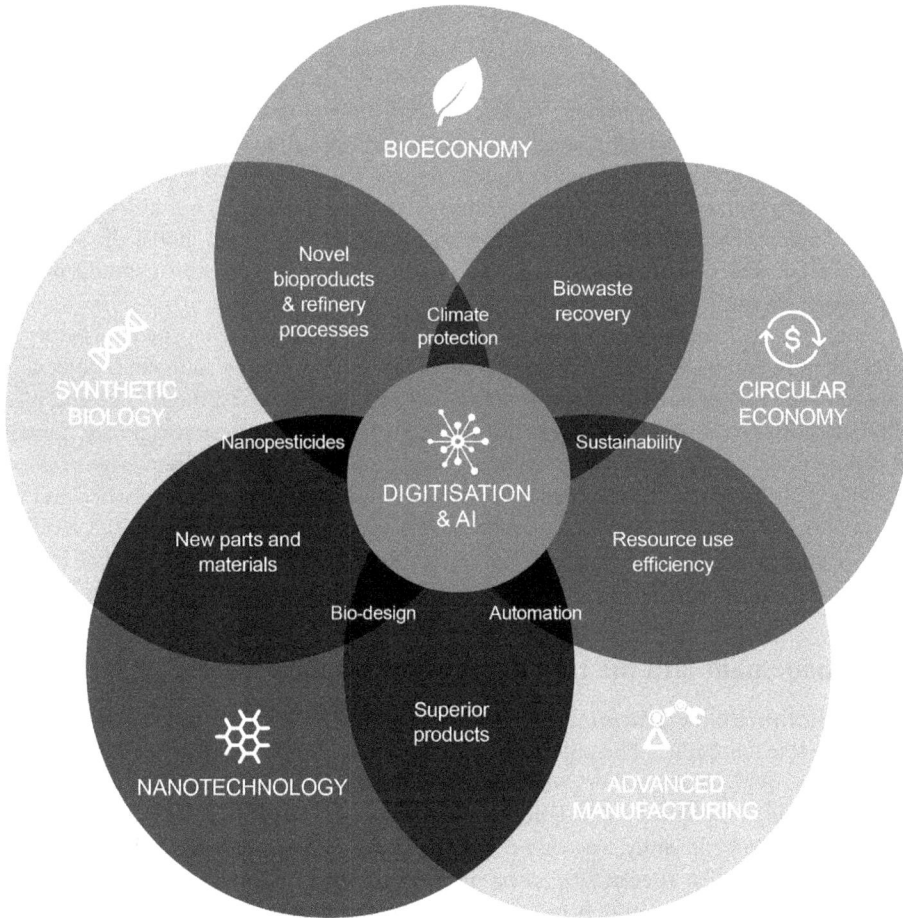

Figure 19.1 The convergence and splitting of current emerging technology and innovation platforms

Source: Adapted from Hussey et al. (2019).

jurisdictions becomes problematic, not least to ensure that environmental problems are addressed holistically, rather than merely shifted elsewhere, but also because of major shifts in *where* technology is created and by *whom*. Until recently, science and technology innovation was dominated by the United States and the EU, under well-developed risk governance regimes ranging from implementation of the precautionary principle to statutory bodies to oversee technology development and risk assessments.

Recently, the locus of technology development has shifted from west to east. China has invested heavily in science and technology and is poised to surpass US investment by 2022; it also overtook the United States in the number of patents filed in 2019 (WIPO, 2020). Likewise, South Korea is becoming a dominant player, with the largest growth in number of patents filed in 2019 and a doubling of GDP spent on R&D over the past decade (Mou 2019; WIPO 2020). Technology development is no longer exclusive to government-funded research institutes or large corporations: a college student can create a tech giant such as Facebook, while do-it-yourself biotechnology, 3D printing and accessible equipment empower individuals to develop innovations that previously required costly infrastructure and development teams (Maynard and Harper 2011). Technologies can be developed and deployed before communities and governments are aware of their potential, limiting the ability to establish safeguards before they become widespread and difficult to control (Collingridge 1980; Thierer 2018). This technology control dilemma has led to calls for a moratorium on certain emerging technologies and field experiments, for example with geoengineering (Shepherd et al. 2009: 37). But such moratoriums create not only an economic, but also a moral and ethical dilemma: if an emerging technology has potential to address a global challenge but possibly create others, should it be given the opportunity to prove itself or should risks be understood rigorously before it is applied and perhaps even developed?

Combined, these features create immense pressure on both the scientific community to rapidly identify and assess risks, and on decision-makers to weigh an abundance of new information and countervailing risks to develop appropriate, timely responses. The end result can be such that risk assessments and policy decisions fail to integrate across policy domains and jurisdictions, or the uptake of technology that proliferates ahead of regulation (Hussey and Pittock 2012). These developments have important implications for how risks and uncertainty are assessed and managed in the context of global environmental politics, not least because rigorous risk management requires a high degree of institutional capacity, but also because it reflects societal values.

Risk management and the treatment of uncertainty

Risk assessments are the primary tools used to assess the level of risk and potential harms posed by new technologies. The objectives are twofold: to determine whether a new process or product should be accepted for commercial release; and to determine the level of regulatory scrutiny (if any) required. The latter helps to ensure that regulatory resources are prioritized and deployed efficiently, and that control measures are appropriate to the level of risk (McHughen 2016). The assessment process involves systematic identification and analysis of risks, including the likelihood of occurrence and level of impact, followed by a *risk evaluation* to appraise the level of tolerable risk (see Figure 19.2). The scope and issues covered in a risk assessment are often subjective and shaped by values and politics (Cothern 2019; Bond et al. 2001; Gibson et al. 2005; Partidário and Clark 2000; Russell et al. 2011). Assessments may consider risks to the environment, human and/or animal health and safety, and/or other social or economic impacts, at a variety of scales. There are also risks in not realizing positive

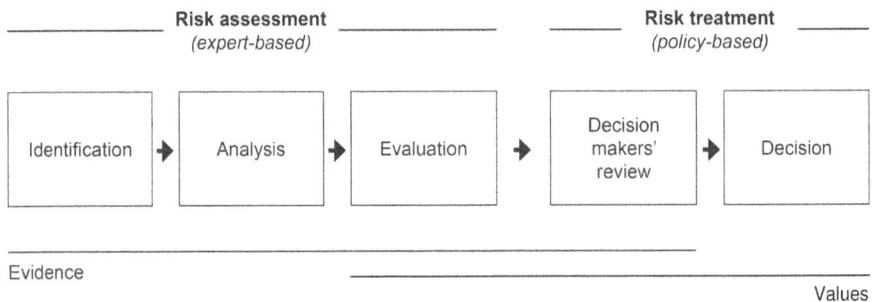

Figure 19.2   Risk management and decision-making process
Source: Adapted from Aven (2016); see also Hansson and Aven (2014).

outcomes (opportunities), particularly for environmental concerns where the risk of doing nothing could be more detrimental (e.g., action on climate change).

Then there are risks related to uncertainty, which is distinguishable from risk itself and is the subject of several examinations (see Cothern 2019; Dovers and Handmer 1995; Hansson and Aven 2014; Hoffman and Hammonds 1994; Peel 2006). Here, we use the definitions of Petersen (2006: 3): "Risk describes situations where there is uncertainty about which outcome will eventuate, but the range of all possible events is known and plausible probabilities can be assigned to each and every possible event". With uncertainty, outcomes are "known but there is insufficient information to permit objective probabilities to be assigned ... Ignorance (or radical uncertainty) is where objective (or sometimes even subjective) probabilities cannot be assigned to outcomes and the full range of possible events cannot be identified".

Uncertainty around emerging technologies arises from many factors. First, there is uncertainty about the consequences of a technology itself: how it might converge with other technologies or split across sectors, how it might be used or misused, unknown harms to environmental or human health, unintended benefits, whether adverse effects may be direct, indirect and/or cumulative, and when such impacts could become evident. Second, there is uncertainty about the nature of the world in which those technologies will be deployed now and in the future (Rip 2006), including social values, changes in the efficacy of the assessment and regulatory regime over time, or a changing operating environment because of demographic, geopolitical, climate, economic or other technological changes. Third, there are knowledge gaps regarding natural processes and phenomena, particularly feedbacks from human activities for which no precedents exist. Because of such complexity, the impacts of emerging technology on the environment often extend to radical uncertainty as we cannot foresee how their application may evolve.

The precautionary principle is a key principle used to deal with uncertainty where there is incomplete information or inconclusive evidence and where potential threats are irreversible, unknown or difficult to estimate (Fisher 2007). Its intention is not to provide a final decision, but rather to emphasize delayed decision-making and to proceed with caution until further knowledge is gained (McHughen 2016). Originally conceived for environmental protection, the principle's scope has extended to human health and safety, following high-profile regulatory failures in the 1980s. Such events, notably the BSE ("mad cow" disease) and dioxin crises, led to greater public scrutiny and fear of food and safety issues, and a general decline in trust of science and government (Anyshchenko and Yarnold 2020). Various formulations of the precautionary principle have been incorporated into international agreements including the Convention on Biological Diversity 1992, and the Framework Convention on Climate Change 1992, and in national law and policy (Peterson 2006). One of the earliest conceptions is found under Principle 15 of the UN Rio Declaration on Environment and Development, which states that "where there are threats of serious or irreversible damage, lack of full scientific certainty shall not be used as a reason for postponing cost-effective measures to prevent environmental degradation" (United Nations 1992).

There is large and contested literature on how the precautionary principle is defined, how it should inform law and policy, and how it should be applied (Fisher et al. 2006). On the premise that it is impossible to eliminate uncertainty, some have argued that the principle is a political construct used to skew public sentiment, rather than one based on science (McHughen 2016). Other discourse relates to its interpretation, including the "strength" of the principle in terms of assigning where the burden-of-proof lies. "Weak" versions put the onus on those advocating precaution to justify such precaution, while "strong" versions place the burden-of-proof on those advocating that the processes will not cause harm. Proponents of "strong" use argue that those introducing a technology generally have more power than those who oppose it (e.g., a multinational in support of a new GM crop versus local farmers opposing it), and must demonstrate its safety to mitigate the power imbalance. Such interpretations have led to regulatory divergence in the approaches taken by jurisdictions (Aldy and Kip Viscusi 2014), even sparking a trade conflict between the United States and the EU over GM crops and food products (Anyshchenko and Yarnold 2020). A further debate concerns vagueness in how the precautionary principle should be applied, leading to a lack of consistencies in precautionary approaches taken toward assessment and approval methods and processes (Dovers 2006; Russell et al. 2011).

While such issues are not easily resolved, one thing is certain: the elimination of all risk and uncertainty is not only impractical, but fundamentally impossible (Stahl, et al. 2003). The question is how to manage risk at a level deemed acceptable? The scientific, policy and political difficulty lies in defining what level of risk is socially acceptable, and how that level is defined. Risk assessments do not provide society and policymakers with "yes/no" answers: the analysis is scientific, but the evaluation and management is subjective (see Figure 19.2). Even prior to risk analysis, subjective decisions are made as to the scope of impacts to be assessed, the thresholds that define the "tolerable" level of risk, the standards of proof required, and the triggers used to determine whether regulatory scrutiny is required (Dovers 2006). A crucial point emerges: uncertainty and risk are value-laden and thus not reducible to simple metrics. Social, scientific, and political arguments over technologies consider impacts beyond the purpose of the technology. They may be founded on different perspectives on the role of science, how much care is afforded to future generations, and whether nature is valued for its own sake. Risk and uncertainty are not simply objective; they are socially constructed and negotiated through politics (Handmer and Dovers 2013; Smithson 1989).

Values, risk perception and attitudes toward technology

Attitudes and perceptions of risk and uncertainty vary widely (Bammer and Smithson 2008; Sjoberg 1979). Different experts, industries, communities and political groups perceive risk differently depending on the technology and its context. Often, those deciding on acceptable risk are not at risk themselves, which may lead to poor outcomes for those that are. Many factors influence technology risk perception, including trust in science, corporations and governments, proximity to the site of application, visibility and/or familiarity of the technology, and one's understanding of the technology. For example, when asked about perceived health impacts of living near wind turbines, those living near them were half as likely to be concerned as those who did not (Baxter et al. 2013). These issues characterize perceptions of many of the technologies listed in Table 19.1, which are shaped by the media, scientists and politicians.

Trust in science and politics has shifted recently and significantly. The role of new media in shaping societal perceptions challenges technology regulation, nationally and internationally (see Rae 2020). A combination of new, expanding information and communication technologies (e.g., social media, smartphones) and the rise of populist political movements and conspiracy theorists have polarized attitudes and popularized opposition to some scientific facts and/or official policy directions. Segments of society have become resistant to rational argument or evidence, well beyond the healthy critical stance required for scrutiny and accountability, as witnessed in climate denialism (see Chapter 32), "anti-vaxxers" (those opposed to vaccinations) and conspiracies surrounding Covid-19 (see, e.g., Gidron and Hall

2017; Kennedy 2019). For risk management, these shifts may be passing or remain marginal, or they may prelude larger moves in social values.

The international response to the Covid-19 pandemic has increased the visibility and arguably the recognized value of well-established assessment institutions and processes. Whether this global visibility will strengthen or weaken acceptance of scientific knowledge and procedures in assessing efficacy and safety remains to be seen.

Prospects for fostering responsible technologies and innovation

Summarizing previous sections, challenges in assessing risks and uncertainties surrounding technologies are complicated by the following, often shared characteristics:

- Risks may be direct, indirect and/or cumulative, and can occur over the long term.
- Perceptions of risk and uncertainty vary widely.
- Applications often spread across multiple sectors and sources of scientific discovery.
- Technological change often outpaces regulation and policy.
- The impacts of a technology may manifest at a distance from where it was developed or first applied.

These characteristics apply to traditional and emergent technologies, and the extent to which risks will be captured by existing regimes depends on governance and institutional regimes. It is clear, however, that the technologies emerging today, and the world into which they are emerging, are different from earlier periods.

To provide robust institutional arrangements and regulatory systems that can keep pace with sweeping technological advancements in the complex world we live in, we must consider: what governance mechanisms currently exist? What suite of approaches to take? And when to regulate, review and revise? These questions require a holistic, systems-based understanding of the technology's processes, applications and supply chains; more proactive, iterative and adaptive tools; and a governance regime that provides appropriate oversight to protect the environment and society without unnecessarily stifling innovation. What might the key elements of such a governance regime be?

In considering options for governing environmental impacts of technology, it is important to consider both *explicit* and *implicit* forms of governance. Explicit governance refers to specific initiatives which are enacted by those with authority to do so, such as formal policy, regulation and law, and publicly funded reviews of emerging technologies by governments or politically neutral agencies. In contrast, implicit governance refers to when other (nongovernment) components of the social/political/economic system act in a way which contributes to governing technology, without necessarily having the authority or intent to do so. This may include voluntary initiatives, market-based mechanisms, organizational culture (Rip 2006: 280), and third-party certification schemes. Applying a suite of complementary explicit and implicit governance mechanisms can help to navigate uncertainty and adapt more rapidly to evolving technology. Some overarching principles and approaches are summarized below.

Anticipatory governance

Anticipatory governance is a broad concept related to managing uncertain events of the future in the present through mechanisms that build adaptive capacity and preparedness.

Anticipatory governance features heavily as an approach to address the rapid transformations of emerging technologies and their consequences to environmental, social and economic systems, promoted widely in the disciplines of sustainability science, climate adaptation and resilience, environmental governance, national security and responsible innovation. A critical analysis of anticipatory governance and related themes by Muiderman et al. (2020) categorized perspectives based on three elements: conceptions of the future (as being probable, plausible, pluralistic or performative); implications for governance and policy in the present (planning, building adaptive capacity, mobilizing diverse stakeholders, or interrogation of existing regimes); and the ultimate purpose of employing anticipatory governance (to reduce future risks, to reflexively navigate uncertain futures, to co-create new futures or to identify implications of current political systems). Critical features of the tools employed focus on foresight, adaptive regulation, stakeholder engagement and policy integration. These are explored further below.

Environmental policy integration

Traditionally, governance processes manage the complexity of society by dividing it into discrete departments and sectors, and applying policy and regulation within those. This can be effective for discrete challenges, but is inadequate for others. It is particularly emerging technologies that split and impact across multiple sectors, stakeholder groups, and generations. One way to overcome regulatory silos is to integrate environmental policy into non-environmental policy areas, as well as into governments' fiscal protocols. This integration can happen at any level of government from whole-of-government policy to specific issues. An example is integrating the impacts of medicines in wastewater effluent on biodiversity and ecosystems into US FDA approval processes (Jordan and Lenschow 2010). Integrated approaches to policy are being applied to sustainability and climate change, and provide a valuable precedent for governing emerging technologies with broad reach such as AI and gene editing.

Adaptive regulation

The traditional approach to the development of new regulation often occurs over several years. New rules are discussed and drafted; the draft is made available for public comment; feedback is reviewed and integrated into the framework; and finally, new regulations are delivered. By the time regulations are implemented, they are often outdated by technology developments and lengthy time frames often deter reviews and reforms needed to address future regulatory gaps. Eggers et al. (2018) suggest shifting from this "regulate once and forget" approach to one based on early and frequent iterative trial-learn-feedback loops integrated with co-design and deep engagement with industry and experts.

One adaptive approach is a regulatory sandbox, which enables a technology to be deployed in the world but within a controlled environment to test products, services and business models arising from new technologies. Sandboxes have been used in the finance sector, by allowing firms to test out payment transfers with cryptocurrencies, with assurances made to a select number of participating clients (Allen 2019). The United States has recently given approval for a regulatory sandbox toward unmanned aerial navigation (Eggers et al. 2018). Of course, tangible, and often irreversible environmental and human health effects require greater containment efforts for such sandboxes, which at times may be inappropriate.

Collaborative design and multi-stakeholder engagement

Technologies impact a wide range of people across society and future generations. To ensure they bring the most good for the most people, the divergent yet legitimate perspectives of various stakeholders must be comprehensively considered to formulate the best approach of technology risk management and to build trust in how they are being used (Peel 2006: 203). Democratic principles instruct that many different groups should contribute to how these technologies are governed. Regulators can benefit from multi-stakeholder engagement, for example by reducing the economic costs associated with regulatory divergence and the risks of potential failures through various types of multi-stakeholder collaboration.

Internationally, coordination between nations through formal agreements can improve regulation of Internet-based technologies, international trade and geographically complex supply chains. Agreements might be to share information between regulators or develop processes to address different regulatory requirements. Intergovernmental and transnational institutions could play a key role in facilitating such agreements, paving the way for standard frameworks and guidelines (Eggers et al. 2018).

Policy laboratories are becoming a popular tool for regulatory co-design (Olejniczak et al. 2020). They foster multiple perspectives to challenges and solutions. This can reduce silos when developing regulation by involving stakeholders from research, corporations, industry bodies, government agencies and communities. When co-designing policy and regulation, care must be taken to manage conflicts of interest and the differing social power of stakeholders through transparency and public accountability. To engage meaningfully, individuals must understand the risks and benefits of new technologies and governments must, in turn, understand their attitudes and concerns. Such feedback enables government to prioritize the areas where technologies are most needed, and reduces the risk of investing in technologies that people do not want or outright oppose.

Fit-for-purpose and well-targeted regulation

To be effective in managing risk and uncertainty, regulation and policy responses must "fit" the characteristics of emerging technologies, rather than apply a one-size-fits-all strategy or extend outdated legacy mechanisms. For example, waste streams such as fly ash, glass, tires and plastics provide cheap and sustainable alternatives for construction and road materials, but require new or reformed construction standards and codes to implement them at scale.

Given the complexity of supply chains and range of impacts that emerging technologies can have on society and the environment, it can be difficult to identify precisely which part of the process is most suited for regulatory measures. The temptation for governments might be to impose risk assessments at some or all stages of that supply chain, but that is a temptation that must be avoided for two reasons: first, the innovation supply chain is more complicated and dynamic than ever before and more assessments are likely to add confusion where clarity is important; and second, the imposition of more assessments imposes unwelcome constraints on competitiveness and potentially on trade (Botterill and Daugbjerg 2011; Hussey and Kenyon 2011). Regulation is most important for components of the technology supply chain and its application which pose greatest risk on the environment. Where supply chains are complex, a whole-of-systems perspective is important to identify these key points of risk and developing harmonious and streamlined regulation. State-based regulation presents a challenge for international trade (see Chapter 24), imposing impediments to trade and investment and altering the competitive position of firms.

While recognizing that regulation is itself a public good, and that the uncertainty and risk surrounding new technologies demands appropriate regulatory oversight, there is a need to find ways of regulating scientific discoveries without unnecessarily stifling them. One relatively urgent need is to achieve as much harmony within and between countries as possible, thus reducing regulatory divergences, administrative redundancy, and barriers to trade and consumer welfare (Hussey and Kenyon 2011). However, this should not lead to the "lowest common denominator" in regulation, if we are to avoid further environmental catastrophes.

Market-based mechanisms and third-party certification

The fast-paced, geographically and temporally disconnected nature of technology research and development means the state is limited in how much oversight it can provide. From the 1990s onwards, in parallel with the institutionalization of the precautionary approach, disappointment with the success of traditional command and control regulation to manage environmental externalities saw the emergence of so-called second- and third-generation environmental policy (see Jordan et al. 2003; Knill and Lenschow 2003; Lenschow 2002). Including policy instruments such as market-based mechanisms, voluntary initiatives and third-party certification schemes, the global environmental politics literature is replete with studies on the efficacy of such instruments in mitigating environmental impacts. Voluntary initiatives from science and industry sectors seek to place checks and balances on their own research. These may include eco-labeling and third-party certification which provide a competitive advantage to firms with more sustainable products; moratoriums on particularly sensitive areas of research; or extensive, government-funded voluntary initiatives to examine the potential risks of a technology (Rip 2006). Initiatives from industry can complement traditional forms of governance, but may also postpone much-needed regulation. A narrow industry framing of risk may exclude important perspectives and issues which, as we have seen, are needed to fully comprehend a technology's impact on society and the environment across time.

Prioritizing technology for environmental and human good

Thus far, we have discussed prospects for governing uncertain environmental impacts of technology while taking the types of technology being developed as a given. However, there is a substantial disconnect between the problems that most need to be fixed and the technologies being developed. Notwithstanding extraordinary advancements made in medical research, problems of hunger, poverty and disease endure as environmental problems escalate, in contrast to the extraordinary proliferation of technological innovations such as smart phones and high-definition television (Bozeman et al. 2011). Technology development may not align with the types and scale of challenges we face. Policy instruments can play an important role in addressing this market failure by incorporating the full cost of environmental externalities into markets, removing subsidies to polluting industries, and providing incentives for responsible innovators. This also prompts questions such as should technologies specifically designed to address significant global challenges (like climate change) be afforded less restrictive regulation than those providing less critical benefits (such as entertainment technology)?

Conclusion

Governments are key in assessing potential risks, defining oversight structures and systems, promoting transparency, protecting citizens, informing the public and steering responsible development of technologies. The theme of stronger, more integrated but also more objective institutions is reiterated in the academic literature, which reinforces our contention that oversight from government or from publicly funded but politically neutral agencies is essential to address market failures. Effectively governing uncertain impacts of emerging technologies on the environment is critical for averting the next global environmental challenge. However, traditional approaches to developing government policy and regulation can be slow to change and often react to emerging technologies which may be too late to manage environmental impacts (as evident of incidents that have become powerful by-words, such as Bhopal, Fukushima and *Exxon Valdez*). It is inevitable that there will be other incidents, and that these will drive further debate and policy change in the future. The fact that we do not know what these will be, how serious, or what changes they will drive underlines both the importance and difficulty of issues surrounding technology, risk, and regulation. Our capacity to address global environmental challenges quickly and responsibly will, to a great extent, rely on humankind's capacity to rethink global technological governance, which, in turn, relies on developing inclusive, adaptive and innovative institutions that reflect modern dynamic circumstances.

References

Ahlm, L., Jones, A., Stjern, C. W., Muri, H., Kravitz, B. and Kristjánsson, J. E. (2017). Marine cloud brightening–as effective without clouds. *Atmospheric Chemistry and Physics* 17(21): 13071–13087.

Aldy, J. E. and W. Kip Viscusi (2014). Environmental risk and uncertainty. In M. Machina and K. Viscusi (eds) *Handbook of the economics of risk and uncertainty*. North-Holland: Elsevier, 1: 601–649.

Allen, H. J. (2019). Regulatory sandboxes. *George Washington Law Review* 87: 579–645.

Anyshchenko, A. and Yarnold, J. (2020). From 'mad cow'crisis to synthetic biology: Challenges to EU regulation of GMOs beyond the European context. *International Environmental Agreements: Politics, Law and Economics*, 1–14. https://doi.org/10.1007/s10784-020-09516-1.

Aven, T. (2016). Risk assessment and risk management: Review of recent advances on their foundation. *European Journal of Operational Research* 253(1): 1–13.

Bammer, G. and Smithson, M. (eds) (2008). *Uncertainty and risk: Multidisciplinary perspectives.* London: Earthscan.

Barnett, J., Matthew, R.A. and O'Brien, K. (2008). Global environmental change and human security. In Brauch, H.G., Oswald Spring, Ú., Mesjasz, C., Grin, J., Dunay, P., Behera, N.C., et al. (eds) *Globalization and Environmental challenges: Reconceptualizing security in the 21st century.* New York: Springer.

Baxter, J., Morzaria, R., & Hirsch, R. (2013). A case-control study of support/opposition to wind turbines: Perceptions of health risk, economic benefits, and community conflict. *Energy Policy* 61: 931–943.

Birkeland, J. (2012). Design blindness in sustainable development: From closed to open systems design thinking. *Journal of Urban Design* 17(2): 163–187.

Bond, R., Curran, J., Kirkpatrick, C., Lee, N. and Francis, P. (2001). Integrated impact assessment for sustainable development: A case study approach. *World Development* 29: 1011–1024.

Botterill, L. and Daugbjerg, C. (2011). Engaging with private sector standards: A case study of GLOBALG.A.P. *Australian Journal of International Affairs* 65(4): 488–504.

Boyden, S. (1987). *Western civilization in biological perspective: Patterns in biohistory.* Oxford: Clarendon Press.

Bozeman, B., Slade, C.P. and Hirsch, P. (2011). Inequity in the distribution of science and technology outcomes: A conceptual model. *Policy Sciences* 44: 231–248.

Brandt, P., Ernst, A., Gralla, F., Luederitz, C., Lang, D.J., Newig, J., Reinert, F., Abson, D.J. and von Wehrden, H. (2013). A review of transdisciplinary research in sustainability science. *Ecological Economics* 92: 1–15.

Clark, W. C. (2007). *Sustainability science: A room of its own.* National Academy of Sciences.

Collingridge, D. (1980). *The social control of technology.* London: Pinter.

Cothern, C. R. (2019). *Handbook for environmental risk decision making: Values, perceptions, and ethics.* Boca Raton, FL: CRC Press.

Cumbers, J. (2020). Synthetic biology startups raised $3 billion in the first half of 2020. *Forbes* [online], 9 September 2020. https://www.forbes.com/

Dovers, S. (2006). Precautionary policy assessment for sustainability. In Fisher, E., Jones, J. and von Schomberg, R. (eds) *Implementing the precautionary principle: Perspectives and prospects*. Cheltenham: Edward Elgar.

Dovers, S. and Handmer, J. (1995). Ignorance, the precautionary principle, and sustainability. *Ambio*. 24: 92–97.

Downes, L. (2009). *The laws of disruption: Harnessing the new forces that govern life and business in the digital age*. New York: Basic Books.

Dunlop, C.A. (2010). The temporal dimension of knowledge and the limits of policy appraisal: Biofuels policy in the UK. *Policy Sciences*. 43: 343–363.

Eggers, W. D., Turley, M. and Kishnani, P. J. D. C. F. G. I. (2018). *The future of regulation: Principles for regulating emerging technologies*. Deloitte Insights. https://www2.deloitte.com/us/en/insights/industry/public-sector/future-of-regulation/regulating-emerging-technology.html.

Esvelt, K. M., Smidler, A. L., Catteruccia, F. and Church, G. M. J. E. (2014). Emerging technology: Concerning RNA-guided gene drives for the alteration of wild populations. *Elife* 3: e03401.

Fisher, E. (2007). *Risk regulation and administrative constitutionalism*. Oxford: Hart Publishing.

Fisher, E. C., Jones, J. S. and von Schomberg, R. (eds) (2006). *Implementing the precautionary principle: Perspectives and prospects*. Cheltenham: Edward Elgar Publishing.

Gibson, R.B., Hassan, S., Holtz, S., Tansey, J. and Whitelaw, G. (2005). *Sustainability assessment: Criteria, processes and applications*. London: Earthscan.

Gidron, N. and Hall, P.A. (2017). The politics of social status: Economic and social roots of the populist right. *British Journal of Sociology* 68 (S1): S58–84.

Handmer, J. and Dovers, S. (2013). *The handbook of disaster and emergency policies and institutions*. 2nd edition. London: Earthscan

Hansson, S.O. and Aven, T. (2014). Is risk analysis scientific? *Risk Analysis* 34(7): 1173–1183.

Hoffman, F.O. and Hammonds, J.S.J.R.A. (1994). Propagation of uncertainty in risk assessments: The need to distinguish between uncertainty due to lack of knowledge and uncertainty due to variability. *Risk Analysis* 14(5): 707–712.

Hennen, L. (1999). Participatory technology assessment: A response to technical modernity? *Science and Public Policy* 26(5): 303–312.

Hussey, K. and Kenyon, D. (2011). Regulatory divergences: A barrier to trade and a potential source of trade disputes. *Australian Journal of International Affairs* 65(4): 381–393.

Hussey, K. and Pittock, J. (2012). The energy–water nexus: Managing the links between energy and water for a sustainable future. *Ecology and Society* 17(1): 31.

Hussey, K., Yarnold, J., McEwan, C., Maher, R., Henman, P., Radke, A., Curtis, C., Fidelman, P., Vickers, C. and Brolan, C. (2019). *Policy futures: Regulating the new economy*. Brisbane: Centre for Policy Futures. https://indd.adobe.com/view/53cfa428-19d1-4a8e-942a-e043c379dc1a.

Jinnah, S. and Nicholson, S. (2019). Introduction to the symposium on 'geoengineering: Governing solar radiation management. *Environmental Politics* 28(3): 385–396.

Jordan, A. and Lenschow, A. (2010). Environmental policy integration: A state of the art review. *Environmental Policy and Governance* 20(3): 147–158.

Jordan, A., Wurzel, R.K.W., Zito, A.R. and Bruckner, L. (2003). European governance and the transfer of 'New' Environmental Policy Instruments (NEPIs) in the European Union. *Public Administration* 18(3): 555–574.

Kennedy, J. (2019). Populist politics and vaccine hesitancy in Western Europe: An analysis of national level data. *European Journal of Public Health* 29: 512–516.

Knill, C. and Lenschow, A. (2003). Modes of regulation in the governance of the European Union: Towards a comprehensive evaluation. *European Integration Online Papers* (EIoP) 7(1).

Kohli, T. (2019). AI's contribution to the global economy will bypass that of China and India by 2030, to reach $15.7 trillion. World Economic Forum [online], 17 September 2019. https://www.weforum.org/agenda/2019/09/artificial-intelligence-meets-biotechnology/

Lenschow, A. (2002). New regulatory approaches in 'greening' EU policies. *European Law Journal* 8(1): 19–37.

Maher, R., Maher, M., Mann, S. and McAlpine, C. (2018). Integrating design thinking with sustainability science: A research through design approach. *Sustainability Science* 13(6): 1565–1587.

Maynard, A.D. and Harper, T. (2011). *Building a sustainable future: Rethinking the role of technology innovation in an increasingly interdependent, complex and resource-constrained world.* A Report for the World Economic Forum Global Agenda Council on Emerging Technologies.

McHughen, A. (2016). A critical assessment of regulatory triggers for products of biotechnology: product vs. process. *GM Crops & Food* 7(3–4): 125–158.

Meadows, D.H. (1999). *Leverage points: Places to intervene in a system.* Hartland VC: Sustainability Institute.

Miller, T. (2013). Constructing sustainability science: Emerging perspectives and research trajectories. *Sustainability Science* 8(2): 279–293.

Mitter, N. and Hussey, K. (2019). Moving policy and regulation forward for nanotechnology applications in agriculture. *Nature Nanotechnology* 14(6): 508–510.

Mitter, N., Worrall, E.A., Robinson, K.E., Li, P., Jain, R.G., Taochy, C., Fletcher, S.J., Carroll, G.Q., Lu, M. and Xu, Z.P.J.N.P. (2017). Clay nanosheets for topical delivery of RNAi for sustained protection against plant viruses. *Nature Plants* 3(2): 1–10.

Mou, X. (2019). *Artificial intelligence: Investment trends and selected industry uses.* Note 71, Sep 19. Washington: International Finance Corporation.

Muiderman, K, Phillips, A, Vervoort, J. and Biermann, F. (2020). Four approaches to anticipatory climate governance: different conceptions of the future and implications for the present. *WIREs Climate Change* 11: e673.

Nassar, N.T. (2017). Shifts and trends in the global anthropogenic stocks and flows of tantalum. *Resources, Conservation and Recycling* 125: 233–250.

Olejniczak, K., Borkowska-Waszak, S., Domaradzka-Widła, A. and Park, Y. (2020). Policy labs: The next frontier of policy design and evaluation? *Policy & Politics* 48(1): 89–110.

Partidário, M.R. and Clark, R. (2000). Introduction. In Partidário, M. and Clark, R. (eds) *Perspectives on strategic environmental assessment.* Boca Raton FL: Lewis Publishers.

Peel, J. (2006). Precautionary only in name? Tensions between precaution and risk assessment in the Australian GMO regulatory framework. In Fisher, E., Jones, J. and von Schomberg, R. (eds) *Implementing the precautionary principle: Perspectives and prospects.* Cheltenham: Edward Elgar.

Peterson, D.C. (2006). *Precaution: Principles and practice in Australian environmental and natural resource management.* 50th Annual Australian Agricultural and Resource Economics Society Conference, Manly, New South Wales 8–10 February 2006.

Rae, M. (2020). Hyperpartisan news: Rethinking the media for populist politics. *New Media and Society* Online 1st: doi.org/10.1177.1461444–820910416.

Rip, A. (2006). The tension between fiction and precaution in nanotechnology. In Fisher, E., Jones, J. and von Schomberg, R. (eds) *Implementing the precautionary principle: Perspectives and prospects.* Cheltenham: Edward Elgar.

Rockström, J., Steffen, W., Noone, K., Persson, Å., Chapin III, F.S., Lambin, E., Lenton, T.M., Scheffer, M., Folke, C., Schellnhuber, H.J.J.E. and society (2009). Planetary boundaries: Exploring the safe operating space for humanity. *Ecology and Society* 14(2): 1–33.

Roco, M.C. and Bainbridge, W.S. (eds) (2013). *Converging technologies for improving human performance: Nanotechnology, biotechnology, information technology and cognitive science.* Springer Science & Business Media.

Rose, R. (1991). What is lesson-drawing? *Journal of Public Policy* 11: 3–30.

Russell, A.W., Vanclay, F.M., Salisbury, J.G. and Aslin, H.J. (2011). Technology assessment in Australia: The case for a formal agency to improve advice to policymakers. *Policy Sciences* 44: 157–177.

Shepherd, J., Caldeira, K., Cox, P., Haigh, J., Keith, D., Launder, B. et al. (2009). *Geoengineering the climate: Science, governance and uncertainty.* London: Royal Society.

Sjoberg, L. (1979). Strength of belief and risk. *Policy Sciences* 11: 39–57.

Smithson, M. (1989). *Ignorance and uncertainty: Emerging paradigms.* New York: Springer-Verlag.

Stahel, W.R.J.N. (2016). The circular economy. *Nature* 531(7595): 435–438.

Stahl, B.C., Lichtenstein, Y. and Mangan, A. (2003). The limits of risk management: A social construction approach. *Communications of the International Information Management Association* 3(3): 15–22.

't Hart, P. (2020). COVID-19 crisis future fault lines: Managing the conflicts we have to have. *The Mandarin*, 15 April. https://www.themandarin.com.au/130611-covid-19-crisis-future-fault-lines-managing-the-conflicts-we-have-to-have/.

Thierer, A. (2018). The pacing problem, the collingridge dilemma & technological determinism, August 2016, 2018. https://techliberation.com/

United Nations (1992). *Rio declaration on environment and development.* Annex 1 of the UN Conference on Environment and Development, Rio de Janeiro, 3–14 June. New York: UN.

Weis, E. (2008). Fundamentals of complex evolving systems: A primer, Inst. of Social Ecology, IFF-Fac. for Interdisciplinary Studies, Klagenfurt.

World Intellectual Property Organization (WIPO) (2012). *World intellectual property indicators, 2011 edition.* Online. http://www.wipo.int/ipstats/en/statistics/patents/.

World Intellectual Property Organization (WIPO) (2020). *Intellectual property statistics.* https://www.wipo.int/ipstats/en/index.html#data.

Wurtzel, E.T., Vickers, C.E., Hanson, A.D., Millar, A.H., Cooper, M., Voss-Fels, K.P., Nikel, P.I. and Erb, T.J. (2019). Revolutionizing agriculture with synthetic biology. *Nature Plants* 5(12): 1207–1210.

Xin, S., You, Y., Wang, S., Gao, H.-C., Yin, Y.-X. and Guo, Y.-G.J.A.E.L. (2017). Solid-state lithium metal batteries promoted by nanotechnology: Progress and prospects. *ACS Energy Letters* 2(6): 1385–1394.

20

Environmental security

International scope, national regimes and the human dimension

Sabina W. Lautensach and Alexander K. Lautensach

Environmental security, despite its multiple contested interpretations, can be broadly defined as the fulfillment of welfare and basic needs being secured against "critical adverse effects caused directly or indirectly by environmental change" (Barnett 2007: 5). This definition gives rise to the following agenda: security *from* environmental threats such as "natural" disasters; security *of* environmental integrity, in the sense that human welfare depends on the contributions of ecosystem functions and services (Zurlini and Müller 2008); the development of resilience as "the capacity of a system, be it an individual, a forest, a city or an economy, to deal with change and continue to develop" (SRC 2015: 1); and security *for* the environment to support the transition to the sustainable survival and wellbeing of human individuals and communities (Arnott et al. 2020). Environmental security guards against a diverse range of environmental changes, from the subtle, gradual kind, such as global climate change, to sudden, acute disasters, such as floods or earthquakes. It includes "natural" and anthropogenic disasters, as well as threats resulting from willful, negligent or inadvertent "behaviour directed against the environment [which] might be seen as a threat to the security of the people or political entities associated with that environment" (Weintraub 1995: 554–555). The definition implies a merely instrumental regard for nonhuman nature, in line with the globally dominant anthropocentric ethic, which could be problematic (Lautensach 2009; see Chapter 27).

Although "environment" usually refers to the natural environment, the definition extends to social, cultural, built and other anthropogenic varieties of environment. Similarly, shortfalls in environmental security inevitably impact our welfare in the social, political, economic, health-related and cultural realms. Because of that multidisciplinary context, the assessment of environmental security goes beyond purely environmental parameters such as resource availability, pollution or climate variability, and takes into account diverse variables concerning the extraction, use, distribution, regulation and administration of natural resources, the culturally contingent values and aspirations that inform a society's treatment of its environment, its susceptibility to adverse environmental change, and its resilience to cope (Barnett 2009). The latter aspect renders it imperative to look at the future, to include into the assessment informed projections about the sustainability of practices in the light of likely changes and trends, from which threats to environmental security might emerge.

DOI: 10.4324/9781003008873-24

Forces that threaten environmental security

What distinguished our species and its immediate ancestors from other animals during the past million years or so was an exceptional proclivity for expanding our habitat, for colonizing diverse environments by adapting to them and by modifying them to our needs (Rees 2004), mainly through the development and use of technology (Dilworth 2010), but reaching back to the earliest examples of planting, fertilizing and controlled burning. As noted by numerous authors (e.g., in Heinberg and Lerch 2010), that proclivity is now for the first time no longer working in our favor. It no longer contributes to our resilience to the same extent. By modifying almost every ecosystem on the planet, by extracting and processing resources in ever more complex ways, and by harnessing diverse energy sources to great effect we succeeded in propagating far beyond the numbers of other medium-sized omnivorous mammals. Humans and domesticated mammals now comprise 36 percent and 60 percent of all mammalian biomass, respectively, while wild mammals make up the remaining 4 percent (Bar-On et al. 2018). As humans introduce competitor species, modify ecosystems, deplete habitats, and modify landscapes and climates, our environmental impact has driven millions of species into extinction and continues to do so at an increasing rate (Kolbert 2014; IPBES 2019). Our limited skills at managing ecosystems could not prevent the "trophic downgrading" of many systems into less complex stable states with fewer species (Estes et al. 2011).

The tragedy in this development lies not just in the irreversible loss of life forms that took millions of years to evolve; because we are still part of the web-like communities of species, subject to dependencies from which no species can be exempt, the loss of biodiversity (UNEP 2020) threatens our very own security (Hawkins 2020; see Chapter 41). Moreover, the loss of biodiversity represents only one way in which the global environmental crisis of the Anthropocene threatens our security. The crisis is also evident in the increasing rates of resource depletion as the global human population and its consumption continue to grow out of control (see Chapter 16). Pollution continues with its disastrous effects on climate, habitat quality and public health. Greenhouse gas accumulation and associated climate change by 2030 may well exceed the IPCC's worst-case scenarios (see Chapter 32). Other manifestations of the crisis include ocean acidification (see Chapter 39), stratospheric ozone depletion (see Chapter 31), phosphorus turnover, freshwater overuse (see Chapter 38), land mismanagement (see Chapter 44), chemical pollution (see Chapter 35) and nitrogen turnover. They correspond to environmental boundaries, describing the safe operating space for humanity, that we are at various stages of transgressing (Griggs et al. 2013). All those factors are not only strongly interconnected, they synergistically threaten our food supply and are thus to be regarded as threats to environmental security (Ehrlich and Harte 2018).

The Anthropocene marks a stage in natural history where many manifestations of environmental change, along with the associated security threats, are anthropogenic, meaning that they are caused by the excessive impacts of human activities on the biosphere (UNEP–MAB 2005; see Chapters 15 and 16). Five causative and self-reinforcing processes have been identified: economic growth, population growth, technological expansion, arms races, and the growing inequity between rich and poor (Coates 1991; Daly and Cobb 1994). Regardless of the technological windfalls we reaped in the past, the capacities of local ecosystems remain limited. Generally, the environmental impact I of a human population on local ecosystems is described by the $I = PAT$ formula, where P means population size, A stands for the affluence or economic means per capita and T represents the technological impact per capita (York et al. 2003; Grossman 2012). The maximum sustainable impact, also referred to as carrying capacity (Curry 2011: 126), is thus described as the product of three variables. It can be reached by small populations with a high-impact lifestyle or by larger populations, where each individual demands less in terms of support services. Cultural differences and historical inequities contribute to that variation.

When animal populations exceed the carrying capacity they enter into ecological *overshoot*, where the services of the local ecosystems are being overtaxed and, depending on their fragility, they may undergo irreversible structural changes (Catton 1980; McMichael 2001; Meadows et al. 2004). Inevitably, the consequence of overshoot for the population is that various biological regulatory mechanisms lead to a decrease in population health, causing an eventual drop in population size, below the system's carrying capacity. Those mechanisms generally include infectious disease, predators, malnutrition, aggressive territorial behavior, outmigration, and infertility. These mechanisms decrease the size of the population to the effect that its impact once more measures below the system's carrying capacity. Numerous case studies from animal populations have allowed ecologists to characterize and predict those dynamics with impressive accuracy. Humans have often, though not always, succeeded in avoiding or escaping such regulatory impacts. However, with the advent of global overshoot in the Anthropocene our choices have become more limited. Global limits are more rigid than regional ones; while local overshoot could be compensated with the help of extraneous resources, off-planet resources remain closed to us. Lasting environmental degradation in many regions, pandemics (which can be triggered by environmental damage, as in the case of HIV and possibly Covid-19), and other manifestations of the global environmental crisis are jeopardizing environmental security, causing, in turn, economic instability, global resource scarcity, social polarization, and cultural conflict (Dobkowski and Wallimann 2002). Overshoot compromises environmental security through both our unprecedented numbers and our counterproductive practices.

The extent of overshoot represents a measure of future environmental security. It can be represented in terms of environmental boundaries being transgressed (Griggs et al. 2013), or using the area of productive land required to support a population's lifestyle. This is referred to as that population's ecological footprint (Wackernagel and Rees 1996). A population whose footprint exceeds the amount of accessible productive land and water is clearly in local overshoot (Chambers et al. 2000). This may not always have immediate negative consequences for their environmental security, as they may obtain the shortfall from other regions that are either underpopulated, defenseless, or otherwise disempowered; empires sustained their overshoot for centuries through their colonialist exploitation of outlying regions. Globally, our shortfall is appropriated from future generations as we continue to draw on the biosphere's "ecological capital" instead of making do with the interest. This has been likened to a family of apple farmers who increasingly rely on the sale of applewood in addition to selling their fruit (Lautensach 2013). A comparison of humanity's collective global footprint with Earth's bioproductive capacity suggests that we first entered overshoot in the mid-1980s and that it has steadily increased since then to a current level above 156 percent, equivalent to at least 1.56 planets (Almond et al. 2020) – (in the apple farm analogy, that would correspond to approximately two-thirds of a year's income coming from fruit and one-third from wood). Globally, the gross inequity in resource demands per person has no bearing on the basic fact that collectively we live "beyond our means" (UNEP–MAB 2005). It should be clear that further growth cannot address the problem.

Animal case studies indicate that overshoot can persist for considerable time spans. With human populations, too, some indicators of wellbeing still increase even though ecosystem services are deteriorating and ecological integrity has been widely compromised (Raudsepp-Hearne et al. 2010). However, in the absence of any scientific evidence to the contrary, it seems safe to assume that no animal species could maintain this extent of bio-mass appropriation, environmental impact, and species displacement for significant lengths of time. Moreover, studies of regional precedents indicated that, as a secondary consequence of overshoot, the ecological carrying capacity gradually decreases because of irreversible damage to ecological support structures through desertification, salinization and erosion of previously productive agricultural areas, resulting in severe environmental insecurity for the inhabitants in later years (Catton 1980; Mancini et al. 2017; Wackernagel and Beyers 2019).

Because ecosystems come in a large variety of sizes, overshoot can assume local (e.g., is-lands), regional (e.g., the Sahel) or global (e.g., the Earth Overshoot Day concept, GFN 2020) dimensions. In many historical cases local overshoot compromised environmental security, resulting in the loss of cultural heritage, invasions, natural disasters or the cultures disap-peared altogether (e.g., the Greenland Norse) (Diamond 2005). Sometimes the crucial factor that drives a population into overshoot is not its excessive growth but a change in the physical environment that lowers the region's carrying capacity. In the case of the Greenland Norse it was cooling of the climate that shortened the local growing season, which contributed to the communities' disappearance (Folger 2017). The threat of "collapse," then, in whatever form it may present itself (McAnany and Yoffee 2009), often includes a decrease in environmental security. It presents an imperative for the culture to adapt quickly enough to the new con-tingencies, taking into account the extent of environmental damage, climate change, hostile neighbors, friendly trade partners and a culture's ability to respond – which was successfully accomplished, for example, by the Icelanders and the Tikopians (Diamond 2005). Icelanders initially destroyed much of their forest cover but then in time learned to regenerate and care for it. The island of Tikopia, situated roughly between the Solomons and Vanuatu, excels in its unique regulation of population size and footprint.

The concept of collapse of a culture or society has been the subject of long-standing debate. It seems clear that Diamond (2005) represented it too narrowly, while some of his critics (e.g., McAnany and Yoffee 2009) tried to deny that it occurs much at all. In our view, the synthesis lies in the heterogeneity of what might feel like collapse to a people – loss of cultural heritage, language, collective identity, health and welfare, self-determination, eco-nomic independence and the like. Defined in this pluralistic way, cultures are seen collapsing all over the world, evident in the examples of First Nations in the Americas that suffer from rampant poverty, poor health, widespread drug use, appalling statistics of suicide and domes-tic violence. Those situations were usually preceded by decreases in environmental security through the loss of ancestral homelands, compulsory residential schooling involving wide-spread abuse of children by organs of the state and churches, violent enforcement of legislated prohibitions, and the obligatory discrimination of colonized peoples (Stromquist 2015). En-vironmental insecurity manifests here as a people's inability to cope with the environmental, social, cultural and political changes imposed on them by modernity.

At the global level, humanity's prospects out of overshoot look rather different. The impact of intensifying climate change and other factors will decrease global environmental security to an extent that generated warnings of a global collapse (e.g., Kuecker 2014). Diamond's five factors are reduced to three: environmental damage, climate change and our capacity for adaptation (Cumming and Peterson 2017). Notable is also that two of the three describe environmental security, and the third describes its improvement. Environmental security demands that regional overshoot be reduced, or at least justly balanced against regions that are still underutilized. This could in theory be accomplished by extending and strengthening the authority of international legal institutions to impose limits to consumption, to regulate development and to protect the ecological integrity of ecosystems in time to prevent their collapse. The current globalized economic and political order represents a threat to global environmental security because it hinders the development of global governance structures that could implement such timely rectification policies (Lautensach 2020).

The international dimension

At the international level, environmental security regimes arising from conventions such as the COP conferences must address threats that act across national boundaries and globally. Those threats potentially constitute reasons for armed conflict, as in the cases of water rights or nuclear disasters, grounds for United Nations (UN) intervention according to the UN Charter (Tinker 2001). Intervention might also be justified when a country destroys the basis of existence for future generations. In 2020 a team of international lawyers and academics convened to gain official recognition by the International Criminal Court (ICC) of the crime of ecocide (Bowcott 2020; Stancil 2020), defined as "extensive damage, destruction or loss of ecosystems of a given territory" and already codified as a war crime under the ICC's Rome Statute (Womack et al. 2014). Their argument follows the one that established the concepts of "crimes against humanity," war crimes and genocide, and is supported by NGOs, some European countries and small island states such as Vanuatu. Initial support came from a 1992 UNSC summit which recognized that "the non-military sources of instability in the economic, social, humanitarian *and ecological fields* have become threats to peace and security" (UNSC 1992; our emphasis), based on the concepts of a planetary trust or common heritage and intergenerational justice (Weiss 1989; Strydom 2020). The UNSC has since held back.

International judicial institutions could contribute to a global legislative groundswell toward progress on that front. They transcend national jurisdictions and potentially influence public opinion worldwide, for example in the cases of indigenous peoples who charged discrimination under national law. The following judicial institutions were established to uphold legal justice at the international level:

- The International Criminal Court (ICC) since 2002 prosecutes individuals. It is based on the Rome Statute, a multilateral treaty with 123 signatory states, not including the United States, Russia, India and China. By prosecuting governments for ecocide and human rights violations, the ICC could make a major contribution toward the environmental security of indigenous minorities and small states. Past opportunities included nuclear testing, forced displacement of tribes, large-scale deforestation and imposed dietary changes.
- The International Court of Justice (ICJ) is an organ of the UN. Since 1947 it hears cases brought by countries, or by the UN, against other countries. Its purview is to interpret, apply and clarify international law; its rulings contribute to case law through a process of "normative accretion."
- International Tribunals, including the International Tribunal for the Law of the Sea, the European Court of Justice, adjudicate disputes between states.

The main problem with international law is that it is binding only for signatory countries, quite unlike the residents in a country or in a local community, where membership is hardly optional. Even though the influence of environmental security applies to all, the duties to respect it do not. A second problem lies in the veto power of the UN Security Council, which as a rule precludes enforcement or intervention even when a clear ruling is made against a country, as in the case of the Philippines and other Southeast Asian countries against China's claims to the resources of the South China Sea. The food security of millions lies in the balance.

The international level also plays an essential role by potentially preventing *corporate* miscreants from escaping local jurisdictions, as in the cases of oil spills that destroy local fisheries, or mining operations that deprive indigenous peoples of their homelands, traditional ways of life and community health. Vast development projects in the Third World often involve ecocide, biodiversity loss, pollution, violent oppression and assassination of local activists that go unpunished (Nolin 2018). In the authors' home province of British Columbia, the environmental security of First Nations and other citizens is threatened by large-scale hydroelectric, mining and pipeline projects operated by transnational corporations.

The potential of international regulation, largely unrealized as of yet, is that it can place natural "resources" beyond the control of any single "sovereign" government, which would serve to enhance environmental security in many regions where health, nutrition and socio-economic welfare are currently jeopardized by unsustainable practices, aggravated by failing governments and kakistocracies such as Brazil, India and the United States under Trump (Spicer 2018; Ehrlich 2020; Lautensach 2020).

Besides numerous bilateral and multilateral aid agreements among governments and NGOs, the United Nations have effectively assumed the responsibility for environmental security at the international level (UN 2000). The UN sets goals and targets for global development, and it plans and implements strategic development programs through subsidiary programs (see Chapters 9 and 10).

A certain commitment to environmental security is evident in the frequent reference to sustainability, most prominent in the UN's 17 Sustainable Development Goals (SDGs) (UN 2015), where the integrity of ecological support structures is invoked in at least three of them. However, their tone belies the daunting picture of global overshoot documented in the UN's own studies (UNEP–MAB 2005; IPBES 2019). Achieving all of the welfare-oriented SDGs would raise the global environmental footprint to between two and five planets – meaning that further development cannot proceed in that direction. Worse, it is overshoot that renders the nine SDGs that depend on physical resources unachievable in principle; it also converts any gains on those goals into losses for SDGs 13,14,15 – goals toward the conservation of physical resources and ecological integrity (Ewing et al. 2010: 21–22; WWF 2012: 61; O'Neill et al. 2018; Shropshire 2019). Those critics indicated that under the balanced priorities of social justice and environmental security, the countries that perform most sustainably are enjoying only a moderate national income, such as Vietnam, Cuba and Guyana. Notwithstanding the merits of the SDGs as the first international effort to tackle the Anthropocene predicament, the fallacy in the UN's approach to development lies in their excessive emphasis on economic growth and their denial of its limits (Rees 2004), their neglect of scientists' warnings about their misinterpretation of "sustainability" (Carter and Woodworth 2018; Ripple et al. 2020), and the strong status quo bias in their views of "development" (Lautensach and Lautensach 2013). Those conceptual constraints reveal little concern for scientifically realistic prospects of the future of global environmental security for all (Wright 2004; Nadeau 2009).

The national dimension

Many of our observations about the international dimension of environmental security apply to the national one as well, especially under the prevailing globalization. The importance of the national dimension arises from the considerable extent to which the principle of sovereignty still overrides international regimes. The primacy of sovereignty is illustrated by the example of the UN Conference on Environment and Development's (UNCED) Agenda 21 action plans that after 28 years largely remain to be implemented at the national level and subject to government approval (see Chapter 25). As long as the cooperation of countries is voluntary, domestic priorities with diverse objectives are likely to override international ones with objectives toward environmental security, and delay their implementation. Voluntary compliance also underlies the SDGs; the UN tried to make compliance as palatable as possible by promising virtually every desirable benefit to all participant countries, still with mixed success.

National initiatives can work in favor of environmental security, as illustrated by the examples of Germany, Switzerland, Belgium and Spain, and (very briefly) Japan forgoing nuclear power after the 2011 Fukushima disaster and citing explicit reasons of environmental security – obviously considering pollution with persistent radionuclides, both in connection with accidents and long-term depositories, a greater security risk than missing national greenhouse gas emission targets, for the time being, in spite of higher energy costs (Cheung et al. 2019). In other contexts, national governments are often more responsive to corporate interests than to the citizenry; nor are the interests of citizens necessarily adequately represented in electoral systems. The Rio Declaration on Environment and Development states: "Environmental issues are best handled with the participation of all concerned citizens, at the relevant level" (Rio Declaration 1992). To that end, citizens are to be guaranteed access to relevant information, and governments are to undertake impact assessments and provide venues for public deliberation. The implication is that effective policies at the national level presuppose a functioning civil society as well as mechanisms to prevent the "tragedy of the commons" (Hardin 1998) – meaning the deterioration of environmental support structures resulting from overexploitation by individual actors in the absence of regulation. In situations where those prerequisites are absent, the active influence of NGOs becomes a vital factor in decisions about environmental security (Tinker 2001). This has become even more significant in cases of failing governance and kakistocracies, where the citizenry is virtually left alone to fend for a modicum of environmental security at the local level.

Even in cases where the will of the citizenry finds adequate political representation, national policies are also influenced by ideological factors and value priorities that do not necessarily prioritize environmental security, such as the "conventional development paradigm" (Hall 2004; Lautensach and Lautensach 2013). It presents development as a fundamentally economic issue and assumes a basic continuity in current market-oriented policies without recognizing the role of physical limits or the consequences of their transgressions for environmental security. The power of these factors reflects the pervasive influence of the media, the entertainment industry, traditional education systems, and other mechanisms that contribute to the reproduction of ideologies (Orr 2004).

Under the UN's agenda for human security, government intervention in favor of environmental security should be mandated on the basis of a human rights theory of fundamental rights to clean air, safe potable water, adequate nutrition, shelter, the safe processing of wastes and adequate health care (UN 2009). However, as in the case of the SDGs, such rights are only grantable within the constraints of ecological support structures; in other words, above a certain total demand, the per capita share is too small to adequately support environmental security for each individual (Lautensach and Lautensach 2013). Moreover, many threats to environmental security, as well as possible solutions, are not confined to single countries but extend over entire regions. Besides the problems with political representation, those two factors severely limit the efficacy of national initiatives. On a more practical level, governments often find their policy latitude toward environmental security to be severely limited by the powerful influence of transnational corporate interests and by trade agreements that are often influenced by those interests (Beder 2006; Piketty 2020; see Chapter 23).

While those contingencies often weaken legislative efforts to promote sustainable environmental security (e.g., through tariffs on resource extraction and trade, rationing of resources and services, taxation on resource use), one area remains where national governments can make a huge difference: education. Despite attempts at privatization, governments are still largely in control of education systems worldwide. This places at their disposal an instrument that can influence the beliefs, values, attitudes and ideals of entire generations, and thus their behavior (Orr 2004; Lautensach 2020). The UN offers its support under SDG #4 (UN 2018). Numerous studies (e.g., Bain et al. 2013) have shown that people's worldviews and values determine their actions toward the environment to a greater extent than their access to information; like no other state-controlled process, education shapes worldviews. In some legislative areas that potential for raising future citizens who respect, appreciate and sustain environmental security is beginning to be realized, but worldwide there remains much to be done.

The human dimension

Security in all its forms, including environmental security, necessarily includes a subjective dimension. Traditional concepts of security paid little attention to it, being focused mainly on the security of states. That changed with the UN's Human Development Report (UNDP 1994), which introduced the concept of *human security* to international and national politics. The UNDP's Human Security Framework (Jolly and Ray 2006) and Human Security Unit (UNHSU 2016) summarize the influence of human security on UN policy. In 2003 the UN Commission on Human Security, chaired by Sadako Ogata and Amartya Sen, reported that the world needed "a new security framework that centres directly on people" (Commission on Human Security 2003). Conceptual reviews of human security have been contributed, for example, by Hampson et al. (2002), Martin and Owen (2013) and Lautensach and Lautensach (2020). The UN's development initiatives such as the SDGs are conceptualized under this shift from state security to human security (UNHSU 2016).

Environmental security is conceptualized as one of four pillars that support human security – the other three being the traditional area of military/strategic security of the state and the rule of law; economic security, particularly as described by heterodox models of sustainable economics; and population health as described by epidemiology and community health; (Lautensach and Lautensach 2020). The four pillars address diverse sources of threats, covering the same ground as the "seven dimensions" preferred by the UN (UNDP 1994): economic, food, health, environmental, personal, community and political security.

One reason for the growing popularity of human security lies in the fact that its value priorities are shared widely, priorities that focus on the continued security and wellbeing of individuals and families. Another strength is its comprehensive, multidisciplinary integration of interdependent sources of insecurity that were traditionally considered under the purview of different academic specialties and were (and still are) thus studied largely in isolation from each other. For a given topic in environmental security, such as the sustainable supply of a healthy diet from local produce in a central African village, the human security approach includes the identification and evaluation of determinant factors and changes in political administration, local governance and land ownership, legislation and the rule of law, income distribution, agricultural exports versus local consumption patterns, land use, demographic profile and migration dynamics, community health, trade and distribution networks of food, provisions for resilience, ethnic profile and intercultural relations. All those factors, and more, constitute possible sources of insecurity that may compromise the dietary provisions of the villagers – in addition to purely environmental factors such as regional climate change, ecological integrity, changes in biodiversity, pollution from local industries, or endemic diseases. Human security integrates them all and describes a multifaceted security profile of the human individual in a given location and point in time.

Environmental security in many respects provides the requisite conditions for the other three pillars of human security, affirming Myers's (1993) thesis of environmental security as the "ultimate security." He predicted that the geopolitics of the twenty-first century will be dominated by environmental changes, such as soil erosion, pollution, deforestation, biodiversity loss and the decline of farmland fertility, as well as the undermining effects those changes exert on economies, public health and regional climates. This does not mean that all human problems are environmental in origin; it merely indicates that many conditions for the security of individuals, their families and communities, cannot be fulfilled unless environmental security, and its sustainability, is ensured (Lautensach and Lautensach 2012). If it is not, secondary effects sooner or later ramify into the socio-political, economic, and health-related areas and compromise human security there. For example, the dependence of population health on ecological integrity has been extensively documented (Garrett 1994; Horwitz 2007; CPHA 2015; Butler 2016) and is being confirmed by the emergence of zoonotic pandemics in recent years, most prominently Covid-19 (Butler 2016, 2020). Resource scarcity is likely to lead to civil strife and violent conflict (Homer-Dixon 1999), which are likely to reduce environmental security further. These interrelations point to the significance of projections into the future, attesting to sustainability: an individual, family or community are only secure if their security can be reasonably assured for the foreseeable future.

That orientation toward future security, as well as the dependence of human security and resilience on environmental prerequisites, has contributed to the growing acceptance of the principles of environmental justice (see Chapter 26). They include sustainable development, intergenerational equity and the precautionary principle (Akins et al. 2019). The rise of environmental justice, and its message that all humans share equal rights to environmental security, to some extent counteracted reactionary efforts to relegate environmental issues to "special interests" of conservation organizations or biologists (Beder 2006).

The multidisciplinary nature of human security lends particular poignancy to the problem of ecological overshoot and its threats to environmental security. As environmental security erodes, any hope of ensuring human security in other areas diminishes. Secondary effects, such as the erosion of the rule of law and of civil society (Myers 1993), as well as the threat of more widespread armed conflict over diminishing resources (Homer-Dixon 1999), add to the urgency of the problem.

Conclusion

The assessment of environmental security at the international, national and human dimension includes multidisciplinary aspects and variables that ramify into the four pillars of human security. Because it focuses largely on resilience to environmental threats, it includes an appraisal of future challenges, as well as opportunities, taking into account current trends and developments. Central to future trends in environmental security is humanity's growing impact on ecological support structures.

In view of the astounding increase of humanity's global resource appropriation (Bar-On et al. 2018), its persistent upward trend despite our overshoot, and the hesitance with which suggestions toward restraint are generally being received, the future of environmental security looks decidedly uncertain. A further complication comes from the growth of the global human population, which is unlikely to cease before we reach at least 10 billion (Roser 2019). Environmental security is being compromised globally from several directions: by the growing socio-economic inequity, by our ecological overshoot and by population growth. The latter has rendered it unlikely that a mere reduction of per capita consumption will suffice to eliminate overshoot; more probably, it would lead to equitable misery all around (Rees 2014; Hedberg 2020). On the other hand, our transition to a sustainable future (including sustainable environmental security) is certain to happen; humanity merely chooses (or not) which kind of transition will eventuate (Raskin 2016; Lautensach 2020). Along the way, we are likely to encounter "transition events" – crises of environmental security, of which the Covid-19 pandemic may be the first example at the global level.

Environmental security is never guaranteed; much of it depends on the practices and conventions of societies interacting with their environment and the ways in which it changes (Diamond 2005). Thus, a culture can be seen as constructing, to some extent, their own environmental security, not just in terms of their perception but through the choices they make. Their extent of environmental security determines their *mode of survival*, aptly described by Potter (1988) as five distinct modes: mere, miserable, unjust, idealistic and acceptable. Each mode is characterized by a distinct state of public health and wellbeing. Sustainable human security on a global scale is identical with the *acceptable* survival of humanity, including an acceptable quality of health for populations at large. This not to be overinterpreted as environmental determinism. The upshot is merely this: Whatever safeguards may be in place to protect the economic security of a population, its public health, its national security, and the rule of law, they seem of little help in the long term unless sustainable environmental security is guaranteed. This resonates with Barnett's (2007) finding of a mutual dependence between environmental security and peace.

Of course, our responsibility for what environmental insecurities may come our way is neither complete, nor is it equitably distributed across humanity. Many poor countries are now experiencing the brunt of climate change and environmental insecurity while having contributed relatively little to their causation. As we discussed above, environmental insecurity is frequently coupled with environmental injustice (see Chapter 25).

References

Akins, A., Lyver, P.O., Alrøe, H.F. and Moller, H. (2019) The Universal Precautionary Principle: New Pillars and Pathways for Environmental, Sociocultural, and Economic Resilience. *Sustainability* 11(2357).

Almond, R.E.A., Grooten M. and Petersen, T. (eds.) WWF (2020) Living Planet Report 2020. WWF, Gland, Switzerland. https://www.wwf.org.uk/sites/default/files/2020-09/LPR20_Full_report.pdf

Arnott, J.C., Mach, K.J. and Wong-Parodi, G. (eds.) (2020) Advancing the Science of Actionable Knowledge for Sustainability. Special Issue of *Current Opinion in Environmental Sustainability* 42: A1–A6, 1–82. https://www.sciencedirect.com/journal/current-opinion-in-environmental-sustainability/vol/42/suppl/C

Bain, P.G., Hornsey, M.J., Bongiorno, R., Kashima, Y. and Crimston, D. (2013) Collective Futures: How Projections about the Future of Society Are Related to Actions and Attitudes Supporting Social Change. *Personality & Social Psychology Bulletin* 39(4): 523–539.

Bar-On, Y., Phillips, R. and Milo, R. (2018) The Biomass Distribution on Earth. *PNAS* 115 (25): 6506–6511.

Barnett, J. (2007) Environmental Security and Peace. *Journal of Human Security* 3(1): 4–16.

Barnett, J. (2009) Environmental Security. In *The International Encyclopedia of Human Geography*, Kitchin, R. and Thrift, N. (eds.), Elsevier, pp. 553–557.

Beder, S. (2006) *Suiting Themselves: How Corporations Drive the Global Agenda*. London: Earthscan.

Bowcott, O. (2020) International Lawyers Draft Plan to Criminalise Ecosystem Destruction. *The Guardian* (30 Nov): https://www.theguardian.com/law/2020/nov/30/international-lawyers-draft-plan-to-criminalise-ecosystem-destruction

Butler, C.D. (2016) Sounding the Alarm: Health in the Anthropocene. *International Journal Environmental Research & Public Health* 13: 665.

Butler, C.D. (2020) Plagues, Pandemics, Health Security, and the War on Nature. *Journal of Human Security* 16 (1): 53–57. http://www.librelloph.com/journalofhumansecurity/article/view/johs-16.1.53

Carter, P.D. and Woodworth, E. (2018) *Unprecedented Crime: Climate Change Denial and Game Changers for Survival*. Atlanta, GA: Clarity Press.

Catton, W.R., Jr. (1980) *Overshoot: The Ecological Basis of Revolutionary Change*. Urbana: University of Illinois Press.

Chambers, N., Simmons, C., and Wackernagel, M. (2000) *Sharing Nature's Interest: Ecological Footprints as an Indicator of Sustainability*. London: Earthscan.

Cheung, G., Davies, P.J. and Bassen, A. (2019) In the Transition of Energy Systems: What Lessons Can Be Learnt from the German Achievement? *Energy Policy* 132 (September): 633–646.

Coates, J. F. (1991) The Sixteen Sources of Environmental Problems in the 21st Century. *Technological Forecasting and Social Change* 40: 87–91.

Commission on Human Security (2003) *Human Security Now*. New York: Commission on Human Security. http://www.humansecurity-chs.org/finalreport/English/FinalReport.pdf

CPHA (Canadian Public Health Association) (2015) *Global Change and Public Health: Addressing the Ecological Determinants of Health*. Ottawa: CPHA/ACSP.

Cumming, G.S., and Peterson, G.D. (2017) Unifying Research on Social-Ecological Resilience and Collapse. *Trends in Ecology & Evolution* 32(9): 695–713.

Curry, P. (2011) *Ecological Ethics: An Introduction*. Cambridge: Polity Press; 2nd edition, 2006.

Daly, H.E. and Cobb, J.B. (1994) *For the Common Good: Redirecting the Economy toward Community, the Environment, and a Sustainable Future*. Boston, MA: Beacon Press, 2nd edition.

Diamond, J. (2005) *Collapse: How Societies Choose to Fail or Succeed*. New York: Viking Penguin.

Dilworth, C. (2010) *Too Smart for our Own Good: The Ecological Predicament of Humankind*. Cambridge: Cambridge University Press.

Dobkowski, M.N. and Wallimann, I. (2002) *On the Edge of Scarcity: Environment, Resources, Population, Sustainability and Conflict*. Syracuse, NY: Syracuse University Press.

Ehrlich, P.R. (2020) Collapse of Civilization is a Near Certainty Within Decades. Interview (9 July), MAHB https://mahb.stanford.edu/blog/paul-ehrlich-collapse-of-civilisation-is-a-near-certainty-within-decades/

Ehrlich, P.R. and Harte, J. (2018) Pessimism on the Food Front. *Sustainability* 10: 1120–1125.

Estes, J.A., Terborgh, J.S., Brashares, M.E., et al. (2011) Trophic Downgrading of Planet Earth. *Science* 333 (July 15): 301–306.

Ewing, B., Moore, D., Goldfinger, S., Oursler, A., Reed, A. and Wackernagel. M. (2010) The Ecological Footprint Atlas 2010. Oakland, CA: Global Footprint Network. https://www.footprintnetwork.org/content/images/uploads/Ecological_Footprint_Atlas_2010.pdf

Folger, T. (2017) Why Did Greenland's Vikings Vanish? *Smithsonian Magazine* (March). https://www.smithsonianmag.com/history/why-greenland-vikings-vanished-180962119/

Garrett, L. (1994) *The Coming Plague: Newly Emerging Diseases in a World out of Balance*. New York: Farrar, Straus and Giroux.

GFN (Global Footprint Network) (2020) About Earth Overshoot Day. https://www.overshootday.org/about-earth-overshoot-day/

Griggs, D., Stafford-Smith, M., Gaffney, O., et al. (2013) Sustainable Development Goals for People and Planet. *Nature* 495 (21 March): 305–307.

Grossman, R. (2012) The Importance of Human Population to Sustainability. *Environment, Development, and Sustainability* 14(3): 973–977.

Hall, C.A.S. (2004) Sanctioning Resource Depletion: Economic Development and Neoclassical Economics. In *Ecojustice – The Unfinished Journey*. Gibson, William, E. (ed.), SUNY Press, 201–212.

Hampson, F.O., Daudelin, J., Hay, J., Reid, H., and Marton, T. (2002) *Madness in the Multitude: Human Security and World Disorder*. Oxford: Oxford University Press.

Hardin, G. (1998) Extensions of "The Tragedy of the Commons". *Science* 280(5364) (1 May): 682–683.

Hawkins, R. (2020) Our War Against Nature: Letters from the Front. In *Human Security in World Affairs: Problems and Opportunities* (op. cit.)

Hedberg, T.G. (2020) *The Environmental Impact of Overpopulation: The Ethics of Procreation*. London: Routledge.

Heinberg, R. and Lerch, D. (2010) *The Post-Carbon Reader*. Heraldsburg, CA: Watershed Media.

Homer-Dixon, T. (1999) *Environment, Scarcity, and Violence*. Princeton, NJ: Princeton University Press.

Horwitz, P. (ed.) (2007) *Ecology and Health: People and Places in a Changing World*. Organising Committee for the Asia-Pacific EcoHealth Conference 2007, Melbourne, Australia. https://www.vichealth.vic.gov.au/~/media/resourcecentre/publicationsandresources/ecology/ecology%20and%20health-people%20%20places%20in%20a%20changing%20world.ashx

IPBES (Intergovernmental Science-Policy Platform on Biodiversity and Ecosystem Services). (29 May 2019) *Summary for policymakers of the global assessment report on biodiversity and ecosystem services of the Intergovernmental Science-Policy Platform on Biodiversity and Ecosystem Service*. New York: UN, UNEP, UNESCO, FAO, UNDP. https://ipbes.net/system/tdf/ipbes_7_10_add.1_en_1.pdf?file=1&type=node&id=35329

Jolly, R. and Ray, B. (2006) *The Human Security Framework*. New York: UNDP.

Kolbert, E. (2014) *The Sixth Extinction: An Unnatural History*. New York: Henry Holt & Co.

Kuecker, G. (2014) The Perfect Storm: Catastrophic Collapse in the 21st Century. In *Transitions to Sustainabilty: Theoretical Debates for a Changing Planet*. Humphreys, D. and Stober, S.S. (eds.) Champaign, IL: Common Ground, pp. 76–92.

Lautensach, A.K. (2009) The Ethical Basis for Sustainable Human Security: A Place for Anthropocentrism? *Journal of Bioethical Inquiry* 6(4): 437–455.

———— (2013). A Story to Remember Us By. *Watershed Sentinel* (Nov./Dec.): 10–11. https://unbc.academia.edu/AlexLautensach

———— (2020) *Survival How? Education, Crisis, Diachronicity and the Transition to a Sustainable Future*. Paderborn, Germany: Schoeningh-Brill.

Lautensach, A.K. and Lautensach, S.W. (2013) Why 'Sustainable Development' is Often Neither: A Constructive Critique. *Challenges in Sustainability* 1(1): 3–15.

———— (eds.) (2020) Introduction. In *Human Security in World Affairs: Problems and Opportunities*. George, Prince (ed.) Canada: UNBC; Victoria, Canada: BCcampus, 2nd edition. https://opentextbc.ca/humansecurity/

Mancini, M.S., Galli, A., Niccolucci, V., et al. (2017) Stocks and Flows of Natural Capital: Implications for Ecological Footprint. *Ecological Indicators* 77: 123–128.

Martin, M. and Owen, T. (eds.) (2013) *Routledge Handbook of Human Security*. New York: Routledge.

McAnany, P.A. and Yoffee, N. (eds.) (2009) *Questioning Collapse: Human Resilience, Ecological Vulnerability, and the Aftermath of Empire*. Cambridge, UK: Cambridge University Press.

McMichael, A.J. (2001) *Human Frontiers, Environments and Disease: Past Patterns, Uncertain Futures*. Cambridge: Cambridge University Press.

Meadows, D., Randers, J., and Meadows, D. (2004) *Limits to Growth: The 30-Year Update*. White River Junction, VT: Chelsea Green Publishing.

Myers, N. (1993) *Ultimate Security: The Environmental Basis of Political Stability*. New York: W.W. Norton.

Nadeau, R.L. (2009) Brother, Can You Spare Me a Planet? Mainstream Economic Theory and the Environmental Crisis. *Sapiens* 2(1): 71–78.

Nolin, C. (2018) Memory-Truth-Justice: The Crisis of the Living in the Search for Guatemala's Dead and Disappeared. In *Human and Environmental Justice in Guatemala*. Henighan, S. and Johnson, C. (eds.) Toronto: University of Toronto Press, pp. 34–55.

O'Neill, D.W., Fanning, A.L., Lamb, W.F. and Steinberger, J.K. (2018) A Good Life for All Within Planetary Boundaries. *Nature Sustainability* 1: 88–95.

Orr, D.R. (2004) *Earth on Mind: On Education, Environment and the Human Prospect*. Washington, DC: Island Press.

Piketty, T. (2020) *Capital and Ideology*. Cambridge, MA: Belknap/Harvard University Press.

Potter, V.R. (1988) *Global Bioethics: Building on the Leopold Legacy*. East Lansing: Michigan State University Press.

Raskin, P. (2016) *Journey to Earthland: The Great Transition to Planetary Civilization*. Boston: Paul Raskin/Telus Institute.

Raudsepp-Hearne, C., Peterson, G., Tengo, M., Bennett, E.M., Holland, T., Benessaiah, K. et al. (2010) Untangling the Environmentalist's Paradox: Why Is Human Well-Being Increasing as Eco-system Services Degrade? *BioScience* 60(8): 576–589.

Rees, W.E. (2004) Waking the Sleepwalkers: A Human Ecological Perspective on Prospects for Achieving Sustainability. In *The Human Ecological Footprint*. Guelph, Canada: Faculty of Environmental Sciences, University of Guelph.

Rees, W.R. (2014) *Avoiding Collapse: An Agenda for Sustainable Degrowth and Relocalizing the Economy*. Vancouver, Canada: Canadian Centre for Policy Alternatives.

Rio Declaration on Environment and Development (1992) Principle 10. UN Doc. A/Conf.151/5/Rev. 1. Reprinted in *International Legal Materials* 31: 874.

Ripple, W.J., Wolf, C., Newsome, T.M., et al. (2020) World Scientists' Warning of a Climate Emergency. *BioScience* 70(1) (Jan): 8–12.

Roser, M. (2019) *Future Population Growth*. Online at OurWorldInData.org. https://ourworldindata.org/future-population-growth#global-population-growth

Shropshire, A. (2019) Human Development & National Eco-Footprints: A Visual Orientation. https://towardsdatascience.com/human-development-national-eco-footprints-a-visual-orientation-9adf86618d4f

Spady, D., and Lautensach, A.K. (2020) Why Human Security Needs Our Attention. In *Human Security in World Affairs: Problems and Opportunities*. (op. cit.)

Spicer, A. (2018) Donald Trump's 'kakistocracy' Is Not the First, But It's Revived an Old Word. *The Guardian* (18 April) – documents use of the word through history. https://www.theguardian.com/commentisfree/2018/apr/18/donald-trump-kakistocracy-john-brennan-us-twitter

SRC (Stockholm Resilience Centre) (2015) *What Is Resilience? An Introduction to Social-Ecological Research*. Stockholm Resilience Centre. https://whatisresilience.org/en/what-is-resilience/

Stancil, K. (2020) Top Lawyers to Define Ecocide to Enforce Environmental Justice. *Ecowatch* (1 December). https://www.ecowatch.com/lawyers-define-ecocide-2649094329.html?rebelltitem=1#rebelltitem1

Stromquist, G. (2015) *Project of Heart: Illuminating the Hidden History of Indian Residential Schools in BC*. Vancouver: British Columbia Teachers Federation (BCTF).

Strydom, H. (2020) Human Security in the Context of International Humanitarian Law and International Criminal Law. In *Human Security in World Affairs: Problems and Opportunities*. Lautensach, A.K. and Lautensach, S.W. (eds.) 2nd edition, ch.21; Prince George, Canada: UNBC; Victoria, Canada: BCcampus. https://opentextbc.ca/humansecurity/

Tinker, C. (2001) Environmental Security: Finding the Balance. In *Adapting the United Nations to a Postmodern Era: Lessons Learned*. W.A. Knight (ed.) New York: Palgrave Macmillan, 2nd ed., 212–228.

UN (United Nations) (2000) *We the Peoples: The role of the United Nations in the 21st century*. New York: United Nations. https://www.un.org/en/events/pastevents/pdfs/We_The_Peoples.pdf

——— (2009) Human Security in Theory and Practice: Application of the Human Security Concept and the United Nations Trust Fund for Human Security. Human Security Unit, Office for the Coordination of Humanitarian Affairs (OCHA). New York: UN. https://www.unocha.org/sites/dms/HSU/Publications%20and%20Products/Human%20Security%20Tools/Human%20Security%20in%20Theory%20and%20Practice%20English.pdf

——— (2015) Transforming our World: The 2030 Agenda for Sustainable Development. NY: United Nations. https://sustainabledevelopment.un.org/index.php?page=view&type=400&nr=2125&menu=1515

——— (2018) The Sustainable Development Goals Report. UN, Department of Economic & Social Affairs, Statistics Division. https://unstats.un.org/sdgs/report/2018/overview/

UNDP (UN Development Program) (1994) *Human Development Report: New Dimensions of Human Security*. New York: UN.

UNEP (UN Environment Program) (2020) *Global Biodiversity Outlook 5*. Montreal: Secretariat of the Convention on Biological Diversity. https://www.cbd.int/gbo5

UNEP-MAB (United Nations Environment Programme–Millennium Assessment Board) (2005) *Living Beyond Our Means: Natural Assets and Human Well-being*. London: UNEP–WCMC. Available HTTP: <http://www.maweb.org/en/Reports.aspx>

UNHSU (UN Human Security Unit) (2016) *The Human Security Handbook*. United Nations Trust Fund for Human Security. https://procurement-notices.undp.org/view_file.cfm?doc_id=212254

United Nations Security Council (UNSC) (1992) *Statement of the President of the Security Council.* UN SCOR 47th Session, 3046th meetg. At 3, UN Doc. S/23500 (31 January).

Wackernagel, M. and Beyers, B. (2019) *Ecological Footprint: Managing Our Biocapacity Budget.* Gabriola Island, Canada: New Society Publishers.

Wackernagel, M. and Rees, W. (1996) *Our Ecological Footprint: Reducing Human Impact on the Earth.* Oxford: John Carpenter.

Weintraub, B.A. (1995) Environmental Security, Environmental Management, and Environmental Justice. *Pace Environmental Law Review 12.* White Plains, NY: Pace University School of Law, 614–618.

Weiss, E.B. (1989) *In Fairness to Future Generations: International Law, Common Patrimony, and Intergenerational Equity.* Tokyo: UN University Press, 17–46, 385–412.

Womack, A., Lelkes, V., and Merz, P. (2014) Ending the Insanity of Ecocide. *The Daly News* and *Mother Pelican* 10(1).

Wright, R. (2004) *A Short History of Progress.* Toronto: House of Anansi Press.

WWF (World Wildlife Fund) (2012) *Living Planet Report 2012.* http://wwf.panda.org/about_our_earth/all_publications/living_planet_report/

York, R., Rosa, E.A., and Dietz, T. (2003) STIRPAT, IPAT and ImPACT: Analytic Tools for Unpacking the Driving Forces of Environmental Impacts. *Ecological Economics* 46(3): 351–365.

Zurlini, G. and Müller, F. (2008) Environmental Security. In Jørgensen, S.E. and Fath, B.D. (Editors-in-Chief), Systems Ecology. Vol. [2] of *Encyclopedia of Ecology.* Oxford: Elsevier, 1350–1356.

21

Earth system governance
World politics in the post-environmental age

Frank Biermann

Earth system governance is a new paradigm in the social sciences to better understand politics in times of socio-ecological system changes at planetary scale. It shares with traditional concepts, such as "global environmental politics," the focus on governance as the collective steering of societal behavior by political actors (see Chapters 1 and 2). Earth system governance radically breaks, however, with traditional notions of environmental policy by replacing their dichotomies of human–nature and human–environment with a decisive socio-ecological systems perspective and by a more explicit focus on planetary interdependence.

This shift from traditional environmental policy to a novel earth system perspective, as reflected in the new paradigm of "earth system" governance, resulted from a new understanding of both global complexity and the rapidly growing planetary role of the human species. Scientific advances of the last decades showed the non-linearity of earth system processes, the threat of rapid system shifts and the complex interrelationships between system components, including the now dominant role of humans. The relative stability of the earth system during the Holocene – the last 12,000 years that brought about the development of human civilization – now seems almost a fortunate exception. This development has been aptly captured in Paul Crutzen and Eugene Stoermer's (2000) call to declare the Holocene ended and to recognize the beginning of a new epoch in planetary history: the "Anthropocene." The Anthropocene, the "age of humans," has become shorthand for the current epoch in our planet's history in which humans are no longer passive elements in planetary evolution but a driving force at earth system scale, comparable in significance to ice ages (Zalasiewicz et al. 2011; see Chapter 15). The Anthropocene notion is not unproblematic and has been criticized by scholars from the humanities and social sciences for what it lacks or tends to blur (Biermann and Lövbrand 2019; Lövbrand, Mobjörk and Söder 2020). For example, the Anthropocene has been critiqued for its universal one-humankind narrative that hides from view centuries of colonialism, patriarchy, exploitation and postcolonial injustices. Yet any alternative terms remain fixated on one aspect only (e.g., Capitalocene, Andropocene or Plantationocene) and fail to fully grasp the breadth of the novel role of people on earth. In short, as a new concept the Anthropocene has caught on (see Chapter 15).

The core argument of the earth system governance approach here is that the traditional mainstream idea of global environmental politics, shaped in the 1970s and 1980s

DOI: 10.4324/9781003008873-25

(see Chapter 2), falls short when dealing with the novel challenges of the Anthropocene. To give just two examples: scientists have suggested a set of "planetary boundaries" as threshold values that would, if transgressed, shift parts of the earth system into a new state (Rockström et al. 2009; Steffen et al. 2015; critically reviewed in Biermann and Kim 2020b). Most of these boundaries, however, fall outside the realm of traditional environmental policy and are understandable only from a complex global-system perspective. Planetary boundaries for the emission of phosphorus and nitrogen (both linked to global food production), the human use of freshwater or global land use can no longer be described as purely "environmental" policy given their huge economic, social and distributive complexities at planetary scale. Global environmental politics is not a useful concept to shape political responses for protecting such complex, interdependent planetary concerns. As a second example, global environmental politics deemphasizes, with its focus on problem-solving and policy effectiveness, important questions of planetary justice, global democracy and postcolonialism. Global environmental politics does not provide much conceptual guidance to study the many novel societal challenges of the Anthropocene, such as providing food, water and energy to 9 billion people while guaranteeing climate stability (see Chapter 32), preserving biological diversity (see Chapter 41) and ensuring intragenerational, intergenerational and interspecies justice (see Chapters 25 and 26). Politically, framing major earth system transformations such as climate change as merely "environmental problems" – to be dealt with by second-tier environmental ministries and agencies – might even have harmed their standing in the policy system and politically marginalized central planetary concerns, which new discourses, such as calls for "climate emergencies," now seek to revert.

As a consequence, several discursive interventions have suggested new framings of governance in the Anthropocene beyond traditional global environmental politics, such as "new earth politics" (Nicholson and Jinnah 2016), "earth governance" (Bosselmann 2015), "planet politics" (Burke et al. 2016), "politics of the Anthropocene" (Dryzek and Pickering 2018), "Anthropocene geopolitics" (Dalby 2020) or, from a legal perspective, "earth system law" (Kotzé and Kim 2019, Stephens 2019) and lex Anthropocenae (Kotzé and French 2018).

The emergence of earth system governance

Earth system governance is an alternative concept that was advanced two decades ago to integrate the new scholarly focus on *integrated socio-ecological systems* at *planetary scale* with an explicit perspective of *governance, institutions and politics* (Biermann 2002 and 2007). The concept of earth system governance is integrative by opening to different disciplines, such as political science, sociology and law. It is conceptually scalable as it allows the study of local- or regional-scale systems within a planetary or earth system perspective. Some scholars have in fact come to use the term in its plural and more scalable version as "earth systems" governance.

Earth system governance relates to other novel concepts of scientific integration, such as earth system analysis, earth system science, sustainability science or resilience theory (Biermann 2014). All these research lines share core characteristics. First, all focus on analyzing interlinked *systems*. Their eventual unit of analysis is the earth as an interdependent system shaped by human and nonhuman agency. Second, all approaches study the *co-evolution* of humans and nature. Boundaries between natural and social systems are broken down by focusing on coupled socio-ecological systems. Third, all approaches break down *disciplinary boundaries* through working toward a new integrating science that combines existing disciplines in joint research. Fourth, all these approaches seek to integrate research *on all scales*. Boundaries between local and global become blurred in both theory and research practice.

Earth system governance stands in the tradition of these novel approaches, with a focus on the study of the political and institutional dimensions.

Importantly, the key concern of earth system governance is not "governing the earth" or the management of the entire process of planetary evolution. Earth system governance is different from technocratic visions of what is referred to as "earth system management" or "geo-engineering." Earth system governance is about the *human impact on planetary systems*. It is about the societal steering of human activities regarding the long-term stability of geobio-physical systems and the flourishing of all species. It is about global stewardship for the planet based on non-hierarchical processes of cooperation and coordination at multiple levels.

Earth system governance is hence not a shorthand for world government but part of a larger academic interest in processes of governance. Governance differs from government and is not confined to states and governments. Governance is marked instead by participation of myriad state and nonstate actors at all levels of decision-making, ranging from networks of scientists and experts, civil society actors and environmentalists to intergovernmental organizations. Governance also often implies the self-regulation by societal actors, private–public cooperation and of new forms of multilevel policy. Governance transcends the more narrow concept of institutions through a dynamic perspective that looks at processes of governing, that focuses on governance systems beyond single institutions, and that brings a stronger emphasis on actors and especially on nonstate actors (see Chapters 13 and 14). In short, governance covers a wide set of phenomena that are all crucial for a complete understanding of politics in coupled socio-ecological systems at planetary scale.

Analytical practice, normative critique and transformative visioning

In the practice of earth system governance research, there are three ways of theorizing about it: earth system governance as analytical practice, earth system governance as normative critique and earth system governance as transformative visioning. The *analytical theory* of earth system governance seeks to explain current politics. This is traditional social science. It is about institutions and regimes, their interlinkages, and the diagnostics of specific institutional designs. It is also about core problems of the social sciences, such as the role of power, ideas, norms, different claims to legitimacy, or about the distributive outcomes of governance and their normative evaluation in terms of justice. Here, earth system governance research can draw on a productive tradition in earlier research on national and international institutions (Young, King and Schroeder 2008).

The *normative theory* of earth system governance, on the other hand, is the critique of current systems of governance. The normative theory does not ask what is, but what should be. It juxtaposes the findings and insights from analytical theory – for instance on the effectiveness of international regimes or national policy – with both the necessities of earth system stability and the needs of social justice. The normative theory is essentially critical theory, focusing on the reform and reorganization of human activity in a way that guarantees to effectively and fairly "navigate the Anthropocene" (Biermann et al. 2012, Biermann 2014).

This combination of analytical theory and normative critique turns earth system governance into a transformative approach in the social sciences. Business-as-usual will not prevent critical transitions in the earth system, and technological revolutions and efficiency gains alone will not suffice. Instead, effective earth system governance must directly address key concerns of societal change. For one, in a highly divided world, earth system governance poses fundamental questions of justice within and among nations. It raises important queries about the legitimacy and accountability of public action and about effective and fair mechanisms of democratic earth system governance. Earth system governance is therefore as much about environmental parameters as about social practices and neoliberal capitalist systems. Its normative goal is not purely environmental protection on a planetary scale – this would make earth system governance devoid of its societal context. Planetary targets – such as control of greenhouse gases – could theoretically be reached through hugely different political means with different costs for actors in different geographies. Earth system governance is about social welfare as well as environmental protection; it is about effectiveness as well as global and local justice.

This foregrounding of normative concerns in earth system governance theory raises new and difficult questions also for scholars and students as agents of change. How do today's universities function as thought leaders, thought provokers, critics and co-creators of transformative normative discourses and novel paradigmatic understandings? Are universities ready for a more active, political role? How can we support engaged scholarship in academia and the generation of novel insights into how to solve the key planetary challenges of the Anthropocene? University communities must critically reflect on their own practices – not the least on the consumption levels of the research community itself. Such discussions about the political and personal roles of scholars in the Anthropocene must undoubtedly become more prominent in the years ahead.

Earth system governance as a research field

Earth system governance as a research field is interdisciplinary, global and complexity-oriented. For one, it transcends traditional concepts of environmental policy and nature conservation and their narrower approaches and questions. The anthropogenic perturbation of the earth system brings research problems that are qualitatively different from those that have traditionally been examined in environmental policy studies. Key questions of earth system governance are, to name a few, the institutional architectures, decision-making procedures and distributive policy impacts in areas as diverse as global adaptation to rising sea levels, the halting of global soil deterioration, the protection of climate migrants or the global implications of speculative carbon dioxide removal technologies – all issues that have barely been covered by traditional environmental policy research.

As a consequence, earth system governance research transcends levels of analysis, from a planetary problem analysis down to national policies and local governance. Earth system governance is more than a problem of global regulation through international agreements. It is first and foremost about people who take decisions in their daily lives and their political struggles. Earth system transformations affect individuals as much as it is driven by individual decisions within social, political and economic structures. As such, earth system governance is informed by planetary complexity while being concerned with local institutions, policies and contestations as well.

Because of this complexity, earth system governance research draws on insights from the full range of the social sciences across the scales, from anthropology to international law. Earth system governance research covers local policies on problems ranging from the preservation of local waters to desertification and soil degradation. At the same time, it includes the study of the hundreds of international regimes that seek to regulate governments and corporations. Earth system governance requires the integration of all these strands of research and bridges scales from global to local.

Importantly, earth system governance research reaches beyond the social sciences. Its problem definition makes it part of the overarching context of integrated sustainability science and earth system science, where social scientists work with natural scientists to advance the integrated understanding of the coupled socio-ecological system that our planet has become (Leemans et al. 2009; Clark and Harley 2020). Both social and natural scientists must collaborate, including in integrated assessments of the state of knowledge. Typical for such global cooperation across disciplinary silos are interdisciplinary research programs, such as the Global Land Programme, the Global Carbon Project, Future Earth Coasts, or the Programme on Ecosystem Change and Society. These global research programs are not unproblematic: often they are driven by frames and problem definitions derived from the natural sciences, lack financial resources and impact, and are dominated by science organizations from the Global North. Yet there is hardly an alternative to global cooperation. Critical social science is called upon to engage and transform such networks to make them more relevant for global sustainability and social justice. Earth system governance serves here as a boundary concept that allows political scientists and institutional analysts to easily operate within larger interdisciplinary research environments.

In short, earth system governance is broader than traditional global environmental politics in its emphasis on the complexity of integrated socio-ecological systems at planetary scale. Key concerns of earth system governance are vast and interdependent challenges, such as ocean acidification, land use change, food system disruptions, climate change, environment-induced migration, species extinction, changing regional water cycles, as well as more traditional environmental concerns. Yet while earth system governance is a broad area of scholarly inquiry, it also has its conceptual boundaries. For instance, global political questions of international security, global communication and digitalization or terrorism are less studied within the earth system governance research community unless there are clear links to the functioning of socio-ecological systems, for example in the nexus of climate disruptions and local conflicts.

The global network of earth system governance researchers

The transformative force of the earth system governance paradigm has given rise to the creation of a global network of scholars working in this field, the "Earth System Governance Project." This network originated in 2008 as a core project of the former International Human Dimensions Programme on Global Environmental Change, based on a ten-year science and implementation plan agreed upon in 2008 (Biermann et al. 2009). After over a decade of operation, the Earth System Governance Project has matured into a global, self-sustaining network of over thousand scientists, with conferences, taskforces, affiliated research centers, regional fellow networks, an affiliated foundation, and a lively presence in social media (see in more detail Biermann et al. 2019). Research fellows organize their own activities, including summer schools, research visits, online training sites and in some regions elect their own representatives. Mid-level and full professors are affiliated with the Project as "senior research fellows." In addition, the Project launched early on its "Lead Faculty," a small globally diverse community of the most influential scholars in the field of earth system governance. This community of people thrives through the annual open science conferences that the Earth System Governance Project organizes. These annual conferences rotate between continents and have ranged from Amsterdam to Fort Collins, Lund, Tokyo, Norwich, Canberra, Nairobi, Utrecht, Oaxaca and in 2021 Bratislava. These conferences are important gathering venues for the earth system governance research community, with many discussions being carried on from conference to conference over the years. As a network organization the Earth System Governance Project has also invested in global community outreach from the outset, with an extensive website, a Facebook presence, several Twitter accounts, a LinkedIn group, a newsletter, the hosting of receptions and dinners at partner alliances, and various other means. Overall, this has allowed the network to further grow and to cumulatively reach several thousand scientists, practitioners and increasingly the public.

Even though the Earth System Governance Project has evolved with a high degree of informality, consensus approaches and bottom-up initiatives, it also developed a portfolio of institutional structures essential for continued scientific exchange and production. The Project maintains three book series with the world's top academic publishers, notably the *Earth System Governance* series with MIT Press launched in 2009; the more recent *Earth System Governance* series with Cambridge University Press focusing on synthesis volumes from the Project; and the new *Cambridge Elements in Earth System Governance*, which provide an outlet for shorter books. In 2019, the Project launched its flagship journal, *Earth System Governance*, as an open access publication with Elsevier. The choice of open access followed the Project's efforts to decolonialize global science and to strengthen its global outreach in a world where millions of scholars lack access to rich university libraries and traditional paywalled journals. Articles in *Earth System Governance* are always freely accessible to anybody anywhere, making the research insights of the community a free knowledge good available to all.

Applications and insights

As earth system governance research can look back at over ten years, several important areas of study have emerged. With a view to international relations, earth system governance research has, for instance, moved from the study of single institutions toward the study of broader *"architectures" of earth system governance*. Whereas traditional studies of global environmental politics had focused on single international regimes (see Chapter 9), increasingly it has become clear that such regimes do not operate in a void but within complex webs of larger governance settings (Young 2017; Zelli, Gerrits and Möller 2020). Earth system governance research has conceptualized such complex institutional settings as "governance architectures," using the metaphor of buildings with copious rooms, lavish apartments, winding staircases, and meandering corridors. The concept of governance architectures has filled a conceptual void in scholarship and shifted the debate to situations in which a governance area is regulated by multiple institutions and norms in highly complex settings at often planetary scale (Biermann and Kim 2020a).

The notion of a governance architecture does not assume an architect. Governance architectures emerge instead incrementally from institutionalization processes that are often decentralized and hardly planned. Many architects shape an architecture, with the resulting configuration then influencing how institutions interact by limiting the choices and opportunities for actors such as governments. A governance architecture is thus in constant flux, evolving through interactions at the micro-level and the dynamic architecture at macro-level. Importantly, the architecture concept allows for structured comparisons between issue areas, regions and over time and to study variant effects of governance architectures. Such architectures have been studied for various socio-ecological systems at planetary scale, including climate governance, ocean governance or biodiversity governance. This research has usually combined analytical and normative debates, theory and practice, scholarly discourse and calls for political reform. Theoretical debates – for instance on the orchestration effects of intergovernmental organizations (see Chapter 8) – go hand in hand with elaborate calls for structural change.

Second, earth system governance research has emphasized from its start the many ways by which *nonstate actors* have gained influence (see Chapter 14). In some areas, such as forest governance, it appears that nonstate actors even play central governing roles. They also often create their own transnational institutions with only a marginal role for governments or without any involvement of governments. All these actors and institutions interact within broader architectures of earth system governance. This new and complex multiplicity of actors in earth system governance has led researchers to emphasize the theoretical question of *agency in earth system governance*. All 11 global conferences on earth system governance that were held so far included a dedicated stream of panels on agency; and many publications have referred to this concept and the detailed research questions that were laid down in the first science plan of the Earth System Governance Project. Earth system governance research advanced here a novel conceptual frame on how state and nonstate actors engage in making decisions from local to global levels, how structure and agency are related and interact, and how governance and agency differs across societies and scales of governance (Betsill, Benney and Gerlak 2020).

Third, the *adaptiveness of earth system governance* has been a core concept in this research community since its very beginning as well. The concept of adaptiveness brings together communities of scholars that have often operated under different conceptual terms, such as resilience, adaptive governance, adaptive management, anticipatory governance or adaptation, as well as scholars of social learning, institutional dynamics or what is known as governance for transformation (Linnér and Wibeck 2019). In earth system governance research, adaptiveness is now an umbrella concept to bring these communities together for a fruitful research program on how societies and governance systems can adapt to a dynamic environment, can anticipate future changes, and learn (Djalante and Siebenhüner 2021).

Fourth, earth system governance research has opened a new debate on the foundations of law, turning from traditional notions of national or international law to a new conceptualization of "earth system law" (see Chapter 10). Earth system law is defined as a new overarching legal phenomenon that accommodates and encapsulates the juridical aspects of earth system governance better than traditional "environmental" law and that can more comprehensively respond to the regulatory challenges presented by a changing earth system (Kotzé and Kim 2019). One important discussion has, for example, been about the legal status of nonhumans in the Anthropocene and the need to "broaden the universe of entities capable of qualifying as legal subjects eligible for legal rights to include both natural and artefactual non-humans, a move integral to obtaining socio-ecological justice under Earth system law" (Gellers 2020).

Fifth, earth system governance research has brought in important questions of *justice, equity and fairness* (see also Chapters 25 and 26). While the problem-solving effectiveness of global environmental policy has been at the center of scholarly attention for long, earth system governance research emphasizes questions of the allocation of the risks and benefits, of power and disempowerment, and more generally on the normative foundations of governance at planetary scale. Earth system governance research seeks here to expand more traditional concepts of "environmental justice" toward a planetary perspective and "planetary justice" (Biermann and Kalfagianni 2020; Hickey and Robeyns 2020). The term planetary justice marks a fundamental departure from traditional approaches toward justice in twentieth-century "Holocene terms." Planetary justice foregrounds that the entire human and nonhuman world is now at stake, not merely a locality. Planetary justice is concerned with justice among humans as well as between humans and the natural world. Planetary justice also transforms older notions of global justice with their focus on international institutions, international relations and global order. Planetary justice instead is equally concerned with the global and the local, with state and nonstate actors, and with individuals and collectives. The notion of planetary justice opens the often localized environmental justice discourse to a more systematic interrogation of its planetary dimensions and the earth system challenges that humanity is facing. It also strengthens the normative critique of the grand designs in earth system science, from planetary boundaries to earth system targets, by questioning their assumptions and the forgotten injustices that are hidden in these meta-narratives.

Sixth, earth system governance research has opened new debates about *global democracy* and about how earth systems could be governed democratically. Much of this work has been linked to theories of deliberative democracy (e.g., Dryzek et al. 2019), with added emphasis on novel ways to strengthen "ecological reflexivity" at planetary scale – the capacity to question and change core commitments while listening and responding to signals from the earth system (Dryzek and Pickering 2018). Yet, the debate on democracy in earth system governance goes further than deliberative democracy (see, e.g., Mert 2019; Schlosberg, Bäckstrand and Pickering 2019). Important debates have addressed the role of human rights in earth system governance (Baber and Bartlett 2020) or representative models of global democracy such as a new "world parliament" within the United Nations (Leinen and Bummel 2018). More specific questions of legitimate and democratic decision-making have been studied by earth system governance scholars as well, including explorations of governance accountability (Park and Kramarz 2019) and transparency (Gupta and Mason 2014).

Seventh, earth system governance research has studied not only global but also local politics. Many studies used frameworks of the Earth System Governance Project to investigate local governance, from floodplain management in Hungary (Werners et al. 2009) to urban climate governance (Van der Heijden 2019; Van der Heijden, Bulkeley and Certomà 2019) or how domestic institutions in climate governance can be (re)made in complex settings (Patterson 2020). At the regional level, a group of African researchers has studied the specific governance implications of earth system transformations in Africa, arguing that the earth system governance approach is useful to guide knowledge generation in this critical area of research. These scholars also underscored the need for robust research capacity and a strong pan-African knowledge network on earth system governance (Habtezion et al. 2015).

After the first ten years of earth system governance research, in 2018 an international group of scholars wrote a new science plan to guide research in the ten-year period until 2028 (Earth System Governance Project 2018; summarized in Burch et al. 2019). This exciting plan develops new approaches to further stimulate a pluralistic, vibrant and relevant research community and advances a new framework with five pairs of research lenses that each offer a distinct perspective on earth system governance, such as "architecture and agency," "democracy and power," "justice and allocation," "anticipation and imagination" and "adaptiveness and reflexivity."

Policy impact

As a novel conceptualization of politics in the post-environmental age, earth system governance is still largely located in the realm of academic discourse. Yet the notion of earth system governance has slowly found its inroad into the public debate as well. For example, in the run-up to the 2012 United Nations Conference on Sustainable Development in Rio de Janeiro, a group of 20 Nobel laureates called for "strengthening of earth system governance" as a priority for coherent global action. Also in reports by the United Nations, references to earth system governance have been made (UNGA 2014).

Moreover, groups of scholars affiliated with the Earth System Governance Project have regularly engaged with policy debates, for example with assessments of the state of knowledge on key policy issues such as the reform of the intergovernmental system (Biermann et al. 2012), better ocean governance (De Santo et al. 2019), the governance implications of the North American Free Trade Agreement (Tienhaara 2019), a proposal for a coal elimination treaty (Burke and Fishel 2020) or, most recently, a collective assessment by over 60 scholars of the political impact of the Sustainable Development Goals (Biermann, Hickmann, and Sénit 2022).

Critique

Earth system governance has met several lines of critique as well over the last years. For one, the conceptual focus on "earth system" governance initially created a misunderstanding that this community would study only global institutions. Among those who follow the Earth System Governance Project only from afar, and hence base their assessment only on their reading of its title, misleading associations are sometimes drawn with planetary engineering, a "proto bio-political regime" (Salleh 2013), technocratic imaginaries (Stirling 2014), and an alleged tunnel vision of global "cockpit-ism" (Hajer et al. 2015). A more recent line of critique has linked earth system governance research with dangers of universal, Northern-based intellectual dominance that marginalizes different epistemologies and in particular actors from the Global South.

The empirical reality of the Earth System Governance Project tells a different story, with much scholarship by the community studying local, national or multi-level governance and often with emphasis on polycentric, networked or experimental governance as opposed to central steering – yet all typically with a planetary concern and a global perspective. Most scholarship in the Earth System Governance Project also is far from top-down managerial approaches but rather focuses on key social concerns or processes, such as justice, power, democracy and legitimacy, and this often from a critical, emancipatory perspective. This position of the community is also extensively reflected in the two science plans of the project from 2009 (Biermann et al. 2009) and 2018 (Burch et al. 2019). Much research on earth system governance has directly criticized ecomodernism, technocracy and postcolonialism, for instance by prioritizing work on "planetary justice," epistemic diversity, decolonializing Western science, or by engaging with ecosocialist and other progressive lines of thinking. In 2017, for example, the (late) Marxist sociologist Erik Olin Wright was invited as the opening keynote speaker at the annual conference of the Earth System Governance Project to lay out his critical insights on capitalism and earth system governance.

Conclusion

The traditional notion of global environmental politics is no longer able to allow for a complete and theoretically satisfying understanding of the human predicament in the Anthropocene. A global environment is impossible to define, and key concerns of global governance cut across traditional binaries of humans versus the environment, or human versus nature. In the twenty-first century, we can understand societies and nature only as integrated socio-ecological systems at planetary scale – systems that are dynamic and instable, interconnected and interdependent, and utterly complex when approached from a global governance perspective. To explore answers to these novel challenges in political science and international relations studies, the earth system governance paradigm has been developed over the past decade as part of a larger family of novel and not mutually exclusive conceptual innovations. Earth system governance brings a new perspective for the theory of global politics that is system-focused as opposed to binary human–environment; integrated across levels instead of being merely inter-governmental or local; and progressive as a research approach by moving from positivist institutional analysis to critical theory and transformative global change.

References

Baber, W. F. and Bartlett, R. V. (2020) *Environmental Human Rights in Earth System Governance: Democracy beyond democracy* (Cambridge Elements in Earth System Governance), Cambridge, UK: Cambridge University Press.

Betsill, M. M., Benney, T. M. and Gerlak, A. K. (eds) (2020) *Agency in Earth System Governance* (Earth System Governance series), Cambridge, UK: Cambridge University Press.

Biermann, F. (2002) "Johannesburg plus 20: From international environmental policy to earth system governance", *Politics and the Life Sciences* 21 (2): 72–77.

Biermann, F. (2007) "'Earth system governance' as a crosscutting theme of global change research", *Global Environmental Change: Human and Policy Dimensions* 17: 326–337.

Biermann, F. (2014) *Earth System Governance. World Politics in the Anthropocene* (Earth System Governance series), Cambridge, MA: MIT Press.

Biermann, F., Abbott, K., Andresen, S., Bäckstrand, K, Bernstein, S., Betsill, M. M., Bulkeley, H., Cashore, B., Clapp, J., Folke, C., Gupta, A., Gupta, J., Haas, P. M., Jordan, A., Kanie, N., Kluvánková-Oravská, T., Lebel, L., Liverman, D., Meadowcroft, J., Mitchell, R. B., Newell, P., Oberthür, S., Olsson, L., Pattberg, P., Sánchez-Rodríguez, R., Schroeder, H., Underdal, A., Camargo Vieira, S., Vogel, C., Young, O. R., Brock, A. and Zondervan, R. (2012) "Navigating the Anthropocene. Improving earth system governance", *Science* 335 (no. 6074): 1306–1307.

Biermann, F., Betsill, M. M., Burch, S., Dryzek, J., Gordon, C., Gupta, A. Gupta, J., Inoue, C., Kalfagianni, A., Kanie, N., Olsson, L., Persson, Å., Schroeder, H. and Scobie, M. (2019) "The Earth System Governance Project as a network organization: A critical assessment after ten years", *Current Opinion in Environmental Sustainability* 39: 17–24.

Biermann, F., Betsill, M.M., Gupta, J., Kanie, N., Lebel, L., Liverman, D., Schroeder, H. and Siebenhüner, B., with contributions from Conca, K., Costa Ferreira, L. da, Desai, B., Tay, S. and Zondervan, R. (2009) *Earth System Governance: People, Places and the Planet* (Science and Implementation Plan of the Earth System Governance Project), Bonn: The Earth System Governance Project.

Biermann, F., Hickmann, T. and Sénit, C.-A. (eds) (2022) *The Political Impact of the Sustainable Development Goals: Transforming Governance Through Global Goals?* (Earth System Governance Series), Cambridge, UK: Cambridge University Press.

Biermann, F. and Kalfagianni, A. (2020) "Planetary justice: A research framework", *Earth System Governance* 6 (December): 100049.

Biermann, F. and Kim, R.E. (eds) (2020a) *Architectures of Earth System Governance: Institutional Complexity and Structural Transformation* (Earth System Governance series), Cambridge, UK: Cambridge University Press.

Biermann, F. and Kim, R.E. (2020b) "The boundaries of the Planetary Boundary framework: A critical appraisal of approaches to define a 'safe operating space' for humanity", *Annual Review of Environment and Resources* 45: 497–521.

Biermann, F. and Lövbrand, E. (eds) (2019) *Anthropocene Encounters. New Directions in Green Political Thinking* (Earth System Governance series), Cambridge, UK: Cambridge University Press.

Bosselmann, K. (2015) *Earth Governance. Trusteeship of the Global Commons*, Cheltenham: Edward Elgar.

Burch, S., Gupta, A., Inoue, C. Y.A., Kalfagianni, A., Persson, Å., Gerlak, A. K., Ishii, A., Patterson, J., Pickering, J., Scobie, M., Van der Heijden, J., Vervoort, J., Adler, C., Bloomfield, M., Djalante, R., Dryzek, J., Galaz, V., Gordon, C., Harmon, R., Jinnah, S., Kim, R. E., Olsson, L., Van Leeuwen, J., Ramasar, V., Wapner, P. and Zondervan, R. (2019) "New directions in earth system governance research", *Earth System Governance* 1: 100006.

Burke, A. and Fishel, S. (2020) "A coal elimination treaty 2030: Fast tracking climate change mitigation, global health and security", *Earth System Governance* 3: 100046.

Burke, A., Fishel, S., Mitchell, A., Dalby, S. and Levine, D. J. (2016) "Planet politics. A manifesto from the end of IR [International Relations]", *Millennium: Journal of International Studies* 44 (3): 499–523.

Clark, W. C. and Harley, A. G. (2020) "Sustainability science: Toward a synthesis", *Annual Review of Environment and Resources* 45 (1): 331–386.

Crutzen, P.J. and Stoermer, E.F. (2000) "The 'Anthropocene'", *Global Change Newsletter* 41: 17–18.

Dalby, S. (2020) *Anthropocene Geopolitics: Globalization, Security, Sustainability,* Ottawa: University of Ottawa Press.

De Santo, E. M., Ásgeirsdóttir, Á., Barros-Platiau, A., Biermann, F., Dryzek, J., Gonçalves, L.R., Kim, R.E., Mendenhall, E., Mitchell, R., Nyman, E., Scobie, M., Sun, K., Tiller, R., Webster, D.G., Young, O. (2019) "Protecting biodiversity in areas beyond national jurisdiction: An earth system governance perspective", *Earth System Governance* 2: 100029.

Djalante, R. and Siebenhüner, B. (eds) (2021) *Adaptiveness: Changing Earth System Governance* (Earth System Governance series), Cambridge, UK: Cambridge University Press.

Dryzek, J. S., Bowman, Q., Kuyper, J., Pickering, J., Sass, J. and Stevenson, H. (2019) *Deliberative Global Governance* (Cambridge Elements in Earth System Governance), Cambridge, UK: Cambridge University Press.

Dryzek, J. S. and Pickering, J. (2018) *The Politics of the Anthropocene,* Oxford: Oxford University Press.

Earth System Governance Project (2018) *Earth System Governance. Science and Implementation Plan of the Earth System Governance Project*, Utrecht: Earth System Governance International Project Office, Utrecht University.

Gellers, J. C. (2020) "Earth system law and the legal status of non-humans in the Anthropocene", *Earth System Governance* 100083.

Gupta, A. and Mason, M. (2014) *Transparency in Global Environmental Governance: Critical Perspectives* (Earth System Governance series), Cambridge, MA: MIT Press.

Habtezion, S., Adelekan, I., Aiyede, E., Biermann, F., Fubara, M., Gordon, C., Gyekye, K., Kasimbazi, E., Kibugi, R., Lawson, E., Mensah, A., Mubaya, C., Olorunfemi, F., Paterson, A., Tadesse, D., Usman, R. and Zondervan, R. (2015) "Earth system governance in Africa: Knowledge and capacity needs", *Current Opinion in Environmental Sustainability* 14: 198–205.

Hajer, M., Nilsson, M., Raworth, K., Bakker, P., Berkhout, F., de Boer, Y., Rockström, J., Ludwig, K. and Kok, M. (2015) "Beyond cockpit-ism: Four insights to enhance the transformative potential of the Sustainable Development Goals", *Sustainability* 7: 1651–1660.

Hickey, C. and Robeyns, I. (2020) "Planetary justice: What can we learn from ethics and political philosophy?" *Earth System Governance* 6 (December): 100045.

Kotzé, L.J. and French, D. (2018) "A critique of the Global Pact for the Environment: A stillborn initiative or the foundation for Lex Anthropocenae?" *International Environmental Agreements: Politics, Law and Economics* 18 (6): 811–838.

Kotzé, L.J. and Kim, R.E. (2019) "Earth system law: The juridical dimensions of earth system governance", *Earth System Governance* 1: 100003.

Leemans, R., Asrar, G., Canadell, J. G., Ingram, J., Larigauderie, A., Mooney, H., Nobre, C., Patwardhan, A., Rice, M., Schmidt, F., Seitzinger, S., Virji, H., Vörösmarthy, C. and Young, O. (2009) "Developing a common strategy for integrative global change research and outreach: The Earth System Science Partnership (ESSP)", *Current Opinion in Environmental Sustainability* 1(1): 4–13.

Leinen, J. and Bummel, A. (2018) *A World Parliament: Governance and Democracy in the 21st Century,* Berlin: Democracy Without Borders.

Linnér, B.-O. and Wibeck, V. (2019), *Sustainability Transformations: Agents and Drivers Across Societies* (Earth System Governance series), Cambridge, UK: Cambridge University Press.

Lövbrand, E., Mobjörk, M. and Söder, R. (2020) "The Anthropocene and the geo-political imagination: Re-writing Earth as political space", *Earth System Governance* 4: 100051.

Mert, A. (2019) "Democracy in the Anthropocene: A new scale", in F. Biermann and E. Lövbrand (eds) *Anthropocene Encounters. New Directions in Green Political Thinking* (Earth System Governance series), Cambridge, UK: Cambridge University Press, 128–149.

Nicholson, S. and Jinnah S. (eds) (2016) *New Earth Politics: Essays from the Anthropocene* (Earth System Governance series), Cambridge, MA: MIT Press.

Park, S. and Kramarz, T. (eds) (2019) *Global Environmental Governance and the Accountability Trap* (Earth System Governance series), Cambridge, MA: MIT Press.

Patterson, J. (2020) *Remaking Political Institutions: Climate Change and Beyond* (Cambridge Elements in Earth System Governance), Cambridge, UK: Cambridge University Press.

Rockström, J., Steffen, W., Noone, K., Persson, Å., Chapin, F. S., Lambin, E. F., Lenton, T. M., Scheffer, M., Folke, C., Schellnhuber, H.-J., Nykvist, B., de Wit, C. A., Hughes, T., van der Leeuw,

S., Rodhe, H., Sörlin, S., Snyder, P. K., Costanza, R., Svedin, U., Falkenmark, M., Karlberg, L., Corell, R. W., Fabry, V. J., Hansen, J., Walker, B., Liverman, D., Richardson, K., Crutzen, P. and Foley, J. A. (2009) "A safe operating space for humanity", *Nature* 461 (24 September): 472–475.

Salleh, A. (2013) *The Idea of Earth System Governance: Unifying Tool? Or Hegemony for a New Capitalist Landnahme?* Working Paper 10/2013 der DFG-KollegforscherInnengruppe Postwachstumsgesellschaften.

Schlosberg, D., Bäckstrand, K. and Pickering, J. (2019) "Reconciling ecological and democratic values: Recent perspectives on ecological democracy", *Environmental Values* 28: 1–8.

Steffen, W., Richardson, K., Rockström, J., Cornell, S. E., Fetzer, I., Bennett, E. M., Biggs, R., Carpenter, S. R., de Vries, W., de Wit, C. A., Folke, C., Gerten, D., Heinke, J., Mace, G. M., Persson, L. M., Ramanathan, V., Reyers, B. and Sörlin, S. (2015) "Planetary boundaries: Guiding human development on a changing planet", *Science* 347 (6223): 736–746.

Stephens, T. (2019) *What is the Point of International Environmental Law Scholarship in the Anthropocene?* University of Sydney Law School, Legal Studies Research Paper Series no. 19/25, Sydney: The University of Sydney.

Stirling, A. (2014) *Emancipating Transformations: From Controlling "the Transition" to Culturing Plural Radical Progress,* Climate Geoengineering Governance Working Paper Series 012, University of Sussex.

Tienhaara, K. (2019) "NAFTA 2.0: What are the implications for environmental governance?", *Earth System Governance* 1: 100004.

UNGA – United Nations General Assembly (2014) *Harmony with Nature. Report of the Secretary-General.* United Nations Doc. A/69/322 of 18 August 2014. URL: https://undocs.org/A/69/322.

Van der Heijden, J. (2019) "Studying urban climate governance: Where to begin, what to look for, and how to make a meaningful contribution to scholarship and practice", *Earth System Governance* 1: 100005.

Van der Heijden, J., Bulkeley, H. and Certomà, C. (eds) (2019) *Urban Climate Politics: Agency and Empowerment* (Earth System Governance series), Cambridge, UK: Cambridge University Press.

Werners, S. E., Flachner, Z., Matczak, P., Falaleeva, M. and Leemans, R. (2009) "Exploring earth system governance: A case study of floodplain management along the Tisza river in Hungary", *Global Environmental Change* 19 (4): 503–511.

Young, O. R. (2017) *Governing Complex Systems. Social Capital for the Anthropocene* (Earth System Governance series), Cambridge, MA: MIT Press.

Young, O. R., King, L., and Schroeder, H. (eds) (2008) *Institutions and Environmental Change. Principal Findings, Applications and Research Frontiers* (Earth System Governance series), Cambridge, MA: The MIT Press.

Zalasiewicz, J., Williams, M., Haywood, A. and Ellis, M. (2011) "The Anthropocene: A new epoch of geological time?", *Philosophical Transactions of the Royal Society A* 369: 835–841.

Zelli, F., Gerrits, L. and Möller, I. (2020) "Global governance in complex times: Exploring new concepts and theories on institutional complexity", *Complexity, Governance and Networks* 6 (1): 1–13.

22

Environmental diplomacy

International conferences and negotiations

Radoslav S. Dimitrov

Diplomacy and negotiations are the principal means of constructing international environmental institutions (Haas et al. 1993; Levy et al. 1995; Hasenclever et al. 1997; Young 1998; Goldstein et al. 2000). Faced with increasing problems that undermine the wellbeing of the planet and human communities, governments have negotiated an astonishing 1,300 multilateral environmental agreements and 2,200 bilateral ones (International Environmental Agreements Database Project 2020). These treaties address a wide range of issues such as ozone depletion (see Chapter 31), chemical pollution (see Chapter 36), biodiversity (see Chapter 41), fisheries (see Chapter 40) and acid rain (see Chapter 30). Collectively, they provide the legal infrastructure of global environmental governance. Most recently, diplomacy produced the 2015 Paris Agreement on Climate Change that now defines the response of the international community to climate change (see Harris 2021: 47–56 and Chapter 32). The sheer number of summits and the political energy invested in preparing for them are staggering. Major global conferences related to only ten of those multilateral agreements took 115 days per year between 1992 and 2007 (Muñoz et al. 2009). When we add other environmental issues as well as the plethora of smaller international meetings, we observe a world in perpetual negotiation over environmental policy.

The empirical magnitude of environmental diplomacy and its importance in forging treaties contrasts with the relatively little academic attention it receives in theory building, compared to other topics in global environmental politics. Once equated with international relations (IR), diplomacy in general is sidelined in modern IR scholarship. Some academic observers dismiss UN conferences as talkshops and media events that create photo opportunities but do little to address global problems (Fomerand 1996; Hoffman 2011). Others interpret "summit theatre" through a Foucauldian perspective, as mechanisms for structuring the conduct of global politics and reifying power relations and political orders (Blühdorn 2007; Death 2011). Indeed, the 1991 Earth Summit reflected and reinforced a normative paradigm of liberal environmentalism that is sympathetic to capitalism (Bernstein 2001). Murray Edelman and his intellectual descendants have argued that high-level decision-making is often a symbolic process of going through the motions, a mere "dramaturgy" to show that the political process is functioning (Edelman 1988; Blühdorn 2007). Carl Death (2011) writes of UN environmental summits as political theater acted out for global audiences to demonstrate

DOI: 10.4324/9781003008873-26

responsible state conduct. Even more worrisome, there is evidence that governments some-times create empty international agreements that are purposefully designed not to deliver. Summits can produce "decoy agreements" that merely hide the failure of international ne-gotiations, create false public impressions of progress, and even legitimize collective inaction and preempt governance by neutralizing calls for real treaties (Dimitrov 2020). And even environmental NGOs (see Chapter 14) could endorse weak agreements in order to secure the participation of the United States (Allan 2019).

Despite academic skepticism or neglect, environmental diplomacy is responsible for the creation of all formal international environmental institutions that exist today. The forma-tion of organizations and international policy regimes (see Chapter 9) occurs through dip-lomatic processes that include conferences, multiple rounds of negotiations and inestimable amount of informal bilateral exchanges. Negotiations typically consist of years of formal and informal discussions on the rules of a treaty, including policy targets, timetables, implemen-tation mechanisms and compliance procedures. The negotiation process unfolds in analyti-cally distinct stages. Oran Young (1994) distinguished between pre-negotiation, negotiating and implementation of international agreements and showed that each stage is affected by different political factors. Pamela Chasek (2001) borrows this insight to provide perhaps the most elaborate discussion of stages in environmental negotiations.

Early research on global environmental politics addressed why some negotiations suc-ceed while others fail to produce policy agreements. One project attempted to identify the determinants of success through a comparison of five empirical cases of successful regime formation and concluded that none of the independent variables under consider-ation could explain the outcomes (Young and Osherenko 1993). Subsequent scholarship scaled down ambition and now typically desists from broad theoretical explanations of negotiation outcomes, focusing instead on explaining particular country positions (Bailer and Weiler 2014) or the political behavior and negotiating tactics of individual actors (Bailer 2012; Michaelowa and Michaelowa 2012; Wu 2013; Cross 2018). For example, Detlef Sprinz and Tapani Vaahtoranta (1994) stress domestic cost-benefit analysis and explain country positions in negotiations with expected policy costs and vulnerability to ecological problems.

Power and leadership

The policy preferences of the United States may help explain the failure of global forestry negotiations (see Chapter 14) to produce a convention (Davenport 2005) or the shape of the 2015 Paris Agreement on Climate Change (see Chapter 32) (Rajamani 2016). Most schol-ars of global environmental politics, however, maintain that structural power matters little in environmental diplomacy (Young 1991; Underdal 1994; Andresen and Agrawala 2002; Weiler 2012; cf. Chapters 4 and 14). In a thorough treatment of the topic, Robert Falkner (2005) shows that hegemony provides an incomplete perspective that explains neither the direction of US policy nor international outcomes. Coalitions of developing countries such as the G77 and BASIC have made the global South successful in shaping climate politics (Al-lan and Dauvergne 2013; Hochstetler and Milkoreit 2013; see Chapter 23). Moreover, even small countries can exercise strong influence in negotiations. The Netherlands have used initiative and shrewd diplomacy to influence both European and global climate negotiations (Kanie 2003). The Alliance of Small Island States are active participants in climate talks who influence the process by "borrowing external power" (Betzold 2010). And a coalition of small Latin American countries became a vocal actor who managed to influence the Paris

Agreement with its ability to hold constructive discussions and build bridges with other countries (Edwards et al. 2017).

The weak relevance of structural power has led to a vibrant body of research on leadership. There are three principal types of leadership: structural, directional and instrumental (Underdal 1994; Gupta and Grubb 2000). (Young [1991] offers an alternative typology and lists three leadership types: structural, entrepreneurial and intellectual.) Structural leadership derives from material resources, including forest cover (Brazil) or share of polluting emissions (China). Directional leaders, such as the European Union in climate change or the United States in the ozone negotiations, lead by example through unilateral domestic policies that demonstrate feasible solutions to other countries (Underdal 1994). Instrumental leadership is a function of political initiative, skill and creativity in the process of negotiations, including submission of policy proposals and persuasive arguments (Betzold 2010; Edwards et al. 2017). It can be further divided into intellectual and entrepreneurial types (Young 1991; Kanie 2003). Intellectual leadership is particularly important at early stages of negotiations (Andresen and Agrawala 2002). The United States played such role in the 1990s by introducing the idea of emission trading in the Kyoto Protocol negotiations. One entrepreneurial leader is the island nation of Tuvalu whose delegation has been remarkably influential in climate discussions by providing concrete proposals, including an elaborate treaty text tabled in 2009 before Copenhagen.

Recent innovative research investigates both the supply and demand side of leadership. Interviews of diplomats in the climate negotiations reveal that the EU and China are most commonly perceived as leaders by their peers (Kilian and Elgström 2010; Karlsson et al. 2011). A crucial variable that affects who countries recognize as a leader is an actor's commitment to the common good rather than exclusive self-interest (Karlsson et al. 2012). The European Union has provided strong leadership in negotiations on various issues (Gupta and Grubb 2000; Vogler 2005; Harris 2007; Schreurs et al. 2007; Oberthür and Kelly 2008; Groen et al. 2012; Torney and Cross 2018). The degree of EU success in achieving its goals varies from summit to summit even within one issue area (Groen et al. 2012). Vogler (2005) considers alternative explanations of European leadership and finds evidence of "normative entrapment," a normative stance on climate change and an enduring self-image that continues to propel strong policies. Others caution against idealism and maintain that political economy and material interests drive the EU (Falkner 2007) and that the EU worked hard to persuade others of the economic benefits of climate policy (Dimitrov 2012).

Jon Hovi and his colleagues argue that the EU persistence in the climate regime is product of the combined effects of domestic institutional inertia and international power-seeking (Hovi et al. 2003). By pulling out of the Kyoto Protocol in 2001, the United States offered the EU and other actors an opportunity to gain political power in an important arena. Similarly, Schreurs and Tiberghien (2007) focus on domestic institutions and argue that "multilevel reinforcement" between key EU states, the European Commission and Parliament vying for power explain leadership. Norichika Kanie (2003) goes even deeper to show that Dutch leadership in climate talks was made possible by domestic political processes and cooperation between the government and Dutch NGOs.

Nonstate actors

The main actors in diplomacy are state delegations (see Chapter 7), yet nonstate actors have access to conferences, participate in large numbers, and affect the process and outcomes (Rietig 2016; Bäckstrand et al. 2017; see Chapter 14). Kal Raustiala (2002) drew a classification of methods of NGO influence and revealed a symbiotic relationship between states and NGOs. Betsill and Corell (2001) developed an influential analytical framework to study systematically the role of civil society and environmental NGOs. Utilizing this framework, a study on forest negotiations reveals that green NGOs can influence negotiations if they get involved early in the process and phrase their policy recommendations in line with neoliberal discourse (Humphreys 2004). We know relatively little about the comparative effectiveness of NGO strategies (Rietig 2016), but NGOs actively seek to influence climate negotiations through awareness raising, coalition building, "corridor politics" and participation on state delegations, yet their actual impact on policymaking is unclear (Gulbrandsen and Andresen 2004; Rietig 2016). Sometimes civil society is altogether disenfranchised (Fisher 2010). Academics also investigate the influence of business and industry groups on environmental negotiations (Levy and Egan 2003; Vormedal 2009; Mecking 2011). Corporate actors rarely manage to prevent international regulation but influence the content of agreements toward market-based policy options such as emissions trading (Mecking 2011).

Domestic-international connections

The interplay between domestic politics and international discussions is another lucrative area of study. Robert Putnam's seminal work established that each delegation plays two simultaneous "games" with domestic constituents and foreign counterparts (Putnam 1988). His concept of the two-level game continues to inform scholars in understanding state behavior (Agrawala and Andresen 2001). In her award-winning work, Beth DeSombre (2000) reveals the domestic sources of foreign environmental policy that can illuminate negotiations, too. Aslaug Asgeisdottir (2008) examines bargaining between Iceland and Norway over fish stocks, and confirms that domestic interest groups strengthen the negotiating position of states vis-à-vis other countries. Iceland's strong fishing industry exerted pressures on the government that helped its delegation win concessions from Norway whose weaker internal pressures left the delegation with more openness for compromise. Other empirical case studies cast doubt on the theory and suggest that state leaders may choose to ignore domestic constraints or, conversely, pursue international strategies without paying close attention to the domestic game. Regarding the Kyoto Protocol, McLean and Stone (2012) argue that the European Union had a principled commitment to climate policy cooperation and developed domestic policy regardless of negotiation outcomes.

Issue linkage

Negotiations on a specific ecological problem rarely unfold in isolation from discussions on other problems. State and nonstate actors often draw connections between issues such as climate change (see Chapter 32), forestry (see Chapter 42), desertification (see Chapter 43), ozone depletion (see Chapter 31) and biodiversity (see Chapter 41). Such linkage has made climate conferences a central hub of global environmental politics. "Indeed, with over 1,200 NGO and IGO observers now accredited to attend the UNFCCC negotiations, representing over 22 issue areas, and drawing over 20,000 observers, it seems that everyone from Mac-Donald's to the Vatican is jumping on the proverbial climate change bandwagon" (Jinnah 2011: 2). Such bandwagoning has the potential to facilitate more effective policy outcomes on climate change (Jinnah 2011). Linking environmental and trade issues made negotiations on ozone depletion easier and contributed to the success of the Montreal Protocol (Barrett 1997). At the same time, linkages increase issue complexity in climate politics and present an obstacle to productive negotiations (Victor 2011; Wapner 2011).

Game theory

All of the above actors and factors come handy in game theory that focuses on modeling of negotiations and utilizes formal logic to derive probable outcomes from fixed actor preferences. Bruce de Mesquita (2009) used computer calculations to predict that the 2009 Copenhagen conference would fail and, indeed, the summit collapsed in a spectacular and highly publicized manner. In more conventional vein, Scott Barrett (1998, 2003, 2008) built a body of work to clarify the obstacles to global environmental cooperation through game theory. Another pioneer in this realm is Hugh Ward who used the game of chicken to illuminate climate negotiations (1993) and later developed a model incorporating divergent national positions of dragger and pusher countries (Ward et al. 2001).

Formal models of bargaining have rarely been applied to actual cases of environmental negotiations (Avenhaus and Zartman 2007). Oran Young (1994) developed a model of integrative bargaining that captures the role of multiple actors, the veil of uncertainty about future costs and benefits, and evolving interest configurations, among other factors. His model is commonly recognized as influential in the discipline but has yet to be applied systematically in comparative empirical studies. A collection of essays used extended game theoretic methods to speculate on potential agreements on the reduction of greenhouse gases and theorized that heterogeneity of state actors enhances the prospects for burden sharing and coalition building (Carraro 1997). Whether this actually occurs remains unclear since studies do not compare models with actual negotiations.

Process and strategy in negotiations

Process variables can affect outcomes in environmental diplomacy. While structural variables such as power distribution and interests remain constant, the management style of organizers who run conferences, for example, helps explain year-to-year differences between climate summits in Copenhagen and Cancun (Park 2015; Monheim 2016). Process is generally under-explored, however, due to lack of data and researcher access to negotiations. More and more scholars attend UN conferences but the number of academics who have access to diplomatic circles and actual negotiations is small. As a result, dynamics around the negotiation table often remain hidden as publications focus on analyzing the legal *outcomes* and evaluate resulting agreements (Clémençon 2016; Falkner 2016)

One particularly problematic gap in academic scholarship is the shortage of studies on communication and dialog in environmental diplomacy. What do delegations actually say to one another? Sweeping literature reviews conclude that the exchange of arguments is the least explored topic in this field of research, partly due to a lack of verbatim records of negotiations (Jönsson 2002; Zartman 2002; Deitelhoff and Müller 2005). Important books by Farhana Yamin and Joanna Depledge rectify the general neglect of process and provide detailed descriptions of logistics in climate conferences but also leave the conversations out (Depledge 2005b; and Yamin and Depledge 2004). Christian Grobe (2010) advanced a rationalist theory of argumentative persuasion that draws on secondary sources, without data on the diplomatic dialog.

Existing research on the micro-dynamics of international conversations show that persuasive arguments can alter policy preferences and affect negotiating outcomes. Soft bargaining strategies that accommodate others increase a delegation's success in negotiations (Weiler 2012). Negotiating strategy helped the EU achieve its goals at the Paris climate conference (Oberthür and Groen 2018). Another study classified argumentation strategies of different countries and identified effective techniques of persuasion (Dimitrov 2012). It found that governments spend considerable efforts to persuade others and engage in purposeful communication aiming to reshape policy preferences in other countries. Vogler (2010) confirms that the British government made efforts to change other countries' perception of the climate problem and their interests in mitigating it. Notably, arguments about the economic benefits of climate action turned the tide in diplomacy and began to transform global climate governance even before the Paris Agreement was reached (Dimitrov 2012; Zhang and Shi 2013; Torney and Cross 2018).

Insider perspectives on environmental diplomacy

Practitioner-scholars who actively participate in environmental conferences can provide rich empirical accounts that improve our understanding of process variables (Benedick 1998; Depledge 2005a, 2005b, 2006; Dimitrov 2005, 2016; Rajamani 2008, 2010; Kulovesi and Gutiérrez 2009; Bodansky 2010; Fry 2011, 2016; Smith 2009). The Earth Negotiation Bulletin (ENB) is widely respected by academics, civil society and policy-makers around the world for providing global transparency of international environmental diplomacy, through daily two-page summaries and political analysis of every round of negotiations on every environmental issue. Writers for the ENB with extensive exposure to negotiations continue to offer valuable insights on various conferences (Wagner 2007; Jinnah et al. 2009; Chasek and Wagner 2012; Allan 2019). These and other works offer palpable taste of environmental diplomacy and in-depth expertise that can inform both theory and practice.

From David Humphreys's (2006) dedicated work to chronicle forest policy negotiations to insiders' perspectives on climate negotiations (Rajamani 2010, 2016; Dimitrov 2016; Fry 2016), scholarship based on direct observations allows readers to get as close to reality as possible. In *Ozone Diplomacy*, US chief negotiator Richard Benedick (1998) provided us with a definitive account of the successful international negotiations that produced the Montreal Protocol (see Chapter 31). A detailed story of the negotiations on the 2015 Sustainable Development Goals was co-written by a chairperson of the UN negotiating process (Kamau et al. 2018). These single-case books are richly informative but did not have scholarly ambitions to engage the academic literature and contribute theoretically informed ideas. Books on climate change negotiations, in particular, also share the problem of ignoring process factors. One volume brought several insiders of climate diplomacy who contribute valuable insights on structural problems and issue complexity, and ignored process variables altogether (Sjöstedt and Penetrante 2013). A more recent volume on the Paris Agreement contains chapters by insiders who focus exclusively on evaluating the legal outcome of climate negotiations, without exploring the political dynamics through which the treaty was negotiated (Klein et al. 2017). As a result, even insider accounts may not help readers learn about what actually occurs during negotiations.

One book from the frontlines of environmental diplomacy avoids both the single-case problem and the process-neglect problem. Richard Smith, an American diplomat who helped negotiate several agreements, illuminates features of environmental diplomacy that often escape academic scholars: real negotiations take place in informal settings and rarely in official Plenary discussions; country delegations sometimes fall silent as a negotiating tactic; and breakthroughs occur during all-night sessions in the final days of conferences. Integrating such insights in academic scholarship would strengthen theory building (Smith 2009).

Climate change negotiations: a special case

Global climate negotiations have been among the top issues on the global political agenda and have attracted particular academic attention (see Harris 2021: 33–59 and Chapter 32). An important body of research analyzes negotiating challenges, debates future policy options and offers policy recommendations (Agrawala and Andresen 2001; Bodansky 2004; Harris 2011; Gupta 2012; Zhang and Shi 2013; Hovi et al. 2016). Various participants in UN climate conferences have illuminated the enormously complex global discussions on post-Kyoto policy with extensive and detailed summaries of issues on the table, positions of main countries, political dynamics and major decision outcomes (Depledge 2006; Rajamani 2008, 2010; Kulovesi and Gutiérrez 2009; Chandani 2010; Dimitrov 2010; Sterk et al. 2010; Fry 2011, 2016; Oberthür 2011). Many analyze existing agreements and discuss future prospects for cooperation (Victor 2001; Yamin and Depledge 2004; Clémençon 2008; Ott et al. 2008; Watanabe et al. 2008). Others focus on national policies and negotiation positions of actors such as the United States (Depledge 2005a), the EU (Vogler and Bretherton 2006; Hovi et al. 2003; Oberthür and Kelly 2008), China (Harris and Yu 2005), developing countries (Najam et al. 2003) and island states (Betzold 2010).

Until recently, climate diplomacy was proverbial for its repeated failures to produce a global treaty to replace the Kyoto Protocol. Formal negotiations on post-2012 climate policy began in 2008 under the Bali Action Plan, a legal mandate to negotiate emission reductions in advanced economies, policy actions by developing countries, and international financial and technological support for climate policy (Clémençon 2008; Watanabe et al. 2008). Two turbulent years of diplomatic activity produced elaborate policy proposals from Japan, Tuvalu, New Zealand and many other countries but made little progress (Bodansky 2010; Rajamani 2010). The talks collapsed in a particularly disastrous summit in Copenhagen, Denmark in December of 2009 (Bodansky 2010). The outside world expected a climate treaty but diplomats had abandoned that option at the previous round of talks in Barcelona (except small island nations that pursued it faithfully to the end). Instead, the plan was to produce a fake "comprehensive decision" that does not contain policy obligations (Dimitrov 2010, 2020). This plot to create an empty agreement failed as did last-minute efforts to "greenwash" the Copenhagen conference with a short political declaration drafted by heads of states and grandly titled the "Copenhagen Accord" when diplomats of six small countries such as Sudan, Cuba and Bolivia decried it as a sham and refused to support it. Another summit in Mexico the following year produced the Cancun Accords whose inadequacy led to yet another legal mandate for negotiations that was adopted in Durban, South Africa in 2011 (Oberthür 2011; Groen et al. 2012).

After 20 rounds of formal negotiations over four years, in Durban states decided to restart negotiations under a new mandate and with a new deadline of 2015 for reaching an agreement that would apply after 2020. Global negotiations were effectively placed on hold, which confirmed Arild Underdal's "law of the least ambitious program" that remains foundational in mainstream scholarship. Underdal (1980) observed that negotiations involving multiple actors tend to produce outcomes that reflect the lowest-common denominator. The large number of actors (194 states) and requirement of global political consensus as a basis for decision-making creates major obstacles to effective multilateralism by giving every actor veto power (Ward et al. 2001).

This poor record of climate diplomacy created a virtual consensus among academics who argued that UN talks cannot possibly succeed (Hoffmann 2011; Victor 2011; Hovi et al. 2003; Andresen 2015). David Victor (2006) and Bruce Bueno de Mesquita (2009) stated with certainty that failure of the current global approach is guaranteed. A veteran diplomat,

Richard Smith (2009) also considered the climate diplomacy process as a manual of how not to negotiate. In an extensively researched piece, Røgeberg, Andresen and Holtsmark (2010) brought charts and numbers to prove that the international community cannot solve the climate problem. Academic observers also debated how to improve the prospects of climate diplomacy. Subject of a cottage industry of academics and think tanks, the proposals for international climate policy were numerous and diverse (Bodansky 2004; Aldy and Stavins 2010). Proposals ranged from incrementalism (Falkner et al. 2010) to negotiating a non-binding agreement by an oligarchy of powerful countries (Victor 2011) to "inclusive minilateralism" involving a global climate council of 8 to 23 countries (Eckersley 2012).

The successful adoption of the 2015 Paris Agreement came as a surprise and became one of the most important events in modern environmental diplomacy. The Twenty-First Conference of the Parties (COP-21) in Paris was a culmination of a four-year diplomatic process that began in 2011. The summit brought 40,000 delegates, including 20,000 government representatives on 195 delegations, and was characterized by a genuine collective effort to reach mutual compromise, skillfully orchestrated by the French presidency. Negotiations addressed a kaleidoscope of diverse issues, with 24 spinoff groups working in the first week. Key contentious issues included: the global long-term goal of the agreement and degree of policy ambition; the legally binding character of national actions; climate finance; and the evolution of the policy regime over time (Rajamani 2016). The result was the first global climate treaty of universal scope and obligations for most countries (see Chapter 32).

Theoretical explanations of the Paris conference by insiders are scant. One account of COP-21 emphasized the role of organizational leadership by the French government and entrepreneurial leadership of the High Ambition Coalition in bridging the North–South divide (Brun 2016). Another study argues that negotiations succeeded because of the skilled use of secrecy, entrepreneurial leadership, and superb organizational tactics by the French government, together with persuasive argumentation by the European Union and island states that altered policy preferences and facilitated social learning (Dimitrov 2016). A more widely shared perspective concerns the pivotal role of the United States. The United States opposed binding commitments concerning mitigation and finance so adamantly that other countries had to choose between a weaker treaty with US participation and a stronger treaty without the United States. The US delegation was singularly responsible for turning national policy "commitments" into "contributions," weakening legal obligations to achieve policy results, and making developed country mitigation policy less legally binding (Rajamani 2016).

While some see the Paris Agreement as a disappointing outcome (Clémençon 2016; Allan 2019), others view it as a major breakthrough in environmental politics (Bodansky 2016; Brun 2016; Dimitrov 2016; Rajamani 2016). Insiders describe it as "a culmination of decades of climate diplomacy" (Brun 2016: 121), and the "most ambitious outcome possible in a deeply discordant political context" (Rajamani 2016: 493-94). Countries demonstrated reciprocal willingness to compromise as most players made sacrifices and gained something in return (Dimitrov 2016; Kinley 2017). The Paris Agreement is a legally binding but ambiguous treaty that gives governments full discretion over domestic policies. It combines the top-down and bottom-up approach to climate governance and relies mostly on voluntary actions by national governments (Andresen 2015; see Harris 2021: 33–59 and Chapter 32). Yet, compliance with the treaty has triggered climate policy developments around the world. A growing pattern of government policies and business practices, particularly in Europe and Asia, converge toward low-carbon development and a global socio-economic shift toward a cleaner economy (Iacobuta et al. 2018). This is a useful reminder of the role of international environmental diplomacy in global governance.

Conclusion

The literature on environmental diplomacy tends to make academic observations from a distant ivory tower, or to provide detailed empirical accounts of single cases. What would strengthen scholarship on this important topic are structured comparative studies that combine the process-oriented insider's vantage point with the theory-oriented academic perspective to provide rich stories that are theoretically informed. Broader comparisons of empirical cases would increase our confidence in the validity of conclusions about the various factors shaping the process and outcomes of environmental negotiations.

Future research should also recognize that environmental diplomacy can affect governance in indirect ways, beyond the production of treaties. Peter Hass (2002) has argued that an important effect of United Nations conferences is the growth of global environmental norms. Its shortcomings notwithstanding, "the Kyoto Protocol" became a household phrase and raised awareness of climate change. In a rich empirical study, Antto Vihma (2010) shows that India's domestic climate policy process has changed as a result of the country's engagement in UN talks. Global climate governance is dramatically different today compared to the 1990s, now a vibrant realm of policy development and implementation (Chapter 22). And while the driving forces behind this shift are multiple, the growing belief in the economic wisdom of green action cannot be separated from the international dialog over the last 20 years. This conversation changed perceptions of national interests even before leading to an agreement. European arguments about the economic benefits of climate policy and "win-win" solutions to global warming affected cost-benefit calculations by other governments and indirectly affected national policies as well as country positions in the UN negotiations (Dimitrov 2010, 2012; Torney and Cross 2018). Thus, global discussions can foster environmental governance even before producing treaties. Students of diplomacy need to reconsider the meaning of "outcome" and recognize the diverse impacts of negotiations on state behavior.

References

Agrawala, Shardul, and Steinar Andresen. (2001). Two-Level Games and the Future of the Climate Regime. *Energy and Environment* 12(2 and 3): 5–11.

Aldy, Joseph E., and R. N. Stavins, eds. (2010). *Post-Kyoto Climate International Climate Policy: Implementing Architectures for Agreements.* Cambridge: Cambridge University Press.

Allan, Jen Iris. (2019). Dangerous Incrementalism of the Paris Agreement. *Global Environmental Politics* 19(1): 4–11.

Allan, Jen Iris and Peter Dauvergne. (2013). The Global South in Environmental Negotiations: the politics of coalitions in REDD+. *Third World Quarterly* 34(8): 1307–1322.

Andresen, Steinar. (2015). International Climate Negotiations: Top-down, Bottom-up or a Combination of Both? *International Spectator* 50(1): 15–30.

Andresen, Steinar, and Shardul Agrawala. (2002). Leaders, Pushers and Laggards in the Making of the Climate Regime. *Global Climate Change* 12: 41–51.

Asgeisdottir, Aslaug. (2008). *Who Gets What: Domestic Influences on International Negotiations Allocating Shared Resources.* New York: SUNY Press.

Avenhaus, Rudolf, and I. William Zartman. (2007). Formal Models of, in and for International Negotiations. In *Diplomacy Games: Formal Models and International Negotiations*, edited by Rudolf Avenhaus and I. William Zartman, pp. 1–22. Berlin: Springer.

Bäckstrand, K., Kuyper, J.W., Linnér, B.O., & Lövbrand, E. (2017). *Non-state Actors in Global Climate Governance: From Copenhagen to Paris and Beyond.* Abingdon, UK: Taylor & Francis.

Bailer, Stephanie. (2012). Strategy in the Climate Change Negotiations: Do Democracies Negotiate Differently? *Climate Policy* 12(5): 534–551.

Bailer, Stephanie and Florian Weiler. (2014). A Political Economy of Positions in Climate Change Negotiations. *Review of International Organizations* 10(1):43–66.

Barrett, Scott. (1997). The Strategy of Trade Sanctions in International Environmental Agreements. *Resource and Energy Economics* 19(4): 345–361.

Barrett, Scott. (1998). On the Theory and Diplomacy of Environmental Treaty-Making. *Environmental and Resource Economics* 11(3–4): 317–333.

Barrett, Scott. (2003). *Environment and Statecraft: The Strategy of Environmental Treaty-Making.* Oxford: Oxford University Press.

Barrett, Scott. (2008). *Why Cooperate? The Incentive to Supply Global Public Goods.* Oxford, UK: Oxford University Press.

Benedick, Richard Elliot. (1998). *Ozone Diplomacy: New Directions in Safeguarding the Planet.* Cambridge, Mass.: Harvard University Press.

Bernstein, Steven. (2001). *The Compromise of Liberal Environmentalism.* New York: Columbia University Press.

Betsill, Michele M. and Elisabeth Corell. (2001). NGO Influence in International Environmental Negotiations. *Global Environmental Politics* 1(4): 65–85.

Betzold, Carola. (2010). Borrowing Power to Influence International Negotiations: AOSIS in the Climate Change Regime. *Politics* 30(3): 131–148.

Blühdorn, Ingolfur. (2007). Sustaining the Unsustainable: Symbolic Politics and the Politics of Simulation. *Environmental Politics* 16(2): 251–275.

Bodansky, Daniel. (2004). *International Efforts on Climate Change Beyond 2012: A Survey of Approaches.* Washington, DC: Pew Center on Global Climate Change.

Bodansky, Daniel. (2010). The Copenhagen Climate Change Conference: A Postmortem. *The American Journal of International Law* 104(2): 230–240.

Bodansky, Daniel. (2016). The Legal Character of the Paris Agreement. *Review of European, Comparative and International Environmental Law* 25 (2): 142–150.

Brun, Aslak. (2016). Conference Diplomacy: The Making of the Paris Agreement. *Politics and Governance* 4(3): 115–123.

Carraro, Carlo, ed. (1997). *International Environmental Negotiations: Strategic Policy Issues.* Cheltenham, UK: Edward Elgar.

Chandani, Achala. (2010). Expectations, Reality and Future: A Negotiator's Reflections on COP15. *Climate Law* 1(1): 207–225.

Chasek, Pamela S. (2001). *Earth Negotiations: Analyzing Thirty Years of Environmental Diplomacy.* Tokyo: United Nations University Press.

Chasek, Pamela S. and Lynn M. Wagner, eds. (2012). *The Roads from Rio: Lessons Learned from Twenty Years of Multilateral Environmental Negotiations.* New York: Routledge.

Clémençon, Raymond. (2008). The Bali Road Map: A First Step on the Difficult Journey to a Post-Kyoto Protocol Agreement. *Journal of Environment and Development* 17(1): 70–94.

Clémençon, Raymond. (2016). The Two Sides of the Paris Climate Agreement: Dismal Failure or Historic Breakthrough? *Journal of Environment & Development* 25(1): 3–24.

Cross, Mai'a K. Davis. (2018). Partners at Paris? Climate Negotiations and Transatlantic Relations. *Journal of European Integration* 40(3): 1–16.

Davenport, Deborah. (2005). An Alternative Explanation of the Failure of the UNCED Forestry Negotiations. *Global Environmental Politics* 5(1): 105–130.

De Mesquita, Bruce. (2009). Recipe for Failure. *Foreign Policy* 175: 76–88.

Death, Carl. (2011). Summit Theatre: Exemplary Governmentality and Environmental Diplomacy in Johannesburg and Copenhagen. *Environmental Politics* 20(1): 1–19.

Deitelhoff, Nicole, and Harald Müller. (2005). Theoretical Paradise-Empirically Lost? Arguing with Habermas. *Review of International Studies* 31: 167–179.

Depledge, Joanna. (2005). Against the Grain: The United States and the Global Climate Change Regime. *Global Change, Peace and Security* 17(1): 11–27.

Depledge, Joanna. (2005). *The Organization of Global Negotiations: Constructing the Climate Change Regime.* London: Earthscan.

Depledge, Joanna. (2006). The Opposite of Learning: Ossification in the Climate Change Regime. *Global Environmental Politics* 6(1): 1–22.

DeSombre, Elizabeth R. (2000). *Domestic Sources of International Environmental Policy: Industry, Environmentalists and U.S. Power.* Cambridge, Mass.: The MIT Press.

Dimitrov, Radoslav S. (2005). Hostage to Norms: States, Institutions and Global Forest Politics. *Global Environmental Politics* 5(4): 1–24.

Dimitrov, Radoslav S. (2010). Inside UN Climate Change Negotiations. *Review of Policy Research* 27(6): 795–821.

Dimitrov, Radoslav S. (2012). Persuasion in World Politics: The UN Climate Change Negotiations. In *Handbook of Global Environmental Politics*, edited by Peter Dauvergne and Jennifer Clapp, 72–86. Cambridge, Mass.: The MIT Press.

Dimitrov, Radoslav S. (2016). The Paris Agreement on Climate Change: Behind Closed Doors. *Global Environmental Politics* 16(3): 1-11.

Dimitrov, Radoslav S. (2020). Empty Institutions in Global Environmental Politics. *International Studies Review* 22(3): 626–650.

Eckersley, Robyn. (2012). Moving Forward in the Climate Negotiations: Multilateralism or Minilateralism? *Global Environmental Politics* 12(2): 24–42.

Edelman, Murray. (1988). *Constructing the Political Spectacle.* Chicago: University of Chicago Press.

Edwards, Guy, Isabel C. Adarve, María C. Bustos and J. Timmons Roberts. (2017). Small Group, Big Impact: How AILAC Helped Shape the Paris Agreement. *Climate Policy* 17(1): 71–85.

Falkner, Robert. (2005). American Hegemony and the Global Environment. *International Studies Review* 7(4): 585–599.

Falkner, Robert. (2007). The Political Economy of 'Normative Power' Europe: EU Environmental Leadership in International Biotechnology Regulation. *Journal of European Public Policy* 14(4): 507–526.

Falkner, Robert. (2016). The Paris Agreement and the New Logic of International Climate Politics. *International Affairs* 92:1107–1125.

Falkner, Robert, Hannes Stephan, and John Vogler. (2010). International Climate Policy after Copenhagen: Towards a 'Building Blocks' Strategy. *Global Policy* 1(3): 252–262.

Fisher, Dana R. (2010). COP-15 in Copenhagen: How the Merging of Movements Left Civil Society Out in the Cold. *Global Environmental Politics* 10(2): 11–17.

Fomerand, Jacques. (1996). UN Conferences: Media Events or Genuine Diplomacy? *Global Governance* 2(3): 361–375.

Fry, Ian. (2011). If a Tree Falls in a Kyoto Forest … *Review of European Community and International Environmental Law* 20(2):123–138.

Fry, Ian. (2016). The Paris Agreement: An Insider's Perspective-The Role of Small Island Developing States. *Environmental Policy and Law* 46(2): 105.

Goldstein, Judith, Miles Kahler, Robert O. Keohane, and Ann-Marie Slaughter. (2000). Legalization and World Politics. *International Organization* 54(3): 385–399

Grobe, Christian. (2010). The Power of Words: Argumentative Persuasion in International Negotiations. *European Journal of International Relations* 16(1): 5–29.

Groen, Lisanne, Arne Niemann, and Sebastian Oberthür. (2012). The EU as a Global Leader? The Copenhagen and Cancun UN Climate Negotiations. *Journal of Contemporary European Research* 8(2): 173–191.

Gulbrandsen, Lars H. and Steinar Andresen. (2004). NGO Influence in the Implementation of the Kyoto Protocol. *Global Environmental Politics* 4(4): 54–75.

Gupta, Joyeeta. (2012). Negotiating Challenges and Climate Change. *Climate Policy* 12:630–644.

Gupta, Joyeeta, and Michael Grubb. (2000). *Climate Change and European Leadership: A Sustainable Role for Europe?* Berlin: Springer.

Haas, Peter M. (2002). UN Conferences and Constructivist Governance of the Environment. *Global Governance* 8(1): 73–91.

Haas, Peter M., Robert O. Keohane, and Marc Levy, eds. (1993). *Institutions for the Earth: Sources of Effective International Environmental Protection.* Cambridge, Mass: The MIT Press.

Harris, Paul G., ed. (2007). *Europe and Global Climate Change: Politics, Foreign Policy and Regional Cooperation.* Cheltenham, UK: Edward Edgar Publishing.

Harris, Paul G., ed. (2011). *China's Responsibility for Climate Change: Ethics, Fairness and Environmental Policy.* Bristol, UK: Policy Press.

Harris, Paul G. (2021). *Pathologies of Climate Governance: International Relations, National Politics and Human Nature.* Cambridge: Cambridge University Press.

Harris, Paul G. and Hongyuan Yu. (2005). Environmental Change and the Asia-Pacific: China Responds to Global Warming. *Global Change, Peace and Security* 17(1): 45–58.

Hasenclever, Andreas, Peter Mayer, and Volker Rittberger. (1997). *Theories of International Regimes.* Cambridge: Cambridge University Press.

Hochstetler, Kathryn and Manjana Milkoreit. (2013). Emerging Powers in the Climate Negotiations: Shifting Identity Conceptions. *Political Research Quarterly* 20(10): 1–12.

Hoffman, Matthew. (2011). *Climate Governance at the Crossroads: Experimenting with a Global Response after Kyoto*. Oxford, UK: Oxford University Press.

Hovi, Jon, Tora Skodvin, and Steinar Andresen. (2003). The Persistence of the Kyoto Protocol: Why Annex I Countries Move on without the United States. *Global Environmental Politics* 3(4): 1–23.

Hovi, Jon, Detlef Sprinz, Håkon Sælen and Arild Underdal. (2016). Climate Change Mitigation: A Role for Climate Clubs? *Palgrave Communications* 2. Available at https://www.nature.com/articles/palcomms201620.

Humphreys, David. (2004). Redefining the Issues: NGO Influence on International Forest Negotiations. *Global Environmental Politics* 4(2): 51–74.

Humphreys, David. (2006). *Logjam: Deforestation and the Crisis of Global Governance*. New York: Earthscan.

Iacobuta, Gabriela Navroz K. Dubash, Prabhat Upadhyaya, Mekdelawit Deribe and Niklas Höhne. (2018). National Climate Change Mitigation Legislation, Strategy and Targets: A Global Update. *Climate Policy* 18(9): 1114–1132.

International Environmental Agreements Database Project. (2020). Available at https://iea.uoregon.edu/.

Jinnah, Sikkina. (2011). Climate Change Bandwagoning: The Impacts of Strategic Linkages on Regime Design, Maintenance, and Death. *Global Environmental Politics* 11(3): 1–9.

Jinnah, Sikkina, Douglas Bushey, Miquel Muñoz and Kati Kulovesi. (2009). Tripping Points: Barriers and Bargaining Chips on the Road to Copenhagen. *Environmental Research Letters* 4(3): 1–6.

Jönsson, Christer. (2002). Diplomacy, Bargaining and Negotiation. In *Handbook of International Relations*, edited by Walter Carlsnaes, Thomas Risse and Beth A. Simmons. 212–234 London: Sage Publications.

Kamau, Macharia, Pamela Chasek and David O'Connor. (2018). *Transforming Multilateral Diplomacy: The Inside Story of the Sustainable Development Goals*. New York: Routledge.

Kanie, Norichika. (2003). Leadership in Multilateral Negotiation and Domestic Policy: The Netherlands at the Kyoto Protocol. *International Negotiation* 8(2): 339–365.

Karlsson, Christer, Charles Parker, Matthias Hjerpe and Björn-Ola Linnér. (2011). Looking for Leaders: Perceptions of Climate Change Leadership among Climate Change Participants. *Global Environmental Politics* 11(1): 89–107.

Karlsson, Christer, Charles Parker, Matthias Hjerpe and Björn-Ola Linnér. (2012). The Legitimacy of Leadership in International Climate Negotiations. *Ambio* 41: 46–55.

Kilian, Bertil, and Ole Elgström. (2010). Still a Green Leader? The European Union's Role in International Climate Negotiations. *Cooperation and Conflict* 45(3): 255–273.

Kinley, Richard. (2017). Climate Change after Paris: From Turning Point to Transformation. *Climate Policy* 17(1): 9-15.

Klein, D. R., Carazo, M. P., Doelle, M., Bulmer, J., and Higham, A. (Eds.). (2017). *The Paris Agreement on Climate Change: Analysis and Commentary*. Oxford: Oxford University Press.

Kulovesi, Kati, and Maria Gutiérrez. (2009). Climate Change Negotiations Update: Process and Prospects for a Copenhagen Agreed Outcome. *Review of European Community and International Environmental Law* 18(3): 229–243.

Levy, David. L. and Daniel Egan. (2003). A Neo-Gramscian Approach to Corporate Political Strategy: Conflict and Accommodation in the Climate Change Negotiations. *Politics and Society* 26(3): 337–361.

Levy, Marc A., Oran R. Young, and Michael Zürn. (1995). The Study of International Regimes. *European Journal of International Relations* 1: 267–330.

McLean, Elena V., and Randall W. Stone. (2012). The Kyoto Protocol: Two-Level Bargaining and European Integration. *International Studies Quarterly* 56: 1–15.

Mecking, Jonas. (2011). The Globalization of Carbon Trading: Transnational Business Coalitions in Climate Politics. *Global Environmental Politics* 11(2): 26–50.

Michaelowa, Katharina and Alex Michaelowa. (2012). India as an Emerging Power in International Climate Negotiations. *Climate Policy* 12: 575–590.

Monheim, Kai. (2016). The "Power of Process:" How Negotiation Management Influences Multilateral Cooperation. *International Negotiation* 21: 345–380.

Muñoz, Miquel, Rachel Thrasher and Adil Najam. (2009). Measuring the Negotiation Burden of Multilateral Environmental Agreements. *Global Environmental Politics* 9(4): 1–13.

Najam, Adil, Saleemul Huq, and Youba Sokona. (2003). Climate Negotiations Beyond Kyoto: Developing Countries Concerns and Interests. *Climate Policy* 3: 221–231.

Oberthür, Sebastian. (2011). Global Climate Governance after Cancun: Options for EU Leadership. *The International Spectator* 46(1): 5–13.

Oberthür, Sebastian and Claire Roche Kelly. (2008). EU Leadership in International Climate Policy: Achievements and Challenges. *The International Spectator* 43(3): 35–50.

Oberthür, Sebastian and Lisanne Groen. (2018). Explaining Goal Achievement in International Negotiations. *Journal of European Public Policy* 25(5): 708–727.

Ott, Herman E., Wolfgang Sterk, and Rie Watanabe. (2008). The Bali Roadmap: New Horizons for Global Climate Policy. *Climate Policy* 8: 91–95.

Park, Siwon. (2015). The Power of Presidency in UN Climate Change Negotiations: Comparison Between Denmark and Mexico. *International Environmental Agreements: Politics, Law and Economics* 16: 781–795.

Putnam, Robert D. (1988). Diplomacy and Domestic Politics: The Logic of Two-Level Games. *International Organization* 42(3): 427–60.

Rajamani, Lavanya. (2008). From Berlin to Bali: Killing Kyoto Softly. *International and Comparative Law Quarterly* 57(4): 909–939.

Rajamani, Lavanya. (2010). The Making and Unmaking of the Copenhagen Accord. *International and Comparative Law Quarterly* 59(3): 824–843.

Rajamani, Lavanya. (2016). Ambition and Differentiation in the 2015 Paris Agreement: Interpretative Possibilities and Underlying Politics. *International and Comparative Law Quarterly* 65(2): 493-514.

Raustiala, Kal. (2002). States, NGOs and International Environmental Institutions. *International Studies Quarterly* 41(4): 719–740.

Rietig, Katharina. (2016). The Power of Strategy: Environmental NGO Influence in International Climate Negotiations. *Global Governance* 22: 269–288.

Røgeberg, Ole, Steinar Andresen, and Bjart Holtsmark. (2010). International Climate Treaties: The Case for Pessimism. *Climate Law* 1(1): 177–197.

Schreurs, Miranda A. and Yves Tiberghien. (2007). Multi-Level Reinforcement: Explaining European Union Leadership in Climate Change Mitigation. *Global Environmental Politics* 7(4): 19–46.

Sjöstedt, Gunnar and Ariel Macaspac Penetrante (eds.). (2013). *Climate Change Negotiations: A Guide to Resolving Disputes and Facilitating Multilateral Cooperation.* New York: Routledge.

Smith, Richard J. (2009). *Negotiating Environment and Science: An Insider's View of International Agreements, from Driftnets to the Space Station.* Washington, DC: Resources for the Future.

Sprinz, Detlef and Tapani Vaahtoranta. (1994). The Interest-Based Explanation of International Environmental Policy. *International Organization* 48(1): 77–105.

Sterk, Wolfgang, Christof Arens, Sylvia Borbonus, Urda Eichhorst, Dagmar Kiyar, Florian Mersmann, Frederic Rudolph, Hanna Wang-Helmreich, and Rie Watanabe. (2010). Something was Rotten in the State of Denmark – Cop-out in Copenhagen. *Policy Paper*, Munich: Wuppertal Institute for Climate, Environment and Energy.

Underdal, Arild. (1980). *The Politics of International Fisheries Management: The Case of the Northeast Atlantic.* New York: Columbia University Press.

Underdal, Arild. (1994). Leadership Theory: Rediscovering the Art of Management. In *International Multilateral Negotiations: Approaches to the Management of Complexity*, edited by I. William Zartman, 178–200. San Francisco: Jossey-Bass.

Victor, David G. (2001). *The Collapse of the Kyoto Protocol and the Struggle to Slow Global Warming.* Princeton: Princeton University Press.

Victor, David G. (2006). Toward Effective International Cooperation on Climate Change: Numbers, Interests and Institutions. *Global Environmental Politics* 6(3): 90–103.

Victor, David G. (2011). *Global Warming Gridlock: Creating More Effective Strategies for Protecting the Planet.* Cambridge: Cambridge University Press.

Vihma, Antto. (2010). Elephant in the Room: the New G-77 and China Dynamics in Climate Talks. *The Finnish Institute of International Affairs.* Available at www.upi-fiia.fi/en/publication/118/elephant_in_the_room.

Vogler, John. (2005). The European Contribution to Global Environmental Governance. *International Affairs* 81: 835–50.

Vogler, John. (2010). The Institutionalisation of Trust in the International Climate Regime. *Energy Policy* 38: 2681–2687.

Vogler, John, and Charlotte Bretherton. (2006). The European Union as a Protagonist to the United States on Climate Change. *International Studies Perspectives* 7(1): 1–22.

Vormedal, Irja. (2009). The Influence of Business and Industry NGOs in the Negotiation of the Kyoto Mechanisms. *Global Environmental Politics* 8(4): 36–65.

Wagner, Lynn. (2007). North-South Divisions in Multilateral Environmental Agreements: Negotiating the Private Sector's Role in Three Rio Agreements. *International Negotiation* 12: 83–109.

Wapner, Paul. (2011). The Challenges of Planetary Bandwagoning. *Global Environmental Politics* 11(3): 137–144.

Ward, Hugh. (1993). Game Theory and the Politics of the Global Commons. *Journal of Conflict Resolution* 37(2): 203–35.

Ward, Hugh, Frank Grundig and Ethan R. Zorick. (2001). Marching at the Pace of the Slowest: A Model of International Climate Change Negotiations. *Political Studies* 49(3): 438–61.

Watanabe, Rie, Christof Arens, Florian Mersmann, Hermann E. Ott, and Wolfgang Sterk. (2008). The Bali Roadmap for Global Climate Policy – New Horizons and Old Pitfalls. *Journal of European Environmental and Planning Law* 5(2): 139–158.

Weiler, Florian. (2012). Determinants of Bargaining Success in the Climate Change Negotiations. *Climate Policy* 12: 552–574.

Wu, Fuzuo. (2013). China's Pragmatic Tactics in International Climate Negotiations: Reserving Principles with Compromise. *Asian Survey* 53(4): 778–800.

Yamin, Farhana, and Joanna Depledge. (2004). *The International Climate Change Regime: A Guide to Rules, Institutions and Procedures.* Cambridge: Cambridge University Press.

Young, Oran R. (1989). The Politics of International Regime Formation: Managing Natural Resources and the Environment. *International Organization* 43: 349–76.

Young, Oran R. (1991). Political Leadership and Regime Formation: On the Development of Institutions in International Society. *International Organization* 45(3): 281–308.

Young, Oran R. (1994). *International Governance: Protecting the Environment in a Stateless Society.* Ithaca, NY: Cornell University Press.

Young, Oran R. (1998). *Creating Regimes: Arctic Accords and International Governance.* Ithaca, NY: Cornell University Press.

Young, Oran R., and Gail Osherenko, eds. (1993). *Polar Politics: Creating International Environmental Regimes.* Ithaca, New York: Cornell University Press

Zartman, I. William. (2002). What I Want to Know about Negotiations. *International Negotiation* 7(1): 5–15.

Zhang, Youngsheng. and He-Ling Shi. (2013). From Burden-Sharing to Opportunity-Sharing: Unlocking the Climate Negotiations. *Climate Policy* 14(1): 63–81.

23

North–South relations

Colonialism, empire and international order

Shangrila Joshi

Conflicts between the Global North and Global South have been one of the mainstays of global environmental politics for well over three decades now. These conflicts result from pervasive differences between North and South over interests, power and socio-economic conditions that are inherently tied to colonialism and imperialism. Amidst recurring questions about the validity of a North–South dichotomy in global environmental politics, international negotiations to successfully mitigate climate change (see Chapter 32) and other global environmental problems have been impeded by apparently irreconcilable differences between North and South. This chapter addresses the origins of North–South conflicts in global politics, their evolution in the context of global environmental change, and theoretical debates over the validity of this frame of reference for understanding global environmental politics.

Origins of North–South conflict and calls for a new world order

North–South politics were borne out of an era of decolonization and have remained focused on challenging the legacies of colonialism. During the Cold War (1945–1990), the dominant binary frame of reference in world politics was East and West, representing the two opposed blocs – the United States and its allies on one side, the Soviet Union and its allies on the other – championing capitalist and communist ideologies respectively. A group of newly decolonized states that first emerged in Asia and then in Africa spearheaded the "Non-Aligned" Movement to mark their independence from these major power blocs. The 1955 Bandung conference in Indonesia is typically seen as the moment that congealed this Third World Project (Prashad 2014). This agenda represented the political position of a coalition of countries self-representing as the Third World that would be institutionalized at the UN as the G77 in 1964. Key aspects of this position included a rejection of colonialism, advancing a security agenda of peace over conflict, and emphasizing the importance of economic growth for the Third World (Gilman 2015; Hickel 2017; Lundestad 2005). The first and third aspects are crucial for understanding contemporary North–South environmental politics (discussed further in the next section).

DOI: 10.4324/9781003008873-27

The Global North usually refers to the core group of developed or industrialized countries, also referred to as Western countries, namely, the United States, Western European countries, Australia, Japan and Canada. These countries have exerted colonial/imperial dominance in the international economic system for several centuries. The term Global South – often used interchangeably with the terms Third World, periphery or developing countries – refers most commonly to countries that have been exploited by colonial/imperial powers in the past (Anand 2004; Isbister 2006). Contemporary relations between these two sets of countries are beset by the legacies of colonialism even in instances where colonial rule may have formally ended, a predicament described as the postcolonial condition (Gandhi 1998), or as coloniality (Grosfoguel 2011). Following a wave of juridico-political decolonial movements that resulted in formal Independence, the power differentials and exploitative relationship have prevailed between the former colonizers and the colonized as a legacy of colonialism (Grosfoguel 2011). This legacy has been further compounded by economic globalization that has further solidified the dominance of the North on an international political economy of neoliberal capitalism that continues to benefit the North (core) at the expense of the South (periphery) (see Chapter 24). Capitalism thus intersects with colonialism to ensure continued prosperity for the North while keeping the South entangled in a perpetual state of dependency and underdevelopment (Blaut 1993).

Three qualifications are necessary here. One is that these categories – North, South, First World, Third World and the West – do not necessarily represent neatly defined geographic regions. These are more accurately understood as geographic imaginaries or political constructs that have formed on the basis of both material and discursive factors (see Chapter 4). Second, not all countries are easily encompassed within these categories, especially when evolving socio-economic conditions are taken into consideration. Scholars of critical geopolitics among others have therefore challenged the validity of these categories (e.g., Toal 1994). The continued self-identification of emerging economies of the Global South in the context of treaty negotiations – and challenges to the same by hegemonic players in the Global North – has been a key feature of North–South environmental politics (Joshi 2021). Third, the North–South distinction often refers to economic differences between countries. As such, some scholars have pointed out that it is too state-centric, thereby failing to capture disparities within countries. Such problematization of the North–South frame of reference will be discussed in greater detail below.

As the Non-Aligned Movement, which was led by the formerly colonized countries, emphasized decolonization and economic development, its members began to demand a new world order (Gilman 2015; Lundestad 2005). It was in the context of trying to create this New International Economic Order (NIEO) that the North–South frame of reference was brought into focus (Jha 1982). The NIEO was designed to challenge the unfair terms of trade between North and South by demanding changes that would enable the South to achieve self-sustaining economic growth and industrialization (Bhagwati 1977; Najam 2004). The impetus of demands for a NIEO was rooted in fears of the economic stagnation of the South if such unequal terms of trade were not overturned. Among the demands were introduction of a range of enabling conditions for developing countries to thrive in international trade, such as removing trade tariffs for exports, stabilization of commodity prices, stepping up loan opportunities and economic development assistance, and regulating the activities of multinational corporations operating in the South (see Chapter 13). Most of these demands, and consequently proposals for a NIEO, were rejected in the United Nations, in large part due to the reluctance of the United States to accept a radically altered status quo in the international political economy. On the part of the South, countries were forced to abandon desires for long-term reform in light of meeting immediate needs because their negotiating position with the North was consistently too weak. Demands for a new economic order consequently fell by the wayside.

The ascendancy of neoliberalism in the post-colonial era – promulgated by the United States through the Washington Consensus and the Bretton Woods institutions – has ensured the continuance of core–periphery dynamics emulating old colonial patterns in the form of Global North imperialism, with the United States as the hegemonic power commanding the global economic system (Peet and Hartwick 2015; Power 2006; Slater 2006). Through economic institutions such as the World Trade Organization (WTO) – ostensibly seeking to strengthen developing countries' economies through incorporation in the global system of trade – the North's domination of the global economic system has been further secured, with a few exceptions offered by China (O'Brien and Leichenko 2003; see Chapter 24). Although the NIEO proposal was thwarted successfully by the United States in the 1970s, the desire to complete the process of decolonization and to challenge the hegemony of the Global North remains, and it is manifested in global environmental politics (Gilman 2015; Joshi 2021).

Emergence of North–South environmental politics

The prospect of unprecedented levels of global environmental change has brought about two major changes in North–South politics. First, it has transformed the negotiating power of the South in global politics, and second, it has called into question the single-minded focus the South has had on pursuing economic growth and development following the wave of decolonization in the 1940s. This became very clear in 1992, when global concern over the environmental crisis culminated in the UN Conference on Environment and Development in Brazil, the "Earth Summit." Two decades prior, in 1972, the international community had gathered in Stockholm for the UN Conference on the Human Environment. Following the seminal publication *Limits to Growth* by the Club of Rome, the key concern of the North at Stockholm was to prevent a drain on global resources that would be inevitable if the dogged pursuit of economic growth by rapidly growing and industrializing countries did not somehow subside. Although the explicit discourse was that the world should put the brakes on pursuit of unlimited economic growth, many believed that the unspoken and perhaps unintended desire was to put the brakes on the South's economic growth. Intellectuals from the South, unsurprisingly, saw this as unfair and a convenient way to maintain the unequal status quo between North and South. Western environmentalism in this guise was therefore seen as a tool for the continued subjugation of the formerly colonized (Agarwal and Narain 1990; Castro 1972).

Whereas industrialization was seen by the newly decolonized states as a valid goal to pursue, it was being portrayed by many in the North as undesirable due to its inevitable ecological consequences. Continued aspirations for development and industrialization were dismissed as something sought only by the South's elites. The South's underdevelopment was therefore viewed as not just an unfortunate outcome of the North's development, but rather deemed a necessity to prevent global ecological catastrophe: "In the name of the survival of mankind developing countries should continue in a state of underdevelopment because if the evils of industrialization were to reach them, life on the planet would be placed in jeopardy" (Castro 1972: 33). Castro and other Southern activists and intellectuals have perceived this as an unfair burden placed on the already disadvantaged South, as it perpetuates "the ostensive imbalance between responsibility for the damage and obligation for repair" (Castro 1972: 35).

North-based environmental advocates continued to argue for a decoupling of justice from development, claiming that "the demand for justice and dignity on behalf of Southern countries threatens to accelerate the rush towards biospherical disruption, as long as the idea of justice is firmly linked to the idea of development" (Sachs 2002: 30). Yet, claims for justice have for the most part continued to be tied to the notion of development (see Chapter 26). Dismissing the neocolonialist overtones of Western environmental prescriptions, the South's position has long emphasized the need to prioritize socio-economic development goals alongside the establishment of ambitious environmental policies. This is evident in the discourses of all major global environmental negotiations and was made explicit by naming the Rio Earth Summit the UN Conference on Environment *and* Development. The importance of development and economic growth for the South, and differential obligations for North and South, were also institutionalized at the Earth Summit in the UN Framework Convention on Climate Change (DeSombre 2002; Williams 2005; see Chapter 32). The emphasis on development in such a global environmental agreement signified that developing countries should not be expected to enact costly environmental policies that would divert limited resources from competing development priorities. Moreover, following the "polluter-pays principle" the responsibility for environmental action is seen to fall squarely on the shoulders of the North, which has made more significant contributions over time to ecological degradation at a global scale.

While many environmental scholars and activists from the South hold this view, it is also acknowledged by some scholars in the North who approach global environmental affairs from ethical perspectives (see Chapter 27). For example, in the context of climate change, Shue articulates the North's historical responsibility as "the acceptance of accountability for the full consequences of an industrialization that relied on fossil fuels" by the countries "that have controlled the process of industrialization, and consequently have tended to benefit the most from industrialization" (Shue 2014: 12–13). Thus, not only is the North seen as obligated to take responsibility for collectively creating ecological harm, but many scholars also claim it owes an ecological debt to the South due to unfair appropriation of ecological space (e.g., Agarwal and Narain 1990; Anand 2004; Goeminne and Paredis 2010; Martinez-Alier 2002; Srinivasan et al. 2008). The North is therefore obligated to do two things to correct this historical and ongoing geographical imbalance: to retreat from the ecological space it has wrongfully occupied to create space for newcomers to development, and to help the South use the space in an ecologically sustainable way (Agarwal and Narain 1990). To create more ecological space, the North is expected to make adjustments to its economy and industry to lessen its ecological footprint. To help the South advance sustainably, it is expected to provide substantial financial and technological support.

The idea of ecological debt is a call for justice in the form of reparations owed collectively by industrialized countries that have benefited from colonialism and imperialism. The demands for reparations are due to past and continuing disproportionate encroachment on ecological space without compensation and without recognition of other countries' entitlements to that space. This appropriation of ecological space can be conceptualized in part as the unidirectional flow of natural resources from the South to the North during the colonial era, which persists in various forms today. This appropriation has a material basis in uneven power in international relations as this has enabled "ecological aggression," or the undeterred exploitation of resources by some states and its corresponding ecological consequences (Goeminne and Paredis 2010; Martinez-Alier 2002; Srinivasan et al. 2008). But the notion of ecological debt has meaning beyond financial resource flows and also includes notions of re-conceptualizing development beyond the exploitative North-centric paradigm (Rice 2009), although this has not been a major emphasis reflected in North–South negotiations. Rather, claims to ecological space have mostly occurred within a neoliberal capitalist paradigm, and have not translated to a viable challenge to the current international political economy, as lamented by those deploring the demise of the Third World as a political project (Berger 2004; Escobar 2004; see Chapter 24). A wide variety of alternatives have been envisioned, articulated, and put into practice but they have yet to be scaled up to the realm of North–South politics which remain firmly ensconced within a neoliberal framework that continues to subjugate Indigenous and non-White people, women, and their knowledge systems.

The institutionalization of North–South environmental politics

In global environmental politics, the Group of 77 (G77) plus China is the negotiating entity representing the Third World or Global South in treaty negotiations (Vihma 2010; Williams 2005). The G77 has been referred to as representing a "new regionalism" in global politics wherein its aim is not only to influence the North's environmental and political agenda but also to demand that the "North confront its responsibilities to the wider world" (Dodds 1998: 729). Scholars have identified a number of specific interests that countries of the South articulate in global environmental negotiations: a concern for explicitly linking environmental concerns to development concerns, seeking additional financial resources and technological assistance for environmental programs, capacity building for negotiation and implementation of environmental agreements, increased time for implementing new regulations, and pushing the North to accept responsibility for the environmental harms it has caused (Dodds 1998; Miller 1995; Najam 2004; Williams 2005). The institutionalization of these interests and their underlying norms of sustainable development and "common but differentiated responsibility" has contributed to the maintenance of the idea of difference between the North and the South (Williams 2005).

These differences have been institutionalized through global negotiations at various points in time. The 1972 Stockholm conference created the conditions that led to substantial discord along North–South lines. The Brundtland Commission's 1987 publication of *Our Common Future* significantly substantiated the North–South discourse in the context of differences in wealth and environmental responsibility. The 1992 UN Conference on Environment and Development (UNCED) further solidified the North–South divide in the context of establishing linkages between environmental and development issues, and it articulated differentiated obligations of developing and developed countries. The international environmental treaties that followed have included significant provisions for realizing North–South equity (DeSombre 2002). Even before UNCED, the 1987 Montreal Protocol on Substances that Deplete the Ozone Layer (see Chapter 31) created a precedent for differentiated treatment of developing countries in the context of negotiations over the atmospheric commons (Rajan 1997). It instituted a financial transfer mechanism, the Multilateral Fund, which developed countries contributed to, and included a time lag to allow large and rapidly industrializing developing countries such as China and India much more time to use ozone-destroying chemicals. The Basel Convention on the Control of Transboundary Movements of Hazardous Wastes and their Disposal (see Chapter 36) also created a space for a North–South frame for debates and negotiations where Third World states articulated a common position despite their heterogeneity in levels of development (Miller 1995).

The 1992 United Nations Framework Convention on Climate Change (UNFCCC; see Chapter 32) has provided the space for a pronounced North–South environmental politics. The UNFCCC articulated several principles to guide the attainment of its overarching objective of greenhouse gas stabilization. These allude to ideas of equity, sustainable economic development for developing country parties, and common but differentiated responsibilities and respective capabilities (CBDRRC). The texts of the UNFCCC and the Kyoto Protocol made two things clear: a clear demarcation between the responsibilities of developed and developing country parties, and the prioritization of economic growth and sustainable development for the latter. Such articulation of differentiated responsibilities and capabilities is compatible with concepts such as ecological debt, as well as other concepts specific to the context of climate change, such as contraction and convergence (GCI 1996) and climate injustice (Roberts and Parks 2007) based on disparities at the level of states (see Chapter 25).

The increased salience of global environmental concerns such as climate change (see Chapter 32), and the role that some countries of the South play in alleviating them – or alternately, their "'power to destroy' a resource" – allowed the traditionally less powerful South relatively more leverage in international environmental negotiations, particularly during the 1990s and 2000s (Anand 2004; DeSombre 2002: 15; Therien 1999). During these decades, demands made by the South, such as for mechanisms to transfer financial and technological resources to developing countries, as well as those for a lag time for developing countries in the implementation of environmental regulations were recognized, at least in theory. Many of the promises made to the South went unfulfilled, leading to a deficit of trust on the part of the South that has long marred treaty negotiations (Roberts and Parks 2007). The increased bargaining power of the South, coupled with the validation of North–South difference through institutionalization in international agreements and frameworks, enabled the South to articulate its claims for justice in increasingly radical terms, as evidenced by the normalization of discourses of ecological debt.

However, the politics leading to the Paris Agreement signal that during the past decade there has been a marked shift in North–South politics reflecting the seemingly dwindling bargaining power of the emerging economies vis-à-vis the North, resulting in differentiated and enhanced obligations for the South while reducing the pressure on the North for taking the lead on mitigation responsibility (Dubash et al. 2018; Hurrell and Sengupta 2012). The apparent weakening of the South's negotiating power in global climate politics may be in part due to the heightened vulnerability of many members of the Global South, and the intransigence of hegemonic powers such as the United States in recognizing North–South equity principles as embedded in the Kyoto Protocol. Paradoxically, these recent shifts in global environmental politics have occurred alongside the ascendancy of some of the emerging economies in the international economic hierarchy, as reflected in the entry of countries such as India and China in the G20, as well as their newly acquired veto power within the WTO (Hurrell and Sengupta 2012). Despite the transition from the Kyoto Protocol to the Paris Agreement, and the subsequent erasure of the explicit binary framework for differentiation of responsibility, though, these countries continue to maintain NIEO-esque desires for North–South equity as evidenced by the Intended Nationally Determined Contributions (INDCs) submitted in response to the Paris Agreement (Joshi 2021).

Debates on the validity of the North–South frame of reference

While North–South politics are inherently about challenging the status quo of the current world order, a number of critiques have been leveled at the legitimacy of such a politics. A variety of such critiques exist. They fall into two broad categories. Scholars falling into one category of critiques have questioned the validity of referencing global inequality on the basis of countries that belong to North or South by way of problematizing the meaning of the terms North and South, with more criticism leveled at the latter. Since the South is seen to originate from and is closely associated with the Third World, the perceived loss of relevance of the Third World as a political project carries over to a corresponding lack of confidence in the relevance of a North–South divide (Berger 2004; Escobar 2004; Isbister 2006; Prashad 2014; Slater 2006; Therien 1999). The Third World is deemed to have lost its meaning due in part to the ending of the Cold War, and because its radical intent is perceived to be all but lost, coopted by the very powers it sought to challenge, so much so that it is considered "intellectually and conceptually bankrupt" (Berger 2004: 31; Prashad 2014).

A second approach to questioning the validity of a North–South divide is based on transformations of the global economy, thus rendering both North and South more heterogeneous than the binary term implies (Eckl and Weber 2007). The understanding is that widespread neo-liberalization of formerly colonized and developing countries has increased the economic differentiation and fragmentation of the Global South, thus destabilizing the traditional core–periphery or North–South configuration of the world (Therien 1999; Williams 2005). A number of explanations have been offered: the gaps between the North and the South are narrowing, countries are graduating from the South to the North (Broad and Landi 1996), and the South is no longer as poor or as dependent on the North, particularly with reference to the stellar economic performance of countries in East Asia (Therien 1999).

Scholars have also pointed out that the North–South divide is too state-centric and rooted in obsolete core–periphery depictions to accurately represent global inequalities. Very often the argument centers on the need to conceptualize global inequality by class rather than on the basis of the development status of countries (Barnett 2007; Newell 2005). More nuanced approaches to complicate the hegemony of a monolithic North are offered by theories of settler colonialism and racial capitalism within countries like the United States, confounding such binary depictions of global inequality (Dunbar-Ortiz 2015; Pulido 2016). Gerard Toal, the so-called father of critical geopolitics, prominently declared that "a critical geopolitics is one that refuses the spatial topography of First World and Third World, North and South, state and state," and that the task of the subdiscipline was to highlight "the precariousness of these perspectival identities" (Toal 1994: 231). This approach has been adopted by some scholars examining climate change (see Chapter 32). Barnett (2007), for example, argued that the geopolitics of climate change has been severely constrained by what John Agnew (1994) called the "territorial trap" – an inaccurate representation of the world as spatially and politically distinct states (Barnett 2007). Therefore, not only has there been a "Third Worldization of certain regions in the developed world" (Toal 1994: 231), and a formation of various Third Worlds representing different sets of collective needs (Escobar 2004), there has also been a rapid burgeoning of a "planetary middle class" that belies the traditional divisions of North and South (Conca 2001: 68; Harris 2010). Scholars have therefore called for the articulation of newer geographies of wealth and poverty (McFarlane 2006).

A manifestation of these questions about both disparities within nations and a move away from a static binary framework for differentiated responsibility particularly for atmospheric pollutants started to become particularly prominent in global environmental negotiations discourse in the build-up to the Paris Agreement (Dubash et al. 2018). Even if the global politics of climate change has begun to entertain prospects of moving beyond the binary framework of differentiation for responsibility, fundamental divisions remain strong in the positions taken about such responsibility, as well as in vulnerability and capability, particularly in terms of financial and technological resources for adapting to the consequences of ecological change (Uddin 2017).

From a postcolonial perspective, the North–South referent is seen by some to perpetuate colonial traditions of "us" and "them" that are predicated on conscious or unconscious forms of Western exceptionalism and superiority (Doty 1996; Nash 2004; Said 1994). With the South representing the developing world and the North the developed world, the binary is seen to connote a teleological relationship where the South is expected to catch up to the North, thus reifying the unequal relationship between North and South through the implicit acceptance of the Western view of the South as well as of itself representing the development model to follow. Such notions are held responsible for obscuring the role of Western imperialism in subjugating the Third World by naturalizing the superiority and the success of the West over the rest of the world (Sidaway 2000). In this mode of thought, international agreements that take on a North–South dimension only serve to institutionalize the North's paternalistic and interventionist role toward developing countries (Eckl and Weber 2007). Meanwhile the subsequent modernization of the South is blamed for the cooptation of the formerly radical Third World agenda (Berger 2004; Escobar 2004). Escobar (2004) therefore saw self-organizing social movements (see Chapter 14) as the only hope for effectively challenging the US-based imperial globality and global coloniality. One of the key contributions of postcolonialism has thus reportedly been to challenge categories that homogenize colonizing and colonized groups through binary representations such as First World/Third World, North/South, developed/developing and core/periphery (Nash 2004).

It has been argued that postcolonial theory should attempt to shed light on the various ways in which the impact of colonialism lingers on in new forms in the formerly colonized countries, such as via internal colonialism or ultra-imperialism (Blunt and McEwan 2002; Sidaway 2000; see Chapter 4). Viewed in this way, the North–South dimension is not the best way to conceptualize global inequality. Much like the argument for a class-based (rather than a state-based) approach to inequality and oppression, such a postcolonial approach seeks to uncover newer relationships of domination and disenfranchisement that feed off older colonial patterns. In this reading of difference, the middle and upper class elites of formerly colonized countries are just as culpable of perpetuating global inequality as former colonial rulers or present-day imperial powers, if not more so. In fact, the "ideological posturing" of Southern elites has been one of the critiques of the North–South framing, with scholars arguing that Third Worldism is a pursuit of Third World elites who stand to gain from North–South politics (Berger 2004; Newell 2005). Here the implication is that identification with Third World solidarity is rhetorical, a charge that others argued has yet to be verified (Roberts and Parks 2007; Williams 2005).

In the context of global environmental politics, the Global South's calls for increased transfers of financial and technological resources from the North and a time lag for accepting stronger environmental regulations might be seen to benefit the state, industrialists, and other elites in the Global South (see Chapter 12). These middle/upper class members of the Global South – typically educated in the West or in Westernized systems of education – are seen to be complicit in perpetuating the legacies of colonialism and imperialism in a postcolonial world (Goldman 2004). From this point of view, only the subaltern members of the "meta-industrial classes" – peasants, Indigenous communities, and women – who are the absolute victims of colonialism and imperialism (Salleh 2011), comprise the true Global South. Anti-capitalist, feminist, decolonial resistance movements led by these groups are in this mode of thinking often seen to be the only valid counterhegemonic challenges to the economic, cultural and political hegemony of the Global North (Acha 2019; Escobar 2004).

Despite these objections to the North–South frame of reference, it has continued to remain a salient framing for representing global inequality, serving as a reminder of the linkages between geopolitics and development, as well as a reminder of abiding core–periphery dynamics and imperialism that are reminiscent of past colonizer–colonized relationships for many scholars (Grosfoguel 2011; Hickel 2017; Ould-Mey 2003; Power 2006; Said 1994; Slater 2004, 2006). In response to the issue of heterogeneity, the argument has been made that the South or the Third World should be seen not as a monolith, but rather as a diverse entity that fluctuates between acting in unity and maintaining plurality according to the geopolitical context (Broad and Landi 1996; Williams 2005). Indeed, given the large number of diverging actors and interests, cohesion particularly for the Non-Aligned Movement has been impressive, especially relative to Western countries (Lundestad 2005).

While some countries of the South may be perceived to have graduated to the North, some caution that such graduation is only partial – usually economic (Hansen 1980; Hurrell and Sengupta 2012) – and that self-identification of nations as members of the South is owing to "a sense of shared vulnerability and a shared distrust of the prevailing world order rather than a common ordeal of poverty" (Najam 2004: 128). For some, the similarities within the North and South outweigh their internal differences. For example, for Anand (2004) the South represents the common experiences of people who have been victimized by a colonial and imperial past. This legacy has not only left countries of the South economically weaker and more vulnerable to the vagaries of a globalized capitalist economy (Williams 2005; see Chapter 24), but also ensured its continued subjugation through an unequal international system where the South's voice wields less influence (O'Brien and Leichenko 2003; Therien 1999). Following Benedict Anderson, Williams (2005: 53) argued that the Third World or Global South represented an "imagined community of the powerless and vulnerable." The Southern bloc or the Third World coalition therefore makes sense when seen in the context of the dominance of industrialized states in global diplomacy and politics because the coalition allows developing countries with relatively marginal influence increased leverage in global negotiations (Hansen 1980; Williams 2005; see Chapter 22). Despite internal differences on a number of issues and despite diverging socio-economic circumstances, therefore, the countries of the Global South continue to have compatible interests, particularly in the context of global environmental politics (Hurrell and Sengupta 2012).

While one approach in postcolonial theory is to view the North–South binary as a colonial framework (Said 1994), another posits that this framing takes advantage of the imaginary of the Global South as a strategic essentialism (Spivak 1993). A Bhabhain approach enables consideration of decolonial strategies that utilize the "disruptive power of hybridity" employing the colonizer's tools of oppression (Jacobs 1996, 14) in a way that achieves "subversive complicity" (Grosfoguel 2011, 24). Dismissing such self-identification as posturing by elites in the Global South might well be a product of a largely Euro-American–centric privileged worldview (Dodds 1998). Such dismissal contributes to silencing voices and forms of resistance emanating from people of formerly colonized countries – even if they might be categorized as the elite – and interestingly functions as a means to preserve or legitimize the North–South status quo in a manner similar to how politics of "color-blindness" serves to uphold White supremacy (Joshi 2021; Omi and Winant 2015).

Even though the diffuse nature of neocolonialism and imperialism, as well as the transnational character of global environmental change, has challenged scholars to question the state-centric nature of analyses of global inequality, methodologically it is difficult to escape using this unit of analysis because states and their representatives are the actors that negotiate international environmental treaties and are held accountable to them (see Chapters 5 and 7). These challenges, however, have pushed scholars to revisit traditional concepts of sovereignty and national security as they are connected to environmental change (DeSombre

2002; see Chapter 20). The increasingly stark exercises of regional hegemonic power – often manifesting alongside extreme right-wing forms of nationalism – on part of some powerful members of the Global South, have further strengthened the legitimacy of these challenges (Joshi 2021).

Prospects for a new international economic order

Debates over the validity of North–South environmental politics seem to be enmeshed in four distinct competing and potentially conflicting interests. One is a normative desire for greater equality among states in terms of economic and material wellbeing. The North–South imaginary is a reminder of the absence of, as well as desire for, such equality. Second is a concern for economic justice that addresses global inequality in a more general sense and transcends concerns at the level of the state (see Chapters 25). The two need not contradict each other, although they are often construed to be mutually exclusive goals. The third interest is to prevent or minimize environmental catastrophe – a normative desire for ecological sustainability (see Chapter 16). Environmental calamities arguably have more immediate and dire implications for the most vulnerable members of the Global South and accordingly an interest in global environmental justice has long been a priority for the South (see Chapter 26). The ways in which these four interests intersect are complicated, not least in part due to the multiple ways in which their pursuit can be envisioned and articulated. Marxist and feminist analyses tend to view all pursuits of economic growth and development within neoliberal capitalist and patriarchal structures as counterintuitive to all four goals, and as tools to perpetuate the legacies of imperialism and colonialism (see Chapter 4). From these perspectives, demands for a new global order constrained by prevailing political economic structures are not adequate to the task: a class-based anti-capitalist movement centered around eco-socialism that takes seriously gendered reproductive justice is deemed essential for addressing global inequality and moving toward a sustainable world. Anti-modernization and anti-development adherents see modernity as inimical to all of the interests outlined above, and they subscribe to a rather romanticized view of Indigenous groups even as their rights are championed. Critical geographers problematize the state-centrism of most analyses of global inequality.

Challenges to transform the international economic order have emanated within a state-centric context and have not problematized neoliberal capitalism as a key impediment to achieving North–South economic and environmental justice. Adherents of North–South equity have sought to capitalize on the strategic essentialism of the Global South as a victim of colonialism, in part due to the global nature of environmental problems. But they have also exhibited a tendency to have blind spots in terms of internal colonialism, thus neglecting to recognize the rights and worldviews of Indigenous groups within Global South territories even as Global North hegemony over the South is problematized. As such, it is doubtful whether erstwhile approaches to North–South equity are compatible with the goal of enabling ecological sustainability and environmental justice because the only known path to international economic equality seems to be industrialization-based economic growth accompanied by deleterious environmental and social consequences. This therefore leads to several questions, including whether states are capable of mediating the forces of global environmental changes and their consequences for their most vulnerable members, how North–South power dynamics are connected to power dynamics within individual countries within the Global South and North (Cox 1993; Joshi 2021), and whether the wide array of neoliberal solutions to problems of both underdevelopment and environmental crisis are conducive to resolving the North–South divide.

While these questions remain unanswered as we wade into the 2020s, the limits of erstwhile North–South politics have become apparent, indicating the need for a shift in approach. An overwhelming focus on parity within prevailing structures of neoliberal capitalism has enabled a handful of Global South powers to ascend in political and economic power, without significantly changing the parameters of the North–South divide. While countries such as China and India have inched closer to the Global North in economic prowess – indicating a shift to a multi-polar world order – under their de facto leadership the South as a whole has not been able to challenge its unequal status relative to the North, nor hold it accountable to global ecological harms. In other words, a North–South politics led by the emerging powers has largely benefited these countries – consolidating their geopolitical and regional hegemony – without benefiting the Global South at large. Meanwhile, internal patterns of neocolonialism have exacerbated within these countries, as evidenced for example by Hindu Nationalism in India and the suppression of Uyghurs in China.

If the North–South politics of the last five decades have largely resulted in the creation of new world hegemons rather than contribute to counterhegemonic possibilities for the greater good, it is a sign that this politics needs to be shelved in favor of a new kind of North–South politics, rather than to abandon the struggle. The success of a revivified NIEO-esque Global South politics may rely on the kind of argument for ecological debt that has emanated from Accion Ecologica – and championed by the Bolivarian Alliance of the Americas at international meetings – that conceptualizes reparations not only in terms of transfer of financial and technological resources, but in changing the paradigm of development (Rice 2009).

Scholars who have studied political and intellectual currents that led to the proposal for a NIEO in the 1970s have noted, on the one hand, that its failure is closely attached to the rise of neoliberalism (Adler 2017; Prashad 2014), and on the other hand, that as an "unfailure" its decolonial possibilities remain dormant (Gilman 2015). It may well be that the key to reviving the radical Third World project lies in the ability of a unified Global South – including the South within the North – to embark on a counterhegemonic politics to challenge the hegemony of a neoliberal capitalist world order in which Global North countries reign supreme. Further, a successful anti-imperial anti-colonial movement at the scale of North–South global politics would need to be built on the foundations of successful anti-imperial anti-colonial movements within countries in the Global South (Cox 1993). A Global South agenda that draws on a plurality of struggles and builds strength in numbers through horizontal relationships of solidarity and accountability – rather than being limited by and subjected to essentialist visions of justice all competing for ascendancy – is needed to create a viable challenge to the hegemony of the Global North in global environmental politics.

References

Acha, M. R. (2019). "Climate justice must be anti-patriarchal, or it will not be systemic." In *Climate Futures*, edited by K.-K. Bhavnani, J. Foran, P. A. Kurian, and D. Munshi. London: Zed Books.

Adler, P. (2017). "'The basis of a new internationalism?': The Institute for Policy Studies and North-South Politics from the NIEO to Neoliberalism." *Diplomatic History* 41(4): 665–693.

Agarwal, A. and S. Narain (1990). *Global Warming in an Unequal World*. New Delhi: Centre for Science and Environment.

Agnew, J. (1994). "The territorial trap: The geographical assumptions of international relations theory." *Review of International Political Economy* 1(1): 53–80.

Anand, R. (2004). *International Environmental Justice: A North–South Dimension*. Burlington, VT: Ashgate.

Barnett, J. (2007). "The geopolitics of climate change." *Geography Compass* 1(6): 1361–75.

Berger, M. T. (2004). "After the Third World? History, destiny and the fate of Third Worldism." *Third World Quarterly* 25(1): 9–39.

Shangrila Joshi

Bhagwati, J. N. (ed.) (1977). *The New International Economic Order: The North–South Debate*. Cambridge, MA: MIT Press.

Blaut, J. M. (1993). *The Colonizer's Model of the World*. New York: Guilford Press.

Blunt, A. and C. McEwan (2002). "Introduction." In A. Blunt and Cheryl McEwan (eds) *Postcolonial Geographies*. London: Continuum.

Broad, R. and C. M. Landi (1996). "Whither the North–South gap?" *Third World Quarterly* 17(1): 7–17.

Castro, J. A. de A. (1972). "Environment and development: The case of developing countries." In K. Conca and G. D. Dabelko (eds) *Green Planet Blues*. Boulder, CO: Westview Press.

Conca, K. (2001). "Consumption and environment in a global economy." *Global Environmental Politics* 1(3): 53–71.

Cox, R. W. (1993). "Gramsci, hegemony and international relations: An essay in method." In *Gramsci, Historical Materialism and International Relations*, edited by Stephen Gill. Cambridge: Cambridge University Press.

DeSombre, E. R. (2002). *The Global Environment and World Politics*. London: Continuum.

Dodds, K. (1998). "The geopolitics of regionalism: The Valdivia Group and southern hemispheric environmental co-operation." *Third World Quarterly* 19(4): 725–43.

Doty, R. L. (1996). *Imperial Encounters*. Minneapolis: University of Minnesota Press.

Dubash, N. K., R. Khosla, U. Kelkar, and S. L. (2018). "India and climate change: Evolving ideas and increasing policy engagement." *Annual Review of Environment and Resources* 43: 395–424.

Dunbar-Ortiz, R. (2015). *An Indigenous Peoples' History of the United States*. Boston: Beacon Press.

Eckl, J. and J. Weber (2007). "North–South? Pitfalls of dividing the world by words." *Third World Quarterly* 28(1): 3–23.

Escobar, A. (2004). "Beyond the Third World: Imperial globality, global coloniality and anti-globalisation social movements." *Third World Quarterly* 25(1): 207–30.

Gandhi, L. (1998). *Postcolonial Theory*. New York: Columbia University Press.

GCI (1996). "Contraction and convergence: Climate truth and reconciliation." Available HTTP: <http://www.gci.org.uk>

Gilman, N. (2015). "The New International Economic Order: A Reintroduction." *Humanity* 6(1): 1–16.

Goeminne, G. and E. Paredis (2010). "The concept of ecological debt: Some steps towards an enriched sustainability paradigm." *Environment, Development, Sustainability* 12: 691–712.

Goldman, M. (2004). Imperial science, imperial nature: Environmental knowledge for the World (Bank)." In S. Jasanoff and M. L. Martello (eds) *Earthly Politics*. Cambridge, MA: MIT Press: 55–80.

Grosfoguel, R. (2011). "Decolonizing post-colonial studies and paradigms of political-economy." *Transmodernity: Journal of Peripheral Cultural Production of the Luso-Hispanic World* 1(1): 1–38.

Hansen, R. D. (1980). "North–South policy – what's the problem?" *Foreign Affairs* 58(5): 1104–28.

Harris, P. G. (2010) *World Ethics and Climate Change*. Edinburgh: Edinburgh University Press.

Hickel, J. (2017). *The Divide*. London: William Heineman.

Hurrell, A. and S. Sengupta. (2012). "Emerging powers, North—South relations and global climate politics." *Royal Institute of International Affairs* 88(3): 463–484.

Isbister, J. (2006). *Promises Not Kept*. Bloomfield, CT: Kumarian Press.

Jacobs, J M. (1996). *Edge of Empire*. New York: Routledge.

Jha, L. K. (1982). *North–South Debate*. Delhi: Chanakya Publications.

Joshi, S. (2021). *Climate Change Justice and Global Resource Commons*. New York: Routledge.

Lundestad, G. (2005). *East, West, North, South*. London: Sage Publications.

Martinez-Alier, J. (2002). "Ecological debt and property rights on carbon sinks and reservoirs." *Capitalism, Nature, Socialism* 13(1): 115–19.

McFarlane, C. (2006). "Crossing borders: Development, learning and the North–South divide." *Third World Quarterly* 27(8): 1413–37.

Miller, M. (1995). *The Third World in Global Environmental Politics*. Buckingham: Open University Press.

Najam, A. (2004). "Dynamics of the Southern collective: Developing countries in desertification negotiations." *Global Environmental Politics* 4(3): 128–54.

Nash, C. (2004). "Postcolonial geographies: Spatial narratives of inequality and interconnection." In P. Cloke, Philip Crang and Mark Goodwin (eds) *Envisioning Human Geographies*. London: Arnold.

Newell, P. (2005). "Race, class and the global politics of environmental inequality." *Global Environmental Politics* 5(3): 70–94.

O'Brien, K. L. and R. M. Leichenko (2003). "Winners and losers in the context of global change." *Annals of the Association of American Geographers* 93(1): 89–103.

Omi, M. and H. Winant. (2015). *Racial Formation in the United States,* 3rd ed. New York: Routledge.

Ould-Mey, M. (2003). "Currency devaluation and resource transfer from the South to the North." *Annals of the Association of American Geographers 93(2)*: 463–84.

Peet, R. and E. Hartwick. (2015). *Theories of Development*, 3rd ed. New York: The Guilford Press.

Power, M. (2006). "Anti-racism, deconstruction and 'overdevelopment'." *Progress in Development Studies 6(1)*: 24–39.

Prashad, V. (2014). *The Poorer Nations.* New York: Verso.

Pulido, L. (2016). "Flint, environmental racism, and racial capitalism." *Capitalism Nature Socialism 27(3)*: 1–16.

Rajan, M. (1997). *Global Environmental Politics: India and the North–South Politics of Global Environmental Issues.* New Delhi: Oxford University Press.

Rice, J. (2009). "North/South relations and the ecological debt: Asserting a counter-hegemonic discourse." *Critical Sociology 35(2)*: 225–52.

Roberts, J. T. and B. C. Parks (2007). *A Climate of Injustice.* Cambridge, MA: MIT Press.

Sachs, W. (2002). "Ecology, justice, and the end of development." In J. Byrne, L. Glover and C. Martinez (eds) *Environmental Justice.* New Brunswick, NJ: Transaction Publishers: 19–36.

Said, E. W. (1994). *Culture and Imperialism.* New York: Vintage Books.

Salleh, A. (2011). "Cancun and after: A sociology of climate change." *Arena 110*: 4.

Shue, H. (2014). "Historical responsibility, harm prohibition, and preservation requirement: Core practical convergence on climate change." *Moral Philosophy and Politics 2(1)*: 7–31.

Sidaway, J. (2000). "Postcolonial geographies: An exploratory essay." *Progress in Human Geography 24(4)*: 591–612.

Slater, D. (2004). *Geopolitics and the Post-Colonial.* Oxford: Blackwell Publishing.

Slater, D. (2006). "Imperial powers and democratic imaginations." *Third World Quarterly 27(8)*: 1369–86.

Spivak, G. C. (1993). *Outside in the Teaching Machine.* London: Routledge.

Srinivasan, U., S. Carey, E. Hallstein, P. Higgins, A. Kerr, L. Koteen, A. et al. (2008). "The debt of nations and the distribution of ecological impacts from human activities." *Proceedings of the National Academy of Sciences (PNAS) 105(5)*: 6.

Therien, J.-P. (1999). "Beyond the North–South divide: The two tales of world poverty." *Third World Quarterly 20(4)*: 723–42.

Toal, G. Ó. (1994). "Critical geopolitics and development theory: Intensifying the dialogue." *Transactions of the Institute of British Geographers 19(2)*: 228–33.

Uddin, K. (2017). "Climate change and global environmental politics: North-South divide." *Environmental Policy and Law 47(¾)*: 106–114.

Vihma, A. (2010). "Elephant in the room: The new G77 and China dynamics in climate talks." Briefing Paper 9. Helsinki: Finnish Institute of International Affairs.

Williams, M. (2005). "The Third World and global environmental negotiations: Interests, institutions and ideas." *Global Environmental Politics 5(3)*: 48–69.

24

Globalization and the environment

Economic changes and challenges

Lada V. Kochtcheeva

Globalization is the intensified interconnectedness and transnational relations, which are accelerating in their speed and scope, and resulting in interdependence of economic, political, social, and cultural spheres (Stearns 2020). Over the last 50 years, the rapidly developing process of economic globalization has continued to pose multiple challenges to reconciling the increasing flows of international trade, production, consumption and investment, with the needs of sustainable development, environmental quality, and domestic and international environmental rights and obligations. Economic globalization has interconnected world economies via improved capital flows, trade openness, innovative opportunities and technology transfer. It has promoted economic growth, development, production and consumption. However, not every country has benefited from economic globalization, and not all national economies benefited equally and evenly. Moreover, global development has brought with it factors that contributed to pollution, increased urbanization, higher energy consumption, industrialization and massive population shifts, disrupting the ecological balancing act and resulting in lower sustainable economic growth and development through welfare impeding channels. Does economic globalization represent the fundamental evolution of the market ensuring previously unconceivable economic development and environmental innovations, which could overcome poverty and sweep away obsolete values? Or, is it a mechanism of increasing economic inequality, environmental degradation, injustice, and inadequate policies?

It is important to emphasize that not only does globalization influence the environment, but the environment, which is inherently global, also affects the speed, course and quality of globalization. Natural environment provides the resources for economic globalization, and policy responses to global environmental challenges restrict and affect the context in which globalization happens. The global Covid-19 pandemic demonstrated the failures of economic globalization and global responses to the crisis. The pandemic undermined the basic tenets of global manufacturing, straining supply chains, and pushing governments, firms, and societies to increase their capacity to cope with prolonged periods of economic quarantine.

There are no perceptions of the linkages between globalization and the environment that can be unequivocal, continuous and impartial. Therefore, this chapter uncovers the nuances and complexities of the economic globalization-environment relationship. It reveals

DOI: 10.4324/9781003008873-28

that economic globalization accelerates market, structural, political, institutional and societal change, thus modifying the economic policies, resource use patterns, and political and institutional responses at the domestic and international levels.

The advance of economic globalization

The world has witnessed how the phenomenon of economic globalization has rapidly gained momentum. An unprecedented integration of the global economy through trade and financial mechanisms has produced catalytic impacts on a range of economic benefits. Intensified trade has become the main instrument that provides an innovative opportunity to improve the process of production, as well as productivity of natural resources. Advanced economic integration and market openness are primary sources of economic development (Sparke 2013; Grossman and Helpman 2015; Shahbaz et al. 2018). The rapid growth in the volume and diversity of world economic relations, accompanied by increasing economic interdependence of countries, is the essence of the economic globalization process. As a result, a relatively integrated economic system is being formed, which influences domestic economic systems and at times dictates its own rules of the game to national economies, societies and polities. The extent and nature of such an impact, and the inclusion of a country in global economic relations, especially international trade, are determined by a number of factors, including the magnitude of its economic capacity, its state of technological development, and its endowment with natural resources (Molle 2013).

Nonetheless, not everyone has benefited from economic globalization and not all countries and societies have benefited evenly, with poverty remaining a major problem around the world (Singh 2015). As argued by the United Nations Environment Program, "Those benefiting from globalization may well do so at the expense of those unable to take part in its essentially short-term, market-led approach... the gap between rich and poor is widening, with no evidence that wealth is 'trickling down'" (UNEP 2007: 1). In the real world of unequal economic opportunity, interdependence does not produce a leveling effect (Sparke 2013). The diversity of domestic reactions to economic globalization is abundant around the world (Hebron and Stack 2017; Postolache et al. 2019; Kochtcheeva 2020). Economic liberalization, trade openness, technological advance and integration in some societies are counterbalanced by severe competition, obstacles to growth, and resentment to global interdependence.

In addition to direct or indirect impacts on the economic and socio-political spheres, globalization affects the state of the natural environment. The environment itself is intrinsically global, with ecosystems, watersheds and species frequently crossing national boundaries, air pollution traveling across continents and oceans, and an atmosphere as a major global commons providing climate protection. Economic globalization has increased resource flows around the planet, raising pressures on the environment (Garmendia et al. 2016; Najam et al. 2007). While human consumption due to economic development and population growth has increased very rapidly, the biological capacity of the world has developed very slowly (Bilgili et al. 2019). The magnitude of ecological change resulting from economic activity over the last several decades has given rise to serious global environmental problems, including human-changed ecosystems (see specific chapters in this volume on different ecosystems), climate change (see Chapter 32), resource exhaustion, species extinction (see Chapter 41), demographic shifts and many other problems.

The UNEP *Global Environmental Outlook-6: Healthy Planet Healthy People* report (2019: 1) points out in key messages that a "healthy environment is both a prerequisite and a foundation for economic prosperity, human health and wellbeing." However, globally environmental health is declining. According to the Intergovernmental Science–Policy Platform on Biodiversity and Ecosystem Services (IPBES) (2019: 10) report, nature and its ecosystem functions and services are deteriorating worldwide. Notwithstanding anticipated reductions in material inputs and pollution made possible by improved technology and increasing efficiency, resource consumption and degradation continue to grow. Under globalization, this growing demand for resources is increasingly being satisfied through international trade, which is contributing to the use of non-domestic ecosystem resources and services (Garmendia et al. 2016).

While economic advances come into conflict with environmental sustainability, abstaining from globalization is generally not considered an answer (Afesorgbor 2019). Technological innovation, policy breakthroughs, trade, investment, and the emergence of new markets are believed to promote more efficient use of resources and spread of knowledge and awareness (Gallagher 2009; Grossman and Helpman 2015). In general, it is believed that trade-generated income growth is important for the environment as raising incomes fuel demand for environmental amenities (Boyce 2012). Global environmental health may present the ultimate opportunity in which interdependencies could come together to shape local and global destinies. No truly autonomous solutions seem to be possible. A critical component of adjustment between economic globalization and environmental quality is needed as a result of changing capacities and preferences, and the failure of current arrangements to cope with the challenges associated with globalization.

Achieving internationally agreed environmental goals on environmental quality, sustainability, and efficiency improvements is essential for transformative change to take place, which can be promoted by coalitions between governments, businesses, researchers and civil society (UNEP 2019). Indeed, many serious attempts at global environmental governance emerged during the 1980s, 1990s and 2000s, paralleling the rise of international economic management, the World Trade Organization (WTO), new information and communication technology (ICT), institution building, and an immense array of policy instruments from regulation to environmental markets and voluntary measures (Timmons Roberts and Newell 2016). The size of the world economy as well as international trade and cross-national pollution have grown and developed into global challenges demanding global cooperation. As a result, various bi-national, regional and multinational environmental agreements and regimes (see Chapter 9) now address the problems of air and water quality (see Chapters 30 and 38), biodiversity and threatened species (see Chapter 41), world fisheries (see Chapter 40), hazardous waste (see Chapter 36), ozone layer depletion (see Chapter 31), climate change (see Chapter 32) and many other environment problems. Depending on the level of domestic development, international environmental policies usually assign different responsibilities for environmental protection to different countries, according to the principle of common but differentiated responsibility (see Chapter 10). Still, many of the developing and newly emerging economies view environmental efforts as a major obstacle to competitiveness and wellbeing, while also experiencing some of the worst environmental problems in the world. Therefore, one of the biggest challenges is to make this differentiated responsibility truly common among countries without significantly hurting their developmental potential. A more comprehensive and detailed understanding is needed to realize the opportunities offered by globalization while fulfilling global environmental responsibilities and advancing equity and justice (see Chapters 25 and 26). Economic globalization may stimulate a new trend, where eco-efficiency is the main principle and where environmental protection is seen not as an obstacle but rather as an opportunity.

To sum up, the world has been moving fast into a global marketplace wherein financial flows, production, technology and policies become more interconnected. This internationalization is leading to an increased transfer of goods and services across borders, increased communication throughout the world, increased importance of trade, innovation, and technology in the economy, and an increase in global policies. Globalization has swelled beyond its economic roots and has proliferated into many facets of human interactions, including in politics, culture and very importantly the natural environment. The understanding is that positive and adverse changes to environmental stability and quality create various conditions for economic growth, and also place new pressures on international cooperation. This has resulted in a policy of power sharing between individual states with a range of new, higher-level political institutions and processes.

The remainder of this chapter describes the scholarly debate on the relationship between globalization and the environment; discusses the positive and negative effects of economic globalization on the natural environment and environmental policy; and reveals the importance of the phenomenon for global environmental politics.

Globalization and the environment: the debate

Over time, globalization has acquired multiple definitions, many of them referring to the realization of extensive linkages and intensified interconnectedness resulting in interdependence of economic, political, social and cultural spheres. Scholars usually emphasize "an intensification of the range and speed of contacts among different parts of the world and an expansion of the kinds of activities intimately involved in global interactions" (Stearns 2020: 6). While literature presents numerous other definitions of the phenomenon, several prominent features of globalization are easily discernible. These features include, among others, modernization, transformation of spatio-temporal and organizational features of the human condition, internationalization, Westernization and economic liberalization, global economic integration, the intensification of communications and technology transfer (e.g., Fukuyama 1992; Scholte 2000, Giddens 2003; Gallagher 2009; Sparke 2013). As such, the very notion of globalization sparks extensive debate on the nature and driving forces of global processes, and on the costs and benefits of an integrated, globalized world driven by economic development, technology breakthroughs, political shifts, cultural influences and communication boost. The only tenet of the globalization argument where most scholars find common ground is that globalization is an unavoidable catalyst of change (Hebron and Stack 2017; Jovane et al. 2017). From that common outlook, several lines of the debate can be detected. One division, which revolves around whether the change associated with globalization yields positive or negative impacts on the environment, frames the whole globalization and environment debate (Shahbaz et al. 2018). Another issue is the emphasis on economy at the expense of the social, cultural and environmental consequences of globalization (Timmons Roberts and Newell 2016). There are concerns about the linkages between economic integration and population growth, consumption and technological change, and human mobility and security (Gallagher 2009; Singh 2015). Still other concerns encompass different ways by which globalization affects various domestic realms, including economy, politics and culture, as well as domestic responses to globalization, including markets, policies, institutions and ideas (Kochtcheeva 2020).

In general, advocates of globalization perceive it, and the associated economic advancements that set conditions for environmental progress, as interconnected and mutually reinforcing. Accordingly, from the advocates' perspective, globalization increases wealth and economic growth, spawns development, and results in rising incomes, all of which are essential elements to allow governments to generate funds for environmental protection and enact sound policies to abate pollution and conserve natural resources (Panayotou 2000; Copeland and Taylor 2004). To effectively address environmental issues, states must first address their basic needs and accumulate sufficient economic capacity to relieve the environment from the burden of development. Economic growth, increased prosperity, and the rising incomes associated with globalization in this case assist in reducing dependence and poverty, raising the standard of living for citizens, allowing governments the ability to focus on the environment, and enabling individuals to be more active in maintaining their rights to a favorable environment.

Globalization's proponents emphasize that it releases economic forces, resulting in enhancement in productivity of domestic firms. It eliminates price distortions and promotes efficient resource allocation in the domestic economy (Shin 2004; Afesorgbor 2019). Advocates claim that the environmental benefits of globalization go beyond economics. The intensified interconnectivity and interdependency facilitate the diffusion of technology that assists less-developed nations to become more environmentally aware and responsible. Enhanced international cooperation in a globalized world aids the assimilation of positive environmental protection standards and norms across the globe (Clapp and Dauvergne 2011; Hebron and Stack 2017). Such expanded cooperation is the basis of any environmental governance aimed at a global solution to environmental issues. Environmental problems transcend political borders and the interdependence and cooperation associated with globalization are necessary in order to confront environmental problems internationally. The collaboration among states due to globalization brings states closer together and is essential in coordinating a global approach to the environment, chartering effective international environmental institutions and regimes, and establishing binding global environmental policy and law. Globalization provides the driving force to facilitate such actions that are necessary to protect the environment.

Critics, however, perceive globalization as disadvantageous, creating increased domination by the highly developed nations over the less developed societies, economic disparities, and eradication of cultural values and traditions. Globalization is seen as destructive force driven by capitalism and its associated consumerism that exponentially increases economic growth at the expense of natural resources and the environment (see Chapter 17). Globalization-fueled consumption exhausts natural resources and, in turn, limits achievable economic growth without intervention and adequate controls (Najam et al. 2007; Garmendia et al. 2016). Economic growth associated with globalization increases "ecological footprints," the stretches of land and sea required annually to support human activities, and the environmental consequences for countries injured by its increasing patterns of trade, finance and consumption (Bilgili et al. 2019; Sethi et al. 2020). Globalization could also result in the marginalization of economies, sectors, and peoples, and produce poverty-related resource depletion and environmental degradation. Instead of growth and convergence, globalization thus could encourage polarization, in the process exacerbating inequalities in environmental quality across countries, especially increasing disparities between the countries of global North and South (see Chapter 23). Environmentally degrading economic activities generally involve winners who benefit from these activities as well as losers who bear their costs (Boyce 2012; Zivanovic and Smolovic 2019).

Globalization may also exaggerate market failures that spread and exacerbate environmental damage. It may also create demands for reform as policies previously perceived as exclusively domestic attract international interest (Panayotou 2000). Negative associations tied to globalization also include the internationalization of decisions made in faraway locations that increasingly impact societies, and the "restructuring of social space" (Patomaki and Teivainen 2002: 40) as a result of technological changes. In many cases, poorer countries are more likely to suffer the negative environmental effects of globalization than the richer nations that gain from its associated economic growth (Boyce 2012). In fact, the triumph of free markets, economic openness, and the spread of liberal principles are not desired or even viable in many places. As Matthew Sparke (2013: 38) claims, "the globe's three richest people have assets that exceed the combined GNP of all of the least developed countries put together." The critics stress that the imposition of free markets produces social dislocation, political and economic instability, and environmental degradation, exemplified in "the race to the bottom" in environmental standards (Hebron and Starck 2017: 148).

The negative effects of globalization are not just limited to the level of the state, however. They also affect communities and individual people (see Chapter 26). As a result, the poor are least likely to benefit from any economic or environmental benefits of globalization, which aggravates existing income disparities and hampers the ability of people to rise to a position from which they can afford environmental concern (Singh 2015).

The debate on globalization and the environment reveals that both positive and negative environmental consequences of globalization have been identified. States must have the willingness, the capacity, the funds and the technology to protect the environment and the international community requires increased state cooperation in order to address environmental issues that eclipse political borders (see Chapters 7 and 8). At the same time, unrestrained consumption, pollution associated with increased economic growth, and the potential disproportionate impact on the poor requires that globalization be sustainable and presents sustainable development as a potential point of embarkation in the quest for a balance between globalization and the environment (see Chapter 16).

Global economic forces and the natural environment: the effects

Positive effects of economic globalization

By promoting competition, trade openness, division of labor, foreign direct investment and technology transfer, globalization has been perceived to be a powerful source of welfare improvement around the world. In the area of trade, economic globalization contributes to an increasing dependence on international commerce as a source of income domestic, economic prosperity, and environmental progress (Molle 2013; Hebron and Stack 2017; Grossman and Helpman 2015; Masoudi et al. 2017; Zivanovic et al. 2019).

The global trade value of goods exported throughout the world in 2019 amounted to around $19 trillion, in comparison to around $6.45 trillion in 2000 (Sabanoglu 2021). China has been the biggest exporter, responsible for almost 13 percent of all trade goods exported around the world in 2019. The United States was the leading importer of goods, with 16.6 percent in imports in 2018 (Sabanoglu 2021). The services share of world trade has grown from just 9 percent in 1970 to over 20 percent in 2019 (WTO 2019: 20). Services already accounted for 76 percent of GDP in advanced economies in 2015, an increase from 61 percent in 1980 (WTO 2019: 15). Emerging economies are becoming more services-based, often at an even faster pace than advanced ones. Despite evolving as the global goods factory in recent decades, China's economy is moving significantly into services, which account for over 52 percent of GDP. In India, services now make up almost 50 percent of GDP, up from just 30 percent in 1970, and in Brazil, the share of services in GDP is even higher, at 63 percent

(WTO 2019: 15). The expanding service sector is believed to be less emission intensive. Also, around "US$ 20 billion of environmental services, including waste disposal, recycling, sanitation and cleaning of pollution were traded in 2017" and rising environmental awareness and subsequent policies are increasing demand for these services worldwide, and their trade is growing (WTO 2019: 31).

A growing number of countries have adopted programs and policies aimed at supporting growth through technological advancement and innovation, including eco-innovation. The global economic downturn linked to the Covid-19 pandemic is especially pushing countries to strengthen innovation policies. Total global R&D expenditures, including both private and public investments, nearly tripled in current dollars since 2000, from US$676 billion to US$2.0 trillion, with emerging and middle-income economies representing 35 percent of total R&D expenditures in 2017 (WTO 2020: 57). Global Internet traffic, a proxy for data flows, grew from about 100 gigabytes per day in 1992 to more than 45,000 gigabytes per second in 2017. Nowadays, 3.9 billion people, or 51 percent of the global population, use the Internet (WTO 2020: 35). For instance, ICTs are used to deliver localized weather forecasts and information on daily market prices to farmers, especially in countries with resource-tight conditions (WTO 2020: 37).

There has been some significant progress at reducing poverty globally, especially in many countries within Eastern and South-eastern Asia. According to the most recent estimates, in 2015, 10 percent of the world's population lived at or below $1.90 a day, which shows a decrease from 16 percent in 2010 and 36 percent in 1990 (UN 2020a). Additionally, ICTs influence the global production pattern and economic development, most significantly providing the opportunity for a group of emerging and developing economies to diversify production activity and become a part of the global value networks. The best-known illustration of economic benefits and tangible gains of globalization is the rise of the East Asian economies, followed by rapid growth and economic integration of Southeast Asian economies and by China's and India's rise as emerging economic powers. These countries experienced sustained GDP growth until the Covid-19 pandemic. South Asia regional growth was expected to contract by 7.7 percent in 2020, after topping 6 percent annually in the past five years (World Bank 2020). After sustained GDP growth that has averaged almost 10 percent a year, more than 850 million people lifted out of poverty, global leadership in clean energy investment and technological innovation the Covid-19 outbreak has led to a significant economic disturbance in China (World Bank 2020). China's economy has started to rebound, as supply restrictions side eased, and as China has continued "to influence other countries through trade, investment, and ideas" (World Bank 2020).

Globalization promotes the increasingly complex inter-state interactions, as the global character of the environment demands global environmental governance. There has been a tremendous rise in global environmental conferences, negotiations, treaties, and action plans (see Chapters 8 and 9); an agenda of fundamental far-reaching environmental concerns has been delineated; there has been an outpouring of significant scientific research (see Chapter 18); national governments and international organizations have created policies and institutions to address environmental challenges (see Chapter 12); and a network of nongovernmental organizations has started multiple environmental projects (see Chapter 14). UNEP is restructuring itself to become a more results-focused and effective organization (UNEP 2019). The third session of the UN Environment Assembly "Towards a Pollution-Free Planet" held in 2017 "expanded the UNEP mandate for work on environment and health issues." World Environment Day 2018, held in India, stimulated extraordinary action on plastic pollution, influencing hundreds of millions of people in over 190 countries (UNEP 2018: 1). The *Emissions Gap*

Report of UNEP (2018: 9) matches the IPCC's warnings and the Paris Agreement's aspirations reignited global attention to the need for action on climate change (see Chapter 32). Research on low-emission growth focuses on decarbonizing energy systems, transport, industries, and infrastructure, and promoting renewable energy. There is also an expanding trend at local and national levels to generate environmental innovations. For instance, more solar photovoltaics panels were installed in 2017 than the combined net capacity additions of fossil fuels and nuclear power (UNEP 2018). Overall, global interactions facilitate the involvement of a growing diversity of participants in addressing environmental threats, accelerate exchange of environmental knowledge, technologies, and best practices, as well as enable environmental consciousness increases with emergence of global environmental networks.

Negative effects of economic globalization

The magnitude and speed of global economic relations vary for different economies and regions, but the common feature is that due to global interconnectedness trade, openness, innovations and financial flows produce benefits for many, but crises produce negative consequences for the globe. The current pandemic further sharpened the unevenness, inequalities and distress around the world. While millions have made it out of poverty and gained access to improved services, including sanitation and medical care, there are still around 1 billion poor people in the world (UN 2020a). Three-quarters of the poorest families live in rural areas and depend in large measure on natural resources for their existence. Almost 42 percent of the population in Sub-Saharan Africa continues to live below the poverty line (UN 2020a). Up to 770 million people in the world, mainly in rural areas, live without access to electricity and 2.6 billion without modern energy services (IEA 2020). Income inequality between countries has improved, yet income inequality within countries has become worse. More than 70 percent of the world's population live in countries where inequality has grown. From 1990 to 2015, the share of income going to the top 1 percent of the global population increased in 46 out of 57 countries, while in more than half of the 92 countries, the bottom 40 percent receive less than 25 percent of overall income (UN 2020b).

There is also growing uneven distribution of environmental burdens within countries and their correlation with disparities in political power (Boyce 2012; Singh 2015). The need for growth and competitive presence leads to extensive air and water pollution, increase in CO_2 emissions, biodiversity stress, and ecosystem degradation because manufacturing and transportation patterns increase to support the economy (Najam et al. 2007). Many developing countries demonstrate enormous hunger for resources to fuel their economies, develop consumption-oriented societies, and undergo urbanization and tremendous social change. The UNEP GEO 6 (2019) report, as well as other authoritative sources, give evidence that the overall environmental situation globally is deteriorating, unsustainable production and consumption developments and inequality, mostly driven by population growth, put at risk the efforts achieve sustainable development (UNEP 2018, 2019). Severe air pollution causes between 6 and 7 million premature deaths per year and is projected still cause between 4.5 million and 7 million premature deaths annually by mid-century (UNEP 2019: 108). Biodiversity loss from land-use change, habitat fragmentation, overexploitation and illegal wildlife trade, and climate change is driving a mass extinction of species. Land degradation locations cover approximately 29 percent of land globally, where some 3.2 billion people reside (UNEP 2019: 203). Numerous pressures on fresh water resulting from the global drivers of environmental change are evident in the rapid deterioration in freshwater quantity and quality in different regions. Pathogens in water remain a major cause of human death and illness

(1.4 million people), particularly in developing countries (UNEP 2019: 78, 638). In 2010, of the 275 million tons of plastic waste generated by 192 countries, an estimated 4.8–12.7 million tons ended up in the oceans because of inadequate solid waste management (UNEP 2019: 186) (see Part IV of this volume).

Some of the major beneficiaries of economic globalization, the emerging world powers, such as China and India, are suffering from severe environmental problems, especially air and water pollution. India is considered to be one of the largest growth engines of the world, yet it is the world's third-largest emitter of GHGs and is expected to account for 10 percent of global emissions by 2035 (Sethi et al. 2020). The situation is aggravated by the fact that about half of the large Indian population is dependent on agriculture or other climate-sensitive sectors with land degradation and deforestation are among the most serious environmental concerns facing India going into the future (Sethi et al. 2020). Some more modest participants in globalization pay a large environmental price as well. Between 1975 and 2015, built-up areas in Africa grew approximately fourfold, and waste generation is expected to double in lower-income African cities by 2030 (UNEP 2019: 214). Poor water and urban air quality still cause substantial problems in some parts of Eastern and Southern Europe. Even the globalizers are not immune to environmental costs. Excessive groundwater withdrawal, nutrient run-off, increasing demands for domestic energy use, and increased human exposure to dangerous chemicals present challenges for the United States (UNEP 2019: 220, 240, 249).

As such, the direct effects of globalization include emissions and environmental damage associated with the physical movement of goods between exporters and importers and increased production and consumption. This includes emissions from fossil fuel use, oil spills, introductions of exotic species, land degradation, waste accumulation and some others (IPBES 2019). Yet, the effects of economic globalization on the environment are anything but uniform and straightforward, and also produce indirect influences. Among indirect effects, which are underpinned by societal values and behaviors, globalization produces changes in scale, composition, and technique of economic activity, allowing for more widely dispersed externalities (Najam et al. 2007). The composition and scale effect may explain changes in pollution arising from the change in a country's industrial composition, following trade liberalization. The technique effect refers to the plethora of channels through which trade openness impacts the rate at which industry and households pollute (McAusland 2008). The complexity of such effects underscores the need to link economic integration with social and environmental policy at the local, national and global level. This will certainly require overcoming multiple persistent challenges. While domestic environmental challenges tend to be perceived by most people as acute, immediate, and rather understandable by the public, the global ones, are more distant, long-lasting, and sometimes difficult to grasp. Global environmental problems cannot be blamed only on economic integration, expanded trade, and liberalization, when individual mismanagement, national routines, and policy failures in the developed and developing countries are clearly at play. Increasingly, environmental degradation reflecting a business-as-usual attitude is being reconsidered, as global interlinkages grow. The fundamental challenge, however, is to better integrate the environment into national development policies, which presents an enormous task that most developing countries cannot tackle due to a lack in basic capacity.

Economic globalization and international environmental governance

Globalization is a highly multidimensional, persistent, vigorous, and intensifying process that increases the linkages between and among actors, as well as structures within which they function, both domestically and internationally (Clapp and Dauvergne 2011; Jovane et al. 2017; Zivanovic and Smolovic 2019). International relations promote a concept of global governance with supranational regimes, institutions, and interests (see Chapter 9). New epistemic communities have developed to provide complex technical information and illuminate emerging problems, which is indispensable for policymaking and norm creation (see Chapter 18). Intergovernmental organizations and businesses have taken a more direct and prominent role in international decision-making (see Chapters 8 and 13). Globalization points to the importance of the global community with global concerns, and it emphasizes the growing importance of transnational actors, organizations, norms and ideas.

Globalization is also altering the fundamental mechanism of global environmental transformations and international responses to them. It influences the way governments, firms, communities, and individuals perceive environmental change (Timmons Roberts and Newell 2016). Due to the increases in the amount and decreases in the cost of communications, globalization also presents extended access to information and data, new channels for policy influence, and the potential of more advanced and effective modes of governance (Masoudi et al. 2017). The processes of global integration and interaction disseminate the principles, norms, codes of behavior, as well as promote environmental markets and organizations and strengthen international law, which affect global environmental governance.

Global environmental governance is a complex and dynamic process. The outpouring of ideas to protect the global environment dates back more than a century, but ground-breaking books of the past half-century, such as *Silent Spring* (1962), *The Limits to Growth* (1972), *The Population Bomb* (1968), and *Our Common Future* (1987), all of which coincided with the advance of globalization, shaped modern conceptions of global environmental action and sustainable development. Similarly, international environmental conferences at Stockholm (1972), Rio (1992), Johannesburg (2002), and Rio+20 (2012) (see Chapter 8); problem-specific meetings and negotiations from biodiversity conservation (see Chapter 41) to hazardous waste management (see Chapter 36), to climate change (see Chapter 32); and local and global activism (see Chapter 14), all took place in a larger economic and political context.

The spread of environmental and health risks on the global scale and the corresponding moves of policy decisions between domestic and global arenas should be seen as important aspects of globalization. The actors, structures, norms, and processes of governance are influenced by wider occurrences in the global political economy. While globalization calls for a growing specialization of labor across countries, it is a model that creates remarkable economic efficiencies but also exceptional vulnerabilities. Shocks to the world's economic system such as the Covid-19 pandemic reveal these vulnerabilities and the recognition that global supply chains and distribution networks are deeply susceptible to disruption (Farrell and Newman 2020). Companies may now reconsider and minimize the multi-country supply chains that have dominated production. Governments may intervene as well, forcing what they consider strategic industries to have national plans and reserves. States may be moving toward selective self-sufficiency, given the perceived need to devote resources to domestic revival and deal with economic consequences of the crisis. Moreover, mounting public and political pressure to reduce carbon emissions has already called into question many companies' reliance on long-distance supply chains (Niblett 2020).

As such, while there is a rich history of formal actions by states and the international community to address global problems, globalization continues to pose challenges for governance, including environmental governance. These include the outpacing of environmental regulations by economic growth, the increasing power of the private sector to shape economic and environmental decisions, the environmental impacts of economic instability and inequality, and questions about the transparency and accountability of international institutions. There are significant polarizing trends, due to disparate impacts across locations and disparities in the extent of governance responses. Such unequal impacts arise from both circumstantial factors, such as geographic location, and greater vulnerability of certain populations, including those of many developing countries (Boyce 2012; Singh 2015). The distribution of causes and effects of environmental problems across space and time contributes to the challenges of identifying those actors failing to cooperate or resolve such problems. Traditional policy remedies, such as standards, taxes, fees, subsidies and especially direct regulation, lose their effectiveness in an international structure of sovereign states and fragmented institutions with overlapping responsibilities, relatively small budgets, and lack of enforcement power (Najam et al. 2007; IP-BES 2019). Contradictions between regimes and organizations in an atmosphere of incomplete information, fragmented policy arenas, and suboptimal transparency exacerbate the situation.

Additionally, domestic environmental policy failures may have international consequences. In a globalizing world, environmental risks, such as uncontrolled air pollution or an oil spill at the local or domestic level may result in regional and even global problems, causing contamination of resources, toxic precipitation, damage to ecosystems, and health disorders. The inability to avoid or alleviate the spillover effects of transboundary pollution creates a risk for the international economic system of being weighed down by market failures. More often than not it is national environmental underperformance that necessitates bringing multiple countries together to produce a common response, which, in turn, represents a much more difficult problem to solve than domestic environmental protection. Vast differences in national regulation style, philosophy and capacity among countries, which are linked economically also create strains in relations (see Chapter 12). The more profound economic integration is between countries, the more sensitive these countries become to the policy decisions and regulatory outcomes of their partners.

Therefore, the challenge for the domestic and global policy communities is to differentiate between the regimes that provide necessary requirements for environmental protection and those regimes that impede economic activity without producing benefits for the natural environment. Yet another challenge for the international environmental system is to overcome the difficulties of dealing satisfactorily with the priorities of both developed, emerging, and developing countries against the background of the proliferation of multinational treaties that place unequal or unfair obligations on different member states. Developing countries need to compel international organizations to induce developed countries to reduce the persistent impact of globalization in the long run (Shahbaz et al. 2018). A growing complexity of global problems require a multifaceted approach that recognizes the dynamism of resource use and pollution abatement and the need for tailored responses with a variety of policy instruments.

Globalization makes it necessary to take a fresh look at the concept of the universal problems, including global climate change, health, conservation and others (Kochtcheeva 2020). Because all countries struggle with some sort of pollution, land degradation, species extinction, waste disposal, and other common environmental problems, dissemination of successful policy experiments and pilot projects, sharing data and research findings globally can be helpful in highlighting issues and illustrating best policy and management practices. The global system serves as an arena and a forum for multiple environmental groups and associations, which help initiate dialogs on trade and environment, direct efforts toward the reform of international institutions, and attempt to reach consensus between various communities. Global reporting of scientific and technical analyses, exchange of ideas, and sharpening of awareness for a multiplicity of actors may benefit from the process of globalization and contribute to the strengthening of interconnectedness around the world. As international policy and problem networks adjust to an increasingly complex global policy environment the goal should be to take advantage of emerging technologies, model institutional responses on relevant existing expertise, examine problems from multiple perspectives, and form new opportunities for cooperation. In the age of globalization, collaboration of states, international organizations, regional institutions, businesses, research centers, and individuals should shape global environmental governance and the debates surrounding it for years to come.

Conclusion

Globalization is an ongoing process that is altering the natural environment. This multifaceted and even controversial phenomenon is shaping trade, production and consumption, as well as promoting new technologies and spreading global norms. Economic globalization specifically has redefined the understanding of and relation to ecosystems, broadened the reach of many environmental decisions, and produced variable environmental outcomes. The interconnectedness of the global economy, investment, trade, and environmental policy, and the nature of their linkages, are often a function of both domestic and international politics (Gallagher 2009; Jovane et al. 2017). The scale, complexity, and the connection of economic globalization and environmental change, however, do not mean that individual countries and the international community are faced with the absolute choice of doing business as usual in the face of complexity and crises (UNEP 2019). The environmental impacts of globalization most importantly depend on how governments and multinational institutions react to the increasing pressure and complexity of economic growth. Environmental globalization requires fundamental changes in approaches, policy and ideas to empower governments, firms, and peoples to respond effectively to different local and regional environmental situations while simultaneously maintaining a global perspective on their environmental impacts. Identifying global economic, political, social and environmental interlinkages offers opportunities for a range of responses at local, regional and global levels. Current challenges and needs relating to existing domestic and international institutions and capacities for integrating the environment into development present an arena where the environmental impacts of globalization not only remain to be seen, but also remain to be determined and solved (Boyce 2012).

The capacity and willingness of many actors to seek solutions to problems of the natural environment are increasing. While efforts and outcomes vary significantly among different countries, environmental initiatives expose aspects of transnational relations that will only grow in significance. It is most likely to happen due to the expanding economic and technological interactions between different jurisdictions, growing linkages between scientific and political communities, the mounting role of non-state actors in producing and disseminating knowledge, and the advent of innovative policy approaches in response to new constellations of actors, institutions, ideas and events that permeate national boundaries. Specifically, the application of alternative and green technologies through innovation, investment and international collaborations can contribute to sustainable environmentally friendly economic growth. The interlinkages between globalization and the environmental highlight the concept of human development and urgency to promote growth and wellbeing in conformity with the laws of nature.

References

Afesorgbor, S.K. (2019). Globalization May Actually Be Better for the Environment. *The Conversation.* HTTP: https://theconversation.com/globalization-may-actually-be-better-for-the-environment-95406 1/.

Bilgili, F., Ulucak, R. and E. Koçak. (2019). Implications of Environmental Convergence: Continental Evidence Based on Ecological Footprint. In M. Shahbaz and D. Balsalobre (eds.), *Energy and Environmental Strategies in the Era of Globalization, Green Energy and Technology*. Switzerland: Springer Nature.

Boyce, J.K. (2012). *Economics, the Environment and Our Common Wealth*. Cheltenham, UK and Northampton, USA: Edward Elgar Publishing Limited.

Clapp, J. and P. Dauvergne. (2011). *Paths to a Green World: The Political Economy of the Global Environment*. Cambridge, MA: MIT Press.

Copeland, B. R. and M. S. Taylor. (2004). Trade, Growth, and the Environment. *Journal of Economic Literature*, 42: 7–71.

Farrell, H. and A. Newman. (2020). Will the Coronavirus End Globalization as We Know It? The Pandemic Is Exposing Market Vulnerabilities No One Knew Existed. *Foreign Affairs*. March 16, 2020.

Fukuyama, F. (1992). *The End of History and the Last Man*. New York: Free Press.

Gallagher, K. P. (2009). Economic Globalization and the Environment. *Annual Review of Environment and Resources*, 34: 279–304.

Garmendia, E., Urkidi, L., Arto, I., Barcena, I., Bermejo, R., Hoyos, D. and R. Lago. (2016). Tracing the Impacts of a Northern Open Economy on the Global Environment, *Ecological Economics*, 126: 169–181.

Giddens, Anthony. (2003). *Runaway World: How Globalization is Reshaping our Lives*. 2nd edn. London: Taylor & Francis.

Grossman, G. M. and E. Helpman. (2015). Globalization and Growth. *The American Economic Review*, 105(5): 100–104.

Hebron, Lui and John F. Stack Jr. (2017). *Globalization: Debunking the Myths*. 3rd Edition. New York: Rowman and Littlefield.

IEA. (2020). SDG7: Data and Projections. HTTP: <https://www.iea.org/reports/sdg7-data-and-projections>

IPBES. (2019). The Global Assessment Report on Biodiversity and Ecosystem Services. Summary for Policymakers. HTTP:< https://www.ipbes.net/sites/default/files/2020-02/ipbes_global_assessment_report_summary_for_policymakers_en.pdf>

Jovane, F., Seliger, G. and T. Stock. (2017). Competitive Sustainable Globalization General Considerations and Perspectives. *Procedia Manufacturing*, 8: 1–19.

Kochtcheeva, Lada V. (2020). *Russian Politics and Response to Globalization*. London, New York: Palgrave Macmillan.

Masoudi, N., Dahmarde, N., and M. Esfandiyari. (2017). The Relationship between Globalization and Openness of the Economy. *Journal of Economic & Management Perspectives*, 11(4): 1278–1287.

McAusland, C. (2008). *Globalization's Direct and Indirect Effects on the Environment. Global Forum on Transport and Environment in a Globalising World*. Guadalajara, Mexico: OECD/ITF.

Molle, W. (2013). *Governing the World Economy*. New York: Routledge.

Najam, A., Runnalls, D. and M. Halle. (2007). *Environment and Globalization: Five Propositions*. New York: International Institute for Sustainable Development.

Niblett, R. (2020). The End of Globalization as We Know It. HTTP: <https://foreignpolicy.com/2020/03/20/world-order-after-coroanvirus-pandemic/>

Panayotou, Theodore. (2000). *Globalization and Environment*. Working Paper N53. Center for International Development at Harvard University.

Patomaki, H. and T. Teivainen. (2002). Critical Responses to Neoliberal Globalization in the Mercosur Region: Roads towards Cosmopolitan Democracy? *Review of International Political Economy*, 9(1): 37–71.

Postolache, A. G., Nastase, M., Vasilache, P. C., and G. Nastase. (2019). Globalization and Environment: Access Success. *Calitatea*, 20: 517–520.

Sabanoglu, T. (2021). Trade: Export Volume Worldwide 1950–2019. Available at: https://www.statista.com/statistics/264682/worldwide-export-volume-in-the-trade-since-1950/#statisticContainer.

Scholte, Jan Aart. (2000). *Globalization: A Critical Introduction*. New York: St. Martin's Press.

Sethi, P., D. Chakrabarti, and S. Bhattacharjee, (2020). Globalization, Financial Development and Economic Growth: Perils on the Environmental Sustainability of an Emerging Economy. *Journal of Policy Modeling*, 42(3): 520–535.

Shahbaz, M., Shahzad, S.J.H., Mahalik, M.K. et al. (2018). Does Globalisation Worsen Environmental Quality in Developed Economies? *Environmental Modeling & Assessment*, 23: 141–156.

Shin, S. (2004). Economic Globalization and the Environment in China: A Comparative Case Study of Shenyang and Dalian. *The Journal of Environment & Development*, 13(3): 263–294.

Singh, P. (2015). Development, Globalization and Environment: The 21st Century Dilemma. *International Journal of Arts & Sciences*, 8(3): 441–453.

Sparke, Matthew. (2013). *Introducing Globalization: Ties, Tensions, and Uneven Integration.* Chichester, UK: Wiley-Blackwell.

Stearns, P. N. (2020). *Globalization in World History.* New York: Routledge.

Timmons Roberts, J. and P. Newell. (2016). Introduction. In Pete Newell and J. Timmons Roberts (eds.), *The Globalization and Environment Reader*, Chichester, UK: Wiley & Sons, Incorporated.

UN. (2020a). *Ending Poverty.* HTTP: <https://www.un.org/en/sections/issues-depth/poverty/>

UN. (2020b). *Inequality – Bridging the Divide.* HTTP: <https://www.un.org/en/un75/inequality-bridging-divide>

UNEP. (2007). *Our Planet: Connected Dreams. Globalization and Environment.* HTTP: https://web.unep.org/ourplanet

UNEP. (2018). *Programme Performance Report.* HTTP: https://www.unep.org/annualreport/2018/index.php#cover

UNEP. (2019). *Global Environmental Outlook: GEO-6.* HTTP: https://www.unenvironment.org/resources/global-environment-outlook-6

World Bank. (2020). *South East Asia.* HTTP: <https://www.worldbank.org/en/region/sar>

WTO. (2019). *World Trade Report 2019. The Future of Services Trade.* Available www.wto.org

WTO. (2020). *World Trade Report 2020.* Available HTTP: www.wto.org

Zivanovic, S., and S. Smolovic. (2019). *Economy and Globalization.* Varazdin: Varazdin Development and Entrepreneurship Agency (VADEA).

25

International justice
Rights and obligations of states

Steve Vanderheiden

Principles of international justice serve as aspirational goals, constraints and evaluative criteria for the development of the institutions and practices of global environmental politics. In this chapter, these three roles will be considered in terms of the way that each shapes political treatment of international environmental issues. First, the notion of international justice itself will be explored, identifying controversies within the scholarly literature that contest the scope of justice as well as its application to environmental issues. Three conceptions of international justice will be described and briefly explored: an older sense based in post-Westphalian norms of state sovereignty; a newer but weaker sense based in the idea of universal human rights and concerned with providing all the requisite minimum resources or protections for those rights to be respected; and a stronger sense, based around the international extension of distributive justice principles and concerned with providing equitable access to key social and economic resources. These conceptions of justice will then be applied to several issues in global environmental politics to illustrate their scope and explore their implications, including global climate change and international fisheries management. Finally, some reflections on the strengths and limits of justice-based analyses of environmental problems shall be offered.

International justice: contested terrain

A preliminary question to its application concerns whether or not justice can defensibly be extended to relations between states, as the notion of "international justice" supposes that it can be. Philosophers and political theorists have long assumed that some ethical norms govern the conduct of nation-states in international politics, occasionally prescribing limits on state actions beyond those inscribed in law or justifying international responses to transgressions of these norms. Just war theory, for example, posits a set of principled limits on conduct within wars between states, or *jus in bello*, as well as upon the decision to resort to war in the first place, or *jus ad bellum*. First articulated in ancient Rome by Cicero's *De Officiis* and further developed by Thomas Aquinas in the thirteenth century, the notion that ideals of justice between sovereign city states govern their conduct, even during wartime, predates the origin of the nation-state itself by centuries, but anticipates some of the ethical challenges

DOI: 10.4324/9781003008873-29

of relations between them. While the concept of justice active in just war theory had yet to acquire the distributive connotations with which it is currently associated, the term has long served as a regulative moral and political ideal in the arena of what would eventually come to be called international relations, and served to inform key principles of international law.

Alongside the nascent ideals of international justice found in just war theory, principles governing the conduct among independent states during peacetime developed with the rise of the modern state. The 1648 Peace of Westphalia established the principles of state sovereignty and territorial integrity, as the parties to those treaties recognized the prerogative of each prince to determine the religion of his own state, subject to provisions guaranteeing toleration of members of other faiths, and generally for each state to control its own territory and people, prohibiting any state from interfering with the internal affairs of others (see Chapter 7). The Westphalian order that emerged among the European powers respected the sovereignty and territorial integrity of states as a matter of international justice, and affirmed the rights of political self-determination as a key internal value toward which such external constraints are oriented. Early notions of international justice thus prohibited certain kinds of conduct – especially aggressive war or interference in the sovereign affairs of other states – but prescribed little or no positive obligations to states within the international order. Despite affirmed commitments to self-determination, which are now viewed as requiring positive obligations that include development aid as well as negative ones against interference, early views of international justice issued primarily negative injunctions against wrongful interference rather than sanctioning mutually beneficial forms of cooperation. They were also collective and statist in that they governed relations between states, but were silent on obligations of states to the plight of sub-state peoples or individual persons, which contemporary notions of international justice have stressed, often against Westphalian norms.

Starting with the 1949 Universal Declaration of Human Rights, the normative ideal of rights that transcend national boundaries and apply regardless of national residence or citizenship took root, modifying the Westphalian order by implicitly limiting what sovereign states could do within their own borders, thus qualifying the older sovereignty-based conception of international justice with a newer rights-based one. The notion of universal human rights implies not only valid moral claims against states or other actors that violate those rights, even if within national borders and so outside the context of war or international relations, but also a legal claim for a remedy to those rights violations in the first place, perhaps including international intervention within those states found to be violating the human rights of their residents. Human rights discourse thus gave rise to the idea of humanitarian intervention, or the international use of force against states or their governments for the purpose of protecting human rights. This idea was later articulated through the 2005 Responsibility to Protect (or R2P) doctrine, which charges each state with the responsibility to protect its resident populations from serious human rights violations, but charges the international community with secondary responsibilities to enforce rights where states fail to do so, through coercive intervention like economic sanctions or force as a last resort. Weighing against Westphalian commitments is now this liberal internationalist imperative of interstate cooperation in advancement of universal ideals, with justice increasingly identified with the goals of protecting human rights rather than Westphalian protections of states from outside interference. International justice has thus shifted from a negative right of states to non-interference and a justification of status quo power relations to a positive case for proactive interference, in some cases, often on behalf of individual persons or marginalized groups whose rights and interests have been neglected by states, either from incapacity or active malice.

As human rights doctrine has gained wide acceptance, another conception of international justice has emerged, issuing a yet more serious challenge to the conventional Westphalian order. Theories of distributive social justice of the kind first developed by John Rawls (1971), and once assumed to apply only within and not between societies, have begun to be extended to apply to at least some aspects of distributive inequality among and between nations. Within justice theory, Charles Beitz's (1975) influential challenge to this national limit in scope of the egalitarian Rawlsian difference principle has given rise to a school of ethical cosmopolitanism, which has argued for the international application of distributive justice principles, against the resistance of a range of scholars that has included Rawls himself in a later work (2001). Among scholars who accept egalitarian justice principles as guiding ideals within societies but reject their application to relations between them, or egalitarian nationalists, the resistance to this wider purview for distributive equity originates in what they take to be the proper Humean circumstances of justice, which govern relations between persons within certain kinds of cooperative schemes and social arrangements. Nationalists typically deny that international society entails such circumstances, and often endorse more limited standards of international justice (human rights, for example) while denying that resource distribution among and between nations is a matter of justice.

For those taking international justice to have the properties vested in it by contemporary ethical cosmopolitans, in which justice principles are applied across national boundaries and international institutions are required to be designed to advance the interests of the world's least advantaged, global environmental politics would focus upon promoting equitable access to resource wealth and other environmental goods and services, and halting environmental harm through pollution or resource depletion, where this results from activities associated with affluence but adversely affects the poor. Note that neither pollution nor resource depletion would be viewed as unjust in itself, but that the injustice of either would depend upon the pattern by which more advantaged parties cause a problem and less advantaged parties suffer its effects, so that either kind of environmental despoliation would exacerbate existing inequalities between the advantaged and disadvantaged. An alternative formulation of egalitarian justice to be considered below sets aside this condition that the agent causing environmental harm be more advantaged than the one suffering its effects, maintaining instead that justice requires that none be made worse off by the polluting or resource depleting acts of others, requiring compensation in those instances when such acts occur. Environmental damage that harms only those parties that cause it is often imprudent and may be morally bad, but would not be unjust, as the injustice lies in the interpersonal or intergroup effects.

Strong environmental protections might be advocated from the weaker sense of justice that is coextensive with human rights doctrine, although with the somewhat more modest goal of ensuring that all meet some threshold of access to environmental goods and services, or are not put at risks that exceed a similar threshold by the acts of others. Since rights are concerned with minimum thresholds below which rights are violated, rather than equality itself, a rights-based approach to water justice, for example, might argue that all persons have access to some quantity of water that is sufficient to meet their basic needs, whereas a distributive justice-based approach might require a higher burden of proof on inequalities in water access, casting such inequalities as unjust unless to the benefit of the least advantaged even if all were above that basic threshold. From the older Westphalian conception of international justice, some prohibitions against certain kinds of environmental harm would issue, as the principle of territorial integrity would prohibit transboundary pollution that originates in one state but has its deleterious effects in another, at least without the former state compensating the latter for its injury. In general, then, the stronger conceptions of international justice serve to justify stronger levels of environmental protection, or can condemn as unjust a wider range of instances of environmental despoliation, but even the earliest conceptions are relevant to some issues in contemporary global environmental politics.

Many local environmental concerns are thus simply outside the purview of international justice, but considerations of justice do weigh on some regional and nearly all global issues, even under the oldest and weakest conceptions. To illustrate, consider the problem of global climate change (see Chapter 32), which shall be examined in more detail below, but which here reveals how international justice in its various senses can be invoked to condemn the status quo and to urge greater levels of climate policy action. The greenhouse gas pollution that causes climate change is not only a kind of transboundary harm in that it crosses one or two national boundaries, as with more local forms of air and water pollution, but it transcends all national boundaries, as its effects are global and are not dependent upon its geographic source. A ton of carbon emitted from anywhere on the earth has the same effect on global climate, and these effects are global, albeit not uniformly or necessarily negative. Hence, all greenhouse pollution violates the terms of international justice established through the Westphalian principle of territorial integrity, insofar as it causes some climate-related harm somewhere outside its territory of origin, requiring compensation to all parties adversely affected by it. Domestic energy, transportation and environmental policies can affect global climate, so they cannot be viewed as protected from international interference by claims to national self-determination. A similar condemnation of climate change is available through human rights-based conceptions of international justice, as Simon Caney (2008) has shown, since climate change is expected to threaten rights to subsistence, health and territory. Applying distributive justice principles, as is done in more detail below, yields a more extensive critique against "business as usual" contributions to climate change as well as providing detailed goals and constraints for the design of international climate policy.

Similar observations could be made about other international or global environmental issues, as transboundary pollution externalities or impacts on the human interests protected by human rights invoke Westphalian and rights-based conceptions of justice, which condemn such effects of environmental despoliation as unjust and require as remedy that the effects be curtailed or their harm be compensated for, while interpersonal or intergroup effects invoke distributive conceptions of justice, as well. As should be evident, all tenable views of international justice can be marshaled on behalf of an active international regulatory regime designed to minimize pollution or promote sustainable resource management practices, even if the various conceptions noted above exist along a continuum of more and less ambitious goals for a just international society. Besides noting the differences between them, these three conceptions might all be contrasted with those which deny that international politics is subject to justice ideals of any kind, such as political realism, which seeks to vindicate self-interested behavior by states, either by positing self-interested action as natural or by denying the binding force of justice outside of its context in domestic society, with its shared political culture and social institutions (see Chapter 3). Since some critics of the application of justice principles to international relations cite realist premises or deny that the circumstances of justice apply in this context, further consideration of skeptical views and the challenge they pose is warranted before considering several applied cases.

International justice and sovereignty

In the context of global environmental politics, considerations of international justice may prescribe a variety of rights and obligations to states, depending upon the issue and justice principles being applied. In some cases, however, these prescriptions would be contested, as those issuing from imperatives of justice may conflict with those arising from other international political norms. For example, conventional views of state sovereignty assign full property rights to natural resources found within territorial borders to the governments or peoples of those states, including rights to use, transfer and profit from resource harvest or extraction. By this account, the oil found within Saudi Arabian oil fields belongs to the Saudi government or people, and its rate of extraction is to be determined by its owners alone. Governments or residents of other states, including international institutions like the United Nations, are not entitled to control the way those resources are used or to any royalties from their extraction. While other states may have a legitimate interest in controlling transboundary pollution that results from oil extraction, refining, or combustion, they have no right to limit what the Saudis can do with the oil resources within their borders. Conventional views of state sovereignty hold likewise with forest, water and other mineral resources, granting ownership and management prerogatives to states alone, in potential conflict with international environmental regulatory imperatives that seek to protect such resources against unsustainable use rates or guarantee access to those resources over time.

This conventional view thus entails what might be termed strong national entitlements to natural resource assets, since the property rights they ascribe are not limited by imperatives for sustainable resource management or considerations of distributive equity, and have as a result several consequences for the notion of international justice as well as for global environmental politics. If justice is taken to occasionally require redistribution between nations in accordance with egalitarian principles, this entitlement claim could undermine the force of its imperatives. For if wealthy nations became wealthy as the result of their resource wealth and they are fully entitled to the proceeds from exploiting those resources, then poor countries have no valid claim to transfers based in wealth to which others are entitled. Ethical cosmopolitans like Thomas Pogge (1994), who argue for a portion of such wealth to be redistributed internationally as a matter of justice, contest this strong national entitlement, grounding poverty relief efforts in the common stake that all share in the world's resources. Indeed, a considerable scholarly literature has developed around the question of resource wealth and development, with Rawls remarking in *The Law of Peoples* (2001) that national development results from its political culture and not from its natural resource wealth, affirming a similar claim made by scholars of the so-called resource curse (Wenar 2008), which postulates that resource wealth can sometimes inhibit forms of development by encouraging corruption and state-sponsored violence. Both claims discount the role that inequitable national resource stocks play in development, and thus tacitly endorse strong national resource entitlements by dismissing challenges to them on grounds of justice, with Rawls arguing for an international duty of assistance that calls for political development aid but not redistributive transfers to be dedicated to poor countries. Those embracing strong national entitlements reject egalitarian redistribution of the proceeds of resource wealth, as well as any system of international development aid that is predicated upon the morally arbitrary nature of natural resource distribution. Significantly for environmental politics, then, they would affirm state sovereignty over natural resources as trumping global concerns for biodiversity, resource depletion, or environmental integrity, save for that range of cases in which environmentally harmful acts within one state can be shown to violate the sovereignty of another.

Ethical cosmopolitans, by contrast, typically reject national entitlement claims to natural resource wealth, often by appealing to Lockean postulates that the world is owned by humanity in common or to the morally arbitrary nature of natural resource distribution, which resembles the Rawlsian "natural lottery" of arbitrarily distributed natural talents and confers no entitlement. According to this analysis, first made by Beitz (1975), the logic of Rawlsian distributive justice entails that persons not be disadvantaged as the result of circumstances like nationality or cultural membership, so advantages in the world that are based on such categories ought to be rectified by international transfers. Beitz first argues for a "resource redistribution principle" that would transfer natural resource wealth from rich to poor nations, following the contractarian analysis of Rawls's original position but supposing that parties are also ignorant of their nationalities, then makes the more expansive case for a fully internationalized difference principle, by which resources would be distributed among all nations and peoples such that inequalities would be justified only insofar as they benefited the most disadvantaged. In making this case for cosmopolitan distributive justice – often referred to as global rather than international justice since applied to individual persons rather than nation-states – Beitz rejects the national resource entitlement premise noted above. It is on this point that Rawls departs from Beitz and other cosmopolitans by insisting that distributive justice applies only within and not among or between nation-states, which are fully entitled to their own resources and thus not obligated to redistribute them in the interest of reducing international inequality, defending instead a set of principles that he terms the "law of peoples" and which includes several basic human rights along with Westphalian commitments to sovereignty and territorial integrity.

The debate between Rawls and Beitz, or more generally between liberal nationalists and those defending global or international justice, has rarely focused upon international regulatory capacity, but the implications of claims concerning control of territory or resources have clear implications for global environmental politics. If land and natural resources are fully owned and controlled by the states in which they are originally located, then international organizations are powerless to promote sustainability imperatives aimed at guarding against resource depletion or environmental degradation that takes place within national territories (see Chapter 8). Besides undermining claims for resource redistribution, the strong nationalist position would prohibit the international imposition of environmental regulations that limit national sovereignty over land or resource use, except where necessary to prevent pollution from crossing national borders. If sovereignty extends to resource use within national borders, international environmental agreements seeking to prevent unsustainable forest management or limit the destruction of species habitat would violate that sovereignty. Only in cases where some kind of transboundary externality arises from internal land management or resource use policies or practices could international regimes trump the authority of national or subnational governments to degrade their environments (see Chapters 9 and 10).

International justice is thus significant in that it posits elements of common concern that transcend national borders and at least occasionally trump national sovereignty, as all members of the international community take on obligations as stewards of their common environment that governments cannot simply annul by invoking sovereign authority. Just as global poverty and human rights are considered to be subjects of international justice, so also can the bases for international justice make possible more extensive international environmental regimes. Once land and resource management decisions are viewed as having myriad external effects, rather than being viewed as properly subject to internal controls alone, the legitimacy of such regimes becomes apparent. Because of its spillover effects on global climate (see Chapter 32), biodiversity (see Chapter 41), and in some cases also international riparian systems (see Chapter 38) and deforestation (see Chapter 42) in Brazil or Indonesia cannot be regarded as strictly an internal matter subject only to the will of the Brazilian or Indonesian governments. Without trumping Westphalian conceptions of international justice, protection of sovereignty can be construed as requiring limits on resource management or pollution policies in some states in order to protect the sovereignty of others against interference, requiring rather than undermining international environmental law designed to protect that sovereignty.

An alternative formulation to distributive justice principles that have developed around the ideal of equity, which lends itself to several problems in global environmental politics, is one that is instead built around the ideal of responsibility (Vanderheiden 2011). While concerned with distribution, responsibility-based conceptions do not necessarily take equal distribution as a default starting point or focus primarily upon the effects of actions or institutions on the least advantaged, as variations upon the Rawlsian difference principle do. Luck egalitarians, for example, typically define justice in distribution in terms of the goods and bads that one acquires as the result of voluntary choices and those acquired by luck, criticizing departures from equal distribution that result from the latter but not those resulting from the former. By this account, responsibility in its descriptive sense is defined both in terms of voluntary control and in its prescriptive sense in terms of desert or entitlement. By this account, persons can be said to deserve or be entitled to those goods or advantages that result from their voluntary choices, such as hard work or willingness to defer gratification, but not those arising from factors beyond their control, such as circumstances of birth or innate talents. Likewise with bads or disadvantages: each is viewed as just insofar as it results from some voluntary (or culpable) choice, but unjust insofar as arbitrarily suffered or imposed by another. Luck egalitarianism is so-called because it maintains that persons should not do better or worse as the result of luck, or factors beyond their control, though it finds nothing unjust about inequality that results from responsible choices.

The implications for environmental politics follow from this key distinction. As applied to climate change, for example, luck egalitarian analysis concludes that persons have no valid complaint against other parties if they suffer climate-related harm for which they are personally responsible, as for example from having caused it through their greenhouse emissions, so no policy response to self-imposed environmental vulnerability would be needed. But insofar as persons or peoples are made vulnerable to climate-related harm caused by others, as is the case for those expected to bear the brunt of climate-related harm, some kind of remedial response to that vulnerability is required as a matter of justice. This policy response could include actions of mitigation (by which vulnerability is minimized by reducing the anthropogenic drivers of climate change), adaptation (by which vulnerability is reduced by proactive efforts to reduce it in the face of expected climatic changes), or compensation (by which wrongfully imposed vulnerability can at least partially be rectified through transfers in the amount of the expected harm). The key for luck egalitarian conceptions of justice is that none suffer harm or disadvantage as the result of phenomena for which they are not responsible, as often happens with pollution problems.

Notice that responsibility-based conceptions of justice are able to capture the nature of many kinds of environmental harm more effectively than equity-based conceptions can. Since equity-based conceptions typically focus upon actions that exacerbate existing inequalities, they cannot identify anything unjust about one relatively affluent polluter exposing another to hazards related to their pollutants, or degrading land or ecosystem services such that future affluent persons will be worse off than they otherwise might be. Unless it affects the least advantaged for the worse, for example, Rawls's difference principle would be unable to condemn pollution or resource depletion as unjust. Merely imposing harm against the will of a vulnerable party or willfully undermining the ecological capacity of a region or people does not in itself violate the terms of equity-based justice, and may in some cases perversely count as advancing justice if those harmed were among the advantaged at the outset. But responsibility-based conceptions, for reasons suggested above, are better able to capture the injustice of imposing avoidable harm through pollution or resource depletion. According to this view, anyone made worse off by the polluting or degrading acts or policies of others is entitled to some form of injunctive relief to mitigate the harm in question, or some form of compensation for losses suffered or serious risks imposed. Since environmental harm is quintessentially of this kind – the imposition of a kind of externality cost, whether through exposure to pollution or exacerbated scarcity from resource depletion, resulting from unsustainable actions, practices and policies – this conception of justice can usefully illuminate its injustice, justifying responses from environmental politics.

Justice and global environmental politics

As an aspirational standard for diminishing wide current disparities in opportunity among persons across national boundaries, or as a set of procedural or substantive constraints on the design of international institutions, justice often plays a peripheral role in global environmental politics. Environmental degradation or resource depletion that exacerbates existing global inequities can tenably and constructively be described as unjust, and international regimes that fail to account for the perspectives and interests of the world's disadvantaged can likewise invoke the same criticism, but in such cases justice is only part of the critique that is typically made against either. The former also concerns the sustainability of the activities causing such problems (see Chapter 16), and the latter the democratic responsiveness and accountability inherent in such regimes (see Chapter 9), with injustice one of several criteria by which bad outcomes can be condemned and justice one ideal toward which good outcomes may aspire. But neither offers a sufficient critique for rendering other normative criteria superfluous, since pollution and resource depletion can be harmful and unsustainable without being unjust, and regimes can follow just procedures but arrive through them at bad outcomes. Justice, that is to say, neither subsumes other ideals as an all-purpose norm nor replaces other criteria for identifying problems – good environmental policies or outcomes should be just, but often must serve other ideals, as well – but justice can nonetheless serve as a valuable concept in critically analyzing several problems toward which political responses are oriented as well as in theorizing remedies to those problems.

Perhaps most notably, justice-based analyses have been constructively applied to the causes and effects of global climate change in order to highlight its nature as in part a problem of justice, as well as to the design of institutions and policies associated with climate change mitigation and adaptation (Caney 2005; Page 2006; Vanderheiden 2008; Harris 2010; Shue 2014). Because the relatively affluent global North is responsible for over half of current greenhouse emissions and for over 70 percent of historical emissions, despite being home to less than 20 percent of global population, climate change has been characterized as resulting from an unjust appropriation of carbon sinks by the world's affluent. At the same time, scientists expect that the world's poor will be most vulnerable to climate-related environmental changes, and have already been born disproportionate impacts from such changes, raising justice concerns in the effects of climate change as well as its causes. Together, critics have aptly cast climate change as both a cause and consequence of global injustice, as wide current disparities in living standards and consumption patterns yield widely disparate national per capita emissions (see Chapter 17), as well as disparities among income groups within all countries, exacerbating the disadvantages of the world's poor by making them most vulnerable to environmental changes toward which they have contributed relatively little. For this reason, "climate justice" has implied a critique against the consumption patterns of the global North and become a rallying cry for stronger action to mitigate the causes and control the effects

Steve Vanderheiden

of climate change, as well as a distributive claim on behalf of the global South in burden-sharing arrangements designed to accomplish those ends.

Norms of international justice are thus violated by climate change, as an environmental externality disproportionately caused by the affluent global North with its costs expected to be borne disproportionately by the poor global South, and those justice norms also oblige states to take action to avoid that outcome. In so doing, it must assign the costs of mitigating climate change justly among various parties. This burden-sharing scheme has been the subject of much work by applied philosophers and political theorists, many of whom have applied various principles of justice to policy problems surrounding mitigation and adaptation. Apart from the Rawlsian difference principle, which is perhaps the most discussed distributive principle in contemporary justice theory and which has been invoked on behalf of equal per capita pollution rights, these have included polluter-pays principles based on current and historical emissions as well as those turning on strict and fault-based liability, beneficiary-pays principles that assign liability for climate change mitigation to those having benefited most from activities associated with carbon pollution, and capacity-based formulae that assign greater burdens to parties most able to afford them. Each aims to define the fair share of burdens to be assigned to various parties, and each arrives at an at least slightly different cost allocation and justification for what each party is required to contribute toward the collective goal of minimizing climate-related harm. Here, distributive justice is used as a principle for resource sharing if assigning shares of global emissions or burden sharing if assigning shares of economic abatement costs, but in each case invokes international justice in its strong distributive sense.

International justice, then, frames the problem of climate change as one of the equitable allocation of either emissions absorptive capacity itself or the costs of mitigation and adaptation, and has served as a key analytical framework for scholars of the normative dimensions of international climate politics. The injustice of unmitigated climate change requires that action be taken to reduce its causes and control its effects, by this analysis, and the manner in which this action is to be taken must likewise follow the constraints of justice. These constraints include procedural elements, by which policy is made on the basis of open and inclusive processes by which less powerful actors may exercise a meaningful role in shaping outcomes, as well as substantive ones, which guard against burdens being assigned in ways that are unfair to relevant parties. One reason that climate change works so well as a case study in applied justice theory is that the problem to which principles are applied is a genuinely international one: greenhouse gas emissions produce their insidious effects regardless of geographical origin, so the climate system functions as a public good that requires collective management, which, in turn, suggests principles by which the terms or costs of its management can be assigned among relevant parties. While some scholars have resisted the idea that relations among states in international society amount to circumstances of justice, which include moderate scarcity and limited altruism in a context of shared resources and common fate, and have thus denied that distributive justice principles apply across national boundaries, the phenomenon of climate change appears to satisfy the conditions for justice to apply internationally. Indeed, global climate change has been heralded as perhaps the paradigm case for the application of international justice analysis for these reasons.

Generally, then, considerations of international justice in its weaker sense require states to recognize and adhere to human rights norms, including those associated with environmental harm such as rights to territory and subsistence. Within the context of climate change, scholars have wielded human rights as instruments for motivating action on climate change by appealing to the threats that climate change poses to life and health, through food shortages,

increasing and more severe storms and floods, and altered disease vectors, as well as to territory, through sea-level rises and land or waterway changes. Insofar as these rights are violated
by significant changes in climate, international justice in its weaker sense may require nations
to take serious and immediate action to mitigate climate change or to assist in adaptation to
its effects, as human rights require international cooperation to ensure that the interests they
represent are protected. Human rights-based justice may also suggest a minimal threshold for
how much the international community as a whole must do in some combination of greenhouse gas abatement and adaptation financing, in order to prevent such rights violations from
occurring, but it cannot prescribe terms by which the obligations of particular states can be
set, as international justice in its distributive sense can. Those vulnerable to climate-related
harm may, for example, advance human rights claims against the United States as a major
emitter and thus culpable party in the violation of their rights, as the 2005 Inuit petition filed
through the Inter-American Commission on Human Rights unsuccessfully attempted to do,
but the international justice framework upon which those rights are based could not in itself
determine what the United States must do in emissions abatement in order to comply with
human rights norms. Prescriptive guidance for the design of international policy action plans
on climate change thus invites the application of justice in its stronger sense, which may help
resolve several of the burden-sharing issues at the core of current policy debates. Insofar as
nationally determined contributions toward international mitigation efforts under the Paris
Agreement still require the fair allocation of national mitigation burdens, as scholars maintain, their calculation and analysis requires the application of justice principles.

Both rights-based and distributive conceptions of international justice can be applied to
other issues in global environmental politics, as well, with the application of principles being
most straightforward in cases that most closely resemble climate change in its international
scope and reliance upon collective management for the maintenance of an international public good. For example, ocean fisheries management involves a shared resource that is subject
to collective action problems, as multiple users threaten to deplete fish stocks, with sustainable management requiring the imposition of catch limits that allocate a scarce resource
among competing claimants according to defensible distributive principles (see Chapter 40).
While overfishing may not itself raise human rights concerns, it may involve distributive injustice insofar as some parties may currently or in the recent past have taken more than their
fair share of existing biomass, and the assignment of individual catch limits certainly suggests
a distributive justice analysis in order to ensure that it follow defensible principles. As with
climate change, international justice principles can help clarify the nature of the problems as
well as prescribe fair solutions to them.

This analysis likewise applies to other common pool resource management issues, where
the resource in question lies outside of national territories or is affected by actions that transcend national borders. In such cases, the need to impose access limits in order to sustainably
manage the resource over time invites the application of distributive justice principles in
order to ensure that burdens associated with collectively managing the resource be fair to
all. While scholars typically do not cast national catch limits in distributive justice terms,
primarily because the relevant units of analysis have been boats rather than nation-states and
sustainable aggregate catches rather than fair individual shares have been the main focus of
management schemes, some have begun to apply such principles to the governance of oceans
and their resources (Armstrong 2020). Such as application follows a two-stage analytical sequence similar to that associated with climate change. Justice between generations would require sustainable fisheries management, as overfishing depletes fish stocks to the detriment of
future persons, and international justice requires that states with claims to limited fish stocks

be assigned defensible shares of them within these parameters. Not all fisheries management issues involve circumstances in which international justice principles can validly be applied, however. Fish habitats may lie largely or wholly within national borders, and so generate no international entitlement claims. Catch limits may be more appropriately allocated according to market forces, as through auctions or fees, rather than being subjected to distributive justice principles. Overfishing and resulting depletion of fish stocks may have no direct and unique impact on the world's disadvantaged. In such cases, prior entitlement claims preempt imperatives for more equitable international access to the resource, or the failure to sustainably manage fisheries is unfortunate but not unjust.

Fisheries management also points to another important limit to justice-based analyses as they apply to global environmental politics, in that justice is typically assumed to be concerned with effects upon human welfare only, and not to govern the human treatment of nonhuman animals or the condition of ecosystems themselves. Although some scholars have proposed notions of ecological justice that apply beyond the human world and are extensions of the sort of principles that are typically found in justice theories (Schlosberg 2007), justice is conventionally viewed as an anthropocentric concept that cannot identify unique wrongs in actions that degrade the environment or harm nonhumans unless those result in harm to humans and this harm has the sort of equity effects noted above. As a result, environmental injustice can identify some but not all wrongs or bad outcomes in human actions or policies with respect to the environment, and other value concepts are needed to fully capture the normativity involved in human relationships with the wider world. It may be wrong but not unjust, for example, to needlessly drive some nonhuman species to extinction or degrade some rare and beautiful landscape, and scholars of environmental ethics have aptly criticized justice for its narrow purview in this regard (see Chapter 27). Since justice as a normative concept requires all other bad effects upon the world to be reduced to equity impacts on human welfare in order to trigger its critique, some justice-based analyses of environmental harm are unavailable, and others so instrumentalize the natural world that they become complicit in the mindset of ecological exploitation that they nominally seek to prevent.

Without a regulative ideal that applies across national borders and thus serves to guide and constrain actions and policies that affect the global environment, nation-states operate within an international context that remains dominated by the Westphalian order, with its commitments to strong versions of state sovereignty and territorial integrity. While these principles remain important to many aspects of contemporary international relations, they fail to adequately address the international and occasionally global scope of some contemporary environmental threats. States alone have the right to control the actions of polluters and resource users within their borders under Westphalian norms of international justice, and those with little or no evident concern with maintaining their ecological support systems retain the rights to degrade and even destroy them. While in one sense this use of sovereignty in the service of environmental damage might be viewed as the epitome of national self-determination, which itself has the status of an important if qualified human right, consideration of the limits of this unconstrained state power of environmental degradation reveals its disjuncture from genuine self-determination, and with that its key flaw. Exclusive state authority over territory and resource allows not only for the degradation of future ecological capacity within that state's borders, imperiling its future people and perhaps violating their rights, but it also threatens the global environment, upon which all states and peoples depend, and in which all sovereign territories are nested and with which all remain interdependent. Analyses based in justice (Kolers 2012; Moore 2015), whether from human rights or distributive equity, provide a useful counterpoint to the Westphalian norms that sometimes allow unsustainable state actions to persist, and which often fail to provide goals toward which the international community might aspire in managing its common environmental challenges.

Conclusion

Although contested, the widespread view that at least some conception of justice applies to international relations has significant implications for global environmental politics. Certain outcomes, whether dangerous anthropogenic interference with the climate system or the collapse of fisheries as the result of overfishing, can be understood as unjust, by one of three conceptions of justice surveyed above. Whether in the relatively weak terms of post-Westphalian norms of state sovereignty, the newer and stronger conception of justice based in human rights doctrine, or the newest and strongest conception rooted in cosmopolitan distributive justice, international justice prescribes rights and obligations to states to care for their common environments, albeit of varying strength and in somewhat different circumstances.

References

Armstrong, C. (2020) "Ocean Justice: SDG 14 and Beyond," *Journal of Global Ethics* 16:2, 239–55.
Beitz, C. (1975) "Justice and International Relations," *Philosophy and Public Affairs* 4:4, 360–89.
Caney, S. (2005) "Cosmopolitan Justice, Responsibility, and Global Climate Change," *Leiden Journal of International Law* 18:4, 747–75.
Caney, S. (2008) "Climate Change, Human Rights, and Discounting," *Environmental Politics* 17:4, 536–55.
Harris, P.G. (2010) *World Ethics and Climate Change: From International to Global Justice*, Edinburgh: Edinburgh University Press.
Kolers, A. (2012) "Justice, Territory and Natural Resources," *Political Studies* 60:2, 269–86.
Moore, M. (2015) *A Political Theory of Territory*, New York: Oxford University Press.
Page, E. (2006) *Climate Change, Justice, and Future Generations*, Cheltenham: Edward Elgar.
Pogge, T. (1994) "An Egalitarian Law of Peoples," *Philosophy and Public Affairs* 23:3, 195–224.
Rawls, J. (1971) *A Theory of Justice*, Cambridge, MA: Belknap Press.
Rawls, J. (2001) *The Law of Peoples*, Cambridge, MA: Harvard University Press.
Schlosberg, D. (2007) *Defining Environmental Justice: Theories, Movements, and Nature*, New York: Oxford University Press.
Shue, H. (2014) *Climate Justice: Vulnerability and Protection*, New York: Oxford University Press.
Vanderheiden, S. (2008) *Atmospheric Justice: A Political Theory of Climate Change*, New York: Oxford University Press.
Vanderheiden, S. (2011) "Globalizing Responsibility for Climate Change," *Ethics and International Affairs* 25:1, 65–84.
Wenar, L. (2008) "Property Rights and the Resource Curse," *Philosophy and Public Affairs* 36:1, 2–32.

26

Environmental justice

Pollution, poverty and marginalized communities

Hollie Nyseth Brehm and David N. Pellow

Environmental hazards disproportionately affect poor communities, communities of color, Indigenous populations and other marginalized communities around the globe. This uneven exposure to environmental risks is variously termed environmental inequality, environmental racism and environmental injustice. Over the past five decades, a body of scholarship and a social movement have emerged in response, and scholars and activists have rallied around the term *environmental justice* (EJ) – the notion that all people and communities are entitled to equal protection of environmental health laws and regulations (Bullard 1996).

While there are numerous ways to define EJ and the problems of environmental racism and inequality, the most important point is that they are not fundamentally *environmental* issues; they are *social* problems. To frame EJ as an ecological problem runs the risk of missing the point that ecological violence is first and foremost a form of social violence, driven by and legitimated by social structures and discourses. According to standard definitions, EJ is the fair treatment of all people with respect to the development, implementation, and enforcement of *environmental* laws, regulations and policies. However, if all environmental laws, regulations and policies were implemented and enforced equally, the globe would still be marked by environmental inequality because the social, political economic, and cultural forces that produce this problem will not have been addressed.

This chapter explores social science EJ scholarship, which has become prominent in many countries. First, we provide a brief history of the political and intellectual movements for EJ followed by a review of key studies that have documented environmental injustices both in the United States and internationally. Next, we review potential causes of environmental inequalities, followed by a consideration of the effects of environmental inequalities and how they interrelate with other social inequalities. Lastly, we examine how social movements, nongovernmental organizations (NGOs), and other key actors respond to the persistence of environmental inequalities.

Origins of environmental justice studies and politics

During the 1970s, several scholars in the United States began to explore the relationship between economic status and exposure to polluted air, revealing troubling correlations between

DOI: 10.4324/9781003008873-30

lower income communities and poor air quality in the United States (Freeman 1972; Zupan 1973; Kruvant 1975). But it was not until protests in Warren County, North Carolina, made national news in 1982 that this emerging focus, soon to be known as environmental justice, became well known. Warren County was the poorest county in North Carolina, and 65 percent of its population was African American (Szasz and Meuser 1997). In the early 1980s, the state decided to build a new hazardous waste landfill in the county and residents organized to protest the proposed landfill and found support from several civil rights organizations (Bullard 2000). These protests were among the first actions that gained national media attention and raised public awareness about the unequal environmental burden that historically marginalized communities confront, although recent research reveals that the news and lessons of the Warren County protests took longer to reach EJ activists in some parts of the country than previously assumed (Perkins 2021). The unique combination of ideas promoting civil rights, social justice, and environmental concern, as well as the growing visibility of hazardous waste, set the stage for the emergence of a new way of thinking about the relationship between ecosystems and humanity.

The protests in Warren County triggered several subsequent events that solidified the place of EJ in the US grassroots political imaginary. In 1983, the US General Accounting Office (GAO) conducted a study of the racial composition of communities near four major hazardous waste landfills in the South. That investigation concluded that in three of the four cases, the communities around the landfills were predominantly African American; and, in the fourth case, the community was disproportionately African American (GAO 1983). This report was followed by a 1987 study by the United Church of Christ (UCC) Commission for Racial Justice, which was the first national-level study of the racial and socio-economic characteristics of communities living near hazardous waste facilities (UCC 1987). Again, a similar pattern emerged – communities of color were much more likely to host hazardous waste facilities (see Chapter 36).

While government and community leaders were studying the situation, scholars turned their attention to the phenomenon as well. In 1990, sociologist Robert Bullard published *Dumping in Dixie*, a book in which he argued that African American communities were being targeted for the location of solid waste facilities throughout the US South. Importantly, Bullard also documented widespread community resistance to these inequitable siting patterns. That same year, environmental studies scholars Bunyan Bryant and Paul Mohai organized a national conference that brought together researchers studying environmental inequality. After reviewing the body of evidence, they concluded that their studies overwhelmingly supported the earlier findings of the GAO and UCC reports (Bryant and Mohai 1992).

While a few studies had explored environmental inequalities before the 1980s, the late 1980s and 1990s saw a flurry of research on environmental injustice. In 1993, sociologist Stella Çapek introduced the EJ frame, which, drawing from Erving Goffman's (1974) idea of framing as a schemata of interpretation, views EJ as a lens that offers a way of constructing meaning for activists. According to Çapek, the EJ frame consists of six key claims, including the right to accurate information from authorities concerning environmental risks; public hearings; democratic participation in decision-making regarding the future of any threatened community; compensation for injured parties from those who inflict harm on them; expressions of solidarity with survivors of environmental injustices; and a call to abolish environmental injustice. Environmental injustice was not just about disproportionate hazards; it was about access to decision-making capabilities, democratic processes, and power. These arguments were later explored by David Schlosberg (2007), who argued that the

EJ literature's focus on justice was limited and that scholars and activists should emphasize the power structures and social systems that give rise to environmental inequalities.

Importantly, the relatively recent development of EJ studies should not be accepted as evidence that this phenomenon is new. European colonization of the Americas, Asia, Africa, and the Pacific was accompanied by many environmental injustices, as people, land, flora, fauna, and waters were exploited for the benefit of colonizers (DuBois 1977; LaDuke 1999; Pellow and Park 2002; Smith 2005). Recent research on settler colonialism takes this observation further. Settler colonialism is the occupation or control over the land, water, aerial space, and peoples of a given territory by an invading power. Indigenous studies scholars argue that settler colonialism itself is a form of environmental injustice because it undermines the ecological conditions necessary for Indigenous peoples to exercise their economic, political, and cultural practices (Whyte 2017: 165, see also Bacon 2018: 6). Moreover, scholars are now building a consensus that settler colonialism is a framework that undergirds all EJ conflicts in the United States (and other settler nations like Australia, New Zealand, Canada, etc.) because these dynamics always reflect entanglements with land and ecological wealth from which Indigenous peoples have been dispossessed (Voyles 2015; Hoover 2017).

Armed with a new lens for viewing environmental injustices, grassroots movements and scholars have worked to document, study, and combat the roots of this social problem. The EJ frame has been extended in numerous ways, and the concept of EJ has spread well beyond the borders of the United States to places as diverse as Australia, Canada, Germany, Hungary, India, Western Africa, South Africa, the former Soviet Union and Mexico (Walker 2009; Agyeman et al. 2010). While EJ studies formally originated in and focused on the United States, scholars are also documenting environmental inequalities around the globe (Pellow 2007; Roberts and Parks 2007; Schroeder et al. 2008). Despite a number of methodological debates concerning the most appropriate tools for documenting environmental inequalities (see Anderton et al. 1994; Been 1995; Mohai 1995; Saha and Mohai 2005), the vast majority of studies conclude that however one measures the phenomenon, communities of color and working class communities often face disproportionate exposure to environmental hazards (Bullard et al. 2007).

Causes of environmental injustice

As scholars and activists document environmental injustices, they seek to situate this problem within the larger context of capitalist production and inequality. Various theories have examined the roots of these dynamics. Here we group the causes of environmental justice into four categories: economic explanations, socio-political explanations, racial discrimination and racial capitalism. While we consider these concepts separately, in reality it is difficult, if not impossible, to disentangle their effects.

Economic explanations

A common explanation for environmental inequality is that hazardous firms do not intentionally discriminate but instead seek to maximize profits and thus place facilities where land is cheap and where there are available labor pools. Often, marginalized communities already live in these same areas. And, once a hazardous facility is present, those who lack the resources to move out remain living in the vicinity (Been 1994). Similarly, Schnaiberg and Gould (2000) use a model known as the treadmill of production to explain environmental injustice. According to this theory, environmental injustices are the byproducts of the routine function of capitalist states and economies. Within treadmill societies, corporations have an ever-growing need to generate goods for sale and make profit (see Chapter 4). This expansion creates wealth but also creates negative byproducts that are not evenly distributed and are disproportionately concentrated among the groups of people with the least ability to resist the location of polluting facilities in their community. Beck (1992, 1995, 1999) adds that modernization contributes to this pernicious cycle. A central aspect of modernity is the application of research to spur economic growth. Industries seek to be frontrunners of development and maximize profits, so they turn to new technologies even though they often do not understand the risks of those technologies. In turn, these risks disproportionately affect marginalized communities.

Socio-political explanations

Socio-political reasons may also explain environmental inequalities. For example, industries and corporations might seek the path of least resistance. They understand that affluent communities, which are often white, have the resources and social capital to oppose the placement of hazardous facilities in or near their neighborhoods and instead place hazards in locations where they will meet little or no local political resistance. Furthermore, communities that are already marginalized are often excluded from participation in policymaking and urban planning. By contrast, industries, corporations, and similar special interests are often highly involved in these processes (Cole and Foster 2001; see Chapter 4). In addition, marginalized communities are relatively invisible in mainstream environmental movements, which has resulted in insidious unforeseen consequences. For example, Andrew Szasz (1994) illustrated that the way in which the mainstream environmental movement negotiated anti-pollution laws led to the shift of certain industries and toxics into low income and minority communities. Similarly, Pellow (2007) has illustrated that, on a global scale, toxic industries and hazardous waste production were shifted to the global South in part due to regulations supported by the mainstream environmental movement (see also Frey 1998).

Racial discrimination

Many scholars have proposed that racism and institutional discrimination are responsible for environmental inequality. Of course, racial discrimination is also embedded in the socio-political and economic explanations. The evidence of racial divides in environmental policymaking is stark and persistent over time, so there is ample documentation of the effects of racism (Bullard 2000). Racial disparities are also mirrored in myriad other aspects of EJ-relevant US institutions, including education, health care, and criminal justice. Often, however, particular acts of racism and discrimination cannot easily be located and measured, as racism is not a specific *thing* whose effects can be neatly isolated or extracted from social life (Pulido 1996).

Racial capitalism

More recently, a number of scholars have embraced a viewpoint that is perhaps a hybrid of the above three explanations: racial capitalism is the term used to describe the ways in which the system of commerce known as capitalism is inescapably racialized. In other words, capitalism is a system that is characterized by class inequalities and by political systems wedded to market logics, but it is also a system that requires and thrives off of racial inequalities, which have provided it with the energy and resources to expand over the centuries (Robinson 1983). Racial capitalism is also a system in which both land and racialized bodies/populations are differentially valued, facilitating the exploitation and abuse of certain communities and territories for the benefit of dominant groups (Pulido 2017).

Intersections of inequality

While the majority of EJ research is devoted to the intersections between race and environmental harm, there are numerous additional social categories of difference that are of critical importance to developing a comprehensive grasp of environmental inequality. These include but are not limited to inequalities surrounding class, gender, sexuality, physical ability, citizenship, indigeneity, space and species (Ray 2013).

As discussed earlier, class inequalities are deeply pronounced within environmental injustices. Class inequality is actually quite overt because market economies publicly embrace the ideology of wealth accumulation and profit for those who are able to achieve these goals over those who cannot. According to this logic, those who remain at or near the bottom of the economic pecking order – and therefore are more likely to live and work in environmentally hazardous conditions – are there because they simply have not availed themselves of what is theirs for the taking. Political economic perspectives embodied in the work of sociologists like O'Connor, Faber, Foster, and Schnaiberg and Gould focus on the devastating effects of capitalism on socio-ecological dynamics. These studies utilize a Marxist viewpoint: when struggles over the means of production tend to favor the capitalist classes, they also produce greater ecological damage and mass social suffering (see Chapter 4). Relatedly, some social scientists have demonstrated that general measures of social and political inequality are correlated with and contribute to greater levels of ecological harm (Downey and Strife 2010). For example, James Boyce (1994, 2008) found societies exhibiting higher levels of economic and political inequality are characterized by comparatively higher overall ecological harm. This body of research is of great importance for linking inequality to ecological harm. Even so, much of it is rather narrowly focused on economic or political measures of inequality that fall short of capturing the complex ways in which inequality also functions across categories of difference.

Gender inequalities are also integrally embedded in environmental inequalities. Men tend to exercise control over states and corporations that produce environmental and economic inequalities, thus gaining the material and social benefits of both the financial and political power that results from and is reflected in environmental injustices. Furthermore, men exercise greater control over national labor and mainstream environmental organizations and enjoy the status and credit for valiantly representing the interests of "the people" in national discourses and campaigns (Seager 1994). Women tend to benefit the least from these struggles, as they are often physically and socially relegated to some of the most toxic residential and occupational spaces in communities and workplaces. In addition, women are less politically visible because they tend to work for smaller, environmental community-based organizations that rarely make headlines and survive on volunteer labor and small grants (Brown and Ferguson 1995; Pellow and Park 2002). Lastly, the very material landscapes being polluted and fought over in EJ struggles are deeply imbued with meanings that are gendered and contained in local and global imaginaries, state policies, corporate practices, and activist resistance campaigns (Adamson et al. 2002; Stein 2004). Several recent studies document the ways women experience and resist discriminatory environmental policies in workplaces, residential communities, and elsewhere (Pellow and Park 2002; Buckingham and Kulcur 2010).

Building on these insights, ecofeminist theory links ecological politics to gender, sexuality, race, class, species, and other social categories of difference, calling for an end to all forms of oppression because any effort to liberate a single oppressed population will only be successful if paired with a parallel attempt to liberate nonhuman natures (Gaard 1993, 2019; Warren 1994). Ecofeminism and EJ discourses and movements have much common ground, but surprisingly few scholars have explored this terrain (Smith 1997; Sturgeon 1997; Taylor 1997).

Citizenship, immigration, indigeneity, and nation also play significant roles in the production of environmental inequalities. Large-scale studies demonstrate that immigrants in the United States are more likely to live in residential communities with high levels of pollution than non-immigrant communities (Hunter 2000; Bullard et al. 2007). Smaller scale ethnographic studies reveal similar dynamics and demonstrate how ideologies of exclusion and nativism support the production and maintenance of such an unequal socio-ecological terrain (Pellow and Park 2002; Park and Pellow 2011). The role of colonial politics weighs heavily in the way that Indigenous peoples fare with regard to environmental outcomes. Specifically, in countries throughout the globe, Indigenous peoples are systematically excluded from participation in environmental decision-making, evicted from their lands, disproportionately exposed to pollution, and restricted from using ecological materials within their territories (Agyeman et al. 2010; Smith 2005; Simpson 2017; Estes 2019; Gilio-Whitaker 2019).

Climate change offers a powerful window into the problem of *global spatial* environmental inequality (see Chapter 32). Though they contribute less to the causes of climate disruption, nations of the global South, people of color, women, and Indigenous communities often bear the brunt of climate disruption in terms of ecological, economic, and health burdens – thereby giving rise to the concept of *climate injustice* (Roberts and Parks 2007). These communities are among the first to experience the effects of climate disruption, which can include "natural" disasters, rising levels of respiratory illness and infectious disease, heat-related morbidity and mortality, and large increases in energy costs. Flooding from severe storms, rising sea levels, and melting glaciers affect millions of people in Asia and Latin America, while sub-Saharan Africa is experiencing sustained droughts. The inequalities associated with climate change are stark. The poorest half of the world's population – 3.5 billion people – are responsible for only 10 percent of global carbon emissions, while the richest 10 percent is responsible for fully half of those emissions (Oxfam 2015). Furthermore, the richest 1 percent of the world's population has consumed twice the carbon as the poorest 50 percent over the last quarter of a century (Oxfam 2021).

Thus, the struggle for social justice is inseparable from any effort to combat climate disruption and the kinds of discourses and actions we witnessed emerging from the EJ movement have spilled over and expanded into a global climate justice movement (Estes 2019; Mendéz 2020). Perhaps because of the clear global and local impacts of anthropogenic climate change, the climate justice movement – which we view as an arm of the EJ movement – has succeeded in mobilizing youth, the elderly, faith-based organizations, Indigenous communities, and communities of color on a global scale in ways that the broader EJ movement has not always achieved (Klein 2014; Bhavnani, Foran, Kurian, and Munshi 2019).

While most EJ scholarship reveals the hardships and suffering associated with environmental inequality and environmental racism, few studies consider the flipside of that reality: *environmental privilege*. Park and Pellow (2011) argue that environmental privilege results from the exercise of economic, political, and cultural power that some groups enjoy, which enables them exclusive access to coveted environmental amenities. While marginalized people living in poor rural towns, in inner cities, and on reservations battle polluting industries and intransigent governments, those living in wealthy enclaves enjoy relatively cleaner air, land, and water and often believe they have earned the right to these privileges. Aspen, the Hamptons, Pebble Beach, and many other exclusive communities are examples of environmental privilege and deserve closer consideration as sites for understanding the roots of EJ struggles (Taylor 2009; Murphy 2016).

Responding to injustice

Scholarship has been a key response to environmental injustices. Scholars have offered definitions of EJ and documented the existence of environmental injustices. However, EJ scholarship has also influenced and been influenced by the broader EJ movements. Movement activists use academic studies to support their claims of injustice, and scholars have collaborated with activists on research and policy projects for decades. Movement activists and scholars jointly articulated the EJ frame as well. The emergence of the EJ frame redefined environmental issues as concerns extending beyond wildlife or wilderness preservation (Bullard 2000). Environmental issues became civil and human rights issues, and the EJ movement combined insights from many causes. As the movement grew, so did its mission, and soon it had developed a broader vision for change centered on the following points: (1) all people have the right to protection from environmental harm; (2) environmental threats should be eliminated before there are adverse human health consequences; (3) corporations and governments, not communities, should be accountable for proving that a given policy or industrial procedure is safe for people and the environment; and (4) grassroots organizations should challenge environmental inequality through political action (Pellow and Brulle 2005).

This vision has inspired diverse actions, ranging from grassroots efforts to United Nations-sponsored conferences. And, though it finds its roots in the United States, the vision has also spread around the globe. While it is difficult to trace the emergence of a global movement, some point to a two-month period in 1984 during which a chemical plant in India and a liquid propane gas plant in Mexico blew up, killing thousands and harming millions (Schroeder et al. 2008). A few years later, press reports of illicit dumping of North American and European toxic waste materials in the global South began surfacing (McKee 1996). Soon, the EJ movement spread "horizontally" to other countries as well as "vertically" to encompass concerns between countries, such as global waste transfers and climate change as discussed earlier (Walker 2009). Although activists around the world have long been fighting environmental injustices, their activism was later redefined through an EJ frame, which has reached places as diverse as South Africa, the United Kingdom, India, and Ecuador (Khagram 2004; Walker 2009). In each case, frames and ideas that originated in the United States are adjusted and recontextualized based on local circumstances, much as research on globalization has illustrated how that phenomenon is negotiated (Lowe and Lloyd 1997).

Definitions of justice also vary both locally and globally. When the call for EJ first rang out, movement activism and scholarship focused on distributive justice. In other words, both focused on issues of equity regarding the distribution of environmental injustices (Schlosberg and Carruthers 2010). However, many activists (and some scholars, see Schlosberg 2004) have also argued for a focus on procedural justice. Arising from the idea of participatory democracy, procedural justice shifts the lens from distributive outcomes to decision-making processes. Proponents of procedural justice maintain that a focus on mere distribution is incomplete and argue for a closer examination of group recognition (see Čapek 1993). This issue has particular salience in the global South, where colonial external powers and internal elites have denied citizens the opportunity to participate in decisions regarding environmental impacts that shape their lives. For example, Al Gedicks (2001) has documented how corporations and governments have threatened the land and culture of Indigenous peoples around the globe. Gedicks points to many examples, such as Nigeria, where oil operations have wreaked havoc on the lives of the Ogoni people, and West Papua, where the Amungme and Komoro peoples have been subjugated by mining companies.

Specific responses

Critical responses to the problem of environmental injustice have come from universities, corporations, governments, and grassroots and transnational movement activists. We review examples of these responses below, though it is important to note that this brief review is not comprehensive, and the actions highlighted are not independent of one another.

Grassroots organizing

Grassroots organizing, driven by small community groups, is at the heart of EJ activism. For generations, electricity for California's Central Coast region has been produced by polluting gas-fired power plants concentrated in Oxnard, a working-class community that is 85 percent people of color and 75 percent Latino. Oxnard already has three power plant smokestacks along its shoreline, more than any other city on the coast of California. In 2014, Oxnard faced a proposal for a 4th power plant that would again produce more greenhouse gases and impose particulate matter pollution on local residents, while making electricity for other cities, meaning that Oxnard would have shouldered the cost and received none of the benefits of this development. Many neighborhoods in Oxnard are above the 90th percentile of asthma rates in the state of California – they are literally gasping for air and choking. But local residents, grassroots organizations, and college students successfully mobilized to oppose this power plant and to call for an end to dangerous and polluting fossil fuel projects that threaten human and environmental health. The proposed plant was rejected by state authorities in 2018, a major victory for environmental and climate justice.

There have also been numerous grassroots responses to the relatively recent problem of transnational waste dumping and trading. For example, groups like the International Campaign for Responsible Technology and the Silicon Valley Toxics Coalition work to ensure that electronic wastes are not being exported from wealthier nations to poorer nations, a practice that has poisoned rivers, air quality, and dramatically impacted the health of residents and laborers in places like China, Ghana, Brazil and India (see Chapter 23). In recent years, the dumping of hazardous wastes – whether electronic or otherwise – has increasingly shifted from being a phenomenon that once flowed from the global North to nations of the global South to a problem being generated from within nations and regions. In other words, while the vast majority of hazardous waste may not be moving from wealthy to poor *nations*; what we find is that the waste is still flowing from wealthier *communities* and *regions* to communities within those nations and regions where low-income and ethnic minority populations are concentrated (Akese and Little 2018).

University responses

Numerous studies reviewed in this chapter are a result of university research centers that have institutionalized EJ studies. This is not surprising, as many EJ activists are scholars, and many EJ scholars maintain close relationships with activists. EJ issues are also becoming institutionalized in college curricula. Bunyan Bryant and Elaine Hockman noted that, in 2002, there were over 60 EJ courses being offered in the United States (see Pellow and Brulle 2005). This number has increased dramatically, and EJ courses are now offered in many other nations around the globe, while research centers and institutes for environmental justice studies have been founded at leading universities around the world. Overall, these scholars and educational institutions have contributed to the movement in a variety of ways, from co-defining the frame to documenting environmental injustices and teaching about them. Furthermore, scholars and research centers have hosted multiple conferences on EJ in order to debate, build consensus, and pressure policymakers to address these issues at the national and global scales.

Corporate responses

A core tenet of EJ holds that corporations and states, not communities, should be responsible for reducing and preventing negative environmental effects. The case of the Patagonia Clothing Company is instructive. Patagonia is widely known for its ability to manufacture lower ecological impact outdoor gear while also maintaining profitable revenue generation. Going further, the company has donated hundreds of millions of dollars to various social causes, including its "Environmental Justice is Racial Justice" campaign. The company's website contains a portal through which anyone can learn about and support community-based organizations fighting for EJ, including Communities for a Better Environment, the Louisiana Bucket Brigade, and various youth movements fighting against oil drilling, for example. The company also hosts a Tools for Grassroots Activists conference every other year in order to bring community leaders together to share and exchange tactics, strategies, and ideas for social change. Patagonia is the exception to the rule, but provides an inspiring example of how a corporation could support environmental and EJ movements (see Chapter 13).

Government responses

As seen with the US GAO report, governments are also key actors in the response to environmental injustice (in addition to frequently being the source of environmental injustice). They often create their own research commissions to study environmental inequalities, and they hold the power to pass regulatory legislation. For example, US Senator Al Gore and Congressman John Lewis introduced the Environmental Justice Act in 1992, which proposed mandatory studies of toxic health impacts of certain facilities. Though the Act never left the committee stage, the Environmental Protection Agency established the Office of Environmental Equity and the National Environment Justice Advisory Council shortly after the Act was introduced. In a major victory for US grassroots movements, the Biden-Harris Administration has made environmental and climate justice a centerpiece of its policy framework, pushing to integrate these goals into the work of all federal agencies and linking this approach to the cause of racial justice as well (Jacobs, King, and Northey 2021).

Governments are also key actors internationally. In fact, countries are the only entities that can be party to international treaties, several of which have explicitly recognized environmental inequality. Importantly, though they cannot sign or ratify treaties, many other actors, such as NGOs and affected communities, are involved in the creation of international treaties. One key convention – the 1992 Basel Convention on the Control of Transboundary Movements of Hazardous Wastes and Their Disposal – currently has 179 parties. Though it first only banned hazardous waste exports to Antarctica, a coalition of global South countries, some European nations, and Greenpeace worked to pass what has become known as the Basel Ban. This 1994 amendment banned hazardous waste exports from 29 wealthy countries of the Organization for Economic Cooperation and Development (OECD) to all non-OECD countries.

The 1992 United Nations Framework Convention on Climate Change also recognizes environmental inequality, though less explicitly. Its main objective is to stabilize greenhouse gas concentrations in the atmosphere. While the Convention did not set explicit goals for reductions in greenhouse gases, the 1997 Kyoto Protocol established legally binding reduction obligations for wealthy countries responsible for the overwhelming majority of carbon emitted into the atmosphere. In 2016, the Paris Agreement – an international climate change treaty – went into effect. This Paris Agreement aims to limit global temperature rise to well below 2 degrees Celsius above pre-industrial levels (see Chapter 32). In a significant reflection of the influence of social movements and engaged scholars, both the preamble and the language of the agreement itself contain specific language reflecting a commitment to climate justice, particularly for vulnerable communities in global South nations (Molesworth 2016).

The 2007 United Nations Declaration on the Rights of Indigenous Peoples (UNDRIP) is another example of government action. Unlike treaties, United Nations Declarations are not legally binding; however, the UNDRIP represents norms that are observed in international law (see Chapter 10). Importantly, the UNDRIP articulates that lands, territories, and resources that indigenous peoples have traditionally owned or occupied are rightfully their property and should be free from hazardous materials. Furthermore, it explicitly states that these territories and lands also must not be slated for "development" by external institutions without informed consent of the Indigenous occupants.

Transnational movement organizing

Many of the responses reviewed thus far include transnational elements. However, the growing prevalence and importance of transnational collaboration deserves further consideration, as numerous EJ transnational advocacy networks (TANs) have emerged (see Chapter 14). TANs comprise actors working internationally on an issue who are bound together by shared values, a common discourse and dense exchanges of information and services (Keck and Sikkink 1998). In this case, TANs have resulted in multiple alliances, conferences and transnational efforts. Several international organizations have been formed around specific issues related to EJ (see Chapter 29). The Basel Action Network (BAN), named after the Basel Convention, is an international NGO dedicated to preventing the dumping of toxic waste and promoting sustainable industrial practices through legislation and voluntary agreements. BAN promotes the Basel Convention and monitors its compliance. Similarly, GAIA (the Global Alliance for Incinerator Alternatives/Global Anti-Incinerator Alliance) is an alliance of over 800 organizations, NGOs, and individuals who work against incinerators and for safe alternatives. GAIA members participated in the International Zero Waste Cities Conference in 2021 as a means of promoting and amplifying the model of urbanization without environmental despoilation (GAIA 2021). Other groups have formed networks around certain causes, such as the International Campaign for Responsible Technology, a network that promotes government and corporate accountability in the global electronics industry, and the Pesticide Action Network, which works to replace hazardous pesticides with ecologically safe alternatives (see Chapters 8 and 21).

The work of these organizations, as well as the work of many grassroots campaigns and academics, is sometimes facilitated through transnational conferences and gatherings (see

Chapter 22), which often support the development and strengthening of international treaties that focus in part on EJ issues (such as the Basel Convention; see Chapter 36). In addition to urging policymakers to embrace EJ norms and principles, these gatherings provide critical opportunities for EJ activist networks to meet, exchange ideas, and build consensus around goals and action plans.

Conclusion

Environmental justice policy and politics are fundamentally social problems, rather than strictly environmental problems, and this framing is critical for advancing both scholarship and policy approaches. EJ studies are primarily concerned with the relationship among race, class, and socio-ecological harm – specifically, the way that marginalized populations are unevenly affected by industrialization and environmental policymaking. Scholars continue to debate the most effective methods for measuring environmental inequality and have offered numerous competing and complementary explanations for the causal roots of this problem. They also continue to expand the depth and breadth of their work by (1) extending the geographic and spatial scope of research beyond the United States to the rest of the globe and, in particular, the global South; (2) embracing interdisciplinarity through an expansion of scholarship beyond the social sciences into law, the humanities and arts, public health, the sciences, and other fields; and (3) including a broader and more complex set of categories of difference through which environmental injustices operate, such as gender, sexuality, nationality/citizenship, indigeneity, physical ability and species. Significant and lasting responses to environmental inequality have come from major stakeholders in grassroots organizations and frontline communities, academia, government, and the corporate sector, resulting in documentation of environmental inequality and institutional policy changes to address this concern at all geographic scales. The study of environmental justice is a thriving and growing field of inquiry, and we expect that trend to continue well into the future.

References

Adamson, J., Evans, M.M., and Stein, R. (eds) (2002) *The Environmental Justice Reader: Politics, Poetics, and Pedagogy.* Tucson: University of Arizona Press.
Agyeman, J., Cole, P., Haluza-DeLay, R., and O'Riley, P. (eds) (2010) *Speaking for Ourselves: Environmental Justice in Canada.* Seattle: University of Washington Press.
Akese, Grace and Little, Peter. 2018. "Electronic Waste and the Environmental Justice Challenge in Agbogbloshie." *Environmental Justice* 11(2): 77–83.
Anderton, D.L., Anderson, A.B., Oakes, J.M., and Fraser, M.R. (1994) "Environmental Equity: The Demographics of Dumping." *Demography* 31: 229–248.
Bacon, J.M. (2018) "Settler Colonialism as Eco-Social Structure and the Production of Colonial Ecological Violence." *Environmental Sociology,* 5: 1–11.
Beck, U. (1992) *Risk Society: Towards a New Modernity.* London: Sage.
Beck, U. (1995) *Ecological Enlightenment: Essays on the Politics of the Risk Society.* Amherst, NY: Humanity Books.
Beck, U. (1999) *World Risk Society.* Cambridge: Polity.
Been, V. (1994) "Locally Undesirable Land Uses in Minority Neighborhoods: Disproportionate Siting or Market Dynamics?" *Yale Law Journal 103*: 1383–1422.
Been, V. (1995) "Analyzing Evidence of Environmental Justice." *Journal of Land Use and Environmental Law 11*: 1–36.
Bhavnani, K., Foran, J., Kurian, P.A., and Munshi, D. (Eds.) (2019) *Climate Futures: Re-Imagining Global Climate Justice.* London: Zed Books.

Boyce, J.K. (1994) "Inequality as a Cause of Environmental Degradation." *Ecological Economics 11* (Dec.): 169–178.

Boyce, J.K. (2008) "Is Inequality Bad for the Environment?" *Research in Social Problems and Public Policy 15*: 267–288.

Brown, P. and Ferguson, F. (1995) "'Making a Big Stink': Women's Work, Women's Relationships, and Toxic Waste Activism." *Gender & Society 9*: 145–172.

Bryant, B. and Mohai, P. (eds) (1992) *Race and the Incidence of Environmental Hazards: A Time for Discourse*. Boulder, CO: Westview Press.

Buckingham, S. and Kulcur, R. (2010) "Gendered Geographies of Environmental Justice." *Spaces of Environmental Justice*. ed. by Holifield, R., Porter, M., and Walker, G. Chichester: Wiley-Blackwell.

Bullard, R.D. (1996) "Symposium: The Legacy of American Apartheid and Environmental Racism." *St. John's Journal of Legal Commentary 9*: 445–474.

Bullard, R.D. (2000) *Dumping in Dixie: Race, Class, and Environmental Quality*. 3rd edn. Boulder, CO: Westview Press.

Bullard, R., Mohai, P., Saha, R., and Wright, B. (2007) *Toxic Wastes and Race at Twenty, 1987–2007*. New York: United Church of Christ.

Čapek, S. (1993) "The 'Environmental Justice' Frame: A Conceptual Discussion and an Application." *Social Problems 40(1)*: 5–24.

Cole, L.W. and Foster, S.R. (2001) *From the Ground Up: Environmental Racism and the Rise of the Environmental Justice Movement*. New York: New York University Press.

Downey, L. and Strife, S. (2010) "Inequality, Democracy, and Environment." *Organization & Environment 23(2)*: 155–188.

DuBois, W.E.B. (1977 [1935]) *Black Reconstruction: An Essay Toward a History of the Part which Black Folk Played in the Attempt to Reconstruct Democracy in America, 1860–1880*. New York: Atheneum.

Estes, N. 2019. *Our History Is the Future: Standing Rock versus the Dakota Access Pipeline, and the Long Tradition of Indigenous Resistance*. New York: Verso.

Freeman, A.M. III. (1972) "The Distribution of Environmental Quality." Pp. 243–278 in *Environmental Quality Analysis: Theory and Method in the Social Sciences*. ed. by Kneese, A.V. and Bower, B.T. Baltimore, MD: Johns Hopkins University Press.

Frey, R.S. (1998) "The Export of Hazardous Industries to the Peripheral Zones of the World-System." *Journal of Developing Societies, 41*: 66–81.

Gaard, G. (1993) "Living Interconnections with Animals and Nature." In *Ecofeminism: Women, Animals, and Nature*. ed. by Gaard, G. Philadelphia, PA: Temple University Press.

Gaard, G. (2019) *Critical Ecofeminism*. Lexington Books: Lanham, Maryland.

Gedicks, A. (2001) *Resource Rebels: Native Challenges to Mining and Oil Corporations*. Cambridge, MA: South End Press.

Gilio-Whitaker, D. (2019) *As Long as Grass Grows: The Indigenous Fight for Environmental Justice, from Colonization to Standing Rock*. Boston: Beacon Press.

Global Alliance for Incinerator Alternatives (2021) "International Zero Waste Cities Conference 2021." https://www.no-burn.org/category/stories/news/

Goffman, E. (1974) *Frame Analysis: An Essay on the Organization of Experience*. Cambridge, MA: Harvard University Press.

Hoover, E. (2017) *The River is in Us: Fighting Toxics in a Mohawk Community*. Minneapolis: University of Minnesota Press.

Hunter, L. (2000) "The Spatial Association between US Immigrant Residential Concentration and Environmental Hazards." *International Migration Review 34(2)*: 460–488.

Jacobs, J., King, P., and Northey, H. (2021) "Biden Climate Plan: Environmental Justice 'Writ Large'." *E&E News*. January 28.

Keck, M.E. and Sikkink, K. (1998) *Activists Beyond Borders: Advocacy Networks in International Politics*. Ithaca, NY: Cornell University Press.

Khagram, S. (2004) *Dams and Development: Transnational Struggles for Water and Power*. Ithaca, NY: Cornell University Press.

Klein, N. 2014. *This Changes Everything: Capitalism vs. the Climate*. New York: Simon & Schuster.

Kruvant, W.J. (1975) "People, Energy, and Pollution." Pp. 125–167 in *The American Energy Consumer*. ed. by Newman, D.K. and Day, D. Cambridge, MA: Ballinger Publishing.

LaDuke, W. (1999) *All Our Relations: Native Struggles for Land and Life*. Cambridge, MA: South End Press.

Lowe, L. and Lloyd, D. (eds) (1997) *The Politics of Culture in the Shadow of Capital*. Durham, NC: Duke University Press.

McKee, D. (1996) "Some Reflections on the International Waste Trade and Emerging Nations." *International Journal of Social Economics* 23(4/5/6): 235–244.

Mendéz, M. (2020) *Climate Change from the Streets*. New Haven: Yale University Press.

Mohai, P. (1995) "The Demographics of Dumping Revisited: Examining the Impact of Alternate Methodologies in Environmental Justice Research." *Virginia Environmental Law Journal* 14: 615–652.

Molesworth, A. (2016) "Climate Justice and its Role in the Paris Agreement." *The Conversation*. April 21.

Murphy, M. (2016) "Mapping Environmental Privilege in Rhode Island." *Environmental Justice* 9(5): 159–165.

Oxfam. (2015) *Extreme Carbon Inequality*. December 2. Nairobi, Kenya: Oxfam International.

Oxfam. (2021) *The Inequality Virus*. January. Nairobi, Kenya: Oxfam International.

Park, L.S. and Pellow, D.N. (2011) *The Slums of Aspen: The War on Immigrants in America's Eden*. New York: New York University Press.

Pellow, D.N. (2007) *Resisting Global Toxics: Transnational Movements for Environmental Justice*. Cambridge, MA: MIT Press.

Pellow, D.N. and Brulle, R. J. (eds) (2005) *Power, Justice, and the Environment: A Critical Appraisal of the Environmental Justice Movement*. Cambridge, MA: MIT Press.

Pellow, D.N. and Park, L.S. (2002) *The Silicon Valley of Dreams: Environmental Injustice, Immigrant Workers, and the High-Tech Global Economy*. New York: New York University Press.

Perkins, T. (2021) "The Multiple People of Color Origins of the US Environmental Justice Movement." *Environmental Sociology* DOI: 10.1080/23251042.2020.1848502

Pulido, L. (1996) "A Critical Review of the Methodology of Environmental Racism Research." *Antipode* 28(2): 142–159.

Pulido, L. (2017) "Geographies of Race and Ethnicity II: Environmental Racism, Racial Capitalism, and State-Sanctioned Violence." *Progress in Human Geography,* 41(4): 524–533.

Ray, S. J. (2013) *The Ecological Other: Environmental Exclusion in American Culture*. Tucson: University of Arizona Press.

Roberts, J.T. and Parks, B. (2007) *A Climate of Injustice: Global Inequality, North–South Politics, and Climate Policy*. Cambridge, MA: MIT Press.

Robinson, C. (1983) *Black Marxism: The Making of the Black Radical Tradition*. Chapel Hill: University of North Carolina Press.

Saha, R. and Mohai, P. (2005) "Historical Context and Hazardous Waste Facility Siting: Understanding Temporal Patterns in Michigan." *Social Problems* 52: 618–48.

Schlosberg, D. (2004) "Reconceiving Environmental Justice: Global Movements and Political Theories." *Environmental Politics* 13(3): 517–540.

Schlosberg, D. (2007) *Defining Environmental Justice: Theories, Movements and Nature*. Oxford: Oxford University Press.

Schlosberg, D. and Carruthers, D. (2010) "Indigenous Struggles, Environmental Justice, and Community Capabilities." *Global Environmental Politics* 10(4): 12–35.

Schnaiberg, A. and Gould, K. (2000) *Environment and Society: An Enduring Conflict*. West Caldwell, NJ: Blackburn Press.

Schroeder, R., Martin, K., Wilson, B., and Sen, D. (2008) "Third World Environmental Justice." *Society & Natural Resources* 21(7): 547–55.

Seager, J. (1994) *Earth Follies: Coming to Feminist Terms with the Global Environmental Crisis*. London: Routledge.

Simpson, L.B. (2017) *As We Have Always Done: Indigenous Freedom Through Radical Resistance*. Minneapolis: University of Minnesota Press.

Smith, A. (1997) "Ecofeminism through an Anticolonial Framework." In *Ecofeminism: Women, Culture, Nature*. ed. by Warren, K.J. Bloomington: Indiana University Press.

Smith, A. (2005) *Conquest: Sexual Violence and American Indian Genocide*. Cambridge, MA: South End Press.

Stein, R. (ed.) (2004) *New Perspectives on Environmental Justice: Gender, Sexuality, and Activism*. New Brunswick, NJ: Rutgers University Press.

Sturgeon, N. (1997) *Ecofeminist Natures: Race, Gender, Feminist Theory, and Political Action*. London: Routledge.

Szasz, A. (1994) *Ecopopulism: Toxic Waste and the Movement for Environmental Justice.* Minneapolis: University of Minnesota Press.

Szasz, A. and Meuser, M. (1997) "Environmental Inequalities: Literature Review and Proposal for New Directions in Research and Theory." *Current Sociology 45(3)*: 99–120.

Taylor, D. (1997) "Women of Color, Environmental Justice, and Ecofeminism." In *Ecofeminism: Women, Culture, Nature.* ed. by Warren, K.J. Bloomington: Indiana University Press.

Taylor, D. (2009) *The Environment and the People in American Cities, 1600s–1900s.* Durham, NC: Duke University Press.

United Church of Christ (UCC). (1987) *Toxic Waste and Race in the United States.* New York: Commission for Racial Justice.

United States General Accounting Office (GAO). (1983) *Siting Hazardous Waste Landfills and their Correlation with Racial and Economic Status of Surrounding Communities.* Washington, DC: GAO.

Voyles, T. (2015) *Wastelanding: Legacies of Uranium Mining in Navajo Country.* Minneapolis: University of Minnesota Press.

Walker, Gordon. (2009) "Globalizing EJ: The Geography and Politics of Frame Contextualization and Evolution." *Global Social Policy 9(3)*: 355–392.

Warren, K.J. (ed.) (1994) *Ecological Feminism.* London: Routledge.

Whyte, K.P. (2017) "The Dakota Access Pipeline, Environmental Injustice, and U.S. Colonialism." *Red Ink*, Spring *19(1)*: 154–169.

Zupan, J.M. (1973) *The Distribution of Air Quality in the New York Region.* Baltimore, MD: Johns Hopkins University.

27

Environmental ethics

Philosophy, ecology and other species

Sofia Guedes Vaz and Olivia Bina

Pressure on the environment has increased in step with economic growth and the mass consumption that fueled unequally distributed benefits and wealth throughout the twentieth century (UNDP 2020; see Chapter 17). Both growth and ecological crises have attained a global reach, challenging our established notions of cause and effect, and our framing of problems and solutions. Accordingly, global environmental politics has witnessed major changes and significant "rescaling" in its "locus, agency and scope" (Andonova and Mitchell 2010: 257; see Chapter 2). Both dimensions of global environmental politics – politics and governance, and the ecological problems that are the subject matter of global environmental politics – are being reinterpreted due to increasing complexity, interconnectedness and interdependence. Accordingly, the range of actors and disciplines that inform global environmental politics and contribute to framing global environmental problems is widening, in an acknowledgment of inescapable pluralism (see the chapters in Part IV of this volume).

This chapter builds on this ontological and epistemological change in the nature of the problems studied in global environmental politics and of the worldviews through which environmental problems are perceived and analyzed. We focus on the (still) dominant Western frames while acknowledging a welcome rise of alternative voices – often captured by the expression of "indigenous and local knowledge" (Díaz et al. 2015) – which will hopefully enrich the depth and breadth of our pathways into the future (see, e.g., Kothari et al. 2019)

This chapter takes its cue from the recognition that the cumulative effects of human behaviors linked to dominant socio-economic systems are both cause and consequence of the complexity of environmental problems (Bina and Vaz 2011). From the now inescapable stage of the "Anthropocene" (Biermann and Lövbrand 2019; see Chapter 15), we explore the strengths, limits and recent developments in Western environmental philosophy and ethics, in informing and shaping global environmental politics. There has been a virtual absence of metaphysical questions in environmental politics, especially since the late 1970s when influential thinkers like Schumacher (1974) sought development models compatible with nature (for an overview of the "classics," see Vaz 2012). This absence helps explain why environmental problems have been framed primarily in scientific, technological and economic terms (see Chapters 18 and 19). If, on the one hand, scientific progress since the 1970s has led to more accurate and comprehensive understanding of the ecosphere, on the other

DOI: 10.4324/9781003008873-31

hand, it has impoverished the epistemology underpinning global environmental politics by avoiding engaging with metaphysics, thereby narrowing the way problems and solutions are identified, debated and implemented (for a reflection on the nature and implications of such impoverishment in society and economics, see Neiman 2009; Sandel 2012; Haraway 2015).

Global environmental politics and environmental ethics

It is the very nature and language of the subject matter of global environmental politics – "environmental problems" – which we wish to problematize in this chapter, suggesting that the *problem* is not so much *environmental* but rather the dominant understanding of the nature of the connection and dependence between humans and nature. By separating environment from its context and from all the causes and effects that interact with it, we reinforce a narrow perception of reality. Metaphysics, and in particular environmental philosophy and ethics, help us clarify the fundamental notions and theoretical principles by which we understand the world, the values that shape the relationship between humans and nature, and the dynamics of cause and effect. The exposure to ethical scrutiny of themes in global environmental politics, such as biodiversity (see Chapter 41), climate change (see Chapter 32) and genetically modified organisms (see Chapter 44), can be uncomfortable because it questions how our societies are evolving, what progress is for, and which values are structuring the relationship between humankind and the natural world (see, for example, the policy implications in IPBES 2019). But failure to do so condemns global environmental politics to narrowly defined problems, and to solutions that achieve little more than postponing an irreversible ecological crisis.

Environmental ethics and its internal debates and tensions can provide precious insights to global environmental politics. Put simply, environmental ethics seeks to determine what is the wrong or right action in relation to the environment and why; that is, it identifies the foundations that best describe and prescribe the moral relationship of human beings to the environment (see Pope and Lomborg 2005). Environmental ethics originates in the recognition that environmental issues, as framed in the West, need an ethical conceptual background. The 1960s and 1970s, with their social movements and public acknowledgment of emerging environmental questions and problems (Carson 1962; Meadows et al. 1972; Schumacher 1974), prompted a series of philosophical debates on environment and development. White (1967), Hardin (1968), Routley (1973) and Næss (1973) published cornerstone papers heralding a philosophical concern for the environmental crisis. The most important question was trying to understand the complexity and the deeper causes of the environmental crisis. The ethical conversation was the most lively and dynamic within environmental philosophy, giving rise to environmental ethics, which became an established discipline.

Environmental ethics can therefore contribute to disciplinary pluralism in global environmental politics by engaging with the philosophical landscape that underpins the meta-narratives that shape our ideas of the human connection and dependence on nature. There are at least three related reasons why this is important. First, global environmental politics aims to set norms, rules and structures to guide behavior with respect to the purpose of sustainable development, and there is a need to re-engage with the ethical dimension of sustainable development to "restructure...our relationship with the Earth and its creatures" (Kothari 1994: 228). Second, we need a radical reconceptualization of humanity's place in nature beyond ideas of duality and separation, as well as of human beings as the sole locus of value – a presumption that excludes all other living and nonliving beings and things. Third, global environmental politics sees human behavior as a major part of the problem, thus it is essential that we also turn to the philosophical landscape and the values that shape it. The following sections outline these meta-narratives, chart the evolution of Western environmental ethics, and link it to the political and policymaking dimension of global environmental politics.

Meta-narratives on the relationship between humankind and nature

Environmental ethics has been investing in identifying and understanding the values that have shaped the relationship humans have with nature, and the roots that determined different types of relationships, including connection and dependence. The way humans understand nature has practical implications. Depending on the value and rights attributed to nature, human actions toward it may or may not be legitimized. Whether humans feel connected and a part of nature, and whether they value this highly, determines how they plan, execute and judge their own ways of life. The humans–nature relationship is characterized by ideas of separation, power relations, domination and exploitation, and by notions of unity, respect, humility and caution. Investigation of different cultures, philosophies and religions helps us understand the meta-narratives of *separation* and *unity*, as we call them throughout this chapter (see Collingwood 1945; Marshall 1992; Pepper 1996; Jamieson 2001).

Most of the ideas and discussions in global environmental politics have, until recently, been framed largely through Western worldviews (the focus of this chapter), but this is only one side of the story, one that is rapidly changing. The major transformations in science and society that occurred during the sixteenth and seventeenth centuries marked the beginning of a new era in which the relationship between humans and nature changed, largely thanks to the shift "from Copernicus to Newton, from Renaissance natural magic to the mechanical worldview, and from the breakup of feudalism to the rise of mercantile capitalism and the nation-state" (Merchant 2006: 517; see Chapters 7 and 18). Galileo distinguished between what could be measured and what could not, establishing ways of knowing what was objective and pertaining to (early modern) science, and what was subjective and thus not pertaining to science (see Chapter 18). This planted the seed for the separation and dualism that came to dominate modern worldviews, interpreted as a rupture in the humans–nature relationship (see Pepper 1996; Merchant 2006).

Descartes reinforced Galileo's idea of the unreality of what is not measurable, and arguably what became known as Cartesian dualism between mind (*Res cogitans*) and matter (*Res extensa*) has marked humankind's relationship with nature to this day. The presumed superiority of the mind and of thought gave human a privileged position toward nature (Pepper 1996), justifying nature's use and eventually abuse by humans, thus failing to heed Schumacher's (1974: 89) warning that humankind "was given 'dominion', not the right to tyrannize, to ruin and exterminate. It is no use talking about the dignity of humans without accepting that noblesse oblige." By the eighteenth century the scientific revolution had all but displaced medieval cosmology. By challenging both medieval theology and science, it opened the way to modernity. This was when the idea of progress became identified with control, domination, manipulation and, thus, loss of respect for nature. Nature existed to serve humankind. Utilitarian and material objectives justified this relation, conceived through empiricist and rationalist perspectives based on assumptions of ontological reductionism. It became natural to think of nature as "something" that is there just for our benefit. We lost fear, then we lost respect, and in recent decades we lost the desire and capacity to connect with nature. Nevertheless, Hansson (2012: 2) notes that, "in our age of globalization and large-scale anthropogenic environmental degradation, the ecological limitations of reductionism are becoming increasingly apparent to both the academic and the global community." For these reasons, the discourse of global environmental politics would benefit from moving away from the vague, and possibly misleading, language of "environmental problems" to one that focuses on the connection and dependence between humans and nature that the narrative of separation has influenced so deeply (exemplified in Pope and Lomborg 2005).

Not everyone had lost the capacity to be fascinated by nature, and thus the narrative of separation was counterposed to one of unity, led by scientists and philosophers who sought and conceived of a positive relationship with nature, respecting, worshipping, loving and admiring it. Hansson explores the early contribution of philosopher Baruch Spinoza (1632–1677) who sought to counter the reductionism promoted by Descartes and Bacon, conceiving of nature as an entity that "subsumes our less inclusive modern-day conception of 'the environment'" (Hansson 2012: 4). Spinoza recognized the contextual interrelation of parts and wholes as key "to properly understand the functional organization of the world," effectively anticipating today's systems thinking (Hansson 2012: 4). Carolus Linnaeus (1707–1778), Friedrich von Humboldt (1769–1859), Charles Darwin (1809–1882) and Ernst Haeckel (1834–1919) are among the scientists who understood the importance of a unified and holistic perspective, one that viewed nature as complex systems, emphasizing the interdependence of all species. Thus the eighteenth and nineteenth centuries witnessed the laying down of modern ecology's foundations and of another view of nature that has yet to permeate Western theory and practice in global environmental politics (see Chapters 3 and 4).

We can therefore see two partially conflicting meta-narratives of separation and unity. In one, science provides an understanding of nature that exposes its holism, complexity and the interdependency and evolution of species (see Chapter 18), which prompts attitudes of respect and admiration. In the other it enhances the dualism between humans and nature as a consequence of the scientific revolution, prompting attitudes of domination and exploitation whose consequences (industrialization, capitalism, progress and technology) are object of analysis in global environmental politics (see Chapters 13, 18, 19 and 24). Environmental ethics was inspired by the first meta-narrative, which is addressed in the following section.

The rise of environmental ethics

Initially, the challenge of environmental ethics was to extend the realm of ethics to future people and to all living beings, ecosystems, nature. Lately, it has been concentrating on applied ethics, such as climate change ethics, sustainable consumption ethics, biodiversity ethics. We will start by presenting the historical debut of environmental ethics, evolving then to the new trends of environmental philosophers worried in dealing with the most pressing environmental questions and even new geographies.

It makes sense to start with Routley (1973), who was exploring the extension of ethics, by asking if we need a new type ethics? He developed the thought experiment of "the last man": "if the last dying man, who barely survived a collapse of the world system, eliminated every living thing, animal or plant – would that be right?" The struggle of environmental ethics to understand the underlying causes of environmental problems pointed to the anthropocentric tradition of the separation meta-narrative explored earlier, enhanced by the power of science and technology, and by an attitude of arrogance toward nature (Carson 1962; see Chapters 18 and 19). A new, non-anthropocentric, ethics was deemed necessary, one that would answer Routley's question negatively, not just for the hypothetical "last man," but also for humanity today. The rationale for a negative answer is that living things have value in themselves, independently of humans. This is why the thought experiment of the "last man" is so important:

if it is not right to destroy all living things even if there are no humans, it must be because living things have intrinsic value.

Early environmental ethics concentrated on attributing an intrinsic value to nature, above and beyond the instrumental one that had dominated the previous few centuries. To be able to extend ethics to other beings, intrinsic value of nature had to be the foundation for this new type of ethics. This led to very complex, sometimes cumbersome, discussions around what would be the value-conferring property uniting humans and nonhumans (De-Shalit 2000; Ball 2001; Light 2002). Different theories claimed different properties for nature, such as interests (Goodpaster 1978), sentience (Singer 1975) or just a good of its own (a *teloi*) that made it a teleological center of life (Taylor 1986). Environmental philosophers developing these ethical theories believed that the intrinsic value of nature would support a different approach to environmental political decision-making. Environmental ethicists viewed non-anthropocentric ethics as fundamental to a proper re-evaluation of the human–nature relationship and as the main added value for a different and wider view of the environmental crisis (Jamieson 2001). Anthropocentrism was therefore rejected as a possible frame for environmental ethics. As Light (2002: 429) put it, "regardless of the early debates over the terminology, the assumption that axiologically anthropocentric views are anti-ethical to the agenda of environmentalists, and to the development of environmental ethics, was largely assumed to be the natural starting point for any environmental ethics."

Discussions on different ways of grounding the intrinsic value of nature dominated environmental ethics for decades, giving rise to different currents, including animal liberation, deep ecologism, biocentrism, land ethics and ecofeminism. These currents evolved during the second half of the twentieth century and had different preoccupations. In addition to the broad theme of "beyond us," scholars sought to deconstruct the separation between humans and nature, between men and women (with whom nature is often identified), and between reason and emotion as artificially opposed ways of solving "environmental problems." They also complemented existing moral rules concerned with the place of individuals in society with a "land ethic," while some actually sought to move beyond moral rules.

Some of the most prominent representatives of these non-anthropocentric schools of thought include Peter Singer's (1975) *Animal Liberation*, which was a seminal work inspiring the movement of animal rights and liberation. There is no moral justification for the mistreatment of animals, as Singer believes in the principle of equal consideration of interests, not only for all human beings but also for nonhuman animals. Sentience, the capacity to suffer or to feel pleasure, which is shared by humans and animals, is used by Singer to justify the equal consideration of interests. This principle of equality also gives ground for Singer to reject and condemn speciesism (nonhuman species are not valued and have no rights). For Singer, it is speciesism that gives the ethical space and justification for causing pain to or killing of animals, disrespecting their existence.

A second non-anthropocentric current, espoused by John Baird Callicott, is land ethics, inspired by the writings of Aldo Leopold (1887–1948), namely, *A Sand County Almanac* (1981, see Callicott in Vaz 2012). This takes the reader through a sequence of concepts that became fundamental for environmental ethics: the extension of ethics; the concept of belonging to an interdependent community; an ecological consciousness that influences what we emphasize intellectually, our loyalties, affections and convictions; the conscience of what it means to use economic and utility arguments to justify the conservation of nature; and the concept of the land pyramid, which makes us understand "the land" not only as soil, but as a fountain of energy flowing through a circuit of soils, plants and animals. Leopold proposes that we should give value to land, not in an economic sense, but in a philosophical sense, anticipating the intrinsic value of nature. Callicott's work (1987, 1989, 1999) sought to develop a philosophical dimension to land ethics, demanded more from humans, than Leopold. He demands an ontological change of the *self*, constructing the thesis of the continuity between human beings and nature, as a whole, as a new being.

A third current in environmental ethics is deep ecology, initially proposed by Arne Næss (1973), who distinguished two different approaches to environment, the shallow ecology and the deep ecology movement the latter characterized by seven *normative* points that provide one unified framework for ecosophical systems. The deeper questioning of the environmental crisis led to a deeper questioning of the self, demanding an ontological effort to understand it. Næss (1973) proposed "ecosophy," believing it should be a broad concept, and later he developed the idea that "ecosophies" should be personal: each person should develop his/her own ecosophy, understood as a philosophy of life oriented to an ecological harmony (Næss 1987, 1989). Næss's own ecosophy is based on the notion of *self-realization*. The self-hood he proposes is based on an active identification with wider and wider circles of being. Self-realization is achieved when this circle of identification is the widest possible. It implies a transition from ego to social self to metaphysical self to ecological self. The upshot is that our self-interest becomes the interest of the rest of life. Næss believed it might also promote a more meaningful life. What makes deep ecology different is its emphasis in ontology, in a realization of a certain status of the self, expanding it as much as possible.

A fourth non-anthropocentric current is ecofeminism, which is divided into two categories: (1) accepting differences between men and women, but seeking to re-evaluate the female characteristics that are undervalued in Western/patriarchal societies; and (2) the idea that masculinity and femininity should both be rejected and we should develop an alternative culture. Dobson (1995) dubs this as "the difference" and the "deconstructive" models. The "difference" model is based on exploring and criticizing the dualisms of human/nature and men/women, basing the discussion on an essentialist argument for a feminine essence that should be universal and common to all women (Mathews 2017). Val Plumwood (1993, 2002) is the main promoter of the "deconstructive" model, believing that dualisms hinder true developments in ecofeminism. Both men and women should challenge the "dualised conception of human identity and develop an alternative culture which fully recognises *human* identity as continuous with, not alien from, nature" (Plumwood 1993: 36). Even though there are many discussions within ecofeminism, the important thing is that it promotes the idea that new ways of thinking in a nonpatriarchal context are needed, and this involves a reconceptualization of knowledge, reality and ethics. Both the value of connections between particular individuals and the value of nature or environment conceived as both material entities and abstractions need to be recognized (Davion 2001). Above all, this approach makes us rethink the relationship of the human being with him/herself and with the world.

These non-anthropocentric arguments are commonly gathered under the umbrella of "ecocentrism," a concept that captures their most relevant themes and promotes rethinking requiring that we proceed with greater caution and humility in our interventions in ecosystems (Eckersley 1992).

There are also anthropocentric strands of environmental ethics, and even if initially despised, they managed to impose themselves in environmental ethics landscape as they also provide support for radical reconsideration of the themes of connection and dependence between humans and nature. These currents are connected with social, political and moral questions, and are represented mainly by environmental virtue ethics and by environmental pragmatism.

Environmental virtue ethics considers the rising importance of wellbeing within development discourses, linking these to the role of virtue in character building, behavior and lifestyles and embracing a perspective of cultivating human character traits that enhance a healthy and harmonious relationship and interaction with nature. It also focuses on protecting future generations and the importance of virtue ethics language in policy responses to the ecological crisis. Van Wensveen (1999) notices that virtue language is present in one way or another in the work of almost all environmental philosophers. Sandler (2005: 7) adds: "virtue language is not only everywhere in the discourse, it is indispensable to the discourse." Hill (1983) realized that there are actions that are not immoral, yet raise some sort of discomfort. So, instead of the traditional question of what is the right or wrong action, Hill (1983) asks "What sort of person would do such a thing?"

Environmental virtue ethics emphasizes the need for thinking about character and behavior of people within environmental ethics, while traditional environmental ethics is more worried about the intrinsic value of nature. People have traits of character, attitudes, habits and dispositions, and it is people who make laws, promote policies and act toward nature (Sandler 2005). Therefore, it makes sense to identify the potential attitudes that constitute environmental virtues, and the role of character in environmental ethics. Furthermore, the rediscovery of the themes of wellbeing and happiness in economic, development and sustainability literature are leading to a growing concern with human flourishing, with what promotes it and what contributes to it (Jackson 2009; see Chapter 16). The idea that nature, living with nature and understanding it are sources of joy, peace, self-knowledge and a feeling of renewal leads one to acknowledge that promoting this openness and sensitivity to nature might be part of a process of one's own flourishing (Bina and Vaz 2011; Vaz 2012). Promoting lifestyles that enhance a balanced and harmonious relationship with nature has been a perennial objective of environmental ethics. Acknowledging the role of virtues to promote this type of lifestyle has been the specific added value of environmental virtue ethics.

Furthermore, as Van Wensveen (1999) observes, ecological virtue discourse, as a distinctive, diverse, dialectical, dynamic and visionary moral language, carries the promise of moral creativity. Such creativity is fundamental for the many problems and dilemmas that environmental ethics is confronted with. For example, questions of the rights of trees, animals or plants might be answered by looking through new moral lenses and by adopting different perspectives. As Van Wensveen (1999) argues, virtue language has pre-modern roots, which is an advantage given that modernity is considered partly responsible for the ecological crisis. We need a new moral language that is independent of such a worldview.

Environmental pragmatism (Light and Katz 1996) contributes with ideas aimed at bridging the gap between the world of ethics and of policymaking, partly appealing to the problem of future generations. Most environmental problems make it clear that future generations are vulnerable to how we develop our policies and therefore it is an inescapable theme for both environmental ethics and environmental policy. Light (2002: 443) argues that environmental ethicists should focus on how best to help the environmental community "to make better ethical arguments in support of the policies on which our views already largely converge." He contends that it is possible to keep the lively philosophical debates and yet be more politically proactive, developing a more public philosophy focused on arguments "that resonate with the moral intuitions that most people carry around with them on an everyday basis" (Light 2002: 444). Obligations to future generations are a powerful intuitive reason that most people easily understand and so might act as a platform of understanding between philosophy and politics.

New trends in environmental ethics

These discussions on the importance of the intrinsic value of nature, of future generations, of anthropocentric versus non-anthropocentric views of ethics have been the building blocks of environmental ethics, but many philosophers have also been exploring new frameworks for answering unavoidable and urgent environmental questions that rose at global, regional and local levels. Anthropocene burst (more or less controversially) into the conversation and even Callicot (2018), recognizing the need for an overhaul of environmental ethics, claimed that Anthropocentric environmental ethics should be anthropocenic *because the looming environmental crisis we face is existential*," adding that it is basically climate change ethics. In fact, even if issues such as biodiversity loss, restoration, sustainable consumption have been in the radar of many philosophers, it is mainly climate change that has been dominating environmental ethics thinking in the last decade. Gardiner´s (2011) description of climate change as "moral storm" hit an accurate key, opening the way for a plethora of papers and books.

These scholars look at climate change from several perspectives, such as that climate change is testing the limits of our current moral systems (Lowe 2019) or that a "complexity ethics" (Lyon 2018) is needed to cope with it, or that there is a need for innovative ethical framings (Palmer 2014) that avoid "post-political" framings (Wetts 2020). They have enriched the conversation and provided a much-needed questioning on our lives, lifestyles, social, political and economic systems. Climate change is a complex issue, in need of finding coherence between the diversity of moral agents (individuals, non-state and state), the diversity of actions at their disposal (mitigation, adaptation, loss and damage), and the diversity of moral settings for finding the best (morally right) responses, knowing that there is no Plan B to deal with it (Pierrehumbert 2019). Some authors have concentrated on supporting moral agents with consequentialism (Nordhaus 2008; Dietz and Asheim 2012), deontology (Milkoreit 2015) or virtue ethics (Sandler 2010; Knights 2019) to deal with climate change. Others have focused on their actions, namely, finding an agenda for ethics and justice in adaptation (Byskov et al. 2019) or in compensation (Jensen and Flanagan 2013) or loss and damage (Mace and Verheyen 2016; Mechler et al. 2019).

Climate change has been dominating much of the agenda (see Chapter 32), but the conceptual strand still kicks with philosophers such as Mathews (2018) and Callicott (2018) who are focusing on suggesting a path capable of reaching beyond the powerful yet narrow frames of rational thought. Plumwood (2002) already had a critical view of our inherited Kantian moral framework of distance from emotion and closeness to reason, criticizing those who ground the need for protecting nature on a rational, cognitive way. Plumwood resented that emotions and care one feels toward nature did not seem to be considered universal, or rational enough, to ground an extended moral theory.

Callicott (2018), turning to the past, namely to David Hume's moral sentiments, follows his intuition that the wellspring of ethics is not reason but feelings. According to Hume, these moral sentiments were informed by reason enabling them to be rightly oriented and engaged. It is this Humean marriage between reason and emotions that inspires Callicott to propose a holistic and affective moral philosophy uniting his new love of an anthropocentric focus on climate stabilization and ecosystem services with his old passion for non-anthropocentric land ethics together with love for self, kith, and kin, still prominent values in any ethical theory.

Mathews's (2018) paper, "We've had Forty Years of Environmental Ethics – and the World's Getting Worse," asks why so many philosophers have failed to influence events, concluded that it is clear that pure reason or argument alone does not mobilize change, neither do blueprints for an ecological society or not even science. What we need, Mathews believes, is social thinking about how value transitions occur and new worldviews arise. Mathews (2018) proposes for the future more than philosophy, more than religion, more than policy: she suggests a cosmology explaining that the future can be both scientific and "mythopoetic." Cosmology, derived from the Greek *kosmos*, means order and therefore is normative. Mathews (2018) says it implies that the physical universe does not merely hang together contingently but is self-conforming to some kind of inner principle of integrity or goodness. Mathews suggests an ecological cosmology based on Earth-based cosmologies of Aboriginal Australia organized around an immanent, normative axis of ecological Law.

This openness to other cultures is denting the dominance of a certain strand of Western thought, suggesting that much more is likely to be achieved through a respectful confrontation and engagement of a plurality of epistemologies and ontologies. We therefore end this section with two illustrations of shifting paradigms, starting with the emergence of African Environmental Ethics.

African environmental philosophy has a considerable number of authors dialoguing directly with diverse philosophical strands, but is mostly focusing on a critical reflection of how could African thought, African traditions and African reality contribute to the body of environmental philosophical and ethical knowledge. Tangwa (2004) alludes to the "live and let-live" attitude that is paradigmatic of African thought and lifestyle, and that justifies many philosophical African theories of a respectful and natural coexistence with nature. Ogungbemi (1997) proposes *"ethics of nature-relatedness,"* Tangwa (2004) investigates *"eco-bio-communitarianism,"* Behrens (2014) develops *"environmental relational theory,"* while Metz (2017) contends that certain traditional values justify animal rights, because all entities form a "chain of being" and those relationships are central to becoming a good person. Ifeakor (2019) based on the African interconnectedness of all beings, considers that African ontology is based on holism. Ibanga (2018) wrote a review paper discussing ten traditional and contemporary methodological paradigms of African environmental ethics showing its richness but the most important thing to have in mind is enriching the field with the synergies in both Western and African environmental ethics like Osuji (2018) did by acknowledging that Pope Francis encyclic *Laudate si* echoes traditional African environmental ethics, namely with the cosmic common good, cosmic harmony and respect for Earth.

Another paradigm is the "Conceptual Framework" of the Intergovernmental Science-Policy Platform on Biodiversity and Ecosystem Services (IPBES 2019), which is intended to contribute directly to global environmental politics by supporting policymakers and different stakeholders in their assessment of complex interactions between the natural world and human societies. One of the main goals of this framework is to strengthen plurality by bringing systematically together different knowledge systems, including Indigenous and Local Knowledge (ILK), and their representations of humans–nature relations, to bear on policy framings (Díaz et al. 2015). Thus, for instance, "Nature" includes both its meaning in Western science (concerning categories such as biodiversity, ecosystems, the biosphere and living natural resources), and in other systems – mainly of indigenous peoples from South American Andes – and their understanding of "Mother Earth" and systems of life. While still facing major challenges, this framework is a bold step in a much-needed direction (Pereira and Bina 2020).

Conclusion

In recent decades global environmental politics has embraced notions of complexity, interconnectedness, interdependencies and, although still tentatively, pluralism, both in terms of the actors in the realm of politics and in terms of disciplines and epistemology.

However, despite clear shortcomings of narrow disciplinary approaches, environmental philosophy and ethics remain marginal in most global environmental politics discourses and literature. Not by chance, the separation between "environmental problems," discussed here as "nature," and development issues, which this chapter treats as humankind and society, continues to be understood through the lenses of dichotomy and reductionism. There is still some way to go before we can discuss global environmental politics themes through a holistic and unified lens, as Baruch Spinoza challenged us to do in the seventeenth century, and as both non-dominant Western and other traditions continue to remind us.

Today's recognition that "ecologically more complex problems" (Andonova and Mitchell 2010: 270) are caused by the combination of various human behaviors requires a more holistic and systemic interpretation. The IPBES (2019) report's pointing to unprecedented losses in biodiversity stands as an emblematic illustration of what is at stake if we persist in viewing the world through narrow frames. Deep ecologists, ecofeminists, biocentrists, land ethicists, defenders of animal rights, environmental pragmatists and environmental-virtue ethicists have different ontological and epistemological perspectives on the environmental crisis. Such diversity is still largely untapped, and to this we must now add the vastly rich and diverse range of perspectives beyond the Western ones. The core preoccupation persists to this day: that the absence on metaphysical questions in environmental politics leads to narrow solutions within global environmental governance. To understand that there is a philosophical landscape behind the way we establish norms, rules, laws and structures that guide our behaviors helps us in the conversation about why we live on one planet as if we had two or three (see Chapters 10 and 17), why we ignore the question of limits, and why we are devoted to such a reductionist understanding of economics (see Chapter 24).

The environmental crisis is linked to the identity crisis of advanced "Western societies," how we relate to ourselves and others near or distant in time and space, and to nature. Environmental philosophy and environmental ethics, in particular, thus have an important role in guiding us to a better relationship between "the other" and ourselves. The currents of environmental ethics have been providing different perspectives aimed at understanding the root causes of the environmental crisis. Both anthropocentric and non-anthropocentric strands defend a need for a radical reconception of humanity's place in nature because there should be no reason to believe that humans are necessarily the most important beings and the sole locus of value in the world. This is an enormous challenge. Global environmental politics cannot overlook the metaphysical questions that are so intrinsic to the place that humanity has in the world.

Sustainable development is a problem-solving strategy shaping much of global environmental politics and related governance norms and structures. Different conceptions of sustainability (see Chapter 16) still reflect the two meta-narratives of separation and unity discussed above. Thus divided, they continue to undermine solutions in political and governance terms (see Pope and Lomborg 2005). The relevant question is which dimensions are constitutive of sustainability. We highlighted the potential of the ethical component of sustainability, in line with Kothari's (1994) appeal for a paradigm shift in sustainability policies, toward an ethical imperative and away from technical fixes (see Chapters 16 and 19). This entails discussing sustainability not only in normative terms but also in terms of purpose, thereby allowing the framing of environmental problems at a metaphysical level – as a set of moral arguments that can justify political action and institutional dynamics.

References

Andonova, L.B. and Mitchell, R.B. (2010). The Rescaling of Global Environmental Politics. *Annual Review of Environment and Resources* 35: 255–282.

Ball, T. (2001). New Ethics for Old? Or, How (Not) to Think about Future Generations. *Environmental Politics* 10(1): 89–110.

Behrens, K.G. (2014). Towards an African Relational Environmentalism. In Imafidon, E., Ayotunde, J. and Bewaji, I. (eds) *Ontologized Ethics: New Essays in African Meta-Ethics*. Lexington Books, pp 55–72

Biermann, F. and Lövbrand, E. (eds.) (2019). *Anthropocene Encounters: New Directions in Green Political Thinking*. Cambridge, UK: Cambridge University Press.

Bina, O. and Vaz, S.G. (2011). Humans, Environment and Economies: From Vicious Relationships to Virtuous responsibility. *Ecological Economics* 72: 170–178. DOI: 10.1016/j.ecolecon.2011.09.029.

Byskov, M.F., Hyams, K., Satyal, P., Anguelovski, I., Benjamin, L., Blackburn, S., Borie, M., Caney, S., Chu, E., Edwards, G., Fourie, K., Fraser, A., Heyward, C., Jeans, H., McQuistan, C., Paavola, J., Page, E., Pelling, M., Priest, S., Swiderska, K., Tarazona, M., Thornton, T., Twigg, J. and Venn, A. (2019). An Agenda for Ethics and Justice in Adaptation to Climate Change. *Journal of Climate and Development*, DOI: 10.1080/17565529.2019.1700774

Callicott, J.B. (ed.) (1987). *Companion to a Sand County Almanac: Interpretive and Critical Essays*. Madison: University of Wisconsin.

Callicott, J.B. (1989). *In Defense of the Land Ethic: Essays in Environmental Philosophy*. Albany: SUNY Press.

Callicott, J.B. (1999). *Beyond the Land Ethic: More Essays in Environmental Philosophy*. Albany: SUNY Press.

Callicott, J.B. (2018). Environmental Ethics in the Anthropocene. *Transtext(e)s Transcultures 跨文本跨文化 Journal of Cultural Studies* 13. DOI : https://doi.org/10.4000/transtexts.1064

Carson, R. (1962). *Silent Spring*. Harmondsworth: Penguin Books, 1971.

Collingwood, R.G. (1945). *The Idea of Nature*. Oxford: Oxford University Press, 1978.

Davion, V. (2001). Ecofeminism. In Jamieson, D. (ed.) *A Companion to Environmental Philosophy*. Malden, MA: Blackwell Publishing.

De-Shalit, A. (2000). *The Environment. Between Theory and Practice*. Oxford: Oxford University Press.

Díaz, S., Demissew, S., Carabias, J., Joly, C., Lonsdale, M., Ash, N., Larigauderie, A., Adhikari, J. R., Arico, S. and Báldi, A. (2015). The IPBES Conceptual Framework—Connecting Nature and People. *Current Opinion in Environmental Sustainability* 14: 1–16.

Dietz, S. and Asheim, G.B. (2012). Climate Policy Under Sustainable Discounted Utilitarianism. *Journal of Environmental Economics and Management* 63: 321–335 doi:10.1016/j.jeem.2012.01.003

Dobson, A. (1995). *Green Political Thought*. 2nd edition. London: Routledge. First edition published in 1990.

Eckersley, R. (1992). *Environmentalism and Political Theory. Toward an Ecocentric Approach*. London: UCL (University College London) Press.

Gardiner, S.M. (2011). *A Perfect Moral Storm: The Ethical Tragedy of Climate Change*. New York: Oxford University Press.

Goodpaster, K.E. (1978). On Being Morally Considerable. *Journal of Philosophy* 75: 308–325.

Hansson, D. (2012). Unpacking Spinoza: Sustainability Education Outside the Cartesian ox. *Journal of Sustainability Education* 3. Available HTTP: <http://susted.com/index.php?sURL=http://www.jsedimensions.org/wordpress/2012-the-geography-of-sustainabilty/>.

Haraway, D. (2015). Anthropocene, Capitalocene, Plantationocene, Chthulucene: Making Kin. *Environmental Humanities* 6: 159–165.

Hardin, G. (1968). The Tragedy of the Commons. *Science* 162: 1243–1248.

Hill, T. (1983). Ideals of Human Excellences and Preserving Natural Environments. Reprinted in Sandler, R. and Cafaro, P. (eds) *Environmental Virtue Ethics*. Boulder, CO: Rowman & Littlefield, 2005.

Ibanga, D-A. (2018). Concept, Principles and Research Methods of African Environmental Ethics. *Africology: The Journal of Pan African Studies* 11(7).

Ifeakor, C.S. (2019). An Investigation of Obligatory Anthropoholism as Plausible African Environmental Ethics. *International Journal of Environmental Pollution and Environmental Modelling* 2(3): 169–176

IPBES. (2019). *Global Assessment Report on Biodiversity and Ecosystem Services of the Intergovernmental Science-Policy Platform on Biodiversity and Ecosystem Services*. Ed. E. S. Brondizio, J. Settele, S. Díaz, and H. T. Ngo. IPBES Secretariat, Bonn, Germany.

Jackson, T. (2009). *Prosperity without Growth? The Transition to a Sustainable Economy.* Sustainable Development Commission (UK). Available HTTP: <http://www.sd-commission.org.uk/publications.php?id=914>.

Jamieson, D. (ed.) (2001). *A Companion to Environmental Philosophy.* Malden, MA: Blackwell Publishing.

Jensen, K.K. and Flanagan, T.B. (2013). Climate Change and Compensation. *Public Reason* 5(2): 21–32. Available at: https://curis.ku.dk/ws/files/154519015/Climate_change_and_compensation.pdf

Knights, P. (2019). Inconsequential Contributions to Global Environmental Problems: A Virtue Ethics Account. *Journal of Agricultural and Environmental Ethics* 32(4): 527–545. 10.1007/s10806-019-09796-x

Kothari, A., Salleh, A., Escobar, A., Demaria, F. and Acosta, A. (eds.) (2019). *Pluriverse: A Post-Development Dictionary.* New Delhi: Tulika Books.

Kothari, R. (1994). Environment, Technology, and Ethics. In Gruen, L. and Jamieson, D. (eds) *Reflecting on Nature. Readings in Environmental Philosophy.* Oxford: Oxford University Press.

Leopold, A. (1981). [1949]. *A Sand County Almanac. And Sketches Here and There.* Oxford: Oxford University Press.

Light, A. (2002). Contemporary Environmental Ethics. From Metaethics to Public Philosophy. *Metaphilosophy* 33(4), 426–449.

Light, A. and Katz, E. (1996). (eds) *Environmental Pragmatism.* London: Routledge.

Lowe, B. S. (2019). Ethics in the Anthropocene: Moral Responses to the Climate Crisis. *Journal of Agricultural and Environmental Ethics* 32: 479–485 https://doi.org/10.1007/s10806-019-09786-z

Lyon, C. (2018). Complexity Ethics and UNFCCC Practices for 1.5° C Climate Change. *Current Opinion in Environmental Sustainability* 31: 38–45. https://doi.org/10.1016/j.cosust.2017.12.008

Mace, M.J. and Verheyen, R. (2016). Loss, Damage and Responsibility after COP21: All Options Open for the Paris Agreement. *RECIEL* 25(2): 197–214 https://doi.org/10.1111/reel.12172

Marshall, P. (1992). *Nature's Web. An Exploration of Ecological Thinking.* London: Simon and Schuster.

Mathews, F. (2017). The Dilemma of Dualism. In MacGregor, S. (ed) *Routledge International Handbook on Gender and Environment.* New York: Routledge.

Mathews, F. (2018). We've had Forty Years of Environmental Ethics – and the World's Getting Worse. *ABC Religion and Ethics,* July 2018.

Meadows, D., Meadows, D., Randers, J. and Behrens, W. III. (1972). *The Limits to Growth: A Report for the Club of Rome's Project on the Predicament of Mankind.* New York: University Books.

Mechler, R., Bouwer, L.M., Schinko, T., Surminski, S. and Linnerooth-Bayer, J. (eds.). (2019). *Loss and Damage from Climate Change. Concepts, Methods and Policy Options.* Springer. https://doi.org/10.1007/978-3-319-72026-5_2

Merchant, C. (2006). The Scientific Revolution and the Death of Nature. *Isis* 97: 513–533.

Metz, T. (2017). How to Ground Animal Rights on African Values: A Constructive Approach. In Chimakonam, J.O. (ed). *African Philosophy and Environmental Conservation.* Earthscan from Routledge, pp. 30–41.

Milkoreit, M. (2015). Hot Deontology and Cold Consequentialism – An Empirical Exploration of Ethical Reasoning Among Climate Change Negotiators. *Climatic Change* 130: 397–409. https://doi.org/10.1007/s10584-014-1170-8

Næss, A. (1973). The Shallow and the Deep, Long-range Ecology Movement. A Summary. *Inquiry* 16: 95–100.

Næss, A. (1987). Self-Realization: An Ecological Approach to Being in the World. *The Trumpeter* 4(3): 35–42.

Næss, A. (1989). *Ecology, Community and Lifestyle.* Cambridge: Cambridge University Press.

Neiman, S. (2009). *Moral Clarity. A Guide for Grown-Up Idealists.* Revised Edition. Princeton, NJ: Princeton University Press.

Nordhaus, W.D. (2008). *A Question of Balance: Weighing the Options on Global Warming Policies.* New Haven: Yale University Press.

Ogungbemi, S. (1997). An African Perspective on the Environmental Crisis. In Pojman, Louis (ed) *Environmental Ethics: Readings in Theory and Application,* 2nd edition. Cengage Learning, Inc: Wadsworth, 330–337.

Osuji, P. (2018). Laudato Si and Traditional African Environmental Ethics in 2018. In Magill, G. and Potter, J. *Integral Ecology: Protecting Our Common Home.* Newcastle upon Tyne, UK: Cambridge Scholars Publishing.

Palmer, C. (2014). Contested Frameworks in Environmental Ethics. In Rozzi, R., Pickett, S., Palmer, C., Armesto, J., Callicott, J.B. (eds.) *Linking Ecology and Ethics for a Changing World: Values, Philosophy and Action.* Dordrecht: Springer, 191–206.

Pepper, D. (1996). *Modern Environmentalism. An Introduction*. New York: Routledge.

Pereira, L. and Bina, O. (2020). The IPBES Conceptual Framework: Enhancing the Space for Plurality of Knowledge Systems and Paradigms. In Pereira, J.C. and Saramago, A. (eds.) *Non-Human Nature in World Politics: Theory and Practice*. Springer Nature, Switzerland, 311–335.

Pierrehumbert, R. (2019). There Is No Plan B for Dealing with the Climate Crisis. *Bulletin of the Atomic Scientists* 75(5): 215–221 DOI: 10.1080/00963402.2019.1654255

Plumwood, V. (1993). *Feminism and the Mastery of Nature*. New York: Routledge.

Plumwood, V. (2002). *Environmental Culture. The Ecological Crisis of Reason*. New York: Routledge.

Pope, C. and Lomborg, B. (2005). Debate: The state of nature: *Foreign Policy*. Available HTTP: <http://www.foreignpolicy.com/articles/2005/07/01/debate_the_state_of_nature?page=full>.

Routley, R. (1973). Is there a Need for a New, an Environmental, Ethic? Reprinted in Zimmerman, M.E., Callicott, J.B., Warren, K.J., Klaver, I.J. and Clark, J. (eds) *Environmental Philosophy. From Animal Rights to Radical Ecology*. 4th edition, (2005). Upper Saddle River, NJ: Pearson Prentice Hall.

Sandel, M.J. (2012). *What Money Can't Buy: The Moral Limits of Markets*. New York: Farrar, Straus and Giroux.

Sandler, R. (2005). Introduction. In Sandler, R. and Cafaro, P. (eds) *Environmental Virtue Ethics*. Boulder, CO: Rowman & Littlefield.

Sandler, R. (2010). Ethical Theory and the Problem of Inconsequentialism: Why Environmental Ethicists Should be Virtue-Oriented Ethicists. *Journal of Agricultural and Environmental Ethics* 23(1): 167–183, DOI: 10.1007/s10806-009-9203-4

Schumacher, E.F. (1974). *Small is Beautiful – A Study of Economics as if People Mattered*. 2nd edition. London: Abacus.

Singer, P. (1975). *Animal Liberation: A New Ethics for our Treatment of Animals*. New York: Avon Books.

Tangwa, G. (2004). Some African Reflections on Biomedical and Environmental Ethics. In Kwasi, W. (ed.) *A Companion to African philosophy*. Oxford: Blackwell Publishers.

Taylor, P.W. (1986). *Respect for Nature: A Theory of Environmental Ethics*. Princeton, NJ: Princeton University Press.

UNDP. (2020). Human Development Report 2020. The Next Frontier: Human Development and the Anthropocene, United Nations Development Programme, New York.

Van Wensveen, L. (1999).The Emergence of Ecological Virtue Language. Reprinted in Sandler, R. and Cafaro, P. (eds) *Environmental Virtue Ethics*. Boulder, CO: Rowman & Littlefield, 2005.

Vaz, S.G. (ed.) (2012). *Environment: Why Read the Classics?* Sheffield: Greenleaf Publishing.

Wetts, R. (2020). Models and Morals: Elite-Oriented and Value-Neutral Discourse Dominates American Organizations' Framings of Climate Change. *Social Forces* 98 (3): 1339–1369, https://doi.org/10.1093/sf/soz027

White, L., Jr. (1967). The Historical Roots of Our Ecological Crisis. *Science* 155(3767): 1203–1207.

28

Participation

Public opinion and environmental action

Sandra T. Marquart-Pyatt

Democracy requires citizens' opinions in the political process and comprises mechanisms through which these can be realized, including participation. According to decades of public opinion research, citizens' concerns about environmental issues, conditions, and topics are wide-ranging and varied, and connect in important ways with political structures. Although recent public opinion polls reveal these environmental concerns are extensive, questions remain about their distribution globally, and how they relate to environmental attitudes and behavior. In this chapter, I provide a review the cross-national literature on public opinion on environmental issues and concerns in recent decades with emphasis on recent work. I then review research on participation with application to the environment, broadly construed, cross-nationally. Throughout the chapter, the goal is to characterize relations among pluralism, participation, and public opinion and how they intersect with environmental issues.

An overview of cross-national public opinion research

Scholars have tracked public opinion on numerous environmental issues and concerns for more than five decades. This scholarship chronicles multiple dimensions of environmental concern – ranging from beliefs about interconnections between the natural environment and humans, trade-offs between economic growth and environmental protection, willingness to pay higher prices for the environment or to give time for environmental causes, including personal involvement in actions like recycling, and engaging in pro-environmental activism, such as signing petitions or being a member of an environmental group. A long-standing definition shows this expansive view, where Dunlap and Jones define environmental concern as "the degree to which people are aware of problems regarding the environment and support efforts to solve them and/or willingness to contribute personally to their solution" (2002: 485).

During the 1990s, cross-national scholarship focused on the conventional wisdom that environmental concern should be present only in wealthy or advanced industrial countries where citizens had their basic economic and material security needs met (for extended discussion, see Dunlap and York 2008). Rather than this proposed geographic concentration among citizens in industrialized nations, however, research pointed to its wide dispersion

DOI: 10.4324/9781003008873-32

globally (Brechin and Kempton 1994; Dunlap ct al. 1993). Ensuing studies examining this observation can be divided into two main research threads. On the one hand, some early research posits that there is a tendency for citizens in wealthier or industrialized countries to express greater degrees of environmental concern based on national material conditions or personally experiencing a baseline of material security (Diekmann and Franzen 1999; Franzen 2003; Inglehart 1995; Kidd and Lee 1997). On the other, a body of scholarship posits the globalization of environmental concern (Brechin 1999; Dunlap et al. 1993; Dunlap and Mertig 1995, 1997; Dunlap and York 2008).

These research streams unite in describing the global character of environmental concern yet differ in explaining it. Numerous intersecting arguments feature prominently in the literature. In accordance with the conventional wisdom, national affluence or prosperity is advanced as a primary influence (Franzen 2003; Inglehart 1995). Other scholars emphasize a measurement framework articulating the multifaceted, multidimensional features of environmental concern to account for its worldwide reach (Brechin 1999; Diekmann and Franzen 1999; Dunlap and York 2008; Marquart-Pyatt 2007, 2008; Xiao and Dunlap 2007). The objective problems–subjective values thesis seeks to explain that environmental concerns may indeed be global, yet they are driven by different factors rooted in contexts (Brechin 1999; Inglehart 1995). Recent research examining environmental topics and concerns across countries describes attitudes on the one hand (Fairbrother 2013; Franzen and Meyer 2010; Gelissen 2007; Franzen and Vogl 2013; Haller and Hadler 2008; Jorgeson and Givens 2014; Marquart-Pyatt 2007, 2008, 2012a; Smith et al. 2017) and behaviors on the other (Doyle 2018; Hadler 2013; Hadler and Haller 2013; Hadler and Haller 2011; Marquart-Pyatt 2012b). Within attitudes, work is moving in new directions to encompass trust (Davidovic and Harring 2021; Fairbrother 2016, 2017; Marquart-Pyatt 2018), political polarization (Birch 2020) and policy preferences (Fairbrother et al. 2019; Harring 2013, 2014; Harring et al. 2019). Questions of how attitudes and actions relate across nations remain germane, particularly (for this chapter) regarding how pluralism, participation and public opinion on environmental issues intersect.

Participation and public opinion

Citizen participation is vital in democracies for the political process, including the representativeness of political institutions and political equality (Verba 1996; Verba et al. 1971). Participation consists of a variety of actions, such as voting, donating time or money to a campaign, signing petitions, or engaging in protest activity that can be grouped into conventional and unconventional forms of participation. For instance, conventional forms of participation include voting, campaign activities, and donating time or money to a political party or candidate. Unconventional forms of participation include actions like signing petitions, engaging in protests, demonstrations, boycotts and participating in strikes (Barnes and Kaase 1979). Recent declines have been observed in conventional forms of participation while, simultaneously, unconventional forms have become more common (Dalton 1996). These have important corollaries in the realm of environmental politics (see Chapter 12).

Public environmental behaviors are an important feature of environmental concern; they are believed to demonstrate an environmental commitment to the environment rooted in institutional structures. These environmental actions involve the realization of individual beliefs, attitudes and actions as applied in formal channels like democratic political structures. There are many dimensions of environmental behavior, including recycling, water and energy conservation, signing petitions, protesting and being a member of an environmental group. Following Stern's (2000) definition of environmental activism, organized participation in environmental issues demonstrates that it is environmentally significant behavior rooted in the political realm and also shows how this differs from routine or everyday behaviors like recycling or conservation, the latter of which are individual environmental actions. That is, environmental activism tends to be expressed in specific activities that are channeled in formal settings and realized through institutional structures like political regimes (Stern 2000). It includes multiple behaviors in the public sphere, such as signing petitions, engaging in protest, and participating in social movements, and comprises one dimension of the broader construct of environmental concern (Dunlap and Jones 2002).

Theoretical frameworks for environmental action

From previous research, two prominent explanatory frameworks can be identified that seek to explain the determinants of environmental actions: the theory of planned behavior (TPB) (Ajzen 1991, Fishbein and Ajzen 2010) and the value–belief–norm (VBN) theory of environmentalism (Stern 2000). Both theories specify values, beliefs, and attitudes as important antecedents to behaviors, and share conceptual frameworks in which they articulate pathways through which individual attributes and attitudes work to affect behaviors. Briefly, the TPB articulates behaviors to be a function of individual beliefs, attitudes and behavioral intentions; VBN theory similarly lays out a causal sequencing of environmental values, beliefs and personal norms as key factors affecting environmental actions. Both approaches also specify key filtering mechanisms: behavioral intention and personal norms, respectively.

The TPB (Ajzen 1991) proposes that individuals are rational actors whose behaviors can be best explained using a path model (Ajzen 1991; Armitage and Connor 2001; Bamberg and Moser 2007; Schwenk and Moser 2009). In the TPB, attitudes related to engaging in particular behaviors, perceptions of others regarding the behaviors (i.e., subjective norms), and perceived behavioral control (i.e., efficacy), or how difficult a particular action may be, affect behavioral intentions, which then influence behaviors (Ajzen 1991; Fishbein and Ajzen 2010). Efficacy is important, as it makes the realization of a behavior possible to perform given an individual's perception of potential barriers in his/her surrounding context. The TPB stresses individual self-interest, outlining a series of cost–benefit calculations in which individuals weigh an array of personal and social normative forces, which then work through behavioral intention and result in a particular behavior.

A subset of TPB applications examines environmentally specific actions. Research has explored recycling (Cheung et al. 1999), energy and water conservation (Bamberg 2003; Harland et al. 1999), reduced car driving (Bamberg and Schmidt 2003; Harland et al. 1999), support for natural resource policy (Routhe et al. 2005), and environmental behaviors (Oreg and Katz-Gerro 2006). Extant work demonstrates attitudes and beliefs influencing behavioral intentions related to recycling, conservation, and natural resource policy (Cheung et al. 1999; Harland et al. 1999; Routhe et al. 2005), behavioral intentions influencing conservation and behaviors (Bamberg 2003; Cheung et al. 1999; Oreg and Katz-Gerro 2006), and efficacy influencing environmentally friendly behavioral intentions (Cheung et al. 1999; Oreg and Katz-Gerro 2006) and some behaviors (Harland et al. 1999). Recent meta-analyses confirm the model across environmentally responsible behaviors including general ecological behavior, public transportation use (see Chapter 31), and recycling. In these reviews, behavioral intention mediates the influences of other predictors in the model, and attitudes, behavioral control, and personal moral norms are key variables affecting behavioral intention (Bamberg and Moser 2007; Schwenk and Moser 2009).

VBN theory posits that the antecedents of environmental behaviors are values, world-views, beliefs and norms (Stern 2000; Stern et al. 1995, 1999; Stern and Dietz 1994). VBN theory proposes cognitive, attitudinal and social factors promoting environmental actions. Briefly, environmental behaviors result from a chain of influences including personal environmental values, beliefs including an ecological worldview, awareness of consequences, assuming personal responsibility, and personal norms linked with environmentally significant behaviors, respectively (Stern 2000). Values and moral norms are central to explaining environmentally relevant behaviors.

Applications of VBN demonstrate effects of norms and values on environmental behaviors. Research demonstrates that values influence individuals' recycling (Guagnano et al. 1995; Milfont et al. 2006; Schultz 2001; Schultz et al. 2005), household conservation behaviors (Black et al. 1985), and some environmental political actions (Stern et al. 1995, 1999). Whereas Steg et al. (2005) show support for VBN theory's causal chain for explaining acceptability of energy policies, Kaiser et al. (2005) demonstrate its significance for conservation behavior. Other research demonstrates personal norms affecting environmental behaviors, with values and attitudes having indirect influences through norms (Nordlund and Garvill 2002, 2003). Support for the role of values and social norms on environmentally significant behaviors is generally confirmed in a meta-analysis (Bamberg and Moser 2007).

Stern (2000) outlines a conceptual framework that posits different sources of environmentally significant behaviors depending on their type, thus anticipating variability in the effects of indicators on particular actions. In other words, the determinants of consumer behavior, environmental citizenship and policy support are likely different, which was confirmed in previous research on environmental citizenship behaviors compared with other actions (Dietz et al. 1998; on environmental citizenship, see Chapter 27). Environmental activism, as one facet of environmental actions or environmentally significant behavior, is influenced by a distinctive set of predictors. Thus, according to Stern (2000), substantial variation is anticipated based on situations, individuals and activities for factors affecting environmental actions.

Prior research reveals differences in the performance of these approaches that link with the type of environmental behavior being studied. For instance, studies suggest that attitudes matter for some behaviors but not others, depending on the degree of effort involved in their realization (Bagozzi et al. 1990) and whether they are routine activities (Schultz and Oskamp 1996). Madden et al. (1992) argue that perceived behavioral control is critical for behaviors in which it is difficult to engage. Research examining VBN theory demonstrates different sets of influential factors depending on the type of environmental behavior considered, where factors affecting contributing to environmental organizations or signing petitions differ from those of consumer behavior and policy support (Dietz et al. 1998; Stern et al. 1999).

Although social–psychological models reveal important pathways to behaviors, they do not account for the role of individual-level characteristics in the same way that other frameworks do. In this regard, it is important to integrate expectations from the literature on political participation. Research on political behavior has consistently shown that individuals of privileged statuses participate more in political activities. While conventional forms of participation (i.e., voting and campaign activity) are essential to democratic regimes, they have waned in popularity in recent years even in countries with established democratic histories. At the same time, citizen activism in unconventional forms of participation, including protests and demonstrations, has increased (Dalton 1996). This activism is essential for the entrenchment of democratic principles in the citizenry. The relationship between privileged status and participation is also important. This model has been extensively studied in the

United States (Brady et al. 1995; Verba et al. 1995). This scholarship demonstrates that privileged status, such as possessing greater amounts of time, money, and skills, as reflected in education and income, influence political participation (Verba et al. 1995). Resource-based explanations articulate how socio-economic status or individual attributes like education and income influence certain behaviors. This relationship is a powerful predictor of political behavior across many decades of published research (Barnes and Kaase 1979; Leighley 1995; Verba et al. 1971, 1995). Socio-demographics and education illustrate individuals' surrounding socio-economic context, which can serve as a constraint on the realization of actions apart from efficaciousness and/or values.

Institutional contexts provide a final piece, providing a structure that might facilitate or constrain individual political actions. Countries around the globe differ on many structural and institutional dimensions including economic, political, and environmental features and contexts, each of which might affect public opinion on a range of issues, including environmental ones (see Chapter 12). Political structures, including democratic governments and international environmental organizational memberships, illustrate institutional structures supportive of an international system of environmental organizations, actors, and treaties that showcase interdependencies among human societies and surrounding natural environments (Frank 1997; Frank et al. 2000; Schofer and Hironaka 2005). For instance, one relation commonly assumed is that citizens of nations with political features like liberal democracy have greater degrees of environmental concern attitudinally and behaviorally. An emerging line of scholarship offers mixed, yet promising, support for these purported relations (Hadler and Haller 2011; Haller and Hadler 2008).

Cross-national research on environmental action

Environmental behaviors are often divided into two domains: public and private environmental action (Hadler and Haller 2011; Hunter et al. 2004). This is an important frame, since signing petitions and participating in demonstrations differ from routine, everyday behaviors like recycling that comprise individual environmental actions or private environmental behavior. Recent cross-national research demonstrates that environmental actions vary widely across nations, even among nations sharing a political regime, and that there is not always a clear correspondence between environmental attitudes and environmentally significant behaviors.

As noted earlier, some recent cross-national research examining environmental topics and concerns describes environmental activism. These studies demonstrate some support for social–psychological models like the TPB and VBN theories (Hadler and Haller 2011; Marquart-Pyatt 2012a; Olofsson and Ohman 2006; Oreg and Katz-Gerro 2006) and mixed support for socio-demographic or positional factors (Freymeyer and Johnson 2010; Hunter et al. 2004; Marquart-Pyatt 2012a, 2012b). More specifically, environmental attitudes, willingness to sacrifice for the environment, and perceived behavioral control (Oreg and Katz-Gerro 2006) affect environmental political behaviors. For instance, Olofsson and Ohman (2006) reveal stable effects for education and general beliefs, yet they also uncover differences in models predicting environmental concerns among four affluent countries. Comparing within regions, Scandinavian countries (Norway and Sweden) were largely similar with one another, and results for the United States and Canada were more ambiguous – some similarities regarding attitudes and environmental political behavior were uncovered but not regarding willingness to make financial sacrifices (Olofsson and Ohman 2006). Hunter et al. (2004) demonstrate that women and men are more likely to engage in private environmental behaviors compared with public ones with variability cross-nationally. Marquart-Pyatt (2012a) revealed that individual resources, awareness of consequences, and attitudes combined to affect environmental activism.

Results further demonstrate that willingness to contribute affects activism across nations directly, yet it also has an important role in shaping how education and efficacy influence activism. Further, studies reveal some support for contextual explanations (Freymeyer and Johnson 2010; Hadler and Haller 2011). Freymeyer and Johnson (2010) investigate how national contexts affect environmental political engagement, showing that national and individual economic wellbeing affects environmental actions. Distinguishing between public and private actions, Hadler and Haller (2011) demonstrate that political opportunity structures and resources drive public behaviors similarly across countries, while the determinants of public behaviors differ and are rooted in local or national contexts rather than necessarily driven by global processes. These studies differ in conceptual frameworks, empirical models, analytical techniques, and number of countries explored. For instance, studies have explored a modified version of the TPB (Oreg and Katz-Gerro 2006), beliefs as key influences (Olofsson and Ohman 2006), an integrated path model (Marquart-Pyatt 2012a), gender differences (Hunter et al. 2004), and national and global dimensions of environmental behaviors (Hadler and Haller 2011).

Recent studies investigating empirical models of activism reveal consistent positive effects for environmental values, attitudes, behavioral intentions, and education, and mixed effects for knowledge and socio-demographics (Dalton 2015; Hadler 2013; Hadler and Haller 2011; Hadler and Wohlkönig 2012; Marquart-Pyatt 2012a; Pisano and Lubell 2015). These studies reveal that national contextual factors like political opportunity structures (e.g., democratization, environmental protection, Green party presence), measures of economic and educational development, and environmental degradation affect public environmental behavior (Hadler 2016; Hadler and Haller 2011; Pisano and Lubell 2015).

Characterizing the multifaceted relations among pluralism, participation and public opinion, and how they intersect with environmental issues, requires brief examination of individual-level public opinion data on environmental and political participation. Data in this section are from the *International Social Survey Program (ISSP) 1993, 2000, 2010: Environment* (International Social Survey Program 2003, 2012) and the *World Values Survey* (WVS) (Haerpfer et al. 2020). The ISSP Environment module has more than 60 questions on environmental values, beliefs, attitudes, actions, issues and topics, and concerns. Although limited regarding country coverage and comprised largely of industrialized nations, since 1993 the ISSP has expanded from 22 to more than four dozen nations. The WVS is a worldwide survey that has conducted national surveys in more than 100 countries containing individual-level information on political participation in a range of countries. The WVS is viewed as the largest source of information on human beliefs and values across a wide range of social, political, and economic topics, particularly regarding how it can be used to study changes in these views. Across both surveys, sample sizes for individual countries range from a few hundred to approximately 1,000 respondents.

Table 28.1 shows data on environmental activism for 22, 27 and 34 countries, respectively, for the 1993, 2000 and 2010 waves of the ISSP Environment public opinion survey (ISSP 1996, 2003, 2012). It presents the percentage of respondents in each country who had signed a petition related to an environmental issue, given money to an environmental organization, joined environmental groups, and engaged in an environmental protest in 1993, 2000 and 2010. The sample composition of countries included in the three waves of data differs, with 16 countries having data for all three waves. Data are shown here to

Table 28.1 Percent of individuals participating in environmental actions, by country

	Petition			Money			Group			Protest		
	1993	*2000*	*2010*	*1993*	*2000*	*2010*	*1993*	*2000*	*2010*	*1993*	*2000*	*2010*
Australia	44	..	28	41	..	23	10	..	7	5	..	5
Austria	..	30	23	..	28	21	..	8	8	..	..	3
Belgium	..	..	22	..	..	13	..	..	8	..	..	4
Britain/UK*	37	30	21	30	24	16	5	6	5	3	3	2
Canada*	44	26	28	41	23	22	7	7	5	7	4	8
Denmark	..	17	17	..	22	20	..	11	10	..	3	3
Finland	..	22	15	..	24	27	..	5	5	..	1	2
France	..	..	27	..	..	11	..	..	6	..	..	**10**
Germany	30	31	21	14	17	13	4	5	5	9	5	4
Iceland	..	..	21	..	..	13	..	..	7	..	..	6
Israel*	15	19	13	8	12	9	6	6	7	5	8	5
Italy	24	..	..	14	..	..	5	..	..	7	..	..
Japan*	25	22	11	11	9	6	2	2	2	2	2	1
Netherlands*	23	22	20	44	**45**	**30**	17	16	12	5	1	2
New Zealand*	**55**	**45**	**35**	**49**	30	24	**17**	11	9	4	4	5
Norway*	19	15	16	29	28	22	4	4	5	5	3	3
Portugal	..	4	..	..	2	..	..	3	..	..	2	..
Spain	15	16	16	10	7	8	2	2	3	6	**8**	**10**
Sweden	..	26	20	..	24	23	..	6	5	..	3	3
Switzerland	..	41	27	..	39	33	..	**19**	**13**	..	7	4
USA*	31	22	17	..	23	18	10	9	2	3	3	6
Bulgaria	9	5	8	4	3	1	2	2	2	6	4	4
Croatia	..	..	16	..	..	4	..	..	3	..	..	2
Czech Rep*	15	15	13	6	9	8	3	3	3	6	3	5
Hungary	5	..	..	4	..	..	3	..	..	2	..	..
Latvia	..	10	5	..	2	5	..	1	1	..	3	3
Lithuania	..	..	6	..	..	1	..	..	2	..	..	3
Poland	10	..	..	18	..	..	4	..	..	4	..	..
Russia*	11	4	4	10	2	3	2	1	2	4	1	2
Slovak Rep	..	..	15	..	..	10	..	..	3	..	..	3
Slovenia*	11	12	11	8	11	7	4	4	4	3	5	3
Argentina	..	..	16	..	..	5	..	..	3	..	..	4
Chile	..	8	6	..	8	10	..	4	2	..	4	4
Mexico	..	12	8	..	12	10	..	5	4	..	8	4
Philippines	6	3	5	15	7	6	11	7	13	5	3	3
South Africa	..	..	4	..	..	5	..	..	9	..	..	4
South Korea	..	..	16	..	..	7	..	..	2	..	..	3
Taiwan	..	..	4	..	..	10	..	..	5	..	..	2

Sources: International Social Survey Program Environment 1993, 2000, and 2010 data (ISSP 1996, 2003, 2019).

facilitate numerous comparisons, including, for instance, overall extent, by action, within regions, over time, and by country over time. Signing an environmental petition is the most frequent environmental political action that survey respondents report in the three waves in all but 4, 6 and 12 countries, respectively. New Zealand has the largest percentage of the population reporting having signed a petition for all three waves and has the largest percentage of the population who had donated money in 1993. The Netherlands has the largest percentage of the population who had donated money in 2000 and 2010. New Zealand has the largest percentage of the population who are members of environmental groups in 1993, with Switzerland having the greatest percentage in 2000 and 2010. Former East and West Germany have the largest percentages of the populations who had engaged in an environmental protest in 1993, and Spain has the largest percentage in 2000, and Spain and France having the largest percentages in 2010. Former state socialist countries have lower percentages of their populations who reported having engaged in environmental actions, although reported levels in 1993 are higher in some countries compared with 2000 and 2010, like Bulgaria and Russia. Within the third set of countries, the reported percentages participating in environmental actions that generally fall between those of the other two groups. However, only two countries have petitions being reported with greatest frequency; instead, donating money was reported by larger percentages for five countries in this group. There is some support for an event-based explanation related to the percentages reporting having engaged in a protest.

Although there is variation regarding environmental activism, in general, advanced industrialized countries, especially in Western Europe and North America, are more environmentally active across the forms of activism included here. Regional differences along with some country-specific differences in activism are shown. This is not entirely unexpected, given previous research and historical legacies like the influence of the former communist regimes in Eastern European countries compared with governmental structures in Western European countries and those in the category including North America and other regions (Dalton 2015; Hadler and Haller 2011; Marquart-Pyatt 2012b). Although these types of activism are individual behaviors reflecting individual efforts, their incidence and thus potential to happen may be realized differently given institutional contexts.

Table 28.2 provides information on political participation by country for three waves of the World Values Survey (Inglehart et al. 2000; World Values Survey 2009, WVS 2020). It presents the percentage of respondents in each country who had signed a petition, joined a boycott, or attended a demonstration for all three waves. The sample of countries included in the three waves of data differs, and data are presented for 41, 30 and 41 countries from Waves 3, 5 and 7, conducted in 1995–1997, 2005–2007 and 2017–2020, respectively. Across all waves, signing a petition is the activity respondents report engaging in with greatest frequency in most countries. New Zealand has the largest percentage of the population reporting having signed a petition for the three waves, and Sweden has the largest percentage of the population who had joined a boycott in all waves. France has the largest percentage of the population who had attended a demonstration across all waves. Although not shown here, France also has the largest percentage of the population reporting having joined an unofficial strike or occupied a building or factory for the 1995–1997 wave, two types of political activity that tend to occur less frequently among individuals than other actions. This is notable when perusing the percentages across the entire table. For instance, in five countries (Belarus, Lithuania, Romania, Russia and Ukraine) demonstrations were the activity for which the largest percentage of respondents reported participating in Wave 3. A perusal of the data for countries in the third group, selected for comparability with the ISSP data, show

Table 28.2 Percent of individuals participating in political activities, by country

	Petition			Boycott			Demonstrate		
	Wave 3	Wave 5	Wave 7	Wave 3	Wave 5	Wave 7	Wave 3	Wave 5	Wave 7
Australia	78	79	76	22	15	16	18	20	19
Austria	57	..	56	10	..	16	17	..	21
Finland	39	51	49	12	16	19	13	10	15
France	68	67	64	13	14	18	**40**	**38**	**41**
Germany	62	50	64	15	9	12	24	31	33
Italy	55	54	37	10	20	8	35	36	29
Japan	56	60	51	9	7	2	11	10	6
Netherlands	59	46	51	21	14	18	31	20	16
New Zealand	**91**	**87**	**78**	19	18	..	21	21	21
Spain	22	23	37	5	7	7	22	36	38
Sweden	72	78	68	**33**	**28**	**24**	30	31	26
Switzerland	64	78	67	12	19	13	17	28	23
United Kingdom	79	68	68	17	17	16	13	17	16
USA	73	70	60	19	20	21	16	15	17
Albania	23	15	12	9	4	4	16	19	7
Belarus	9	8	10	3	4	2	19	16	7
Bosnia/Herz	22	22	31	9	7	12	9	9	11
Bulgaria	8	12	12	3	2	2	11	12	7
Croatia	43	..	**49**	5	..	10	7	..	12
Czech Rep	26	..	39	10	..	6	11	..	14
Estonia	14	..	24	2	..	3	21	..	9
Hungary	15	..	17	3	..	2	5	..	10
Lithuania	3	..	14	5	..	6	17	..	6
Poland	20	24	34	6	5	5	10	10	12
Romania	17	6	13	3	1	2	20	6	9
Russia	11	8	11	2	3	2	21	16	11
Serbia	19	30	26	7	**16**	11	8	..	12
Slovenia	19	31	35	6	6	8	9	13	13
Ukraine	14	7	12	4	4	4	18	17	8
Argentina	29	**28**	16	1	3	2	16	18	12
Chile	17	17	10	2	3	3	15	17	14
Mexico	**32**	21	11	11	3	3	11	16	9
Peru	21	25	11	3	5	3	12	24	12
Philippines	12	3	11	6	3	2	8	7	10

Sources: World Values Survey Wave 3 (1995–1999), Wave 5 (2000–2005), and Wave 7 (2017–2020) data.

quite similar patterns regarding the prevalence of actions by type. Thus, it is important to note evidence of regional and country-specific variation in political participation.

The regional and country-specific variation shown regarding both environmental actions and political participation is instructive for model building in future empirical work.

It can help future research account for patterns of similarities and differences in describing the extent of participation across different modes and in specifying pathways affecting these actions. Frequencies across Tables 28.1 and 28.2 suggest some similarities across countries with different political and economic characteristics, raising questions about other possible factors at work. For instance, does a country's classification as an advanced industrialized country or liberal democracy account for a vibrant civil society that allows for the expression of a range of political *and* environmental actions? Do similar levels of activism across these environmental and political actions by country suggest cultural explanations? How, for instance, can commonalities shown in signing petitions be explained across environmental and political actions? Regional differences could be theorized as offering insights in explaining participation in newly established democracies.

A possible explanation for relatively low percentages who reported engaging in any pro-environmental behaviors or political activities could be explained by the unavailability of such activities or involvement in other activities compared with countries with more entrenched democratic structures. Increasingly, environmental issues are linked with broader concerns like equity and participation (see Chapter 24). Combined with the surge in scholarship on institutional aspects of democratic governance and the role that democratic values play in fostering an active, informed public, a backdrop for examining these interrelations is in place. Pluralism, participation, and public opinion are uniquely poised at the nexus of politics and the environment; despite its importance, however, little research has explored these relations although at least two promising lines of work are emerging related to trust and policy preferences as noted earlier in this chapter.

Conclusion

Recent years have witnessed an expanded array of resources available for addressing these issues including public opinion datasets and analytical techniques. The International Social Survey Program (ISSP) and World Values Survey (WVS), two large-scale, cross-national public opinion surveys, continue to expand their geographic scope in data-gathering initiatives in recent iterations of their surveys. The ISSP, for instance, gathers data on its 48 member countries, as of 2020, and includes topical modules like government, social inequality, religion, work orientations, and the environment. The World Values Survey completed its seventh wave of data gathering in 2020, has amassed survey data from more than 100 countries since it began in the early 1980s as the European Values Survey. This worldwide survey contains information on tolerance, environmental attitudes, democratic values, political participation, and a variety of social and political topics. Analytical techniques like structural equation modeling with latent variables and multilevel modeling stand to make important contributions to this research. A latent variable approach to environmental and political attitudes and behaviors is vital for future scholarship since it specifies abstract, multidimensional constructs that are not directly observable like environmental concern, democratic values, environmental activism and political participation. Multilevel modeling is a technique that enables the investigation of individual-level and aggregate-level characteristics across nations, which has important implications for comparative, cross-national research.

Expanding our understanding of the relations among pluralism, participation and public opinion on environmental and political issues is essential for subsequent research. At least three avenues are especially germane for future scholarship. Researchers should continue to move from exploring to examining how political and institutional contexts shape the expression of political and environmental participation. Future scholarship is also charged with broadening the geographic expanse of research efforts to include more developing/industrializing nations in order to further elucidate these processes across structural contexts. Finally, it is imperative for future work to maintain focus on how key events, historical contexts, mode of industrial development, and political structures relate with expressions of public opinion on social, political and environmental issues. Taking into account institutional factors is essential, as factors affecting participation may be context-dependent. Future scholarship should focus on variation both within and across countries to describe the extent of and determinants of multiple modes of participation across the globe.

References

Ajzen, I. (1991) The theory of planned behavior. *Organizational Behavior and Human Decision Processes* 50: 179–211.

Armitage, C. and M. Connor. (2001) Efficacy of the theory of planned behavior: a meta-analytic review. *British Journal of Social Psychology* 40: 471–499.

Bagozzi, R., Y. Li, and J. Baumgartner. (1990) The level of effort required for behaviour as a moderator of the attitude–behaviour relation. *European Journal of Social Psychology* 20: 45–59.

Bamberg, S. (2003) How does environmental concern influence specific environmentally related behaviors? A new answer to an old question. *Journal of Environmental Psychology* 23: 21–32.

Bamberg, S. and G. Moser. (2007) Twenty years after Hines, Hungerford, and Tomera: a new meta-analysis of psycho-social determinants of pro-environmental behaviour. *Journal of Environmental Psychology* 27: 14–25.

Bamberg, S. and P. Schmidt. (2003) Incentives, morality, or habit? Predicting students' car use for university routes with the models of Ajzen, Schwartz and Triandis. *Environment and Behavior 35(2)*: 264–285.

Barnes, S. and M. Kaase. (1979) *Political Action: Mass Participation in Five Western Democracies*. Beverly Hills, CA: Sage Publications.

Birch, S. (2020) Political polarization and environmental attitudes: a cross-national analysis. *Environmental Politics 29(4)*: 697–718.

Black, J.S., P. Stern, and J. Elworth. (1985) Personal and contextual influences on household energy adaptations. *Journal of Applied Psychology* 70: 3–21.

Brady, H.E., S. Verba, and K.L. Schlozman. (1995) Beyond SES: a resource model of political participation. *American Political Science Review* 89: 271–294.

Brechin, S. (1999) Objective problems, subjective values, and global environmentalism: evaluating the postmaterialist argument and challenging a new explanation. *Social Science Quarterly 84(4)*: 793–809.

Brechin, S. and W. Kempton. (1994) Global environmentalism: a challenge to the postmaterialism thesis?. *Social Science Quarterly* 75: 245–269.

Cheung, S., D. Chan, and Z. Wong. (1999) Re-examining the theory of planned behavior in understanding waste paper recycling. *Environment and Behavior 31(5)*: 587–612.

Dalton, R. (1996) *Citizen Politics: Public Opinion and Political Parties in Advanced Industrial Democracies*. 2nd edition. Chatham, NJ: Chatham House Publishers.

Dalton, R. (2015). Waxing or waning? The changing patterns of environmental activism. *Environmental Politics 24(4)*: 530–552.

Davidovic, D. and N. Harring. (2021). Exploring the cross-national variation in public support for climate policies in Europe: the role of quality of government and trust. *Energy Research & Social Science* 70: 101785.

Diekmann, A. and A. Franzen. (1999) The wealth of nations and environmental concern. *Environment and Behavior 31(4)*: 540–549.

Dietz, T., P. Stern and G. Guagnano. (1998) Social structural and social psychological bases of environmental concern. *Environment and Behavior* 30: 450–471.

Doyle, J. (2018) Institutionalized collective action and the relationship between beliefs about environmental problems and environmental actions: a cross-national analysis. *Social Science Research* 75: 32–43.

Dunlap, R. and R. Jones. (2002) Environmental concern: conceptual and measurement issues. pp. 482–524 in *Handbook of Environmental Sociology*, edited by Riley Dunlap and William Michelson. Westport, CT: Greenwood Press.

Dunlap, R. and A. Mertig. (1995) Global concern for the environment: is affluence a prerequisite? *Journal of Social Issues 51*: 121–137.

Dunlap, R. and A. Mertig. (1997) Global environmental concern: an anomaly for postmaterialism? *Social Science Quarterly 78(1)*: 24–29.

Dunlap, R. and R. York. (2008) The globalization of environmental concern and the limits of the postmaterialist values explanation: evidence from four multinational surveys. *Sociological Quarterly, 49(3)*: 529–563.

Dunlap, R., G. Gallup and A. Gallup. (1993) Of global concern: results of the Health of the Planet Survey. *Environment 35(9)*: 7–40.

Fairbrother, M. (2013) Rich people, poor people, and environmental concern: evidence across nations and time. *European Sociological Review 29(5)*: 910–922.

Fairbrother, M. (2016). Trust and public support for environmental protection in diverse national contexts. *Sociological Science 3*: 359–382.

Fairbrother, M. (2017) Environmental attitudes and the politics of distrust. *Sociology Compass*. https://doi.org/10.1111/soc4.12482

Fairbrother, M., I.J. Seva and J. Kulin. (2019) Political trust and the relationship between climate change beliefs and support for fossil fuel taxes: evidence from a survey of 23 European countries. *Global Environmental Change 59*: 102003.

Fishbein, M. and I. Ajzen. (2010) *Predicting and Changing Behavior: The Reasoned Action Approach*. New York: Psychology Press.

Frank, D. (1997) Science, nature, and the globalization of the environment, 1870–1990. *Social Forces 76(2)*: 409–435.

Frank, D., A. Hironaka and E. Schofer. (2000) The nation state and the natural environment, 1900–1995. *American Sociological Review 65*: 96–116.

Franzen, A. (2003) Environmental attitudes in international comparison: an analysis of the ISSP Surveys 1993 and 2000. *Social Science Quarterly 83*: 297–308.

Franzen, A. and R. Meyer (2010) Environmental attitudes in cross-national perspective: a multilevel analysis of the ISSP 1993 and 2000. *European Sociological Review 26(2)*: 219–234.

Franzen, A. and D. Vogl. (2013) Two decades of measuring environmental attitudes: a comparative analysis of 33 countries. *Global Environmental Change 23(5)*:1001–1008.

Freymeyer, R. and B. Johnson. (2010) A cross-cultural investigation of factors influencing environmental actions. *Sociological Spectrum 30(2)*: 184–195.

Gelissen, J. (2007) Explaining popular support for environmental protection: a multilevel analysis of 50 nations. *Environment and Behavior 39(3)*: 392–415.

Guagnano, G., P. Stern and T. Dietz. (1995) Influences on attitude–behavior relationships: a natural experiment with curbside recycling. *Environment and Behavior 27*: 699–718.

Hadler, M. (2013) Environmental behaviors in transatlantic view: public and private actions in the United States, Canada, Germany, and the Czech Republic, 1993–2010. *International Journal of Sociology 43(4)*: 87–108.

Hadler, M. (2016) Individual action, world society, and environmental change: 1993-2010. *European Journal of Cultural and Political Sociology 3*:341–374.

Hadler, M. and M. Haller. (2011) Global activism and nationally driven recycling: the influence of world society and national contexts on public and private environmental behavior. *International Sociology 26*: 315–345.

Hadler, M. and M. Haller. (2013) A shift from public to private environmental behavior: Findings from Hadler and Haller (2011) revisited and extended. *International Sociology 28(4)*: 484–489.

Hadler, M. and P. Wohlkönig. (2012) Environmental behaviours in the Czech Republic, Austria, and Germany between 1993 and 2010. Macro-level trends and individual level determinants compared. *Czech Sociological Review 48(3)*: 467–492.

Haerpfer, C., Inglehart, R., Moreno, A., Welzel, C., Kizilova, K., Diez-Medrano J., M. Lagos, P. Norris, E. Ponarin and B. Puranen et al. (eds.). (2020) *World Values Survey: Round Seven-Country-Pooled Datafile. Madrid, Spain & Vienna*, Austria: JD Systems Institute & WVSA Secretariat. doi.org/10.14281/18241.1

Haller, M. and M. Hadler. (2008) Dispositions to act in favor of the environment: fatalism and readiness to make sacrifices in cross-national perspective. *Sociological Forum 23(2)*: 281–311.

Harland, P., H. Staats and H. Wilke. (1999) Explaining pro-environmental intention and behavior by personal norms and theory of planned behavior. *Journal of Applied Social Psychology 29(12)*: 2505–2528.

Harring, N. (2013) Understanding the effects of corruption and political trust on willingness to make economic sacrifices for environmental protection in a cross-national perspective. *Social Science Quarterly 94(3)*: 660–671.

Harring, N. (2014) Corruption, inequalities and the perceived effectiveness of economic pro-environmental policy instruments: A European cross-national study. *Environmental Science & Policy 39*: 119–128.

Harring, N. (2016) Reward or punish? Understanding preferences toward economic or regulatory instruments in a cross-national perspective. *Political Studies 64(3)*: 573–592.

Harring, N., S.C Jagers and S. Matti. (2019) The significance of political culture, economic context and instrument type for climate policy support: a cross-national study. *Climate Policy 19(5)*: 636–650.

Hunter, L., A. Hatch and A. Johnson. (2004) Cross-national gender variation in environmental behaviors. *Social Science Quarterly 85(3)*: 677–694.

Inglehart, R. (1995) Public support for environmental protection: the impact of objective problems and subjective values in 43 societies. *PS: Political Science and Politics*, (March): 57–71.

Inglehart, R. et al. (2000) World Values Surveys and European Values Surveys, 1981–84, 1990–93, and 1995–97 [Computer file]. ICPSR version. Ann Arbor: Institute for Social Research [producer] (1999). Ann Arbor: Inter-university Consortium for Political and Social Research [distributor] (2000).

International Social Survey Program (ISSP) (2003) International Social Survey Program (ISSP), 1985–2000 [CD-ROM]. Cologne, Germany: Zentralarchiv fuer Empirische Sozialforschung an der Universitaet zu Koeln [producer], 2003. Cologne, Germany: Zentralarchiv fuer Empirische Sozialforschung/Ann Arbor, MI: Inter-university Consortium for Political and Social Research [distributors].

ISSP Research Group (2019) International Social Survey Programme: Environment III - ISSP 2010. GESIS Data Archive, Cologne. ZA5500 Data file Version 3.0.0, https://doi.org/10.4232/1.13271

Jorgeson, A. and J. Givens. (2014) Economic globalization and environmental concern: a multilevel analysis of individuals within 37 nations. *Environment and Behavior 46(7)*: 848–871.

Kaiser, F., G. Hubner and F. Bogner. (2005) Contrasting the theory of planned behavior with the value-belief–norm model in explaining conservation behavior. *Journal of Applied Social Psychology 35(10)*: 2150–2170.

Kidd, Q. and A. Lee. (1997) Post materialist values and the environment: a critique and reappraisal. *Social Science Quarterly 78(1)*: 1–15.

Leighley, J. (1995) Attitudes, opportunities, and incentives: a field essay on political participation. *Political Research Quarterly 48*: 181–209.

Madden, T., P. Ellen and I. Ajzen. (1992) A comparison of the theory of planned behavior and the theory of reasoned action. *Personality and Social Psychology Bulletin 18*: 3–9.

Marquart-Pyatt, S. (2007) Concern for the environment among general publics: a cross-national study. *Society and Natural Resources 20(10)*: 883–898.

Marquart-Pyatt, S. (2008) Are there similar influences on environmental concern? Comparing industrialized countries. *Social Science Quarterly 89(5)*: 1–24.

Marquart-Pyatt, S. (2012a) Explaining environmental activism across countries. *Society and Natural Resources 25(7)*: 683–699.

Marquart-Pyatt, S. (2012b) Contextual influences on environmental concern cross-nationally: a multilevel investigation. *Social Science Research 41(5)*: 1085–1099.

Marquart-Pyatt, S. (2016) Environmental trust: a cross-region & cross-country study. *Society and Natural Resources 29(9)*: 1032–1048.

Marquart-Pyatt, S. (2018) Trust and environmental activism across regions & countries. *Journal of Environmental Studies and Sciences 8(3)*: 249–263.

Milfont, T., J. Duckitt and L. Cameron. (2006) A cross-cultural study of environmental motive concerns and their implications for proenvironmental behavior. *Environment and Behavior 38*: 745–767.

Nordlund, A. and J. Garvill. (2002) Value structures behind proenvironmental behavior. *Environment and Behavior 34(6)*: 740–756.

Nordlund, A. and J. Garvill. (2003) Effects of values, problem awareness, and personal norm on willingness to reduce personal car use. *Journal of Environmental Psychology 23*: 339–347.

Olofsson, A. and S. Ohman. (2006) General beliefs and environmental concern: transatlantic comparisons. *Environment and Behavior 38*: 768–790.

Oreg, S. and T. Katz-Gerro. (2006) Predicting pro-environmental behavior cross-nationally: values, the theory of planned behavior and value–belief–norm theory. *Environment and Behavior 38*: 462–483.

Pisano, I. and M. Lubell. (2015) Environmental behavior in cross-national perspective: a multilevel analysis of 30 countries. *Environment and Behavior 1–28*.

Routhe, A., R.E. Jones and D. Feldman. (2005) Using theory to understand public support for collective actions that impact the environment: alleviating water supply problems in a nonarid biome. *Social Science Quarterly 86(4)*: 874–897.

Schofer, E. and A. Hironaka. (2005) The effects of world society on environmental protection outcomes. *Social Forces 84(1)*: 25–47.

Schultz, P.W. (2001) The structure of environmental concern: concern for self, other people, and the biosphere. *Journal of Environmental Psychology 21*: 327–339.

Schultz, P.W. and S. Oskamp. (1996) Effort as a moderator of the attitude–behavior relationship: general environmental concern and recycling. *Social Psychology Quarterly 59(4)*: 375–383.

Schultz, P.W., V. Gouveia, L. Cameron, G. Tankha, P. Schmuck and M. Franek. (2005) Values and their relationship to environmental concern and conservation behavior. *Journal of Cross-Cultural Psychology 36(4)*: 457–475.

Schwenk, G. and G. Moser (2009) Intention and behavior: a Bayesian meta-analysis with focus on the Ajzen–Fishbein Model in the field of environmental behavior. *Quality and Quantity 43*: 743–755.

Smith, T., J. Kim and J. Son. (2017). Public attitudes toward climate change and other environmental issues across countries. *International Journal of Sociology 47(1)*: 62–80.

Steg, L., L. Dreijerink and W. Abrahamse. (2005) Factors influencing the acceptability of energy policies: a test of VBN theory. *Journal of Environmental Psychology 25*: 415–425.

Stern, P. (2000) Toward a coherent theory of environmentally significant behavior. *Journal of Social Issues 56*: 407–424.

Stern, P. and T. Dietz. (1994) The value basis of environmental concern. *Journal of Social Issues 50*: 65–84.

Stern, P., T. Dietz and G. Guagnano. (1995) The new ecological paradigm in social-psychological context. *Environment and Behavior 27*: 723–745.

Stern, P., T. Dietz, T. Abel, G. Guagnano and L. Kalof. (1999) A value–belief–norm theory of support for social movements: the case of environmentalism. *Research in Human Ecology 6(2)*: 81–97.

Verba, S. (1996) The citizen as respondent: sample surveys and American democracy. *American Political Science Review 90*: 1–7.

Verba, S., N. Nie and J. Kim. (1971) *The Modes of Democratic Participation*. Beverly Hills, CA: Sage.

Verba, S., K.L. Schlozman and H.E. Brady. (1995) *Voice and Equality: Civic Voluntarism in American Politics*. Cambridge, MA: Harvard University Press.

World Values Survey. (2009) World Values Survey. 2005 Official Data File (v.20090901). World Values Survey Association (www.worldvaluessurvey.org). *Aggregate File Producer*. Madrid: ASEP/JDS.

Xiao, C. and R. Dunlap. (2007) Validating a comprehensive model of environmental concern cross-nationally: a US–Canadian comparison. *Social Science Quarterly 88(2)*: 471–493.

29

Environmental citizenship

De-politicizing or re-politicizing environmental politics?

Derek Bell

The idea of environmental citizenship has "humble origins in a 1990s Environment Canada publication" (Paehlke 2008: 359). It has since become a significant practical and theoretical ideal in global environmental politics. It has been advocated by international organizations, such as the United Nations Environment Programme, as well as national governments and state agencies. The environmental citizen is someone who "does their bit" for the environment. At a minimum, the environmental citizen recycles and installs energy saving light bulbs. More ambitiously, and more generally, the environmental citizen is concerned about sustainability and, especially, about reducing or limiting their impact on the environment. In other words, talking about environmental citizenship is a popular way of reframing discussions of environmental responsibilities.

There is a significant academic literature on environmental citizenship that approaches the idea from various directions. This chapter covers only some of the issues raised in this literature. I divide my discussion of environmental citizenship into four sections: environmental citizenship as a practical ideal; empirical studies of environmental citizenship; environmental citizenship in political theory; and challenges to environmental citizenship. However, there are other important issues that I do not discuss here, including education for environmental citizenship (see, e.g., Schild 2016), environmental citizenship and learning (see, e.g., Gough and Scott 2006), and the psychology of environmental citizenship (see, e.g., Jagers et al. 2016).

In the first section, I introduce the practical ideal of environmental citizenship. I outline several examples of the use of the idea by various political actors. In the second section, I turn to the academic work that has already been done on environmental citizenship. I begin by examining empirical studies of environmental citizenship. I distinguish three kinds of empirical study and I illustrate each of them with examples from the literature. In the third section, I turn from the empirical to the normative. I discuss "liberal" and "post-cosmopolitan" defenses of the practical ideal of environmental citizenship. In the fourth section, I consider two important critical arguments that offer a normative challenge to this practical ideal. The final section is a concluding summary.

DOI: 10.4324/9781003008873-33

Environmental citizenship as a practical ideal

Environmental citizenship has been presented as a practical ideal by a diverse range of actors in environmental politics. For advocates of this ideal, the environmental citizen is to be commended and environmental citizenship is to be promoted. The first use of the term has been attributed to the Canadian environment agency, Environment Canada, in the 1990s (Paehlke 2008: 359; Agyeman and Evans 2006: 199). Environment Canada defined environmental citizenship as "a personal commitment to learning more about the environment and to taking responsible environmental action. Environmental citizenship encourages individuals, communities and organizations to think about the environmental rights and responsibilities we all have as residents of planet Earth. Environmental citizenship means caring for the Earth and caring for Canada" (Environment Canada 2006: 1).

The idea of environmental citizenship was used prominently by the United Nations Environment Programme (UNEP) from the early 1990s, with significant investment, through the Global Environment Facility (GEF), in a US$6 million project called "Global Environmental Citizenship" (Bell 2014; Global Environment Facility undated). The idea of environmental citizenship was also used in other countries and by a diverse range of state and civil society actors (Bell 2014). It is less commonly used by government agencies or intergovernmental organizations today. For example, UNEP has replaced it with more specific campaigns, such as "#SolveDifferent" and "Breathe Life," which promote individual behavior changes by encouraging people to consider "the choice we make in our everyday lives" and "minimize your carbon footprint" (UN Environment Programme, undated). However, there are still a significant number of North American colleges that run courses in environmental citizenship. For example, the University of Guelph in Canada advertises a "Certificate in Environmental Citizenship," which is "beneficial to those who want to learn more about the global environmental issues facing us today and in the future. Participants gain in-depth knowledge about environmental changes, their global impact, and how one can directly contribute to the environment's sustainability" (University of Guelph 2021).

The use of the term "environmental citizenship" might have become less common but the practice of environmental citizenship remains extremely important. We continue to see concerted efforts by governments and nongovernmental organizations to promote pro-environmental consciousness and pro-environmental behavior. For example, the European Commission's *Our planet, our future: Fighting climate change together* rhetorically asks young EU citizens "Do you feel concerned about the threats from climate change? Are you passionate about the need to reduce emissions?" and suggests that "A good place to start is by making changes in everyday actions that reduce your carbon footprint. No action you take is too small" (European Commission 2018: 28). Friends of the Earth UK encourages its supporters to "Save the world in 2021 ... without leaving your sofa" by providing "a list of the lifestyle changes that will have a positive impact on caring for the environment – all of which you can make without having to leave the house" (Friends of the Earth 2021).

The practical ideal of the environmental citizen is someone who is concerned about the global environment and recognizes her responsibility or duty to play her part in preventing environmental harms. She is concerned about the environmental impacts of her individual everyday behavior and seeks to make changes that will reduce her contribution to environmental harms. She may also seek to promote action by government and business leaders by participating in political campaigns. For the environmental citizen, citizenly action is not confined to collective action in the public sphere, such as campaigning (or even voting) for "green" policies. Instead, the private sphere of everyday life is an arena in which each individual can undertake more or less sustainable actions. The environmental citizen chooses more sustainable actions.

Empirical studies of environmental citizenship

The discursive use of the ideal of environmental citizenship by both state and non-state actors in the 1990s and 2000s prompted academic discussion of the concept. This included both theoretical work to develop the idea (which we'll consider in the next section) and empirical studies of environmental citizenship (for an extended discussion of both see Cao 2015).

We might distinguish three types of empirical study. In the first type of study, the researcher attempts to provide a "thick description" of environmental citizens: What do they do? How do they live? What do they believe? What do they value? What are their attitudes? These studies begin from the assumption that the researcher can identify a particular group of people, typically environmental activists, as already being environmental citizens. Therefore, they use the lived experience of environmental activists as a model for environmental citizenship. On this approach, a detailed theoretical elaboration of environmental citizenship is developed from the empirical study of environmental activists.

A good example of this kind of study is Dave Horton's (2006) study of environmental activists as "elite" environmental citizens. Horton (2006: 132) argues that green activists can be understood to be "demonstrating one form of environmental citizenship." Horton's ethnographic study of activists in Lancaster aims to describe the "lifestyles of green activists, examining how these lifestyles are produced and reproduced" (Horton 2006: 133). He argues that the green lifestyles of these "elite" environmental citizens "emerge from a shared green culture," which he characterizes in terms of its "networks, spaces, materialities, and times" (Horton 2006: 127, 133). Horton's environmental citizens/activists participate in "green networks" that are "powerfully productive of green performances" (Horton 2006: 133). They learn green cultural codes and ways of talking through everyday interaction with other environmental activists in particular "green places" (Horton 2006: 136). Horton argues that "specific material objects facilitate the greening of lifestyle," including "bicycles, organic food, and walking boots" as well as Internet and email, while "[other] objects hinder the greening of lifestyle, and so it is their absence that is important," such as "the car and the television" (Horton 2006: 138). Horton argues that the lesson we should learn from his study is that we are unlikely to be able to successfully promote pro-environmental behavior directly. Instead, "broadening environmental citizenship" is only likely to be possible through the promotion of a "green architecture" or a green culture "from which specific behaviours emerge" (Horton 2006: 145).

Kelvin Mason (2014) also proposes an account of "green citizenship" based on the experience of environmental activism. Mason draws on his "10-year engagement with environmental activism" to develop his "auto-ethnographic narratives" through which he elaborates his distinctive theorization of environmental citizenship (Mason 2014: 142). Mason's activism is much more overtly political than the activism studied by Horton. He describes his involvement in "both direct and symbolic confrontations with the coercive forces of states, sometimes placing myself toe-to-toe with riot police but most often preoccupying them and evading them by any means" (Mason 2014: 147). He also describes his involvement in activist groups, participating in meetings, deep-ecology weekends, street singing, mass cycle rides and Climate Camps. Mason acknowledges that "environmental activists do not generally conceive of ourselves as citizens" but he argues that "our prefigurative performance of participatory democracy, our stewardship of the environment, and our creativity can be viewed as constituting a radical green citizenship" (Mason 2014: 155). For him, "Citizen Green is a political being, active in the public realm" where "green citizenship means thinking locally and acting globally as well as the converse" (Mason 2014: 155). Mason's version of environmental citizenship is more radical than Horton's and he rightly recognizes the "extreme challenge of sustaining an effective, material, grounded space of green citizenship ... in a world constituted almost entirely otherwise" and in which that space "will not be readily ceded to us" (Mason 2014: 155–156).

In a second type of empirical study, the aim of the research is to discover whether there are environmental citizens: Do they exist? Where can we find them? Understood in this way, we begin from a theoretical account (or definition) of environmental citizenship and we study a (large or small) group of people to determine whether any of them can be accurately categorized as environmental citizens. We ask whether the subjects of the study hold the values or engage in the behaviors that our theoretical account of environmental citizenship specifies. Some studies of this type focus on "ordinary" members of the public while others look at those engaged in specific practices.

A good example of the second kind of study is Sverker Jagers and Simon Matti's attempt to discover whether the "average [Swedish] citizen is a latent ecological citizen, willing to take on a greater pro-environmental responsibility and responsive to a new set of motivational factors" (Jagers and Matti 2010: 1056). Jagers and Matti define "ecological citizenship" with reference to Dobson's account (see below), emphasizing the three features that we saw in our discussion (in the first section above) of the practical ideal of environmental citizenship: the environmental or ecological citizen is concerned about the global environment; he or she recognizes environmental duties or responsibilities; and he or she is concerned about their individual acts in the private sphere. Building on this conception of ecological citizenship, they attempt to "operationalize" it by drawing on a value–belief–norm (VBN) model from environmental psychology to present an account of the "basic values," "environmental specific beliefs" and "behavioural readiness" that we might expect from the ecological citizen (Jagers and Matti 2010: 1061–1062). They examine the data from a survey of 1,207 Swedish households, conducted as part of a Sustainable Households research program funded by the Swedish Environmental Protection Agency, to see whether they find evidence of values, beliefs and behaviors consistent with ecological citizenship (Jagers and Matti 2010: 1057). They conclude that a "value base consistent with [ecological citizenship], emphasizing non-territorial altruism and the primacy of social justice, already exists among a significant share of Swedes" (Jagers and Matti 2010: 1075). Jagers and Matti's attempt to use large-n survey data to look for evidence of ecological or environmental citizenship has not yet been replicated in other countries (or with a representative sample of Swedes) but there have been (and continue to be) many national and cross-national studies of environmental values, beliefs, attitudes and behaviors, which might be used to test for further evidence of (latent or actual) environmental citizenship.

The use of small-n qualitative studies to look for evidence of environmental citizenship in particular groups has been more common. For example, Rob Flynn and his colleagues carried out nine focus groups with members of the general public in three areas of the United Kingdom – Teesside, south-west Wales and London – "to explore people's understandings of energy and environmental issues and their attitudes towards new hydrogen technologies" (Flynn et al. 2008: 772). Flynn et al. looked for evidence of environmental citizenship among the participants in their focus groups but they found that "Some people indicated that they might try to alter their consumption or approve stricter environmental controls if it was beneficial to their own and their children's health, but their concern for 'global' matters, or even other regions of the country, was more limited or even absent" (Flynn et al. 2008: 780). They found little or no evidence of the research participants' concerns extending beyond the local and little evidence that they recognized a duty or responsibility to change their everyday consumption behavior. Instead, they found that "attitudes seemed to converge on instrumental and privatized outlooks" rather than concern for the "common good" (Flynn et al. 2008: 781).

In other qualitative studies, researchers have looked for evidence of environmental or ecological citizenship among those involved in specific practices. For example, Neil Carter and Meg Huby consider whether either individual or institutional ethical investors are "eco-logical citizens" (Carter and Huby 2005: 262). They also define "ecological citizenship" with reference to Dobson's account – again emphasizing the three features that we identified as common themes in our discussion of environmental citizenship as a practical ideal (Carter and Huby 2005: 262). Carter and Huby draw on data from a large survey of individual ethical investors to argue that individual ethical investors are ecological citizens according to Dobson's definition (Carter and Huby 2005: 262).

Another interesting example of this type of study is Gill Seyfang's discussion of whether sustainable consumption is an act of "ecological citizenship" (Seyfang 2005: 291). On her account, the defining feature of ecological citizenship is the duty "to minimise the size and unsustainable impacts of one's ecological footprints" (Seyfang 2005: 291). She argues that the "mainstream" conception of sustainable consumption as "consumption of more efficiently produced goods" by "green" consumers who "demand sustainably produced goods and exercise consumer choice to send market signals" is not a genuine form of "ecological citizenship" (Seyfang 2005: 294). The ecological citizen's duty to reduce the size of their ecological footprint may "require an absolute reduction in consumption" rather than sustainable consumption (Seyfang 2005: 297).

In the third type of empirical study, the aim of the research is to use the concept of environmental citizenship to interpret government (or nongovernmental organization) campaigns that seek to promote pro-environmental behavior: Does this campaign seek to promote environmental citizens? What kind of environmental citizens does it seek to promote? On this approach, we work between campaign materials and theoretical conceptions of environmental citizenship to try to make sense of the kind of environmental citizens that the state is seeking to create.

A good example of this kind of study is Mirja Vihersalo's (2017) critical analysis of the European Commission's climate change campaign website. Vihersalo argues that the European Commission's climate campaign "portrays the private sphere as the main domain for citizen activity" (Vihersalo 2017: 349). Vihersalo identifies no suggestions of "traditional political activity" on the website, which seems to be seeking to produce a "consumer-citizen" rather than a citizen who engages with policy processes (Vihersalo 2017: 350). Moreover, the "climate citizen … seems to be an individual … performing these private sphere activities by herself, in isolation from others" (Vihersalo 2017: 350). On this conception of environmental citizenship, our response to climate change is the sum of our individual actions. Collective action, co-operation and direct political engagement in the public sphere are not part of environmental citizenship. Moreover, the campaign's behavior change suggestions involve only "minor modifications [to what we do] in the private, domestic sphere" so it implicitly gives "EU citizens permission to maintain a good or even quite luxurious standard of living and consumption level" (Vihersalo 2017: 355–356). For Vihersalo, this is a very minimalist conception of environmental citizenship.

We have seen that the empirical work on environmental citizenship uses (or develops) different conceptions of environmental citizenship. There is a common commitment to the idea that environmental citizens are concerned about the global environment and they recognize that they have some responsibilities to contribute to protecting the environment. However, the empirical work also highlights some disagreements about the extent and nature of those responsibilities. Does environmental citizenship require only modest "lifestyle adjustments rather than deep cultural or value changes?" (Vihersalo 2017: 356). Does environmental citizenship require only changes to what we do in the private, domestic sphere or does it require action in the public sphere?

We will examine these debates in the remaining sections of this chapter. In the next section, I briefly review the initial development of environmental citizenship in political theory before, in the final section of the chapter, considering some important theoretical challenges to environmental citizenship.

Environmental citizenship in political theory

As we might expect, political theorists have offered different accounts of environmental citizenship and different normative justifications or interpretations of those accounts. The most influential theoretical discussion of environmental citizenship is Dobson's book *Citizenship and the Environment* (2003). Dobson distinguishes "environmental citizenship" from "ecological citizenship." He defines "environmental citizenship" as a version of liberal citizenship which "extend[s] the discourse and practice of rights-claiming into the environmental context" (2003: 89). On this account, environmental citizenship simply extends the liberal list of rights beyond civil, political and economic rights to include environmental rights (i.e., rights to environmental goods or to protection from environmental bads). This conception of environmental citizenship is quite different from the notion of environmental citizenship that we saw in our discussion of the practical ideal (in the first section of this chapter). First, it is not global in its scope: the environmental citizen's rights are rights held against his or her own state. Second, it is concerned with rights not duties or responsibilities. Third, because it is not concerned with responsibilities, it does not require either individual action in the private sphere or participation in collective action in the public sphere to protect the environment.

The "gap" between Dobson's "liberal" version of "environmental citizenship" and the practical ideal of it has encouraged some liberals to argue that Dobson mischaracterizes "liberal environmental citizenship." For example, Simon Hailwood argues that political liberalism (as defended, most notably, by John Rawls) can be extended to defend a notion of "reasonable environmental citizenship," which requires citizens to acknowledge duties (as well as rights) to distant strangers (not just those in their locality) that require changes in their individual everyday behavior in the private sphere (as well as in the public sphere) (Hailwood 2005: 204).

I have also previously defended an account of "liberal environmental citizenship" that draws on a cosmopolitan version of Rawlsian political liberalism (Bell 2005). My account begins from the recognition that it is "a common criticism of 'mainstream' liberal conceptions of citizenship that they ignore the fact that members of the political community are embodied individuals living in a physical environment" (2005: 182). However, I argue that this criticism is only partly correct. Contemporary political liberalism does not ignore our embodiment. On the contrary, it is fundamentally (and rightly) concerned with the ability of humans to meet our physical needs for "food, clothing, shelter and health care" (Bell 2005: 182). Moreover, political liberalism does not ignore the fact that we live in a physical environment. However, it does adopt a particular conception of the relationship between humans and the environment: the environment "is conceptualised as property to be owned" by humans (Bell 2005: 182). I argue that this conception of the environment as property is inconsistent with political liberalism's own commitment to the "fact of reasonable pluralism" (Bell 2005: 183).

Basing principles of political justice on a "thoroughgoing conception of the environment as property" is inconsistent with liberal pluralism in the same way that basing principles of political justice on a Buddhist, Christian or secular ecocentric conception of the environment would be inconsistent with liberal pluralism (Bell 2005: 184). Instead, I suggest that we should acknowledge that the relationship between humans and the environment is a "subject about which there is reasonable disagreement" while also acknowledging that human survival is dependent on the environment or, in other words, the environment is the "provider of [our] basic needs" (Bell 2005: 184–185).

I argue that we can draw some substantive conclusions about the rights and duties of "liberal environmental citizens" from this account. More specifically, I defend "substantive environmental rights," such as rights to (adequately) clean air and water, which are necessary to meet our basic needs, as well as procedural rights to defend our substantive environmental rights (Bell 2005: 187). In addition, I argue that liberal environmental citizens will have three kinds of duties: the "duty to obey just [environmental] laws"; the "duty to promote just environmental laws"; and some "non-enforceable…citizens' duties" to undertake individual pro-environmental behaviors in the private sphere (Bell 2005: 189, 191). In sum, I claim to offer a liberal justification of the three key features of the practical ideal of environmental citizenship: the liberal environmental citizen has duties to protect the environmental rights of distant strangers by changing his or her individual behavior in the private sphere (as well as actively seeking to promote just environmental laws in the public sphere).

Unsurprisingly, the attempt to construct a cosmopolitan liberal defense of the practical ideal of environmental citizenship is not satisfying for those who are unconvinced by the merits of liberalism. So, for example, John Barry has argued that the civic republican tradition in political theory is a more attractive starting point for an account of environmental citizenship because it is more explicitly concerned with the "common good" and with "active" rather than "passive" forms of citizenship (Barry 2006: 26). Dobson also suggests that the civic republican tradition has more to offer than the liberal tradition but he argues that both traditions provide an inadequate framework for thinking about citizenship in a "globalizing world" (Dobson 2003: 49). Dobson proposes a new "post-cosmopolitan" account of "ecological citizenship" as an alternative: "At first blush, then, ecological citizenship deals in the currency of non-contractual responsibility, it inhabits the private as well as the public sphere, it refers to the source rather than the nature of responsibility to determine what count as citizenship virtues, it works with the language of virtue, and it is explicitly non-territorial" (2003: 89). Dobson's conception of "ecological citizenship" shares the key features of the practical ideal of environmental citizenship (as identified in the first section of this chapter). However, he offers a novel normative justification of them.

Dobson proposes a "non-contractual" account of responsibility. He claims that "the contractual idiom" is "very common" in discussions of citizenship where "[Citizenship] is regarded as a contract between the citizen and the state," which protects the citizen's rights in return for the payment of taxes (Dobson 2003: 44). He argues that contractual conceptions of citizenship reflect a particular (liberal) "ideological" approach to citizenship (2003: 46). Instead, he proposes that we might have responsibilities that are "*unreciprocated and unilateral* citizenship obligations*": obligations or duties that can be owed by a citizen without either the state or any other agent owing anything in return (Dobson 2003: 47). He suggests that the *"source"* of these new obligations of ecological and post-cosmopolitan citizenship is the capacity of the affluent in the global North "to 'always already' act on others" (rather than a contract between a citizen and a state) (Dobson 2003: 48, 50). In a "globalizing world" the "inhabitants of globalizing nations are *always already* acting on others, as when…our use of fossil fuels causes the release of gases that contribute to global warming. It is this recognition that calls forth the virtues and practices of citizenship" (Dobson 2003: 49). The global North's "antecedent action" of "narrowing the South's options" through our overly large "ecological footprints" and our causal contribution to environmental pollution generates "political [or citizenship] obligations of a non-reciprocal and unilateral type" (Dobson 2003: 50).

Dobson's "post-cosmopolitan" argument offers another way of justifying the key features of the practical ideal of environmental citizenship. First, the ecological citizen is concerned with more than the local. However, Dobson's account is not simply global; it is "non-territorial" (Dobson 2003: 89). Traditional conceptions of citizenship are territorial: the citizens of a state share a territory and non-citizens live (or originate) outside that territory. Dobson distinguishes two ways in which a conception of citizenship might be non-territorial. Cosmopolitan (liberal) citizenship is non-territorial because it invokes the idea of a global political community. In contrast, ecological citizenship is non-territorial because the shared political "community" of ecological citizens is "'produced' by the activities of individuals and groups with the capacity to spread and impose themselves in geographical and diachronic space. This produced space has no determinate size [or territory] (it is not a city, or a state, and nor is it even 'universal') since its scope varies with the case" (Dobson 2003: 81). On this account, ecological citizenship is "a citizenship with international and intergenerational dimensions" and, in a context where our acts have global consequences, it is a citizenship with global reach (Dobson 2003: 49).

Second, ecological citizenship is concerned only with "non-reciprocal and unilateral" duties or obligations; it does not pay attention to environmental rights (Dobson 2003: 50). Dobson suggests that the "principal ecological citizenship obligation" is "to ensure that [one's] ecological footprint does not compromise or foreclose the ability of others in present and future generations to pursue options important to them" (Dobson 2003: 91, 92) Third, ecological citizenship "inhabits the private [sphere] as well as the public sphere" (2003: 89). Dobson endorses the "central feminist point," which is that "the private sphere is a site of the exercise of power" and, therefore, we must "politicize the private sphere" (Dobson 2003: 53). On his "post-cosmopolitan" account, "ecological citizenship" also "inhabits" the private sphere because our "private acts have public implications": if individuals do not "reduce, reuse and recycle in their own homes" or, more generally, reduce the size of their "ecological footprint," their private acts of consumption will causally contribute to the public problem of environmental pollution (Dobson 2003: 56 and 55 quoting Kymlicka and Norman 1994: 360).

Challenges to environmental citizenship

The practical ideal of environmental citizenship, which emphasizes individual lifestyle change in the private, domestic sphere to meet our responsibilities to protect the global environment, has been challenged in two important ways. Critics argue that both Dobson's ecological citizenship and liberal conceptions of environmental citizenship, which seek to defend that practical ideal, are complicit in perpetuating injustice and de-politicizing the struggle against the forces of neoliberalism that support the *status quo*. In this section, I will outline these two challenges.

Critics have argued that the practical ideal of environmental citizenship perpetuates injustice in five important ways. First, if we focus on action in the private, domestic sphere, we place the burden of pro-environmental behavior change disproportionately on women (MacGregor 2006a). In the United Kingdom, for example, women still do "double the proportion of unpaid work when it comes to cooking, childcare and housework" (Office for National Statistics 2016). Therefore, MacGregor argues that advocates of this model of environmental citizenship do not pay sufficient attention to gender inequality and, as a result, their proposals seem likely to perpetuate and even exacerbate existing gender inequalities (MacGregor 2014).

Second, the focus on the responsibilities of affluent individuals in the Global North is exclusionary. This is, perhaps, clearest in Dobson's defense of non-reciprocal obligations. As Hayward (2006: 439) argues the "puzzling result [is] that not everyone is or needs to be an 'ecological citizen': only those responsible for harms need to be, not the victims. ... [The victims] are effectively cast in the role of 'moral patients'." On this account, the (current and future) victims of environmental harms are not regarded as moral agents or active citizens. They are passively waiting for affluent individuals to behave differently, thereby, saving them from the actions of those very same affluent individuals. This is reminiscent of criticisms of neo-colonialism and the "White Saviour Complex" in discussions of volunteer tourism (Wearing et al. 2018: 502). Advocates of this model of environmental citizenship do not pay sufficient attention to the historic – and continuing – exclusion and exploitation of people in the Global South. As a result, their proposals continue to exclude and marginalize by proposing that the solution is to appeal to the Global North's moral sense and paternalistic concern for those that we have exploited.

Third, the practical ideal of environmental citizenship suffers from liberalism's inattentiveness to differences in embodiment. MacGregor (2016: 615) argues that the environmental or ecological citizen is "abstract and disembodied." Dobson and Bell are concerned about the material impact that individuals have on the environment through their everyday actions but they ignore differences in embodiment. As Deborah Fenney Salkeld (2019: 1275) argues "this has led to an implicit embodiment – that of those who have produced it – i.e., white, male, non-disabled – as well as embedded 'ability expectations' which are similarly narrow and exclusive." The result is a model of environmental citizenship that is "exclusive of disabled people" as well as "anyone situated outside the assumed norm" (Fenney Salkeld 2019: 1275). An inclusive conception of environmental citizenship should engage with "new concerns for materiality/corporeality, intersectionality [and] social difference" (MacGregor 2016: 620).

Fourth, environmental citizenship focuses on the voluntary actions of the affluent and ignores the essential role that others play in making those voluntary actions possible. Manisha Anantharaman's (2014) study of emerging pro-environmental behaviors among the new middle classes in Bangalore, India shows how networks of new middle class individuals have worked together to improve waste management through recycling and composting schemes. However, Ananatharaman (2014: 182) argues that the success of this "elite volunteerism" is "dependent on replicating a culture of servitude ... within and outside the home, where waste work is delegated to paid domestic servants, housekeeping staff and other workers." If we focus on the obligations of the "socio-economically privileged" we "[fail] to see the contributions of those actors (the domestic servants and neighbourhood waste workers), who through their livelihood practices play a critical role in producing the systems that make pro-environmental behaviours possible among the elite" (Anantharaman 2014: 182). Environmental citizenship focuses on how "ecological debtors" can change their consumption behaviors; it ignores the contribution of the less privileged and it ignores production behaviors. As a consequence, environmental citizenship fails to challenge the *status quo* in-justices, which enable the world's affluent to choose whether or not to reduce their environ-mental impacts (by moderating their consumption choices) while giving no choice to those who are oppressed by injustice and suffer the environmental effects of over-consumption. In short, environmental citizenship fails to recognize that "Environmental injustice is a result and cause of social, economic and racial inequity" (Agyeman and Evans 2006: 190).

Fifth, environmental citizenship pays insufficient attention to the structural and systemic causes of environmental degradation. Voluntary pro-environmental behavior changes may be welcome but they will not be enough to tackle climate change, biodiversity loss, plastic pollution or many other global environmental problems. Instead, as John Barry suggests, we need a "more ambitious, multifaceted, and challenging" conception of political, social and economic change that "focuses on the underlying structural causes of environmental deg-radation" as well as "other infringements of sustainable development such as human rights abuses or social injustice" (Barry 2006: 24). If we fail to address the structural causes of environmental degradation, we will fail to address both the structural injustices that enable environmental degradation and the injustices that are caused by it.

So far, we have seen that critics argue that environmental and ecological citizenship are problematic ideals because they do not challenge – and, therefore, perpetuate – injustices. The advocates of the practical ideal of environmental citizenship might have a well-intentioned commitment to an egalitarian conception of cosmopolitan (or post-cosmopolitan) justice but their inattentiveness to (or unwillingness to question?) "real world" practices, structures and power relations undermines the egalitarian potential of their theories.

A second major criticism of the practical ideal of environmental citizenship is that it de-politicizes environmentalism. Critics have argued that environmental citizenship de-po-liticizes environmentalism in four ways. First, the "governmentality" critique suggests that the promotion of environmental citizenship, especially through education, can "become a way of disciplining the population to internalize a set of rules for behaviour – to become self-governing – thereby justifying minimal state intervention" (MacGregor 2006b: 115; see also Darier 1996). MacGregor argues that environmental citizenship is "an undesirable ideal for guiding socio-political or ecological movements" because there is a "dangerous dove-tail" between environmental citizenship and the dominant neoliberal discourse that seeks to privatize responsibility for environmental problems (MacGregor 2006b: 116, 113). Envi-ronmental citizenship "place[s] the onus on individuals… to become more educated about environmental issues, to make the necessary changes to their own outlook and behaviours" (MacGregor 2006b: 115).

Second, the ideal of environmental citizenship neglects – and obscures – the role of the state in addressing environmental degradation. If the causes of environmental degradation are structural, voluntary action by consumers will not be enough to resolve the problems. In-stead, co-operation among states will be necessary to re-structure the global political econ-omy. However, environmental citizenship does not challenge neoliberalism's commitment to "reliev[ing] the duty of government to provide goods and services to the population" (MacGregor 2006b: 114). Therefore, MacGregot concludes that environmental citizenship is dangerous because it is "an effective way to 'greenwash' neoliberal resistance to green regu-lation" (MacGregor 2006b: 116).

Third, environmental citizenship focuses on the "ethical or moral realignment of atti-tudes" rather than the political struggle that will be necessary to overcome the power of corporate and other vested interests that defend the *status quo* (Latta 2007: 379). Alex Latta argues that discussion of environmental citizenship in the "existing literature tends to treat ecological citizenship primarily as a normative and institutional tool for promoting a greener future" (Latta 2007: 379). This is an "instrumentalisation of citizenship" (Latta 2007: 385; see also Gabrielson 2008). Advocates of environmental citizenship ask us to believe that we can tackle the problem of environmental degradation by encouraging affluent consumers to fulfill their moral responsibilities when history tells us that the privileged rarely give up their privileges without a struggle. The affluent might be allies in the political struggle for political, social, economic *and environmental* justice but the voice of the oppressed must be central to a transformational form of citizenship that tackles the structural causes and consequences of environmental degradation.

Fourth, the practical ideal of environmental citizenship pre-empts the struggle for the environmental rights of current and future people. It de-politicizes both the idea of citizenship and the human–nature relationship by offering a normative account of environmental citizenship attitudes and behaviors that precedes and is independent from *real* political debates between *real* citizens about environmental rights and responsibilities for sustainability. Instead, Latta argues that we should understand environmental citizenship as "an emergent property of *existing* struggles for sustainability and political–ecological rights" (Latta 2007: 388; original emphasis). On this account, the role of the normative theorist is to interpret the claims of *real* political agents (or activist citizens) and to contribute to their struggles by providing theoretical support for their claims. In this context, "normative theorising must remain provisional and fluid, attentive to the emerging spaces and actors of ecological politics" (Latta 2007: 391).

The critics of environmental citizenship raise important concerns about the practical ideal of environmental citizenship and the theoretical defenses that have been offered to support it. For some, the solution is a more radical re-conceptualization of environmental citizenship, which re-politicizes citizenship and demands attention to injustice (Latta 2007; Cao 2015; MacGregor 2016). For others, it may be better to abandon the concept of environmental citizenship and focus, instead, on other ideals, such as environmental justice (Agyeman and Evans 2006).

Conclusion

Environmental citizenship achieved some prominence as a practical ideal advocated by various state and non-state actors involved in global environmental politics between the 1990s and 2010s. As a result, there are a growing number of empirical studies of environmental citizenship, which aim to discover whether there are any environmental citizens, how environmental citizens live, and what practices are consistent with environmental citizenship. These studies tend to conceptualize environmental citizenship in line with the practical ideal of environmental citizenship familiar from policy and practice. They draw on accounts of environmental citizenship developed by normative political theorists, especially the "post-cosmopolitan" theory of "ecological citizenship" proposed by Andrew Dobson.

As the theoretical literature on environmental citizenship has matured two important lines of criticism have emerged. First, critics argue that the practical ideal of environmental citizenship proposed by governments and developed by liberal and post-cosmopolitan political theorists perpetuates injustices because it pays insufficient attention to the realities of power, structural injustice and the systemic causes of environmental degradation. Second, environmental citizenship de-politicizes the struggle against vested interests that benefit from the *status quo* and is complicit in "greenwashing" neoliberal resistance to state action. It is unclear what the future holds for the concept of environmental or ecological citizenship.

Some critics propose a more radical form of environmental citizenship while others suggest that environmental politics should make use of other ideals. It is, however, clear that environmental degradation will continue unless there is effective political action by citizens across the world to challenge the *status quo*.

References

Agyeman, J. and Evans, B. (2006) "Justice, Governance and Sustainability: Perspectives on Environmental Citizenship from North America and Europe", in A. Dobson and D. Bell (eds) *Environmental Citizenship*. Cambridge, MA: MIT Press.

Anantharaman, M. (2014) "Networked Ecological Citizenship, The New Middle Classes and the Provisioning of Sustainable Waste Management in Bangalore, India", *Journal of Cleaner Production* 63:173–183.

Barry, J. (2006) "Resistance is Fertile: From Environmental to Sustainability Citizenship", in A. Dobson and D. Bell (eds) *Environmental Citizenship*. Cambridge, MA: MIT Press.

Bell, D. (2005) "Liberal Environmental Citizenship", *Environmental Politics* 14: 179–194.

Bell, D. (2014) "Environmental Citizenship: Global, Local and Individual", in P. Harris (ed) *Routledge Handbook of Global Environmental Politics*. Abingdon: Routledge.

Cao, Benito (2015) *Environment and Citizenship*. Abingdon: Routledge.

Carter, N. and Huby, M. (2005) "Ecological Citizenship and Ethical Investment", *Environmental Politics* 14: 255–272.

Darier, E. (1996) "Environmental Governmentality: The Case of Canada's Green Plan", *Environmental Politics* 5: 585–606.

Dobson, A. (2003) *Citizenship and the Environment*. Oxford: Oxford University Press.

Environment Canada (2006) *From the Mountains to the Sea: A Journey in Environmental Citizenship*. Ottawa: Environment Canada.

European Commission (2018) *Our Planet, Our Future: Fighting Climate Change Together*. Luxembourg: Publications Office of the European Union.

Fenney Salkeld, D. (2019) "Environmental Citizenship and Disability Equality: the Need for an Inclusive Approach", *Environmental Politics* 28: 1259–1280.

Flynn, R., Bellaby, P. and Ricci, M. (2008) "Environmental Citizenship and Public Attitudes to Hydrogen Energy Technologies", *Environmental Politics* 17: 766–783.

Friends of the Earth (2021) "Save the World in 2021 … Without Leaving Your Sofa", Available HTTP: <https://friendsoftheearth.uk/sustainable-living/save-world-2021-without-leaving-your-sofa>

Gabrielson, T. (2008) "Green Citizenship – A Review and Critique", *Citizenship Studies* 12: 429–446.

Global Environment Facility (undated) *Global Environmental Citizenship*, Available HTTP: <https://www.thegef.org/project/global-environmental-citizenship-gec>

Gough, S. and Scott, W. (2006) "Promoting Environmental Citizenship through Learning: Toward a Theory of Change", in A. Dobson and D. Bell (eds) *Environmental Citizenship*. Cambridge, MA: MIT Press.

Hailwood, S. (2005) "Environmental Citizenship as Reasonable Citizenship", *Environmental Politics* 14: 195–210.

Hayward, T. (2006) "Ecological Citizenship: Justice, Rights and the Virtue of Resourcefulness", *Environmental Politics* 15: 435–446.

Horton, D. (2006) "Demonstrating Environmental Citizenship? A Study of Everyday Life among Green Activists", in A. Dobson and D. Bell (eds) *Environmental Citizenship*. Cambridge, MA: MIT Press.

Jagers, S., Martinsson, J. and Matti, S. (2016) "The Environmental Psychology of the Ecological Citizen: Comparing Competing Models of Pro-Environmental Behavior", *Social Science Quarterly* 97: 1005–1022.

Jagers, S. and Matti, S. (2010) "Ecological Citizens: Identifying Values and Beliefs that Support Individual Environmental Responsibility among Swedes", *Sustainability* 2: 1055–1079.

Kymlicka, W. and Norman, W. (1994) "Return of the Citizen", *Ethics* 104: 352–381.

Latta, A. (2007) "Locating Democratic Politics in Ecological Citizenship", *Environmental Politics* 16: 377–393.

MacGregor, S. (2006a) *Beyond Mothering Earth: Ecological Citizenship and the Politics of Care*. Vancouver: University of British Columbia Press.

MacGregor, S. (2006b) "No Sustainability without Justice: A Feminist Critique of Environmental Citizenship", in A. Dobson and D. Bell (eds) *Environmental Citizenship*. Cambridge, MA: MIT Press.

MacGregor, S. (2014) "Ecological Citizenship", in H. van der Heijden (ed) *Handbook of Political Citizenship and Social Movements*. Cheltenham: Edward Elgar.

MacGregor, S. (2016) "Citizenship – Radical, Feminist and Green", in T. Gabrielson, C. Hall, J. Meyer, and D. Schlosberg (eds) *The Oxford Handbook of Environmental Political Theory*. Oxford: Oxford University Press.

Mason, K. (2014) "Becoming Citizen Green: Prefigurative Politics, Autonomous Geographies, and Hoping Against Hope", *Environmental Politics 23*: 140–158.

Office for National Statistics (2016) "Women Shoulder the Responsibility of 'Unpaid Work'", Available HTTP: <https://www.ons.gov.uk/employmentandlabourmarket/peopleinwork/earningsandworkinghours/articles/womenshouldertheresponsibilityofunpaidwork/2016-11-10>

Paehlke, R. (2008) "Book Review: 'Andrew Dobson and Derek Bell (Eds) *Environmental Citizenship*. Cambridge, MA: MIT Press, 2006'", *Organization and Environment 21*: 359–361.

Schild, R. (2016) "Environmental Citizenship: What Can Political Theory Contribute to Environmental Education Practice?" *The Journal of Environmental Education 47*: 19–34.

Seyfang, G. (2005) "Shopping for Sustainability: Can Sustainable Consumption Promote Ecological Citizenship?" *Environmental Politics 14*: 290–306.

United Nations Environment Programme (undated) *Campaigns*, Available HTTP: <https://www.un-environment.org/get-involved/campaigns>

University of Guelph (2021) "Certificate in Environmental Citizenship". Available HTTP: <https://www.uoguelph.ca/registrar/calendars/undergraduate/current/c11/c11-certdip-envcit.shtml>

Vihersalo, M. (2017) "Climate Citizenship in the European Union: Environmental Citizenship as an Analytical Concept", *Environmental Politics 26*: 343–360.

Wearing, S., Mostafanezhad, M., Nguyen, N., Nguyen, T., and McDonald, M. (2018) "'Poor Children on Tinder' and their Barbie Saviours: Towards a Feminist Political Economy of Volunteer Tourism", *Leisure Studies 37*: 500–514.

Part V
Key issues and policies in global environmental politics

30

Air pollution and acid rain

Early action and slowing progress

Loren R. Cass

International negotiations on transboundary air pollution present a long and rich set of case studies within the field of global environmental politics. Scholars have produced an extensive literature addressing the successes and failures of international attempts to address these problems. Transboundary flows of air pollution possess a number of characteristics that make these issues particularly interesting for scholars. For example, while smog is clearly visible in urban areas, the most important transboundary air pollutants such as sulfur dioxide, nitrogen oxides, the most dangerous particulate matter, persistent organic pollutants (POPs), and volatile organic compounds (VOCs) are not clearly visible to the naked eye (see Chapter 35). The public is thus unlikely to independently identify these pollutants as problems. Instead, scientists must play a critical role in defining and framing the problems for political debate and policy response (see Chapters 18 and 19). The relationships among science, problem framing and political responses have been a central focus of the international air pollution literature.

The primary sources of air pollution are the burning of fossil fuels for electricity production, transportation, heating and industrial uses (see Chapters 33 and 34). Industrial processes are also a major source of these pollutants. Efforts to address air pollution touch the core of the modern industrial economy and involve substantial costs to those affected by regulations (see Chapter 17). Addressing air pollution will thus be politically divisive and will introduce prominent economic interests into both domestic debates and international negotiations. The high economic costs and politically charged nature of the problem assure that power politics will have a significant effect on the response. The close relationships between international and domestic politics have forced global environmental politics scholars to try to bridge the divide between the subfields of international relations and comparative politics (see Chapters 11 and 12).

Air pollution is also a classic "upstream–downstream" problem common to many environmental issues. In upstream–downstream issues, the perpetrators of the environmental harm do not face the full environmental costs associated with their activities. The upstream perpetrators thus frequently do not have a strong incentive to address the problem while the downstream victims typically have limited leverage to force the upstream polluters to alter their behavior. The solution to the problem will typically require some form of regional agreement and frequently the creation of international institutions to coordinate scientific

DOI: 10.4324/9781003008873-35

research, to facilitate agreements to reduce emissions, to monitor compliance, and perhaps to support financial transfers from downstream states to upstream states to induce compliance (see Chapters 8 and 9). Scholars addressing international air pollution have created a substantial literature analyzing the roles of international institutions in promoting cooperation to address air pollution and in monitoring compliance.

There is a range of potential air pollution problems that could be addressed in this chapter, but the primary focus here will be on the problem of acid rain. Acid rain in Europe was one of the first problems to be addressed through international negotiations after the 1972 Stockholm Convention. It was thus among the first to be studied extensively within the emerging field of global environmental politics. The relatively long history involving nearly 50 years of international efforts to address acid rain offers excellent case studies of the evolution of international institutions designed to address a major transboundary environmental problem and presents an opportunity to explore the effectiveness of these agreements over time.

The chapter provides a brief overview of transboundary air pollution. It traces the origins and political response to acid rain before discussing the global environmental politics literature that has emerged around this issue. Finally, it reviews the evolving international political responses to other forms of air pollution by briefly touching upon particulate matter (PM), volatile organic compounds (VOCs), persistent organic pollutants (POPs), and heavy metals which have been identified more recently as problems deserving international attention. These issues are likely to be the focus of international negotiations in the future.

The problem of transboundary air pollution

Air pollution emerged alongside the processes of urbanization and industrialization. It was thus among the first environmental problems to be acknowledged and addressed with public policy responses. As one of the first countries to experience rapid urbanization and industrialization, it was only natural that England would be the first to identify air pollution as a problem. English royals made various proclamations beginning in 1257 to ban the burning of "sea coal" to address worsening air pollution in London. By the fifteenth century, London's skies were regularly blackened with coal smoke, but its residents recognized the smoky fog or "smog" as more of a local nuisance than a major threat. English diarist John Evelyn published a pamphlet in 1661 entitled "Fumifugium, or the Inconvenience of the Aer and Smoke of London Dissipated" that offered an account of the origins and effects of air pollution on human health and provided a blueprint for improving the air quality of London. However, it would take nearly 300 years before serious domestic efforts would be brought to bear on the problem.

The full extent of the human health and broader ecological effects from exposure to air pollution would not be widely understood and accepted by the public and political leaders until the 1960s, and domestic policies to address the problem were very slow to emerge. The recognition that air pollutant emissions in one country could have ecological implications for other countries would take even longer to be accepted. International air pollution policy thus provides an excellent case study to analyze the forces that shape the identification and framing of a problem for political action at both the domestic and international levels. This is particularly interesting in that the timing of the responses across regions of the world have varied significantly. Acid rain was first identified as a potential cross-national problem in the late 1960s in Europe, but it did not reach the diplomatic agenda of the North American states until the early 1970s, and Asian countries did not begin to address the issue until the mid-1980s (Wilkening 2004: 213). The European response to acid rain and more broadly to air pollution has created the most advanced regional system for addressing air pollution problems. North America and Asia have lagged significantly behind Europe in terms of regional responses, while very little attention has been devoted to transboundary air pollution across the remainder of the world. European responses to air pollution will thus be the primary focus of this chapter.

There is an array of potential environmental and human health problems associated with air pollution. The understanding of these problems has evolved substantially over time. The first of the issues to be addressed internationally was acid rain. The term acid rain was coined in 1872 by Scottish scientist Robert Angus Smith. The basic chemistry of acid rain was well understood by the turn of the twentieth century, but the larger environmental effects of acid precipitation were not well documented until the 1960s. Acids and alkaline (bases) are measured along a logarithmic potential hydrogen (pH) scale from 1 (acidic) to 14 (alkaline) with 7 being neutral. Normal rainwater is slightly acidic with a pH of 5.6. Acid rain would thus be considered rainfall with a pH of less than 5.6. Acid rain is produced when emissions of sulfur dioxide or nitrogen oxides are released during the burning of fossil fuels, when ammonia is released from animal manure, rice paddies, fertilizers, or other sources, or when a range of volatile organic compounds are released from industrial processes. These substances interact with water either while suspended in the air or after falling to the ground to produce a range of acids such as sulfuric acid, nitric acid and ammonium.

The acids produce substantial harm to the natural and man-made environment. In the absence of buffering agents such as carbonate in soils, acid rain enters rivers and lakes and progressively increases the acidity levels of the water, which initially creates stress on fish as pH levels fall below 6.5. Most fish species cannot tolerate acidity levels below a pH of 5.0, and lakes can "die" as fish populations collapse at lower pH levels. At the height of the acid rain problem in Europe in the 1970s and 1980s, there were thousands of lakes across Scandinavia that lost fish species.

Beyond the effects on lakes, acid rain can produce substantial harm to plant life and particularly to forests. Acid precipitation can interact with soils to disrupt the nutrient balance that is critical to the health of plants. Acid precipitation will tend to leach critical nutrients such as magnesium and calcium out of the soil and make them unavailable for plant systems, and it can simultaneously add nitrogen to soils, which is a vital nutrient for plant growth and can an act as a fertilizer. However, the plants cannot effectively utilize the nitrogen in the absence of sufficient quantities of magnesium and calcium. Acid precipitation can also release heavy metals in the soil which can adversely affect root structures and prevent the uptake of water. The overall effect can be widespread forest destruction across vast areas as well as extensive damage to the productivity of agricultural lands (see Chapters 42 and 44). Finally, acid rain also has an effect on the built environment. It degrades metal on buildings and transportation infrastructure and gradually damages a wide range of materials from stone to glass, rubber and ceramics. The costs associated with acid rain can thus be substantial.

While acid rain has been the focus of much of the international response to air pollution, there are many other air pollutants that pose significant environmental and human health risks. While many states have pursued domestic regulation of these problems, international concern has emerged relatively recently (see Chapter 12). For example, airborne particulate matter composed of tiny solid particles and liquid droplets can cause significant problems for human and animal respiratory systems. These particles can be composed of dust, soot, acids, organic chemicals, heavy metals and other toxic substances. Public health officials particularly worry about particles that are under 10 micrometers with the most dangerous particles being under 2.5 micrometers. These tiny particles can be inhaled deep into the lungs where they can cause respiratory problems, aggravate asthma, produce chronic bronchitis, affect heart function leading to heart attacks, and produce premature death due to heart and/or lung disease. The particles can be carried by winds over long distances and thus have the capacity to cross international borders. A World Health Organization (2016) study estimated that 92 percent of the world population live in areas with higher particulate matter levels than deemed safe by WHO standards. While particulate matter has long been the focus of domestic regulatory efforts, there has been much less attention at the international level.

There are many other air pollution problems that have emerged as the focus of international environmental attention. Volatile organic compounds (VOCs) are substances that evaporate easily at room temperature with the chemicals entering the surrounding air where they can be inhaled. Many chemicals in paints, solvents, dry cleaning chemicals, adhesives, and numerous other substances are VOCs. These substances can cause headaches, nausea, respiratory distress, compromised immune systems, cancer, and nervous system damage. Most of these substances pose their greatest threats through indoor air pollution. However, there is growing evidence that they may pose transboundary problems as well. Persistent organic pollutants (POPs) are another group of chemicals that resist degradation and can thus persist for long periods in the environment (see Chapter 35). POPs have recently been the focus of an international agreement under the 2001 Stockholm Convention on Persistent Organic Pollutants. Finally, heavy metals such as cadmium, lead, and mercury also pose potential cross-border threats. Similar to POPs, these metals persist in the environment and can bio-accumulate. Exposure to these metals is associated with kidney and bone damage, cancer, as well as developmental and neurobehavioral problems. While the basic human health threats posed by particulates, heavy metals, POPs, and VOCs are fairly well documented, the specific relationships between exposure levels and biological effects in an international context are less well understood.

Explaining European responses to transboundary air pollution

Analyzing the political response to acid rain requires a theoretical lens through which to evaluate the international negotiations. One approach is to focus on national interests and power politics in dictating international agreement. Levy (1993), McCormick (1997), and Boehmer-Christiansen and Skea (1991) highlight the importance of national interests and power politics in the international negotiations to address climate change. However, while state interests and relative power positions explain aspects of the negotiations, these scholars and others would point to other factors that explain aspects of the international negotiations that power politics cannot explain. Dimitrov (2006), Wilkening (2004), Selin (2006), and VanDeveer (2006) highlight the importance of scientific research and learning in international and national responses. National interests can shift in response to new information about the effects of acid rain and its costs. Political leaders and the public can learn and revise their interests in light of new information or new ways of framing a problem. Underdal and Hanf (2000) and Wettestad (1997, 2002) have utilized regime theory to analyze the evolution of European air pollution policy. Regimes are defined as social institutions that shape actor expectations and associated behavior in a given issue area (see Chapter 9). Regimes create opportunities for interaction among states and help promote the diffusion of information, values and appropriate behavioral norms. In a similar vein, Levy (1993) emphasizes the importance of international institutions in promoting government concern, enhancing the contractual environment for international agreements, and improving the capacity of states to fulfill their commitments. The various theoretical lenses highlight important aspects of the international response to air pollution. There is not sufficient space here to explore each of these perspectives, but the brief discussion of the evolution of the Convention on Long-Range Transboundary Air Pollution (CLRTAP) below will highlight some of the insights from these scholars and provide a starting point for further exploration of this case.

The first step in the emergence of an environmental problem is the recognition that some process is producing an adverse impact on the environment and/or human health. While scientists had understood since the mid-nineteenth century that acid rain could occur, systematic measurements that could actually quantify the presence of acid precipitation were not in place until the mid-twentieth century. The first large-scale project to measure the chemistry of precipitation was established by Hans Egnér in Sweden in 1948 (Wilkening 2004: 214). This project was eventually expanded to become the European Atmospheric Chemistry Network. It was only in the 1960s that scientists began to piece together a more systematic understanding of the origins and impacts of acid rain in Europe. Swedish scientist Svante Odén provided the first model describing the origins and effects of acid rain (Wilkening 2004: 214–215). Increasingly it became apparent that the predominant European winds were carrying acid precipitation from industrial areas of the United Kingdom and Germany north and eastward toward the Scandinavian countries, which due to geological coincidence faced a heavy toll from acid precipitation. Carbonate in soils has the capacity to buffer acidic precipitation and thus reduce acidity levels in runoff. However, Scandinavian soils, similar to soils in the northeast of the United States and large portions of Canada, lack the ability to buffer acidic precipitation and thus face stronger and more immediate effects from acid rain. Because the effects of acid rain were most fully apparent in Sweden, the Swedish government emerged as the primary advocate of international action to address acid rain.

By the late 1960s, acid rain in Europe had been defined as a problem by scientists and was beginning to be addressed domestically in many countries. The next step was for it to become an issue on the international diplomatic agenda. Sweden utilized the 1972 Stockholm United Nations Conference on the Human Environment to raise the problem of acid rain but largely received a skeptical response from the major powers. However, Sweden had some success within the Organization for Economic Cooperation and Development (OECD), which undertook a project to monitor acid deposition in Europe beginning in 1972. In 1977 this research led to the publication of a report that provided substantial evidence of the transboundary transport of acid rain and extensive adverse environmental effects (McCormick 1997: 57). However, the growing scientific evidence that emerged throughout the 1960s and 1970s failed to achieve a significant international impact despite growing calls from scientists and Scandinavian countries to address the problem.

The rise of acid rain to the international diplomatic agenda reflects a fascinating intersection of geopolitics and domestic politics during the 1970s period of Cold War détente between East and West. The success of the 1975 Helsinki Accords to improve relations between the United States and its allies and the Soviet bloc countries produced a renewed effort to find issues on which East and West could make further progress (Darst 2001). The renewed interest in environmental affairs after 1972 made the environment a natural focus for these efforts. The fortuitous presence of the United Nations Economic Commission for Europe (UNECE) provided a forum in which all of the major European countries as well as the United States and Canada were members. The UNECE reviewed a range of possible issues for further negotiation, but air pollution and acid rain emerged as the only issues that diplomats viewed as sufficiently important and had sufficient political backing to launch international negotiations (Levy 1993: 81–83).

Negotiations under the UNECE produced the 1979 Convention on Long-Range Transboundary Air Pollution (CLRTAP), which provided the foundation for European efforts to address acid rain and other major transboundary air pollution problems. While CLRTAP did not require any immediate action to reduce emissions, it created a forum for future efforts to address air pollution and established the principle that cross-border flows of acid precipitation should be reduced as much as was economically feasible. Perhaps most importantly, it also established a program to coordinate scientific research programs across Europe to pursue long-term study of the problem of acid rain and air pollution more generally.

CLRTAP was in many ways a lowest common denominator agreement. Germany and the United Kingdom among other states did not want to undertake the expensive policy changes that would be necessary to actually reduce transboundary flows of pollutants. The Communist Bloc countries were also largely uninterested in actually reducing transboundary flows, but in their case this was not a major problem since the predominant winds took their pollutants further east and away from the other signatories (Darst 2001). As long as states were required to reduce transborder fluxes and not total emissions, the Communist Bloc countries were willing to go along with such a treaty, and the United Kingdom, Germany and several smaller European states were willing to study the problem further as long as they were not required to take immediate actions to reduce emissions. The outcome was a weak agreement that called for additional study but without any immediate requirements for action. This scientific research would provide the foundation for future agreements to address a range of air pollution problems.

As the domestic political situations in the major states of Europe began to shift, the political context within which acid rain was being addressed began to change as well. The Scandinavian countries had sought to include hard targets to force emission reductions of substances producing acid rain in the original CLRTAP agreement or in a protocol to be negotiated immediately after CLRTAP came into force. However, major states such as Germany and the United Kingdom opposed this proposal. As evidence of significant adverse effects from acid rain began to accumulate in Germany and later in the United Kingdom, the domestic political situations in these countries began to shift. As Germany accepted the need to act, the international politics shifted dramatically as momentum built for a protocol to address sulfur emissions under CLRTAP. Boehmer-Christiansen and Skea (1991) and McCormick (1997) provide good discussions of the shifting domestic political responses across the major states involved in the CLRTAP negotiations. There was a combination of forces at work. Domestic political situations were changing, and the larger structure of CLRTAP permitted new scientific evidence to emerge and influence the international negotiations. Dimitrov (2006: 90–91) emphasizes that the scientific working groups created within CLRTAP were structured thematically around types of environmental damage and not specific pollutants so that any substance that could be linked to environmental damage was open to investigation by the working groups. The evolution of CLRTAP to draw ever more pollutants into the negotiations is indicative of the role of these working groups and scientific advice in shaping the evolution of the regime.

CLRTAP provided the forum within which the states of Europe were able to negotiate additional agreements to address air pollution problems as new scientific evidence emerged and the political environment permitted agreement. Members of CLRTAP negotiated eight protocols: 1984 protocol to fund long-term air pollution monitoring; 1985 protocol to reduce sulfur emissions or their transboundary fluxes by 30 percent from 1980 levels by 1993; 1988 protocol to freeze nitrogen oxide emissions at 1987 levels by 1995; 1991 protocol to reduce VOCs emissions by 30 percent by 1998; 1994 protocol to replace the 1985 sulfur protocol with an "effects based" approach that utilized critical loads to minimize the adverse consequences of sulfur precipitation; 1998 protocol to reduce the emission of cadmium, lead, and mercury below their 1990 levels; 1998 protocol to reduce emissions of 16 POPs; and the 1999 protocol to establish ceilings for emissions of a range of substances (sulfur, nitrogen oxides, VOCs, and ammonia) for 2010 and after. While many countries have pursued domestic policies to address these air pollution problems, European countries have been much more aggressive in their international response. There is not sufficient space here to explore the science and politics surrounding each of these protocols. Wettestad (2002), Levy (1993: 91–100), Dimitrov (2006: 73–78), and the edited volume by Lidskog and Sundqvist (2011) provide overviews of the forces shaping the negotiation of the protocols. The evolution of the agreements under CLRTAP illustrates the importance of international institutions in facilitating cooperation. As research emerged regarding the threats posed by various pollutants, the working groups established under CLRTAP placed this information on the political agenda and frequently produced new agreements to expand the scope of CLRTAP.

Wettestad (2006: 290) argues that one of the more interesting examples of the influence of CLRTAP is the development of the "critical loads" concept that became influential in the negotiations leading up to the 1994 protocol on the further reductions of sulfur emissions. The UNECE defines critical loads as "a quantitative estimate of an exposure to one or more pollutants below which significant harmful effects on specified sensitive elements of the environment do not occur according to present knowledge" (UNECE 2012). Prior agreements had been premised upon the need to achieve uniform reductions in transborder flows of acid precipitation. The critical loads approach focused on the effects of the emissions and not exclusively on the total emissions. The concept emerged in the scientific work of CLRTAP and then spread into the air pollution policies of states and the EU.

One of the more interesting debates, however, relates to the actual importance of CLRTAP. While there is evidence of the dissemination of science and policy approaches into broader policymaking, was CLRTAP producing major initiatives that propelled states to undertake new initiatives to reduce emissions or was it merely ratifying the positions that states had already adopted domestically? Addressing this question is complicated by the role of the EU. While the EU had been established to promote economic and political integration, environmental regulation and specifically air pollution policy emerged relatively quickly as areas of EU competence. The expanding responsibility of the EU for air pollution regulation and the concomitant territorial expansion of the EU from six states to 27 states with several applications pending gradually brought the membership of CLRTAP closer to the membership of the EU. Parallel sets of EU regulations and CLRTAP protocols thus emerged with important connections across them.

Underdal and Hanf (2000), and Wettestad (2002) provide very good accounts of the complex relationships among domestic politics, EU politics, and European acid rain policy during the 1980s and into the 1990s. Global environmental politics scholars have increasingly focused on the diverse forms of governance structures that have emerged across environmental issue areas, and acid rain in Europe provides a particularly rich and nuanced set of regulatory structures to study. Selin and VanDeveer (2003) provide an analysis of the complex institutional governance structures and their linkages in European air pollution policy. This is an area of research that remains important (see Chapter 21). How should we understand the complex governance structures that span the local to the national to the EU to the broader regional structures under CLRTAP? How do the various levels influence one another? In which directions do the causal arrows flow? To what extent are grassroots demands for addressing air pollution the critical variables in shaping air pollution policy? How important is scientific evidence in the regulatory process? To what extent do international agreements alter domestic policy debates and regulations? These questions have been the focus of the scholars cited above, and remain important questions going forward.

The ultimate question of course is how effective was CLRTAP? This raises a number of definitional problems. How do you define "effectiveness?" One measure might be to determine whether there was full participation by the most important actors in the agreement. By this measure, CLRTAP could be judged a relative success. Most major European states signed and ratified most of the agreements. Such a measure might be useful in understanding participation in the negotiations, but states could easily undertake agreements that they would ultimately fail to fulfill. Perhaps a better measure of success would be to measure compliance with the agreements by the states that ratified the various protocols. Again, compliance with the protocols under CLRTAP has been relatively high (Wettestad 2002: 197–198). Another level of success would relate to whether CLRTAP and its protocols altered state behavior in ways that led to better environmental outcomes than if CLRTAP had not been present. Wettestad (2002) undertook a review of the "effectiveness" of the CLRTAP regime as part of a larger project seeking to measure the effectiveness of international environmental regimes. He concludes the CLRTAP should be viewed as a "mixed success" (2002: 199–204). There was strong participation, and most states met their commitments; however, the forces that produced the actual reductions in emissions were driven by domestic factors largely unrelated to CLRTAP. For example, national transitions from coal to natural gas in electricity production produced much of the observed reductions in emissions. States were also undertaking policies for domestic reasons that were consistent with CLRTAP but would likely have been pursued even in the absence of CLRTAP. Dimitrov (2006: 72) concurs in the judgment that CLRTAP was a political success, which resulted in a situation of overall compliance. These conclusions raise important questions about what lessons should be taken from the CLRTAP case and whether these lessons can be generalized to other environmental cases.

CLRTAP also presents interesting questions surrounding the functioning of "mature" global environmental institutions. The members of CLRTAP have not added an additional protocol to the agreement since 2005. However, there have been important amendments to some of the protocols that have increased the stringency of some of the commitments and expanded the range of substances covered by the protocols. Many of these amendments are awaiting ratification by the required number of states before coming into force. The 2008 financial crisis appears to have reduced the level of commitment to taking further action. The emphasis of CLRTAP has shifted in recent years to capacity building, seeking geographic expansion to bring in more states from Eastern Europe and Central Asia, and facilitating conversations among countries in other parts of the world who are interested in addressing air pollution. These efforts appear to have achieved limited success. In recent years, momentum toward addressing transboundary air pollution appears to have waned. Byrne (2017) reviews recent initiatives through the UNECE and CLRTAP.

Responses to acid rain outside Europe

No other region of the world has the depth and complexity of the EU's institutional structures with its long history of cooperation on regional issues and ceding of substantial sovereign control over environmental policy to the EU. It should not be surprising that other regions would be slower to address transboundary air pollution issues. While the United States and Canada both participated in the CLRTAP negotiations and are parties to several of the protocols, the United States remained reluctant to address international acid rain issues with Canada until the 1990s. Asian countries did not begin to discuss regional acid rain problems until the 1980s, and air pollution issues remain low on the agendas of most other regions, though many states are pursuing domestic policies.

While the United States and Canada were full participants in the negotiations to create CLRTAP, they lacked the surveillance systems in the 1970s to monitor acid precipitation, and the issue was not perceived as a major problem in the United States until the late 1970s. The United States Atmospheric Deposition Program was only established in 1978 (almost 30 years after a similar program in Europe). By 1980 there was clear evidence of adverse ecological consequence from acid rain in the United States and Canada, but the evidence of widespread damage was still being documented. Canada emerged as the primary advocate of addressing acid rain in North America. The lack of buffering agents in Canadian soils meant that the acid precipitation originating in the industrial heartland of the United States fell across large areas of eastern Canada. Acid rain emerged as a significant irritant in relations between the United States and Canada during the 1980s, but the United States refused to act to address the problem (Schmandt et al. 1988). While Canada signed and ratified the 1985 sulfur protocol under CLRTAP, the United States refused to sign it. However, growing domestic political pressures surrounding US air pollution problems and evidence of effects of acid rain across the northeast of the United States eventually led to the passage of the 1990 Clean Air Act Amendments. The United States undertook significant policies to reduce sulfur, nitrogen oxide, and VOCs emissions, which had the fortuitous effect of reducing acid rain across the northeast of the United States and in Canada. As a result, the United States and Canada signed a transboundary air pollution treaty in 1991, but the treaty essentially affirmed existing US and Canadian policies without requiring additional actions. While European acid rain negotiations may have heightened public interest in acid rain in North America, North American transboundary air pollution policies were overwhelmingly dictated by domestic political considerations and moved no faster than American domestic politics would permit.

Asian responses to acid rain lagged even further behind European initiatives. Japan did not conduct its first acid rain survey until the mid-1980s. A 1985 report noted that acid deposition was occurring in Japan as a result of emissions in China (Wilkening 2004: 230). The evidence of significant harm from acid precipitation was less apparent, but the fact that China was exporting its pollution to Japan led to more political support within the Japanese government for studying the problem at a time when the government was not particularly interested in studying other environmental problems that could impose additional costs on industry. By the early 1990s, there was growing evidence of major plumes of sulfur dioxide emanating from China and crossing over Japan. By the early 1990s, China, South Korea, and Taiwan had joined Japan in studying the problem of acid deposition and trying to measure the overall effects (Wilkening 2004: 231). The growing interest in acid rain eventually produced a Japanese initiative to create the East Asian Deposition Monitoring Network (EANET) that brought together most states from East and Southeast Asia to monitor acid deposition across the region (Shah 2000). While EANET represents progress in addressing air pollution in East Asia, there are currently no larger international institutional structures in place to support the creation of measures to address the problem. Acid rain in Asia is likely to grow in political salience in the coming years.

Other air pollution issues

Beyond acid rain, the international community is only beginning to follow CLRTAP's lead and address some of the other major air pollutants. The United Nations Environment Program (UNEP), as the primary global institution with responsibility for promoting action to address environmental problems, has taken the lead in sponsoring research on other air pollution threats and supporting negotiations to address them. The 2001 Stockholm Convention on Persistent Organic Pollutants represents the leading edge of negotiations to address some of the other pollutants. (Chapter 35 reviews the development of the international response to persistent organic pollutants.)

CLRTAP successfully negotiated a protocol in 1998 to regulate emissions of the heavy metals mercury, lead and cadmium. UNEP has also been promoting global emission reductions of heavy metals and sponsored working groups to assess the global impacts of mercury pollution in 2007 and 2008, which lead to the Governing Council of UNEP to call for the development of a convention to regulate mercury pollution. Bank (2020) provides an overview of the history and evolution of the Minamata Mercury Convention negotiations. An intergovernmental negotiating committee conducted negotiations between 2010 and 2013, and the Minamata Convention on Mercury was signed in October 2013 and entered into force in August 2017. The Convention represents one of the high points in international air pollution negotiations in recent years.

Over the past 20 years UNEP has continued to press for action to address cadmium and lead as well, but there is currently insufficient international support to address these substances at a global level. UNEP has sponsored additional programs to address other potential air pollutants including the Global Chemicals Outlook II, which was launched in 2019 and is intended to provide a framework for evaluating the production, trade, and disposal of a range of chemicals, including VOCs as well as the Strategic Approach to International Chemicals Management (SAICM) which is intended to address a range of emerging threats. While UNEP continues to raise awareness of a range of air pollution and other chemical threats, the discussions involving international regulation of other chemicals remain contentious and are likely to continue for many years.

Finally, the countries of Asia face a number of challenges related to air pollution. The dramatic economic growth in China and India poses several challenges. The problems in Asia extend well beyond acid rain as growing desertification leads to dramatic dust storms spreading from the interior of China toward Korea, Japan, and even the west coast of the United States. The particulate matter in these dust storms combines with other pollutants from China's poorly regulated industries to produce toxic combinations of dust, heavy metals, POPs, VOCs, and other toxics. The dramatic increase in coal consumption to provide electricity and the growth in heavy industry in China are leading to rapid increases in sulfur, particulate, mercury, and other toxic emissions. The OECD (2012: 276) estimates that by 2050 the number of premature deaths from exposure to particulate matter will double to 3.6 million per year with the vast majority of the increased deaths occurring in India and China. The effects of Chinese air pollution across Asia and the Pacific are likely to increase in political salience in the coming years. In addition, the frequent fires in Southeast Asia, particularly in Indonesia, create harmful transnational flows of smoke and haze. The 2002 ASEAN Agreement on Transboundary Haze Pollution was intended to begin to address the haze pollution. While all ASEAN members have now ratified the agreement and a monitoring system has been created, political will to seriously address the problem remains limited. Nguitragool (2011) reviews the origins and challenges associated with the agreement.

India also poses some difficult challenges. Even as India's economy has grown dramatically with increased electricity and transportation fuel consumption, large parts of the country languish in poverty. Increased reliance on charcoal, wood, and animal dung for fuel is dramatically increasing particulate emissions and worsening air quality across the Indian subcontinent. The incomplete combustion of fossil fuels or biomass produces components of soot referred to as "black carbon." Black carbon combines with other chemicals to form harmful particulate matter and is a precursor to smog. Beyond the human health effects, black carbon has been identified by the Intergovernmental Panel on Climate Change as a significant contributor to climate change (Solomon 2007: 163–164). Black carbon has two major effects. It has a direct warming effect by absorbing more sunlight while it is suspended in the air, and it has a secondary effect when it is deposited on ice and snow. By darkening the surface of ice and snow, it increases the absorption of sunlight and accelerates melting which contributes to the accelerating decline in glaciers. Because black carbon remains suspended for only a few days before settling out of the air, reducing these emissions could provide significant benefits in slowing the pace of climate change as well as dramatically improving air quality in India and other developing countries that utilize biomass-based cook stoves.

Asia is likely to emerge as the focus of future air pollution negotiations. The issues are complicated because they involve critical sectors of national economies and will impose significant costs on polluters. The situation is further complicated by the lack of regional institutions to support negotiations to address these problems. While Europe and North America have made progress in addressing their air pollution problems, Asia will pose some of the most difficult future challenges.

Conclusion

International efforts to regulate transboundary air pollution flows have offered a number of insights for global environmental politics scholars. The evolution of the international response to acid rain and other air pollutants has demonstrated the critical roles of scientific research and the framing of environmental problems for the public and policymakers as well as the importance of the larger domestic and international contexts in shaping the political salience of the issues. International institutions have also played important roles in raising awareness, supporting research, and providing forums for international negotiations. However, air pollution negotiations also pose some additional questions. What is the relative importance of international versus domestic forces in shaping air pollution negotiations? Many international agreements merely ratify what states may have done for domestic reasons. How important are the international agreements in propelling more aggressive action and assuring compliance? Do these agreements alter the power positions and material interests of national actors? There is evidence that institutions can facilitate international agreements and promote national implementation, but the degree of influence and the specific conditions under which institutions can significantly affect international agreements are less clear.

The most successful cases of international air pollution regulation are also concentrated in Europe where political and economic integration have created a permissive environment for addressing common problems. To what extent are the European lessons transferable to other regions of the world? This is particularly important to efforts to address Asian transboundary air pollution. How should we understand the complex governance structures that span the local to the national to the EU to the broader regional structures under CLRTAP? How do the various levels influence one another? Which levels are most important in producing meaningful actions to address transboundary air pollution? These are critical questions because China, India, Brazil and other rapidly growing countries are producing ever greater volumes of air pollutants with transboundary effects that will undoubtedly be the focus of conflict and negotiation in the years to come. Are there lessons from prior agreements that can facilitate a more rapid and effective response to emerging air pollution problems in other regions? These questions will likely be the focus of future global environmental politics research. In addition, global negotiations to address POPs, VOCs and heavy metals are likely to continue. The negotiations surrounding some of these pollutants have been less well documented and will undoubtedly be the focus of research in the future.

Attempts to address transboundary air pollution face a number of additional challenges. The growing crisis surrounding climate change has shifted much of the international attention into attempts to address climate change. In addition, the 2008 financial crisis and relatively lethargic economic growth across much of the world since 2008 along with the pandemic of 2020 appear to have drained the energy away from the difficult negotiations that would be necessary to strengthen existing agreements and negotiate new ones to address emerging issues. Scholarship on transboundary air pollution policy is critical to supporting efforts to resolve growing air pollution problems in many parts of the world.

References

Bank, M.S. (2020) "The mercury science-policy interface: history, evolution and progress of the Minamata convention", *Science of the Total Environment* 722: 1–6.

Boehmer-Christiansen, S. and Skea, J. (1991) *Acid Politics: Environmental and Energy Policies in Britain and Germany*, London: Belhaven Press.

Byrne, A. (2017) "Trouble in the air: recent developments under the 1979 convention on long-range transboundary air pollution", *Review of European Comparative & International Environmental Law* 26: 210–219.

Darst, R.G. (2001) *Smokestack Diplomacy: Cooperation and Conflict in East–West Environmental Politics*, Cambridge, MA: MIT Press.

Dimitrov, R. (2006) *Science and International Environmental Policy: Regimes and Nonregimes in Global Governance*, Lanham, MD: Rowman and Littlefield.

Evelyn, J. (1661) "Fumifugium: or the Inconvenience of the aer and smoak of London dissipated. Together with some remedies humbly proposed by J.E. Esq; to his Sacred Majestie, and to the Parliament now assembled". Online. Available HTTP: <http://archive.org/stream/fumifugium00eveluoft#page/n9/mode/2up>

Levy, M.A. (1993) "European acid rain: the power of tote-board diplomacy", in P.M. Haas, R.O. Keohane, and M.A. Levy (eds) *Institutions for the Earth: Sources of Effective International Environmental Protection*, Cambridge, MA: MIT Press: 75–132.

Lidskog, R. and Sundqvist, G. (2011) *Governing the Air: The Dynamics of Science, Policy, and Citizen Interaction*, Cambridge, MA: MIT Press.

McCormick, J. (1997) *Acid Earth: The Politics of Acid Pollution*, 3rd edn, London: Earthscan.

Nguitragool, P. (2011) *Environmental Cooperation in Southeast Asia: ASEAN's Regime for Transboundary Haze Pollution*, New York: Routledge.

OECD (2012) *OECD Environmental Outlook 2050: The Consequences of Inaction*, Paris: OECD.

Schmandt, J., Clarkson, J., and Hilliard, R. (1988) *Acid Rain and Friendly Neighbors: The Policy Dispute between Canada and the United States*, Durham, NC: Duke University Press.

Selin, H. and VanDeveer, S. D. (2003) "Mapping institutional linkages in European air pollution politics", *Global Environmental Politics* 3: 14–46.

Selin, N.E. (2006) "Applying assessment lessons to new challenges: from sulfur to POPs", in A.E. Farrell and J. Jäger (eds) *Assessments of Regional and Global Environmental Risks: Designing Processes for the Effective Use of Science in Decisionmaking*, Washington, DC: Resources for the Future: 84–100.

Shah, J. (2000) "Integrated analysis for acid rain in Asia: policy implications and results of RAINS-ASIA model", *Annual Reviews Energy & the Environment* 25: 339–75.

Solomon, S. (2007) *Climate Change 2007: The Physical Science Basis: Contribution of Working Group I to the Fourth Assessment Report of the Intergovernmental Panel on Climate Change*, Cambridge: Cambridge University Press.

Underdal, A. and Hanf, K. (eds) (2000) *International Environmental Agreements and Domestic Politics: The Case of Acid Rain*, Burlington, VT: Ashgate.

United Nations Economic Commission for Europe (UNECE) (2012) "ICP modelling and mapping critical loads and levels approach". Online. Available HTTP: <http://www.unece.org/env/lrtap/working-groups/wge/definitions.html>

VanDeveer, S.D. (2006) "European politics with a scientific face: framing, asymmetrical participation, and capacity in LRTAP", in A.E. Farrell and J. Jäger (eds) *Assessments of Regional and Global Environmental Risks: Designing Processes for the Effective Use of Science in Decisionmaking*, Washington, DC: Resources for the Future: 25–63.

Wettestad, J. (1997) "Acid lessons? CLRTAP implementation and effectiveness", *Global Environmental Change* 7: 235–49.

Wettestad, J. (2002) *Clearing the Air: European Advances in Tackling Acid Rain and Atmospheric Pollution*, Burlington: Ashgate.

Wettestad, J. (2006) "EU air quality framework directive: shaped and saved by interaction?" in S. Oberthür and T. Gehring (eds) *Institutional Interaction in Global Environmental Governance: Synergy and Conflict among International and EU Policies*, Cambridge, MA: MIT Press: 285–372.

Wilkening, K.E. (2004) "Localizing universal science: acid rain science and policy in Europe, North America, and East Asia", in N.E. Harrison and G.C. Bryner (eds) *Science and Politics in the International Environment*, New York: Rowman and Littlefield: 209–38.

World Health Organization. (2016) *Ambient Air Pollution: A Global Assessment of Exposure and Burden of Disease*, Geneva: WHO Document Production Services.

31

Stratospheric ozone depletion

Elements of success in global environmental politics

David Downie

The science of ozone depletion

The science surrounding the ozone layer problem is complex in detail but relatively easy to summarize. Ozone (O_3) is rare, accounting for about three of every ten million molecules in Earth's atmosphere. Ozone is also highly reactive and a powerful oxidant, properties that also make it toxic. While little naturally occurring ozone exists at ground level, human-produced ozone is a dangerous pollutant and key component of urban smog and long-term exposure can increase the risk of death from certain respiratory and cardiopulmonary problems. This "bad ozone" in air pollution at ground level, which environmental policy seeks to limit, contrasts with naturally occurring "good ozone" in the upper atmosphere, which the Montreal Protocol seeks to protect.

About 90 percent of naturally occurring ozone exists in the stratosphere, which is the section of the upper atmosphere 10–50 km (6–30 miles) above Earth. Stratospheric ozone, commonly called the ozone layer despite its sparse concentration, plays a critical role helping to protect Earth by absorbing or reflecting certain wavelengths of harmful ultraviolet (UV) radiation. Total deterioration of the ozone layer would be disastrous and significant depletion very harmful. Excessive UV exposure can cause skin cancers, eye cataracts, and weakened immune systems in humans and some animals. Higher UV exposure also causes moderate to very severe damage to many kinds of plants including some food crops, to single-cell organisms, and to aquatic ecosystems and also speeds deterioration of certain man-made materials, including plastics (EEAP 2019).

The concern about anthropological impacts on the ozone layer started in 1970 when Paul Crutzen published an article proposing that chlorine atoms released from certain anthropogenic sources could reach the stratosphere where they would interact with and break down ozone molecules (Crutzen 1970). In 1974, F. Sherwood Rowland and Mario Molina published their now famous article showing how chlorofluorocarbons (CFCs), a group of widely used and commercially important chemicals used primarily as coolants, propellants and solvents, can remain intact after being released into the atmosphere (due to their extremely stable molecular composition) until they reach the stratosphere where they break apart due to the higher levels of radiation from the sun. The decomposition of CFCs releases chlorine

DOI: 10.4324/9781003008873-36

atoms that then interact with and break apart ozone molecules (Molina and Rowland 1974). Moreover, each chlorine atom can potentially destroy thousands of ozone molecules because, after it breaks up one ozone molecule, other chemical interactions occur that release the original chlorine atom to start the process all over again. Subsequent research revealed that other chemicals could also release chlorine into the stratosphere while others could release bromine, another atom capable of the catalytic destruction of ozone. In addition to CFCs, other ozone-depleting substances (ODS) include HCFCs (hydrochlorofluorocarbons), which are less ozone damaging CFC-substitutes widely used in air-conditioning and refrigeration; halons, used as fire suppressants; carbon tetrachloride, used primarily as a cleaning agent or solvent; methyl chloroform, used as a cleaning agent; and methyl bromide, a very toxic broad-spectrum pesticide.

Significant scientific debate ensued for years following Rowland and Molina's article (see in particular: Dotto and Schiff 1978; Roan 1989). Many doubts were raised but no firm evidence arose disputing the theories. At the same time, while evidence grew in the lab and consensus broadened on the likely validity of the theory, no observable ozone depletion emerged in nature. This changed in the mid-1980s when reports emerged of an Antarctic "ozone hole," or depletion of stratospheric ozone of as much as 30 to 50 percent above Antarctica during late winter and early spring (Farman et al. 1985). Scientists eventually proved that chlorine atoms released from CFCs were primarily responsible for the ozone hole, although natural causes contributed to its severity. Because the natural chemical reactions that destroy ozone are accelerated in the presence of very cold air, in particular polar stratospheric clouds, ozone depletion from CFCs is most pronounced in the coldest part of the stratosphere, above Antarctic in the winter. (The ozone layer is also naturally "thinner" above the poles and thickest above the equator.) In addition, wind patterns isolate the winter air above the Antarctic, preventing atmospheric mixing with more ozone-rich air until the spring. This combination of factors, starting with CFC emission, creates the ozone hole.

Above Antarctica, and before the Montreal Protocol's measures made a big impact, seasonal ozone depletion often reached 65 percent or higher in spots with significant loss extending to inhabited regions of Argentina, Australia, Chile and Peru. Above the Arctic, it reached 30 percent on a regular basis in some areas and depletion levels over northern Europe measured between 5 and 30 percent (Ozone Secretariat 2012; WMO et al. 2019). These conditions would have worsened significantly without the original scientific work and an effective global policy response. In recognition of their path-breaking and historic discoveries, Crutzen, Molina and Rowland received the Nobel Prize for Chemistry in 1995.

Creating ozone policy

The creation and expansion of the ozone regime is an important case study because of its success and its broader influence on global environmental policy. Many studies delineate and analyze the development and impact of ozone policy (see Haas 1992; Litfin 1994; Downie 1995, 1996, 2020; Benedick 1998; Anderson and Sarma 2002; Canan and Reichman 2002; Parson 2003; Falkner 2005; Ozone Secretariat 2012; Gareau 2012; Chasek and Downie 2021: ch. 3). The 1974 discovery that CFCs posed a serious threat to stratospheric ozone set off a series of intense scientific and political debates, especially in the United States (Dotto and Schiff 1978). The extraordinary versatility, usefulness and economic importance of CFCs made broad controls very difficult to establish. CFCs dominated the markets for coolants in refrigeration and air-conditioning systems, blowing agents for the manufacture of flexible and rigid foam, propellants in aerosol sprays, and other profitable uses. As a result, global production and use of CFCs continued to expand until the late 1980s.

Continuing the environmental leadership that it exhibited in the early 1970s with the Clean Air and Clean Water Acts, creation of the Environmental Protection Agency (EPA), and other legislation, the United States banned the use of CFCs in many "non-essential" uses, including aerosol spray cans in the late 1970s. At the time the United States accounted for more than 40 percent of worldwide CFC production and the banned uses were more than 40 percent of US CFC use. Canada, Finland, Norway, Sweden, Switzerland and a few other countries took similar action. However, the European Community (EC) refused to take meaningful steps, expressing doubt concerning the scientific theory, noting the lack of observed ozone depletion in the atmosphere, and arguing that no substitutes existed or could easily be developed. EC and corporate opposition combined with the elections of anti-regulatory leaders Margaret Thatcher in the United Kingdom in 1979 and Ronald Reagan in the United States in 1980 effectively killed prospects for additional domestic legislation.

The first global discussions occurred in 1976 when the United Nations Environment Programme's (UNEP) Governing Council considered the issue, without significant result. After two small and relatively fruitless international meetings, in 1977 and 1978, the United States and key EC reached a compromise that UNEP and the World Meteorological Organization (WMO) could create a Coordinating Committee of the Ozone Layer (CCOL) to periodically discuss and perhaps assess relevant scientific issues. Despite EC skepticism, and the lack of confirmation in nature, the increasing scientific consensus regarding the CFC-ozone theory led to statements by the CCOL in the early 1980s that enough scientific evidence existed to warrant global concern. Proponents of international action used these statements to propose that UNEP's Governing Council authorize global negotiations. The EC eventually agreed because it supported UNEP and did not want to undercut the new international organization and because the mandate for the negotiations expressly stated that discussions would focus on international cooperation to study the ozone layer rather than on potential controls (Downie 1996).

The agreement that emerged from these negotiations, the 1985 Vienna Convention for the Protection of the Ozone Layer, affirmed the importance of protecting stratospheric ozone, called for international cooperation in conducting research and monitoring the ozone layer and potential threats to it, and instructed Parties to protect human health and the environment from human activities that might impact stratosphere ozone but did not specify what these actions might be. Indeed, the Convention did not even mention CFCs. However, the treaty did include language calling for Parties to convene negotiations on further measures should definitive threats to the ozone layer be identified. Publication of the discovery of an "ozone hole" above Antarctica allowed proponents of CFC controls to argue successfully that new negotiations were needed, despite the lack of firm evidence linking the hole to CFCs. The new negotiations began in 1986 and concluded, relatively quickly in retrospect, with the landmark 1987 Montreal Protocol on Substances that Deplete the Ozone Layer, the centerpiece of global ozone policy.

The Montreal Protocol established binding requirements that industrialized countries reduce their production and use of the five most widely used CFCs by 50 percent from 1986 levels by 2000, and that they freeze the production of three halons. Developing countries had to take the same actions but with ten-year extensions to allow them to increase their use of CFCs for economic development. The Protocol also included important reporting requirements, prohibition on ODS trade with countries that did not ratify the agreement by a certain date, and procedures for reviewing the treaty's effectiveness and strengthening its controls on the basis of periodic reports to be issued by Scientific, Environmental Effects, and Technology Assessment Panels. New chemicals could be added and other changes made to the Protocol by a standard amendment procedure, which required formal ratification to take effect. However, the treaty also allowed the Meeting of the Parties (MOP) to "adjust" the control measure for any chemical already regulated under the Protocol. Such adjustments would take effect immediately, without the need for ratification by the Parties, and were included to give the Protocol the flexibility to respond quickly to future scientific developments.

Since 1987, Parties have used both these mechanisms to strengthen the Protocol significantly in response to new scientific information on the dangers facing the ozone layer and new technological and economic developments regarding substitutes. The first expansion, agreed to at the second MOP, can be considered a historic agreement on its own. The 1990 London Amendment added eight additional CFCs, as well as methyl chloroform and carbon tetrachloride to the Protocol's control measures. Parties also adjusted the existing controls so that countries, rather than meet a 50 percent cut, now had to phase out by 2000 the production and use of the CFCs and halons listed in the original Protocol. This likely represented the first binding global agreement to eliminate specific chemicals that harm the environment (Downie 1996). The 1990 London agreement also created an essentially unprecedented non-compliance procedure (see below).

A third historic achievement of the 1990 London Amendment was creation of the Multilateral Fund for the Implementation of the Montreal Protocol. The fund assists developing countries implement the Protocol by providing funds for capacity building, planning and, in particular, the cost of switching from ODS to alternative chemicals or processes. The Multilateral Fund was the first major assistance fund established under a global environmental agreement. It predated and likely influenced the 1991 creation of the Global Environment Facility (GEF) and was, along with the later phase-out dates granted developing countries, a concrete manifestation of the principle of common but differentiated responsibilities. Creating the fund was also a political necessity as the largest developing countries had refused to ratify the Montreal Protocol unless specific provisions were added to provide financial and technical assistance that would help them access the replacement chemicals. In the late 1980s, China and India were in the process of developing indigenous CFC industries and argued it would be unfair for them to join a global environmental agreement if that meant they would need to pay more for CFC alternatives imported from industrialized countries. Creation of the Multilateral Fund and somewhat vague assurances that HCFC facilities would be built in their countries responded to these concerns (Downie 1996; and personal observations and communications during the 1990 negotiations).

The 1992 Copenhagen Amendment and adjustment introduced binding control measures on HCFCs, methyl bromide, and hydrobromofluorocarbons to the Protocol and further accelerated the phase-out of CFCs and halon. The 1997 Montreal Amendment and adjustment accelerated the phase-out of methyl bromide, earmarked specific Multilateral Fund resources for methyl bromide projects (in a deal with developing countries to get them to accept the faster phase-out), and created a new CFC licensing system to combat illegal trade. The 1999 Beijing Amendment and adjustment mandated the immediate phase-out of bromochloromethane, strengthened controls on HCFCs (including introducing production controls and limits on HCFC trade with non-parties), and increased reporting requirements on methyl bromide to limit unauthorized use. In 2007, Parties significantly accelerated the controls on HCFCs, not only to protect the ozone layer more effectively but also to address climate change, as HCFCs are also potent greenhouse gases (GHGs).

In 2016, Parties took another historic step when they adopted the Kigali Amendment. The agreement added HFCs – which do not deplete stratospheric ozone but are powerful GHGs – to the Montreal Protocol and established a control schedule that will phase-out 85 percent of HFC production in industrialize countries by 2036 and 80 percent in most developing countries by 2045. After years of debate, countries agreed that expanding the treaty to include HFCs was justified because the chemical might not exist if not for the Protocol's controls on CFCs and controlling them under the Protocol would allow developing countries to access the Multilateral Fund to support their reduction activities.

The Montreal Protocol was the first global environmental treaty to enjoy universal participation. A total of 197 countries, including all the major producers and users of ODS, are Parties to the Vienna Convention, the Montreal Protocol and the 1990 London, 1992 Copenhagen Amendments, 1997 Montreal Amendment, and 1990 Beijing amendments. More than 110 countries are Parties to the 2016 Kigali Amendment.

Provisions of the Montreal Protocol and ozone policy

The main elements of global ozone policy are found in the Montreal Protocol and related agreements and decisions of the Protocol's decision-making body, the Meetings of the Parties (MOP). Industrialized country Parties were or are required to phase out their use and production of ODS, and to restrict trade of ODS with non-parties. Developing countries are allowed more time to begin and complete their phase-out schedules. These differentiated obligations were politically necessary to obtain the participation of some large developing countries and also reflect the understanding by all Parties that the industrialized countries had far larger ODS emissions than developing countries when the Protocol and most of its amendments were negotiated, and that developing countries needed access to most of the chemicals for economic development. The control measures include "essential use" exemptions that allow for the production and consumption of CFCs and halon for longer periods, subject to approval by the MOP. A general exemption exists for using very small amounts of ODS in laboratory applications. Perhaps most importantly, there is a large general exemption for the use of methyl bromide for quarantine and pre-shipment applications (e.g., the fumigation of shipping containers and commodities) as well as "critical use" exemptions for using methyl bromide for agricultural purposes.

The Protocol obligates industrialized countries to provide technical and financial assistance to developing countries and countries with economies in transition (CEITs) to help them fulfill their obligations. Since its establishment, the fund has disbursed more than $4 billion to support projects in nearly 150 countries and is widely considered a key ingredient in the success of the ozone regime (Multilateral Fund Secretariat 2020). The existence and effectiveness of the fund has made it easier, both politically and economically, for many developing countries to agree to accelerate ODS phase-out schedules and has resulted in the phase-out of nearly all production of CFCs, carbon tetrachloride, halon and methyl chloroform in developing countries.

Governance and administration of the ozone regime is similar to that in other environmental treaties. The Meeting of the Parties is the supreme decision-making body and meets annually. An Open-Ended Working Group (OEWG) holds discussions in preparation for the MOP, usually four to six months prior. All countries Party to the Protocol can participate in the MOP and OEWG with full decision-making privileges. While the Protocol does allow for supermajority voting, the strong norm is to take decisions by consensus and to date the MOP has held no official vote. The MOP can agree to amend the treaty, changing the text of the Protocol, which then requires ratification by individual Parties to take effect. The MOP can also adjust regulations on chemicals already controlled under the Convention as well as take decisions on other policy matters that do not change the wording of the Protocol. Adjustments and other decisions go into effect immediately. Representatives from international organizations, nongovernmental organizations, industry groups and research institutions can attend MOP and OEWG meetings as observers and participate in Plenary and some contact-group discussions. The Ozone Secretariat, based at UNEP headquarters in Nairobi, performs standard administrative functions.

Successes and challenges of global ozone policy

Arguably, the most important measure of an environmental regime's effectiveness or success is its impact on the problem it was created to address; has the regime produced measurable change in the environment? In addition, because environmental issues exist as consequences of human activity, for a regime to be successful, it must have impacted, in a measurable way, the human activity that produced the environmental harm (Young 2011: 19854; Chasek and Downie 2021).

Other criteria might be relevant as well. For example, measures of state participation, implementation and/or compliance; the strength or quality of regime rules and institutions; cost-effectiveness; the impact of regime norms and principles on actor perceptions of their interests; and its impact on other issue areas (representative discussions include Young 1999, 2011; Sprinz 2000). Along all these measures, the Montreal Protocol and the broader ozone regime should be seen as very successful; not perfect and not without remaining challenges, but highly successful.

Significant reductions in ODS production, use and emissions

The Montreal Protocol has successfully reduced the production, use and emissions of ODS. Nearly all of the production and use of new CFCs, halon, carbon tetrachloride and methyl chloroform have been eliminated. Methyl bromide production has declined drastically and HCFC controls are proceeding according to the control schedule (see Ozone Secretariat 2020a for comprehensive data on ODS production, use, and exemptions). As a result, the atmospheric abundance of all major ODS except halon 1301 and HCFCs is declining, as is the amount of chlorine and bromine in the stratosphere (WMO et al. 2018). Because Argentina, Brazil, China, the EU, India, Indonesia and Thailand, among many other countries, did not take meaningful action to reduce CFCs and other ODS until they joined the Montreal Protocol, and because key ODS alternatives were invented or commercialized in response to controls established by the Protocol, these declines must be attributed to the impact of the ozone regime.

As a consequence of these cuts in ODS, most of the ozone layer will likely return to normal levels near the middle of this century, with recovery of Antarctic ozone following later. Computer simulations show that without the Protocol, ODS emissions would have produced global ozone depletion and solar UV radiation levels far higher than they are now and that much larger levels would have occurred in the future (Newman and McKenzie 2011; WMO et al. 2018).

Impacts on human health and the environment

The Montreal Protocol prevented increases in UV radiation that would have produced large-scale, negative impacts on the environment and human health (Newman and McKenzie 2011; EEAP 2019; WMO et al. 2019). These include the prevention of tens of millions of cases of fatal skin cancer and many more millions of non-fatal skin cancer and eye cataracts (van Dijk et al. 2013; EEAP 2019). The EPA estimates that in the United States alone, "full implementation of the [full amended] Montreal Protocol is expected to prevent approximately 443 million cases of skin cancer, 2.3 million skin cancer deaths, and 63 million cases of cataracts for people born in the years 1890–2100" (Ozone Secretariat 2020b). Globally, from 1987 to 2060, the ozone regime could yield $1.1 trillion in reduced health care costs from skin cancer reductions and $1.8 trillion in overall health care benefits and perhaps "$460 billion in avoided damages to agriculture, fisheries and materials" (Ozone Secretariat 2020b).

Global participation

As noted, the Montreal Protocol and most of its amendments are the only environment treaties to achieve universal ratification. This contrasts with other major environmental treaties, such as the Kyoto Protocol to which Canada and the United States were not Parties; the Stockholm Convention to which Israel, Malaysia, the United States are not Parties; and the Basel Convention Ban Amendment which currently has only 100 Parties.

Most states have met their obligations under the Protocol. This does not mean that all states met all of their phase-out, reporting, financial and other obligations on time. Many national reports are submitted late. A number of Eastern Europe and developing countries missed some of the CFC phase-out targets. However, parties to the Protocol have phased-out 98 percent of ODS production required under Protocol (Ozone Secretariat 2020b).

Strong regime rules and effective institutions

The ozone regime contains strong, clear and binding rules obligating Parties to meet specific obligations to phase-out ODS. Contrast these rules with the weaker provisions of the climate (see Chapter 32), biodiversity (see Chapter 41) and desertification (see Chapter 43) regimes. In addition, the institutions developed under the Protocol, including the Multilateral Fund, Assessment Panels and Implementation Committee, are seen by Parties as operating effectively, albeit not perfectly or without criticism.

The ozone regime is often cited, appropriately and inappropriately, by global environmental policymakers as providing models and lessons in how to design, expand or implement effective global environmental policy (based on this author's observations during more than 70 global environmental negotiations on a variety of issues over the past 30 years). While success should be emulated, cogent analysis argues for careful consideration of what aspects of the ozone regime countries can transfer successfully to other issue areas versus what aspects were products of circumstances somewhat unique to the ozone issue area or the time during which key aspects were developed. At the same time, there is little doubt that experience gained in the ozone regime regarding the design of control measures, reporting requirements, provisions of financial and technical assistance, employment of assessment panels, inclusion of trade provisions, non-compliance procedures, broader participation for NGOs and other issues has positively impacted discussions and developments on other global environmental issues.

Remaining challenges

Successes to date do not automatically necessitate continued success. The scientific analysis that predicts that most areas of the ozone layer should return to pre-ODS levels rests on the assumption that all countries will remain willing and able to fulfill all of their remaining obligations. Indeed, global ozone policy faces important challenges that could delay or even prevent full recovery (Downie 2015).

ODS banks

While almost no legal production and use of new CFCs remain, that does not mean that all the CFCs and other ODS produced in the past have already reached the atmosphere and are no longer a threat. Many millions of tons of CFCs remain in old or discarded refrigerators and air-conditioners, insulating foam and other products and wastes. Large amounts of halons exist in in-use or decommissioned fire extinguisher and suppression systems. Collectively known as ODS Banks, these chemicals will eventually reach the atmosphere unless they are captured and destroyed, delaying recovery of the ozone layer. Indeed, emissions from banks is projected as the largest source of future ODS emissions (WMO et al. 2018). Many governments recognize the seriousness of this issue but coordinated and sustained efforts to rectify it have not yet begun and many developing countries lack the resources to destroy ODS banks in an environmentally sound manner.

Completing the methyl bromide and HCFC phase-out

HCFCs are efficient, cost-effective and far less ozone-depleting alternatives to CFCs used widely in the global air-conditioning and refrigeration sectors. They are also less potent greenhouse gases than another key CFC alternative known as HFCs. Many industrialized and developing countries have based some or even most of their post-CFC refrigeration and air-conditioning infrastructure on HCFCs. Thus, it is possible that as the final HCFC phase-downs arrive, that some countries, particular large developing countries with major HCFC production facilities, might decide that although they have implemented significant reductions, complete elimination is not economically justified or requires more assistance from the Multilateral Fund than donor countries are willing to provide (Downie 2015, 2020).

Methyl bromide presents a different type of obstacle. Industrialized countries phased out most methyl bromide in 2005 and developing countries did the same in 2015. However, methyl bromide remains in use in many countries, including the United States, under the broad exemptions granted for critical agricultural uses and for quarantine and pre-shipment applications. The European Union and others believe that effective, economically viable and environmentally friendly alternatives exist for all uses of methyl bromide. Others argue that alternatives are not sufficiently effective or economically viable for all uses in all countries, especially pre-shipment and quarantine application. While the amounts used under these exemptions has declined significantly, methyl bromide exemptions could mean that some production, use, and emissions of methyl bromide will continue for some time.

Illegal CFC production

Recent measurements suggest there is new, illegal production of CFC-11 in eastern Asia (WMO et al. 2019). While the country, or countries, producing these chemicals is unknown at present, and the relevant national government might not be aware of the production, creation of a global or regional black market in CFCs could pose a threat to the success of the regime and recovery of the ozone layer. There is also the potential for black market HCFC production and use as global phase-outs take effect.

Implementing the HFC reductions and phase-out without a return to CFCs and HCFCs

The Montreal Protocol has made large contributions to mitigating climate change by reducing emissions of ODS that are also greenhouse gases. From 1990 to 2010, ODS reductions under the Montreal Protocol prevented 135 billion tonnes of CO_2-equivalent greenhouse gas emissions, which is about five times larger than the annual emissions reduction target for the first commitment period (2008–2012) of the Kyoto Protocol (Ozone Secretariat 2020b).

At the same time, HCFCs and HFCs, the two most widely used substitutes for CFCs, are also potent greenhouse gases. Thus, the ozone regime has both assisted efforts to mitigate climate change and made it more difficult. Both HCFCs and HFCs are now addressed under the Protocol and their climate impact, while very significant, will decline and eventually be nearly eliminated if countries fulfill all their obligations, However, the full phase-out of HCFCs by all large countries is not guaranteed and the Kigali amendment only mandates about an 85 percent reduction in HFCs (although this could be increased in the future). Addressing HCFCs and HFCs represents an important climate challenge.

Funding and political will

Success can breed complacency. Developing and donor countries face challenges maintaining the political will to eliminate the remaining methyl bromide exemptions, complete the HCFC phase-out, control emissions from ODS banks, and drastically reduce emissions of HFCs. While the political and economic hurdles required for these efforts might be relatively small compared with addressing climate change effectively by drastically reducing carbon-dioxide and methane emission (see Chapter 32), it is possible that the economic difficulties produced by the Covid-19 pandemic, false confidence that the ozone problem has been solved, or broader international political or economic differences could prevent the continued political commitment necessary to ensure recovery of the ozone layer.

Explaining the development and success of the ozone regime

Many factors helped shape the development and extent of the success of the ozone regime. Advancing scientific knowledge played a very important but not a determinative role in the creation and expansion of global ozone policy. (For discussion and analysis of this impact, see Haas 1992; Litfin 1994; Downie 1996, 2020; Benedick 1998; Anderson and Sarma 2002; Canan and Reichman 2002; Parson 2003; Chasek and Downie 2021.) Scientific discovery gave rise to the issue in the first place. Advancing scientific knowledge and consensus in the 1980s undercut European opposition to starting negotiations on a framework Convention. The discovery of the ozone hole galvanized public opinion and policymakers and gave control proponents the platform they needed to restart negotiations aimed at controlling CFCs. Confirmation that chlorine atoms released from CFCs were the ultimate cause of the Antarctic ozone hole eviscerated arguments that further controls should wait until more evidence was found, helped completely reverse the EU negotiating position, and contributed to the strengthening of the ozone regime in London in 1990. Discovery of depletion above the northern hemisphere and the continued worsening of the Antarctic holes contributed to the 1992 Copenhagen Amendment. The 2007 IPPC report and conclusions by the Scientific Assessment Panel that a precautionary approach to protecting the stratospheric ozone layer required further action helped spur the surprising decision in Montreal in 2007 to accelerate the HCFC phase-out.

The complexity of the scientific information also helped a transnational network of experts who supported action, an epistemic community, to influence policymakers who had come to rely on them to interpret the science (Haas 1992; see Chapter 18). It allowed experts who understood the atmospheric science to shape discourse on the issues, to frame discussions, to introduce precautionary and intergenerational time frames, and to influence other policymakers (Haas 1992; Litfin 1994; Canan and Reichman 2002).

From a different perspective, advancing scientific knowledge helped frame the negotiations, constraining and undercutting actors when they supported positions that appeared to go against the consensus knowledge as set forth in the assessment reports (Downie 1996). In the language of simple game theory, advancing knowledge also helped "alter the payoff structure" and "enhance the shadow of the future" (Oye 1985). Cooperation became more likely as countries increased the value they attached to protecting stratospheric ozone (altered payoffs) or came to believe that they would be holding negotiations on the issue for many years (shadow of the future). As the regime grew, scientific information combined with other aspects of the regime to increase the value that actors attached to the ozone regime, further enhancing cooperation. As almost all the analyses of the ozone regime point out, however, while the impact of scientific knowledge and consensus was important, even necessary, it was not sufficient on its own to produce the current regime. A confirming argument is to examine the development of the climate regime at analogous stages of scientific knowledge and consensus in the form of reports by the IPCC (see Chapter 32).

Another set of important causal factors centers on the economic interests of key actors. Perceptions of economic costs, particularly adjustment costs, impacted perceptions of state interests, which, in turn, impacted their policy preferences (Sprinz and Vaahtoranta 1994; Oye and Maxwell 1994; Downie 1996; Falkner 2005; Chasek and Downie 2021). Not surprisingly, economic interests related to ODS production and use often impeded efforts to create stronger controls. Examples include the lack of CFC regulations in most of Europe prior to the Montreal Protocol; Europe preventing the inclusion of control measures in the Vienna Convention; the 50 percent reduction target set in the Montreal Protocol (which allowed the EU to meet much of their obligation through inexpensive controls on the use of CFCs in aerosol sprays); the inclusion of exemptions, especially the critical use exemption for methyl bromide (Gareau 2012); and the relatively lengthy phase-out periods for HCFCs and HFCs, especially in developing countries.

Similarly, at times during the regime's development, countries on both sides of a particular policy debate pushed for policies that would result in low adjustment costs for the relevant industries in their countries, producing policy stalemates. This occurred during creation of the Vienna Convention and the early stages of the Montreal Protocol negotiations when the United States and others advocated banning the use of CFCs in aerosol sprays, which they had already done, while EU countries advocated a cap on CFC production capacity, something they had already enacted and knowing that their companies had significant excess capacity while US CFC producers did not (Downie 1996). A similar situation emerged during the initial efforts to strengthen HCFC controls and the debate on HFCs

However, during several crucial periods economic interests greatly assisted efforts to strengthen the ozone regime. First, the regulation of CFCs in the United States and in the Montreal Protocol created economic incentives for companies to develop substitutes (Downie 1996; Benedick 1998; Falkner 2005). More substantially, the development of effective substitutes, especially for CFCs, altered the economic interests of particular industries, major corporations or governments, lowering the costs associated with eliminating ODS and allowing some actors to profit. By the late 1980s, CFCs had become low-margin chemicals facing imminent competition from large production facilities planned in China and India. Once the major CFC producers in Japan, Europe and the United States were certain that they could produce HCFCs and HCFs, they changed their position and began to support a gradual global CFC phase-out as this would create a market for HCFCs and HFCs (Oye and Maxwell 1994; Downie 1996; Falkner 2005). Along with the new scientific information linking CFCs to the ozone hole, and domestic political realities in the United Kingdom and Germany, this change in long-term economic interests contributed to a rapid change in EU policy (Downie 1996). The Multilateral Fund also impacted economic interests. Some companies that received support from the fund and transitioned away from CFCs also became internal advocates of stronger domestic enforcement actions as they did not want to get undercut by competitors still using CFCs.

Other causal factors involve the neoliberal institutional observation that international institutions can positively impact the development of collective action (e.g., Keohane 1984; Haas et al. 1993). While the ozone regime might have developed exactly as it did, when it did, the presence and actions of UNEP, WMO, the CCOL, and the Vienna Convention and Montreal Protocol themselves significantly assisted efforts by states and other actors seeking effective ozone policy (Downie 1996). UNEP's presence greatly eased the process of initiating international discussions; indeed, the organization was created for that purpose. European actors skeptical or even hostile to the idea of discussing ozone policy had difficulty opposing UNEP's nascent efforts on the issue. Recognition of WMO's function and expertise facilitated creation of the CCOL and the Scientific Assessment Panel. CCOL reports gave greater international weight to the emerging scientific concern, which was still centered in the United States in the early 1980s. If the Vienna Convention had not existed, it likely would have taken far longer to initiate global negotiations on regulating CFCs following discovery of the ozone hole. In addition, the act of negotiating the Convention and Protocol impacted attitudes in some states, helping to raise awareness and concern, lower perceived costs and build trust.

In addition, during the early stages of the scientific debate through creation of the 1990 London Amendment, UNEP played a particular important role (Downie 1995). In the late 1970s, UNEP worked to help initiate international action by organizing the first scientific and political meetings focusing on ozone depletion. UNEP then worked to sustain international attention on the issue when interest in ozone depletion waned significantly during the early 1980s. Once substantive negotiations began, UNEP facilitated regime creation by establishing a procedural foundation and reducing transaction costs. Finally, UNEP – through the efforts of its Executive Director Mostafa Tolba – actively pushed the Parties toward agreements in Montreal in 1987 and London in 1990. Tolba also actively lobbied developing countries in the 1980s and 1990s to join the regime (Benedick 1998).

Finally, the ozone regime could not have strengthened its controls so quickly if Parties had not been able to adjust the control measures on ODS already listed in the Protocol. The controls on CFCs went from requiring a 50 percent cut by 2000 in the original Protocol to a 100 percent phase-out by 2000 in the 1990 London adjustment to a 100 percent phase-out by 1996 in the 1992 Copenhagen adjustment. Controls on methyl bromide and HCFCs were also accelerated in important ways. All these updates became binding immediately. If the agreements had required formal amendments and ratification, the process would have taken years longer, if it would have occurred at all. This is just one aspect of the ozone regime's design identified as important to its success. Others include:

- The concise, clear and obviously binding nature of the obligations to reduce and eliminate ODS production and use.
- The fact that the Protocol did not attempt to specify how countries meet its obligations which allowed each country to set policies appropriate for its circumstances (e.g., controls on specific uses, market-wide reductions, taxes, economic incentives).
- The allowance for exemptions to prevent isolated and relatively small interests from preventing a country from joining the regime while also limiting most exemptions by requiring that countries apply for them annually and receive approval from the MOP.
- Trade restrictions that prohibit Parties from exporting ODS and products containing ODS after a certain date. These provisions acted as a powerful incentive for importing countries, especially smaller countries, to join the regime and act to discourage countries from leaving the regime.
- The Multilateral Fund, which ensured the participation of large developing countries in the regime, assisted developing countries meet and sometimes exceed the phase-out schedules, and created supporters of ODS control among the actors that received funding and had transitioned to alternatives.
- The stated principle that control measures should be guided by scientific understanding of threats to the ozone layer in a precautionary manner and the general although not absolute observance of this principle.
- Requirements for Parties to report annual data on production, imports and exports of the controlled substances.
- The requirement that the MOP review the adequacy of the control measures on the basis of available scientific, environmental, technical and economic information.
- Creation of assessment panels to provide independent, authoritative information directly to the Parties.
- A robust but facilitative non-compliance procedure focused on identifying instances of non-compliance and working with the Parties to seek solutions.

Conclusion

If the terms of the amended Montreal Protocol are fully implemented, the ozone layer should recover. Full recovery could be delayed, however, and ozone depletion could even worsen if countries do not fully implement their remaining obligations, address ODS banks, control exemptions and prevent new black market ODS. Thus, the next few decades will determine if the ozone regime meets its ultimate objective of permanently safeguarding stratospheric ozone.

The ozone regime provides compelling theoretical and empirical evidence that effective global environmental policy is possible. No other regime has combined all the elements that the Montreal Protocol has used to such great advantage: active and influential scientific and technological assessment panels; clear, binding control measures with specific deadlines; strong review requirements; the ability to strengthen controls rapidly; meaningful trade penalties for remaining outside the regime; a dedicated, active, and sufficiently funded financial mechanism; and robust compliance procedures.

While aspects of the issue-area certainly helped produce the ozone regime's success, ODS pollution is not fundamentally different than other global environmental issues. Nothing prevents the global community from taking similar action to address other threats. Thus, perhaps the key question for countries facing other global environmental issues is this: do they have the political will and wisdom to act before it is too late?

David Downie

References

Anderson, S. and K. Sarma. (2002). *Protecting the Ozone Layer: The United Nations History*, London: Earthscan.

Benedick, Richard. (1998). *Ozone Diplomacy*, 2nd edn, Cambridge, MA: Harvard University Press.

Canan, Penelope and Nancy Reichman. (2002). *Ozone Connections: Expert Networks in Global Environmental Governance*, Sheffield: Greenleaf Publishing.

Chasek, Pamela S. and David L. Downie. (2021). *Global Environmental Politics*, 8th edn, Abingdon: Routledge.

Crutzen, Paul. (1970). "The Influence of Nitrogen Oxides on the Atmospheric Content," *Quarterly Journal of the Royal Meteorological Society*, 96(408): 320–325.

Dotto, Lydia and Harold Schiff. (1978). *The Ozone War*, New York: Doubleday.

Downie. David, (1995). "UNEP and the Montreal Protocol," in Robert Bartlett, Priya Kurian and Madhu Malik (eds) *International Organizations and Environmental Policy*. Westport, CT: Greenwood Press.

Downie, David. (1996). "Understanding International Environmental Regimes: The Origin, Creation and Expansion of the Ozone Regime." Unpublished PhD Dissertation. University of North Carolina at Chapel Hill.

Downie, David. (2015). "Still No Time for Complacency: Evaluating the Ongoing Success and Continued Challenge of Global Ozone Policy," *Journal of Environmental Studies and Sciences*, 5(2): 187–194.

Downie, David. (2020). "The Vienna Convention, Montreal Protocol and Global Policy to Protect Stratospheric Ozone," in P. Wexler, et al (eds) *Chemicals, Environment, Health: A Global Management Perspective*, 2nd edn. Boca Raton, FL: Taylor & Francis.

EEAP. (2019). *Environmental Effects of Ozone Depletion and Its Interactions with Climate Change: 2018 Assessment Report*, Nairobi: Environmental Effects Assessment Panel, UNEP.

Falkner, Robert. (2005). "The Business of Ozone Layer Protection: Corporate Power in Regime Evolution," in David Levy and Peter Newell (eds) *The Business of Global Environmental Governance*, Cambridge, MA: MIT Press.

Farman, Joseph, B. Gardiner and J. Shanklin. (1985). "Large Losses of Total Ozone in Antarctica Reveal Seasonal ClO_x/NO_x Interaction," *Nature*, *315*: 207–210.

Gareau, Brian. (2012). *From Precaution to Profit*, New Haven, CT: Yale University Press.

Haas, Peter. (1992). "Banning Chlorofluorocarbons: Epistemic Community Efforts to Protect Stratospheric Ozone," *International Organization*, 46(1): 187–224.

Haas, Peter, Robert Keohane and Marc Levy. (1993). *Institutions for the Earth: Sources of Effective International Environmental Protection*, Cambridge, MA: MIT Press.

Keohane, Robert. (1984). *After Hegemony: Cooperation and Discord in the World Political Economy*, Princeton, NJ: Princeton University Press.

Litfin, Karen. (1994). *Ozone Discourses: Science and Politics in Global Environmental Cooperation*, New York: Columbia University Press.

Molina, Mario and F. Sherwood Rowland. (1974). "Stratospheric Sink for Chlorofluoromethanes: Chlorine Atomic Catalyzed Destruction of Ozone." *Nature*, *249*: 810–812.

Multilateral Fund Secretariat. (2020). "Welcome to the Multilateral Fund for the Implementation of the Montreal Protocol." http://www.multilateralfund.org/default.aspx.

Newman, Paul and Richard McKenzie. (2011). "UV Impacts Avoided by the Montreal Protocol," *Photochemical & Photobiological Science*, 10(7): (1152).1160.

Oye, Kenneth (ed.) (1985). *Cooperation under Anarchy*, Princeton, NJ: Princeton University Press.

Oye, Kenneth and James Maxwell. (1994). "Self-interest and Environmental Management," *Journal of Theoretical Politics*, *64*: 599–630.

Ozone Secretariat. (2012). *Montreal Protocol on Substances that Deplete the Ozone Layer, 2012: A Success in the Making*, Nairobi: UNEP Information Pamphlet.

Ozone Secretariat. (2020). "Data in Tables," https://ozone.unep.org/countries/data-table.

Ozone Secretariat. (2020). "Facts and Figures on Ozone Protection," https://ozone.unep.org/facts-and-figures-ozone-protection

Parson, Edward. (2003). *Protecting the Ozone Layer: Science and Strategy*, Oxford: Oxford University Press.

Roan, Sharon. (1989). *Ozone Crisis: The 15-Year Evolution of a Sudden Global Emergency*, New York: John Wiley & Sons.

Sprinz, Detlef. (2000). "Measuring the Effectiveness of International Environmental Regimes," *Journal of Conflict Resolution, 44*: 630–52.

Sprinz, Detlef and Tapani Vaahtoranta. (1994). "The Interest-Based Explanation of International Environmental Policy," *International Organization, 48*(1): 77–105.

Van Dijk. A., H. Slaper, P.N. den Outer, O. Morgenstern, P. Braesicke, J.A. Pyle et al. (2013). "Skin Cancer Risks Avoided by the Montreal Protocol – Worldwide Modeling Integrating Coupled Climate-Chemistry Models with a Risk Model for UV," *Photochemistry and Photobiology, 89*(1): 234–246.

WMO (World Meteorological Organization) et al. (2019). *Scientific Assessment of Stratospheric Ozone: 2019*, WMO Global Ozone Research and Monitoring Project, report no. 58. Geneva: WMO.

Young, Oran (ed.) (1999). *The Effectiveness of International Environmental Regimes: Causal Connections and Behavior Mechanisms*, Cambridge, MA: MIT Press.

Young, Oran (2011). "Effectiveness of International Environmental Regimes: Existing Knowledge, Cutting-edge Themes, and Research Strategies," *PNAS, 108*(50): 19853–19860.

32

Climate change

International diplomacy and governance from top to bottom

Paul G. Harris

Among the many issues described in this book, climate change – the large-scale, unnatural (anthropogenic) environmental changes caused by emissions of greenhouse gases from the burning of fossil fuels and other human activities – is by far the most scientifically and politically complex. It is an existential threat to people, other species and the biosphere. Major effects of climate change are already being experienced around the world, such as sea-level rise, more powerful storms, intense droughts, unprecedented wildfires and the widening spread of pathogens. Especially since the 1980s, climate change has evolved from a minor issue in international relations to a consistently high-profile global priority. It is the most prominent and difficult challenge in global environmental politics.

The politics of climate change have been tortuous. Despite international agreements brokered by the United Nations (UN), national policies to encourage the use of renewable energy and climate-friendly pledges by governments, countless nongovernmental organizations (NGOs) advocating action and increasing awareness of environmental sustainability among corporations and citizens, global emissions of carbon dioxide (CO_2) and other greenhouse gases continue to rise. This is happening despite repeated warnings from scientists that emissions must be zeroed out by mid-century if the most devastating effects of climate change are to be avoided (see, e.g., IPCC, 2018). Making matters worse, the financial, technological and other resources that are needed for adapting to the many unavoidable impacts of climate change are a fraction of what is required, especially if we consider the needs of poor states, communities and individuals. To be sure, at every level of climate governance – international, national, local, corporate and individual – policies and practices to mitigate greenhouse gas pollution and to address its consequences are increasing. Yet, year after year, decade after decade, they have been inadequate to the task. Put simply, they have amounted to too little, too late.

At all levels of analysis, ranging from the global to the local and the individual, the extent to which climate governance is to be successful is influenced by ongoing international negotiations. For example, agreements reached by diplomats can influence policies at the national level. In turn, what happens at the national level (and other levels) can influence climate diplomacy. With that in mind, this chapter summarizes three decades of international negotiations intended to prevent dangerous climate change, adapt to the change that cannot

 DOI: 10.4324/9781003008873-37

be avoided, and assist those states most affected by its impacts. (For a much more detailed description of the full spectrum of the global politics of climate change, including analysis of climate governance at the international, national and individual levels, see Harris, 2021.)

Climate change: from scientific obscurity to global reality

Until relatively recently, climate change was perceived (when it was perceived at all) to be a problem that would affect future generations and therefore was not a chief concern of current governments. That is no longer the case; climate change is now widely recognized as a problem whose impacts are already being felt around the world, whether they be wildfires in California and Australia, more damaging hurricanes along the shores of the Gulf of Mexico, floods in Europe or prolonged droughts in North Africa. The impacts of climate change on natural ecosystems and on human societies are increasingly severe, particularly in parts of the world where geographic vulnerability and poverty make adaptation to changes difficult or impossible. Importantly for understanding the global politics of climate change, the problem is intimately connected to nearly all economic activity and is particularly wrapped up with modern lifestyles and consumption habits (see Chapter 17).

Since the 1980s, when climate change started to become a prominent political issue, scientists have greatly improved their understanding of its causes and consequences. They have developed an increasingly precise understanding of how greenhouse gas pollution is affecting the environment on land, in the oceans and in the atmosphere. Very importantly, unlike in previous decades, there is no longer credible scientific doubt that human activities are to blame for global warming and myriad manifestations of climate change. The most authoritative official reports have been produced by the Intergovernmental Panel on Climate Change (IPCC). The IPCC's periodic assessments of improving understanding of climate change have been vital to informing international negotiations on what to do about the problem. That said, it is worth bearing in mind that the IPCC is an official international body created for and by governments. Consequently, it tends to reach conclusions based on consensus, and its findings have, at least until recently, tended to downplay the pace and scale of climate change and its likely adverse impacts. Generally speaking, based on reports from the panel over more than three decades, reality is likely to be worse, possibly much worse, than anticipated by the IPCC. For example, many of the effects of climate change that had been predicted for much later in the century, ranging from substantial atmospheric warming to rapid melting of ice sheets, are already occurring. Nevertheless, even the consensus decisions of the IPCC have for decades been routinely challenged by those who deny the reality, or at least the severity, of climate change. The most obvious example of this was US President Donald Trump, the world's most famous climate skeptic. Domestically, his administration denied the significance of the problem and dismantled US policies intended to address it. Internationally, under Trump the United States withdrew from the Paris Agreement (see below) and US diplomats joined with likeminded states (e.g., Australia and Saudi Arabia) to water down efforts toward the effective international governance of climate change.

According to the fifth assessment report of the IPCC, every part of the natural environment has been affected by climate change (IPCC, 2014: 4–13). Water systems around the world have been particularly affected, for example with diminished snowfall reducing runoff in some locations while melting glaciers have increased it in others. The former results in drought while the latter sometimes results in calamitous mudslides. Species on land and in the sea are shifting their ranges. For example, some land animals are moving toward higher elevations and some fish species are moving away from areas of the oceans that are unusually warm. Some creatures have gone extinct, and many others are likely to do so. Crop yields are being adversely affected, reducing food security in places where food security is already lacking. Human health is being undermined by climate change. For example, more people are dying from heat stress and more are being affected by diseases spread by mosquitos and other pathogens whose ranges are expanding alongside environmental changes. Extreme weather events have become more common. Increased heat waves, wildfires, droughts, severe storms and floods are causing suffering to humans and other species. These and other climate-related changes are multiplying the dangers posed by existing risks for communities and individuals, especially the world's poor. People in areas of conflict face added vulnerability due to the impacts of climate change. The tragic reality is that most of these impacts will only become more adverse as time passes; they are "baked into" Earth's climate system due to past greenhouse gas emissions. The challenge for practitioners of global environmental politics will be more profound and difficult with each passing year.

The early years of climate diplomacy: negotiating the framework convention on climate change

Due to increasing scientific knowledge in the 1980s of the potential dangers of climate change, in December 1990 the United Nations created the Intergovernmental Negotiating Committee for a Framework Convention on Climate Change. The objective of the committee was to negotiate a "framework" convention, as was done to address stratospheric ozone depletion (see Chapter 31). As envisioned, the framework convention would be the foundation for subsequent protocols aimed at addressing climate change concretely (again, as done with ozone depletion). The United Nations Framework Convention on Climate Change (UNFCCC), was formally adopted by states at the 1992 UN Conference on Environment and Development, popularly known as the Earth Summit. The official objective of the convention is the "stabilization of greenhouse gas concentrations in the atmosphere at a level that would prevent dangerous anthropogenic interference with the climate system" (United Nations, 1992: article 2). Toward achieving that objective, the convention called on the economically developed states of the Global North to reduce voluntarily their greenhouse gas emissions to 1990 levels by 2000. Developed states also agreed to provide new and additional resources to developing states of the Global South to help them address climate change. Most developed states failed to implement these objectives; collectively, they had not done so even three decades later, and few have done so individually.

Negotiations for the UNFCCC were fraught in many ways, not least due to major disagreements among states of the North and the South (see Chapter 23). These disagreements were manifested in subsequent negotiations toward a protocol and other agreements intended to realize the declared aims of the framework convention. However, disagreements about what to do were, if anything, impetus for further negotiations. During the 1990s those negotiations became regularized. In 1995 the Conference of the Parties to the UNFCCC was established. Its meetings, which are routinely referred to as "COPs," quickly became annual events, or very nearly so. (The 26th COP, scheduled for 2020, was rescheduled for late 2021 due to the global coronavirus pandemic.) The COPs have sometimes involved the participation of heads of government and state.

At COP1 held in Berlin in 1995, developed states acknowledged that they had a greater share of the responsibility for causing climate change and should therefore act to address it first. This acknowledgment became known as the "Berlin Mandate." Central to this declaration was the recurring demand of developing states that the developed states take on greater commitments to reduce their greenhouse gas emissions, assist developing states in achieving environmentally sustainable development (see Chapter 16), and help them to cope with the impacts of climate change. The Berlin Mandate was an affirmation of common but differentiated responsibility (CBDR), an important key principle of climate governance that had been formalized in the framework convention. According to this principle, all states have common responsibility to address climate change, but the developed states have much greater obligation to do so (see Chapter 23).

Building on the framework: the Kyoto Protocol and top-down climate governance

At COP2 in Geneva in 1996, diplomats called for a legally binding protocol to the UNFCCC that would have specific targets and timetables for limiting greenhouse gas emissions from developed states. The COP produced the Geneva Ministerial Declaration, which became the foundation for the Kyoto Protocol. The protocol was negotiated in December 1997 at COP3. It required developed states to reduce their combined greenhouse gas emissions by 5.2 percent below 1990 levels by 2012. In keeping with the CBDR principle, the protocol excluded developing states from its greenhouse gas limitations. In general terms, the Kyoto Protocol was a top-down form of climate governance: it imposed negotiated greenhouse gas limitations on the developed states that accepted it. By the 1990s it was already clear that addressing climate change effectively would require emissions limitations from both developed states and larger developing states, for the simple reason that greenhouse gas pollution of the latter was starting to overtake that of the former. However, for a number of years most attention at COPs was focused on negotiating methods for implementing the very limited objectives of the Kyoto Protocol, despite protestations from the United States and some other developed states calling for limitations on China's emissions. (For details of the climate COPs described in this chapter, see Volume 12 of the *Earth Negotiations Bulletin*, available at https://enb.iisd. org/topics/climate-change.)

Some of the methodology for implementing the Kyoto Protocol was negotiated at COP4, which was held in Buenos Aires in 1998. At COP5 in Bonn in 1999, a timetable for completing outstanding details of the Kyoto Protocol was agreed. Two sessions were needed to complete COP6, starting in 2000 at The Hague and concluding in Bonn in 2001. Around this time, ratification of the Kyoto Protocol by signatories was put into doubt with the election of George W. Bush, who as president of the United States withdrew US support for the protocol – a move that would be mirrored two decades later when President Donald Trump withdrew the United States from the Paris Agreement (see below). COP6 included much debate about several key methodologies for implementing the Kyoto Protocol, including emissions trading, carbon sinks (planting trees and other practices to absorb CO_2 from the atmosphere), compliance mechanisms and aid to developing states.

At COP7 in 2001, diplomats negotiated the Marrakech Accords. The accords comprised a number of proposals for implementing the Kyoto Protocol that might induce enough developing states to ratify the protocol so that it could formally enter into force. Diplomats agreed to increase payments to the UNFCCC's funding mechanism, the Global Environment Facility, and to create new funds for helping least-developed states. These included the Least Developed Countries Fund, the Special Climate Change Fund and the Adaptation Fund. Creation of the Adaptation Fund was recognition by diplomats that efforts to mitigate climate change would not be enough to prevent painful impacts, especially in the developing world. This idea was reinforced in 2002 at COP8 in New Delhi, where diplomats shifted much of their discussions away from limiting greenhouse gas pollution toward finding ways of adapting to unavoidable climate change. At this COP, developed states agreed to provide more assistance for adaptation to developing states. By doing so, developed states might avoid having to reduce their own greenhouse gas emissions as much as would be required if mitigation of climate change remained the UNFCCC's primary objective. At least developing states might benefit from additional assistance from the developed world. Because greenhouse gas concentrations in the atmosphere had by then probably reached the point where painful climate change had become inevitable, adaptation was an obvious near-term priority for those most affected. Because developing states were inevitably to be affected by the impacts of climate change, the shift toward adaptation had some practical logic.

In 2003, at COP9 in Madrid in 2003, diplomats discussed steps for ratification of the Kyoto Protocol and implementation of the Marrakech Accords and other agreements from previous COPs. COP9 saw calls for stronger and more urgent national action on climate change. Such calls have been repeated at every COP since. In 2004, the Kyoto Protocol was ratified by Russia, meaning that it could enter into force in 2005. The same year, COP10 negotiations in Buenos Aires yet again focused more on climate adaptation than on mitigation of greenhouse gas pollution. The focus on adaptation was sufficiently lopsided for the COP to be known as "the adaptation COP." Reflecting what had been and remains a trend in climate negotiations, COP10 resulted in pledges by developed states to provide additional aid to the developing states most affected by climate change. However, there were no clear commitments to enable access to that funding.

COP11, which was also the first formal meeting of the parties to Kyoto Protocol, was held in Montreal in 2005. The COP began the process of formalizing implementation of the protocol, including rules for emissions trading, joint implementation (one country benefiting from emissions reductions in another), crediting of emissions sinks and penalties for non-compliance. Steps to strengthen the Clean Development Mechanism, which was designed to fund projects in developing states, and to establish guidelines for the Adaptation Fund, were also discussed. Several developing states expressed interest in undertaking voluntary measures to limit their greenhouse gas emissions. By doing this they were, in effect (and perhaps in intent), calling the bluff of some developed states, especially the United States, which had been demanding action by developing states before it would agree to taking more action to reduce its own greenhouse gas emissions. However, despite these offers, those same developing states, especially China (which was still interpreted as a developing state in the context of the climate negotiations), remained strongly opposed to *binding* obligations to limit their emissions. Another decade would pass before a more universal formal approach to greenhouse gas limitations, in which all states would take action, was negotiated (see below).

Subsequent climate negotiations resulted in incremental steps, at least in nominal terms, toward action on climate change, in the process highlighting recurring differences among states about how best to achieve the objectives of both the UNFCCC and the Kyoto Protocol. The continued challenges and limited ambitions of most states were exposed at COP12 in 2006, held in Nairobi, during which diplomats concluded that there would be no new commitments under the Kyoto Protocol anytime soon. Similar to other conferences before it, differences among states were apparent at COP13 in Bali in 2007. European states argued for greater international commitments for greenhouse gas cuts, revealing their growing willingness to take more action to address climate change than the United States was willing to do. In contrast, the United States strongly opposed adding new commitments, even as developing states argued that they ought to receive more financial and technological assistance. The discussions at Bali were pushed to a substantial degree by new scientific reports that should have removed any remaining doubt about the main causes and consequences of climate change, notwithstanding lingering influence of climate skepticism, and even denial of the reality of climate change, in a few states, especially Australia and the United States. One significant aspect of COP13 was widespread opposition to efforts by American negotiators to prevent agreement on requiring developed states to go substantially further to reduce their greenhouse gas pollution and to provide much more assistance to developing states for adaptation to climate change. COP13 resulted in the Bali Roadmap, which was intended to guide negotiations toward more robust action, and to an agreement to that effect, over the subsequent two years.

The 2008 COP, COP14 in Poznan, Poland, was a small step toward such an agreement. The COP was noteworthy because the European states, which had for a decade argued in favor of greater action to address climate change, were less supportive of deeper greenhouse gas emissions cuts – not surprising given the global financial crisis that took hold the same year. Nevertheless, by this point in the international climate negotiations – more than two decades since they had begun in earnest – there was realization that much more action was required if the objectives of the UNFCCC, including its overriding objective of avoiding dangerous interference in Earth's climate system, were to be achieved. This realization was reflected in the extent to which states participated in COP15 in Copenhagen in 2009, where 119 national leaders and diplomats representing 192 states were in attendance. The most important outcome of COP15 was the Copenhagen Accord. The accord was agreed on the last day of the conference by a very small group of diplomats and leaders, including US president Barack Obama and Chinese Premier Wen Jiabao, meeting behind closed doors. While this exclusive private meeting may have been necessary to reach agreement, it reflected the persistent influence of powerful states: their consent is needed to reach major climate agreements, but the agreements that result routinely tend to avoid putting pressure on those same states to act robustly to address climate change.

The Copenhagen Accord reaffirmed the science of climate change and acknowledged the need to stop increasing greenhouse gas pollution globally. Importantly, it declared that global warming should be limited to not more than 2°C above the pre-industrial average. This 2°C objective was the construct of negotiators – it was as a political objective that did not emanate directly from climate science (Titley, 2017) – as was demonstrated by the more recent aspiration of limiting warming to 1.5°C, as codified in the 2015 Paris Agreement (see below). The Copenhagen Accord offered a fig leaf of sorts to developing states: the promise of $100 billion in supposedly new annual assistance by 2020 and a new Green Climate Fund to help them cope with the impacts of climate change. At first glance, the Copenhagen Accord appeared to be a significant step toward action on climate change because it appeared to demonstrate commitment by most of the world's governments. However, just as happened with the original 1992 pledge at the 1992 Earth Summit, the accord's provisions were voluntary. Diplomats were willing to pledge action by their national governments, but they still refused to accept and implement major internationally mandated cuts in greenhouse gas emissions. Ultimately, COP15 was so weak in its outcomes, especially compared to expectations and the realities of climate change, that it was routinely deemed to be a failure (see, e.g., Vidal, Stratton and Goldenberg, 2009). Consequently, at COP16, which met in Cancun, Mexico, in 2010, diplomats yet again concluded their negotiations by saying that more effort was needed, even as they were unable to agree on what to do when the Kyoto Protocol was scheduled to expire in 2012.

Supplanting the Kyoto Protocol: the shift away from top-down governance

The international negotiations that produced the Kyoto Protocol and subsequent negotiations on how to implement it exposed a basic predicament of climate governance: as with any collective action problem among states, those that are required to bear the burden are inclined to avoid doing so. Over decades of climate negotiations, most developed states have shown little willingness to take the rapid and extensive action needed to eliminate the causes of the climate change, let alone to provide the financial resources needed by developing states to adapt. The Kyoto Protocol was weak by design. Cutting the greenhouse gas emissions of developed states by barely more than 5 percent, as required by the protocol, was never going to be enough to solve the problem. But even that small cut was too much for many states. Most prominently, the United States used its influence in negotiations to weaken implementation of the protocol, not least because it never liked the premise of differentiated responsibility inherent in both the UNFCCC and the Kyoto Protocol. Ultimately, the top-down approach to global climate governance – negotiations leading to international regulation of states' greenhouse gas emissions, of which the protocol was the prime example – was not working effectively. Too many states were unwilling to compromise their national, usually short-term economic, interests for the common interest of addressing climate change. Diplomats from the United States and several other states had for years advocated an alternative bottom-up approach whereby states would set their individual emissions-reduction targets based on their own capabilities and circumstances. Due to the very limited results of the Kyoto process, those states got their way. This bottom-up approach would guide the next major stage in international climate governance.

When diplomats met in 2011 for COP17 in Durban, South Africa, it was clear that the top-down approach to climate governance would not result in policies that could achieve the objectives of the UNFCCC. The COP produced the Durban Platform for Enhanced Action, which was new insofar as it charted a path toward future greenhouse gas emissions limitations by developing states, specifically "nationally appropriate mitigation actions" (United Nations, 2012: 8). Diplomats at COP17 agreed to keep the Kyoto Protocol alive but also committed their governments to negotiate an entirely new climate agreement that would include pledges for greenhouse gas limitations by all states. Diplomats from the European Union called for the new agreement to include legally binding emissions commitments from large developing states, making essentially the same argument that the United States had been making for many years. Meanwhile, China and India argued that developed states should first implement past agreements, which amounted to a recipe for continued failure given the decades-long reticence of those states to do what they had promised. Despite the title of the Durban Platform for Enhanced Action suggesting that it was about taking much more action, it mostly comprised unenforceable pledges to implement previously negotiated agreements. For example, it called for implementing the Green Climate Fund and finding new sources of financing for developing states, but it did not identify the actual sources of that funding. While COP17 reaffirmed the objective of keeping global warming below 2°C, negotiators simultaneously acknowledged that twice that much warming was likely to occur without new commitments to cut global greenhouse gas emissions much more aggressively. At the conclusion of COP17, there was little prospect that governments would agree to those essential cuts.

Nevertheless, efforts to move in that direction continued, including during negotiations at COP18, which met in Doha in 2012, and at COP19, held in Warsaw in 2013. The outcomes of these COPs were consistent with past conferences: agreement to continue negotiations and to work toward resolving the perennial problem of developed states (most prominently the United States) failing to meet their obligations and developing states (most ominously and dramatically China) rapidly becoming some of the largest sources of global greenhouse gas pollution. At the 2014 COP20 in Lima, Peru, several developed states offered to contribute substantial new money to the Green Climate Fund. This was probably intended as a nudge to developing states to take on greenhouse gas emissions limitations of their own. At a China-US summit, Chinese president Xi Jinping said that China's carbon emissions would level off no later than 2030, and American president Barrack Obama pledged that US emissions would fall by a quarter or more by 2025 (White House, 2014). Significantly, both Xi and Obama used 2005 as the base year for their pledges. This meant that their proposed limits on greenhouse gas pollution were not nearly as impressive as they might have been had both presidents used the 1990 base year that had been the norm since the Earth Summit in 1992.

COP20 produced the Lima Accord, which marked a clear shift away from the traditional differentiation between developed and developing states that had characterized the climate negotiations for decades. The accord was a repudiation, at least in part, of the way that the CBDR principle had guided climate negotiations since the late 1980s. Instead of focusing on the differentiated responsibility of developed states, it emphasized the common responsibility of all states to act on climate change, albeit still expecting historical polluters of the Global North to take on greater responsibility for climate change and to provide aid to developing states affected by it. Developing states would not be expected necessarily to promise future *cuts* in their greenhouse gas emissions (even as the science of climate change clearly indicated that precisely that would be needed from large developing states), but they would be expected to pledge future emissions *limitations*, meaning taking steps to at least restrain increases in their future emissions relative to what they would likely be without acting on such pledges. Tellingly, the Lima Accord did not include any concrete new pledges from states to reduce global greenhouse gas pollution. That was to come the following year as COP21 approached.

Climate governance confronts political reality: the Paris Agreement and nationally determined contributions

The 2015 COP21 conference, held in Paris, was the culmination of efforts to move international climate governance away from top-down international decrees to bottom-up national pledges. It resulted in the Paris Agreement, which requires that developing states join with developed ones to limit their greenhouse gas pollution. What is especially unique about the agreement compared to the Kyoto Protocol is that national emissions limitations are determined by individual states themselves in the form of Nationally Determined Contributions (NDCs). It would have been expected that developed states, not least the United States, would pledge to reduce their emissions substantially – although not all did so – and it was surely hoped that large developing ones, especially China (which by then had overtaken the United States as the largest national source of annual carbon pollution), would begin moving in that direction. The poorest developing states were expected to pledge actions that would mitigate their emissions, although the condition that many of them made was that their actions would depend on long-delayed assistance from developed states. The Paris Agreement's bottom-up approach of NDCs is, in essence, national self-regulation. One advantage is that this approach was almost universally accepted by states. A disadvantage of nationally determined pledges is that they have not gone far beyond what states would have done without the Paris Agreement.

Very importantly, COP21 went beyond previous conferences by affirming that all states would pledge action toward the common goal of limiting global warming to less than 2°C. Importantly, after much lobbying from small-island states and others most vulnerable to climate change, the Paris Agreement went further by calling on states to limit warming to 1.5°C. Furthermore, developed states promised collectively to provide more financial assistance to help developing states address and cope with climate change. Significantly, each state pledged to limit its national emissions in some way, although not necessarily to reduce them. The idea was that the NDCs would become baselines for more robust action in the future. However, it was quickly apparent that the national pledges to implement the Paris Agreement would not be nearly enough. Even if all of them were to be fully implemented by almost every state, global warming was predicted to surpass 3°C (United Nations Environment Program, 2016). That amount of warming was a recipe for dangerous climate change, precisely what the UNFCCC was designed to avoid.

All of the COPs that followed COP21 have focused much of their attention on how to implement the Paris Agreement and increase the "ambition" of states' NDCs enough to make the 1.5°C–2°C global warming objective possible. Meeting in Marrakech, Morocco, in 2016, COP22 negotiations produced the Marrakech Action Proclamation for Our Climate and Sustainable Development. The proclamation contained a laundry list of familiar steps for addressing climate change. Much of it was devoted to the priorities of developing states, such as the provision of additional financing, capacity building and other forms of support. The subsequent COP was hosted by Fiji, a small island developing state highly vulnerable to climate change, in 2017 (although the conference was convened in Bonn). Like many of its predecessors, COP23 was a procedural meeting that largely focused on technical issues and the operationalization of the Paris Agreement. As in all prior COPs, many diplomats aimed to ensure that such technicalities would not undermine their own states' national interests. This was reflected in debates over finance, with developing states calling for more effort by developed states to produce it, and developed states often working to avoid meeting those calls. In the end, they agreed that the Adaptation Fund would serve the Paris Agreement, negotiated aspects of the rules for implementation of the Paris Agreement, and initiated the Talanoa Dialogue for Climate Ambition. The latter was intended to be a facilitative sharing of ideas and information among states and other actors during the period between COP23 and COP24.

COP24 was held in 2018 in Poland, one of the most polluting states in Europe and one that was economically reliant upon, and politically wedded to, its large coal industry. The populist host government did not want to advocate action that would threaten its support among coal miners. It was probably no coincidence that the host city, Katowice, was in the heart of Poland's coal-mining region. Therefore, there was little surprise when, like COP23, COP24 focused on technical matters, notably progress on drafting the Paris rulebook. Once again, the question of CBDR was much debated, with developing states again arguing for their differentiated, flexible commitments to action and developed states generally arguing for equal accountability by all states. As in previous conferences, pledges by developed states to assist developing states with climate finance overcame some differences, as did the fact that many of the most contentious issues were, also once again, pushed to the next COP.

COP25 (the most recent COP as of the time of writing in mid-2021) was held in Madrid at the end of 2019. (The COP was initially planned for Brazil, but it was moved first to Chile after the right-wing politician, and vocal climate-denialist, Jair Bolsonaro was elected president and withdrew Brazil's support for the conference. The COP was then moved from Chile, the formal state host of the conference, to Madrid due to civil unrest in Chile.) COP25 involved tens of thousands of participants. This indicated the growing interest among all actors, not just states but cities, corporations (see Chapter 13), nongovernmental organizations and activists (see Chapter 14), in international attempts to govern climate change. One big area for discussion was whether and under what circumstances developing states would be compensated for loss and damage from climate change and, as in all previous COPs, related questions of climate finance. There was little clarity on where new funds would come from to fulfill the developed states' ageing promise to give $100 billion annually to vulnerable developing states by 2020. A major area of negotiation included a recurring sticking point from previous COPs: market mechanisms and rules for when and how states would be allowed to use them to implement their NDC pledges from COP21. Questions of ambition, specifically strengthening NDCs so that they might achieve the Paris goal of limiting global warming to 1.5°C–2°C, remained largely unresolved. COP25 ended without new emissions pledges, despite a months-long lobbying effort by Antonio Guterres, the UN Secretary General and his Climate Action Summit at the United Nations a few months before the COP. During that summit, 70 states pledged to achieve net-zero greenhouse gas emissions by 2050 – although none of them were the world's largest polluters, and together their greenhouse gas emissions amounted to only one-tenth of the global total (United Nations, 2019).

COP25 was preceded by a large number of scientific reports warning of the dire consequences of climate change if aggressive cuts in greenhouse gas emissions were not forthcoming. Diplomats at the COP faced historically overwhelming pressure from activists, not least Greta Thunberg, the Swedish student who mobilized hundreds of thousands of young people around the world to strike for climate action. Those calling for more action at COP25 may have hoped that states would greatly increase their NDC pledges. However, many of those states seem to have seen things differently. The conference witnessed tensions between states, such as Australia and Saudi Arabia, which perceived their interests to be best served by slowing action on climate change, and those that wanted greater action, including some European states and the most vulnerable states on the frontline of climate change, particularly many of the small-island states. As at previous COPs, the question was which developed states would make big cuts in their greenhouse gas pollution; whether some of the large developing states, not least China, would take greater action; how national greenhouse gas limitations would be measured; and which states would pay, and how much, the poorest states most harmed by climate change.

COP25 demonstrated that increasingly dire warnings from scientists and pleading from activists were not enough to outweigh the national interests of too many states to avoid doing what is necessary to prevent dangerous climate change. It demonstrated, much as COP21 had done, that growing concerns around the world about climate change would not necessarily translate into concerted action. Paradoxically, the Paris Agreement is premised on the very thing that has plagued the climate negotiations for decades: national prerogatives. As experienced observers of COPs over several decades have noted,

> Under the Paris Agreement, the level of states' ambition is determined nationally. There are primarily only soft levers, based on moral suasion, that can convince parties to do more. For COP 25, these limitations were in the agenda, defined by parties themselves. The modest mandates for COP 25 were set years ago, through states' self-interested negotiations.
>
> *(IISD, 2019: 26)*

To be sure, both the Kyoto Protocol's top-down approach to global climate governance and the Paris Agreement's bottom-up approach have achieved some modest progress. All around the world, they are reflected in many new policies intended to address climate change, whether to limit greenhouse gas pollution or to address questions of adaptation to the impacts. However, all of these policies put together have yet to stop the global increase in greenhouse gas emissions, nor have they done much to help the most vulnerable states cope with climate change.

One might argue that it is too soon to assess the effectiveness of the bottom-up approach. Whether the Paris Agreement will be successful will be determined by whether it is a catalyst for states to take far more aggressive action in the very near future to reduce enormously their collective greenhouse gas pollution and to end their economies' dependence on fossil fuels. COP26 would be an opportunity for states to present more ambitious pledges. The future could reveal greater willingness of states to take action to finally give climate the priority it requires – but not if the past three decades of climate negotiations are any kind of guide to what will happen.

National interests versus climate action

When the problem of climate change emerged on the global agenda in the 1980s, addressing it through international cooperation made perfect sense. After all, that was the way that so many transnational problems had been addressed for centuries. Greenhouse gas pollution from one state affects every other state indirectly through global warming and other manifestations of climate change. It is impossible for any individual state, regardless of its economic or military power, to stop climate change without the cooperation of other states. In this context, state autonomy is a fiction. Most states recognize this, reinforcing the desire to deal with climate change collectively. Using international negotiations, COPs and similar approaches has therefore been the obvious approach to the problem. It would have been odd if climate change had been approached any other way. Unfortunately, international negotiations to address climate change have been characterized by much-too-gradual movement toward agreement on responding to the problem in ways that are effective and timely. Scientists have called for greenhouse gas emissions to be reduced very rapidly and eliminated as soon as possible (IPCC, 2018). Yet, the best that developed states together have been able to achieve collectively is less than the meager cut that they promised in the 1997 Kyoto Protocol (comparing 2017 to 1990; IISD, 2019: 25), and emissions from developing states continue to rise. Three decades of top-down attempts to govern climate change, while producing many agreements and policies, failed to produce the one result that is essential: action to cut global greenhouse gas pollution substantially. The success of the framework-protocol approach to stratospheric ozone depletion has not been successful for climate change.

The bottom-up approach to international climate governance that is typified by the Paris Agreement is still relatively new, and it may yet produce better results. However, so far it is yet more evidence of the powerful influence of a system of individual sovereign states vying to protect, and whenever possible to promote, their own national interests. Those interests are often narrowly skewed toward the short-term interests of powerful national political actors and constituencies. The bottom-up approach has encouraged the participation of most states in the climate negotiations because it has allowed them to set their own nationally determined greenhouse gas emissions limitations instead of having to implement policies and regulations for achieving internationally agreed targets. In other words, the Paris Agreement and related bottom-up efforts at international climate governance are affirmations of the national interests that have consistently held back more effective action on climate change. These interests are routinely a function of the sorts of behaviors, not least the burning of fossil fuels to power national economies, that have created the problem of climate change to begin with. Looking at the lengthening history of efforts by states to govern climate change internationally, a pattern emerges. It is a pattern of glacial progress toward an objective – preventing dangerous climate change – even as that objective becomes harder to achieve.

The international climate negotiations are striking in their complexity and comprehensiveness. They have been and remain the most sweeping international negotiations in history, demonstrating the seriousness with which the problem of climate change is viewed by the world's governments. And the international approach has arguably had results. For example, some European states have taken action to reduce their greenhouse gas pollution, and several have pledged to become carbon neutral by mid-century. Even when national governments have sought to thwart international governance of climate change, as in the case of the United States under President Trump, other actors – cities, corporations and so forth – have pointed to the Paris Agreement's objectives as goals that they take seriously when making policies. Yet, looked at holistically in global terms, the actions of most states (and other actors) are still going in the wrong direction. Most importantly, when global greenhouse gas emissions need to be falling rapidly, they are still on the rise.

Conclusion

The very nature of the international system and the states from which it is made explains why both top-down and bottom-up efforts to govern climate change have failed so far and seem very unlikely to bring about sufficiently strong policies anytime soon. By definition, the international system divides the world into supposedly sovereign states whose legitimacy and meaning derive from acting upon their individual, separate interests as perceived at any given time. In such a system, the universal interests of the global climate system are subjugated to the perceived interests of the actors within states that have the most influence over policies, whether those actors be monarchs, as in some petroleum-exporting nations, party officials or oligarchs, as in China and Russia, industry and its lobbyists, as in Japan and many European states, or captured political parties and self-interested politicians, as in Australia and the United States. Diplomats in the climate negotiations have been doing what their governments have told them is in the national interest. This is normal behavior in the international system (see Chapter 3). Thus, like no other topic described in this book, climate change reveals both the challenges for, and apparent limits to, global environmental politics.

Acknowledgments

This chapter draws extensively from the author's earlier work, especially Harris (2013: 41–53, 2018: 129–136, 2021: 39–59), and it benefits greatly from the literature cited therein. A bibliography of nearly 1,000 sources is available at https://paulgharris.net/pathologies-of-climate-governance-bibliography/.

References

Harris, Paul G. (2013) *What's Wrong with Climate Politics and How to Fix It*, Cambridge: Polity.

Harris, Paul G. (2018) "Climate change: science, international cooperation and global environmental politics," in G. Kutting and K. Herman, eds. *Global Environmental Politics: Concepts, Theories and Case Studies*, Abingdon: Routledge, pp. 123–142.

Harris, Paul G. (2021) *Pathologies of Climate Governance: International Relations, National Politics and Human Nature*, Cambridge: Cambridge University Press.

Intergovernmental Panel on Climate Change (IPCC) (2014) *Climate Change 2014: Impacts, Adaptation and Vulnerability*, Cambridge: Cambridge University Press.

Intergovernmental Panel on Climate Change (IPCC) (2018) *Global Warming of 1.5°C*, Geneva: Intergovernmental Panel on Climate Change.

International Institute for Sustainable Development (IISD) (2019) Summary of the Chile/Madrid climate change conference: 2–15 December 2019, *Earth Negotiations Bulletin* **12**(775): 1–28.

Titley, David (2017) How did we end up with a 2C climate limit? *Climate Home News*, August 23, https://www.climatechangenews.com/2017/08/23/end-2c-climate-limit/.

United Nations (1992) *United Nations Framework Convention on Climate Change*, Bonn: United Nations Framework Convention on Climate Change Secretariat.

United Nations (2012) Report of the Conference of the Parties on Its Seventeenth Session, Held in Durban from 28 November to 11 December 2011, https://unfccc.int/resource/docs/2011/cop17/eng/09a01.pdf.

United Nations (2019) *Report of the Secretary-General on the 2019 Climate Action Summit and the Way Forward in 2020*, New York: United Nations, https://www.un.org/en/climatechange/assets/pdf/cas_report_11_dec.pdf.

United Nations Environment Program (2016) *The Emissions Gap Report 2016: A UNEP Synthesis Report*, Nairobi: United Nations Environment Program.

Vidal, John; Stratton, Allegra; and Goldenberg, Suzanne (2009) "Low targets, goals dropped: Copenhagen ends in failure." *The Guardian*, December 19, https://www.theguardian.com/environment/2009/dec/18/copenhagen-deal.

White House (2014) US-China joint announcement on climate change, https://obamawhitehouse.archives.gov/the-press-office/2014/11/11/us-china-joint-announcement-climate-change.

33

Energy

Political–economic strategies

Hugh C. Dyer

Energy has always been central to life, and has now become a significant issue in political discourse, reflecting wider changes in global politics. Energy challenges are linked to other changing circumstances, in particular environmental change, but they have typically been treated separately. This chapter explores the relationship between energy systems and political–economic systems as a strategic issue, given that the existence of that relationship does not imply necessarily complementary policy goals. Each is a challenge for actors with critical roles in setting the global agenda, where incoherence and competing political priorities undermine coordinated, consistent policy. At the same time, there are opportunities for encouraging behavioral modification ("nudging") and social action to support political change toward environmental protection and efficient energy (see Chapter 28). The confluence of energy and economic and environment policy – at the point where carbon is released into the atmosphere from the burning of fossil fuels – may suggest potential benefits of a "win–win" approach by which all policy goals are achieved through "efficiency" and innovation. However, efficiency and innovation alone are not likely to reduce the overall use of carbon-based energy or reduce environmental impacts, and there is seldom room in daily politics for energy sufficiency or urgent policies to address the need for economic change, let alone environmental change. The consequences are uncoordinated tensions rather than coherent solutions, even as economic and environmental change and related energy policies become more central to social and political agendas.

Environmental change is complex and involves many factors, significant among which is human activity; of particular concern is the emission of carbon dioxide through combustion of fossil fuels. As a key source of energy for economic development, fossil fuels are the chief anthropogenic (human) source of carbon emissions that alter the natural balance of Earth's carbon cycle and cause global warming (increased average global temperature). In the natural course of events carbon is released into the atmosphere and extracted from the atmosphere in comparable amounts by natural processes, as carbon sources such as plant respiration and geological activity are matched by sinks such as plant photosynthesis and dissolution in water. When humans burn fossil fuels or reduce plant growth, the sources and sinks are thrown out of balance. The additional carbon dioxide in the atmosphere contributes, along with other "greenhouse gases," to what is sometimes referred to as the "greenhouse effect," whereby

DOI: 10.4324/9781003008873-38

heat that would otherwise escape the atmosphere is reflected and trapped, which leads to global warming and climate change. The use of fossil fuels releases carbon previously sequestered beneath Earth's surface into the atmosphere, overstretching the ability of the planet's natural sinks to recapture it. Hence the relationship between economic activity, fossil-fuel energy consumption and environmental change is direct; if human activity is not the only factor in determining environmental change, it is one that is causing unnatural change.

Recognition of this problem led to a range of policy responses and international agreements, including the 1992 United Nations Framework Convention on Climate Change (UNFCCC) (see Chapter 32). The Intergovernmental Panel on Climate Change, established in 1988 by the United Nations Environment Programme and the World Meteorological Organization, provides a framework for scientific consensus to inform climate policy. The current situation is that policies are generally inadequate to the scale of the problem, and continuing international agreement is troubled by a lack of universal commitment. To varying degrees, countries, cities and citizens have taken steps to reduce their "carbon footprint" (that is, to mitigate the effects of their behaviors and activities), as well as preparing for inevitable change (that is adapting to the effects) (see Chapters 14 and 29). The overall impact of such measures is limited in a world of increasing population and economic growth, with the attendant increase in energy consumption. The only solutions are to develop alternative non-carbon or low-carbon energy sources, or to reduce energy consumption through efficiency or abstinence – that is, to stop burning fossil fuels. As the global economy and human livelihoods are currently heavily dependent on fossil fuels, this is a great economic and social challenge, and thus a significant political issue.

Energy in a political context

Policy debates offer insight into the nature of the "political community" of energy. In the case of environmental change, individual and collective responsibility is an important consideration because it extends the scope of political community beyond the current generation and beyond the human agent (see Chapters 14 and 29). While energy and the environment have received individual attention, the connection between the two issues has come into focus in recent years. The International Energy Agency relates energy markets to "energy security, environmental protection and economic development," and it analyzes related "strategic issues" (International Energy Agency 2012), while in the recent World Energy Outlook 2020 the focus "is firmly on the next 10 years, exploring in detail the impacts of the Covid-19 pandemic on the energy sector, and the near-term actions that could accelerate clean energy transitions," noting that the

> economic downturn has temporarily suppressed emissions, but low economic growth is not a low-emissions strategy – it is a strategy that would only serve to further impoverish the world's most vulnerable populations. Only faster structural changes to the way we produce and consume energy can break the emissions trend for good.
>
> *(International Energy Agency 2020)*

However, the outlook is focused on energy, and still largely if not exclusively on fossil fuels. The connections between energy, environment and economic development goals are such that "an aggressively single-minded pursuit of energy security will compromise these other goals," with current policy a "hotchpotch of measures unlikely to deliver," which suggests policy incoherence (Oxford High-Level Task Force 2007).

Even as energy and environmental change are identified with one another as policy areas, the focus is typically on one or the other without considering the hidden tension between them. There is little discussion of reductions in consumptive lifestyle expectations and declining or altered economic growth. This calls attention to contradictory and complementary aspects of energy provision and environmental protection as strategic goals, and the coherence of policy in these areas. Recognition of the strategic importance of energy and the environment is illustrated, for example, by the inclusion of these issues in the US–China Strategic and Economic Dialogue, with environment and energy cooperation featuring in the strategic track and supported by a "Memorandum of Understanding to Enhance Cooperation on Climate Change, Energy and Environment" (US Department of State 2011; Xinhua 2011). Even in quite recent history such political developments would have seemed unlikely, perhaps unthinkable. An agreement to "enhance cooperation" suggests limited cooperation thus far, which was especially the case for the duration of the recent Trump administration in the United States given its combative relationship with China. A "memorandum of understanding" suggests limited practical significance, yet that it should be deemed necessary at all is suggestive of strategic developments. The strategic content of the political debates emerges most clearly when the underlying characteristics of energy and environment issues are stripped down to potential consequences in terms of conflict and competition. As Shea (2006) points out, "All modern developed economies are dependent upon an abundant supply of energy both in terms of guaranteed supplies and stable prices [making] energy security an issue of strategic importance."

The vulnerabilities in this perspective include lines of communication and transportation, energy distribution infrastructures, difficulty of increasing supplies or finding new energy resources to meet rising demand – particularly in rapidly developing economies – loss of overall energy production due to under-investment in development and infrastructure, and a lack of spare energy supply capacity, making even small decreases in supply significant for areas dependent on imported energy. In all of this, the dominant source of energy is fossil fuels. Where the risks and costs of environmental change are identified, they struggle to acquire the political significance of energy supply (see Chapter 19). Thus energy and environmental issues may be linked, but the economic threats of energy shortage are more immediately obvious in the political domain. If it has now become commonplace to identify energy and its environmental corollaries as significant strategic issues, they are not yet subject to coordinated planning. Even relatively uncontroversial alternative energy sources are not consistently supported, as evidenced by reductions in solar power feed-in tariffs in several countries, and ongoing objections to wind farm development in some areas. Only a handful of countries have *legally binding* net-zero emissions targets (Cavanagh 2012; Murray 2020). Delivering on such commitments remains an economic challenge. Other approaches include trading schemes that try and rationalize and normalize reduced carbon emissions in the economy by imposing a cap on overall emissions while allowing individual emitters to choose and plan their allowed emissions (although too much is still allowed).

Longer-term issues, such as environmental degradation, poverty and underdevelopment, and lack of human rights, which do not yet attract a sense of urgency, will be driven to the margins of the agenda (see Chapter 26). How, then, does a combined energy and environment policy locate itself in the mix of political–economic orientations, and can it help us to appreciate longer-term issues of importance but little apparent urgency? For example, "peak oil" is the historical point – sometime about now – of the maximum rate of extraction of petroleum, beyond which (according to some experts) production declines as reserves are depleted. Yet, historical and ideological debates about the timing and implications of this peak struggle to define the level of urgency in our relationship to petroleum, with the only consensus being that we inhabit a global petroleum-based economy. Some actors appear to defend privileged interests in neoliberal economic policies based on assumptions of plentiful petroleum. Others purport to defend the interests of those who benefit less from the petroleum economy and have even more to lose if no preparation is made for a low-carbon economy (an economy less dependent on fossil fuels and carbon emissions, by policy choice) let alone a "post-petroleum economy" (an economy not dependent on fossil fuels and carbon emissions), whether by policy choice or by lack of fuels. In the world's petroleum-based economy there are direct connections between petroleum and other resource extraction, production and distribution issues, and there is the connection with environmental change and its implications, in turn, for the continued availability of other resources, such as food and water (see Chapters 38 and 44). Consequently, facing up to the importance of energy policy is itself a matter of urgency.

As a commitment to energy issues develops, establishing them as fundamental responsibilities of governance at state and global levels (Downie 2020). As these responsibilities are related to fundamental rights of individuals and communities, a new version of the social contract arises and brings with it a new political style and content (see Chapters 25 and 26). Both producers and consumers of oil have already begun to coordinate as energy markets themselves become a focus of government policy. For example, Saudi Arabia agreed to increase production in the face of an energy price crisis, but it called on consumers to manage demand as well. A Saudi Minister for Petroleum and Mineral Resources emphasized the importance to producers of "access into the markets of oil importing countries, the steady share of oil in total energy consumption over the long term, and fair and stable prices that allow for their sustainable development over the lifetime of the resource" (Fattouh and van der Linde 2011: 61). Industrial states are also coordinating energy policy with varying degrees of success (E3G 2020), but they must do so alongside environmental policy, with energy efficiency being the first step (Holmes and Mohanty 2012). Since fossil fuel energy consumption releases carbon dioxide – a chief source of global warming and long-term climate change – more efficient use of such energy would at least reduce the amount of carbon emitted for the same amount of economic activity. Efficiency would not, of course, reduce overall carbon emissions so long as the global economy continues to expand on the basis of fossil fuel consumption, and in the absence of globally agreed and effective emission limits.

Consequently, achieving the reversal of carbon emission trends to the extent that is necessary is a huge systemic challenge. Furthermore, while addressing energy needs and avoiding damaging environmental change is in itself a gigantic task, these are not the only concerns of people and their governments. However, since energy and the environment are so central to human existence, these issues together are likely to influence the wider pattern of political–economic relations. This suggests that delivering energy and environment policy may involve the scale of cooperation and planning needed to address earlier systemic issues such as global depression, post-war reconstruction and financial crises (and perhaps now, pandemics). To appreciate the political significance of energy policy concerns for global environmental politics, it will be useful to expand on the political context in which they have emerged.

Energy policy

There is little doubt about the centrality of energy in our lives, and yet the implications of this obvious circumstance are perhaps too close to be seen clearly. Kimmins makes the point that all potential solutions to individual energy questions involve a social cost, an ethical dilemma and an impact on the way other problems are resolved. Thus, they can only be looked at within a broader consideration of the functioning of the world system of which energy is but one intimately woven component (Kimmins 2001: 35). This is slightly at odds with the narrow national perspectives of state governments, where energy supply is fundamental to a way of life and national security in that sense. Macfarlane (2007) entitles it, simply, "The Issue of the 21st Century." Kimmins also captures the intergenerational and forward-looking requirements for approaching energy policy in saying that "many ethical issues arise as a result of unequal access to energy and of the environmental repercussions" and this requires "that we consider the consequences for future generations of satisfying the energy needs of the present," while also pointing to the long-term requirement for renewable energy sources: "The only question is how rapidly we should move to such sources and what mix should be used in various parts of the world over time" (Kimmins 2001: 37, 38).

As Shea notes, "tightness in the market has re-ignited the debate over alternative energy supplies such as biofuels or solar power not to mention a renewed interest in nuclear power" (Shea 2006). As we see below biofuels present difficulties, and while there is some political support for the nuclear option it remains very controversial. Nuclear power raises significant issues from both ecological and human perspectives, whatever its short-term appeal as a panacea for addressing the twin challenges of energy and climate (since it provides ample energy and produces no carbon). The negative aspects of nuclear power are amply illustrated by the catastrophic events at Fukushima and previous nuclear disasters, such as Chernobyl, and subsequent policy reactions, such as in Germany, which has now turned away from nuclear energy.

A UNESCO ethics report questioned "whether we could really depoliticize choices about energy," and as "fossil fuel supplies were dwindling and climate change was accepted as a reality, clean renewable energies, like wind energy, geothermal, wave, tidal, hydropower, and photovoltaic were the way of the future" (UNESCO 2007: 5). Already river water supplies about a fifth of all electricity, and over 60 countries meet more than half their electricity requirements from hydropower, but the predictability and long-term future of this energy source is in question as the climate changes (Corley 2010; see Chapter 38). There is also some disparity in exploitation of hydropower potential, with Europe and North America largely developed and Africa hardly tapping this resource at all. In time photovoltaic solar might also generate a fifth of the world's energy (European Photovoltaic Industry Association and Greenpeace 2011). Of course none of these options are without implications, such as land use and demographics, nor do they offer the portability of petroleum fuels for air and sea transport purposes.

Potentially the entire energy system could be renewable in future, and economically beneficial rather than burdensome (Vad Mathiesen et al. 2011). As a practical matter, energy mix depends on "the existing governance and the international sourcing or supply chain of energy" and there is a significant less-developed population which does "not have good access to conventional technology such as electricity and fossil fuels" (UNESCO 2007: 8–9). This perspective challenges any notion that energy is an issue of the future; it is clearly upon some of us now, and will bear more heavily on all of us soon, with human security being "the ultimate goal surrounding the concept of energy equity" (UNESCO 2007: 22–3; see Chapter 20). The perspective of "human security" is usefully linked to energy here, which informs wider debates about human development. Baer et al. identify the basic dilemma in noting that "there is no road to development, however conceived, that does not greatly improve access to energy services" and yet there is "not enough 'environmental space' for the still-poor to develop," thus requiring "a wholesale reinvention of the global energy infrastructure on the basis of low-emission technologies" (Baer et al. 2007: 23, 26). To the extent that this dilemma is now recognized in political debate, there is already evidence of change, with structural implications for global environmental politics.

Energy impacts

Inequities are in respect of both the sources and consequences of any change. If stability is to be achieved, human communities would experience the benefits or burdens according to their location in the ecological and/or industrial structure. Local vulnerabilities, livelihoods and state roles are linked by Barnett and Adger (2007) in noting that "climate change increasingly undermines human security" (see Chapter 20). Elsewhere Barnett (2001: 118) accepts that dealing with climate issues requires wide and deep structural reform, including energy systems. It is likely that political actors will "reinforce their own definitions of 'energy security' and 'energy independence'" (Poruban 2008) which will be limited and instrumental. Singer (2006) points out that "climate change is an ethical issue, because it involves the distribution of a scarce resource," and may not even be readily understood from the conventional economic perspective "given some of the important but often implicit assumptions on which it is based" (Toman 2006), and energy follows the same logical pattern in so far as it remains based on carbon-releasing fossil fuels. The underlying assumptions of our political, moral, economic and social systems (see Chapter 27) do not yet appear to have fully internalized the weight and depth of the issues, even as the challenge is appreciated.

The current state of affairs is even worse than previously anticipated, and given consistent reporting from reliable sources there can be little doubt about the trend in the carbon cycle. Data already indicated carbon dioxide levels up almost 40 percent since the industrial revolution as greenhouse gas levels "continue climbing" (NOAA 2011) and this is confirmed by the trend records (NOAA 2020). There are obviously some limits to what can be done in a relevant timescale, and the little that can easily be done seems woefully inadequate. Emission levels are now so much worse than expected that reaching existing targets for reduction may be unrealistic, since we were already at the upper end of the possible scenarios (Anderson and Bows 2008). Managing the situation will require energy demand management beyond a mere reduction in increases; only economic contraction would have sufficient impact, in the absence of technological intervention. Allen (2008) suggests that addressing critical carbon levels will require technological approaches, such as SaFE (Sequestered at-time-of Fossil Extraction) carbon which allows for sequestration (carbon capture and storage) at an increasing fraction of emissions to stabilize atmospheric content. More striking are proposals for geoengineering of oceans and atmosphere, by developing biological carbon sinks (see Chapter 42), and by introducing solar radiation management technologies to avert the worst-case scenario of sudden climate change. These proposals raise collective action problems, as well as scientific ones, which require coherent governance (Humphreys 2011). Yet such extreme technological interventions could render energy policy extraordinary or exceptional, rather than normalized in a coherent energy policy.

Whatever "efficiency" is achieved by technological means, total emissions continue to rise with production and consumption. Jevons's nineteenth-century paradox still applies: efficiency first does not give frugality second, but rather increased consumption; while on the other hand, frugality first can bring efficiency in response to scarcity (Polimeni et al. 2008; see Chapter 17). Furthermore, if environmental change will be "visited primarily on the globe's most vulnerable populations" it follows that any "response to climate change that hopes to gain international legitimacy must take equity as a central organizing principle"

(Roberts 2007). Sachs argues that equity in regard to fuel access "is about *equality among nations*" while the consequent climate threats suggest that "fundamental rights might be violated" (W. Sachs 2007; see Chapter 25). Baer et al. note the disjointed but overlapping responsibilities of people and nations. In the context of "capacity to mitigate emissions in a global energy regime" they say the main point is obvious: "Recognizing inequality *within* countries is as unavoidable as recognizing inequality *between* countries...If, that is, our goal is a burden-sharing system that actually makes ethical and political sense" (Baer et al. 2007: 31). If "equity in this respect is about human rights," then the "need for low-emission economies in the South and the North is therefore far more than a question of an appeal to morality; it is a core demand of cosmopolitan politics" (W. Sachs 2007; see Chapters 23 and 25).

While this seems patently true, it is not so clear if cosmopolitan politics is a shared aspiration, even if planetary survival is. Equity is surely a political project, as much as a technical or economic one, and success in all other projects in all places may hinge upon its success. A sign of hope emerged in the earlier Bali negotiations on a post-Kyoto consensus when the United States was embarrassed into joining the consensus by the Papua New Guinea delegate (*Newsvine* 2008) – an indication of how structural opportunities in politics may allow a reversal of power dynamics. On the other hand, subsequent failure to agree at Copenhagen and missed opportunities since (in the context of UNFCCC negotiations) suggests business as usual rather than progressive climate policy.

Garvey (2008) identifies three sources for political stances: (1) historically: "the industrialized world has done the most damage"; (2) presently: "the West currently uses more than its fair share of the carbon sinks"; and (3) in future: sustainability creates a general "obligation to leave a hospitable world" behind. Gardiner notes that this involves the convergence of a set of global, intergenerational and theoretical problems. He argues that it is "a perfect moral storm" which "makes us extremely vulnerable to moral corruption" and identifies three characteristics that lead to this "storm": dispersion of causes and effects; fragmentation of agency; institutional inadequacy (Gardiner 2006: 397, 399–400). This leads him to suggest "there is a problem of corruption in the theoretical, as well as the practical, debate" because a focus on political and technical problems of action by nation-states distracts from intergenerational obligations (Gardiner 2006: 408–409). The link to our energy habits demands "rethinking energy options to address climate change" (McGowan 2007), but these options are likely to be ones we do not find convenient or have not taken seriously yet. There are structural assumptions (and corruptions) in our political, social, economic and ecological field of vision that will have to be addressed because the status quo needs to be directly challenged.

The politics of biofuels

The political tensions and ethical issues raised by the confluence of energy and climate policy can be illustrated by a widespread policy response: biofuels. There is considerable controversy around biofuels as an alternative, climate-friendly energy source. Biofuels have a political dimension. While they have a long and checkered history (as old as the internal combustion engine), it seems clear that renewed enthusiasm for biofuels has been dampened by several realizations, including the net energy and environmental consequences, and the impact on food crops (see Chapter 44). As a result, major actors conceded that "plans to vastly increase the amount of fuels such as bio-ethanol and biodiesel might need to be reconsidered" in Europe (Greenpeace UK 2008). US policy wrestles with political commitments on ethanol production. In 2007 Congress approved a fivefold increase in use of biofuels; in 2011 Senate voted to end tax credits and trade protection for ethanol. Low-blend fuel, ordinary automobile fuel containing 5 percent ethanol (E5), is commonplace everywhere. However, all political actors are facing opposition due to a general awareness of the distortions created by this policy, "with political leaders from poor countries contending that these fuels are driving up food prices and starving poor people" (*New York Times* 2008; see Chapter 44).

There are considerable political stakes involved. The EU Commission had to reject claims that biofuels are a "crime against humanity" (Agence France Press 14 April 2008). The situation prompted such protest even while the relative significance of biofuels in the energy mix is quite limited – "only 1% of transport fuels… Oil is still 40% of the global energy mix because of its domination of the transport sector" (Shea 2006; see Chapter 34). There has been a surprising lack of consideration for ecology and sustainability (see Chapter 16): "government agencies said nothing about the degradation of the soil, the nutrients that would be required" nor indeed "about the ridiculously low Energy Returned on Energy Invested (EROEI), the heavy use of water and fossil fuels" (*Energy Bulletin* 2008). Meanwhile, alongside a range of technological initiatives, the US Advanced Research Projects Agency and Department of Energy announced biofuels projects encouraging non-food crop oil production, "Plants Engineered to Replace Oil" (Advanced Research Projects Agency 2011). An alternative assessment of energy return on investment (EROI) may suggest an "acceptable economic EROI" which "allows us to be more appreciative of the fundamental change in energy provision as the energy system transitions from being fossil based to (largely) renewables based and leads to a positive outlook for a post-fossil society" (White and Kramer 2019). In the longer term biofuels will figure in the mix of energy alternatives, and "the key to making sense of these suggestions is for policymakers to re-evaluate biofuels through the prism of rural and industrial development rather than simply employing the somewhat populist food/ fuel framework" (Creamer 2008; see Chapter 44).

Thus, energy and climate policy cannot be treated in isolation from socio-economic policy, let alone in isolation from one another; a holistic perspective is required to capture the complexity (see Dalby 2002 and Chapter 20). Biofuels can serve a range of purposes from substituting petroleum fuels to encouraging agricultural and rural development (see Chapter 44), but this diminishes the energy and climate strategy implied in such initiatives, and completely undermines it if the net use of energy does not actually reduce petroleum dependency and emissions. There are economic motives here, as even old-fashioned energy efficiency ("negawatts") could be significant for energy and climate alike (though the "rebound" or "takeback" effect of increased access and lower prices for fuel leading to greater consumption could cancel 26–37 percent of any gains [The *Economist* 2008]). The ecological motives seem somewhat distant, and the coherence between energy policy and climate policy is weak. Rather than offering an unproblematic quick fix or "free lunch," contra Commoner's (1971) fourth law of ecology, the biofuels debate illustrates the political dilemmas.

Political–economic structure

There can be little doubt now about the cost implications or the likely impact on economic growth of environmental change (Stern 2006), but this could simply lead us to think about the economic opportunities this presents. Any notion of economic change that does not involve growth seems unthinkable given current economic assumptions, but those assumptions could change under the pressure of an energy crisis. It is now common to speak of a "low-carbon economy," or even a "post-petroleum economy," and neither involves the cessation of economic activity; they simply involve change. Yet the balance between states and markets in responses to the energy–climate nexus remains uncertain, often taking the appearance of trade agreements (see Chapter 24). The 2012 Rio+20 summit addresses the prospects for a "green economy," which has been debated since at least Pearce's "Blueprint" suggested that economics is more efficient than traditional "command and control" approaches (Pearce et al. 1989). Market mechanisms involve some element of regulatory intervention to set the boundaries, or to create incentives. The European cap-and-trade system is an interesting example, with controversy and uncertain outcomes around issues of "leaking" emissions outside the system, and total rather than relative emissions (Wråke et al. 2012). Since the policy solutions we seek for energy are so tied up with the cessation of unsustainable practices in both economics and politics it only remains to establish the mechanisms to deliver on that obvious requirement.

There is already considerable cooperative activity around energy policy, but it must cope with predominately structural obstacles. Even in the most developed circumstances of political and economic integration across traditional boundaries (Europe) it is a struggle to establish clear links in energy policy, not least because of a focus on energy supply and markets has distracted from environmental issues (Morata and Sandoval 2012). So there is an intellectual, or attitudinal, hurdle to leap at the outset – we would have to accept that some deeply held assumptions are simply not viable, indeed sustainable, and learn to let them go. A United Nations institutional context illustrates debate about "controversial principles, such as whether to approach from an anthropocentric perspective or from a biocentric approach, or whether the viewpoint was from the individual or community" (UNESCO 2007: 7). There is no progress to be made by thinking that the political significance of energy policy only bears on abstractions. The point is that the underlying principles reflected in political and economic agendas should be flushed out, and the most appropriate ones promoted and acted upon in a pragmatic fashion as political interests. For example, it was noted that "barriers to renewable energy systems were institutional, political, technical and financial," and also that we should be cautious about a "highly centralized and state-controlled source of energy that did not promote participatory democracy" in contrast to "renewable energies such as solar, wind, small hydro, biomass, geothermal and tidal energy are often decentralized and can be used in remote areas without a solid energy supply system" (UNESCO 2007: 8–9). The relevant political structure should not be assumed any more than the economic one, since much action on these issues is driven by non-state actors and local politics, in both the developed and developing worlds (Fisher 2012).

A meaningful energy policy will require anticipation of future post-carbon scenarios. In offering a convincing perspective on "the age of petroleum" as merely a recent blip in the long run of human energy supply (until the late nineteenth century provided by biomass and animate labor, and from the twenty-first century by renewables) the Nuclear Energy Agency argues that the "critical path structure" should include "concurrent risk, economic, and environmental impact analyses…for all technologies and proposed actions for the transition to a post-petroleum economy" (Nuclear Energy Agency 2004: 37). While nuclear power remains under consideration, and hydrogen technology emerges as a potential portable fuel (though electricity-intensive in production), there are many more positive solutions to the challenge. The alternatives to fossil fuels exist, but it is claimed that it "will take a new industrial revolution" (Scheer 2002) or an "energy revolution" (Geller 2002) to develop these more widely. Even a decade ago the *Renewables Global Status Report* indicated that changes in the realm of renewable energy "have been so rapid in recent years that perceptions of the status of renewable energy can lag years behind the reality," with renewables already comprising "one quarter of global power capacity" (REN21 2011). The latest edition of that same report notes: "As disruptive as Covid-19 has been, the crisis does not alter observable trends in the energy sector that have persisted for years. The truth remains: we need to enact a structural shift built on an efficient and renewable-based energy system if we want to decarbonise our economies" and that " renewables lag in other end-use sectors like heating, cooling and transport, and that these sectors suffer a lack of policy support" (REN21 2020).

There is perhaps some promise in the relatively rapid turn toward electric vehicles, in both manufacturing and policy terms (Graham 2021), which if supplied by renewable sources of electricity would suggest some progress. If 100 percent clean renewable energy (Jacobson 2020) is not yet in sight, this offers evidence of continued growth in electricity, heat and fuel production from renewable energy sources, including solar electricity, wind power, solar hot water/heating, biofuels, hydropower and geothermal sources. Reflecting the range of opportunities, the Obama administration in the United States established the White House Office of Energy and Climate Change Policy to promote the president's "all-of-the-above strategy" for the twenty-first century, not surprisingly driven by concern with national security of energy supply (The White House 2012); the Trump administration was less engaged, to say the least, but the Biden administration promises to reverse that trend and go further and faster toward clean energy (joebiden.com) in a rather different political climate (Society of Environmental Journalists 2020).

Heinberg noted that the twenty-first century ushered in an era of declines, in a number of crucial parameters, and he seeks to address "the cultural, psychological and practical changes we will have to make as nature rapidly dictates our new limits" (Heinberg 2007). If Western industrial societies needed a prompt to respond with energy initiatives, it would have found incentives not only in unsettled international energy markets, but also in China's aggressive investment in renewables and clean technologies (ChinaFAQs 2012).

Decades ago, conventional intergovernmental bureaucracies were addressing what may again seem a novel issue, perhaps because a sense of urgency has re-emerged in the confluence of energy and environment policy (FAO 1982). Both producers and consumers of energy have already taken some steps to reflect concern with energy and environment, by experimenting with different practices (improving efficiency, slowly introducing new technologies, attempting to manage the energy situation collectively, etc.), and yet a remaining element of denial is reflected in a slow pace of change limited to the margins rather than the center of planning. On the environment side of the equation, geoengineering solutions could be used in *extremis* (Keith 2000), but this would only prolong our carbon addiction and would likely attract the same level of opposition as biofuels, given some elaborate schemes and the risks of unintended consequences (see Chapter 19). Nevertheless, such technological innovation will necessarily be a part of energy policy debates (Brown and Sovacool 2011). Maintaining economic growth while addressing environmental change will at the very least require prompt development of new technologies and a regulatory and fiscal environment to support them (Sachs 2008). This implies a significant change in current practices, and it remains to be seen whether currently familiar assumptions about economic growth will survive, all requiring long-term energy strategy coordination (Skea et al. 2019).

Conclusion

The policy challenges of energy systems increasingly demand that we make adjustments to our common practices. These will be more than mere instrumental adjustments to meet practical challenges, set within the framework of existing political conceptions and commitments. Our attention should be turned to the systemic and structural implications of this shifting policy area, as it may reflect a substantial underlying change. Furthermore, any opportunity to build on political momentum or economic dynamics that would address the fundamental issues in energy should be identified and capitalized on. This may, in turn, have an impact on opportunity structures and political–economic institutions. While short-term adjustments may advantage some actors, it is of course necessary to go beyond superficial measures and to appreciate the deeper political significance of the energy scenario. In viewing shifts in the surrounding debates as politically significant, we should hold no fixed assumptions about political, economic or social points of reference: this is new political territory, which demands open-mindedness. As a critical report on biofuels concludes, energy security demands a new paradigm, following Einstein's view that we "can't solve problems by using the same kind of thinking we used when we created them" (Santa Barbara 2007).

References

Advanced Research Projects Agency (2011) *Plants Engineered to Replace Oil*. Online. Available HTTP: <http://arpa-e.energy.gov/ProgramsProjects/PETRO.aspx>

Agence France Press (2008) "EU defends biofuel goals amid food crises", 14 April. Online. Available HTTP: <http://afp.google.com/article/ALeqM5gp1nkJeC-IhlYkVtsvPfp3u7mOWQ>

Allen, M. (2008) "Energy and climate: the physics of climate change" (lecture). Online. Available HTTP: <http://climateprediction.net/node/155>

Anderson, K.L. and Bows, A. 2008. "Reframing the climate change challenge in light of post-2000 emission trends", *Philosophical Transactions of the Royal Society – A: Mathematical, Physical and Engineering Sciences*, *366*: 3863–82.

Baer, P., Athanasiou, T., Kartha, S. and Kemp-Bendict, E. (2007) *The Right to Development in a Climate Constrained World: The Greenhouse Development Rights Framework*, Berlin: Heinrich Böll Foundation, Christian Aid, EcoEquity and the Stockholm Environment Institute. Online. Available HTTP: <http://www.ecoequity.org/docs/TheGDRsFramework.pdf>

Barnett, J. (2001) *The Meaning of Environmental Security: Ecological Politics and Policy in the New Security Era*, New York: Zed Books.

Barnett, J. and Adger, W.N. (2007) "Climate change, human security and violent conflict", *Political Geography*, *26*: 639–55.

Brown, M.A. and Sovacool, B.A. (2011) *Climate Change and Global Energy Security: Technology and Policy Options*, Cambridge, MA: MIT Press.

Cavanagh, J. (2012) "Mexico sets legally binding carbon reduction targets". Online. Available HTTP: <https://energycentral.com/c/ec/mexico-sets-legally-binding-carbon-reduction-targets>

ChinaFAQs The Network for Climate and Energy Information (Worldwatch Institute) (2012) *Clean Tech's Rise – Two New Issue Briefs from ChinaFAQs*. Online. Available HTTP: <http://www.chinafaqs.org/blog-posts/clean-techs-rise-two-new-issue-briefs-chinafaqs>

Commoner, B. (1971) *The Closing Circle: Nature, Man and Technology*, New York: Alfred Knopf.

Corley, A.-M. (2010) "The future of hydropower", *IEEE Spectrum*. Online. Available HTTP: <http://spectrum.ieee.org/energy/renewables/future-of-hydropower>

Creamer, T. (2008) "Is the SA biofuels debate properly framed?" Online. Available HTTP: <http://www.engineeringnews.co.za/article/is-the-sa-biofuels-debate-properly-framed-2008-03-28>

Dalby, S. (2002) *Environmental Security*, Minneapolis: University of Minnesota Press.

Downie, C. (2020) Steering global energy governance: who governs and what do they do?. *Regulation & Governance*. Online. Available HTTP: <https://doi.org/10.1111/rego.12352>

E3G (2020) Commission fails first European Green Deal test with energy infrastructure proposals. Online. Available HTTP: <https://www.e3g.org/news/commission-fails-first-european-green-deal-test-with-energy-infrastructure-proposals/>

The Economist (2008) "The elusive negawatt", 8 May. Online. Available HTTP: <http://www.economist.com/node/11326549>

Energy Bulletin (2008) "Food and agriculture", 16 April. Online. Available HTTP: <http://www.energybulletin.net/node/42817>

European Photovoltaic Industry Association and Greenpeace (2011) "Solar Generation 6 – solar photovoltaic electricity empowering the world – 2011". Online. Available HTTP: <http://www.

greenpeace.org/international/Global/international/publications/climate/2011/Final%20So-larGeneration%20VI%20full%20report%20lr.pdf>

Fattouh, Bassam and van der Linde, Coby (2011) "The international energy forum. Twenty years of producer–consumer dialogue in a changing world". Online. Available HTTP: <http://www.clin-gendael.nl/publications/2011/2011_IEF_History_of_IEF_Clinde_BFattouh.pdf>

Fisher, S. (2012) "Policy storylines in Indian climate politics: opening new political spaces?" *Environment and Planning C: Government and Policy*, *30*: 109–127. Online. Available HTTP: <http://www.envplan.com/epc/fulltext/c30/c10186.pdf>

Gardiner, Stephen M. (2006) "A perfect moral storm: climate change, intergenerational ethics and the problem of corruption", *Environmental Values*, *15(3)*: 397–413. Online. Available HTTP: <http://faculty.washington.edu/smgard/GardinerStorm06.pdf>.

Garvey, James (2008) *The Ethics of Climate Change*, London: Continuum. Online. Available HTTP: <http://nigelwarburton.typepad.com/virtualphilosopher/2008/02/james-garvey-in.html>.

Geller, Howard (2002) *Energy Revolution: Policies for a Sustainable Future*, Washington, DC: Island Press.

Graham, John D. (2021) *The Global Rise of the Modern Plug-In Electric Vehicle*, Cheltenham: Edward Elgar.

Greenpeace UK (2008) "Senior EU and Defra figures agree: we were too hasty on biofuel targets". Online. Available HTTP: <http://www.greenpeace.org.uk/blog/climate/we-were-too-hasty-on-biofuel-targets-20080114>

Heinberg, R. (2007) *Peak Everything: Waking Up to the Century of Declines*, Gabriola Island, BC: New Society Publishers. Available HTTP: <http://www.newsociety.com/Books/P/Peak-Everything2>

Holmes, I. and Mohanty, R. (2012) *The Macroeconomic Benefits of Energy Efficiency: The Case for Public Action*. Online. Available HTTP: <http://www.e3g.org/images/uploads/E3G_The_macroeco-nomic_case_for_energy_efficiency-Apr_2012.pdf>.

Humphreys, D.R. (2011) "Smoke and mirrors: some reflections on the science and politics of geoengineering", *Journal of Environment Development*, *20*: 99–120. Online. Available HTTP: <http://jed.sage-pub.com/content/20/2/99.full.pdf+html>

International Energy Agency (2020) *World Energy Outlook 2020*. Online. Available HTTP: <https://www.iea.org/reports/world-energy-outlook-2020>

International Energy Agency (2012) *World Energy Outlook 2012*. Online. Available HTTP: <http://www.iea.org/weo/>

Jacobson, M.Z. (2020) *100% Clean, Renewable Energy and Storage for Everything*, Cambridge: Cambridge University Press.

joebiden.com. Online. Available HTTP: <https://joebiden.com/9-key-elements-of-joe-bidens-plan-for-a-clean-energy-revolution/>

Keith, D.W. (2000) "Geoengineering the climate: history and prospect", *Annual Review of Energy and the Environment*, *25*: 245–84. Online. Available HTTP: <ftp://luna.atmos.washington.edu/pub/breth/PCC/SI2006/readings/Keith_geoengr_AnnRevEnergy_2000.pdf>

Kimmins, J.P. (2001) *The Ethics of Energy: A Framework for Action*, Paris: UNESCO.

Macfarlane, A.M. (2007) "Energy: the issue of the 21st century", *Elements*, *3(3)*: 165–70.

McGowan, A.H. (2007) "Rethinking energy options to address climate change", *Environment*, *49*: 8.

Morata, F. and Sandoval, I.S. (eds) (2012) *European Energy Policy: An Environmental Approach*, Cheltenham: Edward Elgar.

Murray, James (2020) "Which countries have legally-binding net-zero emissions targets?", NS Energy 5 Nov 2020. Online. Available HTTP <https://www.nsenergybusiness.com/news/countries-net-zero-emissions/>

New York Times (2008) "Fuel choices, food crises and finger-pointing", 15 April. Online. Available HTTP: <http://www.nytimes.com/2008/04/15/business/worldbusiness/15food.html>

Newsvine (2008) "Lead or get out of the way! Tension and tears in Bali", Natasha Restrepo, *Style Republic Magazine*. Online. Available HTTP: <http://stylerepublicmagazine.newsvine.com/_news/2008/01/07/1209740-lead-or-get-out-of-the-way-tension-and-tears-in-bali>

NOAA (2011) "NOAA greenhouse gas index continues climbing". Online. Available HTTP <http://www.noaanews.noaa.gov/stories2011/20111109_greenhousegasindex.html>.

NOAA (2020) "Trends in Atmospheric Carbon Dioxide". Online. Available HTTP <https://www.esrl.noaa.gov/gmd/ccgg/trends/>

Nuclear Energy Agency (2004) *Nuclear Production of Hydrogen*, Proceedings of the Second Information Exchange Meeting, Argonne, Illinois, USA, 2–3 October 2003, Paris: OECD.

Oxford High-Level Task Force on UK Energy Security, Climate Change and Development Assistance (2007) *Energy, Politics and Poverty: A Strategy for Energy Security, Climate Change and Development Assistance*, Oxford: University of Oxford.

Pearce, D.W., Markandya, A. and Barbier, E. (1989) *Blueprint for a Green Economy*, London: Earthscan.

Polimeni, J., Mayumi, K., Giampietro, M. and Alcott, B. (2008) *Jevons Paradox and the Myth of Resource Efficiency Improvements*, London: Earthscan.

Poruban, S. (2008) "US presidential politics to brighten spotlight on energy security issue", *Oil & Gas Journal*, 106(1): 24.

REN21 (Renewable Energy Policy Network for the 21st Century) (2020) *Renewables 2020 Global Status Report*, Paris: REN21 Secretariat. Online Available HTTP: < https://ren21.net/gsr-2020/pages/foreword/foreword/>

REN21 (Renewable Energy Policy Network for the 21st Century) (2011) *Renewables 2011 Global Status Report*, Paris: REN21 Secretariat. Online Available HTTP: <http://www.ren21.net/Portals/97/documents/GSR/REN21_GSR2011.pdf>

Roberts, David (2007) "Climate equity: a conversation – introducing an ongoing series on the most undercovered aspect of climate change", *Gristmill*. Online. Available HTTP: <http://grist.org/article/climate-equity-a-conversation/>

Sachs, J. D. (2008) "Keys to climate protection: dramatic, immediate commitment to nurturing new technologies is essential to averting disastrous global warming", *Scientific American*, March. Online. Available HTTP: <http://www.scientificamerican.com/article.cfm?id=technological-keys-to-climate-protection-extended>

Sachs, W. (2007) "Climate change is about equality among nations and fundamental human rights", in *Climate Equity: A Conversation*, edited by David Roberts, *Gristmill*. Online. Available HTTP: <http://grist.org/climate-energy/climate-equity-wolfgang-sachs/>

Santa Barbara, J. (2007) *The False Promise of Biofuels*, A Special Report from the International Forum on Globalization and the Institute for Policy Studies. Online. Available HTTP: <http://www.ifg.org/pdf/biofuels.pdf>

Scheer, Hermann (2002) *The Solar Economy*, London: Earthscan.

Shea, J. (2006) "Energy security: NATO's potential role", *NATO Review*, 3. Online. Available HTTP: <http://www.nato.int/docu/review/2006/issue3/english/special1.html>

Singer, P. (2006) "Ethics and climate change: a commentary on MacCracken, Toman and Gardiner", *Environmental Values*, 15: 415–2 2.

Skea, J., van Diemen, R., Hannon, M., Gazis, E., Rhodes, A. (2019) *Energy Innovation for the Twenty-First Century: Accelerating the Energy Revolution*, Cheltenham: Edward Elgar.

Society of Environmental Journalists (2020) *The 2021 Journalists' Guide to Energy & Environment*. Online. Available HTTP: <https://www.sej.org/publications/sej-news/2021-journalists-guide-energy-environment?emci=50f6aedc-9c56-eb11-a607-00155d43c992&emdi=71049986-ab56-eb11-a607-00155d43c992&ceid=154619>.

Stern, N. (2006) *The Economics of Climate Change ("The Stern Review Report")*, Cambridge: Cambridge University Press.

Toman, Michael (2006) "Values in the economics of climate change", *Environmental Values*, 15: 365–7 9.

UNESCO (2007) *Report of the launch conference for the project Ethics of Energy Technologies in Asia and the Pacific*, Bangkok: UNESCO, in collaboration with the Ministry of Science and Technology and the Ministry of Energy, Thailand. Online. Available HTTP <http://www.unescobkk.org/fileadmin/user_upload/shs/Energyethics/EnergyEthics2007Report.pdf>.

UN Food and Agriculture Organization (FAO) (1982) "Planning for the post-petroleum economy", *Ceres: The FAO Review on Agriculture and Development*, 15(4): 26–3 2.

US Department of State (2011) *US–China Strategic and Economic Dialogue 2011: Outcomes of the Strategic Track*. Online. Available HTTP: <http://www.state.gov/r/pa/prs/ps/2011/05/162967.htm>

Vad Mathiesen, B., Lund, H. and Karlsson, K. (2011) "100% renewable energy systems, climate mitigation and economic growth", *Applied Energy*, 88: 488–501. Online. Available HTTP: <http://www.sciencedirect.com/science/article/pii/S0306261910000644>

The White House (2012) *Energy, Climate Change and our Environment*. Online. Available HTTP: <http://www.whitehouse.gov/energy>

White, Eoin, and Gert Jan Kramer (2019) "The Changing Meaning of Energy Return on Investment and the Implications for the Prospects of Post-fossil Civilization", *One Earth* 1, December 20, 2019. Online. Available HTTP: <https://doi.org/10.1016/j.oneear.2019.11.010>

Wråke, M., Burtraw, D., Löfgren, A. and Zetterberg, L. (2012) "What have we learnt from the European Union's Emissions Trading System?". *Ambio*, *41*: 12–22. Online. Available HTTP: <http://www.springerlink.com/content/yu05415551083434/fulltext.pdf>

Xinhua (2011) "Full text of outcomes of strategic track of 2011 China–US Strategic and Economic Dialogue". Online. Available HTTP: <http://news.xinhuanet.com/english2010/china/2011–05/11/c_13868779.htm>

34

Transport and infrastructure

Toward sustainable mobility

Helene Dyrhauge

Free movements of goods and persons are central to a well-functioning society. Many take their personal mobility for granted without considering their individual transport carbon footprints. However, all transport modes rely on fossil fuels and the continued traffic growth in both goods and persons has contributed to climate change (see Chapter 32), and led to increases in air pollution and congestion. Indeed, driving and flying are two of the highest carbon emission activities individual persons undertake (Wynes and Nicholas 2017). Thus, it is necessary to mitigate transport's climate change footprint and minimize pollution, both locally and globally. This can only be done by breaking the dependence on fossil fuels and changing people's mobility patterns. This change requires technological innovation, government policies and international agreements on mitigating environmental externalities generated by traffic.

Policymakers face two acute challenges: reducing emissions from transport and reducing congestion; both are linked to traffic growth. While technological innovation like electric vehicles can reduce emissions, it will not solve the problem of congestion. Crucially, many national governments have set climate goals for 2030 that entail decarbonizing transport. Policymakers have to make decisions today about future infrastructure and technology investment in order to meet long-term climate goals. Moreover, policymakers have to adopt policies that facilitate more environmentally friendly travel behavior, including everyday and long-distance travel. Changing existing technologies and travel behavior require political vision and commitment to making decisions that reach beyond the next election and set the path for a new and sustainable mobility area. Many transport and mobility problems exist at all levels – from the local to the global. As such, policymakers around the world face similar challenges in changing the existing transport fossil fuel paradigm to a sustainable mobility paradigm. This involves phasing in new technologies and changing existing patterns of mobility, with the aim of minimizing the negative effect of transport on climate and environment.

This chapter focuses on the challenges in transforming the transport sector and personal mobility from a fossil-fuel paradigm into a sustainable-mobility paradigm. I discuss three pathways that originate in the sustainable mobility framework (Banister 2008; Holden 2007; Berger et al. 2014). The first pathway is technological substitution, which focuses on

DOI: 10.4324/9781003008873-39

developing and phasing in low emission fuels, thereby continuing the same growth paradigm, but with new fuel sources. The second pathway, modal shift, aims to reduce traffic
in congested transport modes and redirect it to more environmental modes, like railways
and other public transport. The third pathway, mobility reduction, aims to find alternatives to traveling, thereby radically changing existing principles of free movements. Overall, the chapter analyzes the dynamic between traditional transport policies and pressure to
change due to climate change, demonstrating the problems of changing centuries-old path-
dependent transport and infrastructure policies that are rooted in free movement, fossil fuels
and continued traffic growth.

The chapter starts by discussing how technology-substitution policies facilitate the transition from internal-combustion engine cars to low-emission vehicles. The second section
analyzes modal shift by examining the role of railways as an alternative transport mode.
The third section analyzes the problems of reducing aviation and promoting flying less.
The fourth section discusses how the Covid-19 lockdown affected the transport sector and
our personal mobility. The fifth section shows how we are currently at a critical juncture
in sustainable mobility and analyzes the relation between structure and personal agency in
changing the path from a fossil fueled transport area to the future path of sustainable mobility. Finally, the chapter concludes by summarizing the key challenges in changing transport
and infrastructure policies to a sustainable-mobility era.

Technological substitution and the problems with automobility

Technological substitution focuses on finding new low-emission technologies to substitute
existing fossil-fuel technologies, especially replacing the internal combustion engine with
low emission technologies, for example batteries. Policies that emphasize energy-efficient
transport technology are closely related to industrial policies, which, in turn, aim to facilitate
technological innovation and to protect industrial competitiveness. Often these policies avoid
picking a winner; instead, governments adopt a technology-neutral policy (Meckling and
Nahm 2018 and Azar and Sanden 2011). Technology-neutral policy is a bottom-up approach
to industrial policy representing a multi-level perspective on transition, where niche markets
develop new technologies and where some innovations filter up into wider society, known
as a socio-technical regime, to become mainstream (Geels 2011). Specifically, the emergence
and maturity of battery technologies in cars have led to discussions of how to manage the
transition from internal-combustion engines to low-emission vehicles. This discussion is not
only political but concerns electricity companies that have to deliver (renewable) energy to
cars because the transition changes energy demand away from big oil companies to electricity
companies, thereby creating new challenges requiring restructuring of their energy production to facilitate a new kind of demand and different peak demands.

Simultaneously, governments and private actors need to build a new fuel infrastructure
to recharge electric cars. This requires coordinated efforts by car manufactures, energy providers and governments to develop common standards, so that car owners can recharge their
cars outside their own homes, thereby enabling long-distance mobility. For example, the
European Union is working on both standardization of charging and rolling out new fuel
infrastructure across all 27 Member States. Previously, standardization of electricity plugs in
a country, and later portable adapters, enabled people to take their mobile phones and laptops everywhere within the Union. Indeed, standardization is necessary because it reduces
technical barriers to free movement. Thus, international coordination is central in decarbonizing transport, especially as new actors have to work together with incumbent actors and

policymakers to develop standards and these standards have to be implemented everywhere. Common standards make it easier for consumers to buy electric cars, as they do not have to worry about running out of electricity halfway through their journey.

At the national level, government policies must support the introduction of new low-emission transport technologies, for example via tax exemptions on low-emission vehicles and by increasing taxes on fossil-fuel vehicles, thereby incentivizing consumers to buy low-emission vehicles. Although many countries have promised to phase out fossil-fuel cars, few have adopted legislation; instead several countries are using different policy incentivize to facilitate the transition to low-emission cars. China has signaled a phase-out of fossil-fuel cars and has invested in battery technologies, which it sees as an important industry. These Chinese policies have spurred many European car manufactures to develop low-emission vehicles, especially electric cars, targeted at the Chinese market. Chinese and Asian investments in new technologies are spilling over into the car manufacturers' European markets.

Traditionally, European car manufacturers have lobbied extensively and successfully against stricter emissions standards (Dionigi 2017). Weaker emissions standards and oversight with car manufactures enabled many incumbent European and American car manufactures to deliberately cheat on emissions standards, creating the global Dieselgate Scandal, which subsequently led to stronger emissions regulations, giving regulatory agency powers to check on-road emissions instead of only testing cars in laboratories. The economic role of American and European car manufactures enabled them to influence policymaking, even as governments bailed them out during the last economic crises, which began in 2008. Thus, the transition to low-emission cars becomes more about industrial competitiveness than about climate change, and the road to meeting climate goals are framed through green growth, which until now has not reduced air pollution.

The overall aim of technological substitution is to decouple emissions from mobility (Givoni 2013: 212–215). Technological substitution extends to public transport, where busses with diesel engines are being replaced by those with low- or zero-emission engines that alleviate urban air pollution across the world. In December 2020, the Danish government and regions made an agreement on green public transport that commits the regions to only buy low/zero-emission busses, and only to lease or buy low/zero-emission cars. Public authorities' commitment to replacing their fleets with low-emission vehicles aims to show a pathway to transport decarbonization. However, this pathway does not address the other challenges in transport policy, such as congestion, noise and continued investment in road infrastructure, often including the expansion of existing congested infrastructure.

In China, car ownership has increased and cars have replaced bikes on streets. Consequently, Chinese cities have rebuilt their streets to accommodate automobility instead of bicycling (Notar 2018: 158), thereby contributing to urban congestion and air pollution. In the other side of the world, in Norway, more people buy electric cars. However, these are often the second or third car in a household, which begs the question of whether electrical vehicles are a conservation of the existing mobility paradigm or a transformation of automobility (Kester et al. 2020). The increased numbers of cars on the roads exacerbate existing problems with congestion; electric vehicles are a continuation of the existing car dependence and mobility paradigm. Although smart mobility, autonomous cars and car sharing attempt to address some of the issues with congestion, they only focus on urban areas and neglect rural areas where people are dependent on cars because public transport is not always available. Thus, technological substitution only enhances the existing car-dependent paradigm.

Most societies have developed car dependence that has locked-in mobility patterns and created structures that support the car and car production (Mattioli et al. 2020). This political economy of car dependence is difficult to alter, not only because of the dominant economic role that car companies have in several countries, notably Germany and the United States, but also because policymakers often see road infrastructure construction as the perennial answer to mitigating congestion. However, road building only increases supply, eventually leading to more vehicles on the roads. Thus, road building does not address the root issues in car dependence.

Modal shift and railways as an alternative transport mode

Changing the modal balance in transport entails reducing the number of cars on the roads and increasing availability and accessibility of public transport – busses or trains – and replacing short-haul air travel with long-distance train journeys. In other words, a modal shift in policy redirects existing mobility patterns away from the most polluting modes toward more environmentally friendly modes, for example by increasing the cost of car ownership through higher taxes and charges and simultaneously making public transport more attractive by increasing availability and affordability. The overall aim of modal shift is to decouple transport growth from economic growth (Holden 2007 and Givoni 2013), which will also reduce transport related pollution. However, modal-shift policies can be controversial as they challenge the traditional dominant transport modes, especially road transport. This was evident when the European Commission (2001) in its 2001 Transport White Paper wanted to shift the modal balance away from cars and aviation to public transport and railways. However, the European Commission abandoned this policy due to intense lobbying from other transport actors and reverted to an model that emphasis efficient use of fuel and traffic flows instead of a modal shift (Dyrhauge 2013b: 144).

Modal-shift policies have to be revised regularly to prevent a rebound effect, whereby car owners internalize the extra costs and start to drive more again (Holden 2007). Policymakers continuously have to adopt new measures to promote low-emission public transport and discourage car use. Yet, modal shift policy is a more comprehensive and holistic policy approach to sustainable mobility compared to technological substitution because it requires structural changes to existing mobility patterns whereby people have to use their car less and either have to walk, bike or take public transport like buses or trains. These changes to mobility necessitate changes to investment in cycle lanes and street lights in both rural and urban areas, and better pavements/sidewalks that enable everyone from parents with pushchairs, school children, disabled persons and the elderly to use them. These low-emission mobility activities have been neglected in some cities, for example in the United States where out-of-town shopping centers often can only be accessed by car, compared to European city centers with pedestrian areas accessible by public transport, whereas many Asian cities experience heavy road congestion due to lack of planning and investment. Thus, planning policies are important for creating a sustainable low-emission mobility paradigm that makes cities greener and more livable for the people residing and working in them.

By comparison, leisure travel often involves traveling long distances. Inexpensive airfares challenge long-distance rail. City pairing with competitive rail and aviation links does not facilitate modal shift; instead they are more likely to increase demand for services. London-Paris and Sydney-Melbourne are examples of successful inter-modal competition. Unsurprisingly, both case studies show that aviation has a higher negative impact for climate and air pollution compared to high-speed railways (Givoni 2007; Robertson 2018) because the latter often use electricity, which can come from both fossil fuels and renewables. Emissions from electricity production happen at the power plant instead of the railways, thereby moving transport emissions to the electricity sector. Importantly, even fossil-fuel electricity in high-speed rail services is better for the climate and environment than flying (Givoni 2007). Thus, investment in renewable energy from a climate perspective benefits high-speed rail and regular railways compared to other transport modes. Indeed, the decarbonization of transport modes often means electrification, which, in turn, necessitates cooperation between policymakers, the transport sector and the electricity sector in order for public authorities to meet their low-emission targets.

Fuel substitution is only one element in creating a sustainable mobility paradigm. Most policymakers agree that railways have an important role in making transport more sustainable; this necessitates that they are competitive vis-à-vis other modes. The European Commission has tried to liberalize national railways and make European railways competitive, including by trying to promote international rail travel. This has been an uphill struggle because many EU member states and incumbent national railways with monopolistic market positions have not supported the idea (Dyrhauge 2013a). Moreover, night-train services have, over the past decade, closed because fewer people have used them, instead favoring cheaper and faster aviation that enable people to go on short weekend breaks. Nevertheless, recent years have seen new national and European grass root movements emerging that are lobbying for sustainable travel, especially pushing for the return of international night trains and more connected international rail services as an alternative to air travel. These grass root movements are part of the broader climate movement. Importantly, consumer demands have led several European governments and railway companies to discuss reopening night-train services. Today, more people have started to appreciate slow travel and seeing the journey as part of the holiday. Indeed, 2021 was designated the European Year of Rail during which the European Commission aimed to relaunch the EU's railways.

Governments in other parts of the world have successfully invested in rail infrastructure, especially high-speed railways to facilitate mobility within their countries. For example, Chinese and Japanese high-speed passenger services connect major cities and successfully compete with short-haul flying, whereas older and slower rail infrastructure cater to shorter distances, daily commuting and feeder traffic to high-speed rail services. Building new infrastructure, including high-speed rail is costly. Most countries do not have high-speed railways, instead relying on older infrastructure. Indeed the construction of new rail infrastructure projects negatively influences land-use and biodiversity because such projects often go around built up areas and instead go through farmland, woodlands and bio-sensitive natural areas, in some cases even protected nature reserves. For example, High-Speed 2 in the United Kingdom aims to alleviate congestion on both the existing rail network and motorways connecting the north and south of the country, thereby creating better territorial cohesion. However, the project is disputed, not only due to the cost of building a new rail network to complement the existing network, but also because the planned route will go through old woodlands and protected natural areas. This creates a dilemma of protecting nature versus facilitating territorial cohesion, and thus economic growth, through transport infrastructure, a dilemma that is often resolved in favor of the latter, despite protests from environmental groups.

Overall, modal shift policies that favor public transport and a shift to low-emission transport are preferable to increased car usage. However, modal shift that involve building new infrastructure affects local land use and biodiversity, thereby creating new environmental problems. Modal shift policies coupled with infrastructure building increases the supply of transport between cities, which lead to traffic growth instead of modal shift. While traffic growth itself is not bad as long as it is low-emission travel, the inter-modal competition between rail and air travel has negative consequences for air pollution generated from flying.

Reducing transport growth by flying less

Reducing traffic growth and restricting mobility require socio-economic changes to societies at both the local and global levels. This approach aims to "reduce travel demand through increasing positive environmental attitudes," for example through planning policies that encourage low emission mobility like biking, and through developing communication tools that minimize the need for travel (Holden 2007: 72). Reducing mobility necessitates radical lifestyle changes (Givoni 2013: 218–222). It might entail restrictions to free movement by introducing a personal annual carbon footprint allowance, yet such a policy is controversial and taboo for policymakers (Gössling and Cohen 2014). Policymakers around the world value the principle of free movement, and the principle is integral to democratic societies across the globe. Transport is energy-intensive and almost exclusively fossil-fuel based; free movement therefore sets human needs and satisfactions above climate and environmental protection. The question of a transport taboo is particularly evident in the public debate about flying.

In 2019, aviation accounted for 2.8 percent of global carbon dioxide (CO_2) emissions, but it is a growth industry and it has grown by 5 percent per year since 2000 (IEA 2020). The continued growth in aviation is problematic, especially as much of the emissions are not accounted for in the international climate agreements, including the Kyoto Protocol and the Paris Agreement (Gössling and Humpe 2020: 9; on climate change, see Chapter 32). Indeed, the Chicago Convention that governs international aviation through bilateral agreements exempts aviation from fuel duties. This means that national governments unilaterally cannot impose environmental taxes on aviation fuel for international flights. Moreover, the International Civil Aviation Organization (ICAO), which regulates the Chicago Convention, has rejected the EU's lobbying attempts to implement an international emissions-trading system for aviation (Lindenthal 2014). Instead, ICAO and the International Air Transport Association have worked toward finding technological solutions such as increased fuel efficiency and use of biofuels to mitigate the sector's impact on the climate (Larsson et al. 2019). These technological solutions are still in their infancy and unlikely to become mainstream before 2030. Therefore, governments will have to agree on other policy mechanisms to mitigate climate change derived from flying. Crucially, governments have few incentives to act because aviation represents a small share of global emissions and simultaneously is a mobility activity disproportionately used by the world's more-affluent people.

Only a minority of people on the planet can afford to fly, but those that can do so are high emitters. Among the people who fly, frequent flyers account for the majority of emissions. Indeed, frequent-flyer programs encourage their members to fly by rewarding them with different types of bonuses the more they fly. These "super emitters, i.e. the 10% of the most frequent fliers emitting more than half of global CO2 emissions from commercial air travel[,] as well as the users of private aircraft[,] cause emissions of up to 7,500 t[ons] CO2 per year" per person (Gössling and Humpe 2020: 9). In short, a tiny number of people contribute toward 2.8 percent of annual global CO_2 emission. This makes much of global aviation an elite mobility practice.

Public debate about how to tax those who fly is becoming more mainstream as climate change has moved up the political agenda, especially as social movements like "Flying Less" and "Fridays for Future" campaign for people to fly less and stay on the ground by taking trains instead. Flight shaming is central to these campaigns. The concept originated in Sweden in 2017 and a "vivid debate about individual responsibility and air travel exploded and spread across mainstream and social media" (Jacobson et al. 2020: 3). For the first time, there was a reduction in flying in Sweden in 2019 (Jacobson et al. 2020: 3). Instead, people campaigned for new night-train services and international railway services as substitutes for flying, thereby promoting modal shift. Famously, the Swedish environmental activist Greta Thunberg took a low-carbon route to North America in autumn 2019 to attend a climate conference. Her crossing the Atlantic both ways in yachts increased awareness of aviation's negative impact on the climate. Thunberg called out several celebrities who campaigned for the environment yet who took private jets (one example being Emma Thompson, who flew from the United States to the United Kingdom to participate in a climate change demonstration). The issues of flying less and flight shame are not just about reducing emissions among a few people. It is also about social justice in relation to those who fly and those who do not. This has led to debates about environmental taxes on flying. However, the intergovernmental Chicago Convention framework prevents governments from taking unilateral action. Instead, the onus is on individuals to change their travel behavior and take responsibility for their emissions.

The discussion about individual responsibility and agency is especially evident in the academic debates about flying, where academics have openly questioned the need for flying. Academics are part of the mobile elite that often fly. They recognize the need for going on fieldwork, which might be in places that can only be reached by flying. Instead of taking multiple smaller trips, one can go on longer field trips, to the extent that family life permits this, thereby taking responsibility for one's personal carbon footprint. Moreover, academics often participate in several international conferences and workshops per year. Environmental and climate change scholars are no different than scholars in other fields; they also fly to international conferences to discuss climate change. Holden et al. (2017) has called out this contradictory practice, which some see as hypocritical. These activities give individual academics the opportunity to present and discuss their research with their peers, extend their networks and, for some, it is an opportunity to extend the conference with a holiday. However, conference participation and high personal emissions due to travel do not equal enhanced academic outcomes (Chalvatis and Ormosi, 2021).

A study of six European Consortium for Political Research conferences shows that the group's 2018 conference in Hamburg had the lowest emissions because of its central location in Europe and its connectivity to international railway services (Jäckle 2019: 640). Jäckle (2019: 648) concludes that Frankfurt is the best location for European conferences because of its central location in Europe and hub for international railway services. This is unrealistic in, for example, the Asia Pacific area where countries are separated by the sea instead of a landlocked border, thus here participants will have to fly if they must meet in person. A question is whether such in-person conferences are needed. At the very least, this study raises important questions about academics' practices and the need to travel across the global to participate in conferences. In 2020, almost all international academic conferences were cancelled due to Covid-19, and many of them went online. Online conferences only require a good Internet connection and thus allow academics at different stages of their careers and from all corners of the world to participate. Thus, online conferences are better for social equity than physical conferences, which can be expensive to attend.

The Covid-19 pandemic: lockdowns and restrictions on movement

Covid-19 was not only a pandemic; governments' restrictions aimed at stopping infection rates fundamentally changed people's daily mobility patterns, as more people were forced to work from home. Indeed, during spring 2020 "road transport in regions with lockdowns in place … dropped between 50% and 75%, with global average road transport activity almost falling to 50% of the 2019 level by the end of March 2020. …. As a result of declines in mobility, in March alone world oil demand plummeted by a record 10.8 mb/d [million barrels per day] year-on-year" (IEA 2020: 18). In short, the worldwide lockdown in spring 2020 had a positive impact on climate change.

Simultaneously, lockdowns and border closings effectively grounded the world's airlines. During the spring 2020 lockdown, "aviation activity in some European countries declining more than 90%. Aviation activity in China … rebounded slightly from the low at the end of February, as lockdown measures … eased slightly. Nonetheless, as lockdowns spread, global aviation activity declined a staggering 60% by the end of Q1 [first quarter] 2020" (IEA 2020: 18). While the grounded airplanes have benefited the environment, the lock-down and restrictions on personal mobility had severe personal consequences for those people working in transport, especially at airports and in aviation, who in many cases lost their jobs. During the pandemic governments across the world focused on supporting the health sector and making emergency financial packages for the sectors most affected by the lockdowns, for example airlines, airports and tourism. While transport emissions dropped due to lock downs, policies to facilitate transport decarbonization were put on hold.

Moreover, Covid-19 lockdowns reduced road fatalities as people worked from home, just as normally busy commuter trains were less crowded. Congestion almost vanished as roads lay quiet and all major cities in the world, from Paris to New Delhi, experienced clean and clear skies as air pollution dropped and their inhabitants for the first time in many years could see clear blue skies. Similar several cities closed streets to make space for pedestrians and bikers, some cities kept the streets closed after lockdowns, while others reverted them back into busy congested streets with high levels of air pollution.

Overall, it is difficult to predict what the transport sector and our personal mobility will look like in a post-pandemic world. From a climate change perspective, one hopes that daily commuters will return to public transport instead of using cars to prevent catching Covid-19. Some people are likely to continue to work from home more than previously. Nevertheless, over time the daily mobility patterns are likely to revert to normal. The real question will be how people will plan their leisure travel. Will they return to cruise ships and airlines? There are unanswered question about to what extent people will take to the skies again once they have been vaccinated and lockdowns finally have eased. Some people are likely to do so once restrictions have lifted so that they can satisfy their longing for travel after extended time at home, while others will have discovered the joys of staying on the ground and exploring places closer to home. The lockdowns have shown how governments' restrictions on free movements can fundamentally change people's personal mobility. From a climate perspective, this shows a double-edge sword that simultaneously protects the environment and restricts individual free movement.

Facilitating transition, free movement and personal travel choices

The transport sector as a fossil-fuel sector needs a radical change to achieve the goals set out in the Paris Agreement. Yet, government transport policies are often fragmented, focusing on one mode instead of developing coherent strategies, thus making it difficult to create a sustainable mobility paradigm with low emissions. Politicians in democracies only look to next (re)election while investments in transport infrastructure are long-term commitments. The discrepancy between governments' short-termism and the need for long-term commitment to transport decarbonization is stalling the phase-in of low-emission technologies for road transport. Car buying is a long-term commitment because cars bought today will still be on the road for a decade or more. Incentives for people to switch to low-emission cars have to be in place immediately. This entails investing in alternative-fuels infrastructure and adjusting environmental taxes and charges to make fossil-fuel cars more expensive and low-emission cars cheaper. However, taxes and charges from transport activities are revenues for governments that use the income to finance welfare, education, health and more. Thus, the structure of national budgets can make it difficult for governments to radically change transport taxes and charges. In many countries, a plan for long-term adjustment to public finances that will facilitate the transition to a sustainable mobility paradigm will require agreements between the government of the day and their opposition to make sure that the next government does not reverse decisions, preventing the countries from meeting their climate commitments.

Government policies can create structural incentives for people to change their personal travel behavior, such as choosing public transport over car ownership and making life-style changes to reduce their personal transport carbon footprints. Indeed, personal agency is an important element in breaking fossil-fuel dependency, thereby achieving sustainable mobility. Modal-shift policies are not only concerned with persuading more people to leave their cars behind in favor of public transport or taking the train instead of flying; they are also concerned about making public transport more attractive compared to other transport modes, for example through timetabling, pricing and accessibility. While governments can use different policy instruments to incentivize changes by creating structural changes, green attitudes to transport at the individual level must exist before it is possible to establish a sustainable mobility paradigm.

Personal choices are important, especially as transport is *the* activity that contributes the most to personal carbon footprints. Wynes and Nicholas (2017) "identify four recommended actions which, [they] believe to be especially effective in reducing an individual's greenhouse gas emissions: having one fewer child, living car-free, avoiding airplane travel, and eating a plant-based diet." Making the choice to live a low-carbon life entails radically changing travel behavior in terms of both basic transport *needs* and transport *wants*. Basic transport *needs* refer to meeting daily transport requirements, such as commuting to work and going to the shops and health care facilities. In developed countries, these transport activities are often carbon- and energy-intensive, and they have increased over time (Mattioli 2016). Transport *wants* refer to leisure travel and long-distance traveling, often by flying, and contribute toward "excess travel" (Mattioli 2016). Moving from a fossil-fuel transport paradigm to a sustainable-mobility paradigm requires changing how people meet their transport *needs* and adjust their understanding of transport *wants*. Here policies and infrastructure provide the structure that facilitates changes to personal choice about mobility. Importantly, the transition needs to be just.

Transport justice has an important role in preventing social exclusion and ensuring that everyone can participate in the transition to a sustainable-mobility paradigm (Markovich 2013; Martens 2017). Transport justice entails policies that support accessible and affordable infrastructure. On a personal level, "a person's accessibility depends on both context (transportation systems and land use patterns) and personal attributes (such as vehicle ownership, income levels, and abilities" (Martens 2017: 13). To make sure that everyone is able to take part in the transition to sustainable mobility, it is necessary to develop coherent policies at all levels, including local urban planning, national transport policies for infrastructure and public transport, environmental taxes and charges, and international agreements on climate mitigation in civil aviation. These structures can incentives consumer choice and travel behavior.

Individual transport choice is also determined by where people live. While those living in urban areas might not need a car and instead use public transport or bikes, people living in other places depend on cars to fulfill their transport needs. Here government policies are important to ensure that all car owners can afford to switch from fossil-fuel powered cars to low-emission cars. A long-term policy strategy for phasing in low-emission vehicles enables a stable transition and transparency for individuals in deciding what their next cars should be and how to change their transport activities.

Policy incentives can structure consumer choice, for example by taxing SUVs higher than electric cars. The car is a symbol of freedom and individualism. The choice of car signals one's socio-economic status and political perception of climate change. Moreover, "the cultural underpinning of car dependence go beyond just car travel and are connected to the very idea of travel as an intrinsically desirable practice" (Mattioli et al. 2020: 10). The car gives people freedom to go places whenever they want because they do not have to wait for a bus or airplane. Many people are reluctant to give up their cars and the freedom that comes with them (Mattioli et al. 2020: 11), even though they might spend a considerable time stuck in congested traffic. Similarly, many perceive free movement and traveling as essential to human flourishment (Goodwin 2010: 70–75). This perception, together with economic development, has contributed to both local and global traffic growth. Reducing transport emissions necessitates that people take personal responsibility for both their transport needs and wants. While personal decisions of not being a high emitter and reducing travel, especially leisure travel to faraway places, is crucial for a successful transition to a sustainable-mobility paradigm, currently personal agency to change is constrained by a fossil-fuel dependent paradigm.

Path dependence in transport exists at multiple levels, including in transport policy, infrastructure and individual travel behavior. The climate change agenda has placed transport at a critical juncture that requires radical change to meet long-term climate goals. While policymakers across the world face similar issues of decarbonizing transport, national transport policies and infrastructure differ, ranging from the focus on personal car ownership in the United States to high-speed rail services in China and Japan. Transport policies create long-term structures and infrastructure that guide personal mobility, and thus personal agency.

The policies to facilitate the transition to a sustainable-mobility paradigm need to combine the three pathways discussed earlier: technological substitution, modal shift and reducing transport growth. Without agency to change behavior and better policy coordination, transport will continue along a fossil-fuel pathway with continued traffic growth. The transition from a fossil-fuel transport paradigm to a sustainable mobility paradigm is complex, but at the heart of the transition lies the relation between structure (i.e., physical, spatial and technological infrastructure) and agency, particularly policy instruments that guide and incentivize personal agency. Personal agency is also important for changing structure. This is evident in the campaigns mentioned above that advocate flying less and having more night trains. Individual transport choice can create demand for different forms of transport, thereby shifting the modal balance. In other words, the relation between structure and agency is crucial for a just transition to sustainable mobility.

Conclusion

Sustainable mobility and transport decarbonization are essential for mitigating climate change and meeting the Paris Agreement's objectives (see Chapter 32). However, sustainable mobility requires more than technological solutions. It requires replacing a fossil-fuel infrastructure with a low-emission infrastructure and making people change their travel behavior. This entails better policy coordination across policy fields and long-term planning. While phasing in low-emission technologies is slowly happening around the world, thereby changing infrastructure, it is more difficult to convince people to change travel behavior and take responsibility for their personal transport-related carbon footprints. The change from a fossil-fuel transport paradigm to a sustainable-mobility paradigm requires a long-term policy vision that combines all three pathways discussed above and which supports the transition to low-emission travel behavior. Without dramatic new transport policies and widespread behavioral change, instead of creating a sustainable mobility paradigm the world is likely to continue along the path of traffic growth that has been experienced over the past almost 200 years.

References

Azar, C. and Sanden, B.A. (2011). "The elusive quest for technology-neutral policies". *Environmental Innovation and Societal Transitions* 1: 135–139.

Banister, D. (2008). "The sustainability mobility paradigm". *Transport Policy* 15, pp. 73–80.

Berger, G., Feindt, P.H, Holden, E. and Rabik, F. (2014). "Sustainable mobility – challenges for a complex transition". *Journal of Environmental Policy and Planning* 16(3), pp. 303–320.

Chalvatis, K. and Ormosi, P.L. (2021). "The carbon impact of flying to economic conferences: is flying more associated with more citations?" *Journal of Sustainable Tourism* 29(1), pp. 40–67.

Dionigi, M.K (2017). *Lobbying in the European Parliament*. Bakingstoke, Palgrave MacMillan.

Dyrhauge, H. (2013a). *EU railway policy-making: On track?* Bakingstoke, Palgrave Macmillan.

Dyrhauge, H. (2013b). "EU sustainable mobility – between economic and environmental discourses" in P.M. Barnes and T.C. Hoerber (eds.) *Sustainable development and governance in Europe: The evolution of the discourse on sustainability*. Abingdon, Routledge.

European Commission (2001). *WHITE PAPER European transport policy for 2010: Time to decide*. COM (2001) 370 final. 12 March 2001.

Geels, F.W. (2011). "The multi-level perspective on sustainability transitions: Responses to seven criticisms". *Environmental Innovation and Societal Transitions* 1, pp. 24–40.

Givoni, M. (2007). "Environmental benefits from mode substitution: Comparison of the environmental impact from aircraft and high-speed train operations". *International Journal of Sustainable Transportation* 1(4), pp. 209–230,

Givoni, M. (2013). "Alternative pathways to low carbon mobility" in M. Givoni and D. Banister (eds.) *Moving Towards Low Carbon Mobility*. Cheltenham, Edward Elgar.

Goodwin, K.J. (2010). "Reconstructing automobility: The making and breaking of modern transportation". *Global Environmental Politics* 10(4), pp. 60–78.

Gössling, S. and Cohen, S. (2014). "Why sustainable transport policies will fail: EU climate policy in the light of transport taboos". *Journal of Transport Geography* 39, pp. 197–207.

Gössling, S. and Humpe, A (2020). "The global scale, distribution and growth of aviation: Implications for climate change". *Global Environmental Change* 65 (2020), p. 102194.

Holden, E. (2007). *Achieving sustainable mobility. Everyday and Leisure-time travel in the EU*. Farnham: Ashgate.

Holden, M.H., Butt, N., Chauvenet, A.M. Plein, M. Stringer and Chadès, I. (2017). "Academic conferences urgently need environmental policies". *Nature Ecology Evolution* 1(19), pp. 1211–1212.

International Energy Agency (IEA) (2020). *Tracking Transport 2020*. IEA, Paris. https://www.iea.org/reports/tracking-transport-2020

Jäckle, S. (2019). "WE have to change! The carbon footprint of ECPR general conferences and ways to reduce it". *European Political Science* 18, pp. 630–650.

Jacobson, L., Åkerman, J., Giusti, M. & Bhowmik, A.K. (2020). "Tipping to staying on the ground: Internalized knowledge of climate change crucial for transformed air travel behaviour". *Sustainability* 12, p. 1994.

Kester, J., Sovacool, B.K., dw Rubens, G.Z. & L. Noel (2020). "Novel or normal? Electric vehicles and the dialectic transition of Nordic automobility". *Energy Research & Social Science* 69, p. 101642. https://www.sciencedirect.com/science/article/abs/pii/S2214629620302176

Larsson, J., Elofsson, A., Sterner, T. & Åkerman, J. (2019). "International and national climate policies for aviation: A review". *Climate Policy* 19(6), pp. 787–799.

Lindenthal, A. (2014). "Aviation and climate protection: EU leadership within the International Civil Aviation Organization". *Environmental Politics* 23(6), pp. 1064–1081.

Markovich, J. (2013). "Accessibility, equity and transport" in M. Givoni and D. Banister (eds.) *Moving towards low carbon mobility*. Cheltenham, Edward Elgar.

Martens, K. (2017). *Transport justice: Designing fair transportation systems*. Abingdon, Routledge.

Mattioli, G. (2016). "Transport needs in a climate-constrained world. A novel framework to reconcile social and environmental sustainability in transport". *Energy Research & Social Science* 18, pp. 118–128.

Mattioli, G., Roberts, C., Steinberger, J.K. and Brown, A. (2020). "The political economy of car dependence: A systems of provision approach". *Energy Research & Social Science* 66, p. 101486.

Meckling, J. and Nahm, J. (2018). "When do states interrupt industries? Electric cars and the politics of innovation". *Review of International Political Economy* 25(4), pp. 505–529.

Notar, B.E. (2018). "Car crazy: The rise of car culture in China" in A. Hansen and K.B. Nielsen (eds.) *Cars, automobility and development in Asia. Wheels of change*. Abingdon, Routledge.

Robertson, S. (2018). "A carbon footprint analysis of renewable energy technology adoption in the modal substitution of high-speed rail for short-haul air travel in Australia". *International Journal of Sustainable Transportation* 12(4), pp. 299–312.

Wynes, S. and Nicholas, K. (2017). "The climate mitigation gap: education and government recommendations miss the most effective individual actions". *Environmental Research Letters* 12, p. 074024.

35

Persistent organic pollutants

Managing threats to human health and the environment

David Downie and Jessica Templeton

Since the 1960s, more than 100,000 synthetic chemicals have been registered for commercial use (Selin 2010: 40). Many of these chemicals are both extremely useful and harmless; others are extremely harmful to human health and the environment. In the early 1970s, governments in many industrialized countries began enacting national legislation to regulate the production and use of certain pesticides and industrial chemicals, with most starting with DDT and PCBs. However, as chemical production and use expanded globally in the 1970s and 1980s, governments realized that no country acting alone could effectively address transboundary chemical pollution.

Consequently, over the past 40 years, governments have pursued a series of increasingly ambitious international initiatives to address the negative impacts of toxic chemicals (Lönngren 1992; Downie and Fenge 2003; Selin 2010; Abelkop et al. 2017; Chasek and Downie 2021). These efforts have included the creation of: international organizations and programs such as the Intergovernmental Forum on Chemical Safety and the United Nations Environment Programme (UNEP) Chemicals Branch; voluntary global guidelines such as the International Code of Conduct on the Distribution and Use of Pesticides; regional treaties, including Protocols adopted under the Convention on Long-Range Transboundary Air Pollution (CLRTAP, see Chapter 30); and several legally binding global treaties, including the 1989 Basel Convention on Hazardous Wastes (Chapter 36), the 1998 Rotterdam Convention on the Prior Informed Consent (PIC) Procedure for Certain Hazardous Chemicals and Pesticides in International Trade, the 2001 Stockholm Convention on Persistent Organic Pollutants (POPs) and the 2013 Minamata Convention on Mercury. (The 1987 Montreal Protocol addresses chemicals that deplete stratospheric ozone, and while several are toxic to human health, the Protocol is usually considered to be an atmospheric treaty and not part of the chemicals and waste cluster. See Chapter 31.)

This chapter examines the Stockholm Convention on POPs. This agreement is particularly important not only because of the dangerous chemicals it addresses, but also because it is the first global, legally binding treaty that specifically seeks to eliminate the production and use of chemicals that are directly toxic to the environment and human health.

DOI: 10.4324/9781003008873-40

Persistent organic pollutants: global threats to the environment and human health

POPs possess four key characteristics that necessitate collaborative international action. First, POPs are toxic to humans and wildlife. While impacts vary, exposure to these chemicals is associated with disease, cancer, birth defects, developmental impairment, reproductive difficulties, autoimmune problems and other significant harms to aquatic mammals, fish, birds and humans. Second, POPs are persistent. Once released into the environment, they remain intact and toxic for extended periods. While all POPs have relatively long half-lives, some are amazingly persistent and resist degradation for decades or longer. For example, perfluorooctane sulfonic acid (PFOS), used in applications such as firefighting foams, upholstery and cleaning products, degrades only in high-temperature incineration.

Third, POPs bioaccumulate, or increase in concentration, within the tissues of individual animals and as they pass through food webs. Concentrations in animals and people can be thousands of times greater than the levels found in their local environment. Humans and animals can also absorb high concentrations of POPs quickly if they eat multiple organisms in which POPs have already accumulated. Thus, POPs have been found in high concentrations in top predators, such as seals and polar bears, as well as humans who rely on fatty aquatic animals such as seals or whales for food. Mammals can then pass POPs to their offspring during prenatal development and through lipid-rich breastmilk.

Fourth, POPs can travel thousands of kilometers from their emission sources via water and air currents (Hung et al. 2016). This crucial characteristic, referred to as long-range environmental transport, means that POPs pose risks even to humans and wildlife living far from where the substances are produced or used. POPs tend to travel northward, where colder temperatures cause them to fall to Earth. Some of the world's highest concentrations of POPs are found in countries bordering the Arctic Circle (AMAP 1998).

The transboundary dangers posed by POPs, resulting from the combination of toxicity, persistence, bioaccumulation, and especially long-range environmental transport, came to light in the mid-1980s when Canadian government researchers discovered surprisingly high concentrations of PCBs in territory near the Arctic Circle (see Downie and Fenge 2003). Because industrial chemicals are neither produced nor widely used in the Arctic, scientists had planned to use the area as a pristine reference standard against which they could compare data from other areas known to be contaminated. The discovery led to further research that revealed how POPs were affecting a variety of species including fish, birds, seals and bears.

The research also revealed the impact of POPs on people living in the region. In particular, tests of the umbilical cord blood of newborns, the blood of infants, and mothers' milk in areas near the Arctic were found to contain many times the amount of POPs found in the blood of babies born in more southern regions (Hillman 1999; Downie and Fenge 2003). The diets of indigenous Arctic people tend to rely heavily on fish, mammals and birds, species with particularly high levels of POPs in their fatty tissues due to the persistence, bioaccumulation and long-range transport of POPs.

From science to negotiations

The discovery of substantial contamination throughout the Arctic, and evidence of bioaccumulation and long-range transport, led to growing recognition among scientists and policymakers of the need for international action to control POPs (Downie and Fenge 2003; Selin 2010). Particular attention was paid to the risks faced by sensitive groups such as small children, pregnant women, and indigenous peoples. Canada and Sweden were instrumental in supporting this research and pushing the issue on to the international agenda. During the 1990s concern also grew in the southern hemisphere about local health and environmental risks from the unregulated use of certain pesticides and other hazardous substances (Kohler and Ashton 2010).

Scientific assessments conducted by the regional CLRTAP Task Force on POPs also enhanced awareness of the need for international action. The Task Force played a crucial role in defining POPs scientifically and politically by agreeing on the general physical and chemical characteristics of POPs, and divided POPs into three categories: industrial chemicals, pesticides and unintentionally produced by-products created as a result of other industrial processes, such as waste incineration (Selin 2010). These assessments not only informed the development of a regional POPs Protocol under CLRTAP, but also greatly assisted calls to consider formal global negotiations. (The discussion below draws on previous work by the authors, including Downie and Fenge 2003; Downie et al. 2005; Templeton 2011, 2020; Allen et al. 2018; Chasek and Downie 2021; contributions to United Nations documents and Earth Negotiation Bulletin reports; and direct observations made during negotiation of the Stockholm Convention and subsequent meetings of the Conference of Parties.)

In May 1995, UNEP's Governing Council requested that the Intergovernmental Forum on Chemical Safety (IFCS) and the Inter-Organization Programme for the Sound Management of Chemicals (IOMC), two international programs created by governments to address chemicals, conduct a global scientific assessment of 12 POPs known as the dirty dozen: eight pesticides (aldrin, chlordane, DDT, dieldrin, endrin, heptachlor, mirex and toxaphene); two industrial chemicals (PCBs and hexachlorobenzene, which is also a pesticide) and two by-products (dioxins and furans). These substances were chosen both for the dangers they posed and because they were already subject to significant regulation, and in some cases outright bans, in many industrialized countries.

In response, the IOMC summarized the scientific literature on POPs, consolidating available information on their chemistry, toxicity, environmental dispersion and other relevant properties, and IFCS established a working group to conduct the assessment. In June 1996, IFCS concluded that sufficient evidence existed to warrant significant international action to reduce the risks posed by POPs. In February 1997, the Governing Council (in decision 19/13c) endorsed the IFCS report and formally authorized UNEP to convene negotiations with a mandate to draft a global legally binding agreement that would address the dirty dozen and include procedures for identifying, reviewing and listing additional POPs in the future.

Meanwhile, Canada and Western European governments were concluding their regional POPs agreement under CLRTAP. Signed in June 1998, the Aarhus Protocol on Persistent Organic Pollutants seeks to eliminate or reduce releases of 16 POPs. Covering Europe, Russia, Canada and the United States (although not ratified by the United States or Russia), the agreement provided diplomatic momentum to the global effort, established additional scientific justification, and supplied certain templates that assisted development of a global convention (Downie and Fenge 2003; Eckley and Selin 2003).

Formal negotiations on a global POPs treaty began in June 1998. Despite the Montreal Protocol, Aarhus Protocol, and Rotterdam PIC Convention serving as successful models, and extensive preparations by UNEP (including a series of regional awareness-raising and pre-negotiation workshops), governments had a difficult time resolving several important issues, including how to structure control measures, add new POPs to the treaty in future, and provide financial and technical assistance to developing countries.

The Stockholm Convention

Adopted in May 2001, the Stockholm Convention seeks to protect human health and the environment from POPs (Article 1) by: eliminating or reducing the production, use, release and trade of specific POPs listed in the treaty; establishing specific criteria and procedures for placing controls on additional POPs; providing financial and technical assistance to developing countries to support their implementation of the Convention; monitoring POPs in the environment and humans; facilitating information exchange; requiring national reports on implementation plans and activities; and providing for reviews of the treaty's effectiveness and adjusting it accordingly. A Conference of the Parties (COP), made up of all states that have ratified the Convention, is the supreme decision-making body and meets every two years (see Chapter 22). This section summarizes four critical parts of the treaty.

Control measures

Countries faced two broad questions with regard to the control measures: what POPs to include in the Convention, and how to structure the controls. No country objected to placing significant restrictions on the dozen POPs that had received attention during the agenda-setting phase, except for DDT, which some countries insisted was necessary for use against malaria and other vector-borne diseases. Some delegations, including Canada, Norway, Switzerland and the EU, supported considering additional POPs (e.g., those addressed in the regional Aarhus Protocol), but this position never gained sufficient traction.

Questions about the design of the control measures proved more intriguing. Opinions differed on the most effective way to reduce the production, use, and emissions of POPs; what type of controls would allow the flexibility that some countries required to join the Convention; and what design features would allow new POPs to be added to the treaty with the fewest legal and procedural complications.

In the end, negotiators created a multi-tiered set of control measures. POPs regulated by the Stockholm Convention are divided into three annexes, depending on their source and the control measures placed on them. New POPs can be added to each annex by amending the annex, without amending the main text of the treaty. In summary, the key control measures of the Stockholm Convention require that all Parties:

- *Prohibit the production and use of POPs listed in Annex A.* The Convention initially listed all the pesticides and industrial chemicals among the dirty dozen except DDT. To date, since the agreement's entry into force, Parties have added 17 new substances, including both pesticides and industrial chemicals, to Annex A.
- *Restrict the production and use of POPs listed in Annex B.* Annex B currently includes DDT and PFOS. Parties can continue to produce and use DDT only for disease vector control, especially against malarial mosquitoes, and when "locally safe, effective and affordable alternatives are not available" (Annex B). PFOS, added to the Convention in 2009, can still be used production of the insect bait sulfluramid.
- *Minimize, and where feasible eliminate, releases of the POPs listed in Annex C.* While Annex A and B list POPs intentionally produced for use as pesticides or in industry, the POPs listed in Annex C are unintentionally created by-products of waste incineration, chlorine production, pulp bleaching, metallurgy, and other industrial processes. In some cases, eliminating their production is physically impossible; hence their placement in

Annex C, which currently lists dioxins, furans, hexachlorobutadiene, pentachloroben-
zene, polychlorinated naphthalenes and other compounds.

- *Take efforts to prevent the commercial development of new POPs.*
- *Ban the import or export of POPs controlled under the Convention*, except for narrowly defined purposes or environmentally sound disposal.
- *Reduce or eliminate releases of POPs from existing POP stockpiles or wastes containing POPs.*
- *Promote the adoption and use of the best available technologies (BAT) and best environmental practices (BEP)* for reducing emissions of POPs, managing and disposing of POP wastes, and replacing POPs with alternatives. Detailed annexes to the Convention, as well as subsequent decisions by the Conference of Parties, delineate guidelines with respect to different POPs and different activities.

Exemptions

Despite broad consensus on the need to control POPs, many countries argued for provisions that would allow continued use of chemicals which were of significant economic impor-tance. Mindful of the need to ensure broad participation in the Convention without sacri-ficing the long-term effectiveness of the regime, negotiators agreed on a system of general and specific exemptions that allow for the continued use of specified POPs, for specified purposes, for limited time periods. Thus, general exemptions were granted for certain uses of DDT and, later, for PFOS (and thus their placement in Annex B), but these exemptions are subject to review and reporting requirements. Another general exemption allows countries to use and maintain existing equipment containing PCBs until 2025, which allows gradual replacement of the hundreds of thousands of tons of existing equipment that contain PCBs.

Parties are also allowed "country-specific exemptions" that permit five years of continued use of a POP for specific uses listed in a *Register of Specific Exemptions*. Parties must tell the Secretariat, which maintains the *Register*, if they intend to use a particular exemption, and must reapply every five years to the COP to keep using the exemption. Once no Party is using a particular exemption, it is removed from the Register, allowing no future use. Thus, over time, fewer Parties will use these loopholes and certain exemptions will disappear. When the COP agrees to list newly identified POPs, it usually needs to include exemptions as part of the agreement.

Adding new chemicals

From the beginning, negotiators understood that the Convention would include criteria and procedures for considering the addition of new POPs to Annexes A, B, and C. The Gov-erning Council's mandate for the negotiations called for such provisions, and proponents of a strong regime, including Canada, the EU, Norway and Switzerland, considered creation of a robust system critical given that the Convention would initially cover only the dirty dozen and would likely include exemptions. However, creating an agreement required bal-ancing the views of governments that preferred an active approach based on the precaution-ary principle and delegated rule making that would allow substances to be added relatively quickly (e.g., the EU), with the views of governments that preferred an approach based on extensive evidence of existing harm, procedures that emphasize sovereign control over decision-making (e.g., Australia, China and the United States), and full consideration of the socio-economic impacts of banning particular substances (e.g., many developing countries). In the end, delegates reached a compromise.

Any Party can nominate a substance to add to the Convention. Nominated chemicals are then reviewed by the POPs Review Committee (POPRC), a subsidiary scientific advisory body composed of 31 experts affiliated with Parties to the Convention. First, POPRC reviews a nominated substance to determine whether it meets basic screening criteria (i.e., certain levels of toxicity, persistence, bioaccumulation and potential for long-range environmental transport) indicating that the substance is a POP. POPRC members and observers then gather more information, which is compiled in a risk profile. If POPRC decides the evidence indicates that a substance meets the thresholds for listing in the Convention, it drafts a risk management evaluation and forwards this document to the COP, along with its recommendation whether to place controls on the chemical by listing it in one or more annexes to the Convention and whether to include potential exemptions. Each stage in the POPRC review process (screening criteria, risk profile, risk management evaluation) is subjected to review by the COP and typically takes one year, although some controversial chemicals have progressed more slowly.

The COP reviews POPRC's recommendation, considers socio-economic issues associated with potential listing, and makes the final decision regarding controls and exemptions. Thus, the Convention calls for POPRC to create a science-based foundation for action and for the governments that comprise the COP to make the actual policy decision.

Once adopted by the COP, the decision to add a chemical to Annexes A, B and/or C represents an amendment to the relevant annex(es). During negotiations, many governments argued that such amendments should take effect immediately, similar to "adjustments" under the Montreal Protocol (see Chapter 31), so that the Convention could respond quickly to new threats. Others argued that adding chemicals represents a substantive amendment that requires formal ratification before a Party could be required to comply. In the end, neither view was conceded. The final compromise allows Parties to choose, when ratifying the Convention, which rule will apply to them. Thus, some Parties, including countries in the EU, chose to comply with additions to Annexes A, B, and C immediately unless they formally "opt-out" by a specified time. Other countries, like Australia, Canada, China and India, are not bound by an amendment to Annex A, B, or C unless they "opt-in" by formally ratifying the amendment. This dual structure is rare in multilateral environmental agreements (MEAs).

Financial and technical assistance

Following negotiation of the 1990 London Amendment to the Montreal Protocol (see Chapter 31), global environmental agreements are widely expected to include mechanisms to provide financial and technical assistance to help developing countries implement their obligations. Such assistance is seen by developing countries as an operational manifestation of the principle of common but differentiated responsibilities.

Reaching agreement on a financial mechanism proved difficult (Downie and Fenge 2003). Developing countries strongly supported creation of a stand-alone financial mechanism for POPs, similar to the Montreal Protocol's Multilateral Fund. Developed countries opposed the creation of a new dedicated institution, arguing instead that utilizing the Global Environment Facility (GEF) would provide important economies of scale, expertise and opportunities to leverage additional bilateral, multilateral and private sector co-financing while eliminating the operational expenses associated with a new institution. Developing countries argued that such a system would not provide sufficient and guaranteed financial resources and that the GEF would not respond to the needs of the Convention or wishes of the COP in the same way as a dedicated fund. These arguments were repeated in the negotiations of the Minamata Convention on mercury (authors' observations of the mercury negotiations).

In the final compromise, the Convention states that developed country Parties will provide new and additional financial resources to enable developing country Parties, and Parties with economies in transition, to fulfill their obligations under the treaty. The GEF was designated as the principal entity of the financing mechanism, although officially only on an interim basis, with provisions for the COP to review its effectiveness at regular intervals. Since then, the GEF has created an official funding window for POPs, donors earmarked specific amounts for addressing chemicals in their collective funding of the GEF, and the GEF allocated well over US$1.2 billion to support Stockholm Convention implementation (GEF 2020). However, the amount of financial assistance to be provided under the Convention remains unspecified, and disagreement remains about whether developing country obligations should be contingent upon provision of sufficient funds.

General obligations

The Stockholm Convention includes many other provisions relevant to its implementation, review and operation. Most of these follow standard patterns established in previous MEAs. These include the obligation of all Parties to report on their production, use, import and export of POPs, use of exemptions, and other issues. Parties must also develop national implementation plans detailing their strategy for implementing the Convention and to share information and raise public awareness concerning POPs, threats posed by exposure, and alternatives. Additionally, Parties must collectively establish systems to monitor POPs in the environment and to evaluate the effectiveness of the Convention on a regular basis.

Successes and challenges

The Stockholm Convention has had a successful start. To date, 185 countries have ratified the Convention. Thirty POPs are now subject to binding global controls aimed at eliminating or significantly restricting their production, use, and emission. The GEF distributes funds to assist Parties implementing POPs-related projects. POPs stockpiles in many developing countries are being identified and steps taken for their environmentally sound management or disposal. The COP designated regional centers to help provide technical assistance to developing countries and created the PCBs Elimination Network and the DDT Global Alliance to speed transitions to POP-free alternatives. Although, as with the Basel and Rotterdam conventions, the US Senate has not ratified Stockholm, the United States has complementary domestic chemicals controls and generally supports the global POPs regime. At the same time, significant challenges exist, including some long-standing divisions among Parties, which could impede the effectiveness of the Convention and the broader international agenda on chemicals.

The dirty dozen

Even though most of the original 12 POPs were no longer intentionally produced by the time the Stockholm Convention was signed, listing these substances represented a significant step toward protecting human health and the environment because it created a treaty framework to address their remaining uses, the deterioration of POPs-containing products (e.g., PCBs in old refrigerators, transformers and capacitors), and management of POP stockpiles and wastes. For example, the COP established the BAT/BEP committee to provide techni-cal guidance to help governments control dioxin and furan emissions, address POPs wastes, replace current-use POPs with safer substances, and prevent development of new POPs.

Similarly, in 2009, the COP established the Global Alliance to help promote interna-tional action to deploy cost-effective alternatives to DDT for use in combating malarial mosquitoes. The Alliance seeks to identify gaps, promote coordination, enhance information sharing among existing initiatives, catalyze new action and raise public awareness. Also in 2009, the COP established the PCB Elimination Network (PEN) to facilitate information exchange on the environmentally sound management of PCBs, help stakeholders cooper-ate and promote development of improved techniques for managing PCBs, particularly in developing countries. The Global Alliance and PEN demonstrate how global agreements can go beyond control measures and be used to develop initiatives that coordinate and sup-port multi-sector action among governments, corporations, nongovernmental organizations (NGOs) and other stakeholders to achieve common goals.

Adding new chemicals

As of January 2022, Parties have added 18 new chemicals to the annexes of the Conven-tion since the treaty entered into force. More chemicals are under review by POPRC and could be added in the future. Equally important, decisions to list some of the new chemicals marked a critical shift in the regime's focus from addressing largely "dead chemicals," or substances no longer widely produced or used, to tackling "live chemicals," which continue to be of socio-economic importance in many parts of the world. Addressing live POPs is obviously more difficult, as evidenced by the successful but contentious discussions on endo-sulfan, an agricultural chemical in widespread use during its review (Templeton 2011). The listing of several socio-economically important POPs and the ongoing consideration of new substances by POPRC represent major successes.

To address the need for information about possible risks associated with POP alternatives, the COP has asked POPRC to assess chemical and non-chemical alternatives to some listed substances, including DDT and endosulfan. This request expands POPRC's mandate and demonstrates the potential for the Convention to develop a mechanism to evaluate alterna-tives, much like TEAP does under the Montreal Protocol.

At the same time, the Stockholm Convention is facing significant challenges. The shift to live chemicals has introduced more economic considerations into COP debates, making listing chemicals more difficult. Socio-economic concerns have also begun to intrude earlier and more prominently into POPRC deliberations, threatening the intended independence and scientific focus of the body. It has also become necessary to ensure that the alternatives introduced for the banned chemicals do not create other, equally significant environmental problems.

Similarly, expanding POPRC's workload to consider more alternatives might detract from POPRC's core mission of examining chemicals proposed for addition to the Conven-tion. It could also undermine POPRC if members draw conclusions on issues outside their areas of expertise or are seen as unfairly favoring one commercial alternative over another. Indeed, some observers have questioned the technical ability of POPRC to assess environ-mental hazards that differ substantially from those the committee must consider in evaluating POPs (Allan et al. 2011).

Adding new chemicals also creates potential implementation and even legal challenges. For example, some "opt-in" Parties have not ratified all of the amendments that added chemicals to the annexes. Widespread refusal to ratify such amendments could lead to uneven implementation of the Convention, as some Parties would be bound by controls on a new substance while others were not. Such disparities could serve as a disincentive for the COP to list more live substances, as doing so could put some countries at a competitive disadvantage. Uneven patterns of ratification could also create legal uncertainties in cases in which some countries have banned the use of a particular POP while other countries wish to include products made with or containing that POP in international trade.

On the other hand, the decision to list the pesticide endosulfan suggests that the regime may be able to negotiate agreement on chemicals that are of significant economic value to some Parties. Endosulfan was widely used in some countries, particularly India, in both domestic and export-related applications. The Indian government negotiated for many years to prevent its addition to the control measures (Ashton et al. 2011). However, the final COP decision to list endosulfan, albeit with exemptions won by India, contributed to domestic pressure within India (from both environmentalists and manufacturers of alternatives) to eliminate the production and use of this pesticide, as well as to a decision by the Indian Supreme Court to ban its production, use, and sale. Although several chemicals currently under review by POPRC present similar or even more challenging tests of the COP's ability to reach agreement to control live substances and of opt-in Parties' willingness to ratify such listings, the endosulfan case suggests that these challenges are surmountable and that the Stockholm Convention can play an important agenda-setting role and influence domestic decision-making even in the face of significant economic interests.

Managing exemptions

Allowing time-limited exemptions creates the flexibility necessary to reach agreement on listing some chemicals. Without them, some Parties might block a listing or refuse to ratify the amendment to the annex. However, the success of the exemption system rests on Parties' willingness to transition to alternatives. This requires domestic motivation, possibly financial and technical assistance, and, arguably, the expectation that the COP will not renew exemptions indefinitely. As more live chemicals are listed, it is possible that countries will find it difficult to stand in opposition to renewal of another's exemption if they themselves are hoping to find support for continuing an exemption. This problem is magnified for broad exemptions, like those granted to DDT, or for live chemicals with significant commercial applications, such as PFOS and PFOA (BRS Secretariat 2020). The long-term effectiveness of the Convention depends on closing individual exemptions within a reasonable time frame.

Technical and financial assistance

The effectiveness of the Convention requires provision of sufficient financial support for developing countries, or at least certain developing countries, to build capacity, regulate chemicals, transition to specific alternatives, manage stockpiles of obsolete substances, raise awareness about risks, and implement other obligations. Important successes in providing financial assistance have been achieved. The GEF took the critical step of establishing a formal focal area for POPs funding and developed a memorandum of understanding with the COP to guide its activities. The most recent replenishment of GEF resources by donor countries included another increase in funding for the POPs focal area. By January 2021, the GEF had allocated more than US$1.2 billion to projects in more than 135 countries that support Stockholm Convention implementation (GEF 2020).

However, some developing countries have expressed frustration with the amount of funding available, given the complexities of managing toxic chemicals and the challenges many have experienced with securing funding for certain projects, meeting co-financing requirements, or navigating the process required to access GEF funds (authors' observations during COPs; see also the official negotiation reports on the Secretariat's website: www: http://chm.pops.int).

The addition of new POPs to the Convention has exacerbated concerns about the availability of project funding. Without adequate technical and financial assistance to help implement regulatory action and to find and deploy affordable and effective alternatives, the listing of a large number of new chemicals in the Convention will become a largely aspirational exercise for some Parties. Moreover, many developing countries, including some with sufficient resources, have stated in regime negotiations that it will be difficult for them to agree to list additional chemicals without confidence that funds to support implementation will be available.

Compliance mechanism

Article 17 requires the COP to develop procedures to identify and determine appropriate responses to Parties found to be in non-compliance. However, the text provides no firm deadline, and the COP has made little progress beyond a general consensus that the mechanism should focus on identifying and addressing obstacles to compliance rather than on judging and punishing individual countries for non-compliance. Some developing countries argue that financial and technical assistance must be increased before a compliance mechanism is created, especially as countries struggle to keep pace with the addition of new chemicals to the Convention. They also argue that provision of specific amounts of financial and technical assistance should be a central compliance requirement for developed countries. Meanwhile, some developed countries insist that because a compliance mechanism would not be punitive, there is no reason to link the issues.

The lack of a compliance mechanism potentially undermines the COP's ability to monitor the global effectiveness of the agreement. A compliance mechanism that reviews whether and how Parties meet their commitments under the Convention could help identify domestic or regional problems and facilitate action to help Parties respond to implementation challenges. However, while many countries see this as a positive mechanism to increase the effectiveness of the Convention, as has been the case under the Montreal Protocol, others perceive the potential mechanism as a threat to national sovereignty.

Synergies

Two other major initiatives have significantly affected the development of the Stockholm Convention: the "synergies" initiative and the Strategic Approach to International Chemicals Management (SAICM). Both sought to coordinate and streamline different aspects of international chemicals and hazardous waste management to enhance national implementation of global agreements and to achieve, as an ultimate goal, the environmentally sound management of chemicals during all stages of their life cycles in all regions of the world.

The synergies process was a UNEP-led effort to enhance cooperation and coordination among the Stockholm, Basel, and Rotterdam (BRS) conventions in areas where the conventions overlap or complement each other (Allan et al. 2018). Since 2010, the BRS Conventions have moved to reduce administrative costs and improve the overlapping implementation and effectiveness of the BRS conventions by integrating their Secretariats, initiating or expanding joint implementation activities – including regional centers that provide technical assistance for all three conventions, and coordinating overlapping actions such as reporting requirements and other activities. The COPs of each Convention now hold joint and back-to-back biannual meetings in order to facilitate greater cooperation on issues of mutual concern; however, each Convention maintains individual legal autonomy over its activities. Supporters of the synergies process believe it has and will continue to improve the efficiency and effectiveness of all three conventions, and allow Parties, the Secretariat, the financial mechanism, and international organizations to enhance information exchange, direct more resources to implementation activities, and otherwise achieve advantageous synergies that would not have been possible in the absence of such significant coordination.

SAICM is a broader policy initiative. Initiated by governments at the International Conference on Chemicals Management in 2006, SAICM seeks to achieve a multi-stakeholder, multi-sector policy framework for the global, sound management of chemicals throughout their entire life cycle so that chemicals are produced and used in ways that minimize significant adverse effects to human health and the environment. SAICM's more specific objectives are grouped into five themes: risk reduction, knowledge and information, governance, capacity building and technical cooperation, and illegal international traffic. SAICM is structured to engage a range of actors involved in different aspects of chemicals production, use and disposal, including civil society, industry, national and local governments, and intergovernmental agencies. Thus, the initiative has been particularly welcomed by actors who support multi-stakeholder engagement and recognize the importance of focusing on managing the entire life cycle of chemicals.

Achieving the goals of these initiatives would represent huge successes, but their pursuit also carries some risks. Joint COP meetings allow for productive consideration of cross-cutting issues but also opportunities for counterproductive cross-treaty bargaining and `issue hostage-holding' (Allan et al. 2018). The language and operation of the Stockholm Convention gives more prominence to the precautionary principle and joint COPs and operations might augment its inclusion the Basel and Rotterdam deliberations, and at times this appears to have been the case (personal observations), but the explicit consideration of socio-economic issues that is an acknowledged part of the process of listing chemicals in the Stockholm Convention has now become part of debates under the Rotterdam Convention – which runs counter to the intent, design and past operation of this information and entirely non-regulatory regime. The International Pollutants Elimination Network (IPEN), an umbrella organization representing public health and environmental advocacy groups, and others have also raised concerns that integration under the synergies initiative could reduce resources available for capacity building related to unique obligations under each agreement (IPEN 2010; see Chapter 14). Thus, the challenge of the synergies and SAICM initiatives is to ensure that integration and coordination do not inadvertently undermine key strengths of individual agreements or take time and money away from important implementation activities.

The role of non-state actors

The Stockholm Convention is an inclusive forum in which nongovernmental delegates (including advocacy groups and industry associations, among others) make contributions to a degree not found in many other MEAs, and its development provides examples of the valuable role that different types of non-state actors can play in global environmental policymaking (see Chapters 13 and 14). Advocacy groups played a key agenda-setting role in framing the proposed regime as an essential tool for protecting human health, especially the health of indigenous peoples living near the Arctic Circle (Downie and Fenge 2003). Non-state actors with various forms of expertise play an active role in the work of the Convention, participating in the work of the COP and POPRC both during meetings and as part of official intersessional groups that work on a range of issues (although only Parties take part in formal decision-making). On broad policy issues, IPEN and other civil society groups advocate for aggressive precautionary action to protect human health and the environment from POPs, providing data and first-hand experiences of the impact of chemicals on vulnerable populations. Industry associations such as CropLife International, which represents manufacturers of pesticides, play what they consider to be a "watchdog" role, ensuring that implementation of the Convention is legally rigorous (Templeton 2011). As producers and users of some of the chemicals and pesticides reviewed under the Convention, industry observers also contribute data and technical expertise regarding both chemicals under review and the efficacy of possible substitutes and alternatives.

Conclusion

The Stockholm Convention offers an effective mechanism for reducing the threats to human health and the environment posed by POPs. The control measures, process for adding new chemicals, provisions for providing financial and technical assistance, reporting requirements, and mandates for effectiveness evaluations demonstrate that governments can create comprehensive and sustainable agreements to address complex global environmental issues. Furthermore, activities during the regime's first two decades, including work to reduce emissions of the original dirty dozen and the many new POPs, funding activity by the GEF, creation of national implementation plans, designation of regional centers, and reviewing the financial mechanism, show that governments are capable of implementing these global agreements in specific national contexts.

More broadly, the positive impact of UNEP's activities prior to the negotiations, especially the series of regional workshops, illustrates how effective action by international organizations during the agenda-setting phase can pave the way for successful negotiations by catalyzing and sustaining action, raising awareness, augmenting capacity, and improving the contractual environment (e.g., Haas et al. 1993; Downie 1995). The role played by the combination of advancing scientific knowledge and an epistemic network of scientists and aligned policymakers (Downie and Fenge 2003) confirms arguments regarding how, in some situations, these factors can enhance (but certainly not guarantee) prospects for successful agenda-setting and reaching an initial agreement (Haas 1990). The initial success of POPRC, the use of exemptions to overcome lowest common denominator problems and ensure participation, and the development of national implementation plans, confirm that "regime design matters" (Mitchell 1994), in that each of these treaty components was at some point configured differently (and likely less effectively) during negotiation of the Convention. The relatively speedy negotiations that produced the Convention, which included heavy reliance on productive textual and organizational precedents set in the ozone, CLRTAP and Rotterdam Convention negotiations, demonstrate how cumulative knowledge and experience in creating effective MEAs can positively influence outcomes.

The success of the Stockholm Convention also speaks to the importance of multi-layered governance systems. The governance of toxic chemicals involves global, regional, and domestic initiatives, formal and informal governance systems, and a range of actors that address different aspects of chemicals production, use and disposal; these include IFCS, UNEP

Chemicals, SAICM, voluntary initiatives and guidelines developed under UNEP and FAO, the Basel and Rotterdam conventions, environmental NGOs, industry groups and guidelines, domestic laws, among many others. The development, composition, impact and effectiveness of multi-layered governance systems deserve more attention in the study of global environmental politics.

At the same time, the long-term success of the Convention faces major challenges (for a similar discussion regarding another successful regime, see Downie 2015). Will more countries fully implement the BAT and BEP guidelines to control POPs emissions, especially of dioxins and furans? Will Parties be able to add more live POPs? Will the effectiveness of new listings be compromised by extensive exemptions? Will key Parties choose not to "opt-in" to the controls on new chemicals, creating de facto POPs havens? Will POPRC avoid further politicization? Will the GEF perform effectively as the regime's financial mechanism? Will sufficient funds be available to meet the implementation needs or political demands of developing countries? Will funding uncertainties arise that keep key countries from agreeing to list additional chemicals? Will sufficient monitoring take place so Parties can determine the true effectiveness of the control measures? (For discussion, see Hung et al. 2016.) Will the ongoing shifts within the developing country coalition between the interests of the fastest growing economies of Brazil, China, India and South Africa and those of the least developed countries hinder operation of an effective, politically acceptable and economically viable financial mechanism? Will financial difficulties due to the Covid-19 pandemic limit regime implementation and expansion? Will the United States ratify the Convention and, if not, how will this impact its effectiveness over time? Will policy disputes in other issue areas, such as climate, help or hurt regime developments related to chemicals?

The Stockholm Convention is the centerpiece of the only global environmental regime designed to eliminate substances directly toxic to human health and the environment (see Chapter 9). The challenges facing the Convention magnify this importance because of their relation to, and potential impact on, other aspects of global environmental politics. How events unfold over the next decade could have a significant impact on how the world regulates toxic chemicals, and perhaps on the study and practice of global environmental politics.

References

Abelkop, D.K., Graham, J., and Royer, T. (2017) *Persistent, Bioaccumulative, and Toxic (PBT) Chemicals: Technical Aspects, Politics and Practices*, Boca Raton, IL: Taylor & Francis.
Allen, J., Downie, D., and Templeton, J. (2018) "Experimenting with TripleCOPs: Productive Innovation or Counter-productive Complexity?" *International Environmental Agreements: Politics, Law and Economics*, 18(4): 557–572.
Allan, J., Kohler, P., and Templeton, J. (2011) "Summary of the Seventh Meeting of the Persistent Organic Pollutants Review Committee of the Stockholm Convention: 10–14 October 2011," *Earth Negotiations Bulletin*, 15: 189.
AMAP (Arctic Monitoring and Assessment Programme) (1998) *AMAP Assessment Report: Arctic Pollution Issues*, Oslo: AMAP.
Ashton, M. et al. (2011) "Summary of the Fifth Meeting of the Conference of Parties to the Stockholm Convention on Persistent Organic Pollutants: 25–29 April 2011," *Earth Negotiations Bulletin*, 15: 182.
BRS Secretariat (2020) "Exempted use for PFOS, SCCP, DecaBDE and PFOA." Available <http://chm.pops.int/Implementation/Publications/BrochuresandLeaflets/tabid/3013/Default.aspx
Chasek, P. and Downie, D. (2021) *Global Environmental Politics*, 8th edn, New York: Routledge.
Downie, D. (1995) "UNEP and the Montreal Protocol: New Roles for International Organizations in Regime Creation and Change," in R. Bartlett, P. Kurian, and M. Malik (eds) *International Organizations and Environmental Policy*, Westport, CT: Greenwood Press.

Downie, D. (2015) "Still No Time for Complacency: Evaluating the Ongoing Success and Continued Challenge of Global Ozone Policy," *Journal of Environmental Studies and Sciences*, 5(2): 187–194.

Downie, D. and Fenge, T. (eds) (2003) *Northern Lights against POPs: Combating Toxic Threats in the Arctic*, Montreal: McGill-Queens University Press.

Downie, D., Krueger, J., and Selin, H. (2005) "Global Policy for Toxic Chemicals," in R. Axelrod, D. Downie, and N. Vig (eds) *The Global Environment: Institutions, Law and Policy*, 2nd edn, Washington, DC: CQ Press.

Eckley, N. and Selin, H. (2003) "Science, Politics, and Persistent Organic Pollutants: Scientific Assessments and their Role in International Environmental Negotiations," *International Environmental Agreements: Politics, Law and Economics*, 3(1): 17–42.

GEF (2020) "Chemicals and Wastes," https://www.thegef.org/topics/chemicals-and-waste.

Haas, P. (1990) "Obtaining International Environmental Protection through Epistemic Consensus," *Millennium*, 19: 347–364.

Haas, P., Keohane, R., and Levy, M. (1993) *Institutions for the Earth: Sources of Effective International Environmental Protection*, Cambridge, MA: MIT Press.

Hillman, K. (1999) "International Control of Persistent Organic Pollutants: The UN Economic Commission for Europe Convention on Long-Range Transboundary Air Pollution, and Beyond," *RECIEL*, 8(2): 105–112.

Hung, H. Katsoyiannis, A., and Guardans, R. (eds.) (2016) "Persistent Organic Pollutants (POPs): Trends, Sources and Transport Modelling," special issue of *Environmental Pollution*, 217: 1–158.

IPEN (2010) "An NGO View on Synergies and the ExCOPs." Prepared by Dr. Mariann Lloyd-Smith. Available HTTP: <http://ntn.org.au/wp/wp-content/uploads/2010/04/synergy_excops.pdf>

Kohler, P. and Ashton, M. (2010) "Paying for POPs: Negotiating the Implementation of the Stockholm Convention in Developing Countries," *International Negotiation*, 15: 459–484.

Lönngren, R. (1992) *International Approaches to Chemicals Control: A Historical Overview*, Stockholm: National Chemicals Inspectorate/KemI.

Mitchell, R. (1994) "Regime Design Matters: International Oil Pollution and Treaty Compliance," *International Organization*, 48(3): 425–458.

Selin, H. (2010) *Global Governance of Hazardous Chemicals: Challenges of Multilevel Management*, Cambridge, MA: MIT Press.

Templeton, J. (2011) "Framing Elite Policy Discourse: Scientists and the Stockholm Convention on Persistent Organic Pollutants." Unpublished PhD Dissertation, London School of Economics and Political Science. Available HTTP: <http://etheses.lse.ac.uk/361/1/Templeton_Framing%20elite%20policy%20discourse.pdf>.

Templeton, J. (2020) "Persistent Organic Pollutants Convention," in J. Frédéric and A. Orsini, eds., *Essential Concepts of Global Environmental Governance*, 2nd edn, New York: Routledge.

36

Hazardous waste

Fragmented governance and aspirations for environmental justice

Katja Biedenkopf

Hazardous waste is not only a local but also an international environmental problem. It has a strong international dimension since it can be traded among countries. Especially, hazardous waste can cause severe environmental and health damage when it is not processed with appropriate methods and care. Hazardous waste trade can have a pronounced environmental justice dimension when the waste is shipped from high-income to low-income countries where it can cause serious damage if it is not appropriately handled. In the 1980s, countries recognized that this environmental injustice should be addressed through an international treaty: The Basel Convention on the Control of Transboundary Movements of Hazardous Wastes and Their Disposal was adopted in 1989 and promotes environmentally sound management of hazardous waste and restricts its transboundary movements. The Convention, and especially its export ban amendment and liability protocol, have however experienced challenges in their ratification. Moreover, the Convention lacks global coverage since the United States is not a Party to it. While it has improved international governance of hazardous waste, the Basel Convention faces some significant effectiveness challenges.

Waste is defined as "substances and objects which are disposed of or are intended to be disposed of or are required to be disposed of by provisions of national law" (Basel Convention, Article 12). With a growing global population and in conjunction with rising consumption, the production of waste has soared over past decades. In 2016, global household waste amounted to 2 billion tonnes and is expected to grow by 70 percent by 2050. High-income countries generate about 34 percent of all global waste although they account for only 16 percent of the global population. It is estimated that about 5 percent of global carbon dioxide emissions came from the treatment and disposal of waste in 2016 (Kaza et al. 2018). It is difficult to estimate the amount of hazardous waste more specifically, since reliable figures exist only for some countries but not for many others that do not collect and report such data. Nonetheless, the number of incidents that occur due to inappropriate disposal and handling of hazardous waste, especially in low-income countries, are a testimony to the severity of the problem.

Waste is hazardous when it "exhibits characteristics or qualities that make it dangerous to humans or the environment" (Saunier and Meganck 2009: 166). Hazardous waste results from a number of industrial manufacturing processes in sectors such as chemicals, iron,

DOI: 10.4324/9781003008873-41

steel and fertilizers. Household waste can also be hazardous. For example batteries, paints, detergents and electronics need to be disposed and handled with special methods. When hazardous waste is poorly managed, it can cause tremendous harm to humans and the environment. For example, when waste electric and electronic equipment (e-waste) is landfilled, hazardous substances such as lead and mercury can leech into the environment and entre drinking water. This problem has received some media attention in the past with documentaries and articles, for example about the Agbogbloshie area in Ghana's capital Accra (e.g., McElvaney 2014; Yeung 2019). An estimated 10,000 workers recover precious metals from old electronics, using primitive methods that damage their health. Many electronics stem from high-income countries and are traded legally but often illegally to such dump sites in low-income countries. This illustrates that ensuring the safe management of hazardous waste is crucial for our health and that of our planet.

Yet, it is not always easy to define whether something is waste or a resource. While in the 1990s and early 2000s the waste policy debate focused on collection and recycling, it now has turned to the concept of circular economy. In a circular economy, products are designed for reuse and recycling while waste is avoided. The conventional economic model is linear starting with resource extraction, moving to production, then consumption and finally disposal. A circular economic model however closes the circle and reuses or recycles used products so that they become materials for new production. In the ideal type of this model, there is no waste but rather everything is fed back into the system. In this conception, waste becomes a resource. Fully implementing this model is not easy. For example, e-waste contains precious metals such as copper and gold that can easily be reused. Some other parts, however, are less valuable and incur costs when being recycled. Not all materials are recyclable or they lose quality after recycling so that they cannot be used for the same purpose multiple times. These considerations illustrate the complexity of defining what waste is and how to handle it. Those challenges are at the core of the debates within the context of the Basel Convention.

This chapter provides an overview of international hazardous waste governance and politics with a focus on the 1989 Basel Convention. It first sets the scene by explaining why hazardous waste is an international policy problem, followed by a description of the Basel Convention's origins. The next four sections discuss key features of the global hazardous waste regime. The first feature pertains to how the Basel Convention addresses environmental (in)justice. The second feature is the Convention's slow progress and implementation, which is marked by a slow pace of ratification or even non-ratification. The third and fourth features relate to fragmentation. International hazardous waste governance is fragmented horizontally at the international level (third feature) and vertically across different levels of governance (fourth feature). The concluding section highlights some of the remaining core challenges of international hazardous waste politics.

The problem of hazardous waste

Hazardous waste can pose health and environmental risks when it is not handled with appropriate methods and technologies. It has a strong international angle since hazardous waste is shipped across borders, which transforms it from a local to a global problem. The nuisance that is at the core of the Basel Convention is the incentives for high-income countries to ship hazardous waste to low-income countries because disposal costs a fragment of what it would cost at home. This is the case because low-income countries tend to have laxer environmental policy and limited capacity to process hazardous waste. As a consequence, international trade of hazardous waste to low-income countries has led to multiple accidents in which people's health and the environment were compromised.

Traditionally, waste was a domestic policy problem. It concerned questions about building appropriate infrastructures for collecting and disposing of waste. With technological progress and growing consumption, waste production increased and handling the waste became more complex. Recycling became an ever more prominent alternative to landfilling or otherwise disposing of waste. In a number of countries, a waste hierarchy was developed to guide policy: reduce, reuse, recycle, recover, dispose. Reducing waste is the best option and tops the hierarchy. Environmentally sound disposal is the least optimal solution at the bottom of the hierarchy. Designing products so that they can be used longer, are less hazardous, and are easy to recycle has become a policy objective advocated by many environmentalists and policymakers. Yet, this is a difficult objective since it requires changing consumer behavior and reversing the economic trend of producing ever more products with ever shorter life spans. National policymakers have grabbled with those challenges for years and decades. The problem, however, reaches beyond the national realm. A number of hazardous waste problems are global. This links to international supply chains (see Chapter 24) and, more directly, to the international trade in hazardous waste.

Hazardous waste governance at the international level was triggered by a number of incidents linked to health and environmental damage that occurred due to trade practices. For example, in 1986–1988 the garbage ship *Khian Sea* that was loaded with ash from a Philadelphia-based waste incinerator remained at sea for two years since no country allowed it to unload its cargo. Originally, the shipping company planned to dump the ash in the Bahamas, which however refused the ship. After this, a number of other countries also denied the ship entry into their harbors and returning to Philadelphia failed. Eventually and after two years at sea, most of the ash was illegally dumped in the ocean. Another example from the same time period is polychlorinated biphenyl (PCB) containing waste that was shipped from Italy to Nigeria in 1987. Disguised as building material, the hazardous waste was stored in a warehouse from where it leaked and affected the local population's health. After a diplomatic dispute, the Italian government took responsibility for the waste in 1988 (Krueger 1999: 10–11; Puthucherril 2012: 296).

E-waste illustrates the severity of international hazardous waste-related problems and its environmental justice dimension. Old electronics contain hazardous substances that require handling with the appropriate methods and technologies. In the early 2000s, NGOs such as Greenpeace and the Basel Action Network drew attention to the disastrous circumstances in countries such as China (Basel Action Network 2002) and Ghana (Greenpeace 2008) where unregistered businesses – which generally are known as the informal sector – extract valuable metals from the e-waste (see Chapter 14). For this process, unprotected workers use primitive methods such as open-air incineration and acid baths, which emit dioxins, furans and other pollutants that are harmful to human health and the environment. The incentive to ship e-waste to such locations is high since the costs are only a fraction of the recycling costs in high-income countries. As this example demonstrates, the problem of hazardous waste has a clear and pronounced North–South dimension (see Chapter 23). Environmental (in)justice among countries (see Chapter 25) was one of the driving factors that led to the adoption of the Basel Convention. The problem is a combination of international trade with low-income countries' lack of capacity to process hazardous waste in an environmentally sound manner.

Most of the legal hazardous waste trade occurs among high-income countries, typically within Europe and North America (Baggs 2009). Nonetheless, many low-income countries are net importers of hazardous waste. In an analysis of official country reports submitted by the Parties to the Basel Convention, its Secretariat concluded that 86 percent of hazardous waste trade occurs between countries that are a member of the European Union, the OECD and Liechtenstein. Only 5 percent of hazardous waste trade occurs between those countries and the other Parties to the Convention (Secretariat of the Basel Convention 2010: 15). Yet, the officially reported figures do not reflect the totality of all activities that occur. Actual amounts of hazardous waste trade are difficult to determine since illegal trade is significant (Bisschop 2012). This observation can explain why the EU Network for Implementation and Enforcement of Environmental Law (IMPEL) comes to different conclusions than the Basel Convention's Secretariat in their assessment of hazardous waste shipments. According to one of their reports, more than 50 percent of the shipments that were checked at ports did not comply with EU and Basel Convention regulations (IMPEL 2006: 20). Misdeclaration of products as reuse or recycling material although they actually are waste is a frequent illegal activity (OECD 2012). The Basel Convention bans most trade in hazardous waste while exports for reuse are permitted since the products are not classified as waste. Due to misdeclarations and other illegal activities, exports from high-income to low-income countries are problematic and likely much higher than the official statistics.

The Basel Convention's origins

The first official hazardous waste shipment took place in the 1970s, but undocumented trade is likely to have occurred much earlier (O'Neill 2000). With the adoption of domestic waste legislation in a number of high-income countries, which increased the costs of waste disposal, the incentive for shipping it to countries with lower health and environmental standards became greater. As a consequence, trade in hazardous waste increased and was a largely unregulated activity (Clapp 2001). In the 1980s, a number of high-profile incidents, such as the *Khian Sea* odyssey described above, attracted public attention to the severe implications such practices can have if safe handling is not ensured. This led to the negotiations of what became the 1989 Basel Convention on the Control of Transboundary Movements of Hazardous Wastes and Their Disposal.

Public attention and pressure as a result of the high-profile incidents was one of the triggers of a number of developments that eventually led to the negotiation and adoption of the Basel Convention. The incidents motivated not only low-income countries to call for an international treaty, but also high-income countries to recognize their responsibility and support the negotiations. In the 1980s, the Organization for Economic Cooperation and Development (OECD) developed voluntary guidelines for managing transnational movements of hazardous waste among the OECD countries. This already included a procedure called Prior Informed Consent (PIC), which later became a core element of the Basel Convention. Its details are discussed in the next section. The voluntary guidelines were followed by the drafting of legally binding rules on the transboundary movement of hazardous waste. The process was however suspended since meanwhile the negotiation process of the Basel Convention had started focusing on similar rules (Selin 2010).

At around the same time, the United Nations Environmental Programme (UNEP) had started addressing the problem of hazardous waste trade between high-income and low-income countries (Krueger 1999). In 1982, the UNEP Governing Council established a working group on new technical guidelines and policy recommendations on the improved management of hazardous waste for the purpose of better human health and environmental protection. This led to the first global standard on the transnational transport of hazardous waste in 1987, the so-called Cairo Guidelines and Principles for the Environmentally Sound

Management of Hazardous Waste. The Cairo Guidelines included a PIC procedure, similar to the OECD voluntary guidelines for the transnational transport of hazardous waste (Selin 2010). PIC and environmentally sound management – two core provisions of the Basel Convention – were already developed previously in other fora and fed into the negotiations by some of the Parties. This illustrates how policy, in a number of instances, is developed in different fora based on incentive structures specific to those fora and through this builds up momentum and support for an international agreement and shapes its design.

A growing number of actors – states and NGOs – found the Cairo Guidelines insufficient and pushed for a legally binding treaty so as to provide better protection for low-income countries. The Organisation of African Unity (the African Union's predecessor) and Greenpeace strongly advocated a complete trade ban, which gained support by the Scandinavian countries. A number of other high-income countries were, however, not prepared to go that far. They only accepted a PIC procedure similar to the OECD rules and the Cairo Guidelines. This coalition was led by the United States and included countries such as Germany, the Netherland and Switzerland. They argued that trade could ensure environmentally sound management at lower costs than at home. Eventually, this coalition was stronger than the ban advocates and the original Base Convention did not include a trade ban (Selin 2014). Yet, this changed quickly. As described below, an amendment to introduce a trade ban between high-income and low-income countries was adopted shortly after the Convention entered into force. This demonstrates how, in some instances, building political support from a number of sources can be slow when faced with strong opposition.

The Basel Convention was signed in 1989 and entered into force in 1992. Its objectives are to minimize hazardous waste generation in terms of quantity and hazardousness, to manage hazardous waste in an environmentally sound manner and to minimize transboundary movements of hazardous waste (Basel Convention, Article 4). Whether and to what extent trade in hazardous waste should be banned was controversial not only during the negotiations but it has remained a bone of contention until today.

Environmental justice and the Basel Convention

Redressing environmental injustice was one of the main motivations for adopting the Basel Convention. Specifically, the Convention aims to redress the imbalance that resulted from hazardous waste exports from high-income to low-income countries and alleviate the disproportionate burden on low-income countries (Puthucherril 2012: 305–306). This is done in a number of ways, most prominently through the PIC procedure, the ban of exports from OECD to non-OECD countries, the not-yet-ratified liability protocol, and capacity building.

The Basel Convention establishes a procedure that allows waste exports only when the importing country has received detailed information about the characteristics of the waste, the proposed means of transport, the disposal methods and the contract between the parties to ensure environmentally sound management. The importing country must give its explicit consent prior to the transaction, a provision known as Prior Informed Consent (PIC). It is also used in other international environmental agreements such as the Rotterdam Convention and the Nagoya Protocol on Access and Benefit-sharing (see Chapter 41). The duty to ensure that hazardous waste treatment complies with rules on environmentally sound management remains with the exporting country and cannot be transferred (Basel Convention, Article 4.10). Exporting countries must ensure that hazardous waste is taken back in case the export transaction is not completed (Basel Convention, Article 8) and in case of illegal exports (Basel Convention, Article 9). Hazardous waste exports to countries that are not a Party to the Convention must be subject to an agreement that is at least as strict as the Convention's requirements.

While it aims to enable countries to make informed choices, the PIC procedure has been criticized because it places the responsibility to verify the information about hazardous waste shipments on the importing country. In the case of low-income countries, the lack of sufficient administrative capacity to assess the information can hamper the process. In some cases, corruption has proven to be a problem (Andrews 2009: 173–174). Those two factors render the PIC procedure inefficient and do not achieve the anticipated goal. This illustrates one of the factors that can explain why the Convention faces effectiveness challenges. It also highlights the importance of capacity building to enable countries design and implement effective procedures and policies. At the same time, it is used as one of the arguments in favor of more stringent provisions, especially a complete ban of trade between high-income and low-income countries.

As a result of the controversial debate about the stringency of the Basel Convention's provisions and, more specifically, between proponents and opponents of a trade ban, Parties adopted an amendment to the Convention that bans all hazardous waste exports from OECD countries, the European Union and Liechtenstein to all other countries in 1994. The amendment became known as the Basel Ban, which eventually was ratified in 2015. Since then, the Basel Convention only allows hazardous waste exports between OECD and non-OECD countries when the waste serves as raw material for recycling and recovery (Basel Convention, Article 4.9). This aims at protecting low-income countries from imports and forces high-income countries to take responsibility for their hazardous waste. Yet, some actors do not consider the amendment as stringent enough. The exemption of exports for reuse and recycling has been criticized by NGOs and other actors as a loophole that is prone to abuse due to factors such as the ones mentioned above: lack of capacity and corruption.

Redressing environmental injustice by strengthening the provisions on exporters' liability for the possible damages is another contentious aspect of the Basel Convention negotiations. In order to provide more protection and support to low-income countries, in 1999 Parties adopted the Protocol on Liability and Compensation, which aims to create a fund to compensate for damage from transboundary movements of hazardous waste. This responds to low-income countries' concerns that they are not able to cope with illegal dumping and accidents, and that capacity building alone is unlikely to remedy the problem in the near future. In 2021, however, the Protocol still had not been ratified, which demonstrates the contentiousness of the issue. Although the protocol would add more protection, it is criticized because it holds exporting countries liable only up until the sale of the hazardous waste. Liability ends once the possession of the hazardous waste has been transferred to a local waste handler. This is judged as problematic since at the stage of local handling most incidents occur through leeching of hazardous substances and basic waste handing methods (Andrews 2009: 176–177).

Capacity building and training in low-income countries is yet another part of the Basel Convention that addresses environmental justice concerns. To this end, a network of Regional Centres for Training and Technology Transfer has been created. This is a rare kind of center established by an international environmental convention. Only the Stockholm Convention has similar regional centers, some of which are combined with the Basel Convention Regional Centres. Their purpose is to encourage regional cooperation, attract resources for implementing the Convention and support implementation projects. In pursuing their objectives, the Regional Centres raise awareness, strengthen administrative capacity and provide scientific information and technical support. These activities aim to help countries develop systems and procedures for the safe handling and processing of hazardous waste. Yet, they lack sufficient resources and political support to effectively achieve their objectives. One of the reasons for the low financial endowment is that the Convention fails to prescribe mandatory contributions from high-income countries. As a consequence, the Regional Centres tend to be underfinanced for their task to provide capacity building and technology transfer to low-income countries (Andrews 2009: 177; Selin 2012).

In sum, the Basel Convention contains a number of provisions that aim to ensure environmental justice and alleviate the burden on low-income countries, which tend to suffer most from the consequences of hazardous waste imports. The PIC procedure, the ban of exports from OECD to non-OECD countries, the (not yet ratified) liability protocol and capacity building aim to strengthen the position of low-income countries but as shown in this section, they are not without criticism. Especially the Basel Ban and the Liability Protocol are highly contentious issue, which can explain why they have made very slow progress in ratification and implementation, which is discussed in more detail in the following section.

Slow progress in ratification and implementation

The Basel Convention has been marked by slow ratification and implementation. At the 1989 diplomatic conference at which the Basel Convention was adopted, only 33 Parties signed the treaty. All African countries refused to sign because of the controversy over whether or not to ban hazardous waste exports. In 1992, the Convention entered into force with only 20 ratifications. Over time, ratifications increased and nowadays the Basel Convention has almost universal geographic coverage with 188 Parties in 2021, which however does not include the United States. At the third Conference of the Parties (COP) in 1994, the advocates of a trade ban succeeded in pushing for the adoption of the Basel Ban amendment. They succeeded despite strong opposition. Given its controversy, it took until 2015 for the Basel Ban to be ratified. Some difficulties to ensure compliance have emerged. The voluntary nature of some technical guidelines allows for the close involvement of non-state actors but is criticized for heterogenous application across countries and non-state actors. Overall, the Basel Convention has moved slowly toward its goals, which have not been achieved yet.

As Figure 36.1 shows, ratification of the Basel Ban amendment was slow; only somewhat more than half of the Convention's Parties have ratified it, and the threshold for its entry into force was only passed in 2015. Besides the slow pace of ratification, a disagreement on the ratification threshold complicated the process. The amendment itself refers to a three-fourth threshold for ratification but does not detail how this should be calculated. While some argued that it should be calculated based on the number of Parties at the point in time when ratification occurs, others argued that the calculation should be based on the number of Parties at the time of adopting the Basel Ban, namely 1994. The disagreement finally was settled in 2011 when Parties decided that the three-fourth of Parties that needed to ratify the Basel Ban for it to enter into force should be interpreted as three-fourth of the 82 countries that were a Party to the Basel Convention at the time of adoption of the Basel Ban (Selin 2014: 435). This threshold was reached in 2015 after many years of intensive advocacy by actors such as the NGO Basel Action Network.

Ratification of the 1999 Protocol on Liability and Compensation has been even slower; it had still not been achieved by 2021. The Protocol responds to the concern by many low-income countries that high-income countries are not willing to repatriate illegally dumped hazardous waste and pay for the clean-up costs. This is a very politically sensitive topic. As

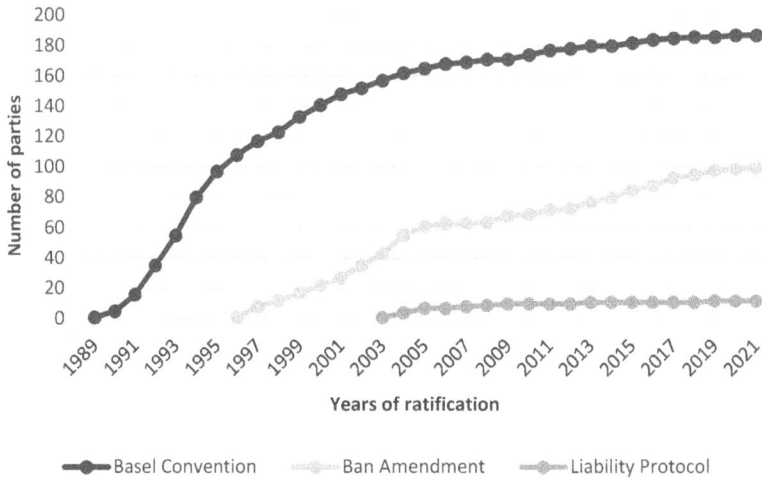

Figure 36.1 Ratification over time
Source: Author's own elaboration.

such, high-income countries have taken a number of measures to take their responsibility for their hazardous waste, as explained in the previous section, but formally accepting liability for damages to human health and the environment as a result of illegal trade is considered as a too far-reaching measure (Selin 2014: 435). Controlling illegal activities and exports has proven to be a great challenge for high-income countries as the previously mentioned IM-PEL report shows. Countries are reluctant to being held liable for possible damage resulting from those activities. This can explain the extremely slow pace of ratification.

Despite being one of the major hazardous waste producers and exporters, the United States is not a Party to the Basel Convention. The United States signed the Convention in 1990 and, in 1992, the US Senate voted in favor of ratification. Nonetheless, the final ratification step has not been taken. The United States needs to adopt additional implementing legislation that provides the necessary statutory authority. As of 2021, this had not happened. The main reason for the United States' non-ratification seems to be opposition to a complete ban of hazardous waste exports as enshrined in the Basel Ban amendment (Dreher and Pulver 2008). For this reason, the entry into force of the Basel Ban amendment makes US ratification of the Convention even more unlikely than previously.

Not all rules that have been adopted as part of the Basel Convention regime are binding on the Parties. A number of non-binding technical guidelines have been adopted. These define, for example, environmentally sound management of a number of waste streams for use by state and non-state actors. The Convention also includes some private voluntary initiatives. For example, technical guidelines for the management of e-waste were jointly developed by governments and industry. While the non-binding nature of these guidelines makes it easier for Parties to agree and facilitates the involvement of and application to non-state actors, their voluntary character also can have downsides. The main shortcoming is that not every Party applies the guidelines and that uniform application is not necessarily ensured. Moreover, to improve compliance with the Convention's rules in general, a compliance mechanism was established in 2002. Most countries are however wary of interference in their domestic affairs and want their sovereignty safeguarded so that the compliance mechanism mainly concerns the collection of data. Its effectiveness remains uncertain (Selin 2010).

The Basel Convention has been criticized for not achieving its objectives since hazardous waste trade has not been reduced and hazardous waste dumping still occurs. Especially illegal trade poses a huge challenge (Bisschop 2012: 227–231). The 2006 incident of a ship chartered by a Dutch company that dumped hazardous waste in Abidjan, Ivory Coast, illustrates that dumping still takes place. Hazardous substances leaked from the waste into the groundwater and polluted the drinking water, soil and fish, which led to at least ten deaths and over 100,000 people in need of medical care (Widawsky 2008: 579). Another example is the 2000 incident in which 122 containers of hospital waste from Japan were illegally exported to the Philippines (Krueger 2001: 43). Overall, the Basel Convention's track record in terms of ratification and implementation is mixed.

Horizontal fragmentation

At the international level, the Basel Convention is part of a regime complex (see Chapter 9) that includes the 1998 Rotterdam Convention, the 2001 Stockholm Convention and the 2013 Minamata Convention. These different Conventions overlap and interact horizontally but also leave wide gaps in the governance of the broader field of chemicals. In recognition of their close interrelation, the Basel, Rotterdam and Stockholm Conventions coordinate much of their work. The three Conventions' Secretariats are joined in the same location in Geneva, and they hold their Conferences of the Parties (COPs) at the same time and location. Their first joint extraordinary COPs were held in 2010 and the first regular COP in 2013. Since then, the so-called Triple COPs have taken place in conjunction.

The Basel, Rotterdam and Stockholm Conventions are closely related since they address similar and complementary practices and substances/products. Both the Rotterdam and Basel Conventions cover international trade in hazardous materials but at different stages of the life cycle. The Stockholm Convention addresses the production and use of certain chemical substances. Yet, the three Conventions do not necessarily address the same chemicals. The Stockholm Convention focuses on Persistent Organic Pollutants (POPs). In 2020, 29 POPs were covered (see Chapter 35). The Rotterdam Convention covers a larger number of chemicals. The Basel Convention focuses on hazardous waste, which can include chemical substances not necessarily covered by the other Conventions. For example, lead is covered by the Basel Convention but not by the others. The Minamata Convention addresses one substance only – mercury – but all of its life-cycle stages, from mining to the end-of-life stage. The Conventions are thus closely related but not a coherent set of rules that neatly fit with each other.

As illustrated, the chemicals regime complex is fragmented in terms of life-cycle stages. While the interaction is complementary rather than conflictive, there are a number of gaps since not all stages are covered for all substances. This links to the fragmentation in the different Conventions' coverage of substances. Another fragmentation relates to membership. Similar to a number of other international environmental agreements, the United States is not a Party to the Basel Convention, neither to the Rotterdam and Stockholm Conventions though it is a Party to the Minamata Convention. For the Basel Convention, this geographical fragmentation reduces its effectiveness since the waste export ban, PIC and other provisions do not apply to the United States, which is a large (hazardous) waste generator. Since the Basel Convention's provisions only allow Parties to trade hazardous waste with countries with which they have an agreement that ensures environmentally sound management at least at the Convention's level of stringency, the United States concluded a multilateral agreement with the OECD countries on trade in hazardous waste for recovery. It also has bilateral agreements with Canada and Mexico.

Ships that have reached the end of their useful lives constitute a problematic waste stream that led to the adoption of a separate Convention. The Hong Kong International Convention for the Safe and Environmentally Sound Recycling of Ships was adopted in 2009 to focus exclusively on this issue. The shipbreaking industry asserts that ships are not waste, and a large share of all ship dismantling takes place in one country, Bangladesh, with significant health and environmental problems. The Convention covers aspects such as hazardous substances that may be contained in ships – including asbestos and heavy metals – and working and environmental conditions of ship recycling. It addresses the design and construction of ships to make them better recyclable and the operation of recycling facilities to ensure environmentally sound management. As of 2021, the Convention had not entered into force since it had not passed the threshold of 15 ratification, representing 40 percent of the world merchant shipping by gross tonnage. Even when the Hong Kong Convention enters into force, the Basel Convention provisions and its technical standards on end-of-life ships will continue to apply (Puthucherril 2012: 307–309). In sum, at the international level the Basel Convention is part of a patchwork of semi-matching international agreements. The gaps of this horizontally fragmented regime complex led to the adoption of a number of measures at lower levels of governance, adding vertical fragmentation to the complex picture.

Vertical fragmentation of hazardous waste governance

Given the political backlog and slow progress at the international level, a number of regional and non-state initiatives have been adopted to fill gaps and move faster than the Basel Convention process. Article 11 of the Basel Convention encourages Parties to conclude bilateral, multilateral and regional agreements to help achieve the Convention's goals. When it became clear that the Basel Ban would not be ratified quickly, a group of African states, a group of Pacific islands and the European Union adopted their own regional agreements to compensate for the slow international progress. Also some non-state actors have launched initiatives to fill some of the governance gaps.

The Member States of the African Union (AU, previously Organisation of African Unity) signed in 1991 the Bamako Convention on the Ban of the Import to Africa and the Control of Transboundary Movement and Management of Hazardous Wastes within Africa, which entered into force in 1998. Out of fear that hazardous waste would continue to be dumped in Africa, the Bamako Convention introduces a complete ban on all hazardous waste imports. Contrary to the Basel Convention, the Bamako Convention also covers radioactive waste. It does not only ban imports from outside Africa but also aims to reduce transboundary movements of hazardous waste among African countries, prohibits dumping in the oceans and inland waters as well as hazardous waste incineration. Moreover, the Convention aims to ensure that hazardous waste is handled in an environmentally sound manner and promotes cleaner production. In 2021, the Convention had 29 Parties.

Analogously, in 1995 the Pacific Island Forum adopted the Convention to Ban the Importation into Forum Island Countries of Hazardous and Radioactive Wastes and to Control the Transboundary Movement and Management of Hazardous Wastes within the South Pacific Region (Waigani Convention), which entered into force in 2001. Similar to the Bamako Convention, the Waigani Convention bans the import of hazardous and radioactive waste into the Pacific Island states that are Party to the Convention. In 2021, 12 countries had ratified the Convention.

Not only low-income countries initiated agreements to compensate for the Basel Conventions shortcomings and slow progress. The European Union (EU) also adopted a law that implemented the Basel Ban amendment years before its eventual ratification. In 2006, the EU adopted Regulation No. 1013/2006 on Shipments of Waste, which prohibits waste and hazardous waste exports for disposal to all non-OECD countries. It also transposes OECD rules that date from 2001 by establishing a control system for waste shipments for recovery within the OECD. In 2013, the EU moreover unilaterally adopted rule on ship recycling, which however only apply to vessels sailing under EU flag. As these three examples show, individual groups of countries and Regional Integration Organizations can take the lead by adopting certain rules. If the organization is a significant trade actor such as the EU or the AU, those unilateral measures can exert a significant effect, however, not as impactful as a decision at the international level. Those national and regional activities bear the risk of fragmentation and making compliance with the diverse rules highly complex for companies and other actors.

Not only state but also some non-state actors have adopted initiatives that, in parts, aim to address some of the Basel Convention's shortcomings, in particular the United States' non-ratification, persisting illegal trade and dumping. In the United States, a number of actors attempted to fill the void of the country's non-ratification. For example, the NGO Basel Action Network developed a private certification scheme for e-waste that includes an export ban. Other standards also provide rules on environmentally sound management but allow for international trade (Renckens 2015). NGOs such as Greenpeace have engaged in active advocacy and public awareness raising on trade in hazardous waste. The Solve the E-waste Problem (StEP) initiative is a public-private partnership that strives to find solutions to the global e-waste problem. The partnership comprises electronics manufacturers, recyclers, public administration, NGOs and research institutions. These multiple initiatives demonstrate the multifaceted nature of state and non-state initiatives that emerged in response to the slow progress of the Basel Convention. While they contribute to filling governance gaps and increase the level of ambition, they have led to a fragmented patchwork of initiatives.

Conclusion

Achieving environmental justice is at the core of the Basel Convention that aims to minimize the severe damages that can occur from the inappropriate handling and processing of hazardous waste. Trade between high-income and low-income countries has been the main cause of direful incidents. For this reason, the Basel Convention regulates the international trade dimension, environmentally sound management and some other aspects of hazardous waste management. While many initiatives have been taken, which have improved the situation, more work is needed. For example, capacity building and enabling low-income countries to establish environmentally sound processes is underfunded but also the United States' non-ratification poses challenges. Overall, the Basel Convention has given rise to some tense political discussions. Many of these political tensions are illustrative of global environmental politics more generally. For example, the fragmentation of governance initiatives can be witnessed in many environmental areas. The United States' non-ratification and the corresponding challenges with regard to global coverage and effectiveness are also shared by most other international environmental treaties. The North–South dimension, and more specifically, debates about how and to what extent high-income countries should take responsibility for their global footprint transcends global environmental politics and is at the core of many political dynamics.

One of the open and controversially discussed questions in regard to the Basel Convention is whether banning hazardous waste trade is the best solution to address the problem. An alternative option that is advocated by a number of countries and observers is allowing hazardous waste trade in conjunction with the insurance of environmentally sound waste management practices. This controversy is at the root of the Basel Ban discussion and can explain why it took more than two decades to ratify the amendment. This links to the economic value of some types of waste. Through recycling, valuable materials such as copper and gold can be recovered (Puthucherril 2012: 304). Recycling according to high environmental and health standards can replace mining of raw materials because they are recovered from waste instead and avoid dumping. This could generate both environmental and health gains. For this reason, opponents of a trade ban argue that a viable recycling sector in low-income countries could enhance their economic productivity. Yet, reality shows that so far illegal exports persist and much of the recycling in low-income countries is done in the informal sector, causing health and environmental harm.

The characteristics of the hazardous waste problem have changed since the 1980s when the Basel Convention was adopted. At the time, it was a clear North–South problem. Hazardous waste was shipped to low-income countries where it was dumped or treated with harmful methods. Nowadays, the problem is more complex. While the North–South dimension remains, it has become to a large extent a problem of illegal trade. Another big change is the domestic production of hazardous waste in the Global South. Low-income countries increasingly produce hazardous waste domestically, for example, in the form of e-waste but also industrial and other kinds of waste. The problem therefore is not necessarily one of trade anymore or, if hazardous waste is traded, it is among low-income countries, a transaction to which the Basel Ban does not apply.

The Basel Convention's adaptability to technological, economic and societal changes is thus important to ensure that it still addresses the core activities that are at the root of hazardous waste problems. Besides changes in hazardous waste generation and trade, the move toward the paradigm of a circular economy is another change that was not anticipated by the drafters of the Basel Convention and that requires adjustment. Waste is increasingly perceived as a resource, which changes incentive structures for trade and recycling/recovery. This can create opportunities, but the environmental and health dimension remains an important consideration and risk. Making processes circular needs to ensure environmentally sound management. For this reason, the Basel Convention remains relevant and has already been adjusted over the course of time to accommodate some of the recent technological, economic and societal changes.

Acknowledgments

The author would like to thank Paul G. Harris for his helpful comments on an earlier draft of this chapter.

References

Andrews, A. (2009) 'Beyond the Ban – Can the Basel Convention Adequately Safeguard the Interests of the World's Poor in the International Trade of Hazardous Waste?' *Law, Environment and Development Journal* 5: 167–184.
Baggs, J. (2009) 'International Trade in Hazardous Waste'. *Review of International Economics* 17: 1–16.
Basel Action Network (2002) *Exporting Harm. The High-Tech Trashing of Asia*, Seattle: Basel Action Network.
Basel Convention. 1989. Basel Convention on the Control of Transboundary Movements of Hazardous Wastes and their Disposal. Basel, 22 March 1989.

Bisschop, L. (2012) 'Is It All Going to Waste? Illegal Transports of E-Waste in a European Trade Hub'. *Crime, Law and Social Change* 58: 221–249.

Clapp, J. (2001) *Toxic Exports: The Transfer of Hazardous Wastes from Rich to Poor Countries*, Ithaca: Cornell University Press.

Dreher, K. & Pulver, S. (2008) 'Environment as 'High Politics'? Explaining Divergence in US and EU Hazardous Waste Export Policies'. *Review of European Community & International Environmental Law* 17: 306–318.

European Union Network for the Implementation and Enforcement of Environmental Law (2006) *International Cooperation in Enforcement Hitting Illegal Waste Shipments*, Project Report. IMPEL-TFS Seaport Project II.

Greenpeace (2008) *Chemical Contamination at E-Waste Recycling and Disposal Sites in Accra and Korforidua, Ghana,* Amsterdam: Greenpeace.

Kaza, S., Yao, L., Bhada-Tata, P. & Van Woerden, F. (2018) *What a Waste 2.0. A Global Snapshot of Solid Waste Management to 2050*, Washington, DC: International Bank for Reconstruction and Development/The World Bank.

Krueger, J. (1999) *International Trade and the Basel Convention*, London: Earthscan.

Krueger, J. (2001) 'The Basel Convention and the International Trade in Hazardous Waste', in Schramm Stokke, O. & Thommessen, O.B. (eds.) *Yearbook of International Cooperation on Environment and Development 2001/2002*, London: Earthscan Publications.

McElvaney. (2014). 'Agbogbloshie: The World's Largest E-waste Dump – in Pictures'. *The Guardian*, 27 February 2014. (available at: https://www.theguardian.com/environment/gallery/2014/feb/27/agbogbloshie-worlds-largest-e-waste-dump-in-pictures).

OECD (2012) *Illegal Trade in Environmentally Sensitive Goods. OECD Trade Policy Studies,* Paris: OECD Publishing.

O'Neill, K. (2000) *Waste Trading among Rich Nations. Building a New Theory of Environmental Regulation*, Cambridge, MA and London: The MIT Press.

Puthucherril, T.G. (2012) 'Two Decades of the Basel Convention', in Alam, S., Bhuiyan, J.H., Chowdhury, T.M.R. & Techera, E.J. (eds.) *Routledge Handbook of International Environmental Law*, London and New York: Routledge, 295–311.

Renckens, S. (2015) 'The Basel Convention, US Politics, and the Emergence of Non-state E-waste Recycling Certification'. *International Environmental Agreements: Politics, Law and Economics* 15: 141–158.

Saunier, R.E. & Meganck, R.A. (2009) *Dictionary & Introduction to Global Environmental Governance*, London: Earthscan.

Secretariat of the Basel Convention (2010) *Waste Without Frontiers*, Geneva: The Secretariat of the Basel Convention.

Selin, H. (2010) *Global Governance of Hazardous Chemicals. Challenges of Multilevel Management*, Cambridge, MA and London: The MIT Press.

Selin, H. (2012) 'Global Environmental Governance and Regional Centers'. *Global Environmental Politics* 12: 18–37.

Selin, H. (2014) 'Hazardous Wastes', in Harris, P.G. (ed.) *Routledge Handbook of Environmental Politics*, London and New York: Routledge, 427–438.

Widawsky, L. (2008) 'In My Backyard: How Enabling Hazardous Waste Trade to Developing Nations can Improve the Basel Convention's Ability to Achieve Environmental Justice'. *Environmental Law* 38: 577–625.

Yeung, P. 2019. 'The Toxic Effects of Electronic Waste in Accra, Ghana'. *Bloomberg CityLab*, 29 May 2019. (available at: https://www.bloomberg.com/news/articles/2019-05-29/the-rich-world-s-electronic-waste-dumped-in-ghana).

37

Plastics

From resources to waste and back again

Josephine van Zeben and Violet Ross

Plastics enhance the safety of electronic devices and hygiene of both food and medicine; they are durable, relatively cheap and can be molded into a variety of forms from hard to mailable, depending on their function. The widespread use of plastics, and their slow and incomplete degradation, have created a vast, and seemingly permanent, environmental problem. The presence of microplastics – plastics smaller than 5 mm in diameter – in the world's rivers and oceans, and in human bodies, is a worrying example of the consequences of plastics usage, which are hard to reverse or remedy. Though it is difficult to imagine a complete phasing out of plastics at this stage, the extent of current plastic usage is increasingly called into question.

This chapter highlights the physical and regulatory aspects of plastics. It considers how the physical aspects of plastics inform our regulatory responses regarding production, consumption, waste management and pollution. It also considers the most important shift in regulatory thinking, namely the potential of a circular economy (CE) for plastics, and considers the geopolitics of plastics production, consumption and waste. We discuss the shift toward a more coordinated international approach to plastics that has been developing in recent years, which has cumulated in the inclusion of certain plastics in the Basel Convention on the Control of Transboundary Movements of Hazardous Wastes and Their Disposal (commonly referred to simply as the Basel Convention). The chapter concludes with reflections on the future outlook of plastics regulation.

Physical aspects of plastics

The term "plastics" refers to a diverse group of synthetic materials known as polymers. The varying chemical structures of the polymers and chemical additives used in production lead to a range of physical properties. Broadly, plastics are either thermosoftening, meaning they will soften when heated and can be reshaped, or thermosetting, which means they do not soften with heat (e.g., silicone used in car engines and kettles) (Plastics Europe, 2020: 6). Table 37.1 shows the plastic resins, fibers and additives that dominate the plastics market, the typical uses for them, and their annual production.

Highly versatile plastics that can be made strong or mailable are most common and include PP, HDPE and LDPE, some of which are more sensitive to heat increases than others (i.e.,

 DOI: 10.4324/9781003008873-42

Table 37.1 Dominant plastic materials produced globally by the plastics industry with typical use and the estimated annual tonnage

Plastic Resin/Fiber/Additive	Typical Use	Annual Production (Million Tons)
Polypropylene (PP)	Packaging, reusable food containers, car parts	68
Low-density polyethylene (LDPE)	Drink container liner, reusable bags, food packaging film	64
Polyphthalamide fibers (PP&A fibers)	Polyester fabrics	59
High-density polyethylene (HDPE)	Toys, housewares, containers for drinks, hazardous liquids (e.g. bleach)	52
Polyvinyl chloride (PVC)	Soft as packaging, inflatable pools or hard as flooring, pipes, windows	38
Polyethylene terephthalate (PET)	Bottles for drinks, cleaners	33
Polyurethanes (PUR)	Wheels, insulation panels	27
Additives	Stabilizers, plasticizers, lubricants, flame retardants	25
Polystyrene or styrofoam (PS)	Expanded as insulation and packaging, or rigid as disposable cutlery and CD cases	25

Geyer, Jambeck and Law (2017).

LDPE). PVC and PS can also be made both rigid or thin and flexible, but this depends on the additives used, while buoyancy is a key property of PUT. The additives used in plastic production make up approximately 7 percent of non-fiber plastics by mass (Geyer, Jambeck and Law, 2017: 1) and, as shown in Table 37.1, 25 million tons of it are produced each year. While some plastics more often fulfill one-time use (e.g., PET and LDPE), others are designed to last several decades (e.g., hard PET and PP). Plastics are almost invariably extremely durable. This also means they do not decompose but accumulate in the environment. Traditionally, plastics are made from the monomers extracted from fossil hydrocarbons: ethylene from natural gas or propylene from crude oil. Bioplastics – plastics made from bio-based (renewable) feedstock, such as corn starch or wood, and/or plastics that are biodegradable, even if they are made from fossil hydrocarbons – currently represent only 1 percent of market share. However, the use of bioplastics is increasing and diversifying (European Bioplastics, 2019).

Regulatory aspects of plastics

In order to appreciate the environmental impacts associated with plastics, this section adopts a lifecycle approach that follows plastics through the stages of production, consumption, waste management and, eventually, pollution.

Production

The main environmental impact arising in the plastics production stage concerns the use of polymers in conventional plastics. The monomers extracted from oil and gas to make polymers are residues from natural gas processing and crude oil refining, which explains their over-supply and low cost (CIEL, 2017; Sicotte, 2020). This creates a link between the petrochemical and plastics industry, leading plastics markets to fluctuate with oil and gas markets (Milios *et al.*, 2018: 186), which have been historically low over the past years. It also means that the phasing out of virgin plastics is complicated by the economic contribution and lobbying power of the petrochemical sector (CIEL, 2017; The Pew Charitable Trusts and Systemiq, 2020: 31). Recycled plastics often cannot compete with the low prices of virgin plastics (Milios *et al.*, 2018; Gu *et al.*, 2020). To resolve this, many jurisdictions have introduced targets for recycled content of plastic used in production, including the EU and various Association of Southeast Asian Nations (ASEAN) countries including Vietnam, Thailand, Indonesia and Malaysia (UNEP, 2019).

The monomers that make plastics and additives achieve desired functions carry potential health, safety and environmental impacts (Gallo *et al.*, 2018; Hahladakis *et al.*, 2018; Beaumont *et al.*, 2019). Various public and private, national and transnational regulations exist that govern the use, trade and information concerning chemicals. This includes national legislation in 15 countries, the Global Harmonized System of Classification and Labelling of Chemicals (GHS) by the UN, and international conventions such as the Stockholm and Basel Conventions (Chem Safety Pro, n.d.). Though bioplastic production does not suffer from the same issues as conventional plastics, it has high water demands and creates potential competition over land use when crops are grown for bioplastics instead of agricultural goods (Spierling *et al.*, 2018).

Finally, hundreds of types of plastic resins, fibers and additives exist and their categorization is completed in the production stage through the use of coded identification systems. These aim to ease the separation of different polymers for recycling. The dominant system of seven Resin Identification Codes (RICs) was drafted in 1988. However, the expansion of polymer types since then means that RICs fail to capture this growing diversity. New systems aim to resolve this issue, such as China's system of 140 identification codes (*Standardization Administration of the People's Republic of China (SAC) GB16288*, 2008). Poor enforcement leaves producers responsible to ensure the inclusion of these codes, which often results in them being missing or incorrect (Coltro, Gasparino and Queiroz, 2008). These elements contribute to confusion on the consumer side, ineffective sorting by both consumers and waste sector, and ultimately lower recycling rates.

Consumption

In many sectors, the use of plastics has become a de facto necessity insofar as it would be excessively costly, or technically impossible, to replicate the same properties in non-plastic products. In light of health and hygiene benefits, the reduction of plastics in such sectors is considered extremely unlikely, and even undesirable, despite the negative consequences. In these sectors, bioplastics are considered the most promising alternative, though they are also imperfect. However, there are equally many sectors where the consumption of plastics is a question of convenience and low cost rather than necessity. The dividing line between convenience and necessity is at times clear cut – for example, the choice between a re-usable cloth bag and a plastic carrier bag – but at times, surprisingly difficult – such as benefits from super absorbent polymers in such products as nappies/diapers compared to reusable or biodegradable alternatives. In making such decisions, a product's utility is weighed up against its environmental impact and available alternatives. Environmental impacts have previously been compared via life cycle analyses (LCAs), but parameters are not standardized and weight is not often given to the societal need of certain products. "Utility" also has a degree of subjective variation, which has been considered in the EU's new single-use plastic directive whereby "medical purposes" allow for the lifting of restrictions on certain plastic products (e.g., water bottles) (European Union, 2019). More recently, utility has been weighed against growing data on plastic production and waste generation. Restrictions now increasingly target single-use items and packaging, with the latter comprising approximately 40 percent of all plastics produced and discarded (Geyer, Jambeck and Law, 2017). A range of measures have been put in place to change behavior and drive demand away from these to promote reuse in the consumer stage. Regulations have taken shape as bans (e.g., to straws and microplastics in cosmetics), charges (taxes) on certain plastic items such as plastics bags, and green procurement plans (UNEP, 2019).

Increased media attention to the negative environmental impacts of plastics consumption has led to higher consumer awareness, and can be linked to the growth in bioplastics and the reuse and repair market. Despite greater concern by consumers, a recent survey of attitudes in Australia shows that many consumers still place primary responsibility on regulators and industry to solve the plastics problem (Dilkes-Hoffman et al., 2019).

Waste management

Most plastics regulation focuses on plastic *waste*, which has been seen as less controversial than regulating production and consumption (Nielsen et al., 2019). Regulation of production and consumption has led to strong pushback by plastic producers and retailers, while increased regulation on waste has created business opportunities in the waste sector without undermining the existing economic model of "take-make-use-throw." Different plastics require different waste management techniques. Conventional plastics are either landfilled, permanently destroyed via incineration (i.e., energy recovery) or recycled. These waste management options are also true for bioplastics, though they could also biodegrade in an anaerobic digester or a (home/industrial) compost facility. In the mid-1990s, spatial limitations sparked a range of policies deterring landfill usage, including bans, taxes and extended producer responsibility (EPR) schemes (OECD, 2018: 47). These debates also led to the promotion of recycling and recovery technologies and targets, as well as CE policies, by most governments.

Waste management options with regard to technology vary from country to country and income level is largely indicative of the options available. These levels are defined as high-income country (HIC), upper middle income (UMI), lower middle income (LMI) and low income (LI), and are based on 2015 gross national income (Brooks, Wang and Jambeck, 2018). Massive investment is required for the high-tech facilities more typical in high- and middle-income countries (The Pew Charitable Trusts and Systemiq, 2020: 114–116). Though providing minimal job opportunities, these facilities provide returns through gate fees, whereby incoming waste deposits are charged by weight.

Traditional landfills (dumps) and a large informal waste sector are more characteristic of LMI and LI countries. As an urban common, the informal waste sector is an important contributor to the livelihoods of people in such economies (Zapata and Zapata Campos, 2015; Hartmann, 2018). However, the health impacts from dangerous working environments (e.g., on mismanaged landfills) bring high public health costs due to chemicals, poor hygiene, disease, accidents and psychological harm (Yang et al., 2018). With the absence of formalized and centralized waste management systems, recycling of plastics is estimated to be lower than 1 percent in these countries (Woldemar d'Ambrières, 2019).

Technically speaking, all plastics can be recycled, but the extent that they are depends on the collection, sorting and processing infrastructure in a location. Neither mechanical nor chemical recycling can cope with excess levels of dirt, salt (from marine plastics), chemical additives (added during production) or other polymers, including bioplastics (from poor sorting) (Alaerts, Augustinus and Van Acker, 2018; Clift *et al.*, 2019). These contaminants result in "downcycling" whereby the original physical properties (e.g., durability, flexibly) degrade following the recycling process and the plastics can only be used in "lower value" products. In order to prevent downcycling, the sorting and handling of plastic to minimize contamination are crucial. Consumers are primarily responsible for this sorting.

In HICs, household recycling systems can be hindered by improper sorting and/or contamination (e.g., dirt) caused by consumer non-compliance or confusion regarding which plastics should go into recycling or regular waste streams. This confusion can be compounded by differing waste management systems between and within countries; insufficient recycling resources given to consumers (e.g., information, trash bins) to facilitate the desired sorting; and unclear or confusing labeling on plastic products (OECD, 2018: 26). Debate exists as to whether household separation of plastics should be avoided altogether through the use of mechanical biological waste treatment plants that are used around Europe and retrieve plastic waste directly from municipal solid waste (Feil *et al.*, 2017). In UMIs, we see increased household sorting of plastic waste. For instance, the Domestic Waste Classification System Implementation Plan requires 46 urban areas in China to implement a mandatory household waste separation system by 2021. Deposit return schemes for plastic bottles have been effective in some countries, such as Germany and the Netherlands, in increasing waste data (Zhou *et al.*, 2020). Increasing efforts are also being made on the industry side to improve tracking technology and consumer sorting (P&G, 2020).

Pollution

Plastics become pollutants when they occur in "higher-than-normal concentrations […] in the air, water or soil, which may have effects on humans or non-human organisms" (van Zeben and Rowell, 2021: 92). Examples of this include plastics in rivers and oceans, and poisoning or harming birds or fish, but also high concentrations of microplastics in food and/or humans. Plastic pollution is caused by the mismanagement of waste, either in transit or at a landfill or other facility, and waste crimes such as littering and illegal dumping (also called fly tipping), especially in countries that lack centralized collection and disposal systems. While littering and illegal dumping are largely regulated domestically, marine plastics are a global problem, receiving increased attention in research and on the global political stage (Nielsen *et al.*, 2019: 5; see Chapter 39). The undesirability but large quantities of contaminated plastics have also caused much illegal trade in plastic waste. Plastics and other materials are often mixed with hazardous waste under the guise of plastics recovery to export the regulatory costs associated with hazardous waste (Baird, Curry and Cruz, 2014: 99–100; see Chapter 36).

Plastics' resistance to degradation makes them likely to survive in open air and water for centuries, making them a particularly problematic pollutant. Research shows that plastic pollution negatively impacts almost all marine ecosystem services as well as human wellbeing and local economies by affecting tourism and fisheries (Beaumont *et al.*, 2019; see Chapter 39). There are also growing concerns around the uncertain impacts of microplastics (De-la-Torre, 2020). These tiny particles leak into the environment through cosmetic products and textiles (primary microplastics) or are formed by the breakdown of larger pieces of plastic (secondary microplastics). Existing, research suggests that they can act as toxic sinks, causing harm to humans and other organisms (Gallo *et al.*, 2018; Lam *et al.*, 2018). Apart from these waste impacts, the plastics lifecycle is responsible for approximately 4 percent of current global carbon emissions (Zheng and Suh, 2019)

Toward a circular economy for plastics

The idea of a circular economy aims for a lifecycle approach to all aspects of the economy, including the plastics sector. At its core the CE calls for systems change to production and consumption processes to reduce reliance on natural resources. It promotes keeping "products and materials in use" and places greater emphasis on the "reduce" and "reuse" of materials, before choosing "recycling" and energy "recovery" (Ellen MacArthur Foundation, 2017). A CE specifically for plastics is controversial due to the finite nature of fossil fuels – the main ingredient for conventional plastics. The Ellen MacArthur Foundation is a leading civil society organization on CE research and private regulation. It highlights distinctive issues hindering circular plastics, such as the chemical additives present within plastic products, or on waste plastic and downcycling. Its depiction of "The New Economy for Plastics" deviates from the standard CE by emphasizing that only those necessary plastics should remain in the economy and should retain as high a utility value as possible (The Ellen MacArthur Foundation, 2019). It is thus a big step away from merely focusing on recycling to push for the reduction of unnecessary plastics.

There is diversity in definition of CE and the use of the concept (Kirchherr, Reike and Hekkert, 2017). For plastics, an end-of-pipe focus dominates both legislation and academic material because earlier stages of the lifecycle are not politicized (Nielsen *et al.*, 2019: 11). For instance, early CE policies in China centered on industrial ecology principals, yet more recent policies such as the "National Sword" ban, and a law for mandatory household waste separation in urban areas, center around waste (Brooks, Wang and Jambeck, 2018). In the United States and EU, a focus on recycling and recovery, rather than reduce and reuse, has dominated (Nielsen *et al.*, 2019; Sicotte and Seamon, 2020). However, in the EU, increasing emphasis is being placed on the early stages in the lifecycle, with the new Circular Economy Action Plan (CEAP) and related laws, such as the single-use plastics directive (European Union, 2019). Also, law and policy for plastics is in place and increasing in a number of ASEAN countries. Similar to the EU, Japan, Indonesia and Malaysia have introduced legislation on EPR, which shifts some financial burden of waste management onto the producer, and many ASEAN countries, including Cambodia, Malaysia, Thailand and Vietnam, have policies on minimum recycled content for packaging (UNEP, 2019: 8). Questions are also being raised regarding the export of recyclables, and whether this should be considered pollution transfer or part of the global circular economy (Liu, Adams and Walker, 2018).

Geopolitics of plastics

Having set out the main characteristics of plastics as a resource and pollutant, this section will discuss key factors in the geopolitics of plastics. As mentioned, some plastics are seen as necessary based on the benefits provided to society. However, the politics of how to define what is "necessary," who produces and consumes this plastic, and who then receives the plastic at the end of its lifecycle, are complex. Moreover, the power dynamics in these relationships are changing, particularly regarding the responsibility for plastics pollution.

Responsibilities of producers, consumers and receiver countries

Producer countries are those involved in the production of polymer resins and fibers (i.e., virgin plastics) from oil and gas monomers – and countries converting these resins into plastic products. In 2018, approximately 359 million tons of plastics were produced globally which increased to 368 million in 2019 (Plastics Europe, 2020). Currently, virgin plastic is predominantly made in North America (19 percent), Europe (16 percent) and Asia (51 percent, with China accounting for 31 percent) (Plastics Europe, 2020: 17). Specifically, China produces 28 percent of global resin and 68 percent of global PP&A fibers (Geyer, Jambeck and Law, 2017: 1). The sheer amount produced in recent years is highlighted in a study by Geyer *et al.*, who found that half of the total amount of plastic resins and fibers manufactured from 1950 to 2015 were produced in the past 13 years (2017:1).

Oil and gas extraction is an essential step that precedes virgin production. As a result large petrochemical companies dominate this sector, including some of the major oil and gas extraction companies (CIEL, 2017). This makes it hard to disconnect key oil and gas supplier countries including Norway, Russia, Saudi Arabia and Venezuela, from linkage to plastics supply chains. Polymer resins, often in the form of pellets, go to converters around the world – there is limited data on their specific geographical diffusion – who manufacture products. The most common final products are produced by the packaging industry (42 percent of all non-fiber plastics, primarily PE, PP and PET), followed by the building and construction sector (19 percent), consumer and institutional products (12 percent) and "other" (13 percent) which includes transportation (8 percent), furniture, industrial machinery and textiles (Geyer, Jambeck and Law, 2017: 1).

Consumer countries are locations where plastic products are used and, when disposed, contribute to waste generation figures. The dominant regions consuming plastics in 2016 were North America, consuming approximately 21 percent, followed by China (20 percent), Western Europe (18 percent), Latin America and the Caribbean (8 percent) and the rest of Asia (8 percent) (Statistica 2021, 2020). The generation of plastic waste globally has been estimated at 300 million tons in 2015 (Geyer, Jambeck and Law, 2017) and of this, only 4 percent was exported legally (OECD, 2018: 9). However, the differences between countries vary significantly; countries such as the United Kingdom and Germany export a high proportion (40 percent) of their waste, while other countries, like Bulgaria, export far less (5 percent) (OECD, 2018, suffix 47).

The ten top exporting countries are listed in Table 37.2. These exporters correspond with the dominant plastic consuming countries mentioned above, with North America, China and Western Europe dominating. The majority of exporters are HIC which is also true for the importers, with the exception of Mexico, China and Hong Kong SAR, which are UMI countries, and India, which is an LMI country. It must be noted that shipments of illegal plastic waste are not counted in these figures, which, according to best estimates from 2017, amount to 56,000 tons ('Plastic Waste Trafficking', 2020: 1187).

The receiver countries are those importing plastic waste. Table 37.2 shows the top ten cumulative importers from 1988–2016 with China and the SAR of Hong Kong taking an overwhelming majority at 72.4 percent combined. China's infamous ban, known as "The National Sword," drastically pivoted the direction of global plastic waste imports when it came into force in 2018. To demonstrate this, in 2017 between 5.8 and 8.3 Mt of plastic waste entered China while in 2018 this was reduced to 52 kt (Liang *et al.*, 2021). Instead, total global exports decreased and imports to various other countries increased, including Malaysia, Thailand, Vietnam, Turkey and Taiwan (Principles for Responsible Investment (PRI), 2019: 20).

The mismanagement of plastic waste and its negative environmental impacts have become increasingly documented and promulgated. Though data limitations exist, current information shows that dominant regions that mismanage plastic waste are East Asia and the Pacific (60.1 percent), South Asia (12.1 percent) and Sub Saharan Africa (10.6 percent) and

Table 37.2 Top ten cumulative exporters and importers of plastic waste from 1988 to 2016

Exporters	
Reporter in Order of Rank	*Percent of Global Exports*
China, Hong Kong SAR[a]	26.1
United States	12.4
Japan	10.3
Germany	8.22
Mexico	4.90
UK	4.31
Netherlands	3.59
France	3.52
Belgium	2.99
Canada	1.81
Importers	
Reporter in order of rank	*Percent of global imports*
China	45.1
China, Hong Kong SAR[a]	27.3
United States	3.60
Netherlands	2.72
Germany	2.27
Belgium	1.76
Canada	1.62
Italy	1.41
India	1.31
Other Asia, nes[b]	1.01

[a] Special Administrative Region.
[b] Other Asia, not elsewhere specified (nes) is 1 of 16 UN areas nes. These areas are used (i) for low-value trade or (ii) if the partner designation was unknown to the country or if an error was made in the partner assignment. The reporting country does not send details of the trading partner in these cases, sometimes to protect company information.
Brooks, Wang and Jambeck (2018).

the main sources of marine pollution are mis-managed and un-regulated landfills in Asian countries (Jambeck *et al.*, 2015; Qu *et al.*, 2019). However, global environmental problems with plastic pollution are a collective concern as they are affecting ecosystems and biodiversity crucial to human existence on the planet (Villarrubia-Gómez, Cornell and Fabres, 2018; Barnes, 2019).

Much denial and debate exists on the sources of plastic pollution and who should shoulder the financial and environmental responsibility regarding the management of this waste (Nielsen *et al.*, 2019: 5–6). Following the 2018 China ban, where waste was redirected to various other countries, strong correlations have been shown between import-export and the income level of a country and plastic waste acceptance, with LMI and LI countries more likely to act as importers of plastic waste, and HMI and UMI countries as exporters (PRI, 2019: 20). Many South-East Asian countries, whose imports have increased due to the China ban, have become critical of this situation and strengthened their regulations because of it

(Brooks, Wang and Jambeck, 2018; Qu *et al.*, 2019: 72; Sasaki, 2020). Though receivers benefit financially from importing plastic waste, it has become increasing clear that they do not have sufficient capacity with regard to space, technology or infrastructure to process plastics in an environmentally sound manner.

This raises the question as to why consumer countries export so much waste. First, it is due to the sheer amount of plastic. Globalization has reinforced the "take-make-use-throw" culture despite stricter waste regulations. Second, waste is exported due to the poor quality of most plastic waste and the inability of recycling infrastructure to handle high contamination levels complicate waste management. Rather than cleaning, sorting and pre-treatment, it has been easier and cheaper to offset the problem and export the low quality waste. Linked to this is the aforementioned illegal trade of contaminated plastic waste which continues to burden importers. The China ban was pivotal in illuminating such contamination as it only banned low quality recyclables, including eight types of plastic (Brooks, Wang and Jambeck, 2018). The sustainability of plastic waste exports has come under severe public scrutiny, especially as the mis-management of plastic waste in LMI and LI income countries has led to dire global problems. Questions of responsibility are not dissimilar to those in climate justice debates whereby more developed nations are larger historical contributors of greenhouse gases, have reaped more benefits and arguably have more responsibility in resolving related issues (see Chapter 26). With this perspective, not only do the consuming countries have greater resources and higher quality infrastructure to process plastics in an environmentally sound manner, but arguably, "disposal" else ware is not in keeping with their emphasis on sustainability and circularity. Facts pertain that both the trade and pollution of plastics is complex and transnational and recognition of the need for an integrated global approach has developed.

Toward an international approach to plastics

In recent years, recognition of the need for an integrated global approach to solutions around plastics, and plastic waste, has grown. This is first due to the increased awareness and understanding of the plastics crisis through increased media coverage of poorly constructed landfills in LMI and LI countries and the great pacific garbage patch. This raised concern from consumers in HIC and UMI countries who before then, had largely been shielded from the extent of the plastics problem. With this, the global nature of plastics supply chains has become increasingly researched and recognized as linked to plastics pollution (Nielsen *et al.*, 2019). Alternately, the illegal trafficking of plastics is still not widely known or understood ('Plastic Waste Trafficking', 2020).

International law (see Chapter 10) plays a role in such global approaches to manage plastics. Nielsen *et al.* identify 11 global agreements related to plastics (2019: 12–14). Aside from the Stockholm Convention, which regulates persistent organic pollutants (see Chapter 35), these agreements predominantly focus on marine plastic pollution (see Chapter 39). All have thus far proven unsuccessful in reducing the negative effects from the over production and consumption of plastic, and in reducing the trade of hazardous waste (Yang, 2020: 314). However, international law is not the only mechanism that can be used in achieving a global approach. Non-binding coordinated approaches in global private regulation such as the Global Commitment and Global Plastic Action Partnership have come about since 2018. These mechanisms emphasize the need to address systems change on a global and local scale by integrating efforts from governments, industry and community actors (Nielsen *et al.*, 2019: 10–11).

Plastics as hazardous waste under the Basel Convention

The 1989 Basel Convention is an international treaty on hazardous waste aiming to min-imize the trade and enhance the proper management of hazardous waste (see Chapter 36). Stemming from concern over the transfer of waste pollution problems to developing coun-tries, it recognizes "the right of a country to ban the entry or disposal of foreign hazardous wastes and other wastes in its territory" (UNEP, 2011: 6). Annex I lists the 48 types of wastes originally included and categorized as Y codes.

In 2017, China made notifications to the World Trade Organization (WTO) and the Ba-sel Convention banning 24 categories of waste imports, including some plastics, and placed restrictions on the minimum acceptable level of contamination for imports, including plastics (OECD, 2018: 10). These and previous undocumented restrictions in 2013 (Brooks, Wang and Jambeck, 2018) were enforced against a backdrop of decreasing landfill capacity and CE policy in China, and targeted improvements to the quality of waste imports and stopping illegal waste trafficking and smuggling (Brooks, Wang and Jambeck, 2018: 1). The implications of the ban were three-fold. First, it significantly reduced the amount of waste being exported globally with the big exporters increasing domestic stockpiling and strengthening domestic waste policy (OECD, 2018: 10). For instance plastic waste exports from the EU decreased 39 percent from 2016 to 2018 (Plastics Europe, 2020: 30). Second, it caused a complete reshuffle of the global trade in plastic waste, highlighting "the fragility of global dependence on a single importer" (Brooks, Wang and Jambeck, 2018: 1; Huang *et al.*, 2020). And third, it resulted in increased il-legal trafficking and smuggling of plastic waste ('Plastic Waste Trafficking', 2020; Sasaki, 2020).

Following the Chinese ban, the UN Environment Assembly on plastic pollution was held in March 2019. As a result of this meeting, the 14th meeting of the Basel Convention an-nounced amendments to three annexes, leading plastic waste to be included in its provisions on hazardous waste. The amendments have come into effect on 1 January 2021 and they stipulate which plastic waste requires the completion of a PIC procedure. If a PIC is required, exporting countries must receive the consent that importing countries accept the waste, and they must ensure that the importing countries have the capacity to manage the plastic waste in an environmentally sound manner (Basel Convention, n.d.). The three amendments are as follows:

- Annex II: Y48, plastic waste, including mixtures of such wastes will be subject to the Prior informed Consent (PIC) procedure (excluding those that would fall under A3210 or B3011).
- Annex VIII: A3210, clarifies the scope of plastic waste presumed to be hazardous and therefore subject to the PIC procedure.
- Annex IX: B3011, plastic waste destined for recycling and almost free from contami-nation and other types of waste that remain excluded from the PIC procedure (certain single polymers or mixture of PE, PP and/or PET).

These amendments stipulate that plastic waste to be subject to PIC are mixes of various polymers or are hazardous, and that excluded plastic waste is that which is made from single polymers or mixtures of PE, PP and/or PET. The amendments will be the first international legally binding mechanism to date related to plastic waste and will likely increase the trans-parency and availability of data on the global plastics trade and strengthen regulations on ex-ported waste (Nielsen *et al.*, 2019: 11). As such, the amendments affect the illegal waste trade, as plastic waste that was purposefully covered in hazardous substances to hide the illegal trafficking of such substances, are anticipated to be easier to track and prevent. Moreover, the amendments only affect contaminated or not easily recycled plastics, which means that easily recycled plastics in support of a CE are minimally affected.

In addition, these amendments are likely to have positive repercussions for a number of other areas not originally targeted. For instance, the new processes are likely to enhance collaboration between international organizations on the problem of plastics as organizations like the WTO and the International Criminal Police Organization (Interpol) expand and harmonize their jurisdiction over plastic waste. Second, the amendments have the potential to reduce the generation of waste in the first place and boost circularity. The China ban was shown to reduce the amount of plastic waste exports globally (OECD, 2018: 10), and matched with an increase in "start-of-pipe" policies already mentioned, future plastics policies will no doubt increase steer toward "reduce" and "reuse" measures and encourage trade toward capable countries. In relation to this, the amendments may also accelerate enhancements to domestic waste management technologies and infrastructure – one means to reduce plastic pollution (Barnes, 2019). In HIC countries, where collection and sorting infrastructure tends to be more advanced, the amendments and need to reduce exports may push further development. While in LMI and LI countries, the amendments may lead to developments in collection, sorting infrastructure as well as technologies. The amendments also come with guidance on how best to reduce and manage plastic waste further aiding these developments (Basel Convention, 2019). And finally, the amendments may improve the harmonization and effectiveness of identification codes for plastic resins. Currently, no coherent set of resin or fiber codes are used and the predominant one (RICs) are too limited for plastics' diversity and not enforced effectively. Coordination brought by these amendments have potential to influence the codes' development, and therein, improve data collection and processing to better manage and separate waste polymers and increase recyclability.

The Basel Convention, however, is not without its flaws and critique exists regarding the limitations of its remit and country categorization, lack of incentives for capable handlers, and lack of assistance in building effective regulation in capacity poor areas (Yang, 2020; see Chapter 36 for more detail). Nevertheless, the amendments have led to international legal recognition of the plastics problem and pave the way for more regulatory development.

Conclusion

The outlook for plastics regulation is fast-changing, with many uncertainties. Awareness of the global nature of plastic supply chains and dangers to human health and the environment caused by mismanagement of plastic waste has increased impetus to resolve the plastics crisis. As consuming countries increase domestic capacity to better manage plastic waste, receiving countries are also looking for improvements. The growing number of CE policies for plastics, including LMI and LI countries, are sure to see continued infrastructural and technological development in the already fast-changing field. But the conflicting use of CE concepts and diverse material properties of plastics raise questions about what kind of changes will be seen. Information gaps and a lack of data transparency in each stage of the life cycle are evident and calls for measures to remedy these are getting louder. Such information is vital to enhance communication and circularity and reduce the illegal trafficking of plastic waste.

The aims of the Basel Convention amendments relate to these broad issues by supporting recycling and boosting global corporation on plastics and hazardous waste to reduce illegal waste trafficking and increase information transparency. Furthermore, the amendments increase pressure on nations to improve waste management infrastructure and technologies.

Yet this is still overshadowed by uncertainty about whether a systems perspective on reduce and reuse will be enhanced.

What is clear is that holistic, international solutions are needed to amalgamate the disjointed strategies around plastics from both governments and private or civil society actors (Dauvergne, 2018). Despite the growing involvement and interest of the international community (Nielsen *et al.*, 2019), national, local, and community levels continue to play a central role in transforming the plastics economy (Vince and Hardesty, 2018).

References

Alaerts, L., Augustinus, M. and Van Acker, K. (2018) "Impact of bio-based plastics on current recycling of plastics", *Sustainability*, 10: 1–15.

Baird, J., Curry, R. and Cruz, P. (2014) "An overview of waste crime, its characteristics, and the vulnerability of the EU waste sector", *Waste Management and Research*, 32: 97–105.

Barnes, S. J. (2019) "Understanding plastics pollution: The role of economic development and technological research", *Environmental Pollution*, 249: 812–821.

Basel Convention (2019) *Guidance and awareness raising*. Online. Availabale HTTP: <http://www.basel.int/Implementation/Plasticwaste/Guidance/tabid/8333/Default.aspx>.

Basel Convention (n.d.) *Plastic waste amendments FAQs*. Online. Available HTTP: <http://www.basel.int/Implementation/Plasticwaste/PlasticWasteAmendments/FAQs/tabid/8427/Default.aspx>.

Beaumont, N. J. *et al.* (2019) "Global ecological, social and economic impacts of marine plastic", *Marine Pollution Bulletin*, 142: 189–195.

Brooks, A. L., Wang, S. and Jambeck, J. R. (2018) "The Chinese import ban and its impact on global plastic waste trade", *Science Advances*, 4: 1–7.

Center for International Environmental Law (CIEL) (2017) *Fueling plastics: How fracked gas, cheap oil, and unburnable coal are driving the plastics boom*. Online. Available HTTP: <https://www.ciel.org/wp-content/uploads/2017/09/Fueling-Plastics-How-Fracked-Gas-Cheap-Oil-and-Unburnable-Coal-are-Driving-the-Plastics-Boom.pdf>.

Chem Safety Pro (n.d.) *REACH and chemical control laws*. Online. Available HTTP: <https://www.chemsafetypro.com/topics.html>.

Clift, R., Baumann, H., Murphy, R. J. and Stahel, W. R. (2019) "Managing plastics: Uses, losses and disposal", *Journal of Law, Environment and Development*, 15: 93–107.

Coltro, L., Gasparino, B. F. and Queiroz, G. de C. (2008) "Reciclagem de materiais plásticos: A importância da identificação correta", *Polímeros*, 18: 119–125.

Dauvergne: (2018) "Why is the global governance of plastic failing the oceans?" *Global Environmental Change*, 51: 22–31.

De-la-Torre, G. E. (2020) "Microplastics: An emerging threat to food security and human health", *Journal of Food Science and Technology*, 57: 1601–1608.

Dilkes-Hoffman, L. S., Pratt, S., Laycock, B., Ashworth, P. and Lant, A. (2019) "Public attitudes towards plastics", *Resources, Conservation and Recycling*, 147: 227–235.

Ellen MacArthur Foundation (2017) *What Is a Circular Economy?* Online. Available HTTP: <https://www.ellenmacarthurfoundation.org/circular-economy/concept>.

European Bioplastics (2019) *Bioplastics market data 2019. Global Production Capacities of Bioplastic 2019–2024*. Online. Available HTTP: <https://docs.european-bioplastics.org/publications/market_data/Report_Bioplastics_Market_Data_2019.pdf>.

European Union (2019) *Directive (EU) 2019/904 of the European Parliament and of the Council of 5 June 2019 on the reduction of the impact of certain plastic products on the environment*.

Feil, A., Pretz, T., Jansen, M. and Thoden Van Velzen, E. U. (2017) "Separate collection of plastic waste, better than technical sorting from municipal solid waste?", *Waste Management and Research*, 35: 172–180.

Gallo, F. *et al.* (2018) "Marine litter plastics and microplastics and their toxic chemicals components: The need for urgent preventive measures", *Environmental Sciences Europe*, 30: 1–14.

Geyer, R., Jambeck, J. and Law, K. (2017) "Production, use, and fate of all plastics ever made", *Science Advances*, 3: 1–5.

Gu, F., Wang, J., Guo, J. and Fan, Y. (2020) "Dynamic linkages between international oil price, plastic stock index and recycle plastic markets in China", *International Review of Economics and Finance*, 68: 167–179.

Hahladakis, J. N., Velis, C. A., Weber, R., Iacovidou, E. and Purnell: (2018) "An overview of chemical additives present in plastics: Migration, release, fate and environmental impact during their use, disposal and recycling", *Journal of Hazardous Materials*, 344: 179–199.

Hartmann, C. (2018) "Waste picker livelihoods and inclusive neoliberal municipal solid waste management policies: The case of the La Chureca garbage dump site in Managua, Nicaragua", *Waste Management*, 71: 565–577.

Huang, Q. et al. (2020) "Modelling the global impact of China's ban on plastic waste imports", *Resources, Conservation and Recycling*, 154: 1–12.

Jambeck, J. R. et al. (2015) "Plastic waste inputs from land into the ocean", *Science*, 347: 768–771.

Kirchherr, J., Reike, D. and Hekkert, M. (2017) "Conceptualizing the circular economy: An analysis of 114 definitions", *Resources, Conservation and Recycling*, 127: 221–232.

Lam, C. S. et al. (2018) "A comprehensive analysis of plastics and microplastic legislation worldwide", *Water, Air, and Soil Pollution*, 229: 1–19.

Liang, Y., Tan, Q., Song, Q. and Li, J. (2021) "An analysis of the plastic waste trade and management in Asia", *Waste Management*, 119: 242–253.

Liu, Z., Adams, M. and Walker, T. R. (2018) "Are exports of recyclables from developed to developing countries waste pollution transfer or part of the global circular economy?", *Resources, Conservation and Recycling*, 136: 22–23.

Milios, L. et al. (2018) "Plastic recycling in the Nordics: A value chain market analysis", *Waste Management*, 76: 180–189.

Nielsen, T., Hasselbalch, J., Holmberg, K. and Stripple, J. (2019) "Politics and the plastic crisis: A review throughout the plastic life cycle", *Wiley Interdisciplinary Reviews: Energy and Environment*, 9: 1–18.

Organisation for Economic Co-operation and Development (OECD) (2018) "Improving plastics management: Trends, policy responses, and the role of international co-operation and trade", *OECD Policy Paper No. 12*. Online. Available HTTP: <https://doi.org/10.1787/c5f7c448-en>.

P&G (2020) *P&G continues support of HolyGrail with AIM test market*. Online. Available HTTP: <https://us.pg.com/blogs/HolyGrail/>.

Plastics Europe (2020) *Plastics – the facts 2020*. Online. Available HTTP: <https://www.plasticseurope.org/en/resources/publications/4312-plastics-facts-2020>.

Principles for Responsible Investment (2019) *The plastics landscape: Risks and opportunities along the value chain*. Online. Available HTTP: <https://www.unpri.org/plastics/risks-and-opportunities-along-the-plastics-value-chain/4774.article>.

Qu, S. et al. (2019) "Implications of China's foreign waste ban on the global circular economy", *Resources, Conservation and Recycling*, 144: 252–255.

Sasaki, S. (2020) "The effects on Thailand of China's import restrictions on waste: measures and challenges related to the international recycling of waste plastic and e-waste", *Journal of Material Cycles and Waste Management*, 1–7, DOI: 10.1007/s10163-020-01113-3.

Sicotte, D. M. (2020) "From cheap ethane to a plastic planet: Regulating an industrial global production network", *Energy Research and Social Science*, 66: 1–6.

Sicotte, D. M. and Seamon, J. L. (2020) "Solving the plastics problem: Moving the U.S. from recycling to reduction", *Society & Natural Resources*, 1–10, DOI: 10.1080/08941920.2020.1801922.

Spierling, S. et al. (2018) "Bio-based plastics – a review of environmental, social and economic impact assessments", *Journal of Cleaner Production*, 185: 476–491.

Standardization Administration of the People's Republic of China (SAC) GB16288 (2008).

Statistica 2021 (2020) *Distribution of plastic consumption worldwide in 2016, by region*. Online. Available HTTP: <https://www.statista.com/statistics/1002005/distribution-plastic-consumption-worldwide-by-region/>.

The Ellen MacArthur Foundation (2019) *A vision of a circular economy for plastic*. Online. Available HTTP: <https://www.newplasticseconomy.org/assets/doc/npec-vision.pdf>

The Pew Charitable Trusts and Systemiq (2020) *Breaking the plastic wave*. Online. Available HTTP: <https://www.pewtrusts.org/en/research-and-analysis/articles/2020/07/23/breaking-the-plastic-wave-top-findings>.

UNEP (2019) *The role of packaging regulations and standards in driving the circular economy.* Online. Available HTTP: <https://www.unenvironment.org/resources/report/role-packaging-regulations-and-standards-driving-circular-economy>.

United Nations Environment Programme (UNEP) (2011) *The Basel Convention on the control of transboundary movement of hazardous wastes and their disposal (Text and Annexes).*

Van Zeben, J. and Rowell, A. (2021) *A guide to EU environmental law.* Oakland: University of California Press.

Villarrubia-Gómez, Cornell, S. E. and Fabres, J. (2018) "Marine plastic pollution as a planetary boundary threat – The drifting piece in the sustainability puzzle", *Marine Policy*, 96: 213–220.

Vince, J. and Hardesty, B. D. (2018) "Governance solutions to the tragedy of the commons that marine plastics have become", *Frontiers in Marine Science*, 5: 214.

Woldemar d'Ambrières (2019) "Plastics recycling worldwide: Current overview and desirable changes", *Field Actions Science Reports*. Online. Available HTTP: <http://journals.openedition.org/factsreports/5102>.

Yang, H., Ma, M., Thompson, J. R. and Flower, R. J. (2018) "Waste management, informal recycling, environmental pollution and public health", *Journal of Epidemiology and Community Health*, 72: 237–243.

Yang, S. (2020) "Trade for the environment: Transboundary hazardous waste movements after the Basel Convention", *Review of Policy Research*, 37: 713–738.

Zapata, P. and Zapata Campos, M. J. (2015) *Producing, appropriating and recreating the myth of the urban commons.* London: Routledge.

Zheng, J. and Suh, S. (2019) "Strategies to reduce the global carbon footprint of plastics", *Nature Climate Change*, 9: 374–378.

Zhou, G. *et al.* (2020) "A systematic review of the deposit-refund system for beverage packaging: Operating mode, key parameter and development trend", *Journal of Cleaner Production*, 251: 1–13.

38

Water, rivers and wetlands

Governance paradigms and principles

Edward Challies and Jens Newig

Water is essential to human activity across multiple sectors. Agriculture, energy generation, industry, and rural and urban settlements, for example, are highly reliant on water. Even more fundamentally, the ecosystem services that underpin human collective wellbeing depend on clean water and the integrity of the hydrological cycle (UNEP 2009). Without the protection of water resources and associated ecosystems there can be no sustainable development or economic growth within ecological limits. However, water resources are increasingly under pressure as a result of human activity (United Nations 2018; UN Water 2020).

Recognition of the centrality of water to human wellbeing, and the complexity of interacting drivers of change, which often escape the reach of local or national governing institutions, has led increasingly to calls for integrated management and multilevel governance of water resources (Pahl-Wostl 2015). Perhaps most prominently, the 2030 Agenda for Sustainable Development, through inclusion of Sustainable Development Goal (SDG) 6 on "Water and Sanitation," and in convening the International Decade for Action on "Water for Sustainable Development" (2018–2028), positions integrated water resource management as vital to "peace and prosperity" at the planetary scale (United Nations 2018).

This chapter examines policy and governance institutions that seek to address the pressing water issues facing the world. It describes major paradigms of water governance before discussing international expert networks, key actors, formal agreements, and the substance of policy and governance.

Paradigms of water governance

Ideas on how to collectively govern water have been crystallizing in what have become known as major paradigms of water governance. Informed by debates on sustainability, democracy and development, labels such as "integrated water resources management," "adaptive water governance," "collaborative water governance" and "water security" have served to channel often-competing discourses and form guiding principles of water policy and management. The respective paradigms simultaneously imply a particular problem framing and suggest a particular set of solutions. As such, they have served as important agenda-setters for political action at different scales. Paradigms thus function as symbolic, "normative–cognitive ideas"

DOI: 10.4324/9781003008873-43

and focal points for joint action, whose circulation may even explain the enactment of certain local policies better than functional necessity or strategic considerations of the involved parties (Blatter and Ingram 2000).

Arguably the most influential, global paradigm in water governance has been *Integrated Water Resources Management (IWRM)*, defined by the Global Water Partnership as "the coordinated development and management of water, land and related resources, in order to maximize the resultant economic and social welfare in an equitable manner without compromising the sustainability of vital ecosystems" (GWP-TAC 2000: 22). From an IWRM perspective, fragmented and disjointed management of water in "sectors" is the primary barrier to sustainable water governance. IWRM therefore addresses underlying drivers of change across key water-interdependent social and economic sectors, as well as ecosystems. As a normative concept, IWRM draws on the "Dublin principles" of water and sustainable development, which emerged out of the 1992 International Conference on Water and the Environment, and stress: the importance and vulnerability of fresh water, its economic value, gender dimensions of water management, and the need for participatory approaches (WMO 1992).

Recognizing the recurrent failure of isolated and technology-oriented "solutions" to water governance issues, IWRM advocates a more holistic approach. Waterbodies are to be understood as "elements of broader and more complex socioecological systems" (Conca 2006), and their management should address ground and surface water systems, integrate planning across sectors and ecosystems, and engage multiple scales and levels of government (i.e., local, regional, national and transnational institutions) (Conca 2006). More generally, IWRM incorporates the principles of social equity see Chapter 26), economic efficiency and environmental sustainability (Butterworth et al. 2010; see Chapter 16).

Meanwhile, IWRM is said to have "become *the* discursive framework of international water policy – the reference point to which all other arguments end up appealing." Like the concept of sustainability, "IWRM combines intuitive reasonableness, an appeal to technical authority, and an all-encompassing character of such great flexibility that it approaches vagueness" (Conca 2006: 126–7). In the latter sense, IWRM has been called a "nirvana concept" (Molle 2008). Unresolved conflicts revolve around issues of public versus private governance (Bakker 2010; Kammeyer et al. 2020), water as an economic good versus a (human) right (Tremblay 2011), and the operationalization and monitoring of IWRM indicators (Petit 2016). Notwithstanding its ambiguities, there have been numerous attempts to implement IWRM. While its impact on water management in practice has been questioned for quite some time (Biswas 2004), the UN maintains an active program to monitor IWRM implementation toward SDG6, which indicates considerable variability across countries, and "medium low" implementation on average globally (see UNEP-DHI 2021).

Over the past two decades, *adaptive water management* has been increasingly promoted as a novel paradigm of water governance. Adaptive management (AM) was developed in the context of ecosystem management (Holling 1978) in response to limited understanding of the complex dynamics of ecosystems on the one hand, and their widespread degradation on the other. A key strategy is to implement policies as systematic experiments whose outcomes are monitored and evaluated to inform policy adaptation. Adaptive approaches, which aim to increase institutional capacity to respond to uncertainty and change, have been applied to water governance in the face of climate change (Pahl-Wostl 2007; Bruch 2009; Mysiak et al. 2010; see Chapter 32), figuring prominently in the United Nations World Water Development Report 4 on "managing water under uncertainty and risk" (WWAP 2012; see Chapter 19). Adaptive water management has been defined as "a systematic process for continually improving management policies and practices by learning from the outcomes of implemented management strategies" (Pahl-Wostl 2007: 51). Unlike IWRM, adaptive water management has been developed in a research context. "[I]t is concerned with organizational learning, whereas IWRM is concerned with transforming governance arrangements" (Medema et al. 2008). Extending from the core ideas of AM (especially systematic policy experimentation), adaptive water governance essentially adds public participation, polycentric governance arrangements and the river basin approach (Huitema et al. 2009; Cosens and Gunderson 2018).

Collaborative water governance has gained momentum over several decades (Lubell et al. 2009; Koontz and Newig 2014), and focuses on opening up decision-making to a wide range of stakeholder groups and the public. Collaborative approaches respond to the diversity of values and interests that must typically be accommodated in water management, and are expected to deliver better-informed decisions through (local) stakeholder input; education and capacity building among stakeholders and the public; conflict resolution; enhanced legitimacy of decisions; and thus more widely accepted decisions and improved implementation (Carr 2015; Newig et al. 2018). The effects of collaborative and participatory approaches, however, remain disputed, as decision-making may be ineffective or delayed, or it may be captured or manipulated by powerful interests or groups (Brisbois and de Loë 2016), and produce a wide range of outcomes depending on how processes are designed and run (Bell and Scott 2020; Yoder et al. 2020).

Although *water security* has been an issue for at least as long as IWRM, it has only emerged as a paradigm over the past decade (Cook and Bakker 2012; on environmental security, see Chapter 20), in response to the social and economic consequences of water resource depletion and degradation. Now regarded as a dominant concept in water resources management (Opperman et al. 2020), water security was introduced prominently at the Second World Water Forum in The Hague with the aim of

> ensuring freshwater, coastal and related ecosystems are protected and improved; that sustainable development and political stability are promoted; that every person has access to enough safe water at an affordable cost to lead a healthy and productive life; and that the vulnerable are protected from the risks of water-related hazards.

This essentially mirrors the aims of IWRM. However, a shift toward a more "neoliberal" as opposed to a "social-democratic" political model (Mollinga 2008) has been observed. In a recent review, Gerlak et al. (2018) find that although the quickly growing paradigm of water security is predominantly linked to issues of water quantity, it also embraces a diversity of governance frames across different regional contexts.

All of these paradigms have been subject to critique. In general critics have warned that adherence to any one paradigm of water governance risks promoting "universal remedies," which are bound to fail in real (local) settings (Ingram 2011). Calls are therefore made for context-tailored, perhaps "messy" solutions (Butterworth et al. 2010; Ingram 2011), and addressing water governance in diverse local settings or "problem-sheds" (Woodhouse and Muller 2017).

Actors and networks of experts

Actors in (global) politics of fresh water comprise international organizations such as the World Bank and United Nations (UN) organizations, nation-states (national governments), subnational governments and agencies, (transnational) corporations, nongovernmental organizations (NGOs), scientific experts, professional associations, and a wide range of other, partly intermediary actors (see chapters in Part III of this volume). In the field of water politics, bilateral or multilateral agreements on transboundary surface waters are commonplace (De Stefano et al. 2010), whereas truly international regimes are largely non-existent (as discussed below; on regimes, see Chapter 9). The principal actors in water politics are therefore not necessarily unitary states, but rather private companies, (subnational) agencies, river basin organizations, NGOs, or scientific communities, such that policymaking is a "complex web of interactions…without a central actor or arena for decisionmaking processes" (Blatter and Ingram 2000: 470).

At the global level, water-related claims and competencies are spread across a multitude of international bodies. More than 20 UN and related sub-organizations claim authority in freshwater matters, including the World Health Organization, the Food and Agriculture Organization, the World Meteorological Organization, and the UN Development and Environment Programmes, each with a different focus (Conca 2006). Commissioned in 2003, UN-Water has been established as a coordinating body in the UN framework (Baumgartner and Pahl-Wostl 2013).

Arguably, the most influential players in global water politics and policy in a development context are the large donor organizations, first and foremost the World Bank (Goldman 2007). It has funded major infrastructure projects in developing countries (e.g., in hydropower and irrigation), including on international watercourses. By means of loan conditionality, the World Bank has effectively been issuing water policy for over 50 years, starting with its 1956 policy for "Projects on International Inland Waterways" (Salman 2011: 596), and has pushed privatization as a precondition for lending. Moreover, the World Bank has been driving institutional reforms on participation, mandating water user associations in Mexico, India and China, for example (Wilder and Romero Lankao 2006; Wang et al. 2010). Other influential donor organizations include the European Union and its "EU Water Initiative" (Fritsch et al. 2017) and the Asian Development Bank with its "Water Financing Program," which has, alongside the World Bank, been influential in implementing IWRM (Allouche 2016).

More recently, scholarly attention has turned to the role of policy intermediaries (Moss et al. 2009) and social entrepreneurs (Meijerink and Huitema 2009; Partzsch and Ziegler 2011), highlighting how a myriad of different actors exercise agency beyond the nation-state. As units of governance shift from the nation-state (e.g., to basin-wide governance or supranational structures, as is the case with the EU), new actors emerge, and extant actors adapt their "scalar" strategies to a restructured multilevel governance landscape (Moss and Newig 2010).

Freshwater policy and politics have to a considerable degree been shaped by international expert networks – perhaps more so than via intergovernmental collaboration and codified regimes (Conca 2006; see Chapter 18). Many of these water policy networks formed at a time when ecological concerns were less salient and debates were dominated by issues of access to safe drinking water, sanitation and irrigation (Conca 2006). Following the 1977 UN Water Conference in Mar del Plata, Argentina, a global network of water experts emerged that has since gradually brought increasing attention to issues of (surface) water quality and biodiversity of water-related habitats.

Global expert networking has been institutionalized in essentially two different ways: routinization of global water-related conferences, and organization-building. Much of this professional networking activity has been related to the development, rise and consolidation of the IWRM concept (Conca 2006).

Influential in the global exchange of technical expertise have been the International Commission on Large Dams (ICOLD) and the International Commission on Irrigation and Drainage (ICID). Founded in 1928, ICOLD addresses issues of dam construction and safety, but increasingly also environmental impacts of large dams, holding triennial international congresses and other workshops. ICID was founded in 1950, initially to promote large-scale irrigation. It has since broadened its focus to include sustainable irrigated agriculture, flood management, economics, and ecological and social issues (Conca 2006: 85; see Chapter 44). Following these earlier entities, the non-profit International Water Resources Association (IWRA) has, since its foundation in 1971, promoted an integrated perspective on water issues, including ecological, social and economic aspects.

On the initiative of the IWRA, the World Water Council (WWC) was founded in 1996, following an initial call in the 1992 Dublin Declaration for a forum to unite "private institutions, regional and non-governmental organizations along with all interested governments" (WMO 1992). The WWC has played a key role in structuring the field of global water politics, and currently comprises more than 400 member organizations from 60 countries, including ministries, international organizations, private enterprises and professional networks (Subramaniam 2018). Its mission is to provide an "umbrella organization to raise the profile of freshwater issues globally, provide expertise and authoritative recommendations, and undertake periodic assessments of the world water situation" (Conca 2006: 146).

Also in 1996, the Global Water Partnership (GWP) was founded by the World Bank, the UNDP, and the Swedish International Development Agency. It seeks to promote the Dublin principles, and in particular IWRM. In 2002, the GWP was split into a GWP *network* and a GWP *organization*. In a 2004 joint memorandum of understanding, the GWP and WWC committed themselves to promoting governance of global water issues, with private sector participation, privatization and cost recovery as important principles.

These influential associations have contributed to institutionalizing global water conferences as fora for expert and political exchange. Notably, the WWC initiated the triennial World Water Forum (WWF), which began in 1997 in Marrakech (see Conca 2006; Subramaniam 2018 for comprehensive overviews). These conferences attract increasing numbers of participants, with more than 100,000 attendees from 172 countries in Brasília in 2018. Another notable event is World Water Week, which has been held annually in Stockholm since 1991, attracting around 2,000 experts. It is not easy to assess what these mega-events have actually accomplished. Clearly, they have facilitated exchange of ideas among experts, policymakers and civil society (Subramaniam 2018). They have also expressed shared perspectives on freshwater governance via issuance of ministerial declarations. For instance, an ecosystem-centric view of water governance is relatively prominent in the WWF declarations from Istanbul and Marseille. Moreover, these fora have arguably served to legitimize the work of expert networks. However, the sheer size of the events does not imply that they are representative. In fact the WWF is paralleled by an Alternative World Water Forum, organized by social movements, trade unions, NGOs, indigenous groups, citizens and elected representatives, in opposition to what is seen as hegemonic and industry-dominated global water governance (Maganda 2010; Subramaniam 2018); both the WWF and the Alternative WWF were last held in Brasília in 2018.

A notable episode in global expert networking was the work of the World Commission on Dams (WCD). In response to increasing criticism of large dam-building projects, it was founded in 1998 by the World Bank and the International Union for Conservation of Nature (IUCN), and in 2000 produced a widely received (critical) report on large dams. It consisted of representatives from the scientific community, civil society, the private sector, government agencies, and river basin authorities (WCD 2000). However, its recommendations were only partially taken up, and were in part outright rejected, for example by China (Schulz and Adams 2019).

International networks of experts on water management have played a complex role in water politics. As epistemic communities or "social learning networks" (Conca 2006: 125–6) they have contributed to a shift from a technocratic paradigm of water diversion and damming, toward a more integrated, and process-oriented approach to governing water. On the level of paradigms, these networks have been important in agenda-setting and diffusing concepts around IWRM into political arenas. However, the actual implementation of these concepts, and thus improvements in water management on the ground, has been limited (Conca 2006). Water expert networks as "institutionalized site[s] of normative struggle" (Conca 2006: 160) would require a certain political legitimacy in order to effectively impact on political agendas. Although these networks are unlikely to attain full democratic legitimacy, they could attain a degree of legitimacy if they were to fairly represent relevant stakeholders and achieve consensus and/or effective problem-solving. These criteria, however, are only partly achieved; participation of a broader public and non-technical stakeholders remains a desideratum.

Transnational and international collaboration: water law and formal agreements

Formal rules on water – water law – are almost as old as human civilization itself. Indigenous peoples have long employed sustainable forms of water resources management, often integrated within holistic belief systems, customary law, and traditions of environmental stewardship. Ancient civilizations emerged in part because they were able to tame floods and manage irrigation through centralized bureaucracies – the "hydraulic state" (Wittfogel 1957). Modern water law has mainly developed within territorial states, with occasional cross-border policy transfers. Whereas formal transboundary collaboration has a long tradition, global water law in terms of international agreements is almost non-existent (on international environmental law, see Chapter 10). The EU, however, has developed a consolidated supranational water governance regime.

Dellapenna and Gupta (2008) have identified five major trends in national water law:

> National water law systems (1) have long histories and are contextual in nature…; (2) are more coherent and integrated in developed countries and more pluralistic in developing countries; (3) cover similar subjects; (4) have increasingly integrated environmental issues since 1972; and (5) have gradually welcomed stakeholder participation as well as private sector participation since the 1980s.
>
> *(Dellapenna and Gupta 2008: 440)*

Diffusion of regulatory concepts is, however, not uncommon. National water law can adopt broad concepts (such as IWRM) from the international arena, but (elements of) water law may also be exported by force (e.g., colonization) or through collaboration or active demand on the part of the "importer." For the importing jurisdiction, such "legal transplants" may have the advantage that they are already coherent regulations, but they also often fail to fit to given local contexts.

Transboundary collaboration includes common rules arising out of bilateral and regional negotiations, joint river basin institutions, and multilateral law. Worldwide, some 260 major international rivers exist (Wolf et al. 2002). The Transboundary Freshwater Dispute Database (http://transboundarywaters.science.oregonstate.edu) lists more than 600 agreements, covering the years 1820 to 2007. Notable examples of transboundary river basin collaboration include the Great Lakes Regime, the Great Lakes–St. Lawrence River Basin Water Resources Compact (Schulte 2012), the International Commission for the Protection of the Rhine (e.g., Mostert 2009) and the Mekong River Commission (Dore and Lebel 2010; Suhardiman et al. 2012). Transboundary collaboration, it has been observed, is increasingly the result of network governance, involving sub-national agencies and non-state actors, rather than unitary states (Blatter and Ingram 2000; Sayles and Baggio 2017).

As to the effects of transboundary collaboration, research suggests that the resultant information sharing and capacity building among government agencies are even more important determinants of pollution reduction than specific transboundary measures (Bernauer and Moser 1996). In the case of upstream–downstream water pollution, Bernauer and Kuhn (2010) found that while international treaties do help prevent transboundary pollution, free-riding remains prevalent and cooperation is limited.

Global and international law in general certainly has had important impacts on national water law and transboundary agreements (Dellapenna and Gupta 2008), but global or international law on water is largely missing. One notable exception is the Ramsar Convention on Wetlands (1975), which promotes "the conservation and wise use of all wetlands through local and national actions and international cooperation" (Ramsar Convention Secretariat 2016: 2). Covering a wide range of wetland systems, the convention has formally identified 2,416 sites for protection across its 171 signatory states (RSIS 2021). Furthermore, states undertake to collaborate in the management of transboundary wetlands and shared species.

One attempt at a global water regime is the United Nations Watercourses Convention of 1997, which entered into force in 2014. As an early piece of global water law, it is rather "conservative" and hardly considers ecological issues, but rather reiterates the 1996 "Helsinki Rules" on the Uses of the Waters of International Rivers (Dellapenna and Gupta 2008). In this context, the International Law Commission (ILC) has played a key role, in that it drafted the 1997 convention. It also produced the 2008 draft articles of a Law of Transboundary Aquifers. While these draft articles have been formally annexed to a United Nations General Assembly Resolution (Eckstein and Sindico 2014), there is still debate as to whether they should be legally adopted by the UN, and literature has so far remained sparse on the topic (Devlaeminck 2021). So despite their paramount importance to water provision and ecosystems, transboundary aquifers have hardly been treated by international law, with some regional exceptions, for example, the multipartite transboundary collaboration on the South American Guarani Aquifer System (Sindico et al. 2018) or three agreements on the Nubian Sandstone Aquifer System in North Africa (Burchi 2018). In the absence of an enforceable global framework for sustainable water governance, the International Law Association (an international NGO) drafted the "Berlin Rules on Water Resources" in 2004. According to Dellapenna and Gupta (2008), the Berlin Rules, which integrate domestic and international water law, are comprehensive in their coverage of the aquatic environment, also involving one of the first attempts to specifically address groundwater and its distinct characteristics. The key ingredient in (attempts at) codified international water law, such as in the Watercourses Convention and the Berlin Rules, is the "equitable use principle," maintaining that international natural (water) resources be shared equally, and that the rights of riparian states to a share of the water resource be respected (Dellapenna 2019).

A prominent example of an international "regime-like" structure is the EU supranational water policy, embodied by the EU Water Framework Directive (WFD), passed in 2000.

The WFD introduced a Europe-wide ecological goal of good water status (which was to be attained by 2015, with possible extensions), governance principles of river basin management and participation, and economic principles of pricing and cost recovery. Indeed these provisions had notable impacts on national water policy in the member states as regards river basin management and public participation (Jager et al. 2016). Pursuing a "mandated participatory planning" approach, the EU requires its member states to produce river basin management plans and programs of measures, which are to be updated every six years, drawing on the participation of citizens and organized stakeholder groups (Newig and Koontz 2014).

Given the heterogeneity of its member states, implementation of the WFD has varied considerably across the EU (Liefferink et al. 2011; Jager et al. 2016). In order to achieve coherent WFD implementation, a EU-wide "Common Implementation Strategy" was established, which, with the participation of national administrations and non-state organizations, produced dozens of "guidance documents" (DG Environment 2006). Attempts have even been made to transfer elements of the WFD to other contexts such as China (Ravesteijn et al. 2009) (see above on policy transfer).

While hopes and expectations associated with the WFD and its governance approach had been enormous, actual achievements appear to have been at best moderate. The European Commission's expectation that citizen participation is imperative to achieve environmental objectives has, by and large, not been confirmed (Rimmert et al. 2020). In substantive terms, the Directive has fallen short of its potential (Voulvoulis et al. 2017). In particular, pressing issues of nitrate pollution have not yet been tackled effectively (Wiering et al. 2020).

Principles: the substance of policy and governance

Facing issues of water depletion and degradation, the impairment of water-related ecosystems, and the compounding impacts of climate change (UN Water 2020), political programs are expected to respond effectively. On the substantive level, the following strategies and instruments are being employed:

- *Review of baselines, and monitoring.* Policies such as the WFD demand a review of the status of waters and the extent to which they are at risk; the assessment of environmental flow requirements (Arthington et al. 2018); and the monitoring of improvement (or deterioration). Such strategies are typically employed in developed countries with well-functioning public administrations and the technical means for gathering the requisite data.
- *Pollution control.* Initially, industrialized countries mostly employed an emissions-control approach, addressing pollution at source. This was effective for point sources such as industrial discharges to rivers and lakes. Likewise, wastewater treatment plants were built and/or upgraded. Increasingly, the emissions-based approach has been complemented with an *immission-* or quality-based approach, focusing on the actual status of water bodies, and taking into consideration possible antagonistic effects of multiple polluters.
- *River and wetland restoration.* Increasingly, the ecological value of freshwater systems is recognized (Darwall et al. 2018). While in many modern industrialized countries, water quality has much improved in the past decades, many rivers remain in a highly modified state, so river restoration – and also wetland conservation and restoration – remains an important issue (Nilsson et al. 2005). Where water quantity is an issue, a variety of *water saving* strategies are pursued, mostly through targeting efficiency gains in industry, private households or agriculture.

- *Economic instruments.* These include water pricing, pollution taxes and water quality trading. The first two are common practice in many countries. However, water pricing in the sense of full recovery of costs related to water supply and purification and wastewater treatment is still not common practice. As water is so undervalued in many places, metering of household consumption is also not common globally. Water quality trading is a relatively new development. Promoted in particular by the United States Environmental Protection Agency, it has expanded considerably since the late 1990s (Morgan and Wolverton 2008), alongside related instruments like wetland mitigation banking (Galik and Olander 2018) and payments for freshwater ecosystem services (Hawkins et al. 2009).
- *Planning instruments.* These provide the opportunity for broader consideration of freshwater resources within geographic areas, possibly in conjunction with other sectoral demands (as per IWRM). Land-use planning regarding agriculture or settlements (e.g., use of water for irrigation, draining of wetlands, deforestation; see Chapter 42), energy (e.g., dam construction; see Chapter 33) and navigation/transportation (canal and river regulations) has enormous impacts on water resources and related ecosystems.

Governance principles relate to the ways in which, and the scalar levels at which, substantive policy decisions are taken. They include:

- *River basin planning and management.* The river drainage basin, along with hydrological subunits such as watersheds and catchments, has long been advocated as the optimal spatial unit by which to govern water. The main rationale is that cross-border pollution spillovers – upstream–downstream in particular – can best be addressed if water management decisions are jointly taken by those who inhabit the basin. This has been described as achieving institutional "fit" between the natural (hydrological) scale and the governance scale (Moss and Newig 2010). Collaboration across borders within basins is a particularly important issue (Sabatier et al. 2005; Lubell et al. 2009). However, the principle of river basin management has also been challenged, because for some of the most pressing problems of diffuse (groundwater) pollution, hydromorphological issues or wetland quality, the surface drainage basin is not necessarily definitive (Benson and Jordan 2010; Ingram 2011).
- *Decentralization.* Whereas river basin management implies a spatial governance unit large enough to integrate possible pollution spillovers, decentralization calls for smaller, more local units of governance. This is assumed to increase ownership by local stakeholders, thereby allowing for local self-governance (e.g., in water user associations), and more flexible technical infrastructure such as decentralized wastewater treatment (WWAP 2012). In spite of widespread decentralization over the past three decades, there is evidence of a re-centralization of control over certain water governance functions in many countries in recent years (Birkenholtz 2015; Stuart-Hill and Schulze 2017).
- *Privatization.* This mostly refers to putting water services in the hands of private companies in order to achieve more efficient water service provision. Privatization has been strongly advocated by the World Bank and private companies (Finger 2005; Bakker 2010). As a prerequisite, privatization also involves water pricing (see above). The effects of privatization have been much discussed and much criticized, partly from more general anti-capitalist and anti-globalization perspectives (see Chapter 24), and partly on the basis of empirical evidence that private companies have not succeeded in more efficient and equitable water provision, especially in developing countries where the concept has been most strongly promoted (Page 2005; Braadbaart 2007; Bakker 2010).

Conclusion

Bogardi and colleagues have noted that "the global 'water crisis' is ultimately a 'governance crisis' extending from the local to the planetary scale" (Bogardi et al. 2012). Clearly, the status of the world's ground and surface waters crucially depends on the way governance institutions are crafted and implemented, as well as the processes that lead to the creation of these institutions (Garrick et al. 2017). National water laws have matured to a large extent, in many places now incorporating ecological and sustainability aspects of water management. Transboundary collaboration on shared surface waters and wetlands is now commonplace. However, the development and implementation of such institutions in global comparison largely depend on the cultural and economic context. Moreover, global law on fresh water is still relatively underdeveloped.

Over the years, a multitude of guiding principles for (transboundary) water management and governance has been proposed, originating typically from research and expert networks. The crucial question, though, is whether IWRM, adaptive water governance, privatization, decentralization and collaborative governance actually live up to the promise they hold. The promotion of "universal remedies" has drawn criticism from many sides (Ingram 2011), and prompted calls for more context-tailored governance solutions that "fit" the respective local context (Conca 2006; Moss 2012). Yet knowledge of how to tailor to local contexts is sparse as well. Often the key contextual factors that impact on success or failure of water governance institutions are simply not known. Comparative research on the shaping and implementation of transboundary, multi-, supra- and international freshwater governance should aim to identify the boundary conditions under which key governance principles and approaches work. Given the prevailing ambiguities surrounding many of the concepts of water governance discussed here, research should work toward clear and commonly shared conceptual understandings, such that knowledge cumulation is facilitated. Careful analysis that is attentive to local contexts will help develop evidence-based best practice in water governance, and in doing so contribute to more socially just and ecologically sustainable water management.

References

Allouche, J. (2016) "The birth and spread of IWRM: a case study of global policy diffusion and translation." *Water Alternatives* 9(3): 412–433.

Arthington, A.H., A. Bhaduri, S.E. Bunn, S.E. Jackson, R.E. Tharme, D. Tickner, . . . S. Ward (2018) "The Brisbane Declaration and Global Action Agenda on Environmental Flows (2018)." *Frontiers in Environmental Science* 6(45): 1–15. https://doi.org/10.3389/fenvs.2018.00045

Bakker, K. (2010) *Privatizing Water: Governance Failure and the World's Urban Water Crisis.* Ithaca, NY: Cornell University Press.

Baumgartner, T. and C. Pahl-Wostl (2013) "UN-Water and its role in global water governance." *Ecology and Society,* 18(3): 16. http://dx.doi.org/10.5751/ES-05564-180303

Bell, E. and T.A. Scott (2020) "Common institutional design, divergent results: a comparative case study of collaborative governance platforms for regional water planning." *Environmental Science and Policy* 111: 63–73.

Benson, D. and A. Jordan (2010) "The scaling of water governance tasks: a comparative federal analysis of the European Union and Australia." *Environmental Management* 46(1): 7–16.

Bernauer, T. and P. Kuhn (2010) "Is there an environmental version of the Kantian peace? Insights from water pollution in Europe." *European Journal of International Relations* 16(1): 77–102.

Bernauer, T. and P. Moser (1996) "Reducing pollution of the River Rhine: the influence of international cooperation." *Journal of Environment and Development* 5(4): 389–415.

Birkenholtz, T.L. (2015) "Recentralizing groundwater governmentality: rendering groundwater and its users visible and governable." *WIRES Water* 2: 21–30.

Biswas, A.K. (2004). "Integrated water resources management: a reassessment." *Water International* 29(2): 248–56.

Blatter, J. and H. Ingram (2000) "States, markets and beyond: governance of transboundary water resources." *Natural Resources Journal 40*: 439–73.

Bogardi, J.J., D. Dudgeon, R. Lawford, E. Flinkerbusch, A. Meyn, C. Pahl-Wostl, et al. (2012) "Water security for a planet under pressure: interconnected challenges for a changing world call for sustainable solutions." *Current Opinion on Environmental Sustainability* 4(1): 35–43.

Braadbaart, O. (2007) "Privatizing water: The Jakarta concession and the limits of contract." In P. Boomgaard (ed) *A World of Water: Rain, Rivers and Seas in Southeast Asian Histories*, 297–320. Leiden: KITLV Press.

Brisbois, M.C. and R.C. de Loë (2016) "State roles and motivations in collaborative approaches to water governance: A power theory-based analysis." *Geoforum* 74 (1): 202–212.

Bruch, C. (2009). "Adaptive water management: strengthening laws and institutions to cope with uncertainty." In A.K. Biswas, C. Tortajada, and R. Izquierdo (eds) *Water Management in 2020 and Beyond*, 89–113. Berlin: Heidelberg.

Burchi, S. (2018) "Legal frameworks for the governance of international transboundary aquifers: pre- and post-ISARM experience." *Journal of Hydrology: Regional Studies* 20: 15–20.

Butterworth, J., J. Warner, P. Moriarty, S. Smits, and C. Batchelor (2010) "Finding practical approaches to integrated water resources management." *Water Alternatives* 3(1): 68–81.

Carr, G. (2015) "Stakeholder and public participation in river basin management—an introduction." *WIRES Water* 2(4): 393–405.

Conca, K. (2006) *Governing Water. Contentious Transnational Politics and Global Institution Building*. Cambridge, MA: MIT Press.

Cook, C. and Bakker, K. (2012) "Water security: debating an emerging paradigm." *Global Environmental Change 22*(1): 94–102.

Cosens, B. and L. Gunderson (Eds.) (2018) *Practical Panarchy for Adaptive Water Governance*. Cham: Springer.

Darwall, W., V. Bremerich, A. De Wever, A.I. Dell, J. Freyhof, M.O. Gessner, . . . O. Weyl (2018) "The Alliance for Freshwater Life: A global call to unite efforts for freshwater biodiversity science and conservation." *Aquatic Conservation: Marine and Freshwater Ecosystems* 28(4): 1015–1022.

De Stefano, L., J. Duncan, S. Dinar, K. Stahl, K. Strzepek, and A.T. Wolf (2010) *Mapping the Resilience of International River Basins to Future Climate Change-Induced Water Variability*. Water Sector Board Discussion Paper Series No. 15. Washington, DC: World Bank.

Dellapenna, J.W. (2019) "The work of international legal expert bodies." In S.C. McCaffrey, C. Leb and R.T. Denoon (eds.), *Research Handbook on International Water Law*, 26–43. Cheltenham: Edward Elgar.

Dellapenna, J. and Gupta, J. (2008) "Toward global law on water." *Global Governance* 14(4): 437–453.

Devlaeminck, D.J. (2021) "Reassessing the draft articles on the law of transboundary aquifers through the lens of reciprocity." *International Journal of Water Resources Development* 37(1): 162–177.

DG Environment (2006) *Common Implementation Strategy for the Water Framework Directive (2000/60/ EC). Improving the Comparability and Quality of the Water Framework Directive Implementation. Progress and Work Programme for 2007 and 2009*. Environment Directorate-General, European Commission.

Dore, J. and L. Lebel (2010) "Deliberation and scale in Mekong region water governance." *Environmental Management* 46(1): 60–80.

Eckstein, G., and F. Sindico (2014) "The law of transboundary aquifers: many ways of going forward, but only one way of standing still." *Review of European, Comparative & International Environmental Law* 23(1): 32–42.

Finger, M. (2005) "The new water paradigm: the privatization of governance and the instrumentalization of the state." In D.L. Levy and P.J. Newell (eds) *The Business of Global Environmental Governance*, 275–304. Cambridge, MA: MIT Press.

Fritsch, O., C. Adelle and D. Benson (2017) "The EU water initiative at 15: origins, processes and assessment." *Water International* 42(4): 425–442.

Galik, C.S., and L.P. Olander (2018) "Facilitating markets and mitigation: a systematic review of early-action incentives in the U.S." *Land Use Policy* 72: 1–11.

Garrick, D.E., J.W. Hall, A. Dobson, R. Damania, R.Q. Grafton, R. Hope, C. Hepburn, R. Bark, F. Boltz, L. De Stefano, E. O'Donnell, N. Matthews and A. Money (2017) "Valuing water for sustainable development." *Science* 358(6366): 1003–1005.

Gerlak, A.K., L. House-Peters, R.G. Varady, T. Albrecht, A. Zúñiga-Terán, R.R. de Grenade, C. Cook and C.A. Scott (2018) "Water security: a review of place-based research." *Environmental Science and Policy* 82: 79–89.

Goldman, M. (2007) "How 'water for all!' policy became hegemonic: the power of the World Bank and its transnational policy networks." *Geoforum* 38(5): 786–800.

Hawkins, S., H. Murray and T. Greiber (2009) "Understanding water-related ecosystem services." In T. Greiber (ed) *Payments for Ecosystem Services: Legal and Institutional Frameworks*, 5–10. Gland: IUCN.

GWP-TAC. (2000) "Integrated water resources management." *TAC Background Papers*. Stockholm: Global Water Partnership.

Holling, C.S. (ed.) (1978) *Adaptive Environmental Assessment and Management*. Chichester: Wiley.

Huitema, D., E. Mostert, W. Egas, S. Moellenkamp, C. Pahl-Wostl, and R. Yalcin (2009) "Adaptive water governance: assessing the institutional prescriptions of adaptive (co-)management from a governance perspective and defining a research agenda." *Ecology and Society* 14(1): Art. 26.

Ingram, H. (2011) "Beyond universal remedies for good water governance". In A. Garrido and H. Ingram (eds) *Water for Food in a Changing World*, 241–261. Routledge: London.

Jager, N., E. Challies, E. Kochskämper, J. Newig, D. Benson, K. Blackstock, . . . Y. von Korff (2016) "Transforming European water governance? Participation and River Basin Management under the EU Water Framework Directive in 13 Member States." *Water* 8(4): 156.

Kammeyer, C., R. Hamilton and J. Morrison (2020) "Averting the global water crisis: three considerations for a new decade of water governance." *Georgetown Journal of International Affairs* 21: 105–113.

Koontz, T.M. and J. Newig (2014) "From planning to implementation: Top-down and bottom-up approaches for collaborative watershed management." *Policy Studies Journal* 42(3): 416–442.

Liefferink, D., M. Wiering, and Y. Uitenboogaart (2011) "The EU Water Framework Directive: a multidimensional analysis of implementation and domestic impact." *Land Use Policy* 28: 712–22.

Lubell, M., W.D. Leach, and P.A. Sabatier (2009) "Collaborative watershed partnerships in the epoch of sustainability." In D.A. Mazmanian and M.E. Kraft (eds) *Toward Sustainable Communities: Transition and Transformations in Environmental Policy*, 255–88. Cambridge, MA: MIT Press.

Maganda, C. (2010) "Water management practices on trial: the Tribunal Latinamericano del Agua and the creation of public space for social participation in water politics." In K.A. Berry and E. Mollard (eds) *Social Participation in Water Governance and Management: Critical and Global Perspectives*, 265–287. London: Earthscan.

Medema, W., B.S. McIntosh, and P.J. Jeffrey (2008) "From premise to practice: a critical assessment of integrated water resources management and adaptive management approaches in the water sector." *Ecology and Society* 13(2): Art. 29.

Meijerink, S. and D. Huitema (2009) "Water transitions, policy entrepreneurs and change strategies: lessons learned." In D. Huitema and S. Meijerink (eds) *A Research Companion to Water Transitions around the Globe*, 371–91. Cheltenham: Edward Elgar.

Molle, F. (2008) "Nirvana concepts, narratives and policy models: insights from the water sector." *Water Alternatives* 1(1): 131–156.

Mollinga, P.P. (2008) "Water, politics and development: framing a political sociology of water resources management." *Water Alternatives* 1(1): 7–23.

Morgan, C. and A. Wolverton (2008) "Water quality trading in the United States: trading programs and one-time offset agreements." *Water Policy* 10(1): 73–93.

Moss, T. (2012) "Spatial fit, from panacea to practice: implementing the EU Water Framework Directive". *Ecology and Society*, 17(3): 12. http://dx.doi.org/10.5751/ES-04821-170302

Moss, T. and J. Newig (2010) "Multilevel water governance and problems of scales: setting the stage for a broader debate." *Environmental Management* 46(1): 1–6.

Moss, T., W. Medd, S. Guy, and S. Marvin (2009) "Organising water: the hidden role of intermediary work." *Water Alternatives* 2(1): 16–33.

Mostert, E. (2009) "International co-operation on Rhine water quality 1945–2008: an example to follow?" *Physics and Chemistry of the Earth* 34(3): 142–149.

Mysiak, J., H.J. Henrikson, C. Sullivan, J. Bromley, and C. Pahl-Wostl (eds) (2010) *The Adaptive Water Resource Management Handbook*. London: Earthscan.

Newig, J., E. Challies, N.W. Jager, E. Kochskaemper, and A. Adzersen (2018) "The environmental performance of participatory and collaborative governance: a framework of causal mechanisms." *Policy Studies Journal* 46(2): 269–297.

Newig, J. and T.M. Koontz (2014) "Multi-level governance, policy implementation and participation: the EU's mandated participatory planning approach to implementing environmental policy." *Journal of European Public Policy* 21(2): 248–267.

Nilsson, C., C.A. Reidy, M. Dynesius, and C. Revenga (2005) "Fragmentation and flow regulation of the world's large river systems." *Science* 308(5720): 405–408.

Opperman, J.J., S. Orr, H. Baleta, D. Garrick, M. Goichot, A. McCoy, A. Morgan, R. Schmitt, L. Turley and A. Vermeulen (2020) "Achieving water security's full goals through better integration of rivers' diverse and distinct values." *Water Security* 10: 100063.

Page, B. (2005) "Paying for water and the geography of commodities." *Transactions of the Institute of British Geographers* 30(3): 293–306.

Pahl-Wostl, C. (2015) *Water Governance in the Face of Global Change: From Understanding to Transformation.* Cham: Springer.

Pahl-Wostl, C. (2007) "Transition towards adaptive management of water facing climate and global change." *Water Resources Management* 21(1): 49–62.

Partzsch, L. and R. Ziegler (2011) "Social entrepreneurs as change agents: a case study on power and authority in the water sector." *International Environmental Agreements: Politics, Law and Economics* 11(1): 63–83.

Petit, O. (2016) "Paradise lost? The difficulties in defining and monitoring Integrated Water Resources Management indicators." *Current Opinion in Environmental Sustainability* 21: 58–64.

Ramsar Convention Secretariat (2016) *An Introduction to the Ramsar Convention on Wetlands* (7th ed.). Gland: Ramsar Convention Secretariat.

Ravesteijn, W., X. Song, and R. Wennersten (2009) "The 2000 EU Water Framework Directive and Chinese water management: experiences and perspectives." *River Basin Management V* 124: 37–46.

Rimmert, M., L. Baudoin, B. Cotta, E. Kochskämper and J. Newig (2020) "Participation in river basin planning under the Water Framework Directive: Has it benefitted good water status?" *Water Alternatives* 13(3): 484–512.

RSIS (2021) Ramsar Sites Information Service. Gland: Ramsar Convention Secretariat, https://rsis.ramsar.org/

Sabatier, P.A., W. Focht, M. Lubell, Z. Trachtenberg, A. Vedlitz, and M. Matlock (eds) (2005) *Swimming Upstream. Collaborative Approaches to Watershed Management.* Cambridge, MA: MIT Press.

Salman, S.M.A. (2011) "The World Bank policy and practice for projects affecting shared aquifers." *Water International* 36(5): 595–605.

Sayles, J.S. and J.A. Baggio (2017) "Social-ecological network analysis of scale mismatches in estuary watershed restoration." *Proceedings of the National Academy of Sciences of the United States of America* 114(10): E1776–E1785.

Schulte, P. (2012) "The Great Lakes Water Agreement." In P.H. Gleick (Hrsg.) *The World's Water Volume 7: The Biennial Report on Freshwater*, 165–70. Washington, DC: Island Press.

Schulz, C., and W.M. Adams (2019) "Debating dams: The World Commission on Dams 20 years on." *WIREs Water*, 6(5), e1396.

Sindico, F., R. Hirata and A. Manganelli (2018) "The Guarani Aquifer System: From a beacon of hope to a question mark in the governance of transboundary aquifers." *Journal of Hydrology: Regional Studies* 20: 49–59.

Stuart-Hill, S. and R. Schulze (2017) "Reflections on the framework of water governance in South Africa." *New Water Policy & Practice* 3(1–2): 46–65.

Subramaniam, M. (2018) *Contesting Water Rights: Local, State, and Global Struggles.* Cham: Palgrave Macmillan.

Suhardiman, D., M. Giordano, and F. Molle (2012) "Scalar disconnect: the logic of transboundary water governance in the Mekong." *Society and Natural Resources* 25(6): 572–586.

Tremblay, H. (2011) "A clash of paradigms in the water sector? Tensions and synergies between integrated water resources management and the human rights-based approach to development", *Natural Resources Journal* 51: 307–356.

UNEP (2009) *Water Security and Ecosystem Services: The Critical Connection*, edited by World Water Assessment Programme (WWAP). Nairobi, Kenya.

UNEP-DHI (2021) *IWRM Data Portal: Tracking Global Progress on Implementation of Integrated Water Resources Management.* Hørsholm: UNEP-DHI Centre on Water and Environment, http://iwrm-dataportal.unepdhi.org/

UN Water (2020) *The United Nations World Water Development Report 2020: Water and Climate Change.* Paris: United Nations Educational, Scientific and Cultural Organization (UNESCO).

United Nations (2018) *Sustainable Development Goal 6: Synthesis Report on Water and Sanitation.* New York: United Nations.

Voulvoulis, N., K.D. Arpon and T. Giakoumis (2017) "The EU Water Framework Directive: from great expectations to problems with implementation." *Science of the Total Environment* 575: 358–366.

Wang, J., J. Huang, L. Zhang, Q. Huang, and S. Rozelle (2010) "Water governance and water use efficiency: the five principles of WUA management and performance in China." *Journal of American Water Resources Association* 46(4): 665–85.

WCD (2000) *Dams and Development: A New Framework for Decision-making.* London: World Commission on Dams (WCD); Earthscan.

Wiering, M., D. Liefferink, D. Boezeman, M. Kaufmann, A. Crabbé and N. Kurstjens (2020) "The wicked problem the Water Framework Directive cannot solve: The governance approach in dealing with pollution of nutrients in surface water in the Netherlands, Flanders, Lower Saxony, Denmark and Ireland." *Water* 12(5) 1–22. https://doi.org/10.3390/w12051240

Wilder, M. and P. Romero Lankao (2006) "Paradoxes of decentralization: water reform and social implications in Mexico." *World Development,* 34(11), 1977–1995.

Wittfogel, K.A. (1957) *Oriental Despotism: A Comparative Study of Total Power.* New Haven, CT: Yale University Press.

WMO (1992) *Dublin Statement on Water and Sustainable Development.* World Meteorological Organization, Dublin, Ireland, January 31. Available HTTP: <http://www.wmo.int/pages/prog/hwrp/documents/english/icwedece.html>.

Wolf, A.T., K. Stahl, and M.F. Macomber (2002) "Conflict and cooperation within international river basins: the importance of institutional capacity." *Water Resources Update* 125. Universities Council on Water Resouces.

Woodhouse, P. and M. Muller (2017). "Water governance—an historical perspective on current debates." *World Development* 92: 225–241.

WWAP (2012) *The United Nations World Water Development Report 4: Volume 1 – Managing Water under Uncertainty and Risk.* Paris: World Water Assessment Programme (WWAP); United Nations Educational, Scientific and Cultural Organization (UNESCO).

Yoder, L., A.S. Ward, S. Spak, and K.E. Dalrymple (2020) "Local government perspectives on collaborative governance: a comparative analysis of Iowa's Watershed Management Authorities." *Policy Studies Journal*: 1–23. https://doi.org/10.1111/psj.12389

39

Pollution and management of oceans and seas

Challenges in an unresponsive international system

Peter J. Jacques

The World Ocean System (WOS) is the ensemble of interconnected oceans and seas around the world. The WOS contains not only the sea surface – which makes up over 70 percent of Earth's surface, but also the water column, the sea floor and the sub-sea floor, all of which together provide the largest overall living space, habitat, on the planet (see, e.g., Haedrich, 1996). Indeed, most of Earth's "ecosystem services" come from marine systems (Costanza, 2000). Ecosystem services are direct and indirect benefits of environmental systems to people (Hassan et al., 2005) in four categories: provisioning, regulating, cultural and supporting services. For example, the World Ocean provides fish for us to eat (see Chapter 40), it regulates the climate (see Chapter 32), provides the basis of the hydrologic cycle that brings freshwater to us, and provides spiritual, educational, economic and recreational opportunities for people around the world. Perhaps most importantly, the World Ocean cycles nutrients and elements, such as carbon, thereby supporting all the other ecosystem services. Figure 39.1 shows all four ecosystem services, including supporting services and their benefits to humanity.

The World Ocean system

The World Ocean is a system of systems. Within the WOS, there are subsidiary ecological systems. These involve coastal systems, inter-coastal and inter-tidal systems, open-ocean and deep-ocean systems that all work together, driven by solar energy that warms the upper ocean layers and fuels primary production of green phytoplankton – the base of the marine food chain and the source for 40–50 percent of Earth's oxygen (Bigg, 2003). The WOS is also a space where ecological systems and social systems are integrated together. These marine social–ecological systems have stable conditions, but when they are disturbed enough, these stable conditions can be pushed beyond breaking points, or thresholds, that move that system into a different state (Gunderson and Holling, 2002). Clearly, the WOS cannot sustain an infinite amount of disturbance, such as pollution, before critical life support systems are undermined (on sustainability, see Chapter 15).

DOI: 10.4324/9781003008873-44

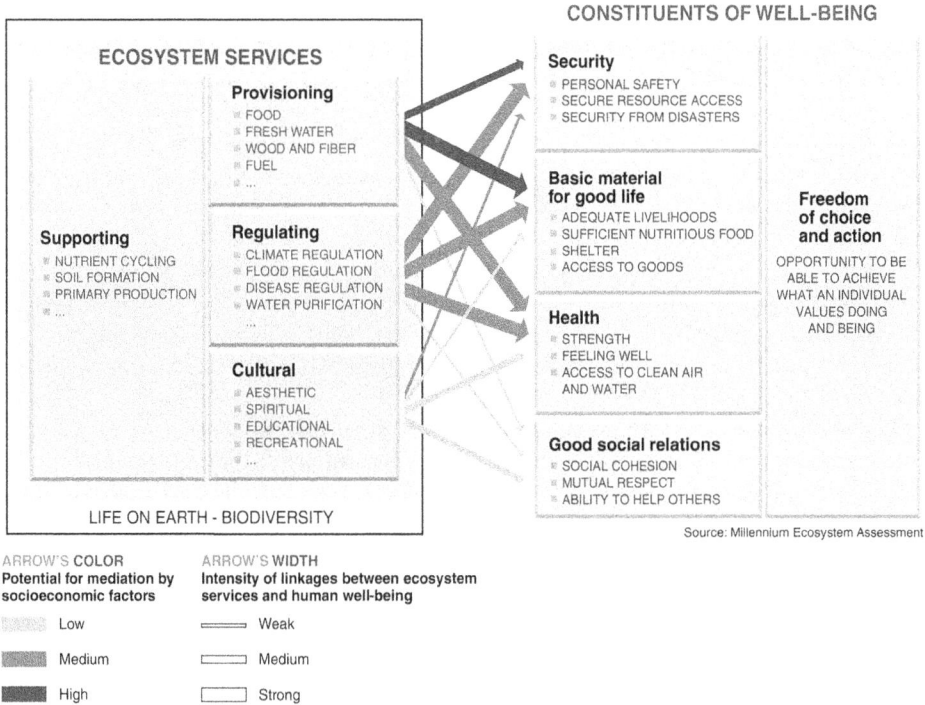

Figure 39.1 Ecosystem services and constituents of wellbeing

Source: Millennium Ecosystem Assessment (2005), https://www.millenniumassessment.org/en/GraphicResources.html.

There are three main explanations for ocean pollution. One is the institutional setting, or the types of regimes in place that provide pollution deterrence, incentives, and opportunities. Second, are values, or constellations of values – paradigms – that determine how people perceive the ocean. The "old coastal cultures" saw the ocean as a primal, living being with its own agency and will that affected their lives and to which they were responsible (Acheson and Wilson, 1996). It is less culturally acceptable to pollute "Mother Ocean" than an ocean that is, as the English jurist John Seldon described, "som[e] dull heap of matter that Nature could not bring to perfection" (Selden, 1972: 172). In contrast, today the WOS is perceived mainly as a highway and store for commerce, and certainly not as a "treasure of Mother Earth" (Jackson, 1993, Steinberg, 2001).

Combined with rules and paradigms are political economic causes of pollution, which have changed a great deal since World War II. Industrial production and consumption have since multiplied several times (see Chapter 17), which has magnified and expanded marine pollution in ways that still are not accounted for in governing regimes. From here, two kinds of marine pollution are explained. First are the "usual suspects" which are the pollutants with the longest histories, such as human sewage and which can be regulated in simple ways; then there are the "invisibles" which are relatively new pollutants stressing the WOS which are all but impossible to keep out of the WOS once they are produced. All of these pollutants, except sewage, are direct results of industrial production.

The "usual suspects" of marine pollution are oil, toxic and hazardous waste (see Chapter 36), and human sewage. This is only the beginning, though, because a whole new class of pollutants have become major concerns: carbon, nitrogen, phosphorous, heat and plastics (see Chapter 37). Most marine pollution originates on land, but this is amplified in this new class of pollutants, which are much harder, if not impossible, to control once they are produced. These trends fit three arguments in this chapter: (1) If protections are actually instituted they are almost always reactive, not anticipatory, even when the hazard is clear. (2) Most marine pollution is best explained by political economic structure, such as through global market forces and economic power (see Chapter 24). (3) The governing capacity of the international system has substantially weakened in the face of neoliberal capitalism starting in the 1980s.

Argument one is evident in that pollution regimes (see Chapter 9) only come after problems emerge. Some theorists might expect exactly this in rational actors maximizing their interests (see Chapter 3), not wanting to limit activity before there is a compelling reason. However, these reasons are clear well before the agreements are made and much damage, and cost, is incurred.

The second argument is seen in the increased burden on environmental systems by successive expressions of capitalism. While there are more environmental protections in place than ever before in history, we also have the most extractive, and polluting, political economic system in history, neoliberal capitalism, which abhors limits and regulation of economic activity. Neoliberal capitalism has not only produced more pollution (see Chapter 24), but more categories of pollution than ever before, for example as seen in relatively new pollutants, such as marine plastic (see below and Chapter 37). Consequently, marine pollution is at an unprecedented level, which threatens the WOS.

Argument three is demonstrated by the increasing failure of states to commit to adequate regulatory limits on pollution. This is consistent with "new governance theory" (Ruggie, 2014). Ruggie (2014) notes that "new governance" occurs where an "already weak system of global governance apparently becomes more so." Dimitrov (2005) argues that states purposefully create "hollow" institutions that appear to be governing environmental problems as a "decoy" *actually meant to impede governance*, an example being the United Nations Forum on Forests. Dimitrov (2019: 2) reflects that other "empty institutions," like the climate-related Copenhagen Accord, which have no "policy obligations for any country" and are "deliberately designed to not deliver substantive policy," are a growing problem for marine pollution, too.

Classical marine politics

The WOS serves as a link between people and time (Borgese, 1998), and the major periods of ocean politics also form major political–economic eras (Jacques, 2010). Political economy refers to the structural conditions that form markets, organize property and labor, direct trade, and shape power through a specific mode of production. A mode of production is a combination of forces that determine how *and why* things are produced and consumed. For example, subsistence modes of production focus on everyday needs, but capitalist modes focus on the accumulation of wealth.

The first era of world politics starts at least as far back to 8,000 BCE and extends to 1492, when the use of the oceans was primarily determined by local customs, rituals, taboos, and Indigenous knowledge. Human civilization consisted of villages, chiefdoms, principalities, some larger cities, and empires, but people mostly used the ocean at an artisanal or small scale (see, e.g., Acheson and Wilson, 1996). In this era, there were no global institutions for regulating the use of the oceans. During the second era, from 1492 to 1945, European countries used the oceans as highways to build empires that removed natural resources and people from other lands to enrich and empower European kingdoms (see Chapters 17 and 24). During this period of mercantile capitalism (sometimes called economic nationalism), the colonial system created the first global oceanic changes, and the first global ocean institution was established. The first human-caused global oceanic change was through whaling, as European empires relied on whale oil and other parts for energy and food. Colonial whaling fleets collapsed several whale populations and many whale populations have not been the same since (see Chapter 40). In addition to whaling, colonial fishing and hunting, significantly affected populations of green and hawksbill turtles, manatees (including the extinct Steller sea cow), and the extinct Caribbean monk seal, and entire estuary systems, such as the Atlantic North American coastline that saw serial collapses of oysters (Jackson, 2008). This period saw much damage to marine systems, but pollution was not a main driver until the post-war period (Lotze and McClenachan, 2014)

The third and current era is the post-World War II period of global market capitalism operating in a system of nation-states. Within this period, a "neoliberal" political economy developed in the 1980s. Neoliberal economics place more social power in the hands of industry and business and regulations that limit economic efforts are increasingly relaxed (Centeno and Cohen, 2012).

During the period of mercantile capitalism and colonialism, the principle of *mare liberum* (Latin for "freedom of the seas") was established by Hugo Grotius (1916) in the seventeenth century. *Mare liberum* posited that since no sovereign owned the seas, they were open to all nations to use for commerce and navigation. While it took some time for the proposition to be adopted by the international community and there was debate around the idea, ultimately the concept worked in favor of colonial ambitions and it would then come to dominate ocean politics for almost 400 years (Vieira, 2003, Russ and Zeller, 2003). This principle was based on the European system of natural law and the belief that true law was permanent and from God, and the belief that all peoples around the world should know and observe the tenets of natural law, such as the Law of Nations, regardless of where they are in the world or their cultural norms. The Law of Nations asserted that all national leaders, and ultimately ships with a national flag, had the right to travel and trade with other peoples.

As such, *mare liberum* was the first global ocean institution. *Mare liberum* created an "open regime" that lacked rules limiting access and use. It was not until the 1950s that any international limits to polluting the oceans were developed, and *mare liberum*, in part, explains why it took so long. *Mare liberum* is the reason why on the "high seas" (those areas beyond national jurisdiction) little international law existed before the post-War period (on international environmental law generally, see Chapter 10). Since *mare liberum* set up this open regime, there were not any substantial rules to limit what people did on the ocean, and thus there were no serious limits to polluting the ocean.

This open regime changed with the United Nations Convention on the Law of the Sea (UNCLOS), which eventually closed much of the open ocean by extending state control. UNCLOS went into force in 1994 (United Nations, 1982). The convention involved decades of negotiation through three United Nations Conferences on the Law of the Sea (CLOS) (on environmental diplomacy, see also Chapter 22). CLOS I was convened in 1958 to establish limits to national claims and boundaries because, after World War II, some countries announced broad marine claims that had no international consensus (Juda, 1996). The 1958 meeting codified *mare liberum* and the freedom to use and navigate the oceans. CLOS II met in 1960 because CLOS I did not settle the boundaries of coastal state jurisdictions, but failed to achieve any agreement at all. CLOS III, which was negotiated over several years (1973–82), produced a landmark treaty that established national sovereign control of coastal seas out to 12 nautical miles, and management of resources – but not sovereign control – out another 188 nautical miles in exclusive economic zones (EEZs). Beyond the EEZs is the "Area" where resources developed on and below the sea floor are considered to be the "Common Heritage of Mankind," the development of which were to be used to raise money to aid the world's least-developed countries.

Marine pollution: the "usual suspects"

Oil

The first of the usual suspects in ocean pollution is oil. While oil pollution makes headlines during a catastrophic shipwreck or spill, most oil pollution in the ocean is emitted from natural seeps in the ocean, and between the 1980s and the 1990s, oil pollution from tanker accidents declined from 140 million gallons per year to 17 million gallons per year (National Research Council, 2003). And, of the human sources of oil in the ocean, 90 percent comes from low-level leaks and runoff from extracting and refining oil (National Research Council, 2003). Thus, while media may focus on a "tanker on the rocks" the more important oil problems come from continuous contributions, such as an oil spill that lasted for about 40 years unabated on the coast of California (Beamish, 2002). In this case, workers, the oil company, *and regulators* knew the oil was being spilled into the ocean from an oil rig onto the beach, but no one identified it as a problem until a whistle-blower notified those same regulators.

This does not minimize the damage of oil spills because the oil released can stay in the seawater, coastal waterways, and shorelines for a very long time. For example, toxic oil from the 1989 *Exxon Valdez* spill in Alaska was found on local beaches decades later (Barron et al., 2020). In another case, oil from a spill off Massachusetts's coast was found in sediment a few centimeters under the surface in concentrations like those just after the "accident" more than 30 years afterward (Rabalais, 2003). Oil tanker spills and other releases into the ocean occur in an economic system that relies on oil. They are a persistent and "normal" part of that system and therefore should be anticipated by policymakers and diplomats.

The first attempts to regulate ocean oil pollution began in the 1920s but did not succeed until after the establishment of the International Maritime Consultative Organization (now simply the IMO), the international governmental organization responsible for coordinating shipping agreements. In 1954, the London Oil Pollution Convention (OILPOL) established rules to curb the release of oil by ships. At this point, ships carrying oil would purposefully release "slops," or the mixed oil and water from seawater stored in ballast tanks. Prior to OILPOL, an average of 300 tons of these slops per trip were released directly into the ocean; afterward, ships could not release slops within 50 miles of shorelines (Mitchell, 1994). Still, none of this addressed tanker accidents, and after the wreck of the *Torrey Canyon* in 1967, a new regime was negotiated. Indeed, the 1973/78 International Convention for the Prevention of Pollution from Ships (MARPOL) exemplifies how ocean pollution politics is reactive rather than proactive to known hazards, waiting for a crisis to precipitate political action. MARPOL required a permanent and physical change to tankers that was hard to remove once installed, making compliance much more common (Mitchell, 1994).

Toxic, hazardous and nuclear waste

In 1971, the *Stella Maris* left the Rotterdam port with the intention to dump 650 tons of chlorinated toxic waste straight into the ocean, which was perfectly legal at that time. Citizen and foreign diplomatic protests forced the ship back to port and led to the 1972 Convention for the Prevention of Marine Pollution by Dumping from Ships and Aircraft, known as the Oslo Convention. This convention was merged with the 1974 Paris Convention, becoming OSPAR, that regulated harmful substances discharged from platforms, rivers, and the atmosphere into the North Atlantic. This treaty only has 16 Western European parties, but it does demonstrate the feasibility of an international regulation of inland sources of ocean pollution.

The 1972, the Convention on the Prevention of Marine Pollution by Dumping of Wastes and Other Matter, or the London Dumping Convention, now has 87 parties. It was negotiated to regulate intentional waste at sea and is thought to be quite successful because it became increasingly restrictive and precautionary. At first, there were three levels of waste that could only be dumped if the flagging country issued a permit. Now, under a related instrument called the London Protocol, with 53 parties, waste can only be dumped from ships and ocean platforms if they are on a "safe" list, and these materials still require a permit. Nuclear waste has been more difficult to regulate, but the Convention bans "high level" radioactive waste, with low levels of radioactive waste allowed through a national permit (Hunter et al., 2006: 737). Note that MARPOL, OSPAR and the London Convention – all active regulatory regimes – are established in the 1970s prior to the neoliberal period.

Urban runoff and wastewater

Of all the usual suspects, urban runoff is most concerning because it is growing. It is not being reduced like oil spills or radioactive waste – and, in fact, runoff is the major source of oil pollution in coastal waterways. By 2020 the human family was almost 8 billion strong, and since 2008 most people have lived in urban areas, making urban sewage and runoff an imperative. These urban centers have both point-source (such as pipes) and non-point-source runoff. Point-source pollution comes from outlets of human sewage, which is mainly untreated in poor countries, where mega-cities have multiplied. Non-point sources include all the contaminants that get washed into the ocean, such as soil that harms coral (Fabricius, 2005) and fertilizer (below).

Coastal pollution includes organic chemicals, heavy metals, pharmaceuticals and sludge that contribute to deformity, disease and death of marine organisms (Decker et al., 2002). Despite efforts to holistically manage coastal pollution through integrated coastal management schemes that have been widely adopted to promote sustainable development of coastal areas, runoff has continued to increase. This runoff threatens key coastal ecosystems – coral reefs, sea grass beds, mangroves, coastal fisheries and food webs.

Sewage fouls coastal areas and harms coral communities, mangroves, and marine life through harmful bacteria, sludge, increased nitrogen and phosphorous, harmful hormone-mimicking chemicals in the sewage, known as endocrine disrupters, among other problems. In areas like east Asia, rapidly developing countries with growing urban populations and industry are also rapidly and cumulatively polluting marine and coastal environments (Khim et al., 2020). Further, pollution from nutrients, metals, and persistent organic pollutants (POPs) combine with warming coastal and marine waters from climate change to significantly damage ecosystems across the globe (Lu et al., 2018). Wealthier industrialized countries have addressed the baseline sewage output through environmental policies and improved infrastructure. However, even in the United States the infrastructure of sewage pipes and treatment facilities were built during a time with a smaller population, less development, and less urbanization, so many of these industrialized systems still shut down after short intervals of, say, 20 minutes of heavy rainfall, not to mention rural septic-tank leakage into waterways (Lapointe et al., 2015, Duhigg, 2009). These sewage systems normally feed into treatment facilities that become overwhelmed. When this happens, valves are shut off, sending feces, storm water, debris, and other pollution into rivers, bays, and coastal waterways. This is less a problem if storm-water systems are separate from sewer systems and sewage is adequately treated before it is released.

In poorer countries, pipes and sewers, or drainage canals, often drain untreated sewage into rivers and coastal waterways. The demographic change resulting from massive urbanization has dramatically increased the amount of sewage flowing into the sea, even as environmental policies have improved in many places (Jiang et al., 2001).

Threats to the World Ocean: the "invisibles"

Many of the usual suspects in ocean pollution have been addressed using controls at the source – the ship, the pier, the ocean platform. And this seems to have had some margin of success in *controlling that point source*. However, new types of ocean pollution are made up of much more difficult contaminants that create overlapping stresses on the WOS. Unfortunately, there is no single point source to look to when we think of heat, carbon, nitrogen, or plastic. The usual institutions, then, will probably not be effective, forcing us to think about the global political economic system itself. States demonstrated little interest in significant regulatory limits after the 1980s.

Ocean warming

The Earth system has a heat balance that is produced by heat from the sun minus the heat that escapes the atmosphere into space. Since the last Ice Age, humanity has enjoyed a rather comfortable inter-session called the Holocene, during which all of civilization – agriculture, writing, and of course everything in the Industrial Revolution – have occurred. Before the Industrial Revolution, humanity produced only small amounts of carbon dioxide (CO_2) through things like forest clearing, when compared with the rest of nature, such as when plants decay. However, now human emissions make up about 55 percent of all carbon emissions in the Earth system (Yue and Gao, 2018). At the time of writing, emissions of carbon dioxide have increased by 140 parts per million (ppm), from the pre-industrial level of 280 ppm to about 420 ppm (National Oceanic and Atmospheric Administration, 2020). Because CO_2 absorbs heat, more CO_2 has resulted in global warming (Stone et al., 2009).

Importantly, about 93 percent of the heat from global warming has been absorbed by the WOS (Johnson and Lyman, 2020). Water at the bottom of the Atlantic and the Pacific has warmed, as has the upper 3,000m of the Southern Ocean, while the whole cryosphere (ice)-atmosphere system has been destabilized (Stone et al. 2009). Heat pollution, in combination with other factors, is disturbing the WOS *much* more than episodic oil, radioactive, or toxic dumping ever did. Richardson (2008: 291) writes, that "Some of the current leading conceptual thinkers in marine ecology… have warned that marine systems are undergoing abrupt shifts to unwanted stable states dominated by microbes, flagellates, bacterial mats, and jellyfish. These opportunistic species are capitalizing on ecosystems stressed by overfishing, eutrophication, pollution, and climate change."

Between 1955 and 2010, the water from the surface to 2,000 meters warmed by 0.18°C because it absorbed 24×10^{22}J, or 240,000,000,000,000,000,000,000J, of heat. According to Levitus et al, "If this heat were instantly transferred to the lower 10 km of the global atmosphere it would result in a…warming of this atmospheric layer by approximately 36°C (65°F)"(Levitus et al., 2012: L10603). Thus, the oceans have softened the impacts of global warming, but in the process they have absorbed a giant amount of energy. Warming has already had profound impacts on the WOS, including shifts in entire food webs in fast, non-linear ways across large marine ecosystems through changing the abundance of plankton, which is the productive base of all marine organisms (Kirby and Beaugrand, 2009).

What is more, this added heat causes seawater to expand, leading to sea-level rise (SLR). SLR is caused more and more, however, by the addition of water from inland glaciers. Early in the twentieth century, SLR was ~2 mm/year, but this accelerated to 3.4 ±0.3 mm/year after the 1960s (Dangendorf et al., 2019). Even small amounts of SLR do more than simply raise the shoreline, having also profound consequences inland, causing erosion, loss of salt marshes, shifting and eroding barrier islands, exposing coastal residents to storm dangers, and putting many small island states in an existential crisis (Yamamoto and Esteban, 2010). Even small amounts of SLR will affect *hundreds of millions* of people in low-lying deltas, such as the densely populated Nile delta, both physically and economically (FitzGerald et al., 2008).

Further, the way that seawater mixes is changing. This mixing is a fundamental aspect of the Meridional Overturning Current (MOC), or the thermohalene current, that drives currents around the world. This current is caused by varied densities of water based on heat and salinity, which circulates the water column vertically and then around the planet. The MOC is a fundamental regulatory ecosystem service that helps maintain a stable climate by moving cold water to the equatorial zones and warm water to the poles. However, the MOC began slowing after 1980, a likely effect of climate change, and is expected to continue slowing across the twenty-first century (Liu et al., 2020). This stress may force the WOS into regional or global abrupt shifts, which, in turn, will affect weather and climate patterns, such as monsoons, which are tied to abrupt climactic changes in the past, as well as unexpected and undesirable changes to biological systems from plankton and fish to apex predators (Kirby and Beaugrand, 2009).

One tipping point in Earth's climate system is the loss of Arctic sea ice. The Arctic has experienced non-linear (slow, then fast) thinning and loss of ice since satellite measurements started in 1979 compared to the 1981–2010 long-term average across all months (Stroeve and Notz, 2018). Loss of Arctic sea ice is a major problem in at least two ways. First, ice acts as a reflector of heat entering the Earth system (this process is called the albedo effect), and when it melts the darker water absorbs more heat, ironically causing more ice melt resulting in a "positive feedback" where ocean warming ultimately causes yet more ocean warming. Second, ice loss is causing major biological shifts because plants and animals that have evolved in the Arctic niche now face existential challenges (Solan et al., 2020), such as polar bears that need the sea ice to hunt and without which they may starve. In 2007, the Northwest Passage opened for the first time in human memory, allowing ships to pass through the Arctic. This system may have already entered a tipping point of no return – at least in human times scales.

Carbon dioxide

Another invisible pollutant is carbon dioxide (CO_2). Not only does CO_2 cause greenhouse warming, but it also affects the chemistry of the oceans. Absorption of carbon into the water is a normal process by which carbonate or bicarbonate is created. However, as more and more carbon has been taken up by the oceans as carbonic acid, the oceans' ability to convert the carbonic acid into carbonate and bicarbonate has slowed, lowering marine pH. This acidification has already affected the development of marine life and will continue to challenge the basic physiology of plankton, mussels, coral, and other organisms that depend on calcifying carbon (Doney et al., 2009). Acidification is occurring faster than at any other time in the past 300 million years and will have "serious consequences for the millions of people who are dependent on coastal protection, fisheries and aquaculture" (Hall-Spencer and Harvey, 2019: 197). Like the other "invisibles" there is no way to stop CO_2 from mixing with sea water once it is emitted into the atmosphere.

Elemental pollution: nitrogen and phosphorous

Nitrogen and phosphorous are both elements in agricultural fertilizers and are ironic contaminants because through spurring on life they create death through algae blooms (Wurtsbaugh et al., 2019). When those algae die, microbial decomposition depletes oxygen in the water through a process called eutrophication. Eutrophication creates "dead zones" that are depleted of enough oxygen to kill local organisms. Between nitrogen and phosphorous, the former is a more important marine pollutant, although phosphorous is more damaging to systems with lower salinity (Altieri and Diaz, 2019). Dead zones have increased with the industrialization of agriculture, sometimes called the Green Revolution (see Chapter 44). In the Green Revolution, petro-chemicals and biocides are added to crops to increase yield. In the United States, 10 out of 12 teragrams of applied fertilizer filter into fresh water and then marine systems (Robertson and Vitousek, 2009). Globally, there are over 400 known dead zones mainly in areas that receive industrial N runoff, but scientists believe there are "hundreds" more in tropical areas that have not yet been discovered (Altieri et al., 2017). Further, climate change is expected to increase the severity and prevalence of dead zones globally (Altieri et al., 2017).

Some scientists believe that humanity exceeds the "planetary boundary" for nitrogen and phosphorous pollution, meaning that these pollutants harm the "safe living space" for humanity (Steffen et al., 2015). The consequences of nitrogen also include the release of a powerful greenhouse gas, nitrous oxide (see Chapter 32), the release of reactive nitrogen in the troposphere, nitrogen deposition in natural areas like forests (see Chapter 42), biodiversity loss (see Chapter 42), compromised air and water quality (see Chapters 30 and 38) and "threats to human health across large areas of Earth" (Robertson and Vitousek, 2009: 98).

Plastics

The amount of plastic pollution has risen dramatically in recent decades (Thompson et al. 2009; see Chapter 37). Plastics are different from the "invisibles" discussed above in that heat, CO_2 and nitrogen occur in nature (albeit not in the concentrations emitted by humanity), but plastics are entirely synthetic products. Unfortunately, plastics are very difficult to control once produced; they are not simply dumped into oceans, but are first released on land and then emigrate to rivers and then to coasts and beyond. Plastic pollution is ubiquitous with an estimated 19 to 23 million metric tons entering waterways, with this number expected to double to 53 million metric tons by 2030 (Borrelle et al., 2020).

Plastic is often buoyant and may last hundreds or thousands of years before it breaks down into simpler minerals. However, plastic can photodegrade and fragment into pieces, as well as absorb water and other elements as it floats in the sea. Plastic in the ocean is often in the form of micro-pieces that are dispersed across large areas and float often unseen under the surface of the water. It is not always feasible to scoop it up.

Plastic poses serious problems for marine organisms, which may become ensnared in discarded fishing lines or may confuse it for food. In autopsies of regal leatherback turtles, which mistake floating plastic bags for their main food source, jellyfish, almost 40 percent of the animals had plastic in their gastrointestinal tract, yet even with many studies on plastic damage to individual animals we still do not know what the impact is on whole populations of these turtles (Senko et al., 2020). In addition, synthetic additives in plastics, such as flame retardants and bisphenol A (BPA), are harmful to human and animal health.

Recent developments in marine governance

The most important developments in global marine governance relevant to ocean pollution since the first edition of this volume (i.e., since 2013) are a legally binding treaty under development regarding Areas Beyond National Jurisdiction (ABNJ) to protect Biodiversity Beyond National Jurisdiction (BBNJ) under UNCLOS. The United Nations began the treaty process in 2015 and called for an International Conference in 2017. As of this writing, the Conference had met in three sessions. The need for biodiversity protections in ABNJ is clear, but there are immediate problems the BBNJ treaty face already. For example, the opening lines of the proposed new treaty, which are found across all meetings and in a non-committal "zero drafts," set up a failure to implement serious protections:

> This Agreement shall be interpreted and applied in a manner that [respects the competences of and] *does not undermine* [existing] relevant legal instruments and frameworks and relevant global, regional and sectoral bodies, and that promotes coherence and coordination with those instruments, frameworks and bodies, provided that they are supportive of and do not run counter to the objectives of the Convention and this Agreement (Article 4.3, emphasis added).

In other words, the Agreement cannot be used for oversight in relation to other issues, such as fishing, some pollution, or shipping (De Santo et al., 2020). Tiller and Nyman (2018) note that plastic pollution should be included in the treaty, but the United States and other states have worked "to ensure that member states would not be bound by any decision not made in the context of these negotiations" (De Santo et al. 2020: 2–3). The end result is that the promise of UNCLOS-based marine governance is diminished as states continue to evade significant regulatory limits.

Conclusion

Everything that humans do on land has consequences for the ocean, and we cannot effectively divorce our energy, agricultural or manufacturing policies from marine systems. In several of these areas, there are now agreements to regulate water quality and pollution, and new dumping protections, but these are foiled by the larger scale and structures of a globalized economy, industrialized consumption that is growing not only in Northern countries, but in rising powers like Brazil, India, and China, and a very large growing urbanized human population (Clapp and Helleiner, 2012). Based on the work of Clapp and Helleiner (2012) and others, it appears that these structural concerns overwhelm many efforts to protect oceans, including better farming techniques to limit nitrogen pollution or efficiency gains in fossil fuel use.

The complex adaptive systems of the oceans, as part of the Earth system, have overshot several limits that provide critical life supports and ecosystem services required for human and nonhuman wellbeing. One of the largest assemblies of scientists in history, the

Millennium Ecosystem Assessment, concluded with "a stark warning. Human activity is putting such strain on the natural functions of Earth that the ability of the planet's ecosystems to sustain future generations can no longer be taken for granted. The provision of food, fresh water, energy, and materials to a growing population has come at considerable cost to the complex systems of plants, animals, and biological processes that make the planet habitable" (Millennium Ecosystem Assessment, 2005: 5).

At the global scale, marine pollution has fundamentally changed from the "usual suspects" to the "invisibles," most of which are completely natural elements, but which are over-whelming oceanic systems and cycles. None of the pollutants in this new class are being addressed in a systematic way in marine regimes. This is not surprising because, as noted above, the international system only reacts to pollution problems after the fact, and then when it does, these actions are subservient to global economic interests that have now weakened the capacity for effective international regulatory action that could make a difference. Treating the invisibles requires a major overhaul of the global economic system, which also might require a paradigmatic shift in the way we see the ocean.

References

Acheson, J. M. & Wilson, J. A. (1996). Order out of Chaos: The Case for Parametric Fisheries Manag. *American Anthropologist,* 98, 579–594.

Altieri, A. H. & Diaz, R. J. (2019). Dead Zones: Oxygen Depletion in Coastal Ecosystems. *In:* Sheppard, C. (ed.) *World Seas: an Environmental Evaluation, 2nd ed.,* London: Academic Press.

Altieri, A. H., Harrison, S. B., Seemann, J., Collin, R., Diaz, R. J. & Knowlton, N. (2017). Tropical Dead Zones and Mass Mortalities on Coral Reefs. *Proceedings of the National Academy of Sciences,* 114, 3660–3665.

Barron, M. G., Vivian, D. N., Heintz, R. A. & Yim. (2020). Long-Term Ecological Impacts from Oil Spills: Comparison of Exxon Valdez, Hebei Spirit, and Deepwater Horizon. *Environmental Science & Technology,* 54, 6456–6467.

Beamish, T. (2002). *Silent Spill: The Organization of Industrial Crisis,* Cambridge, MA: MIT Press.

Bigg, G. (2003). *The Oceans and Climate, 2nd ed.,* Cambridge, Cambridge University Press.

Borgese, E. M. (1998). *The Oceanic Circle: Governing the Seas as a Global Resource,* New York, United Nations University Press.

Borrelle, S. B., Ringma, J., Law, K. L., Monnahan, C. C., Lebreton, L., Mcgivern, A., Murphy, E., Jambeck, J., Leonard, G. H., Hilleary, M. A., Eriksen, M., Possingham, H. P., De Frond, H., Gerber, L. R., Polidoro, B., Tahir, A., Bernard, M., Mallos, N., Barnes, M. & Rochman, C. M. (2020). Predicted growth in plastic waste exceeds efforts to mitigate plastic pollution. *Science,* 369, 1515–1518.

Centeno, M. A. & Cohen, J. N. (2012). The Arc of Neoliberalism. *Annual Review of Sociology,* 38, 317–340.

Clapp, J. & Helleiner, E. (2012). International Political Economy and the Environment: Back to the Basics? *International Affairs,* 88, 485–501.

Costanza, R. (2000). *The Ecological, Economic, and Social Importance of the Oceans,* New York, Pergamon Press.

Dangendorf, S., Hay, C., Calafat, F. M., Marcos, M., Piecuch, C. G., Berk, K. & Jensen, J. (2019). Persistent acceleration in global sea-level rise since the 1960s. *Nature Climate Change,* 9, 705–710.

De Santo, E. M., Mendenhall, E., Nyman, E. & Tiller, R. (2020). Stuck in the middle with you (and not much time left): The Third Intergovernmental Conference on Biodiversity Beyond National Jurisdiction. *Marine Policy,* 117, 103957.

Decker, E. H., Elliott, S. & Smith, F. A. (2002). Megacities and the Environment. *The Scientific World Journal,* 2, 374–386.

Dimitrov, R. S. (2005). Hostage to Norms: States, Institutions and Global Forest Politics. *Global Environmental Politics,* 5, 1–24.

Dimitrov, R. S. (2019). Empty Institutions in Global Environmental Politics. *International Studies Review,* 22(3), 626–650. https://doi.org/10.1093/isr/viz029

Doney, S. C., Fabry, V. J., Feely, R. A. & Kleypas, J. A. (2009). Ocean Acidification: The Other CO2 Problem. *Annual Review of Marine Science,* 1, 169–192.

Duhigg, C. (2009). Sewers at Capacity, Waste Poisons Waterways. *New York Times,* p.A1, A18.

Fabricius, K. E. (2005). Effects of Terrestrial Runoff on the Ecology of Corals and Coral Reefs: Review and Synthesis. *Marine Pollution Bulletin,* 50, 125–146.

Fitzgerald, D. M., Fenster, M. S., Argow, B. A. & Buynevich, I. V. (2008). Coastal impacts due to sea-level rise. *Annual Review of Earth and Planetary Sciences,* 36, 601–647.

Grotius, H. (1916). The Freedom of the Seas or the Right Which Belongs to the Dutch to Take Part in the East Indian Trade. *In:* Scott, J. B. (ed.). New York, Carnegie Endowment for International Peace/Oxford University Press.

Gunderson, L. H. & Holling, C. S. (eds.) (2002). *Panarchy: Understanding Transformations in Human and Natural Systems,* Washington, D.C: Island Press.

Haedrich, R. L. (1996). Deep-Water Fishes: Evolution and Adaptation in the Earth's Largest Living Spaces*. *Journal of Fish Biology,* 49, 40–53.

Hall-Spencer, J. M. & Harvey, B. P. (2019). Ocean acidification impacts on coastal ecosystem services due to habitat degradation. *Emerging Topics in Life Sciences,* 3, 197–206.

Hassan, R. M., Scholes, R. & Ash, N. (2005). Ecosystems and Human Well-Being: Current State and Trends: Findings of the Condition and Trends Working Group of the Millennium Ecosystem Assessment. *Millennium Ecosystem Assessment* [Online], 1.

Hunter, D., Zaelke, D. & Salzman, J. (eds.) (2006). *International Environmental Law and Policy,* Boulder CO: West Group.

Jackson, J. (2008). Ecological Extinction and Evolution in the Brave New Ocean. *Proceedings of the National Academy of Sciences,* 105, 11458–11465.

Jackson, M. (1993). *Indigenous Law and the Sea,* Washington, DC, Island Press.

Jacques, P. J. (2010). International Regulation of Ocean Pollution and Ocean Fisheries. *In:* DENEMARK, B. (ed.) *International Studies Encyclopedia.*

Jiang, Y., et al. (2001). Megacity Developments: Managing Impacts on the Marine Environment. *Ocean and Coastal Development,* 44, 293–318.

Johnson, G. C. & Lyman, J. M. (2020). Warming Trends Increasingly Dominate Global Ocean. *Nature Climate Change,* 10, 757–761.

Juda, L. (1996). *International Law and Ocean Use Management: The Evolution of Ocean Governance,* New York, Routledge.

Khim, J. S., Wang, T., Zhang, X. & Hong, S. (2020). Coastal Ecosystem in East Asia: Pollution and Management. *Environment International,* April 149, 106185.

Kirby, R. R. & Beaugrand, G. (2009). Trophic Amplification of Climate Warming. *Proceedings of the Royal Society B,* 276, 4095–4103.

Lapointe, B. E., Herren, L. W., Debortoli, D. D. & Vogel, M. A. (2015). Evidence of Sewage-Driven Eutrophication and Harmful Algal Blooms in Florida's Indian River Lagoon. *Harmful Algae,* 43, 82–102.

Levitus, S., Antonov, J. I., Boyer, T. P., Baranova, O. K., Garcia, H. E., Locarnini, R. A., Mishonov, A. V., Reagan, J. R., Seidov, D., Yarosh, E. S. & Zweng, M. M. (2012). World Ocean Heat Content and Thermosteric Sea Level Change (0–2000 m), 1955–2010. *Geophysical Research Letters,* 39, L10603.

Liu, W., Fedorov, A. V., Xie, S.-P. & Hu, S. (2020). Climate Impacts of a Weakened Atlantic Meridional Overturning Circulation in a Warming Climate. *Science Advances,* 6, eaaz4876.

Lotze, H. & Mcclenachan, L. (2014). Marine Historical Ecology: Informing the Future by Learning from the Past. *Marine community ecology and conservation,* 165–200.

Lu, Y., Yuan, J., Lu, X., Su, C., Zhang, Y., Wang, C., Cao, X., Li, Q., Su, J., Ittekkot, V., Garbutt, R. A., Bush, S., Fletcher, S., Wagey, T., Kachur, A. & Sweijd, N. (2018). Major Threats of Pollution and Climate Change to Global Coastal Ecosystems and Enhanced Management For Sustainability. *Environmental Pollution,* 239, 670–680.

Millennium Ecosystem Assessment, M. (2005). *Ecosystems and Human Well-Being: Our Human Planet: Summary for Decision-Makers,* Washington, D.C., Island Press.

Mitchell, R. (1994). *International Oil Pollution at sea: Environmental Policy and Treaty Compliance,* Cambridge, MA: MIT Press.

National Oceanic and Atmospheric Administration, N. (2020). Trends in Atmospheric Carbon Dioxide. *In: Earth System Research Laboratory,* G. M. D. (ed.). Mauna Loa, Hawaii, United States Department of Commerce.

National Research Council. (2003). *Oil in the Sea III: Inputs, Fates, and Effects*, Washington, D.C.: The National Academies Press.

Rabalais, N. (2003). Oil in the Sea. *Issues in Science & Technology,* 20, 74–78.

Richardson, A. J. (2008). In Hot Water: Zooplankton and Climate Change. *ICES Journal of Marine Science,* 65, 279–295.

Robertson, G. P. & Vitousek, P. M. (2009). Nitrogen in Agriculture: Balancing the Cost of an Essential Resource. *Annual Review of Environment and Resources,* 34, 97–125.

Ruggie, J. G. (2014). Global Governance and "New Governance Theory": Lessons from Business and Human Rights. *Global Governance,* 20, 5–17.

Russ, G. R. & Zeller, D. C. (2003). From mare liberum to mare reservarum. *Marine Policy,* 27, 75–78.

Selden, J. (1972). Of the Dominion, Or, Ownership of the Sea. *In:* SILK, L. (ed.) *The Evolution of Capitalism.* New York: Arno Press.

Senko, J. F., Nelms, S. E., Reavis, J. L., Witherington, B., Godley, B. J. & Wallace, B. P. (2020). Understanding Individual and Population-Level Effects of Plastic Pollution on Marine Megafauna. *Endangered Species Research,* 43, 234–252.

Solan, M., Archambault, P., Renaud, P. E. & März, C. (2020). The Changing Arctic Ocean: Consequences for Biological Communities, Biogeochemical Processes and Ecosystem Functioning. London: The Royal Society Publishing.

Steffen, W., Richardson, K., Rockström, J., Cornell, S. E., Fetzer, I., Bennett, E. M., Biggs, R., Carpenter, S. R., De Vries, W. & De Wit, C. A. (2015). Planetary Boundaries: Guiding Human Development on a Changing Planet. *Science,* 347, 1259855.

Steinberg, P. E. (2001). *The Social Construction of the Ocean,* Cambridge, Cambridge University Press.

Stone, D. I. A., Allen, M. R., Stott, P. A., Pall, P., Min, S.-K., Nozawa, T. & Yukimoto, S. (2009). The Detection and Attribution of Human Influence on Climate★. *Annual Review of Environment and Resources,* 34, 1–16.

Stroeve, J. & Notz, D. (2018). Changing State of Arctic Sea Ice across All Seasons. *Environmental Research Letters,* 13, 103001.

Thompson, R. C., Swan, S. H., Moore, C. J., & Vom Saal, F. S. (2009). Our plastic age. *Philosophical Transactions of the Royal Society B: Biological Sciences,* 364(1526), 1973–1976. https://doi.org/10.1098/rstb.2009.0054

Tiller, R. & Nyman, E. (2018). Ocean Plastics and the BBNJ Treaty—Is Plastic Frightening Enough to Insert Itself into the BBNJ Treaty, or Do We Need to Wait for a Treaty of Its Own? *Journal of Environmental Studies and Sciences,* 8, 411–415.

United Nations. (1982). *United Nations Convention on the Law of the Sea.* United Nations, Division of Ocean Affairs and the Law of the Sea.

Vieira, M. B. (2003). Mare Liberum vs. Mare Clausum: Grotius, Freitas, and Selden's Debate on Dominion over the Seas. *Journal of the History of Ideas,* 64, 361–377.

Wurtsbaugh, W. A., Paerl, H. W. & Dodds, W. K. (2019). Nutrients, Eutrophication and Harmful Algal Blooms along the Freshwater to Marine Continuum. *WIREs Water,* 6, e1373.

Yamamoto, L. & Esteban, M. (2010). Vanishing Island States and sovereignty. *Ocean & Coastal Management,* 53, 1–9.

Yue, X.-L. & Gao, Q.-X. (2018). Contributions of Natural Systems and Human Activity to Greenhouse Gas Emissions. *Advances in Climate Change Research,* 9, 243–252.

40

Fisheries and marine mammals

The complexities of collective management

Elizabeth R. DeSombre

Protecting or conserving the animal resources – be they fish or mammals – that live in the ocean is among the most difficult of global environmental issues. Large portions of the oceans are international space, owned by no one and accessible by many. Fisheries are common pool resources, meaning that it is difficult, either practically or legally, to exclude others from access to them, and that use by some diminishes the ability of others to use the resource. Uncertainty is rampant, about the condition of the resource and about the behavior of those who make use of it (and, sometimes, about the relationship between the two) (see Chapter 19).

But fishery resources create their own incentive for management: if enough members of a given species remain in the ocean, they will reproduce and thereby create the next genera-tion, and at least some of them do so on a reasonably short time frame for natural resources. If those who harvest them can successfully cooperate to ensure that sufficient numbers remain, they will be able to continue harvesting these resources indefinitely. This logic applies as well to the harvesting of marine mammals when they are seen as resources. Yet even with this incentive structure it has been difficult to manage international fisheries or marine mammal populations successfully.

The history of efforts to protect marine fisheries and mammals suggests that the diffi-culties can overwhelm the incentives for cooperation, especially when there is some level of uncertainty about stock sizes or behavior of fishing vessels. Increased national control of some ocean resources failed to protect fish stocks, because states instead worked to increase fishing capacity. Even the most successful cooperative efforts regionally to conserve some fish stocks have suffered from the global nature of the fishing industry, in which capacity can move from one area to another in response to restrictions or opportunities. Some of the most successful protection of ocean resources came in the case of marine mammals, by changing the reason they are protected – moving them from a resource to be harvested sustainably to individuals protected from harm. Even then, this redefinition is contentious and thus fragile, and the future protection of fisheries and marine mammals remains in question.

DOI: 10.4324/9781003008873-45

Elizabeth R. DeSombre

Resource overharvesting

Ocean fish and mammals have provided protein and other resources for human populations for millennia. As long as the human population was sufficiently low and fishing technology simple, there was little human impact on the availability of fish resources globally. Technological innovation made fishing, and the distribution of fish and marine mammal resources worldwide once caught, much more efficient over time, and thus contributed to the human ability to affect marine resources at a global level.

These technological advances included storage and transportation; refrigeration (and freezing) both on land and at sea made it possible for ships to stay out for long periods and travel far to catch previously inaccessible fish or whales, and to transport them to new markets (inland or overseas) once they are landed. The decreasing cost and increasing speed of transport also meant that fresh fish could be shipped long distances. Increasingly available and decreasingly expensive air travel allowed high-value fish like bluefin tuna to reach Japanese markets where the freshness of fish mattered if it were to be eaten raw (Cushing 1988; Kurlansky 1997).

Some of the most important technologies included those that allowed fish to be found, and harvested, more efficiently. Mechanization allowed enormous nets or lines to be deployed and hauled in. Radar and then sonar were used to locate large schools of fish. Global Positioning System (GPS) technology aided ships in navigation to fishing grounds. Increased sophistication of technology makes it possible for a smaller number of fishers to catch more fish (Cushing 1988; Roberts 2007).

This technology has frequently been underwritten by national subsidies, as states encouraged development of domestic fishing fleets, both for reasons of food security and to have ready access to naval capacity in times of war. Almost every country in the world with a fishing fleet subsidizes it in one way or another, and there are many different ways that fishing can be subsidized, from low-interest loans or grants for ship modernization or purchasing, to assistance with operating costs (such as exemptions from fuel taxes). Subsidies ultimately lead to overcapitalization, an increase in investment in fishing capacity beyond what the resource can support (UNEP 2004).

This technological innovation and subsidization led to ever-increasing catches, from a global average of approximately 2 million tonnes in 1850, to 10 million tonnes annually by 1930 and reaching 20 million tonnes per year by 1950. These catches continued upwards, reaching a peak of almost 90 million tonnes per year in the late 1990s. Since then they have declined somewhat, holding steady at around 80 million tonnes per year. Current catches are likely not sustainable, and certainly do not allow for continued growth in fishing. The United Nations Food and Agriculture Organization estimates that 34.2 percent of global marine fish stocks are overfished and another 59.6 percent are fished at maximally sustainable levels, meaning that they are incapable of sustaining any additional fishing pressure (UN Food and Agriculture Organization 2020).

Managing fisheries

As soon as it became clear that human fishing efforts were having an effect on fish stocks, either regionally or globally, states understood the necessity of cooperating to attempt to prevent overfishing. Because most of the ocean has, historically, been un-owned space, there was no governmental entity with the authority to prevent overfishing. There was also serious likelihood of fishers or states free-riding on the efforts of others to protect the resource;

although fishers would gain collectively by fishing at a level that would allow fish stocks to replenish over time, individual fishers would be better off if others conserved and they continued to catch at preferred levels. Even those hoping to act collectively realize that if others do not, their good behavior will be undermined and the stocks might collapse. Fishery conservation can thus only be achieved collectively, and on the open ocean these issues could only be addressed internationally via voluntary cooperative efforts.

Underlying fisheries management globally is the United Nations Convention on the Law of the Sea (UNCLOS), a sweeping global treaty completed in 1982 with the intention of covering all aspects of ocean regulation. Although its breadth decreases its ability to formulate specific rules, the agreement did create two important structures for fisheries regulation. The first was the expansion and codification of what came to be known as Exclusive Economic Zones (EEZs) – areas extending from coastlines that states have jurisdiction over. States have always had a territorial sea – a small stretch of ocean (previously three nautical miles) extending from the shore that they have been allowed to control in the same way they control their territory. But beginning in the 1940s some Latin American states started to declare jurisdiction of 200 miles or more, which created a patchwork of unilateral and sometimes conflicting claims of control over ocean spaces and resources. UNCLOS extended the territorial sea to 12 miles, and created EEZs out to 200 miles. In these areas states were permitted to control access to resources. This process put more than 35 percent of ocean spaces (Sanger 1986: 67), and the majority of commercially caught fish (Colson 1995: 100), under some form of national jurisdiction.

A large part of the logic for this extended zone of control was that it would help protect fisheries resources. Because the common pool resource nature of fisheries led to cooperation problems – and problems when states failed to cooperate – it was logical to assume that putting the resources of large parts of the oceans under state control could ameliorate that problem. Fishers from different nationalities would no longer be competing for the same fish, and in nationally controlled spaces governments actually had the jurisdiction to impose and enforce management priorities.

Although states did, indeed, force foreign fishers out of their newly controlled waters, the broader principles of successful management were not implemented. Many states responded to the new control over resources by subsidizing domestic fishing fleets, thereby increasing global fishing capacity, and threatening to overwhelm the fish stocks newly released from foreign fishing pressure. Moreover, few states responded by strict regulation of national fishing inside their EEZ. To the extent that there were regulations they were often collective (rather than individual) catch limits, thereby increasing the incentive for overcapitalization as fishers now turned to competition with other vessels of the same nationality where they had previously competed with foreign fishers (Scheiber 2001). Fish stocks in these newly nationalized areas frequently declined, sometimes precipitously, as was the case with cod in Canadian Atlantic waters. This new global capacity then also competed in international waters as vessels moved elsewhere in search of increasingly scarce fish stocks, and sometimes illegally within the EEZs of other states (Englander 2019).

Other issues were not resolved by UNCLOS, most notably the fish that straddled the EEZs of multiple states or those that swam long distances across the ocean, perhaps crossing from EEZs to the high seas and back. Although UNCLOS directed states to cooperate in addressing these stocks, in practice states resisted such cooperation, and these species were the subject of conflict (Mack 1996). The United Nations convened a conference in 1993 to negotiate a way to address this situation; the result was the Agreement for the Implementation of the Provisions of the United Nations Convention on the Law of the

Sea relating to the Conservation and Management of Straddling Fish Stocks and Highly Migratory Fish Stocks (1995). It does not lay out an actual management process for dealing with these fish stocks, but rather makes clear the necessity to cooperate to address them and the obligation of states to join the relevant regional fishery management organizations or comply with their regulations, if fishing for species is regulated by these organizations, and allow for the enforcement of rules to manage these fish stocks (*Agreement for the Implementation of the Provisions of the United Nations Convention on the Law of the Sea relating to the Conservation and Management of Straddling Fish Stocks and Highly Migratory Fish Stocks* 1995: Articles 8, 13 and 19–23).

Regional fishery management organizations

Most international efforts to protect fish stocks in international waters are undertaken through Regional Fishery Management Organizations (RFMOs). These are organized by species, region, or some combination of the two, like the Indian Ocean Tuna Commission, the Northwest Atlantic Fisheries Organization or the Commission for the Conservation of Antarctic Marine Living Resources (CCAMLR).

The regional and species-based organization of international fisheries management is, in part, a historical accident, as the first agreements arose to address specific management issues. The International Convention for Regulating the Police of the North Sea Fisheries Outside Territorial Waters, for example, was negotiated in 1882 to harmonize domestic fishing regulations (and enforcement) among Great Britain, Germany, France, Denmark and Belgium. The Agreement for the Establishment of the Asia-Pacific Fishery Commission (1948) created a research and regulatory process for fishing in that region that still exists today.

The proliferation of RFMOs may in part have been path-dependent: once some organizations had been created, this management model was easy to adapt for the next stock or region in need of cooperation. Management by species or region also avoided the difficulties of having to gain the agreement of large numbers of states over multiple issues (a problem UNCLOS itself later demonstrated). And although RFMOs have had varying degrees of success, initial regional cooperation at least helped to address coordination problems among states and allow for collective research and decision-making (on global and regional environmental cooperation more generally, see Chapter 8).

Prior to the expansion of EEZs, these regional agreements addressed fisheries management outside of territorial seas; in the wake of the new nationalization of management of resources some RFMOs changed their mandates (or reconstituted themselves altogether) to reflect their new jurisdiction; some, however – especially those protecting highly migratory fish stocks – maintained their regulation of the relevant stocks wherever they were found (Peterson 1993). Although each RFMO runs in its own way there are some common elements across them generally. Many have open membership – any state with vessels fishing in the region can join. Most RFMOs have two central components: a scientific committee (sometimes called a council) and a commission. The scientific committees are charged with either conducting, or, more frequently, aggregating the relevant scientific research on stock health to inform the political decisions; often these committees make recommendations about sustainable catch levels (on science and expertise in global environmental politics, see Chapter 18).

Any actual regulations are passed by the commissions. RFMOs set out in their founding documents what types of rules they can adopt, but these most frequently involve catch limits, opened or closed areas or seasons for fishing, size limitations on the fish caught, gear restrictions and (less frequently) bycatch limits. The process of passing these rules varies by organization, but generally each member state has one vote. Commissions usually meet annually (or at least every two or three years), and regulations are passed at each meeting. These procedures combine to allow for rapid changes in regulations that can take account of new scientific information and the potential impacts of previous regulatory decisions. But they also allow for political factors, such as the interests states have in supporting their domestic fishers, to trump scientific ones in the setting of quotas.

Although recently created RFMOs are more likely to require unanimous voting to pass rules, those created in previous decades generally require a majority or supermajority (such as three-quarters) vote. RFMOs that use non-unanimous voting almost always include an objections procedure by which states that did not vote in favor of a rule that passed can opt out of being bound by it. The process used by the Northwest Atlantic Fisheries Organization is illustrative. Rules can be passed by simple majority. Within 60 days after a regulation has been passed, any state may lodge an objection, which means that it chooses not to be bound by that obligation. Because a state that might have been willing to follow the rule if every state agreed to take it on might feel differently if it knew that some states were objecting, an additional 40 days is allowed after a state has lodged an objection to allow other states to do so as well. If by the end of this period a majority of states has objected to the measure, it does not become binding. Otherwise, it does, but only for those that have not objected to it. In addition, states may remove their objections at any time (*Convention on Future Multilateral Cooperation in the Northwest Atlantic Fisheries* 1978: Article XII).

This objections process is controversial and has occasionally resulted in fishing seasons in which none of the major fishing states in an RFMO were bound by some of the rules the organization had passed (Schiffman 2008). But without such a rule non-unanimous voting would not be possible, and regulation could not move forward without the agreement of all states, including those most reluctant to accept regulation. Situations in which most major states opt out of rules are counterbalanced by those in which many do agree and the process of regulation can move forward. In the best situations, states that have initially objected to a rule can be persuaded to remove their objections.

Other potential difficulties plague efforts within RFMOs to pass regulations sufficiently stringent to ensure the sustainability of the fish stocks they oversee. Frequently there is a difference between the catch levels recommended by scientific committees and the catch limits agreed to by fisheries commissions. Not surprisingly, that difference usually involves the rules allowing for larger catches than the scientific process recommended (Barkin and DeSombre 2013). Once rules are passed, there is also the potential for non-compliance. Discussions of problems with global fisheries regulation often use the acronym "IUU" – Illegal, Unreported, and Unregulated fishing – but actively illegal fishing (much of which is also unreported) is different conceptually than unregulated fishing. Although ships and states find many ways to avoid being bound by fishing regulations (hence counting as "unregulated"), non-compliance involves fishing in a manner that contravenes regulations that the vessels are unquestionably obligated to follow.

It is remarkably easy to avoid international rules. The ocean is vast; fish move around in it and individual ships are difficult to keep track of at any moment. Each fishing vessel has the opportunity to obey, or disobey, existing rules every time it catches fish, and there are too many commercial fishing vessels for anyone to realistically check at all moments whether a given ship is obeying the rules. Vessels can catch fish of the wrong size or species or in the wrong location, and they can use prohibited gear; in some cases ships even falsify log books so that their records show compliance with rules that they are nevertheless breaking. Some efforts, discussed below, have managed to decrease the ability of ships to break fishing regulations that they are obligated to uphold, but non-compliance is a more difficult problem for international fisheries regulation than for many other issues of international cooperation (see Chapters 8 and 9).

Problems with existing regulatory approaches

There are some advantages to small numbers of organizations that focus on a certain area or set of species. Uncertainty is a major issue in fisheries management, and the scientific analysis can focus on the information relevant to the actors and the species in the region the organization oversees. Agreement on management is difficult enough with a small number of actors; increasing the number of those participating in any decisions might make agreement more difficult. But there are difficulties created by this approach as well. Although regulation is regional, the high-seas fishing industry is, for the most part, global. The largest fishing fleets or companies, when faced with regulation in one species or region, can switch to another species or region to continue catching fish. In this type of situation, successful fisheries management in one region may help address problems there, but this only shifts the capacity elsewhere, depleting additional stocks and causing problems in new regions (Barkin and DeSombre 2013).

The other way that fishing vessels can escape regulation is by changing their registration. All ships are required by international law to be registered with a state; this registration determines the nationality of the vessel and hence the domestic and international regulations it must follow. The phenomenon of open registration (also derisively called "flags of convenience," FOC) involves states offering registration to vessels owned by non-nationals, as a revenue-generating mechanism (because taxes and registration fees can be collected). One way to lure ships to register is to keep regulation levels low, which can include remaining outside of international cooperative agreements, since ships are only bound by international rules their registry states have taken on (DeSombre 2006).

This use of flags of convenience in order to escape international regulation has been a major problem for RFMOs. In the early 2000s it was estimated that between 10 and 21.5 percent of the major commercial fishing vessels were registered in open registries (ICTFU et al. 2002; DeSombre 2005). RFMOs saw the effects of this open registration. Estimates of the impact of fishing in the regulatory area by vessels flagged in non-member states (most of which are FOCs) during this period ranged from 10 percent in some RFMOs (such as International Commission for the Conservation of Atlantic Tunas and the Indian Ocean Tuna Commission; OECD 2004) to between 15 and 35 percent for the Commission for the Conservation of Southern Bluefin Tuna (CCSBT) (Swan 2002) and considerably higher than that in the Commission for the Conservation of Antarctic Marine Living Resources (Lack and Sant 2001).

Recently this situation has improved, due to carefully crafted action by RFMOs. Although they often cannot practically, or in many cases legally, exclude vessels that are not members of the organization from fishing in a given management area, they have created a tracking process to require vessels fishing within the rules to document their catches and allow – or sometimes require – states that permit these catches to be landed or transshipped to do so only if it can be demonstrated that they were caught within the regulatory process. In other words, member states in the RFMOs can exclude unregulated catches from their markets, thereby decreasing the value to the vessels of catching fish outside the regulatory process (DeSombre 2005).

While this approach has certainly not eliminated the problems of flag-of-convenience fishing, it has persuaded some FOCs to join RFMOs where their vessels are fishing and in other cases caused registry states to remove fishing vessels from their rolls. And although those who catch fish outside these RFMOs can continue to find markets on which to sell them, the price that these fish fetch is considerably lower than when sold in the major fish markets from which fish caught outside the regulatory process are excluded. For example, in CCAMLR, undocumented catches of regulated species fetch a price that is 20 to 40 percent lower than documented catches (Stokke and Vidas 2004).

The problem of FOC-fishing, while improved, has not been eliminated. More recent efforts to track the use of FOCs for illegal fishing examines lists that individual RFMOs keep of vessels accused of engaging in IUU fishing in their regulatory areas. The vast majority of ships listed on RFMO IUU lists are registered in open registries, most prominently Panama, Belize and Georgia, and nearly 30 percent of IUU-designated vessels have no state of registry, a violation of international law. Moreover, vessels engaged in IUU fishing are likely to shift registries regularly in a likely effort to avoid regulations or detection. More than 25 percent of the IUU-listed vessels had been registered in at least five different states in recent years (North Atlantic Fisheries Intelligence Group and INTERPOL 2017).

Transhipment – the practice of off-loading fish on the ocean from the vessel that caught it to a refrigerated vessel that takes it to its destination – introduces another option for illicit fish to reach market. A 2017 study found that up to 40 percent of transhipment takes place on FOC-registered vessels, allowing additional avenues for unregulated ships in the fishing value chain (Skyteam and Global Fishing Watch 2017).

Another difficulty with the RFMO regulatory focus on species is the problem of bycatch. While global bycatch is estimated to have been cut in half in the past 20 years, current estimate suggests that at least 10 percent of fish caught are non-target species (Gilman et al. 2020). Often these fish are returned to the ocean, but mortality from discarded catches is extremely high (up to 100 percent), and the methods, like trawling, that produce the greatest bycatch are also responsible for the greatest mortality from bycatch (Pascoe 1997). Bycatch also poses a problem for scientific analysis; neither the discards nor their levels of mortality are generally recorded (Hilborn and Walters 1992). Scientific efforts to estimate catches thus undercount actual fishing mortality, and the extent of depletion may be much greater than expected for a given catch limit, making recovery or sustainability estimates inaccurate.

RFMOs have attempted to deal with bycatch issues, often through gear restrictions, fishing techniques, or other prohibitions. Mandating or prohibiting certain types of fishing gear is frequently used to address the bycatch of non-fish species, such as turtles, marine mammals (discussed below), or birds. Shrimp fishers in many areas are now required to use "turtle excluder devices" (TEDs), escape hatches in shrimp trawl nets that allow sea turtles pulled in with shrimp to escape before the nets are hauled in. In order to prevent seabirds from being caught in swordfish longlining, some RFMOs mandate the use of circle hooks that are less likely to ensnare birds. In addition, some RFMOs have bycatch limits, closing the fishery (regardless of the catch of targeted species) once they are reached (Lodge et al. 2007). A related regulatory problem is the issue of high-grading, or catching more fish than allowed, keeping those that are highest value, and discarding the rest overboard; as with bycatch, most fish discarded in this manner do not survive (Chopin et al. 1996).

In addition, RFMOs continue to face the problem of non-compliance with existing rules. Most RFMOs rely on self-reported data. States are obligated to report catches by the vessels they register, and to ensure that their ships are following the rules. Ship crews have an incentive to underreport catches and states have little incentive to police their own ships. While it has been difficult to estimate the extent to which ships bound by international regulation are evading it, there is circumstantial evidence that some are (such as improbably distributed catch sizes or locations); the Australian Scientific Committee to the CCSBT, for example, concluded from an analysis of Southern bluefin tuna at the Japanese wholesale fish markets that the number sold exceeded the number reported as caught by Japan and the other states that provide tuna to those markets by at least 100 percent (Polachek and Davies 2008). There is also the occasional spectacular example of ships caught with illegal catches and illegal gear, falsified logbooks, or hidden storage areas (Springer 1997).

Efforts to address this non-compliance are intrusive and controversial, but have been increasingly accepted in contexts where non-compliance is rampant. One early approach was the exchange of observers on fishing vessels so that nationals from one state would observe and report on fishing behavior on another state's vessels. This practice was initiated in the regulatory process for whaling. Other fishery commissions experimented with observer exchanges, including CCAMLR and a robust current one in the Inter-American Tropical Tuna Commission, ensuring that dolphins are not harmed in the course of tuna fishing (as discussed below). Observer schemes are expensive, however, and observers regularly report harassment or bribery in efforts to prevent them from reporting illegal behavior (Rojas 2008).

A more recent approach is a vessel monitoring system (VMS) in which a satellite (or sometimes radio) tracking system is mounted on a fishing vessel. It relays real-time information about its position and speed either to regulatory agencies within registry states or to regional fishery management organization headquarters. This technology is used by some regional fishery management organizations, such as in the Patagonian toothfish fishery in Antarctic waters and for Southern bluefin tuna caught under the auspices of the CCSBT. The information provided by this technology can be used to determine whether fishing happened within the proper regulatory area.

Catches can also be inspected at the point of landing, and frequently are; vessel monitoring or observer reports contribute to this process. But it may nevertheless be impossible to determine at that point whether the fish were caught using the proper procedures, and truly illegally caught fish are likely landed in difficult-to-monitor locations. In short, because of the vast and distant spaces and large number of actors involved in ocean fishing, it is much more likely than in other resource extraction or environmental issues that international rules are regularly broken.

Marine mammal conservation

Historically marine mammals were simply considered another "fishery," although they are, of course, not actually fish. They were addressed in the same way as any other resource to be harvested. In that capacity, they suffered many of the same problems that face fisheries conservation currently, with overharvesting rampant, and difficulties setting and enforcing catch limits. Some of the earliest international environmental cooperation efforts emerged in the context of marine mammal conservation. One of the earliest international resource agreements of any sort was the Fur Seal Treaty of 1911 (officially the Convention between the United States, Great Britain, Russia, and Japan for the Preservation and Protection of Fur Seals), which attempted to use biological indicators to ensure that seals (and other marine mammals, such as otters) were not overharvested. It outlawed hunting of these species in the open ocean and protected the most endangered species from being hunted on land (*Convention between the United States, Great Britain, Russia, and Japan for the Preservation and Protection of Fur Seals* 1911: Articles I and III).

Similarly, there have been many international efforts to conserve whale stocks globally. The first of these was the Geneva Convention for the Regulation of Whaling, negotiated under the auspices of the League of Nations in 1931. It limited the taking of especially depleted species, such as right whales, and prohibited the taking of whale calves. Because this agreement was not signed by some of the important whaling states at the time, such as Germany, Japan, and the Soviet Union, its effectiveness was hindered (Francis 1990: 209–210). A second global agreement, the International Agreement for the Regulation of Whaling, was attempted in 1937. It attempted to regulate through the setting of season limits rather than catch limits, and ultimately had little success in protecting whale stocks (Francis 1990: 210). Negotiations of additional protocols continued in an effort to make the agreement more effective, but the best protection for whale stocks came during World War II, when whaling largely ceased because of the fighting, and many whale stocks recovered.

The agreement that currently oversees international whaling efforts is the International Convention for the Regulation of Whaling, which was negotiated in 1946. As with other early whaling (and marine mammal agreements more generally) it was intended as a conservation treaty, restricting catches so that "increases in the size of whale stocks will permit increases in the number of whales which may be captured without endangering these natural resources" (*International Convention for the Regulation of Whaling* 1946: preamble). It created an International Whaling Commission (IWC) to make decisions, with advice from a Scientific Committee, annually on regulations about whale catches. The decision process for rules about catch limits requires a three-quarters majority vote. The same type of process described above for RFMOs allows states to "object" to (opt out of) rules by following a particular process within 90 days after they are passed (and then, if some do object, others are given an additional period of time during which they can also object). This process resulted in some whaling seasons in which the majority of the whaling states opted out of the regulations that were passed.

The IWC creates regulation through a "schedule," passed every year. Over time the approaches it takes to conservation have evolved. The Commission began by following pre-IWC practice of regulating catches in "Blue Whale Units" rather than by specific numbers and species – a comparison based on how much oil each whale contained (since whale oil was the primary product that came from whales at the time) designated in comparison to the largest whale species. Quotas during this period were global, rather than being allocated by state or vessel, leading to a form of competition for who could catch whales the fastest, which came to be known as the "Whaling Olympics." Since the whaling season closed after the annual quota had been caught, those who caught whales fastest would bring in the greatest numbers. Owners of whaling vessels had an incentive to increase size and technological sophistication of their ships in order to be able to find and catch whales before others could. As ships became faster and more sophisticated, the length of the whaling season decreased dramatically, moving from 112 days in 1946 to only 64 days by 1951 (Clark and Lamberson 1992: 107–109).

By 1972 the IWC had moved to regulating by species (or sometimes subspecies) and region, with the catching of some especially endangered species prohibited altogether. It was difficult, however, to gain agreement to decrease quotas to a low enough level, and by the 1970s whale stocks were unquestionably depleted. The IWC began debating the possibility of a moratorium on commercial whaling to allow whale stocks the chance to recover; this moratorium (officially an annual quota set at zero) was finally passed in 1982, to take effect beginning with the 1986 whaling season.

Early marine mammal conservation efforts had mixed success, and often demonstrated in a particularly stark way the difficulties of international cooperation to protect ocean resources. Some of the clearest problems of non-compliance were witnessed in the case of whaling. In the early years of international cooperation efforts there were minimum catch sizes required, so that juvenile whales would not be caught before they had the chance to reproduce. As with almost all international agreements, information on catches was self-reported, and there were years in the 1960s when 90 percent of whale catches were reported at within one foot of the legal minimum, a statistical impossibility (Birnie 1985: 338).

Whaling also provides the clearest example of intentional state-level non-compliance (see Chapters 9 and 10). After the collapse of the Soviet Union, scientists in Russia released the official state-level statistics that were kept on whale catches, which were different from those submitted to the IWC. The Soviet Union had reported catches during this era that were entirely in line with its obligations, while its actual catches were not within legal limits and often included catches of species that were under moratorium and should not have been caught at all (Brown 1994). In addition to demonstrating the difficulties of monitoring and implementing international agreements, this episode demonstrated the relationship between scientific research and compliance. During this period some of the species – such as humpback, right and blue whales, which were under moratorium – were failing to rebound as scientific models predicted they should have if, as reported, none were being caught. This misreporting made scientific estimates of stock recovery or future sustainable harvesting unreliable ("Call Me Smiley" 1994).

From conservation to preservation

What is particularly interesting about efforts to protect marine mammals is that the reason for protecting them has changed over time: from seeing them as a resource to be hunted sustainably so that they can continue to be used over time, to seeing them as entities deserving of protection as individuals. This view is not universally accepted. There are enormous political fights in regulatory contexts among states that believe these species should be considered in the same way as any other resource and harvested in a sustainable manner, and those that argue that they should not be harmed for their own sake. But the latter view has played an increasingly important role in how they have been protected.

This context has been clearest in the case of dolphins. Although there are some places where small cetaceans are hunted as a resource, most frequently by indigenous populations, dolphins are often killed as bycatch in the process of fishing for other species. That has been particularly true of fishing for yellowfin tuna in the Eastern Tropical Pacific (ETP) Ocean. In that region, dolphins school with tuna, for reasons that are not entirely understood. Because dolphins, as mammals, have to surface to breathe, they are much easier to locate than are schools of tuna, which remain under the water. A fishing technique developed in the 1950s to take advantage of this association: fishers would encircle dolphins with purse seine nets to catch the tuna that would be schooling below. As the nets were drawn closed to gather in the fish, the dolphins would be held under water and would drown.

By the early 1970s, dolphin deaths in the ETP from this practice were above 300,000 annually (Bonanno and Constance 1996: 127). An exposé of these dolphin deaths by an environmental organization helped change public opinion in the United States; the United States was the biggest fishing state in this region at that point (DeSombre 2000). This public outrage helped lead to the passage in the United States in 1972 of the Marine Mammal Protection Act, which imposed restrictions on human behavior that might harm marine mammals. Beginning in 1974 the National Marine Fisheries Service created quotas for how many dolphins could be killed by fishers.

Complicated political struggles followed, in which the United States, in part to appease its domestic tuna fishing industry, imposed economic sanctions on states that did not follow the same fishing practices (and prevent the same number of dolphin deaths) as US tuna fishers (DeSombre 2000). These sanctions, poorly designed from a trade perspective, were twice ruled illegal by the dispute-resolution procedure of the General Agreement on Tariffs and Trade (DeSombre and Barkin 2002), but nevertheless led many fishing vessels to change their tuna-fishing practices to protect dolphins sufficiently that their tuna could be exported to the United States. At the same time, many US tuna vessels left the region, either moving to fish elsewhere or changing registration to other states with laxer dolphin-protection regulations.

More importantly, the Inter-American Tropical Tuna Commission (IATTC) took up the mantle of dolphin protection. As the RFMO that regulated fishing behavior in this region, it was the most obvious international organization to get involved in addressing this issue. The IATTC led the negotiation of the Agreement for the Conservation of Dolphins (also known as the La Jolla Agreement) in 1992, which created continually decreasing mortality limits on dolphins in the context of tuna fishing, and required observer coverage on all member vessels to ensure that rules about protecting dolphins were followed. Dolphin mortality in tuna fishing has dramatically decreased from this combination of endeavors.

This shift of approach to protecting marine mammals has also contributed to the way whales are protected in the International Whaling Commission. "Save the Whales!" became a cry of the early environmental movement, and whales were seen as worthy of saving not because they are a resource, but because they can feel pain, are intelligent, and form strong family relationships (D'Amato and Chopra 1991). Some traditional whaling states, like the United Kingdom, the United States, Australia and New Zealand, shifted to oppose whaling, driven largely by domestic pressures. In the United States, the transition officially came with passage of the Marine Mammal Protection Act in 1972.

There was no question that the process of regulating whaling as a resource conservation effort had failed to successfully conserve whale stocks. When the moratorium was initially passed, it was through the combined agreement between whaling states that recognized that whaling had to pause in order to allow stocks to regenerate, and those states that had concluded – due largely to domestic animal-rights pressure – that whaling for any reason was unethical. Since then the moratorium has held, despite evidence that some whale stocks have recovered sufficiently that they could be harvested in a sustainable manner. To states or domestic populations that oppose whaling for ethical reasons, the sustainability argument is no longer persuasive.

The battle over the basis for regulating whaling persists, with Japan, the predominant whaling state, deciding at the end of 2018 to withdraw from the IWC and resume commercial whaling as of 2019 (Wold 2020). Other states – such as Norway – continue to hunt whales commercially through a previously existing objection to the zero-catch limits initially passed. This political problem is in some ways intractable, since the two different bases for protecting whales differ conceptually, with no clear compromise possible.

Conclusion

The efforts to protect ocean fisheries and marine mammals demonstrate that, despite collective incentives to use these resources sustainably, actual protection of them has been extremely difficult. The broader problems of uncertainty about resources (and behavior of

Elizabeth R. DeSombre

actors) in an international space combine with a common-pool resource structure that means that it is difficult to exclude actors from access to the resource and that overfishing by some affects the state of the resource for others.

The specific form of regulation has also experienced difficulties, with regionally based organizations unable to deal with the shifts of capacity to new regions or species in the context of a global fishing industry, and collective action problems faced within RFMOs as the political process results in choosing catch levels that are higher than recommended by scientific advice. Even when regulations are created, fishing vessel owners find ways around them by registering their ships in states that are not members of the relevant RFMOs or occasionally through outright – but difficult to detect – non-compliance.

Marine mammal protection has in some ways had a more successful recent history, and it has changed the approach from one of sustainable use to complete preservation. But that perspective is not universally accepted, and may thus not be a more broadly applicable model, especially with respect to fisheries. Overall, although there have been some signs of progress in better management of resources, efforts to address marine fisheries and mammals demonstrate how difficult cooperation to protect international environmental resources can be.

References

Agreement for the implementation of the provisions of the united nations convention on the law of the sea relating to the conservation and management of straddling fish stocks and highly migratory fish stocks (1995).
Barkin, J.S. and DeSombre, E.R. (2013) *Saving global fisheries: reducing fishing capacity to promote sustainability*, Cambridge, MA: MIT Press.
Birnie, P. (1985) *International regulation of whaling: from conservation of whaling to conservation of whales and regulation of whale watching*, New York: Oceana Publications.
Bonanno, A. and Constance, D. (1996) *Caught in the net: the global tuna industry, environmentalism, and the state*, Lawrence: University Press of Kansas.
Brown, P. (1994) "Soviet Union Illegally Killed Great Whales," *Guardian* (12 February): 14.
"Call me Smiley" (1994) *New York Times Magazine* (13 March): 14.
Chopin, F., Inoue, Y., Matsushita, Y., and Arimoto, T. (1996) "Source of Accounted and Unaccounted Fishing Mortality," in Alaska Sea Grant College Program, ed., *Solving bycatch: considerations for today and tomorrow*, Report 96–03, Fairbanks: University of Alaska.
Clark, C.W. and Lamberson, R. (1992) "An Economic History and Analysis of Pelagic Whaling," *Marine Policy* 6(2): 107–109.
Colson, D.A. (1995) "Current Issues in Fishery Conservation and Management," *US Department of State Dispatch* 6(7). Available HTTP: <(http://dosfan.lib.uic.edu/ERC/briefing/dispatch/1995/html/Dispatchv6no07.html)>.
Convention between the United States, Great Britain, Russia, and Japan for the preservation and protection of Fur Seals (1911).
Convention on future multilateral cooperation in the Northwest Atlantic Fisheries (1978).
Cushing, D.H. (1988) *The provident sea*, Cambridge: Cambridge University Press.
D'Amato, A. and Chopra, S.K. (1991) "Whales: Their Emerging Right to life," *American Journal of International Law 85*: 21–62.
DeSombre, E.R. (2000) *Domestic sources of international environmental policy: industry, environmentalists, and US power.* Cambridge, MA: MIT Press.
DeSombre, E.R. (2005) "Fishing under Flags of Convenience: Using Market Power to Increase Participation in International Regulation," *Global Environmental Politics* 5(4): 73–94.
DeSombre, E.R. (2006) *Flagging standards: globalization and environmental, safety, and labor standards at sea*, Cambridge, MA: MIT Press.
DeSombre, E.R. and Barkin, J.S. (2002) "Turtles and Trade: The WTO's Acceptance of Environmental Trade Restrictions," *Global Environmental Politics* 2(1): 12–18.
Englander, G. (2019) "Property Rights and the Protection of Global Marine Resources." *Nature Sustainability* 2(10): 981–987.

Francis, D. (1990) *A history of world whaling*, Ontario: Viking.

Gilman E., Roda A.P., Huntington T., Kennelly S.J, Suuronen P., Chaloupka M., Medley P.A. (2020) "Benchmarking global fisheries discards," *Scientific Reports* 10(1):1–8.

Hilborn, R. and Walters, C.J. (1992) *Quantitative fish stock assessment: choice, dynamics and uncertainty*, New York: Chapman and Hall.

ICTFU, Trade Union Advisory Committee to the OECD, ITC, and Greenpeace International (2002) "More Troubled Waters: Fishing Pollutions and FOCs," Major Group Submission for the 2002 World Summit on Sustainable Development, Johannesburg.

International convention for the regulation of whaling (1946).

Kurlansky, M. (1997) *Cod: a biography of the fish that changed the world*, New York: Walker.

Lack, M. and Sant, G. (2001) "Patagonian Toothfish: Are Conservation and Trade Measures Working?" *Traffic Bulletin* 19(1): 1–19.

Lodge, M.W., Anderson, D., Løbach, T., Munro, G., Sainsbury, K., and Willock, A. (2007) *Recommended best practices for regional fishery management organizations*, London: Chatham House.

Mack, J.R. (1996) "International Fisheries Management: how the UN conference on straddling and highly migratory stocks changes the law of fishing on the high seas," *California Western International Law Journal* 26(Spring): 318–321.

North Atlantic Fisheries Intelligence Group and INTERPOL (2017) *Chasing red herrings: flags of convenience and the impact on fisheries crime law enforcement*, Oslo: NA-FIG.

OECD (2004) "Draft Chapter 2 – Framework for Measures Against IUU Fisheries Activities," OECD, Fisheries Committee, Directorate for Food, Agriculture and Fisheries, AGR/FI/IUU(2004) 5/PROV.

Pascoe, S. (1997) "Bycatch Management and the Economics of Discarding," *FAO Technical Paper 370*, Rome: FAO.

Peterson, M.J. (1993) "International Fisheries Management," in Peter M. Haas, Robert O. Keohane, and Marc A. Levy (eds) *Institutions for the Earth*, Cambridge, MA: MIT Press, 249–305.

Polachek, T. and Davies, C. (2008) "Consideration of the Implications of Large Unreported Catches of Southern Bluefin Tuna for Assessments of Tropical Tuna, and the Need for Independent Verification of Catch and Effort Statistics," *CSIRO Marine and Atmospheric Research Paper* 23.

Roberts, C. (2007) *The unnatural history of the sea*, Washington, DC: Covelo Press.

Rojas, E. (2008) "Fisheries Observer Harassment and Interference – A Global Challenge," *APO Mail Buoy* 2(3): 6–11.

Sanger, C. (1986) *Ordering the oceans: the making of the law of the sea*, London: Zed Books.

Scheiber, H.N. (2001) "Ocean Governance and the Marine Fisheries Crisis: Two Decades of Innovation and Frustration," *Virginia Environmental Law Journal* 20: 119–137.

Schiffman, H.S. (2008) *Marine conservation agreements: the law and policy of reservations and vetoes*, Leiden: Martinus Nijhoff Publishers.

SkyTruth and Global Fishing Watch (2017) *The global view of transshipment: revised preliminary findings* Washington, DC: Global Fishing Watch https://globalfishingwatch.org/data/.

Springer, A.L. (1997) "The Canadian turbot war with Spain: unilateral state action in defense of environmental interests," *Journal of Environment and Development* 6(1): 26–60.

Stokke, O.S. and Vidas, D. (2004) *Regulating IUU fishing or combating IUU operations?* OECD, Fisheries Committee, Directorate for Food, Agriculture and Fisheries, AGR/FI/IUU, Rome: FAO.

Swan, J. (2002) "Fishing Vessels Operating under Open Registers and the Exercise of Flag State Responsibilities – Information and Options," *FAO Fisheries Circular No. 980*, FITT/C980, Rome: FAO.

UNEP (2004) *Analyzing the resource impact of fisheries subsidies: a matrix approach*, Geneva: United Nations Environment Programme.

UN Food and Agriculture Organization (2020) *The state of world fisheries and aquaculture*, Rome: FAO.

Wold, C. (2020) "Japan's Resumption of Commercial Whaling and Its Duty to Cooperate with the International Whaling Commission," *Journal of Environmental Law & Litigation* 35: 87–143.

41

Biodiversity, migratory species and natural heritage

Global challenges ahead

Volker Mauerhofer and Felister Nyacuru

This chapter introduces the overlapping issues of biodiversity, migratory species and natural heritage. In the context of global environmental politics, these issues have somewhat different meanings in theory and in practice. This chapter describes each of the issues and focuses on the practical governance of each one, and it explores their relationships to one another. The chapter applies a distinction made by North (1990), who pointed to organisations as 'players of the game' and formal institutions as the "rules of the game" in terms of formal regulatory rules that govern individual behavior and structure social interactions. The chapter aims to provide a focused overview of these formal global institutions, with a lesser emphasis on the relevant organizations (see also Chapters 8 and 9). Where deemed to be appropriate, governance-relevant information from the regional, national and subnational levels has also been included (see also Chapters 7, 12 and 14).

International (global) formal institutions are understood in this chapter as a subset of multilateral environmental agreements (MEAs). Specific MEAs regarding biodiversity, migratory species and natural heritage emerged early in the last century. They include, for instance, the 1900 London Convention for the Protection of Wild Animals, Birds and Fish in Africa (replaced by the 1933 London Convention Relative to Preservation of Flora and Fauna in their natural state), the 1902 Convention for the Protection of Birds Useful for Agriculture, the 1911 Treaty for the Preservation and Protection of Fur Seals, and the 1923 Convention for the Preservation of the Halibut Fishery of the Northern Pacific Ocean (see Brown-Weiss 1992: 479; Sands 1995: 338; see also Chapter 40).

Biodiversity

The three issues examined in this chapter are characterized by a considerable degree of overlap. Biodiversity, for example, covers all migratory species. Nevertheless, some significant differences in the conceptual basis and further understanding can also be found, as discussed below. There are numerous frameworks governing biodiversity (see Morgera 2017, Kütting and Herman 2018). Distinctions can be made quantitatively and qualitatively. Basic quantitative differentiations can concern the number of species covered while qualitative distinctions address the different level of political commitment or the different structure of

DOI: 10.4324/9781003008873-46

the framework itself (e.g., Memorandum of Understanding versus Convention) or its organizational units (e.g., frameworks with or without enforcement mechanisms).

The database of United Nations Environment Programme (UNEP) currently contains 11 MEAs that are in some way relevant to biodiversity in the global sense (UNEP 2021). Numerous bilateral environmental agreements also exist wherein two parties, in most cases countries, agree on issues related to biodiversity. Several definitions of biodiversity exist; the term is often used synonymously with biological diversity. The most widely agreed definition is enshrined in the Convention on Biological Diversity (CBD) concluded in 1992: "'Biological diversity' means the variability among living organisms from all sources including, inter alia, terrestrial, marine and other aquatic ecosystems and the ecological complexes of which they are part; this includes diversity within species, between species and of ecosystems" (Article 2 CBD). The CBD's definition of biodiversity does not expressly cover abiotic issues, but these also provide a continuous contribution to shaping species and ecosystems. In comparison, another definition explicitly includes abiotic factors: "The term biodiversity encompasses all of the species that currently exist on Earth, the variations that exist within each species, and all of the interactions that exist among all of these organisms and their biotic and abiotic environments as well as the integrity of these interactions" (Gowdy 1997: 186).

There are numerous politically agreed-upon frameworks for governing biodiversity. They range from binding international treaties to informally institutionalized networks among nations and other stakeholders. International treaties, which are largely steered by the nation-states themselves, set up rules and regulations to which the state parties commit (Jardin 2010). Such formal frameworks can be very broadly formulated thematically. They can address an ecosystem or habitat type, or they can focus exclusively on species (see Table 41.1). Several formal MEAs listed in Table 41.1 address the issue of conservation areas (Mauerhofer et al. 2015; Beresford et al. 2016).

Related to all of these agreements, boundary organizations facilitate the uptake of scientific information through the policy process by enabling a bidirectional flow of communication between scientists and policymakers (Morin et al. 2020). In the context of biodiversity conservation, UNEP administers IPBES, a boundary organization that was established in 2011 with a global mandate (Hrabanski and Pesche 2017). Reports produced by IPBES have now been acknowledged as authoritative through the governance processes of international environmental agreements, such as the CBD, despite, for example, critical references to

Table 41.1 Examples of biodiversity multilateral environmental agreements and their themes at global and regional levels

Themes	Geographic level	
	Global	*Regional*
General	• Convention on Biological Diversity	• European Alpine Convention (Alpine Convention 2021)
Habitat/ ecosystem-related	• World Heritage Convention	• EU – Habitat and Birds Directives
Species-related	• Convention on the Conservation of Migratory Species of Wild Animals	• Lusaka Agreement on Cooperative Enforcement Operations Directed at Illegal Trade in Wild Flora and Fauna
	• Convention on the International Trade in Endangered Species of Wild Fauna and Flora	• Agreement on Conservation of African Eurasian Migratory Water Birds

timing and scope of future global assessments, as well as to overlaps and synergies with future editions of the Global Biodiversity Outlook (CBD COP 2018). The work of IBPES has also not been free from other criticism. Controversies have arisen about the insufficient involvement of social scientists (Vadrot et al. 2018), as well as about – despite the unchanged name of IPBES – the focal switch from "ecosystem services" to "nature's contributions to people" (see, e.g., Braat 2018), as the new term still retains a largely instrumental perception of nature that widely ignores other world views.

Notably, nongovernmental organizations (NGOs) (see Chapter 14) have, during the past decade, strongly entered into the field of implementation of these conventions, as part of negotiation teams of Parties based on the Almaty Guidelines of the regional Arhus Convention (UNECE 2005; Prideaux 2015) and as enforcers of "Natures' rights" (Chapron et al. 2019). There has also been improved standing for NGOs in administrative and juridical procedures granted by, for example, the Arhus Convention (Mauerhofer 2016) and the Escazú Agreement, which entered into force on 22 April 2021 (Stec and Jendrośka 2019; CEPAL 2021).

Informal governing frameworks establish more loose networks and platforms for initiating a broad range of enterprises and policy initiatives. One of the informal frameworks is the International Union for Conservation of Nature (IUCN), which is the world's oldest and largest global environmental organization. It is a leading authority on the environment and sustainable development, and has many members drawn from both governmental and nongovernmental organizations. The IUCN has a network of more than 11,000 voluntary scientists and experts who have established global standards in their respective fields by setting definitive international standards, such as the IUCN Red List of Threatened Species (Betts et al. 2020), the IUCN Green List of Protected Areas (Akcakaya et al. 2018) and the IUCN Red List of Ecosystems adopted in 2014 (Bland et al. 2019). Another example of such a network is the Global Biodiversity Information Facility, an international initiative created and funded by governments that is focused on making biodiversity data available to all for scientific research, conservation and sustainable development (Zizka et al. 2020; GBIF 2021).

Global governance of biodiversity

A number of key international agreements have been reached in recent decades to address questions of biodiversity, including the Convention on Biological Diversity (CBD), the Convention on International Trade in Endangered Species of Wild Fauna and Flora, and the International Treaty on Plant Genetic Resources for Food and Agriculture (see Chapter 44). The CBD, which was concluded in 1992 during the Rio Earth Summit, has been ratified by 196 parties. Its wide scope covers ecosystems, species and genetic resources, and it aims at the conservation of biological diversity, sustainable use of the components of biodiversity as well as the fair and equitable sharing of benefits arising from the utilization of genetic resources (see Chapter 25). The Conference of the Parties (COP) is the most important decision-making structure of the CBD. It governs the Convention and furthers its implementation through the decisions taken at its periodic conferences (CBD 2021). The CBD is administered by a Secretariat, which is embedded in a larger organizational structure (Siebenhüner 2007; Mauerhofer 2019a). The parties play a key role toward implementation of the Convention; at their respective national levels, they use National Focal Points (NFPs) to prepare national reports (Ette and Geburek 2020).

As a framework convention, the CBD has enabled the negotiation of additional institutional arrangements. The Cartagena Protocol on Biological Safety was concluded in 2000 under the CBD's aegis. The protocol, which had 173 parties as of January 2021, aims to ensure the safe handling, transport and use of living modified organisms resulting from modern biotechnology (Falck-Zepeda and Zambrano 2011; CBD 2021). The Protocol currently has an intrinsic political weakness: the United States has not ratified the CBD. The United States prefers the WTO framework (Young 2008; see Chapter 24). In 2010 at COP10, the Nagoya Protocol was developed as an instrument for the implementation of provisions on Access and Benefit Sharing of genetic resources under the CBD framework (Aviles-Polanco et al. 2019; CBD 2021). This protocol, which is open to all CBD parties, had 128 parties as of December 2020 (CBD 2021). The Nagoya Protocol has already been critically assessed, with praise for its positive aspects, including its clear definition of "utilization of genetic resources" on the one side, and criticism of its negative aspects, such as weak language ("endeavor," "encourage" and the like), as well as the lack of a self-standing obligation for user states to ensure benefit sharing, on the other side (Kamau et al. 2010; Harrop 2011; Kariyawasam and Tsai 2018).

In October 2010 the Nagoya–Kuala Lumpur Supplementary Protocol on Liability and Redress to the Cartagena Protocol on Biosafety was adopted, parallel to the Nagoya Protocol. Its parties include 47 countries and the European Union, and it has been in force since May 2018 (CBD 2021). It is recognized that this new Supplementary Protocol concretizes the norms of the Cartagena Protocol, while, on the other hand, the concrete outcome has been considered quite disappointing as it primarily focuses on binding international rules rather than nationally defined civil liability solutions (Sands et al. 2012).

Judging by the number of ratifications, the CBD and its protocols can already be considered one of the most successful biodiversity-related international frameworks, with no other convention coming near them. This number can also be seen, however, as a direct reflection of how many of the Convention's articles are formulated, leaving much up to the discretion of its ratifying parties. Doubtless, the CBD has shown its merits by bringing biodiversity closer to the center of global environmental policy over the past three decades (Glowka 2000; Mauerhofer 2019a; Keiper and Atanassova 2020). Numerous countries have started to produce national and local strategies and action plans (McCay and Lacher 2021) and, based on Article 26 of the CBD, countries with weak or non-existent policies have had to continuously and publicly report on biodiversity issues. It also generated several more binding formulations of efforts to stop the loss of biodiversity, notably within the EU, and to reduce the rate of that loss (Maes et al. 2016). Several of these aims, in particular the more ambitious ones, failed spectacularly due to insufficient political efforts and too many countervailing interests (Harrop and Prichard 2011; Gamero et al. 2017). It is yet to be seen whether the outcomes of recent and future COP meetings can contribute to more focused efforts and increased accountability by global environmental policy stakeholders.

In general, all the current protocols of the CBD focus on trade and other use-related issues. Topics that are more protection-oriented appear to gain less attention when it comes to further specifying the effects of the CBD in a legally binding way within global environmental policy forums.

Another outcome of CBD COP10 in 2010 was the Strategic Plan for Biodiversity 2011–2020, which had 5 goals and 20 (only partly) quantifiable targets (the "Aichi Targets"), such as "Target 5": "By 2020, the rate of loss of all natural habitats, including forests, is at least halved and where feasible brought close to zero, and degradation and fragmentation is significantly reduced" (UN 2010). Parallel to this, the UN introduced the Sustainable Development Goals (SDGs), which include biodiversity-related targets (UN 2012, 2015; Kanie et al. 2017). This happened alongside an upgrade of UNEP toward universal membership in its governing body and securing improved funding from the UN regular budget (UN 2012 para. 88). Goal

14 ("Life under water") and goal 15 ("Life on Land") of the SDGs are specifically focused on biodiversity conservation and its sustainable use, with 10 and 12 related targets respectively (UN 2015). Thus far, the SDGs show substantial weaknesses related to their environmental indicators and their performance seems to be largely trumped by socio-economic goals, leaving little hope that they will really live up to the promise of halting biodiversity loss (see, e.g., Elder and Olsen 2019; Zeng et al. 2020). Given this outlook, together with the recent wide failure to meet the Aichi Targets (Buchanan et al. 2020), it is not surprising that the post-2020 Global Biodiversity Framework is intended to contribute to the implementation of the SDGs (UN 2020, para. 4). This Framework was to be negotiated and considered for approval at the 15th CBD COP (UN 2020; Williams et al. 2020). It includes an even more ambitious goal of global environmental net improvements by 2050 (UN 2020, echoed by the EU 2020, as "net gain principle"). This poses tremendous challenges for the Global North to achieve a net gain in governance for biodiversity by 2030 (Mauerhofer 2021).

The CBD also has two biodiversity-relevant sister conventions, both of which were also concluded at the 1992 Rio Summit, namely the Framework Convention on Climate Change (see Chapter 32) and the Convention to Combat Desertification (see Chapter 43). The significance of these two sister conventions rests in the fact that they complement one another to help address two of the most important driving forces of global problems that are directly or indirectly affecting biodiversity.

The Convention on International Trade in Endangered Species of Wild Fauna and Flora (CITES) was concluded in March 1973. Politically it can be seen as an outcome of the 1972 United Nations Conference on the Human Environment (also known as the Stockholm Conference). Adoption of a resolution at that conference led to another large conference in Washington where CITES was finally adopted (Wijnstekers 2018). This Convention entered into force in July 1975 and currently has 183 parties (CITES 2021). CITES aims to improve international cooperation on the conservation of species of wild animals and plants while officially reported trade in these species has quadrupled since 1975 (Harfoot et al. 2018). It functions by subjecting international trade in specimens of selected species to certain controls. All import, export, re-export and introduction by sea of species covered by the Convention must be authorized through a licensing system. The COP, as the supreme governing body of CITES, decides on changes within the Appendices. The COP's meetings are organized by the CITES Secretariat, which is located in Geneva, Switzerland, and is administered by UNEP. The Secretariat primarily distributes information to the parties through meeting documents and notifications while the function to sanction Member States is quite unique in the global biodiversity convention setting (Mauerhofer 2019a). Within CITES, NGOs have recently increased influence in highly contested COP-issues (Gaffney and Evensen 2020) and, externally, CITES-cooperation with other conventions, such as Convention on Migratory Species (CMS), has gained additional momentum (Hellinx 2020). A stronger monitoring of demand and supply, particularly related to poaching, has been recommended to CITES to increase its effectiveness (Challender et al. 2015). Improvement in information compliance of management authorities was also recommended by a study that found a lack of data on live seizures among 70 percent of country parties for 2010 to 2014 (D'Cruze and Macdonald 2016).

The International Treaty on Plant Genetic Resources for Food and Agriculture was negotiated by the United Nations Food and Agriculture Organization (FAO) and entered into force in 2002. It replaced a previous voluntary agreement – the 1983 International Undertaking on Plant Genetic Resources for Food and Agriculture. The treaty seeks to increase recognition of the importance of diversity among certain crops, to establish a global system of access sharing for plant genetic materials, to ensure benefit sharing with countries of origin of these genetic materials, and to further conservation and sustainable use (Tsioumani 2018; see Chapter 40). Since 2006 it has had its own Governing Body under the aegis of the FAO. This Body is the highest organ of the Treaty as established in Article 19 and composed of representatives of all 148 contracting parties (as of January 2021; FAO 2021).

Regional governance of biodiversity

Several regional multilateral agreements related to biodiversity have been reached, including the Lusaka Agreement on illegal species trade and the European Union's Habitats Directive. The Lusaka Agreement on Cooperative Enforcement Operations Directed at Illegal Trade in Wild Fauna and Flora was concluded in 1994; it entered into force in 1996 (Lusaka Agreement 2021). There are currently seven parties to the Agreement, namely the Republic of the Congo (Brazzaville), Kenya, Liberia, Tanzania, Uganda, Zambia and the Kingdom of Lesotho (Lusaka Agreement 2021). It has its headquarters at the Kenya Wildlife Service in Nairobi. The Agreement establishes a three-tier institutional mechanism with a Governing Council functioning as a policy- and decision-making organ, a Task Force as a permanent law enforcement institution, and a National Bureau as a governing body. The Lusaka Agreement aims, in particular, to strengthen the enforcement capacity among its members, for example through a Wildlife Enforcement Monitoring System (WEMS). WEMS addresses information and reporting processes as well as analyzing capabilities regarding the monitoring of the illegal wildlife trade at both the national and the regional levels (Chandran et al. 2015; Lusaka Agreement 2021).

Cornerstones of the EU's nature conservation policy include the 1979 Birds Directive and the 1992 Habitats Directive forms. EU policy is built around two pillars: the so-called Natura 2000 network of protected sites and a strict system of species protection (Bouwma et al. 2017). In total, the Directives protect over 1,000 animal and plant species, and over 200 so-called "habitat types" (i.e., special types of forests, meadows, wetlands and so forth). The EU Commission supervises the implementation of these Directives by the EU's 27 member states. Binding rulings on infringement, preliminary rulings, and penalty procedures made at the European Court of Justice (ECJ) can be seen as the world's most progressive enforcement mechanism for MEAs (Mauerhofer 2019b). The central aim of the Habitat Directive is to maintain or restore a favorable conservation status for species and habitat types (and similarly for the Birds Directive). The ECJ was very progressive in, for example, the application of the precautionary principle for a case dealing with annual fishery activities within a Natura 2000 site in the Netherlands (Mauerhofer 2008; Lees 2016). In this case, the ECJ twice shifted the burden of proof beyond scientific doubt on to the shoulders of those proposing an activity: first concerning the question of whether an appropriate assessment of the implications of these activities is necessary and, second, concerning the question of whether these activities adversely affect the integrity of the site. This and similar decisions "in *dubio pro nature*" of the ECJ contributed to the Natura 2000 site network aiding the increase in populations of wild birds in Europe (Beresford et al. 2016). This has been due to the network's important role in bird migrations and its crucial influence on bird populations in other regions of the world.

International grades

"International grades" are non-binding certificates of excellence regularly used for site conservation. They often play an important role in influencing biodiversity-related decisions by policymakers and other stakeholders. On the global level, the nomination of biosphere reserves, established as a network in 1977 by UNESCO's Man and Biosphere program, is among the best-known examples of such international grades (Price et al. 2010; UNESCO 2011, Reed 2019; UNESCO 2021). The 714 sites, including 21 transboundary ones, in 129 countries (as of March 2021) include Serengeti-Ngorongoro, Archipiélago de Colón (Galápagos), Danube Delta and the Rocky Mountains (UNESCO 2021). Entry into the IUCN's list of protected areas is also a type of international grade. This list contains a classification into different categories according to management objectives (IUCN 2021). One example of a classification category II is "National Park," which has some quantified management criteria and has played an important role – although a non-binding one – in national designations of protected areas, for example in Austria. Other examples of such international grades include designation of areas as Globally Important Agricultural Heritage Systems (GIAHS) by the FAO (Santoro et al. 2020) and the European Diploma of Protected Areas granted by the Council of Europe to protected areas based on their outstanding scientific, cultural or aesthetic qualities (CoE 2021).

Migratory species

Migration of wild species can be classified according to different factors. These can be broken down into factors influenced by humans and those that are not. With the projected impacts of global climate change (Chapter 32), the issue of migratory species is likely to gain increased prominence. In the broadest sense, "migratory species" can be defined as "the entire population or any geographically separate part of the population of any species or lower taxon of wild animals or plants, a significant proportion of whose members autonomously cross one or more national jurisdictional boundaries" (CMS Article 1). This definition includes plants that are able to migrate, for example due to climate change (LeDee et al. 2021), as well as alien species in general (Essl et al. 2020).

Global governance of migratory species

The Convention on Wetlands of International Importance – the Ramsar Convention – was concluded in 1971 in the city of Ramsar in Iran (see Chapter 34). It has 171 parties (Ramsar 2021). The development of this Convention was largely driven by the efforts of the nongovernmental International Waterfowl Research Bureau (Sands et al. 2012). The Ramsar Convention focuses in particular on the conservation and sustainable maintenance ("wise use") of internationally important wetlands (Davidson et al. 2020). Sites are selected based on certain criteria, including among other things the occurrence of a specified number of migratory bird species (Ramsar 2021). The effectiveness of its implementation is steered in different continents by various governance regimes (Mauerhofer et al. 2015). Currently, the Ramsar List covers 2,414 wetlands with a total surface area of 254,540,512 hectares (as of January 2021; Ramsar 2021). The preamble to the Convention states that waterfowl migration is seasonal and not caused by humans, although there is likely to be a large influence from human-caused climate change (see Chapter 32). The Ramsar Convention is organizationally administered by a Secretariat based in Gland, Switzerland, which has organized 13 COPs and also administers the Montreux Record. The latter is a sort of watchlist that includes 45 sites (by March 2020) that are in some way endangered by human pressures (Ramsar 2021). However, this process has not been effective in preventing the decline of wetlands, and proposals for improvements have been advanced (Davidson et al. 2020).

Table 41.2 List of agreements and examples of memoranda of understanding concluded under the Convention on Migratory Species

Agreements (species listed in CMS Appendix II)	Memoranda of Understanding
• African–Eurasian Waterbirds Agreement	• African Elephant – West Africa
• Albatrosses and Petrels	• Aquatic Warbler
• Small Cetaceans – Mediterranean and Black Seas	• Bukhara Deer
• Small Cetaceans – Baltic and North Seas	• Cetaceans – Pacific
• Wadden Sea Seals	• Great Bustard
• European Bats (Eurobats)	• Marine Turtles – Indian Ocean/Southeast Asia
• Gorillas and their Habitats	• Birds of Prey
	• Saiga Antelope
	• Siberian Crane
	• Slender-billed Curlew

The Convention on Migratory Species was signed in 1979 in Bonn under the aegis of UNEP. It entered into force in November 1983 and had 132 parties as of December 2020 (Hensz and Soberon 2018; CMS 2021). The CMS is fully focused on the conservation of wild animals throughout their range, with an emphasis on terrestrial, marine and avian migratory species (Koester 2002). It aims to conserve migratory species and their habitats by providing strict protection for species through multilateral agreements, Memoranda of Understanding (MoU) and cooperative research activities. The COP is the decision-making organ of the Convention. It reviews the implementation of the Convention, can adopt recommendations, and provides an overview of information from all the agreements concluded under the CMS (CMS Articles IV, V and IX). For many countries, the Convention on its own does not have much immediate influence because its Appendix 1 list of species often does not contain many species that occur on their national territory. Furthermore, the impacts of climate change have rendered migratory patterns increasingly unpredictable (Essl et al. 2020; LeDee et al. 2021; on climate change, see Chapter 32). In general, more intense cooperation and coordination with other conventions is strongly recommended (Glowka 2000; Kuunal et al. 2020). Under the aegis of the CMS, seven legally binding agreements and 19 MoUs were concluded up to mid-2021 (CMS, 2021; see Table 41.2).

Regional governance of migratory species

Apart from the internationally agreed upon CMS, there are several regionally concluded treaties dealing with migratory species. The reason these treaties were not concluded under the CMS is partly because one or more of the parties are not members of the Convention (Boardman 2006), Several bilateral migratory bird agreements exist, for example between Australia and other countries, namely the 1974 Japan–Australia Migratory Bird Agreement, the 1986 China–Australia Migratory Bird Agreement, and the 2006 Republic of Korea–Australia Migratory Bird Agreement (Hamman 2019). Examples of regional agreements from other continents include the 1937 US–Mexico Convention for the Protection of Migratory Birds and Game Animals, and the 1972 US–Japan Convention for the Protection of Migratory Birds in Danger of Extinction and their Environment (EPA 2021). A common factor among these conventions, treaties and agreements is that they seek to conserve species as "common heritage of mankind" on behalf of future generations (as noted in the preambles of the CMS and Agreement on the Conservation of African-Eurasian Migratory Waterbirds).

Outside formal intergovernmental processes, hybrid arrangements have been completed regionally, too. That is the case with the East Asian-Australasian Flyway Partnership, a voluntary agreement for conserving migratory waterbirds in the Asia-Pacific that includes state and non-state actors. Beyond these actors, various cities have entered non-binding institutional arrangements for conserving migratory shorebirds (Gallo-Cajiao et al. 2019).

Natural heritage

The term "natural heritage" can, on the one hand, be understood to include both biotic and abiotic environments, and therefore be broader than "biodiversity." On the other hand, the term has a clear anthropocentric and unidirectional orientation, putting humanity into the role of a sort of supervisor or protector of something we allegedly have "inherited" from past generations and which we are to maintain in the present and bestow to the benefit of future generations. The World Heritage Convention (WHC) was adopted in 1972 by the General Conference of UNESCO in Paris. It has been ratified by 194 States Parties (as of October 2020; UNESCO 2021). Article 2 of the WHC defines the term "natural heritage" in three different ways: (1) natural features consisting of physical and biological formations or groups of such formations, which are of outstanding universal value from the aesthetic or scientific point of view; (2) geological and physiographical formations, and precisely delineated areas, which constitute the habitat of threatened species of animals and plants of outstanding universal value from the point of view of science or conservation; and (3) natural sites or precisely delineated natural areas of outstanding universal value from the point of view of science, conservation or natural beauty.

The WHC contains procedures and criteria, such as legal and integrity checks, for areas to be listed as cultural, natural or mixed World Heritage Sites (Albert and Ringbeck 2015; Cameron and Rössler 2016). The list included 869 cultural heritage sites, 213 natural heritage sites, and 39 sites containing mixed objects as of January 2021 (UNESCO 2021). Delisting is possible, as was the case with the former Dresden Elbe World Heritage Site, which was delisted due to the serious impact on the integrity of the site's landscape due to the construction of a bridge across the Elbe. If there is a threat to a protected site, it is put on the "WHL in Danger" list until substantial renovations have been carried out and the site is out of danger (Morrison et al. 2020). Additional dynamics toward stronger conservation could be developed by national courts. The High Court of Australia judged in 1983 that the conservation obligations of the WHC regarding a listed riverine landscape in Tasmania are legally binding for Australia (in the Tasmania Dam Case, High Court of Australia 1983). If other courts would take a similar stance, the impact of the WHC would increase tremendously in global environmental policy.

Conclusion

Proposed and concretely envisaged global environmental policies can foster and strengthen biodiversity, migratory species and natural heritage. Some policy approaches are aimed at strengthening mandates for existing institutions, such as expanding the mandate of the United Nations Security Council (Elliott 2005), creating new institutions, such as a World Environment Court (Pauwelyn 2005). Another proposal aims at clustering MEAs in order to overcome their fragmentation (von Moltke 2005). These proposals and ongoing efforts can all be seen as largely addressing the biodiversity and species protection from a top-down perspective. New and innovative bottom-up implementation mechanisms are additionally needed (Mauerhofer 2011; Otero et al. 2020). Indeed, they are gaining increasing importance, in part due to a lack of adequate existing enforcement mechanisms (Mauerhofer 2012).

It is obvious that this governance system is largely not effective, as the status of biodiversity, migratory species and natural heritage continues to deteriorate (CBD-Secretariat 2020; Allan et al. 2017). Conversely, it is understandable that those three elements – biodiversity, migratory species, and natural heritage – would be in an even much worse condition should such a system never have emerged (Beresford et al. 2016). At least as a strong political signal, leaders participating in the UN Summit on Biodiversity in September 2020, representing 84 countries from all regions and the European Union, have committed in a 'Leaders' Pledge for Nature" to reversing biodiversity loss by 2030, which has also been endorsed by multiple non-state actors (Leaders Pledge for Nature 2020). A steep, stony and urgent way of implementation still lays ahead.

Acknowledgment

The authors are grateful to Chiharu Takei for her comments on the first edition of this chapter and Eduardo Gallo-Cajiao for his extensive input to former versions of this second edition of this chapter. The usual disclaimer applies.

References

Akcakaya, H.R., Bennett, E.L., Brooks, T.M., Grace, M.K., Heath, A., Hedges, S. et al. (2018). Quantifying species recovery and conservation success to develop an IUCN Green List of species. *Conservation Biology* 32, 1128–1138.

Albert, M.-T. and Ringbeck, B. (2015). Introduction. In: Albert, M.-T. and Ringbeck, B. (eds) *40 years World Heritage Convention Popularizing the Protection of Cultural and Natural Heritage*, Berlin: Walter de Gruyter, pp. 1–9.

Allan, J.R., Venter, O., Maxwell, S., Bertzky, B., Jones, K., Shi, Y. et al. (2017). Recent increases in human pressure and forest loss threaten natural world heritage sites. *Biological Conservation* 206, 47–55.

Alpine Convention (2021) Alpine Convention https://www.alpconv.org/en/

Aviles-Polanco, G., Jefferson, D.J., Antonio Almendarez-Hernandez, M. and Felipe Beltran-Morales, L. (2019). Factors that explain the utilization of the Nagoya Protocol framework for access and benefit sharing. *Sustainability* 11, 1–18.

Beresford, A.E., Buchanan, G.M., Sanderson, F.J., Jefferson, R. and Donald, P.F. (2016). The contribution of the EU nature directives to the CBD and other multilateral environmental agreements. *Conservation Letters* 9, 479–488.

Betts, J., Young, R.P., Hilton-Taylor, C., Hoffmann, M., Rodriguez, J.P., Stuart, S.N. et al. (2020). A framework for evaluating the impact of the IUCN Red List of threatened species. *Conservation Biology* 34, 632–643.

Bland, L.M., Nicholson, E., Miller, R.M., Andrade, A., Carre, A., Etter, A., et al. (2019). Impacts of the IUCN Red List of ecosystems on conservation policy and practice. *Conservation Letters* 12, 1–8.

Boardman, R. (2006). *The International Politics of Bird Conservation*. Northampton: Edward Elgar Publishing.

Bouwma, I., Arts, B. and Liefferink, D. (2017). Cause, catalyst or conjunction? The influence of the Habitats Directive on policy instrument choice in Member States. *Journal of Environmental Planning and Management* 60, 977–996.

Braat, L.C. (2018). Five reasons why the Science publication "Assessing nature's contributions to people" (Diaz et al. (2018). would not have been accepted in ecosystem services. *Ecosystem Services* 30, A1–A2.

Brown-Weiss, E. (1992). Appendix B: Chronological Index of selected international environmental legal instruments. In Brown-Weiss, E. (ed) *Environmental Change and International Law: New Challenges and Dimensions*, Tokyo: United Nations University Press, pp. 479–490.

Buchanan, G.M., Butchart, S.H.M., Chandler, G. and Gregory, R.D. (2020). Assessment of national -level progress towards elements of the Aichi biodiversity targets. *Ecological Indicators*, 116. doi: 10.1016/j.ecolind.2020.106497.

Cameron, C. and Rössler, M. (2016). *Many Voices, One Vision: The Early Years of the World Heritage Convention*, London: Routledge.

CBD. (2021). *Convention on Biological Diversity.* http://www.cbd.int/

CBD COP. (2018). *Decision Adopted by the Conference of the Parties to the Convention on Biological Diversity*, 14/36 Second work programme of the Intergovernmental Science-Policy Platform on Biodiversity and Ecosystem Services CBD/COP/DEC/14/36.

CBD Secretariat. (2020). *Global Biodiversity Outlook 5*, Montreal, Canada, https://www.cbd.int/gbo5

CEPAL. (2021). *Regional Agreement on Access to Information, Public Participation and Justice in Environmental Matters in Latin America and the Caribbean*, https://www.cepal.org/en/escazuagreement

Challender, D.W.S., Harrop, S.R. and MacMillan, D.C. (2015). Understanding markets to conserve trade-threatened species in CITES. *Biological Conservation* 187, 249–259.

Chandran, R., Hoppe, R., de Vries, W.T. and Georgiadou, Y. (2015). Conflicting policy beliefs and informational complexities in designing a transboundary enforcement monitoring system. *Journal of Cleaner Production* 105, 447–460.

Chapron, G., Epstein, Y. and Lopez-Bao, J.V. (2019). A rights revolution for nature. *Science* 363, 1392–1393.

CITES. (2021). *Convention on International Trade in Endangered Species of Wild Fauna and Flora* https://cites.org/eng

CMS. (2021). *Convention on the conservation of migratory species of wild animals.* https://www.cms.int/

CoE. (2021). *Council of Europe: European Diploma.* https://www.coe.int/en/web/bern-convention/european-diploma-for-protected-areas

D'Cruze, N. and Macdonald, D.W. (2016). A review of global trends in CITES live wildlife confiscations. *Nature Conservation-Bulgaria*, 15, 47–63.

Davidson, N.C., Dinesen, L., Fennessy, S., Finlayson, C.M., Grillas, P., Grobicki, A. et al. (2020). A review of the adequacy of reporting to the Ramsar Convention on change in the ecological character of wetlands. *Marine and Freshwater Research* 71, 117–126.

Elder, M. and Olsen, S.H. (2019). The design of environmental priorities in the SDGs. *Global Policy* 10, 70–82.

Elliott, L. (2005). Expanding the mandate of the United Nations Security Council. In Chambers, B.W. and Green, J.F. (eds) *Reforming International Environmental Governance*, Tokyo: United Nations University Press, pp. 204–226.

EPA. (2021). Migratory bird initiatives. https://www.epa.gov/wetlands/migratory-bird-initiatives

Essl, F., Lenzner, B., Bacher, S., Bailey, S., Capinha, C., Daehler, C., et al. (2020). Drivers of future alien species impacts: An expert-based assessment. *Global Change Biology* 26, 4880–4893.

Ette, J.S. and Geburek, T. (2020). Why European biodiversity reporting is not reliable. *Ambio* doi: 10.1007/s13280-020-01415-8

EU. (2020). *EU Biodiversity Strategy for 2030 Communication from the Commission to the European Parliament, the Council, the European Economic and Social Committee and Committee of the Regions*, Brussels, 20.5.2020, COM(2020) 380 final

Falck-Zepeda, J.B. and Zambrano, P. (2011). Socio-economic considerations in biosafety and biotechnology decision making: The Cartagena Protocol and National Biosafety Frameworks. *Review of Policy Research*, 28, 171–195.

FAO. (2021). *International Treaty on Plant Genetic Resources for Food and Agriculture.* http://www.fao.org/plant-treaty/en/

Gaffney, A.C.B. and Evensen, D. (2020). Addressing the elephant in the room: Learning from CITES CoP17. *Global Environmental Politics* 20, 3–10.

Gallo-Cajiao, E., Morrison, T.H., Fidelman, P., Kark, S. and Fuller, R.A. (2019). Global environmental governance for conserving migratory shorebirds in the Asia-Pacific. *Regional Environmental Change* 19, 1113–1129.

Gamero, A., Brotons, L., Brunner, A., Foppen, R., Fornasari, L., Gregory, R.D., et al. (2017). Tracking progress toward EU biodiversity strategy targets: EU policy effects in preserving its common farmland birds. *Conservation Letters* 10, 395–3402.

GBIF. (2021). *Free and Open Access to Biodiversity Data.* https://www.gbif.org/

Glowka, L. (2000). Complementarities between the convention on migratory species and the convention on biological diversity. *Journal of International Wildlife Law and Policy* 3, 205–252.

Gowdy, J.M. (1997). The value of biodiversity: Markets, society, and ecosystems. *Land Economics* 73, 25–41.

Hamman, E. (2019). Bilateral agreements for the protection of migratory birdlife: The implementation of the China-Australia Migratory Bird Agreement (CAMBA). Asia Pacific *Journal of Environmental Law* 22, 137–159.

Harfoot, M., Glaser, S.A.M., Tittensor, D.P., Britten, G.L., McLardy, C., Malsch, K. and Burgess, N.D. (2018). Unveiling the patterns and trends in 40 years of global trade in CITES-listed wildlife. *Biological Conservation* 223, 47–57.

Harrop, S.R. (2011). Living in harmony with nature? Outcomes of the 2010 Nagoya Conference of the Convention on Biological Diversity. *Journal of Environmental Law* 23, 117–128.

Harrop, S.R. and Pritchard, D.J. (2011). A hard instrument goes soft: The implications of the convention on biological diversity's current trajectory. *Global Environmental Change: Human and Policy Dimensions* 21, 474–480.

Hellinx, E. (2020). The CMS-CITES African Carnivore Initiative as an Illustration of Synergies Between MEAs. *Frontiers in Ecology and Evolution* 8(Article 10), 1–9.

Hensz, C.M. and Soberon, J. (2018). Participation in the convention on migratory species: A biogeographic assessment. *Ambio* 47, 739–746.

High Court of Australia. (1983). *Commonwealth v Tasmania ("Tasmanian Dam case")* [1983] HCA 21; (1983) 158 CLR 1 (1 July 1983) http://www6.austlii.edu.au/cgi-bin/viewdoc/au/cases/cth/HCA/1983/21.html

Hrabanski, M. and Pesche, D. (2017). Introduction, analyzing IPBES functioning within the biodiversity regime complex and beyond. In: Hrabanski, M. and Pesche, D. (eds) *The Intergovernmental Platform on Biodiversity and Ecosystem Services (IPBES) Meeting the Challenge of Biodiversity Conservation and Governance*. London: Routledge, pp. 1–17.

IUCN. (2021). *World Database on Protected Areas*. https://www.iucn.org/theme/protected-areas/our-work/world-database-protected-areas

Jardin, M. (2010). Global biodiversity governance: The contribution of the main biodiversity related conventions. In Billé, R., Chabason, L., Chiarolla, C., Jardin, M., Kleitz, G. and Le Duc, J.P. (eds) *Global Governance of Biodiversity: New Perspectives on a Shared Challenge*, Paris: IFRI, pp. 6–44. https://en.calameo.com/read/000437626b0ecd176a3f9f

Kamau, E.C., Fedder, B. and Winter, G. (2010). The Nagoya Protocol on access to genetic resources and benefit sharing: What is new and what are the implications for provider and user countries and the scientific community? *Law, Environment and Development Journal* 6, 246–262.

Kanie, N., Bernstein, S., Biermann, F. and Haas, P. (2017). Introduction: Global Governance through Goal Setting. In Kanie, N. and Biermann, F. (eds). *Governing through Goals, Sustainable Development Goals as Governance Innovation*. Cambridge, MA: The MIT Press, pp. 1–28.

Kariyawasam, K. and Tsai, M. (2018). Access to genetic resources and benefit sharing: Implications of Nagoya Protocol on providers and users. *Journal of World Intellectual Property*, 21(5–6), 289–305.

Keiper, F. and Atanassova, A. (2020). Regulation of synthetic biology: Developments under the convention on biological diversity and its protocols. *Frontiers in Bioengineering and Biotechnology*, 8. doi:10.3389/fbioe.2020.00310

Koester, V. (2002). The five global biodiversity-related conventions: A stocktaking. *Review of European Community and International Environmental Law* 11, 96–103.

Kütting, G. and Herman, K. (2018). Introduction. In Kütting, G. and Herman, K. (eds) *Global Environmental Politics: Concepts, Theories and Case Studies*, London: Routledge, pp. 1–6.

Kuunal, S., Mair, L., Pattison, Z. and McGowan, P.J.K. (2020). Identifying opportunities for improving the coherence of global agreements for species conservation. *Conservation Science and Practice* 2(12) 1–11 doi:10.1111/csp2.294.

Leaders' Pledge for Nature (2020). *United to Reverse Biodiversity Loss by 2030 for Sustainable Development*. https://www.leaderspledgefornature.org/Leaders_Pledge_for_Nature_27.09.20.pdf

LeDee, O.E., Handler, S.D., Hoving, C.L., Swanston, C.W. and Zuckerberg, B. (2021). Preparing wildlife for climate change: How far have we come? *Journal of Wildlife Management* 85, 7–16.

Lees, E. (2016). Allocation of decision-making power under the habitats directive. *Journal of Environmental Law* 28(2), 191–219.

Lusaka Agreement. (2021). Lusaka agreement task force. https://lusakaagreement.org/about-us/

Maes, J., Liquete, C., Teller, A., Erhard, M., Paracchini, M.L., Barredo, J.I., et al. (2016). An indicator framework for assessing ecosystem services in support of the EU biodiversity strategy to (2020). *Ecosystem Services* 17, 14–23.

Mauerhofer, V. (2008). 3-D Sustainability: An approach for priority setting in situation of conflicting interests towards a sustainable development. *Ecological Economics, 64*, 496–506.

Mauerhofer, V. (2011). A bottom-up "Convention-Check" to improve top-down global protected area governance. *Land Use Policy, 28*, 877–886.

Mauerhofer, V. (2012). A "Legislation-Check" based on "3-D Sustainability" – addressing global precautionary land governance. *Land Use Policy, 29*, 652–660.

Mauerhofer, V. (2016). Public participation in environmental matters: Compendium, challenges and chances globally. *Land Use Policy* 52, 481–491.

Mauerhofer, V. (2019a). Activities of environmental convention-secretariats: Laws, functions and discretions. *Sustainability* 11(11), 3116. doi: 10.3390/su11113116.

Mauerhofer, V. (2019b). Ignorance, uncertainty and biodiversity: Decision making by the court of justice of the european union. In Voigt, C. (ed.) *The Environment in International Courts and Tribunals: Questions of Legitimacy*, Cambridge: Cambridge University Press, pp. 146–164.

Mauerhofer, V. (2021). Introduction into the role of net gain law for governance of a sustainable development. In Mauerhofer, V. (ed.) *The Role of Law in Governing Sustainability*, London and New York: Routledge, pp. 1–20.

Mauerhofer, V., Kim, R.E. and Stevens, C. (2015). When implementation works: A comparison of Ramsar Convention implementation in different continents. *Environmental Science & Policy* 51, 95–105.

McCay, S.D. and Lacher, T.E., Jr. (2021). National level use of International Union for Conservation of Nature knowledge products in American National Biodiversity Strategies and Action Plans and National Reports to the Convention on Biological Diversity. *Conservation Science and Practice*, 3(11). doi:10.1111/csp2.350

Morgera, E. (2017). Introduction to Volume III: The research challenges of international biodiversity law. In Morgera, E. and Razzaque, J. (eds) *Biodiversity and Nature Protection Law*. Volume III in Faure, M. (ed) *Elgar Encyclopedia of Environmental Law*, Cheltenham: Edward Elgar, pp. 1–9.

Morin, J.F., Orsini, A. and Jinnah, S. (2020). *Global Environmental Politics*, Oxford: Oxford University Press.

Morrison, T.H., Adger, W.N., Brown, K., Hettiarachchi, M., Huchery, C., Lemos, M.C. et al. (2020). Political dynamics and governance of World Heritage ecosystems. *Nature Sustainability* 3, 947–955.

North, D.C. (1990). *Institutions, Institutional Change and Economic Performance*, Cambridge: Cambridge University Press.

Otero, I., Farrell, K.N., Pueyo, S., Kallis, G., Kehoe, L., Haberl, H. et al. (2020). Biodiversity policy beyond economic growth. *Conservation Letters*, 13(4): e12713.

Pauwelyn, J. (2005). Judicial mechanisms: Is there a need for a world environment court. In Chambers, B.W. and Green, J.F. (eds) *Reforming International Environmental Governance*, Tokyo: United Nations University Press, pp. 150–177.

Price, M., Park, J.J. and Bouamrane, M. (2010). Reporting progress on internationally designated sites: The periodic review of biosphere reserves. *Environmental Science and Policy* 13, 549–557.

Prideaux, M. (2015). Wildlife NGOs: From adversaries to collaborators. *Global Policy* 6, 379–388.

Ramsar. (2021). Ramsar Convention. https://www.ramsar.org/

Reed, M.G. (2019). The contributions of UNESCO Man and Biosphere Programme and biosphere reserves to the practice of sustainability science. *Sustainability Science* 14, 809–821.

Sands, P. (1995). *Principles of International Environmental Law Vol. 1: Frameworks, Standards and Implementation*, Manchester: Manchester University Press in Association with IUCN.

Sands, P., Peel, J., Fabra, A. and MacKenzie, R. (2012). *Principles of International Environmental Law*, 3rd edition, Cambridge: Cambridge University Press.

Santoro, A., Venturi, M., Bertani, R. and Agnoletti, M. (2020). A Review of the role of forests and agroforestry systems in the FAO Globally Important Agricultural Heritage Systems (GIAHS) Programme. *Forests* 11(8), 860. doi: 10.3390/f11080860.

Siebenhüner, B. (2007). Administrator of global biodiversity: The secretariat of the convention on biological diversity. *Biodiversity and Conservation* 16, 259–274.

Stec, S. and Jendrośka, J. (2019). The Escazú agreement and the regional approach to rio principle 10: Process, innovation, and shortcomings. *Journal of Environmental Law* 31(3), 533–545.

Tsioumani, E. (2018). Beyond access and benefit-sharing: Lessons from the law and governance of agricultural biodiversity. *Journal of World Intellectual Property* 21, 106–122.

UN. (2010). *Strategic Plan for Biodiversity 2011–2020, Including Aichi Biodiversity Targets*, decision X/2, tenth meeting of the Conference of the Parties, held from 18 to 29 October 2010.

UN. (2012). *The Future We Want*, Outcome Document of the United Nations Rio+20 Conference, New York. https://sustainabledevelopment.un.org/futurewewant.html

UN. (2015). *Transforming Our World: The 2030 Agenda for Sustainable Development*, General Assembly, Resolution A/RES/70/1, 25 September.

UN. (2020). *Updated United Nations Zero Draft of the Post-2020 Global Biodiversity Framework* (CBD/POST2020/PREP/2/1).

UNECE. (2005). *Almaty Guidelines on Promoting the application of the principles of the Aarhus Convention in international forums* https://unece.org/fileadmin/DAM/env/documents/2005/pp/ece/ece.mp.pp.2005.2.add.5.e.pdf

UNEP. (2021). *InforMEA Treaties and MEAs in Biological Diversity.* https://www.informea.org/en/topics/biological-diversity

UNESCO. (2011). *Strategic Action Plan for the Implementation of the World Heritage Convention 2012–2022.* WHC-11/18.GA/11

UNESCO. (2021). *Man and the Biosphere (MAB) Programme.* https://en.unesco.org/mab

Vadrot, A.B.M., Rankovic, A., Lapeyre, R., Aubert P-M. and Laurans, Y. (2018). Why are social sciences and humanities needed in the works of IPBES? A systematic review of the literature. *Innovation: The European Journal of Social Science Research*, 31:sup1, 78–100.

Von Moltke, K. (2005). Clustering international environmental agreements as an alternative to a world environment organization. In Biermann, F. and Bauer, S. (eds) *A World Environment Organization. Solution or Threat for Effective International Environmental Governance?* Aldershot: Ashgate, pp. 175–204.

Wijnstekers, W. (2018). *The Evolution of CITES*, 11th edition. https://cites.org/sites/default/files/common/resources/The_Evolution_of_CITES_2018.pdf

Williams, B.A., Watson, J.E.M., Butchart, S.H.M., Ward, M., Brooks, T.M., Butt, N. et al. (2020). A robust goal is needed for species in the post-2020 global biodiversity framework. *Conservation Letters.* e12778.

Young, O. (2008). Deriving insights from the case of the WTO and the Cartagena Protocol. In Young, O., Chambers, W.B., Kim, J.A. and ten Have, C. (eds) *Institutional Interplay Biosafety and Trade*, Tokyo: United Nations University Press, pp. 131–158.

Zeng, Y.W., Maxwell, S., Runting, R.K., Venter, O., Watson, J.E.M. and Carrasco, L.R. (2020). Environmental destruction not avoided with the sustainable development goals. *Nature Sustainability.* doi:10.1038/s41893-020-0555-0

Zizka, A., Carvalho, F.A., Calvente, A., Baez-Lizarazo, M.R., Cabral, A., Ramos Coelho, J.F. et al. (2020). No one-size-fits-all solution to clean GBIF. *PeerJ*, 8, e9916. doi: 10.7717/peerj.9916.

42

Forests

The political ecology of international environmental governance

Constance L. McDermott and David Humphreys

This chapter surveys the complex international politics of forest conservation. Forests emerged as a global environmental concern in the latter half of the twentieth century due largely to the acceleration of tropical deforestation. Debates over what constitutes a fair and equitable means of addressing tropical forest loss, as well as the degradation of other forest ecosystems worldwide, continue to be central drivers of forest geopolitics to this day. Multiple attempts to negotiate a legally binding international forest convention have failed, and instead both state and non-state actors have sought to create a new forest politics in which the forces for conservation are greater than those for degradation. Many of these new initiatives are market-based. They range from sustainability and legality certification to corporate and state pledges for "zero deforestation" commodity supply chains and payments for forest carbon to mitigate the impact of forest loss on climate change. Yet, progress across all of these initiatives has been limited and the benefits are unevenly distributed, leading some to argue for more transformative, equitable and inclusive approaches to global forest governance.

The quest for a global forest convention

The acceleration of tropical forest loss since the 1960s (Rudel et al. 2009) has sparked, and continues to propel, the idea of forests as a "global commons" in need of coordinated international governance (Humphreys 2006). However, the challenges of achieving such coordinated action are complex, diverse and dynamic. The lack of globally shared interests on forests has been highly apparent since the 1992 United Nations Conference on Environment and Development (UNCED), when governments first tried, and failed, to reach agreement on a global forest convention (United Nations 1992b).

The UNCED conference, also known as the Rio Earth Summit, saw the launch of two core environmental conventions – the UN Framework Convention on Climate Change (UNFCCC) and Convention on Biodiversity (CBD) – and the initiation of negotiations for a third on combatting desertification – the UN Convention to Combat Desertification (UNCCD). Initially the CBD held the most direct relevance to forests, with its strong emphasis on expanding the global network of protected areas. The proposal at UNCED to establish a fourth convention on forests, in contrast, pursued a broader mandate to arrest

DOI: 10.4324/9781003008873-47

forest loss and ensure the sustainable management of both protection and production forests (McDermott et al. 2007). The resulting forest convention debate continues to be an undercurrent of forest politics and policy to this day, although the demand for a convention has weakened over the past 30 years as states have pursued alternative international forest policy measures (MacKenzie 2012).

An international forest convention could have various advantages, depending on how it is designed. First, it could demonstrate political commitment from states, serving as a statement of intent that governments are serious about addressing deforestation. Second, a convention could complement other international environmental conventions, including the UNFCCC and CBD established in Rio, and the Convention on International Trade in Endangered Species of Wild Fauna and Flora (CITES; see Chapter 41), with a forest-related mandate. Third, a forest convention could provide a coordination function, integrating in one comprehensive instrument all forest-related provisions in international environmental law, and in the process eliminating gaps, duplications and uncertainties (see Chapter 21). In so doing, a forest convention could re-energize international forest policy by providing strategic leadership on forests. Finally, a forest convention could clarify how some established principles of international environmental law should apply to forests. These principles include sovereignty, the precautionary principles and common but differentiated responsibilities (see Chapter 10).

However, there are some arguments against these claimed advantages. A convention is not a precondition for political action to address deforestation; states and other actors can take meaningful action on forests without such an instrument. Indeed a convention could deflect political attention and resources away from implementation on the ground. A convention could also generate jurisdictional complexities. For example, were a forest convention to be agreed the question would inevitably arise over which instrument should take the lead on conserving forest-related carbon sinks: the UNFCCC or the forest convention. There is certainly no legal basis for a forest convention acting as the lead mechanism on forests, with other forest-related instruments, such as the UNFCCC, having a secondary role. Conferences of parties to other instruments would be under no obligation to take their lead from parties to a forests convention. On this view, far from rationalizing international forest-related law a forest convention could add another layer of political and legal uncertainty to international forest policy.

Conflicting national interests

In addition to the above generalized arguments for and against a forest convention, different states may hold different value-based conceptions of the national interest. These interest-based perspectives can be broadly summarized through five different ideal-type arguments. The first relates to the anticipated environmental and social benefits. For example, a state may desire a convention because it wishes to contribute to intergenerational equity, promoting long-term forest conservation for future generations. However, this appears to be a motivating factor for very few states. For example, countries from the global South have often asserted in international forest negotiations that states should address current issues of equity, poverty and economic inequality before turning to issues of intergenerational equity.

A second set of arguments revolves around sovereignty and forest ownership. Some tropical countries resist a convention, arguing that such an instrument could erode their sovereign rights to exploit their natural resources. Brazil in particular has taken a persistently strong line on sovereignty, making clear it will accept no international regulation of its policies for the Amazon. During the UNCED forest negotiations, Malaysia, which at that time led for the

Group of 77 Developing Countries (G77), spoke out against a convention on the grounds that it would infringe national sovereignty over forest use. Related to sovereignty is the question of forest ownership and privatization. There is a suspicion from some countries in the global South that some countries in the global North seek a forest convention to promote forest privatization and further trade liberalization in the forest sector (on North–South relations, see Chapter 23). Most of the world's forests are publicly owned by the state or regional forest authorities (White and Martin 2002). Some developed countries, in particular the United States, have argued that forests are more effectively and sustainably managed when under private ownership. The G8 countries endorsed a forest privatization agenda in 1998, presumably anticipating that forest-based corporations in G8 countries would be among the main beneficiaries (Humphreys 2006). However, tropical forest countries are unwilling to privatize their forests, as under international trade and investment rules their forests could end up under foreign ownership (on international trade and the environment, see Chapter 24).

A third set of arguments revolve around forest management standards. Countries that have adopted high forest management standards may incur additional costs and thus find themselves at a disadvantage in international trade relative to states with lower standards. The former may favor a convention in order to promote high international standards that erode any such disadvantages. The long-term support of Canada, a country with relatively prescriptive forest regulations, can be explained in this light (McDermott et al. 2010). Until 2015, Canada was the most persistent advocate for a forest convention in international forest negotiations. Higher global forest management standards would improve the international competitiveness of Canadian timber. Many Canadian forest businesses have voluntary signed up to the International Organization of Standards (ISO), and the Canadian Pulp and Paper Association has argued that ISO standards should apply worldwide (Lipschutz 2001; on corporate influence in global environmental politics, see Chapter 13). However, states with low management standards tend to oppose a forest convention that might impose additional costs on their forest industry, thus eroding the international competitiveness of their forest products.

A fourth set of arguments also focuses on trade. States with a strong forest industry may seek a forest convention as a mechanism to promote the international trade in forest products and gain access to new markets, thus realizing economic gains. The desire to achieve market openings for domestic industries has driven some previous international environmental instruments (see Chapter 22). Davenport (2006) argues that the United States was a leader in negotiating the Vienna Convention and Montreal Protocol on ozone depletion because US industry foresaw market opportunities in the production of ozone-friendly chemicals for aerosols and refrigeration (see Chapter 31).

A fifth set of arguments rest on bargaining between forest conservation and other issues. A recurring thread to all international forest negotiations is the G77's linkage of forest conservation with the transfer of financial and technological aid. Some tropical countries support a convention that would provide increased flows of official development assistance (ODA) to tropical forest countries vulnerable to deforestation. Similarly, some European and North American states have opposed a convention as this could increase the expectations of tropical countries of increased ODA. Preferring to deal with aid issues on a bilateral rather than a multilateral basis, countries such as the United States, the United Kingdom and Sweden have argued that merely agreeing a convention would not increase forest-related aid transfers, emphasizing that increased funding can come from a variety of sources, including the private sector.

North–South differences are often overemphasized in analyses of international relations (see Chapter 23). However, in the UNCED forest negotiations there was a clear North–South polarization. Broadly speaking, all the countries of the global North favored a forest convention while all those of the global South opposed one. No country from the global North opposed a convention which was supported by the United States, Canada, Japan and the European Union (or European Community as it was then called). Meanwhile, no country from the global South supported a convention, with many vocally against, in particular India and Malaysia.

Sovereignty was a major issue for the global South at Rio. Part of the reason for this was a shift in World Bank forest policy during the lead up to the UNCED. In 1991 the World Bank introduced a ban on financing logging projects in tropical forests, the key provision being that "the Bank Group will not under any circumstances finance commercial logging in primary tropical moist forests" (World Bank 1991). This was in response to criticisms from nongovernmental organizations (NGOs) and community groups that the Bank had financed destructive developments in the Amazon, with the Grande Carajás iron ore and mining project and the Polonoroeste highway construction and colonization project attracting strong criticism (on NGOs in global environmental politics, see Chapter 14). But the logging ban did not apply to non-tropical forests, fueling the suspicions of tropical forest governments that developed countries and international development organizations such as the World Bank were seeking to deny them the right to exploit their forests.

During the UNCED process the United States promoted a convention, although without particular enthusiasm (Humphreys 2006). The shift from a pro- to an anti-convention position in the United States can be attributed to the change of administration from George H.W. Bush to Bill Clinton, the latter arguing it would impose higher costs for the forest industry (Davenport 2006: 131). The subsequent administration of President George W. Bush continued Clinton's opposition to a convention, although for different reasons; the Bush administration was ideologically opposed to any measure that could be construed as environmental regulation either nationally or internationally. President Barrack Obama also maintained an anti-convention position, while President Donald Trump opposed any international commitment on environmental issues.

Malaysia – the most outspoken critic of a convention during the UNCED negotiations – now supports a convention. The Malaysian change of position can be explained by a shift in the lead agency for international forest policy from the Ministry of Foreign Affairs, which framed forestry as an issue of national security and sovereignty, to the Ministry of Primary Resources, which sees forests as an issue of trade and market access (Kolk 1996: 162). Another Southeast Asian forest power, Indonesia, has also shifted to support a convention after earlier opposition. Here the reason lies in the change of regime following the resignation of President Suharto in 1998. Suharto had a policy of aggressively exploiting Indonesia's timber resources, including extensive clear felling in Sumatra and Kalimantan. Since 1998 Indonesia has favored a forest convention and adopted a pro-conservation stance, including leading a regional Southeast Asian initiative against illegal logging in 2001.

States have revisited the convention question four times since UNCED: at the UN Intergovernmental Panel on Forests (IPF), a temporary forest body that existed for two years between 1995 and 1997; at the UN Intergovernmental Forum on Forests (IFF), the IPF's successor that existed for three years until 2000; and in 2006–2007 and then again in 2015 at the United Nations Forum on Forests, a body created in 2001 that reports directly to the UN Economic and Social Council. On each occasion there was no consensus. Countries arguing most strongly for a convention were Canada (until its change of position in 2015), Malaysia, most of the EU countries (in particular Germany and the Netherlands), Russia, Norway, Finland, many African countries (including Nigeria) and the Central American countries (in particular Costa Rica). Meanwhile opponents have included the United States, the United Kingdom, Sweden, China, Japan, New Zealand and almost all South American countries.

The South American countries have been led by Brazil. As the tropical forest state with the largest expanse of tropical forest cover, Brazil has issue-specific power that it uses to good effect in international forest negotiations. Brazil regularly attracts forest-related development assistance from bilateral and multilateral donors, and at the same time is a major global economic player with lessening dependence on foreign aid. The Central American countries, by contrast, are much smaller and some have more difficulty attracting forest aid, which could help explain their greater support for a convention.

With no agreement for a convention, states have opted for soft law on forests, a key recent example being the 2007 Non-legally Binding Instrument on All Types of Forests (United Nations 2008).

A turn to the market

Around the same time as the 1992 Rio Earth Summit, another coalition was forming outside the confines of the United Nations. Frustrated with the inadequacy of government action, a core group of NGOs and companies joined together to create forest certification as a voluntary, non-state, market-driven labeling scheme (Cashore et al. 2004; Klooster 2005). Certification of this type awards green labels to companies that have met an agreed upon set of environmental and social standards. Companies can then use these labels to add market value to their products, or to improve their access to green markets.

The first major forest certification scheme to emerge was the Forest Stewardship Council (FSC). The FSC was established in 1993 after the International Tropical Timber Organization – an intergovernmental organization that aims to promote the trade in tropical timber while also taking action to ensure the long-term conservation of the resource base – decided against introducing a labeling scheme for timber from sustainable sources (Gale 1998). FSC is a non-state membership-based organization with a decision-making structure designed to balance power among different interest groups. Decision-making authority in the FSC is divided equally between environmental, economic and social chambers, with equal representation between developed and developing countries (Cashore et al. 2004). This chambered structure was designed to counterbalance the perceived dominance of economic concerns over environmental and social interests in forestry decision-making, as well as to re-balance power between the global North and South.

The FSC awards its label to forest concessions that meet the FSC criteria for well-managed forests that protect biodiversity and the rights of workers, indigenous peoples and local communities. The FSC relies on both supply-side measures (the willingness of forest producers to comply with FSC standards) and demand-side measures (retailers and consumers showing preference for FSC-certified timber). As of December 2020 the FSC had certified 229,849,033 hectares of forests in 89 countries (FSC 2020). The success of the FSC led to the creation of industry-backed competitor schemes, most of them now consolidated under the Programme for the Endorsement of Forest Certification (PEFC). The PEFC endorses national certification schemes. As of December 2020 the PEFC had certified 320,000,000 hectares of forest in 55 countries (PEFC 2020). The PEFC has thus certified more forest area than the FSC, reflecting industry preference for a scheme more strongly led by industrial interests (Judge-Lord et al. 2020).

From the early 1990s to the mid-2000s forest certification and labeling was the dominant international policy response to promoting sustainable forestry. Both the FSC and PEFC remain important initiatives (Chan and Pattberg 2008; Gulbrandsen 2012). However, it soon became clear that certification was growing much more rapidly in the global North than in the tropics, and that it was doing little to address the underlying causes of tropical forest loss.

A growing body of scientific research has uncovered a number of reasons for certification's uneven growth, as well as its limited relevance to deforestation. Critically, research on the drivers of land use change has identified the expansion of commercial agriculture, rather than timber production, as the primary driver of tropical deforestation (e.g., Rudel et al. 2009). At the same time, market demand for certified timber comes largely from the global North, and Northern markets are supplied primarily by Northern timber. Tropical wood producers not only have fewer market incentives for certification, they also face extra barriers. The governance of industrial wood production in the tropical forest frontier is generally weak and plagued by unclear or disputed land and resource rights, while many tropical producers lack the finance and capacity to meet certification requirements (Ebeling and Yasué 2009; McDermott 2013).

Reinserting the state

By the mid-2000s, a growing number of state and civil society actors began to shift their attention to the issue of weak governance, and in particular illegal logging, as the chief barrier to more sustainable tropical forest management. This refocusing on legality proved highly effective in expanding the political traction of international forest governance. Instead of pressuring Southern states to change how they managed forests, Northern governments could focus attention on strengthening the power of Southern governments to enforce their own rules. Forest industries in the United States and Europe also stood to gain by exposing the high levels of illegality in the tropical wood trade, as did civil society groups focused on tropical forest protection (McDermott 2014). A series of government- and industry-funded reports emerged estimating that informal, and hence illegal, logging amounted to as much as 80 percent of total timber production in some tropical countries (SCA and WRI 2004; Lawson and MacFaul 2010). The result has been a string of initiatives to address illegal logging, ranging from World Bank-supported Ministerial Declarations in Europe and North Asia, East Asia and Africa, to new legislation introduced in the United States, the European Union, Australia and elsewhere aimed at prohibiting the import of illegally produced timber.

Several US and EU unilateral and bilateral actions on illegal logging are particularly notable for having set new legal precedents and new policy coalitions. In 2008 the United States expanded the existing Lacey Act on illegal wildlife trade to encompass wood products, thereby prohibiting the import of wood products produced in violation of the laws of the country of origin. This use of sovereign power to enforce foreign forest laws received broad-based support, including from the American Forest and Paper Association, a leading US industry association (Cashore and Stone 2012). Similar multi-stakeholder support coalesced around the European Union's 2003 Forest Law Enforcement, Governance and Trade (FLEGT) Action Plan. A core mechanism under the Plan are Voluntary Partnership Agreements (VPAs) between the EU and participating developing countries. VPAs are based on the development of FLEGT legality licensing systems within partner countries, and the subsequent exclusion of all un-licensed wood from entering the EU. Complementary to this, the EU Timber Regulation (EUTR) came into force in 2013. The EUTR requires "first placers" of wood products in EU markets to show proof of due diligence that these products were produced in accordance with the laws of the country of origin (McDermott and Sotirov 2018).

While this growing array of illegal logging initiatives speaks to their political traction, they have yet to be successful in addressing forest loss and degradation. The narrow focus on law enforcement and international trade overlooks more systemic problems such as government corruption and lack of resources, conflicts among ministries, and highly complex forest laws that exclude domestic and local actors from legal access to forest resources (Myers et al. 2020). Such challenges help explain why 15 years after the launch of the VPA program only Indonesia has received approval to award FLEGT licenses (EU FLEGT Facility 2020), and the validity of even those is subject to debate (Acheampong and Maryudi 2020). Furthermore, just as with forest certification, the focus on timber production overlooks the role of agriculture as the leading driver of tropical forest loss.

It was not until the late 2000s that international forest politics began to break out of the forest sector and generate strategies aimed more directly at deforestation. The catalyst was climate change, and the growing body of scientific evidence linking forest loss to global carbon emissions.

Forests and climate

Recent scientific advances have played a central role in galvanizing global attention to the negative impacts of human-induced climate change (see Chapters 18 and 32). While initially global attention focused on fossil fuels as the leading source of greenhouse emissions, studies of global carbon budgets combined with new remote sensing technologies soon uncovered tropical forest loss as another significant factor that, according to the influential Intergovernmental Panel on Climate Change (IPCC), amounted to as much as 17 percent of total human-induced emissions (IPCC 2007). The consequent framing of forest loss as a climate change problem has led to a shift in the political center of gravity in intergovernmental negotiations from the United Nations Forum on Forests to the United Nations Framework Convention on Climate Change (UNFCCC). Forests are now increasingly valorized for the role that they can play in sequestering and storing carbon from the atmosphere.

In December 2007 the 13th conference of parties to the UNFCCC agreed in Bali a decision on reducing emissions from deforestation in developing countries (United Nations 2007a). This followed a proposal from the Coalition of Rainforest Nations led by Costa Rica and Papua New Guinea, and evolved over time into a mechanism known as "Reducing Emissions from Deforestation and forest Degradation and enhancing forest carbon stocks" (REDD+). The idea of REDD+ is consistent with Article 2 of the UNFCCC on "stabilization of greenhouse gas concentrations in the atmosphere at a level that would prevent dangerous anthropogenic interference with the climate system" (United Nations 1992a). REDD+ can be seen as part of a drive from the global South to shape international climate and forest policies in a way that is consistent with developing country interests and understandings of fairness and climate justice (see Chapters 25 and 26). The underlying rationale of REDD+ is that countries in the global North should pay those in the South for reducing deforestation by attaching monetary value to the carbon stored in their standing forests.

A core principle underlying the global South's support for REDD+ is the idea of compensation for opportunity costs forgone, a claim made by the G77 during the UNCED forest negotiations. Simply stated, the principle means that if a country is to desist from exploiting forests or other natural resources for the common good of humanity, then that country should expect to receive some compensation in lieu of the money it would have received from exploiting the resources (Humphreys 1996; Grainger 1997). The G77 did not succeed in inserting this principle into the UNCED outputs and has not secured its inclusion in any post-UNCED multilateral legal and political declarations on forests. However, it remains a key guiding principle for developing countries in international forest politics. No forest country has indicated that it is prepared to make significant forest conservation pledges without some financial *quid pro quo* from donor countries.

Initial proposals for REDD+ finance were ambitious. An influential report written in 2008 recommended financial flows of between US$17 and 33 billion per year "to halve emissions from the forest sector to 2030" (Eliasch 2008: xvi). Carbon offsetting, coupled with ambitious national commitments to emissions reduction, were to form a core source of such funds. Whether or not such finance will materialize remains to be seen, and this was a key sticking point in early negotiations over REDD+.

The use of carbon offsetting to reduce forest emissions was highly controversial from the outset. For example, Brazil's official position under the UNFCCC has been strongly opposed to offsetting based in part on arguments also made by NGOs, that is, that offsets are in effect a "license to pollute" that reduces pressure on developed countries to curb their emissions (Seymour and Busch 2016). Concerns about offsetting also relate to issues of "permanence," "additionality" and "leakage." "Permanence" refers to the risk of reversing back to forest loss after the delivery of REDD+ payments, "additionality" to the difficulty of proving that deforestation would have been higher without REDD+, and "leakage" to the displacement of deforestation from one area to another. All of these factors affect the credibility of forest carbon credits, particularly as a legitimate means to offset fossil fuel emissions.

Concerns about REDD+, and more generally about market-based approaches to forest carbon accounting, also extend to its potential effects on other environmental and social values. Environmental NGOs have stressed that REDD+ should not just focus on carbon but incorporate biodiversity and other environmental services. Likewise community and indigenous peoples' groups have argued that REDD+ must respect traditional land rights. Indigenous groups, in particular, have claimed that their approval should be sought before REDD projects are implemented on their traditional, customary land. These groups often cite the principle of Free, Prior and Informed Consent (FPIC) that appears in the United Nations Declaration on Indigenous Peoples (UNDRIP) adopted by the UN General Assembly in 2007 (United Nations 2007b). This principle holds that consent should be free (with no coercion or intimidation), prior (before the authorization and commencement of any project or development activities) and informed (by full knowledge of what any proposed project or development activities will entail) (Anderson 2011).

These and other debates around REDD+ design have been at least partially addressed in a series of key UNFCCC decisions adopted between 2010 and 2014. In 2010 the Cancun Agreement (UNFCCC 2011) established REDD+ as a "nationally-driven process," enabling individual countries to opt in or out of REDD+ and decide for themselves whether to allow offsetting. Likewise, it was decided that the level of reduced forest emissions would be measured at the national level, based on a country's own proposed "baseline" of emissions under a business-as-usual scenario. The Cancun Agreement also introduced a set of seven "safeguards" addressing a range of environmental and social concerns. These include requirements that REDD+ actions align with "relevant international conventions and agreements," protect biodiversity and not promote the loss of natural forests, generate social benefits, respect the rights of indigenous and local communities (with reference to UNDRIP), and address the risk of reversals and leakage. By the time of the 2013 Warsaw Framework (UNFCCC 2014), which elaborated on arrangements for REDD+ finance and "safeguard reporting systems," much of the key architecture for REDD+ was formally decided.

Throughout this process of formal REDD+ negotiations, a wide range of donor agencies, private investors, governments and NGOs were already implementing a variety of REDD+ actions on the ground. As outlined in the 2011 Cancun Agreement, these early activities might be seen as part of three recognized "phases" of REDD+. Phases I and II consist of REDD+ "readiness" actions in preparation to reach the final Phase III of "results-based" payments, when national governments would be paid for reduced forest-based emissions. However, and as noted by McDermott et al. (2012), all of the organizations involved have their own priorities and constraints, and they have contributed to operationalizing REDD+ via diverse and overlapping rules and safeguards for REDD+. While the complexity of REDD+ implementation continues to grow, the corresponding finance has fallen far short of the tens of billions first envisioned. The UNFCCC has since formalized the role of the Green Climate Fund (GCF) as a core funder of REDD+, governed by UNFCCC rules. But in practice, the reliance of REDD+ on diverse funding sources guarantees a much more complex and expansive governance regime (Skutsch 2019).

By the time of the UNFCCC Paris Agreement in 2015, frustration was growing over the shortfalls in REDD+ finance and the slow progress of countries in reaching REDD+ Phase III, involving national results-based payments. The Paris Agreement helped reinvigorate global attention to forests by establishing the ambitious target of limiting global warming to 1.5 degrees (see Chapter 32), and by directly mentioning the role of forests in reaching that target. This, combined with a series of intergovernmental and multi-stakeholder pledges to protect and restore hundreds of millions of hectares of forests around the globe (e.g., the 2014 New York Declaration on Forests and the 2011 Bonn Challenge (NYDF 2019)), contributed to more expansive consideration of the tools to address forest-cover change, of which REDD+ was just one.

As of 2020, 9 countries have reported Phase III results of reduced emissions on the UNFCCC's REDD+ platform (UNFCCC 2020). Brazil, which was among the first to reach Phase III, has been lauded as a particular success in dramatically reducing its rates of deforestation between 2004 and 2012 by about 80 percent (West and Fearnside 2021). However, this reduction had little directly to do with REDD+ and carbon payments, and much more to do with a suite of regulatory, incentive- and market-based measures instituted by the Brazilian government (West and Fearnside 2021). Among these were public-private partnerships in the cattle and soy sector aimed at removing beef associated with deforestation from Brazilian supply chains.

Since 2012, Brazil's deforestation rate has begun to climb again. Illustrating the critical role of political leadership, the election of President Bolsonaro in October 2018 has resulted in the rapid dismantling of environmental regulations and incentives, and a 30 percent rise in deforestation the following year (West and Fearnside 2021). Nevertheless, and despite these recent setbacks, Brazil's initial success in reducing the deforestation embedded in its commodity supply chains has helped build momentum for similar efforts at the international level, known collectively as "zero-deforestation" initiatives.

Tackling the forest-agriculture nexus: "zero-deforestation" commodities

While REDD+ injected forest politics into the climate sphere, a growing suite of "zero deforestation" initiatives have ushered it into agriculture as well. These efforts attempt to directly address the production of agricultural commodities for urban and international trade as the leading driver of forest loss (DeFries et al. 2010). To date, the alternative economic incentives generated by global forest governance initiatives, from certified timber to forest carbon credits, have failed to compete with the value of clearing land to grow commodities such as beef, soy, palm oil and cocoa. This has driven a push for a more "landscape-level" approach to forest governance that recognizes the interactions of forests with surrounding landscapes and land uses (Ros-Tonen et al. 2018).

By 2017, over 470 consumer goods and producer, processor and trader companies in the agricultural sector had enacted some form of "zero deforestation" pledge (Haupt et al. 2018: 6), drawing on existing tools such as certification to prove their compliance. That number continues to grow and is matched by demands from hundreds of investment firms, responsible for over US$16 trillion in assets, for companies to verify that they have eradicated deforestation from their supply chains (Forest 500 2020). While the majority of these efforts have been criticized for their lack of speed, comprehensiveness and ambition (NYDF 2019), there are signs of growing political momentum backed by the increasing participation of state actors. Mirroring strategies embedded in FLEGT and the EUTR, both the EU and the United Kingdom are pursuing new regulations requiring that companies demonstrate due diligence to remove deforestation from key commodity imports (EC 2019; GOV.UK 2020). Just as with timber, there is a strong focus on legality in these initiatives. In this way, Northern states aim to use markets to reinforce the power of Southern states to enforce their own laws. This helps bypass claims of interference with Southern state sovereignty.

Despite the widespread support that the concept of "zero deforestation" has garnered from diverse stakeholders, the narrow focus of many state actors on the illegality of forest cover change has also been challenged by civil society organizations. For example, and echoing similar critiques of FLEGT, critics have argued that states and legal systems in many producing countries may perpetuate human rights abuse, reinforce "land grabbing" – as in the corporate consolidation of landholdings and displacement of local communities or violation of indigenous peoples' rights (D'Odorico et al. 2017; Davis et al. 2020) – and may fail to protect biodiversity and other environmental values (Saunders 2020). In other words, "legal" does not necessarily mean "sustainable" or "ethical." Aggressive law enforcement risks the violation of human rights for large rural populations with insecure tenure who are reliant on informal markets and economies for their livelihoods.

Conclusion

A great many environmental and social values are at stake in forest politics, and the question of how to prioritize them is a matter of ongoing debate. The rapid rise of tropical deforestation since the 1950s has been instrumental in the reframing of forests as a global commons in need of international collaboration and agreement. This has driven attempts to agree on a global forest convention framed broadly around "sustainability," which would potentially encompass not only the protection of forest cover, but also ensure biodiversity conservation, indigenous peoples' rights and sustainable forest use. However, conflicting interests and priorities have made countries reluctant to relinquish sovereignty to a set of global rules on forests. Meanwhile, nongovernmental actors, frustrated with government action, have stepped in to create market-based certification schemes to incentivize sustainability. Yet, the concept of sustainability is disputed among certification schemes as well. While forest certification has grown rapidly in the global North, it has not addressed the leading drivers of forest loss in the global South.

More recently, many actors have narrowed their focus from sustainability down to the seemingly simpler and more universal goals of "legality," "carbon mitigation" and "zero deforestation." This narrowing of focus has helped broaden political support across a diversity of stakeholders. Yet it also, ironically, creates a very complicated governance landscape of overlapping rules, standards, procedures and initiatives, and does little to address underlying debates over social and environmental priorities. An increasing number of actors, therefore, have called for more "transformative" and "emancipatory" approaches to forest governance that are more socially inclusive and resilient (Scoones et al. 2020). This would require recognizing the inherently political nature of forest conservation, and prioritizing a more equitable balance of power in international forest governance. While current approaches are strongly market-based, and driven by expert knowledge, large-scale actors and the global North, more "emancipatory" governance would draw on diverse knowledge systems and more horizontal collaboration to create more inclusive and effective stewardship of the world's forest resources.

References

Acheampong, E., Maryudi, A. (2020). Avoiding legality: Timber producers' strategies and motivations under FLEGT in Ghana and Indonesia. *Forest Policy and Economics.* 111, 102047–102048.

Anderson, P. (2011). *Free, Prior and Informed Consent in REDD+: Principles and Approaches for Policy and Project Development,* RECOFTC (Centre for People and Forests)/Deutsche Gesellschaft für Internationale Zusammenarbeit (GIZ) GmbH. Available online at: <http://www.recoftc.org/site/uploads/content/pdf/FPICinREDDManual_127.pdf>

Cashore, B., Auld, G., Newsom, D. (2004). *Governing through Markets: Forest Certification and the Emergence of Non-state Authority,* New Haven, CT: Yale University Press.

Cashore, B., Stone, M.W. (2012). Can legality verification rescue global forest governance? Analyzing the potential of public and private policy intersection to ameliorate forest challenges in Southeast Asia. *Forest Policy and Economics 18,* 13–22.

Chan, S. and Pattberg, P. (2008). Private rule-making and the politics of accountability: Analyzing global forest governance. *Global Environmental Politics 8(3),* 103–21.

Davenport, D.S. (2006). *Global Environmental Negotiations and US Interests,* Basingstoke: Palgrave Macmillan.

Davis, K.F., Koo, H.I., Dell'Angelo, J., D'Odorico, P., Estes, L., Kehoe, L.J., Kharratzadeh, M., Kuemmerle, T., Machava, D., Pais, A.d.J.R., Ribeiro, N., Rulli, M.C., Tatlhego, M. (2020). Tropical forest loss enhanced by large-scale land acquisitions. *Nature Geoscience 5,* 433–413.

DeFries, R., Rudel, T., Uriarte, M., Hansen, M. (2010). Deforestation driven by urban population growth and agricultural trade in the twenty-first century. *Nature Geoscience: Letters 3,* 178–181.

D'Odorico, P., Rulli, M.C., Dell'Angelo, J., Davis, K.F. (2017). New frontiers of land and water commodification: Socio-environmental controversies of large-scale land acquisitions. *Land Degradation & Development 28,* 2234–2244.

Ebeling, J., Yasué, M. (2009). The effectiveness of market-based conservation in the tropics: Forest certification in Ecuador and Bolivia. Journal of environmental management 90, 1145–1153.

EC. (2019). Stepping up EU Action to Protect and Restore the World's Forests, Brussels, pp. 1–22.

Eliasch, J. (2008). Eliasch Review: Climate change: financing global forests. UK Office of Climate Change.

EU FLEGT Facility. (2020). The FLEGT License Information Point. Available online at: https://www.euflegt.efi.int/es/flegt-licensed-timber

Forest 500. (2020). Fueling the fires: Why investors need to do more to protect the Amazon. Global Canopy Programme. Oxford. Available at: https://globalcanopy.org/insights/publication/fuelling-the-fires-why-investors-need-to-do-more-to-protect-the-amazon/

Forest Stewardship Council (FSC). (2020). Facts & figures. Available online at: https://fsc.org/en/facts-figures.

Gale, F. (1998). *The Tropical Timber Trade Regime,* New York: St Martin's Press.

GOV.UK. (2020). Government sets out world-leading new measures to protect rainforests. Available online at: https://www.gov.uk/government/news/government-sets-out-world-leading-new-measures-to-protect-rainforests

Grainger, A. (1997). Compensating for opportunity costs in forest-based global climate change mitigation. *Critical Reviews in Environmental Science and Technology 27*(S001): 163–76.

Gulbrandsen, Lars H. (2012). International forest politics: Intergovernmental failure, non-govern-mental success? In Steinar Andresen, Elin Lerum Boasson and Geir Hønneland (eds) *International Environmental Agreements: An Introduction*, London: Routledge.

Haupt, F., Streck, C., Bakhtary, H., Behm, K., Kroeger, A., Schulte, I. (2018). Zero- deforestation Commodity Supply Chains by 2020: Are We on Track? International Sustainability Unit, TFA 2020, CDP, Climate Focus, pp. 1–33.

Humphreys, D. (1996). *Forest Politics: The Evolution of International Cooperation*, London: Earthscan.

Humphreys, D. (2006). *Logjam: Deforestation and the Crisis of Global Governance*, London: Earthscan.

Judge-Lord, D., Cashore, B.W., McDermott, C.L. (2020). Do Private Regulations Ratchet Up? How to Distinguish Types of Regulatory Stringency and Patterns of Change. *Political Studies* 33, 96–125.

IPCC. (2007). *Climate Change 2007: Synthesis Report*. Geneva: Intergovernmental Panel on Climate Change (IPCC).

Klooster, D. (2005). Environmental certification of forests: The evolution of environmental gover-nance in a commodity network. *Journal of Rural Studies* 21, 403–417.

Kolk, A. (1996). *Forests in International Politics: International Organizations, MGOs and the Brazilian Ama-zon*, Utrecht: International Books.

Lawson, S., MacFaul, L. (2010). *Illegal Logging and Related Trade: Indicators of the Global Response*. Lon-don: Chatham House, p. 154.

Lipschutz, R.D. (2001). Why is there no international forestry law? An examination of international forestry regulation, both public and private. *Journal of Environmental Law* 19(1): 153–79.

MacKenzie, C.P. (2012). Future prospects for international forest law. *International Forestry Review* 14(1): 1–9.

McDermott, C.L. (2013). Certification and equity: Applying an "equity framework" to compare cer-tification schemes across product sectors and scales. *Environmental Science and Policy* 33, 428–437.

McDermott, C.L. (2014). REDDuced: From sustainability to legality to units of carbon—The search for common interests in international forest governance. *Environmental Science and Policy* 35, 12–19.

McDermott, C.L., Cashore, B., Kanowski, P. (2010). *Global Environmental Forest Policies: An Interna-tional Comparison*. Earthscan, London.

McDermott, C.L., Coad, L., Helfgott, A., Schroeder, H. (2012). Operationalizing social safeguards in REDD+: Actors, interests and ideas. *Environmental Science and Policy* 21, 63–72.

McDermott, C., O'Carroll, A., Wood, P. (2007). International Forest Policy – the instruments, agree-ments and processes that shape it. United Nations Forum on Forests, UN Department of Economic and Social Affairs.

McDermott, C.L., Sotirov, M. (2018). A political economy of the European Union's timber regulation: Which member states would, should or could support and implement EU rules on the import of illegal wood? *Forest Policy and Economics* 90, 180–190.

Ministry of the Environment of Norway (2009) "Guyana–Norway partnership in climate and for-ests". Available online at: <http://www.regjeringen.no/en/dep/md/Selected-topics/climate/the-govemment-of-norways-intemational-/guyana-norwaypartnership.html?id=592318>

Myers, R., Rutt, R., McDermott, C.L., Maryudi, A., Acheampong, E., Camargo, M., Cam, H. (2020). Imposing legality: Hegemony and resistance under the EU Forest Law Enforcement, Governance, and Trade (FLEGT) initiative. *Journal of Political Ecology* 27, 125–146.

NYDF Assessment Partners. (2019). Protecting and restoring forests: A story of large commitments yet limited progress. New York Declaration on Forests Five-year Assessment Report. Climate Focus (coordinator and editor). Accessible at forestdeclaration.org, pp. 1–96.

Programme for the Endorsement of Forest Certification (PEFC). (2020). What's new: Facts & figures. Available online at: <https://www.pefc.org/>

Ros-Tonen, M.A.F., Reed, J., Sunderland, T. (2018). From synergy to complexity: The trend toward integrated value chain and landscape governance. *Environmental Management* 42, 439–415.

Rudel, T., DeFries, R., Asner, G.P., Laurance, W. (2009). Changing drivers of deforestation and new opportunities for conservation. *Conservation Biology* 23, 1396–1405.

Saunders, J. (2020). Tackling Deforestation is Balance of Local and Global | Chatham House – International Affairs Think Tank, pp. 1–7. Available at: https://www.chathamhouse.org/2020/11/tackling-deforestation-balance-local-and-global

SCA and WRI (Seneca Creek Associates and Wood Resources International). (2004). Summary: "il-legal" logging and global wood markets: The competitive impacts on the U.S. Wood Products Industry. American Forest and Paper Association, pp. 1–20.

Scoones, I., Stirling, A., Abrol, D., Atela, J., Charli-Joseph, L., Eakin, H., Ely, A., Olsson, P., Pereira, L., Priya, R., van Zwanenberg, P., Yang, L. (2020). Transformations to sustainability: Combining structural, systemic and enabling approaches. *Current Opinion in Environmental Sustainability*, 20, 1–11.

Seymour, F., Busch, J. (2016). *Why Forests? Why Now? The Science, Economics, and Politics of Tropical Forests and Climate Change*, Center for Global Development.

Skutsch, M.M. (2019). *The Evolution of International Policy on REDD+, Oxford Research Encyclopedia of Climate Science*. Oxford: Oxford University Press, pp. 1–30.

United Nations. (1992a). *United Nations Framework Convention on Climate Change*, New York: United Nations.

United Nations. (1992b). *Rio Declaration on Environment and Development*. Available online at: <http://www.unep.org/Documents.Multilingual/Default.asp?documentid=78&articleid=1163>

United Nations. (2007a). FCCC/SBSTA/2009/L.19/Add.1. Reducing emissions from deforestation in developing countries: approaches to stimulate action. Available online at: <http://unfccc.int/resource/docs/2009/sbsta/eng/l19a01.pdf>

United Nations. (2007b). *United Nations Declaration on the Rights of Indigenous Peoples*. Available online at: <http://www.un.org/esa/socdev/unpfii/documents/DRIPS_en.pdf>

United Nations. (2008). *Non-Legally Binding Instrument on All Types of Forests*. Available online at: <http://daccess-dds-ny.un.org/doc/UNDOC/GEN/N07/469/65/PDF/N0746965.pdf?OpenElement>

UNFCCC. (2011). *The Cancun Agreements Dec 1/CP.16*. United Nations Framework Convention on Climate Change, pp. 1–31.

UNFCCC. (2014). Report of the Conference of the Parties on its nineteenth session, held in Warsaw from 11 to 23 November 2013: Part two. UNFCCC, pp. 1–43.

UNFCCC. (2020). REDD+ Web Platform. Available online at: https://redd.unfccc.int/info-hub.html

West, T.A.P., Fearnside, P.M. (2021). Brazil's conservation reform and the reduction of deforestation in Amazonia. *Land Use Policy* 100, 105072–105012.

White, A., Martin, A. (2002). *Who Owns the World's Forests? Forest Tenure and Public Forests in Transition*, Washington, DC: Forest Trends and Center for International Environmental Law.

World Bank. (1991). *The Forest Sector: A World Bank Policy Paper*, Washington, DC: World Bank.

43

Desertification

Competing knowledge claims and land-management agendas

Meri Juntti

Desertification, like climate change (see Chapter 32) and biodiversity loss (see Chapter 41), has been deemed a global environmental challenge that merits its own multilateral convention to achieve coordinated action to combat and mitigate its impact (see Chapters 9 and 10). The United Nations Convention to Combat Desertification (UNCCD), established in the wake of the 1992 Rio Earth Summit on sustainable development, orchestrates action on desertification at a global level. However, like many other environmental issues, desertification is a nebulous concept, where scientific knowledge, political opinion and operative-level experience and know-how converge and sometimes conflict. Over a hundred definitions of desertification are identifiable from the literature, but most relate it to the loss of an area's resource potential, through depletion of soil cover, vegetation cover or loss of useful plant species (Middleton 2013). Desertification is seen as a serious threat to food security in specific dryland regions and one that renders global poverty and food security efforts such as the UN Sustainable Development Goal 2 (SDG 2 Zero Hunger) difficult to reach.

Desertification poses a so-called "wicked" challenge (Turnpenny et al. 2009) to environmental managers and legislators. Due to its complex and interdisciplinary nature, characterized by high uncertainty and ambiguous relations of cause and consequence that vary by context, the allocation of responsibility and identification of possible solutions in practice is hard.

This chapter outlines the debates surrounding the definition and extent of desertification globally and the evolution of global and local governance efforts focused on addressing it. Particular attention will be given to the evidence base on which efforts to diagnose and address desertification are founded and some of the context specific political and socio-economic processes that are seen to drive desertification. The chapter concludes by summing up on the challenge of desertification and the extent to which policy and mitigation practices are beginning to embrace its complexity.

Extent of the problem: defining and governing desertification

While desertification remains a debated term, the UNCCD defines it as "land degradation in arid, semi-arid and dry sub-humid areas resulting from various factors, including climatic variations and human activities" (UN 1994). Land degradation, in turn, is defined as:

DOI: 10.4324/9781003008873-48

the reduction or loss, in arid, semi-arid and dry sub-humid areas, of the biological or economic productivity and complexity of rainfed cropland, irrigated cropland, or range, pasture, forest and woodlands resulting from land uses or from a process or combination of processes, including processes arising from human activities and habitation patterns, such as: (i) soil erosion caused by wind and/or water; (ii) deterioration of the physical, chemical and biological or economic properties of soil; and (iii) long-term loss of natural vegetation.

(UN 1994: 5)

As with other similar global governance regimes centering on climate change and biological diversity for example, the UNCCD has been a powerful force in institutionalizing this definition globally and with that, the notion that desertification is a universal threat to localized productivity of land and subsequently, livelihoods and food security in already vulnerable dryland regions (IPCC 2018).

But despite the global convention and the consensus that desertification concerns the loss of productivity of land, considerable discord remains around how to measure the extent of desertification, its exact drivers (human and natural) and its impacts (Behnke and Mortimore 2016; Cherlet et al. 2018; Briassoulis 2019). The World Atlas of Desertification (Cherlet et al. 2018) for example, has omitted any precise areal mappings of the occurrence and extent of desertification from its latest edition, and suggests that desertification cannot be represented by a comprehensive global model, as this would inevitably be inaccurate due to complexity and uncertainty. Blaikie and Brookfield (1987) suggest that the degree of degradation can only be meaningfully defined in relation to actual or potential uses of a specified area of land, and is therefore contingent. It makes sense that some deserts exist naturally, and are not by definition an adverse phenomenon. For example, Behnke and Mortimore (2016; see also Thomas and Middleton 1994) point to the "Sahel desertification crises" which stemmed from reports of a rapidly encroaching desert and was widely researched and publicized in the late twentieth century, but has since been refuted. These authors suggest that the notion of desertification is misleading altogether and, while the loss of productivity of land is a real problem in certain regions, the Sahel serves as an example where the characteristic fluctuation of conditions in dryland environments has been misinterpreted as desertification (see UNCCD Knowledge Hub 2016). This is not to say that communities in the Sahel region do not suffer from vulnerability in terms of access to resources and nutrition, but that rather than encroaching deserts, these crises are driven by political and socio-economic causes (e.g., UN News 19th of October 2020). Nevertheless, it is indisputable that human actions and climatic changes are increasingly colluding to corrode the biological and economic functions of land. But this is driven and manifest in different ways depending on context. For example, land degradation can be associated with large intensive monocultures and heavy usage of chemical nutrients with detrimental impact on soil organic matter and the loss of these key nutrients from the natural circulation of matter on the planet (see, e.g., Steffen et al. 2015). But conversely, degradation may also be driven by the abandonment of beneficial land management practices due to the declining viability of traditional farming methods and urban migration (e.g., Briassoulis 2019).

All things considered, the overwhelming consensus in the scientific community is that desertification in terms of natural and human-induced loss of productivity of land remains a growing threat in specific dryland regions (IPCC 2018). The Millennium Ecosystem Assessment along with several other global reports (MEA 2005; WRI 2005; IPBES 2018; FAO et al. 2020) collate evidence on the constituting processes and manifestations of desertification and portray various aspects of its impact on human wellbeing, pointing out that while the local magnitude of the impact varies in relation to degree of aridity and population pressure, desertification occurs on all continents except Antarctica, and affects millions of people, the majority of whom already live in poverty and can be classified as vulnerable. Many reports also place the problem of desertification in the context of the need to feed a global population of an estimated 9 billion people by 2050 on available and diminishing land resources, where desertification, in many areas exacerbated by climate change, poses a threat to food security (IPCC 2018: 273).

The global convention (UNCCD) identifies 197 countries as parties to the Convention and thus as affected by processes of desertification (UNCCD 2020). The Convention constitutes the most significant legally binding international agreement linking environment and development to sustainable land management. It is committed to a bottom-up approach, encouraging the participation of local people in combating desertification and land degradation, as well as knowledge and technology transfer from North to South. Its principal aims are to improve the living conditions for people in drylands, to maintain and restore land and soil productivity, and to mitigate the effects of drought (UNCCD 2020). Figure 43.1 details the different institutions involved in the UNCCD.

The 2018–2030 strategic framework makes an explicit commitment to global "zero net degradation" or Land Degradation Neutrality (LDN) by 2030 (UNCCD 2017). This aim is also expressed in the UN Sustainable Development Goals (SDGs), where goal 15, and specifically target 15.3, is to, "by 2030, combat desertification, restore degraded land and soil, including land affected by desertification, drought and floods, and strive to achieve a land degradation-neutral world" (UN 2015: 27). LDN is hailed as a paradigm shift in addressing desertification as it places great emphasis on remediation. In practice, LDN is defined as involving (a) managing land more sustainably, which would reduce the rate of degradation; and (b) increasing the rate of restoration of degraded land, so that the two trends converge to give a zero net rate of land degradation (UNCCD 2017:44).

Perhaps because of the ambiguity surrounding the term, global policy efforts to address desertification have been accompanied by meager economic means, however. The Global Mechanism (GM, see Figure 43.1) was established in 1998 with the remit to support developing countries in increasing investment in land as a resource at the national and international levels. The GM also helps countries to identify national and international, private and public sources of finance for sustainable land management practices. In 2010 the Global Environment Facility (GEF) finally adopted the mandate to finance the UNCCD. It directs funds toward monitoring and reporting on the national Action Plans by Conference Parties, but in order to release any significant funding, it has been necessary to link desertification to other cross-cutting issues, such as climate change, and the Secretariat of the Convention plays a key role in this (Conliffe 2011). Scientific evidence linking climate change and incidences of desertification is widespread, and desertification has been hailed by many as a noteworthy potential contributor to carbon emissions – particularly in terms of loss of soil carbon sequestration capacity (Conliffe 2011; IPCC 2018). Since 2017, the GM-led initiative termed the Land Degradation Neutrality Fund (LDNF; UNCCD 2020) has worked to channel money from public and private impact investors – those, including national governments, wishing to invest with the specific goal of achieving beneficial outcomes – toward sustainable land management and restoration projects implemented by the private sector in areas affected by desertification. These financially viable projects funded through the LDNF are intended to create sustainable jobs and improve food and water security in their respective locations.

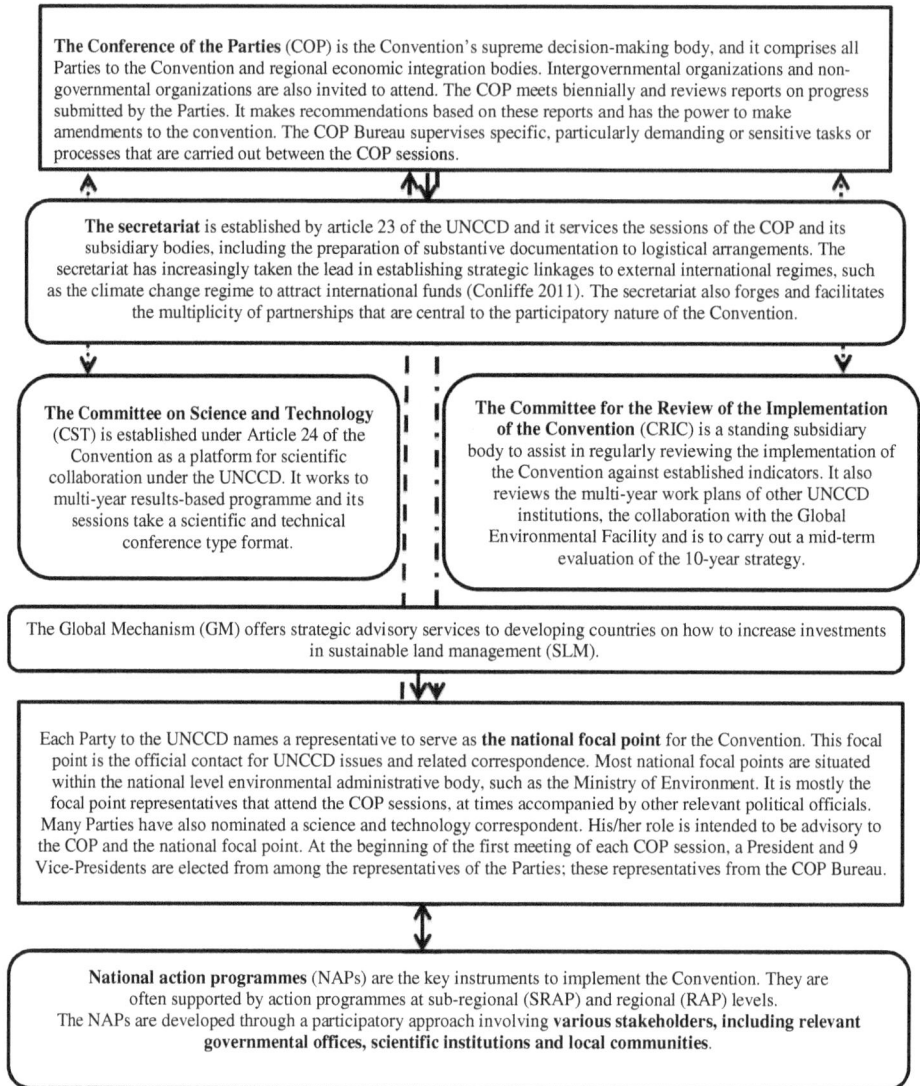

The Conference of the Parties (COP) is the Convention's supreme decision-making body, and it comprises all Parties to the Convention and regional economic integration bodies. Intergovernmental organizations and non-governmental organizations are also invited to attend. The COP meets biennially and reviews reports on progress submitted by the Parties. It makes recommendations based on these reports and has the power to make amendments to the convention. The COP Bureau supervises specific, particularly demanding or sensitive tasks or processes that are carried out between the COP sessions.

The secretariat is established by article 23 of the UNCCD and it services the sessions of the COP and its subsidiary bodies, including the preparation of substantive documentation to logistical arrangements. The secretariat has increasingly taken the lead in establishing strategic linkages to external international regimes, such as the climate change regime to attract international funds (Conliffe 2011). The secretariat also forges and facilitates the multiplicity of partnerships that are central to the participatory nature of the Convention.

The Committee on Science and Technology (CST) is established under Article 24 of the Convention as a platform for scientific collaboration under the UNCCD. It works to multi-year results-based programme and its sessions take a scientific and technical conference type format.

The Committee for the Review of the Implementation of the Convention (CRIC) is a standing subsidiary body to assist in regularly reviewing the implementation of the Convention against established indicators. It also reviews the multi-year work plans of other UNCCD institutions, the collaboration with the Global Environmental Facility and is to carry out a mid-term evaluation of the 10-year strategy.

The Global Mechanism (GM) offers strategic advisory services to developing countries on how to increase investments in sustainable land management (SLM).

Each Party to the UNCCD names a representative to serve as the national focal point for the Convention. This focal point is the official contact for UNCCD issues and related correspondence. Most national focal points are situated within the national level environmental administrative body, such as the Ministry of Environment. It is mostly the focal point representatives that attend the COP sessions, at times accompanied by other relevant political officials. Many Parties have also nominated a science and technology correspondent. His/her role is intended to be advisory to the COP and the national focal point. At the beginning of the first meeting of each COP session, a President and 9 Vice-Presidents are elected from among the representatives of the Parties; these representatives from the COP Bureau.

National action programmes (NAPs) are the key instruments to implement the Convention. They are often supported by action programmes at sub-regional (SRAP) and regional (RAP) levels. The NAPs are developed through a participatory approach involving various stakeholders, including relevant governmental offices, scientific institutions and local communities.

Figure 43.1   United Nations convention to combat desertification: institutions and decision-making structure

Source: Collated from UNCCD (2020).

The politics of desertification

The complexity of desertification – the plethora of processes and factors that drive the loss of land productivity – and their variation across geographic contexts, as well as the socially constructed nature of the phenomenon have left ample space for its politicization. Literature identifies several modes of the politicization of desertification and explores and identifies different manifestations, ranging from the political motives behind the framing of the UN Desertification Convention itself (Adger et al. 2001) to unraveling some of the "myths" associated with desertification and revealing the underpinning uncertainties (Thomas and

Middleton 1994). Piers Blaikie and Harold Brookfield suggested already in 1987, that land degradation (that in many cases amount to desertification) is driven by certain social, economic and cultural processes of resource use and modification that produce highly politicized patterns of appropriation and link to uneven relationships of power. The link between environmental degradation resulting from marginalization, where land users have been forced to move to areas that are not fit for cultivation for example, is well established (e.g., Benjaminsen 2015). While this may be driven by "land grabbing" in a context of poorly established property rights or insecure tenancy conditions (Benjaminsen and Bryceson 2012), expansion of cultivation onto unsuitable land may also happen as an unintended consequence of policy, such as the spread of durum wheat cultivation to "badlands" in southern Europe (Clarke and Rendell 2000). It is now commonly accepted that desertification needs to be understood as a resource use issue taking place within a bigger picture of socio-economic structures, processes, needs and disparities (Halbac-Cotoara-Zamfir et al. 2020).

Blaikie and Borookfield (1987) also highlight the need to think carefully about how we define and conceptualize the society–degradation relationship – not only what are the underlying drivers of harmful land management practices, but what potential land uses and benefits that are being lost to desertification – because this is significant for how we address the problem. The varying interpretations of the extent and manner in which desertification is actually posing a problem for livelihoods in specific areas are rife among stakeholders, and this has a significant impact on how natural resource management policies and programs are implemented (Wilson and Juntti 2005; Briassoulis 2019). Therefore, the local perspective is crucial in understanding what desertification means in context and how it has come about.

But according to Adger et al. (2001), policies orchestrating local solutions to desertification and other major global environmental issues are often informed by powerful international knowledge regimes and are the result of multilevel actions and interactions that rarely reflect the multiplicity of local contexts in a realistic way (see Chapter 18). Global discourses tend to be guided by a managerialist ideal, where the issues at hand are presumed to be somehow resolvable via global action, and incorporate shared "myths" and blueprints (Thomas and Middleton 1994; Behnke and Mortimore 2016). Adger et al. (2001) identify a strong discourse of crisis on which much of the international desertification policy is based but often does not resonate with local level experiences of desertification. The WRI report (2005) provides an explanation of how this crisis discourse might have come about by suggesting that a political motivation can be identified for the founding of the UNCCD. The United States acted as an unexpected proponent of the African demands for a global convention on desertification because it harbored hopes that African states would support the broader Rio Declaration in return. The United States may also have been responding to criticism regarding its lack of action on the other multilateral conventions. The subsequent establishment of a desertification control unit within the United Nations Environment Programme (UNEP) further institutionalized the crisis discourse, whereby it became purposeful in order to maintain the existence of the unit (Warren and Agnew 1988, cited in Middleton 2013; see also Behnke and Mortimore 2016). Whether more politically motivated or deriving from valid evidence, the crisis discourse shapes much of the managerial approach to combating desertification, embodied by the UNCCD, which binds all Parties to the Convention to establishing National Action Plans (NAPs) to combat desertification.

This managerial approach obscures what many term the deeper drivers of desertification processes, embedded in global trade relations, weak property and tenancy rights and unequal access to resources such as water, resulting marginalization of smallholders who are thus forced to exploit lands that are prone to desertification processes (Adger et al. 2001;

Benjaminsen 2015). A powerful interpretation stemming from the crisis discourse is that desertification is the catalyzer of underdevelopment and that local level human actions are the root causes of desertification (for prevalence of this, see WRI 2005). But more critical local perspectives have begun to gain recognition in the past decades and this is manifest in the emphasis on participatory approaches and indigenous and local knowledge (ILK) in more recent global initiatives addressing desertification (IPCC 2018). Increasing number of authors highlight that desertification processes need to be considered as taking place in a scalar and often unequal socio-economic context of land use drivers, motivations and needs (Halbac-Cotoara-Zamfir et al. 2020). Better representation of local stakeholders in desertification discourses and remedial actions may be a route to uncovering the deeper drivers of harmful land management practices.

It is important to understand that decertification can be a highly politicized issue at the local scale in particular. What should be a simple matter of agreeing what desertification means in practice in a specific location frequently uncovers quite disparate stakeholder understandings of the term, its extent and the processes involved. Perceptions regarding the driving forces of desertification processes at local level may vary greatly, with some stakeholders emphasizing climatic drivers and others placing more focus on human causes such as intensive land management practices and irrigation. Confusion prevails between the concepts of desertification, drought and progressive desiccation (Goudie 1990). Desertification is an evocative and misleading concept: it does not necessarily manifest itself in the spreading of desert-like conditions and it certainly does not consist of a single process but rather of a combination of mutually reinforcing and often cyclical developments (Wellens and Millington 1992). It may therefore be difficult to provide exact evidence of desertification even in a well-defined location. Indeed, over the years, it has proved very difficult, even though empirical scientific exploration, to define whether an area is desertified or not.

While the acceleration of land degradation processes may be both climatically induced and anthropogenic (as per the UNCCD definition), the relative influence of these drivers is difficult to determine. Juntti and Wilson (2005) point out that the difference in emphasis detectable in the above definitions can lead to very different ways of conceptualizing and diagnosing the problem and the appropriate remedial methods, and crucially, that different emphases can also be deployed to serve different stakeholder interests. For example, from an analysis of desertification discourses in four northern Mediterranean countries, Juntti and Wilson (2005) identify five different ways of defining desertification at the "operative" level, by stakeholders such as farmers, water and environmental officials and citizens living and working in environments that the UNCCD identifies as affected by desertification. While the five categories overlap to some extent, each holds a different interpretation of the role of the natural resources in the economy, the justifiable ends toward which these resources are to be used and, hence, a different morality according to which the extent and nature of desertification has been defined and is influencing how natural resources are managed. For example, in areas where intensive agriculture and the use of water resources for irrigation was linked to resource scarcity but also to significant economic growth, a reductive understanding of desertification as a water provisioning issue was prevalent. This reflects a morality where the rural population and the existing forms of land management (mainly irrigation farming and horticultural production) are regarded not only as necessary economically productive functions, but also as examples of good management of the natural resources of the locality, especially the productivity of the soil (Oñate and Peco 2005). Economic development is often seen to justify overexploitation or natural resources and technological solutions such as desalination or water transfers are seen as the solution, with much of the responsibility for providing these falling on the state and water officials.

Both of the above interpretations shift responsibility away from land managers and sanction the continuation of resource management practice as usual. In this sort of a normative context, the lack of broader political will to undertake economically detrimental policy decisions at the local level poses a significant problem for any desertification mitigation efforts questioning land use intensity (Juntti and Wilson 2005). This has meant that water management and the implementation of land use policies in these water-scarce regions has since been guided by the aspirations of the irrigation farming industry rather than by resource availability and sustainable use (Oñate and Peco 2005; Ripoll et al. 2010).

This plurality of understandings of desertification and its drivers poses a challenge for the design and implementation of the National Action Plans that the UNCCD requires from all designated parties, as well as any individual efforts to combat desertification in practice. The following section looks critically at the role of evidence in the emergence of desertification into the global arena as a significant environmental problem and considers options for managing the complexity and competing knowledge claims in policy and practice.

Desertification drivers, competing knowledge claims and implications for global policy

Adger et al. (2001) outline two types of global desertification discourses that allocate the blame for desertification to either international power relations or locally induced resource depletion, respectively. While this can be seen as a polarized debate, it is likely that both interpretations apply to a varying extent in different contexts. Understanding desertification and identifying its extent and impact in a locality requires a contextualized understanding of the biophysical resource base, land and water management practices, key sectors and actors such as farmers, other land owners and the local business sector, and their linkages to broader socio-economic scales through processes such as trade and tourism. As discussed above, acquiring a sound scientific understanding of ecology, soil morphology and the hydrological conditions and relevant cause-consequence relationships underpinning possible desertification alone, is a challenge. But despite the complexity, science has made significant progress in understanding and identifying desertification in the last decades and there are now some well-established assessment practices and indicators (Briassoulis 2019).

Many still complain of the domination of uncertain scientific evidence and obscure modeling in understanding and assessing the extent of desertification risk, and most indices that are commonly used focus solely on biophysical variables (D'Odorico et al. 2013). However, science based approaches are increasingly beginning to emphasize the need to understand the role of socio-economic structures such as land use subsidies, the structure of landownership and the existence and implementation of mitigating policies (Kosmas et al. 2013), the omission of a broader range of contextually informed socio-economic divers of desertification risk, such as processes of trade and price setting, intensification and marginalization portray desertification as a de-politicized concept and hinder the identification of relevant, workable and legitimate solutions.

To really pin down the manifestations, drivers and impacts of desertification in a locality, it is necessary to look at local land-use needs and practices, nested socio-economic structures and processes that drive these as well as demographic and social factors that modulate human–nature interactions. Based on a secondary analysis of a broad range of case study data, Geist (2017) suggests that in the majority of cases globally, desertification is caused by a combination of intensifying agricultural practices (particularly livestock and to a lesser extent crop production), increasing aridity (often associated with climate change), extension of infrastructure (including that associated with urban sprawl but mainly irrigation infrastructure such as the establishment of dams and canals), and extraction of wood or other materials. Underpinning these processes, are institutional and socio-economic drivers that need to be understood at regional and local levels (Geist 2017; Halbac-Cotoara-Zamfir et al. 2020). For example, in some cases, agricultural intensification is driven by low prices and the need to extract more value from scarce land or to expand farming into vulnerable territories, and in others, policies aimed at modernization and increasing the competitiveness of agricultural practices have underpinned unsustainable expansion, intensification and the adoption of harmful production practices (Juntti and Wilson 2005; D'Odorico et al. 2013; Blake et al. 2020).

Understanding what is driving desertification in a given context requires careful consideration. Adger et al. (2001) refer to a case study of subsistence farmers in Mali (Benjaminsen 2000 cited in Adger et al. 2001) to illustrate how the presumption that poor subsistence farmers are forced to overexploit forest resources for firewood is over simplistic and overlooks the methods whereby local subsistence farmers have managed to sustainably use local forests for firewood for centuries (see Gray and Moseley 2005 on many similar myths related to the poverty–environment relationship). This renders many of the managerial solutions to deforestation overly coercive and inefficient or even detrimental to the existing actually sustainable livelihoods. As Thomas (1997) points out, scientific solutions are rarely easily transferrable from one context to another and, where this is done, it is the small variations in physical and environmental factors as well as the socio-economic context that often lead to unexpected and inefficient outcomes and downright rejection by local land managers. For instance, Middleton (2013) describes how the diversity of processes whereby overgrazing encourages desertification, not just by removal of biomass but also through trampling and resulting erosion and changing of soil chemical components inviting an invasion of desert shrubs, are well understood, but nevertheless, the impacts of measures taken to curb overgrazing may vary unexpectedly in context. Where settlement of nomadic pastoralists has in some instances proven a good way to curb growing herd sizes, forced settlement has in many cases just become an alternative underpinning driver of accelerated desertification processes particularly near watering holes and in the best grazing lands (Geist 2017).

In a positive development, the engagement of the experiential understanding of many nomadic tribes of how their environments work and respond to different management options is now recognized as crucial in global policy (IPCC 2018; UNCCD 2017). This message remains relevant even for the proponents of the more critical analytical approaches that recognize power relations and scalar drivers of desertification. As Adger et al. (2001) point out, those assign the blame for desertification to cash crops and marginalization through resource appropriation, may also underestimate the resource management knowhow of local people that has accumulated through generations. Adger et al. (2001) cite a further example from Benjaminsen's research in Mali to show that sustainable cash-crop farming is possible and can even reinforce the ability of the local farmers to undertake sustainable management practices of food crops.

What scientists and practitioners involved in combating desertification have come to realize, is that little can be done to reverse desertification processes without the complete involvement of those farmers, pastoralists and other natural resource users being directly affected by desertification in its many forms. But while the UNCCD is explicitly supportive of bottom-up action, deeming participation of stakeholders as crucial for achieving workable

NAPs, genuine and equitable participation is difficult to achieve. Seely (1998) points to the significance of conductive policy and planning framework, environmental framework and socio-economic framework conditions for achieving full participation. It is often, however, the policy and planning frameworks themselves that have led to land use changes accelerating desertification and triggered the institutionalization of unsustainable resource management practices, and policies often have contradictory impacts (Juntti and Wilson 2005; Oñate and Peco 2005; Briassoulis 2019). Insecure land tenure conditions and lack of land policy which permits land-grabbing by those who have the means are often blamed for degrading land management (Bugri 2008). Briassoulis (2019) suggest that while land use planning is in a pivotal position in curbing the human drivers of desertification, as with environmental policies and individual efforts to address desertification, its ideal functioning is also hindered by the highly politicized stakes and complex drivers of the phenomenon. Where unsustainable land and water uses have become deeply embedded in resource management institutions and practices as well as local economic fortunes, the implementation of any mitigating measures is extremely difficult as these are perceived by many locals to be detrimental to their livelihoods. Subsequently, for example the development and implementation of the Spanish National Action Plan to Combat Desertification, which was passed in 2008 after considerable delay, have been slow and ineffectual (Oñate and Peco 2005; MMA 2008).

The Spanish example illustrates how policies and other socio-economic factors can lead to conditions where local management structures as well as resource managers themselves are highly resistant to desertification mitigation measures, but broader "social" factors can play a role also. Bradley and Grainger (2004) discuss the role of what they term "social resilience" in aiding the mitigation of and adaptation to desertification in a developing country context where technological fixes are not available. Exploring a case study of two nomadic pastoral tribes in Senegal, Bradley and Grainger demonstrate how land managers have historically adapted their management systems in response to repeated eco-climatic fluctuations or social constraints. While learning from past experience or indigenous knowledge passed from previous generations is significant for continuous sustainable management, it becomes even more crucial when a constraint, such as resource depletion or appropriation, is imposed. In this case land managers must choose between substituting an activity wholly or partially by others that function better under more constrained conditions; adopting new non-extractive or non-land-based activities under severe or prolonged survival conditions; traveling longer distances to avoid place-based constraints; or becoming more reliant on support from other households or on income generated by household members who have moved elsewhere (Bradley and Grainger 2004: 454; see also Dorward et al. 2009 for livelihood strategies). The choice between these alternatives can be made partly through trial and error, and Bradley and Grainger conclude that the social resilience of communities, built through a cyclical model of learning, could help explain the lack of widespread desertification in the silvopastoral zone of Senegal where the two tribes operate. It is important to understand these kinds of existing strategies in order to devise sustainable desertification mitigation measures for any context.

Many other social factors influence the ability of communities to make sustainable decisions regarding resource management and, as Seely (1998) points out, it is also important to identify any weaknesses in these. Bugri (2008) highlights a broader range of factors as significant. One of these is the marginalization of women in decision-making relating to land resources, although women are often in charge of agricultural work. Although the roles of women and men in land management differ widely between regions, women tend to have lower levels of education, and lower participation in community-based organizations (OECD 2002). Pirmoradi et al. (2011) found that there was a positive relationship between the level of literacy, income and participation in training courses, and participation in plans to combat desertification among farmers in Iran; and, indeed, women's position is seen by many as requiring attention in the pursuit of successful desertification mitigation measures. Effective participation requires local-level gender-sensitive understanding of livelihood roles (OECD 2002; FAO 2003). Building on these understandings, Blake et al. (2020) emphasize the need to "co-design" desertification mitigation activities with local land managers, as this facilitates the engagement of scientific findings and techniques while not overlooking local knowhow, needs and aspirations.

While the UNCCD does not define exactly what is meant by participation, it makes several references to local populations, communities and nongovernmental organizations (NGOs) and thus to more empowering and democratic decision-making (Stringer et al. 2007; see Chapters 14 and 28). NGOs and community-based organizations (CBOs) play a key role in the UNCCD and have been drafted in to establish a link with land users who, through NGO initiatives, are supposed to help identify and assess cases of desertification and contribute ideas for mitigation measures (see Figure 43.1). However, NGO accountability and ability to enable equal participation opportunities for all communities regardless of socio-economic status have been found to be challenging in many case studies (Stringer et al. 2007; Blake et al. 2018). This is particularly problematic where existing inequalities between communities and within communities exist and are linked to the root causes of desertification, as discussed above (Bugri 2008). However, Reed et al. (2008) report on successful collaboration between ecologists and pastoralists in identifying and evaluating a set of indicators suitable for assessing land degradation in three field locations in Botswana by applying a scientific methodology to validate local indicator knowledge. Reed et al. also note that the use of focus groups in the study methodology increased the exchange of information also among the pastoralists, which addressed the initially "thinly spread" indigenous knowledge relating to species and land uses linked to degradation (see Chapter 41).

While Redclift (2005) argues that traditional scientific definitions and categories obscure local meanings of nature and natural resources, the study by Reed et al. (2008) appears to serve as an example of participatory research where, crucially, community members' experience-based expertise is engaged in the whole research process, from the identification of analytical units to defining the use of findings (Fischer 2002). The notion of co-design implies a genuinely empowered position for community members in the identification of the local nature of the desertification problem and the design of mitigation measures (Blake et al. 2018). In participatory planning literature, Booher and Innes (2002) advocate marshaling of "network power" through "authentic communication" which empowers all parties to communicate and contribute to an equal extent so that collaboration is as innovative as possible. This requirement highlights the need to pay attention to the purpose and form of participation; ideally participation is seen not as instrumental for compliance with management prescriptions but intrinsic to the creation of new knowledge, better contextual understanding and achieving innovative new management options. How much of the participation under the UNCCD aspires to these kinds of aims and objectives seem to have increased in the past decade or so.

Conclusion

This chapter has illustrated the complex nature of the phenomenon of desertification itself and the implications this has for efforts to address it. It is evident that the barriers and challenges related to managing desertification are as varied as the contexts where it manifests itself. A wicked or highly politicized issue, desertification must be addressed through approaches that embrace the complexity of the scientific evidence and the methodologies used to derive it, appreciate the significance of physical and environmental variations for the workability of any scientific solutions and remain awake to the implicitly and inevitably politicized nature of knowledge production, adoption and application in complex and varied socio-economic conditions (see Chapter 18). This is of course a challenging task and the different sections of this chapter illustrate the fault lines but also some positive examples in mitigating desertification.

The extent to which mitigation initiatives are willing and able to address the underlying causes of desertification such as product price driven intensification or marginalization due to insecure land ownership and tenancy conditions, remains limited. This suggests that as with climate change resilience and adaptation, there is an evident need to embrace transformational change that addresses power imbalances at the grassroots level as well as other scalar drivers of unsustainable land uses. While technical solutions are important, these must not create dependency or be used to justify land grabbing. Scientific evidence is of course important but measures addressing desertification need to be co-designed to integrate indigenous and local knowledge.

The land degradation neutrality target and the accompanying fund (LDNF) are new developments that suggest that the international community, including funders, is beginning to take desertification and the state of soils seriously. The UNCCD submission to the Rio+20 conference (UNCCD 2011) saw addressing land degradation and desertification as key to achieving the kind of adaptation capacity and resilience that the impending challenges of population growth combined with climate change will pose to food security and water provision globally (see also CGIAR 2011; see Chapter 32). The strategy outlines payments for ecosystem services as a good means of addressing the short-term economic losses that landowners will incur from introduction of the kind of rates of sustainable land management that enable meeting the zero net degradation aim. The failure to establish legally binding instruments for soil management, for example the inability to pass the EU soil directive (CEC 2012), is regrettable, but the emphasis given to decertification in the 2018 IPPC report on Climate Change and Land and the FAO et al. (2020) report on soil biodiversity as well as the EUs new soil observatory (CEC 2021) provide a more optimistic picture. These initiatives arguably respond to the UNCCD (2011: 5) call for "scientifically credible, transparent and independent assessment of existing, policy-relevant but not policy-prescriptive knowledge. This assessment should be provided by a globally recognized, strong and effective science–policy interface, similar to those established for climate and biodiversity (IPCC and IPBES respectively)."

It remains to be seen how well the new funding instruments, mainly the LDN Fund, function within the context of the deeper drivers of desertification such as institutionalized land-use interests, uncertain tenancy conditions, price competition that drives intensification and urban sprawl. As these are increasingly recognized in literature, perhaps the actions needed to address them will slowly gain traction also.

References

Adger, N.W., Benjaminsen, T.K., Brown, K., and Svarstad, H. (2001) Advancing a political ecology of global environmental discourses. *Development and Change 32*: 682–715.
Behnke, R. and Mortimore, M. (2016) Introduction: The end of desertification. In same (eds.) *The end of desertification: Disputing environmental change in drylands*. Springer Nature: Switzerland, pp. 1–34.

Benjaminsen, R.A. (2015) Political ecologies of environmental degradation and marginalization. In Perreault, T., Bridge, G., and McCarthy, J. (eds.) *The Routledge Handbook of Political Ecology*. London: Routledge, pp. 354–365.

Benjaminsen, T.K. (2000) The Malian cotton zone: Economic success but environmental failure? In Benjaminsen, T.K. and Lund, C. (eds.) *Politics, Property and Production in the West African Sahel*. Uppsala: Nordic Africa Insitute.

Benjaminsen, T.A. and Bryceson, I. (2012) Conservation, green/blue grabbing and accumulation by dispossession in Tanzania. *The Journal of Peasant Studies 39*(2): 335–355.

Blake, W.H., Rabinovich, A., Wynants, M., Kelly, C., Nasseri, M., Ngondya, I, Patrick, A., Mtei, K., Munishi, L., Boeckx, P., Navas, A. Smith, H.G., Gilvear, D., Wilson, G., Roberts, N., and Ndakidemi, P. (2018) Soil erosion in East Africa: an interdisciplinary approach to realising pastoral land management change. *Environmental Research Letters* 13: 124014.

Blake, W.H., Kelly, C., Wynants, M., Aloyce, P., Lewin, S., Lawson, J., Nasolwa, E., Page, A., Nasseri, M., Marks, C., Gilvear, D., Mtei, K., Munishi, L., and Ndakidemi, P. (2020) Integrating land-water-people connectivity concepts across disciplines for co-design of soil erosion solutions. *Land Degradation and Development*. https://doi.org/10.1002/ldr.3791

Blaikie, P. and Brookfield, H. (1987) *Land Degradation and Society*. New York: Methuen.

Booher, D.E. and Innes, J.E. (2002) Network power in collaborative planning. *Journal of Planning Education and Research 21*(2): 221–236.

Bradley, D. and Grainger, A. (2004) Social resilience as a controlling influence on desertification in Senegal. *Land Degradation and Development* 15: 451–470.

Briassoulis, H. (2019) Combating land degradation and desertification: The land-use planning quandary. *Land 8*(2): 1–27. http://dx.doi.org/10.3390/land8020027

Bugri, J.T. (2008) The dynamics of tenure security, agricultural production and environmental degradation in Africa: Evidence from stakeholders in north-eastern Ghana. *Land Use Policy 25*: 271–285.

CEC (2012) Report from the Commission to the European Parliament, the Council, the European Economic and Social Committee and the Committee of Regions: The implementation of the soil thematic strategy and on-going activities. COM (2012) 46 final. Available online at: http://eur-lex.europa.eu/LexUriServ/LexUriServ.do?uri=COM:2012:0046:FIN:EN:PDF.

CEC (2021) Concept Note for the EU Soil Observatory. Ispra, September 2021 JRC.D/GDS. Available online at: https://ec.europa.eu/jrc/en/eu-soil-observatory

CGIAR (2011) *Farming's Climate Smart Future: Placing Agriculture at the Heart of Climate Change Policy*. Wageningen: Technical Centre for Agricultural and Rural Cooperation (CTA). Available online at: http://ccafs.cgiar.org/sites/default/files/assets/docs/farmings_climate-smart_future.pdf.

Cherlet, M., Hutchinson, C., Reynolds, J., Hill, J., Sommer, S., and von Maltitz, G. (2018) *World Atlas of Desertification*. Publication Office of the European Union, Luxembourg. Available online at: http://wad.jrc.ec.europa.eu

Clarke M.L. and Rendell H.L. (2000) The impact of the farming practice of remodeling hillslope topography on badland morphology and soil erosion processes. *Catena* 40: 229–250.

Conliffe, A. (2011) Combating ineffectiveness: Climate change bandwagoning and the UN Convention to Combat Desertification. *Global Environmental Politics 11*(3): 44–63.

D'Odorico, P., Bhattachan, A., Davis, K.F., Ravi, S., and Runyan, C.W. (2013) Global desertification: Drivers and feedbacks. *Advances in Water Resources 51*: 326–344. http://dx.doi.org/10.1016/j.advwatres.2012.01.013

Dorward, A., Anderson, S., Bernal, Y.N., Vera, E.S., Rushton, J.. Pattison, J., Paz, R. (2009) Hanging in, stepping up and stepping out: Livelihood aspirations and strategies of the poor. *Development in Practice 19*(2): 240–247.

FAO (2003) *Gender and Sustainable Development in Drylands: An Analysis of Field Experiences*. Rome: FAO.

FAO, ITPS, GSBI, SCBD and EC (2020) *State of Knowledge of Soil Biodiversity – Status, Challenges and Potentialities, Report 2020*. Rome: FAO. https://doi.org/10.4060/cb1928en

Fischer, F. (2002) *Citizens, Experts and the Environment*. Durham, NC: Duke University Press.

Geist, H. (2017) *The Causes and Progression of Desertification*. London: Routledge.

Goudie, A.S. (1990) *Techniques for Desert Reclamation*. Chichester: John Wiley.

Gray, L.C. and Moseley, W.G. (2005) A geographical perspective on poverty–environment interactions. *Geographical Journal 171*(1): 9023.

Halbac-Cotoara-Zamfir, R., Colantoni, A., Mosconi, E.M., Poponi, S., Fortunati, S., Salvati, L., and Gambella, F. (2020) From historical narratives to circular economy: De-complexifying the "desertification" debate. *International Journal of Environmental Research and Public Health 17*: 5398. http://dx.doi.org/10.3390/ijerph17155398.

IPBES (2018) The IPBES assessment report on land degradation and restoration. In Montanarella, L., Scholes, R., and Brainich, A. (eds.). *Secretariat of the Intergovernmental Science-Policy Platform on Biodiversity and Ecosystem Services*, Bonn, Germany, 744 pages. https://doi.org/10.5281/zenodo.3237392

IPCC, 2018: Global Warming of 1.5°C. An IPCC Special Report on the impacts of global warming of 1.5°C above pre-industrial levels and related global greenhouse gas emission pathways, in the context of strengthening the global response to the threat of climate change, sustainable development, and efforts to eradicate poverty [Masson-Delmotte, V., P. Zhai, H.-O. Pörtner, D. Roberts, J. Skea, P.R. Shukla, A. Pirani, W. Moufouma-Okia, C. Péan, R. Pidcock, S. Connors, J.B.R. Matthews, Y. Chen, X. Zhou, M.I. Gomis, E. Lonnoy, T. Maycock, M. Tignor, and T. Waterfield (eds.)]. In Press.

Juntti, M. and Wilson, G.A. (2005) Conceptualizing desertification in southern Europe: Stakeholder interpretations and multiple policy agendas. *European Environment 15*: 228–249.

Kosmas, C., Kairis, Or., Karavitis, Ch., Ritsema, C., Salvati, L., Acikalin, S., Alcala´, S., Alfama, P., Atlhopheng, J., Barrera, J., Belgacem, A., Sole´-Benet, A., Brito, J., Chaker, M., Chanda, R., Coelho, C., Darkoh, M., Diamantis, I., Ermolaeva, O., Fassouli, V., Fei, W., Feng, J., Fernandez, F., Ferreira, A., Gokceoglu, C., Gonzalez, D., Gungor, H., Hessel, R., Juying, J,. Khatteli, H., Khitrov, N., Kounalaki, A., Laouina, A., Lollino, P., Lopes, M., Magole, M., Medina, L., Mendoza, M., Morais, P., Mulale, K., Ocakoglu, F., Ouessar, M., Ovalle, C., Perez, C., Perkins, J., Pliakas, F., Polemio, M., Pozo, A., Prat, C., Qinke, Y., Ramos, A., Ramos, J., Riquelme, J., Romanenkov, V., Rui, L., Santaloia, F., Sebego, R., Sghaier, M., Silva, N., Sizemskaya, M., Soares, J., Sonmez, H., Taamallah, H., Tezcan, L., Torri, D., Ungaro, F., Valente, S., de Vente, J., Zagal, E., Zeiliguer, A., Zhonging, W., and Ziogas, A. (2013) Evaluation and selection of indicators for land degradation and desertification monitoring: Methodological approach. *Environmental Management 54*: 951–970.

MEA. (2005) *Ecosystems and human well-being. Synthesis report.* Washington, DC: Island Press.

Middleton, N. (2013) *The Global Casino.* London: Hodder Education.

MMA (2008) *Programa de Acción Nacional contra la Desertificación.* Madrid: Ministerio de Medioambiente.

OECD (2002) Poverty–Environment–Gender Linkages. *Off-print of the DAC Journal 2001 2*(4).

Oñate, J. and Peco, B. (2005) Policy impact on desertification: Stakeholders' perceptions in southeast Spain. *Land Use Policy 22*: 103–114.

Pirmoradi, A.H., Hosseini, S.M., Hosseini, S.F.J. (2011) Effective parameters on farmers' participation in plans to combat desertification (PCDs). *African Journal of Agricultural Research 6*(25): 5582–5590.

Redclift, M. (2005) Sustainable development (1987–2005) an oxymoron comes of age. *Sustainable Development 13*: 212–227.

Reed, M.S., Dougill, A.J., Baker, T.R. (2008) Participatory indicator development: What can ecologists and local communities learn from each other? *Ecological Application 18*(5): 1253–1269.

Ripoll, S., MacMillan, T., Muños, M.J.B., Velázquez, E., López, C.M., and Levidov, L. (2010) *WP3: Water Scarcity and its Virtual Export from Spain to the UK.* Final Report of the Cooperative Research on Environmental Problems in Europe (CREPE) Project (coordinated by the Open University, UK).

Seely, M.K. (1998) Can science and community action connect to combat desertification? *Journal of Arid Environments 39*: 267–277.

Steffen, W., Richardson, K., Rockström, J., Cornell, S.E., Fetzer I, Bennett, E.M., Biggs, R., Carpenter, S.R., de Vries, W., de Wit, C.A., Folke, C., Gerten, D., Heinke, J., Mace, J.M., Persson, L.M., Ramanathan, V., Reyers, B., and Sörlin, S. (2015) Planetary boundaries: Guiding human development on a changing planet. *Science 347*(6223) DOI: 10.1126/science.1259855

Stringer, L.C., Reed, M.S., Dougill, A.J., Seely, M.K., and Rokitzki, M. (2007) Implementing the UNCCD: Participatory challenges. *Natural Resources Forum 31*: 198–211.

Thomas, D.S.G. (1997) Science and the desertification debate. *Journal of Arid Environments 37*: 599–608.

Thomas, D.S.G. and Middleton, N. (1994) *Desertification: Exploding the Myth.* Chichester: Wiley.

Turnpenny, J., Lorenzoni, I., and Jones, M. (2009) Noisy and definitely not normal: Responding to wicked issues in the environment, energy and health. *Environmental Science and Policy 12*: 347–358.

UN (1994) *The Convention to Combat Desertification.* United Nations: New York.

UN (2015) Transforming our world: The 2030 agenda for sustainable development. A/RES/70/1.

UN News 19th of October 2020 Burkina Faso 'one step short of famine', warns UN food relief agency. Available online at: https://news.un.org/en/story/2020/10/1075712

UNCCD (2011) *Land and Soil in the Context of a Green Economy for Sustainable Development, Food Security and Poverty Eradication*. Bonn: UNCCD Secretariat. Available online at: http://www.unccd.int/Lists/SiteDocumentLibrary/Publications/Rio%206%20pages%20english.pdf.

UNCCD (2017) *2030 Agenda for Sustainable Development: Implications for the United Nations Convention to Combat Desertification. The Future Strategic Framework of the Convention*. Conference of the Parties. Thirteenth session. Ordos, China, 6–16 September 2017.

UNCCD (2020) About the convention. Available online at: https://www.unccd.int/convention/about-convention

UNCCD Knowledge Hub (2016) The Sahel: Land of Opportunities, Land with a Future. Web page available online at: https://knowledge.unccd.int/knowledge-products-and-pillars/unccd-e-library/sahel-land-opportunities-land-future

Warren, A. and Agnew, C. (1988) *An Assessment of Desertification and Land Degradation in Arid and Semi-arid Areas*. Drylands Paper 2. London: International Institute for Environment and Development.

Wellens, J. and Millington, A.C. (1992) Desertification. In Mannion, A.M. and Bowlby, S.R. (eds.) *Environmental Issues in the 1990s*. London: Wiley.

Wilson, G.A. and Juntti, M. (2005) *Unravelling Desertification: Policies and Actor Networks in Southern Europe*. Wageningen: Wageningen Academic Publishers.

WRI (2005) *Ecosystems and Human Wellbeing: Desertification Synthesis. A Report of the Millennium Ecosystem Assessment*. Washington, DC: World Resources Institute.

44

Food and agriculture

Global dynamics and environmental consequences

Jennifer Clapp and Sarah J. Martin

The ways in which food and agriculture are organized around the world have enormous implications for the global environment. The industrial organization of agriculture that feeds the dominant food system – including large-scale production methods and intensive livestock operations – is associated with soil degradation, biodiversity loss (see Chapter 41), pollution (see Chapter 35), climate change (see Chapter 32) and the depletion of water supplies (see Chapter 38). At the same time, international economic forces in the food system – including the international trade in food and global financial activities that add to tensions between food, fuel and land – also contribute to environmental problems including greenhouse gas emissions (see Chapter 32) and deforestation (see Chapter 42). The environmental effects of industrial agricultural production and the integration of that production system into global food and agricultural commodity markets extend far beyond national borders. The environmental challenges of the global food system issue, however, have not been dealt with effectively at the global scale.

Food and agriculture have a grounded quality because of their intimate relationship with the soil, and because individuals consume food on a daily basis. Individual food choices of course have important implications for how agriculture relates to the environment and politics. But it is not just individual choice that matters on this issue. Political choices about how societies collectively organize agricultural systems are of overriding importance because these choices shape individual food choices in many ways. These issues thus require consideration not just at a local or national scale, but also at the international level.

The environmental dimensions of large-scale industrial agriculture and the rise of a globally organized food system are widely understood, but there is no clear agreement on the pathway forward toward a more sustainable way to feed the world – at least as organized on a global scale. There are vastly different interpretations of what exactly constitutes a more environmentally friendly international organization of agricultural and food systems, and how those systems should be structured and governed. Some argue that agroecological methods and more locally oriented food systems will reduce the environmental damage caused by industrial agriculture and global food markets. But others are skeptical that such methods will be sufficient to feed the world's growing population. Instead, they argue for the use of more sophisticated technologies, such as agricultural biotechnology, genome editing techniques,

DOI: 10.4324/9781003008873-49

and new digital farming software and equipment. These new technological approaches are often paired with a call for the development of more globally integrated markets, along with voluntary market-based sustainability governance schemes, to provide the required food at the least environmental cost.

In this chapter, we examine the linkages between agriculture, food and the environment on an international scale, and assess the state of the political debate over how best to make global food systems more sustainable. We begin by outlining the development and operation of industrial agriculture and livestock production, which have serious environmental impacts. We then highlight some of the less visible effects of global food and agriculture market dynamics, such as trade and finance. Next, we explore how the linkage between the organization of food and agriculture systems and the natural environment has been debated at the international level, mapping out the main competing perspectives. Finally, we conclude the chapter with some reflections on how the polarized debate over how to vision sustainable agriculture has stalled progress on international cooperation in this arena of global environmental politics.

Industrial agriculture

The sustainability of agriculture has long been a concern, but the widespread adoption of industrial agricultural methods has generated new environmental issues to the extent that some analysts now assert that food systems as they currently operate have crossed several of the "planetary boundaries" that delineate a safe operating space within which humanity must stay to ensure long-term sustainability (Springmann et al. 2018; Willett et al. 2019). The uptake of a more "scientific" approach to agriculture in North America and Europe in the late 1800s and the first half of the 1900s was supported by laboratory research. The industrial agricultural model relied on a system of new specialty seeds, chemical fertilizers and pesticides, irrigation, and machinery for both planting and harvesting. Machines saved labor and chemical inputs, which, in turn, proponents argued, saved land. The industrial model is premised on energy and capital-intensive production with large tracts of land with a single crop, known as "monocropping." The early post-war adoption of this kind of agriculture resulted in enormous production increases per acre for certain crops. The United States and Canada, for example, quickly produced massive surpluses of grain that began to be exported around the world, both commercially and as food aid in the 1950s (Barrett and Maxwell 2005).

The industrial agricultural production model was promoted not just in North America and Europe, but also globally. The Union of Soviet Socialist Republics adopted a program of "chemicalization" of agriculture in the 1960s (McNeill 2000) and the developing world in particular was encouraged to adopt a technical and managerial model of industrial production as part of the broader push for a "green revolution" in the 1960s and 1970s (Clapp 2020; on industry generally, see Chapter 13). In the late 1970s, China also adopted a model of industrial agriculture. International aid programs promoted research, development and adoption of new seeds designed for tropical climates, pesticides, fertilizers, machinery, irrigation and monocropping in many developing countries. While at first the green revolution was hailed as a success in the countries that embraced it fully, including large parts of South Asia, Southeast Asia and Latin America, its negative environmental effects soon became apparent. The industrial agricultural system as it has manifested around the world is deeply entrenched and difficult to change.

A major consequence of the expanding reach of the industrial agricultural model has been the loss of biodiversity. Widespread adoption of specialty varieties of hybrid seeds raised in monocultural fashion meant that fewer traditional crops and crop varieties were planted around the world. The Food and Agriculture Organization of the United Nations (FAO) has estimated that between 1900 and 2000, around three-quarters of the world crop diversity was lost, with the most rapid decline occurring between 1950 and 2000 (Commission on Genetic Resources for Food and Agriculture 2010). Since the 1960s, there has been a significant decline in the number of crop varieties planted. Just three crops – rice, wheat and maize – constitute over half of the world's food energy, and only 12 crops and 5 livestock species make up over three-quarters of the world's food supply (Bioversity International 2017, p.54). Diversity is necessary for maintaining species viability and if it is lost, agricultural systems become more precarious and vulnerable to pests and diseases. For this reason, diversity is vital for long-term resilience in agriculture and food systems.

The rise of an industrial agricultural model around the world has also contributed to a decline in nature's ecosystem services and functions (FAO 2020). For example, industrial farm practices have contributed to a reduction in soil biodiversity (the living organisms in the soil that contribute to its fertility and other services) and some argue that the crucial functions of soil, such as water filtration, carbon fixing, climate regulation and nutrient cycling, are being compromised as a result. These problems are on top of the fact that declining soil fertility is a constraint on increasing food production. Land and resource degradation associated with the spread of industrial agriculture also threatens habitats for pollinators on which over 75 percent of global food and agriculture crops rely for pollination (IPBES 2019).

Early in the development of scientific forms of farming, agricultural scientists encouraged the use of synthetic inputs – chemical fertilizers and pesticides – to make up for these lost soil and ecosystem functions caused by intensive agricultural practices. Synthetic chemical inputs can temporarily make up for the loss of soil fertility and can protect against pests. But over time, as modern agriculture has become ever more reliant on these chemical inputs, their effectiveness has come into question. The reliance on synthetic fertilizer use increased five-fold between 1960 and 2010 (Foley et al. 2011: 338). Although yields overall have increased with the rising use of synthetic fertilizers, the yield of grain as a ratio of fertilizer application has declined significantly, and the responsiveness of crops to these chemicals has been highly uneven (Keating et al. 2010). For example, the use of synthetic fertilizers has contributed to increased crop yields per acre for highly responsive crops, such as maize, while for other crops the increases have not been nearly so dramatic.

Synthetic herbicides and pesticides often accompany the use of synthetic fertilizers in industrial farm operations, and their use has increased even more sharply than that of synthetic fertilizers. In the United States, for example, the use of insecticides and herbicides increased by a factor of 40 between the mid-1940s and the mid-1970s. Globally, pesticide use increased in the 1970s and 1980s at a rate of around 5 percent per year (Pimentel et al. 1993) and the most recent figures place annual global use at 2 million tonnes per year (Sharma et al. 2020). Because pests – both insects and weeds – can build up resistance to chemicals designed to control them, there has been an increase in the use of those chemicals, and the use of even more harmful chemicals, just to keep pests at a manageable level. The dramatic rise in chemical use intensified the ecological side-effects of industrial agriculture (see Chapter 36). Given that up to 75 percent of pesticides do not ultimately reach their target, especially in aerial spraying, it is not surprising that pollution of soils, water and air resulted (Pimentel and Burgess 2014). The spread of these chemicals into the environment has far-reaching human and ecosystem health impacts.

Agricultural intensification has also resulted in greater use of energy and water, with implications for climate change (see Chapters 32, 33 and 38). Modern industrial agriculture and food systems more broadly are particularly reliant on fossil energy (see Chapter 33) – especially petroleum products – not only to fuel farm machinery but also because they are key ingredients in fertilizers and pesticides and are extensively used in transport of food items in globalized markets (Pellegrini and Fernández 2018). The Intergovernmental Panel on Climate Change (IPCC) estimates that between 21 and 37 percent of greenhouse gas emissions are associated with food systems, including both pre- and post-production activities (IPCC 2019). Irrigation accounts for some 70 percent of global freshwater withdrawals and intensive irrigation has led to significant drawdown on groundwater (Campbell et al., 2017; see also Chapter 38). Climate change and growing water scarcity, in turn, threaten future food production (FAO 2016).

New technological developments since the 1980s and 1990s – especially the development of genetically modified (GM) seeds – have brought additional concerns and controversy to studies on the environmental implications of industrial agriculture. The main kinds of GM crops are engineered either to produce their own pesticide, or to be resistant to herbicides. Increasingly, seeds are being engineered to do both. Since they were first commercialized in the mid-1990s, there has been a dramatic increase in the planting of genetically modified crops globally. From the mid-1990s to 2019, the number of hectares planted with GM crops grew to 190.4 million hectares (ISAAA 2019). Increased GM crop acreage has been associated with a growing use of herbicides with GM crops, such as glyphosate, which has generated a growing number of herbicide tolerant weeds that require more toxic herbicides to control (Bonny 2016). GM crops have also brought concerns about the potential for genetically altered plants to cross with traditional varieties and crop wild relatives, which could impact genetic diversity.

Industrial meat production

A transformation in livestock production models paralleled the transformation to industrial agricultural methods. The dramatic rise of surplus grains that accompanied industrial crop production in developed countries such as the United States in the 1950 and 1960s spurred an increase in grain-fed livestock production. This surplus, along with new pharmaceuticals and new animal breeds, enabled the growth of large-scale industrial animal operations known as intensive livestock operations (ILOs) or concentrated animal feeding operations (CAFOs). Through these developments, large-scale grain production and large-scale intensive livestock farming became closely linked (Weis 2013). As the organization of meat production began to change, so too did consumption of meat. These developments have had enormous environmental impacts on a global scale.

Prior to the adoption of industrial livestock production and processing methods, livestock was seasonal, and constrained by weather and the availability of surplus crops and forage. At the turn of the last century, the spread of railways, refrigeration and centralized slaughterhouses and grain storage contributed to industrializing livestock processing in the United States (Cronon 1992). After World War II, seasonal limitations on livestock production were eased in the United States when poultry began to be raised indoors in CAFOs. In the 1980s, poultry CAFOs became a model for the expansion of US hog production (Drabenstott 1998; Rhodes 1995). Indoor confinement allowed animals to be raised faster with less feed and the rise of new veterinary pharmaceuticals suppressed diseases previously associated with large-scale animal confinement (Rhodes 1995). A growing proportion of antibiotics are used for livestock around the world, as the industrial livestock model has been increasingly established and replicated in countries around the world, giving rise to concerns about antimicrobial resistance (Kirchhelle 2018). The growth of China's hog industry has outstripped all other countries and now accounts for just under half of the world's production (FAO 2019). Genetically specialized breeds have been developed for industrial animal farming which, similar to industrial agriculture, has contributed to a narrowing of farm animal diversity. The FAO estimates that around 17 percent of livestock breeds are at risk of extinction (FAO 2015).

The rise of industrial livestock farming has increased the availability of inexpensive meat and poultry, which is now globally produced, traded and increasingly available. Unlike the global car that is assembled from parts sourced globally, animals are disassembled and their parts are distributed to specialty markets around the world. But the dumping of cheap meat from the global North can undercut local markets in the global South (Chemnitz et al. 2014). In addition, an associated increase in per capita consumption of meat and poultry, or "meatification" of diets (Weis 2020), has accompanied the growth of industrial livestock production, although there are indications that meat consumption in some of the wealthier states is declining (on consumption generally, see Chapter 17).

Livestock has a significant "ecological hoofprint" (Weis 2013) with global effects and huge climate implications (Katz-Rosene 2020). Livestock is the single largest user of agricultural land and uses a whopping 77 percent of it (UNEP 2019). An indirect result of livestock production is that clearing of land for cattle grazing has resulted in deforestation that contributes a significant amount of carbon dioxide (Weindl et al. 2017; see McDermott and Humphreys). More directly, livestock raising itself is credited with being responsible for 14.5 percent of anthropogenic greenhouse gas emissions and these gases are considered more problematic than carbon dioxide (e.g., methane, nitrous oxide and ammonia which contributes to acid rain and acidification of ecosystems; on acid rain, see Chapter 30) (Grossi et al. 2019). Finally, livestock's "hoofprint" has a significant impact on energy consumption. Whereas agriculture is estimated to use energy at a 3:1 ratio, the energy use ratio for meat is 35:1 ratio (Pew Commission 2009: 9). Not only are there consequences to the production of feedgrains as chronicled above, livestock, especially ILOs, impose significant external costs in the form of manure and waste, which pose risks of excessive nutrient loads that causes eutrophication, pollution from antibiotics as well as the spread of diseases.

Intensive livestock operations have come under criticism by scholars examining environmental justice (see Chapter 26). In particular, the location of ILOs is often in rural areas where the residents are predominantly poor (Stull and Broadway 2004) and/or are unable to gain political support to resist the establishment of ILOs in their communities (Novek 2003). As a result, conflict often accompanies the establishment of ILOs. For workers, there are serious health concerns caused by the gas and dust produced in confined areas (Thu and Durrenberger 1998) and CAFOs are associated with high rates of respiratory disease (Pew Commission 2009). More recently, in the context of Covid-19, ILOs are associated with high rates of infection for slaughterhouse workers in the United States and Canada (Douglas 2020).

Globalized food and agriculture market dynamics

As the world economy has become more globalized in recent decades (see Chapter 24), there has been an accompanying rise in the physical trade of food and agricultural products across borders as well as a sharp increase in the trade in financial derivatives linked to food and agriculture. The globalized food market, as well as its growing ties to financial markets, has sparked increased investment in a new complex nexus of finance, food, biofuel and land that has been associated with myriad ecological effects around the world.

The international trade in food has grown significantly over the past 30 years, and 20–25 percent of all food produced is traded internationally (D'Odorico et al. 2014). In 2017, global food trade reached over US$1.5 trillion, up from just US$315 billion in 1990 (Clapp 2020). Global food trade has grown more rapidly than production, signaling the growing significance of global markets in the food system. The global trade in food has been associated with significant greenhouse gas emissions (Schmitz et al. 2011). These emissions can arise from land use changes for trade-based production systems, production methods themselves, fuels required to transport food, and the storage and processing infrastructure required for global food trade (Garnett 2013). These impacts are not always easy to tease out, and they are often unevenly distributed, depending on the specific foods and their origins and production systems (Dalin and Rodríguez-Iturbe 2016). These uncertainties have left ample room for debate over the ecological implications of food trade, with proponents often arguing that food trade is more resource efficient because it capitalizes on comparative advantages, while critics note the ecological destruction it can cause in regions heavily reliant on the export of certain food crops, like palm oil (Clapp 2017).

It should be noted that the growing global food market is highly reliant on large-scale industrial agriculture, and as such a growing global food market and associated trade reinforces an energy intensive system. Moreover, the carbon emissions and pollution from trade are in addition to greenhouse gases from industrial agricultural production. It may be that in some cases more environmentally sound production methods reduce the climate impact of certain traded foods – for example, an imported organic tomato may be responsible for fewer carbon emissions than a local hot-house tomato. But, at the same time, locally traded foods do not necessarily produce high levels of greenhouse gases, whereas internationally traded foods certainly produce additional carbon emissions because of the fossil fuels used in long-distance transportation.

The rise of a globally integrated food market is also linked to the rise in the trade in complex financial derivatives based on agricultural and food commodities. According to a number of analysts, the food system has become increasingly "financialized," meaning that financial actors – investors and financial institutions in particular – have become a significant influence on the sector (see Chapter 13). Their behavior, although geared primarily to financial profit, has an important impact on food system outcomes – including the environment. Given the global nature of agriculture and food markets, this financialization has important implications for global environmental politics (Clapp 2014).

The increasingly important role of financial actors was highlighted in the aftermath of the 2007–2008 food price crisis, as many began to point to the role of financial speculators in driving up global food prices. This concern was sparked by a sharp increase in the speculative trade in agricultural commodity futures contracts and other agricultural commodity derivatives following financial deregulation that allowed them to gain more exposure to these markets. There is a heated debate over whether the growth in financial speculation on agricultural commodity futures markets was a cause or a response to rising food prices (see Clapp and Isakson 2018). But whether cause or response, financialization has certainly facilitated further financial investment – both in biofuels and in land – both of which have profound ecological implications, as mapped out below.

Since the early 2000s, there has been growing interest in crop-based biofuels as a renewable fuel source. The financialization of commodities more broadly, including not just food and agricultural products, but also petroleum, has facilitated this investment. Like food, oil price rises have also been associated with financialization, which has made biofuels a competitive product in energy markets. In turn, this has led to a further push to invest in land for their production. As food prices remained elevated in the 2007–2012 period, some say in part due to rising investment in biofuels, investment in biofuel operations became even more attractive because both food and oil prices were also rising, making investments in production facilities attractive as a hedge against further rising grain and fuel prices (Clapp 2014). As a result, biofuel investment is used as a hedge against both agricultural commodity volatility and fuel price volatility. There is a complex nexus of these investments that have enormous environmental implications at the global level.

Stoking the role of biofuel investment in this nexus has been the fact that the largest producers of biofuels, including the United States and the European Union, enacted policies in the early 2000s that required a certain percentage of fuels to be from renewable sources. Along with subsidies, these policies attracted an increased interest in biofuel investments. For example, in the United States, maize production diverted to ethanol production increased sharply after 2000, rising from less than 10 to nearly 40 percent by 2020 (USDA 2020).

Although requiring the use of renewable fuels is aimed at reducing a reliance on the very fossil fuels that create greenhouse gas emissions, biofuel production is far from evenly "green" in its environmental credentials (McMichael 2010). The energy return on investment for corn-based ethanol, for example, is very low when compared with that from sugar-based ethanol. Palm oil, a common feedstock for biofuels in developing countries, is notoriously inefficient as a biofuel. Critics generally agree that replacing food crops with biofuels will likely worsen climate change (see Chapter 32).

Rising food and fuel prices and the financialization of food have also encouraged and facilitated foreign land acquisition, a phenomenon that has increased sharply since 2006. Critics have raised warning bells about the possible negative effects of some of these land deals (Fairbairn 2020). In most cases, these land investments involve the import of large-scale industrial farming methods and the establishment of infrastructure to export the crops that are planted on them, including biofuel crops. The link between global capital and investment, and agricultural land and biofuel development is occurring globally, from Brazil, to Ethiopia and areas in Southeast Asia such as Malaysia.

The ecological impacts of large-scale land investments can be significant, especially if they involve the import of large-scale industrial farming operations for both food and biofuels. In such circumstances, large-scale industrial agriculture can pose serious risks to the ecosystem as noted above. Tropical forests have already been cleared, for example, in many parts of Asia and Africa in order to establish palm oil plantations for the production of biofuel (Dauvergne 2018). These ecological risks associated with large-scale foreign land acquisition are a particular concern since these investors are seeking short turnaround on returns and longer-term environmental impacts are often left off the balance sheet (Clapp 2014).

The international politics of more sustainable food and agriculture

The environmental impact of food and agricultural systems has received less attention than it deserves within the field of global environmental politics (Clapp and Scott 2018). This is likely due to the fact that, unlike the issues of climate (see Chapter 32), stratospheric ozone (see Chapter 31) or hazardous waste (see Chapter 36), there is no one international "regime" or agreement that seeks to promote sustainable agriculture on a global scale (on regimes, see Chapter 9). A number of existing regimes do touch on some of the problems – such as biodiversity loss and genetically modified organisms (see Chapter 41) and hazardous pesticides (see Chapter 35). But these are not coordinated as "agriculture" agreements; agriculture is but one source of these particular problems. Part of the reason for the lack of a global agreement specifically focused on the environmental dimensions of food and agriculture is the lack of an international consensus on what exactly a more sustainable agriculture should or could look like in practice (Swinburn et al. 2019). There is much agreement that the current situation is not sustainable over the long run. While this view is widely supported, there are differing views on how to move forward in a way that provides sufficient food for the world's population with the least damage to the environment.

Two views have become especially prominent in the debate over the topic, and these have shaped the political discussion at the international level on sustainable agriculture. The first is more mainstream in its approach. It takes as a starting point that some patterns in the global food and agriculture system are a given: for example, that meat consumption is likely to increase over time as the world population grows and as the size of the middle class in developing countries grows; an outgrowth of this assumption is a resolve that increased food production is essential. They remain committed to industrial technologies (see Chapter 19) and globalized markets (see Chapter 24) for food, but in new ways that they see as both more productive and more sustainable.

On the technology side, this mainstream view promotes new digital agricultural technologies and sophisticated computer-assisted genome editing of food crops. The aim of these technologies is to increase food production while also minimizing its environmental impact (Rotz et al. 2019). Digital farming technologies utilize digital sensors, drones, and global positioning systems with the aim of improving the efficiency of farm inputs, such as reducing the amount of agrochemicals sprayed on crops and matching appropriate seeds to soil types and weather patterns (Weersink et al. 2018). Genome editing enables more precise editing of the genetic code of plants to turn on and off different traits that can, in theory, make them more productive or resistant to stresses like water scarcity and greater heat caused by climate change (Chen et al. 2019; Henry 2020). These technologies, while they have gained a great deal of attention for their promises of sustainability, have also been critiqued because they are largely controlled by large transnational corporate actors, whereas data privacy and access concerns of small-scale producers are at risk (Rotz et al. 2019; Clapp and Ruder 2020).

On the markets side, this approach promotes an extension of globalized food trade and financial investment systems, as a means by which to further promote efficiency of food distribution and financing, especially in developing countries. Promoters of this view see the international trade as an important tool in adapting to climate change, by ensuring countries whose agricultural production is negatively affected by a changing climate can still import the food they need from countries that are less impacted (Baldos and Hertel 2015). This view also promotes market-based governance initiatives, which are often voluntary measures that rely on third party certification, to ensure that production of those food items is sustainable. The Roundtable on Sustainable Palm Oil (RSPO) and the Global Roundtable for Sustainable Beef, are examples of these kinds of initiatives (Carlson et al. 2018; Buckley et al. 2019). Responsible investment initiatives are also promoted, to ward against unsustainable outcomes linked to financial speculation and land grabbing (World Bank 2014). These kinds of voluntary responsibility and/or sustainability initiatives have been critiqued by many analysts, however, as being weak and ineffective (e.g., Clapp and Isakson 2018; Dauvergne 2018).

Standing in opposition to this mainstream view is a more radical vision that draws on broader ideas of "food sovereignty," a social movement that prioritizes rights of peasant and small-scale producers to determine the shape of their own food systems (Wittman et al. 2010). This view calls for production systems based on the science of agroecology, which is based on diverse, regenerative and resource efficient food production practices that do not relay on external inputs such as genetically altered seeds and synthetic chemicals. This approach addresses many of the ecological problems associated with modern industrial agriculture, because it utilizes techniques such as intercropping and carbon sequestration in soils that builds natural climate cooling and pest resistance. It also restores and enhances ecosystem resilience by minimizing pollution and building biological diversity, including that of soils, animals and plants (Altieri 2019; Wezel et al. 2020).

This vision also calls for more diversity in systems for food processing and distribution. Food sovereignty, for example, is skeptical of globalized food trade that is reliant on large-scale transnational corporations, and instead calls for more "territorialized" food markets that are centered on local and regional food supply chains (e.g., Van der Ploeg et al. 2012; CSM 2016). More localized food and agriculture markets can work to strengthen ecological and social resilience because they require a more localized and diverse production base that is geared to local food demand, as opposed to large-scale monocultural production for distant global markets, and directly reduces carbon emissions from long-distance trade. Food sovereignty also promotes land rights for small-scale food producers, thus reducing environmentally destructive practices associated with large-scale land acquisitions that are focused on the production and export of industrial crops, such as biofuels (Wittman et al. 2010).

Beyond the debate over the global organization of agriculture and food systems is the question of diets. A number of scholars and reports have called for a shift toward "sustainable diets" as a way to reduce the environmental impacts of food systems (Mason and Lang 2017; IPCC 2019). Reduced meat consumption could, for example, play a role in future climate change mitigation policies by reducing livestock's contribution to greenhouse gases (Gilligan 2019). The concept of sustainable diets, and the role of meat within them, however, is highly complex and subject to debate, opening opportunities for powerful corporate actors to shape the agenda (Scott 2020).

Alternatively, some scholars have called for "de-intensification" of meat production – because pastoral systems and mixed crop – livestock systems have significant carbon sequestration potential and the fact that a majority of the global population are dependent on these systems for their livelihoods (Herrero et al. 2009, 2010). Many pastoralists, especially in developing countries, live in extreme poverty (HLPE 2016). Human-animal relationships are complex and heterogenous, and suggests that policy-makers would benefit from changing their focus from agricultural and livestock *specialization* to greater support of *diversification*. Inter-species synergies, including mixed crop and livestock operations that enhance environmental improvements such as soil quality, can provide a buffer against changing climate conditions and volatile markets (Katz-Rosene and Martin 2020).

The 2021 UN Food System Summit presented an opportunity for a more explicit global dialog on sustainability in agriculture. However, there was controversy around the planning for the event, as civil society groups complained about a lack of representation and fear that the mainstream vision will take precedence in policy recommendations. It is as yet unclear how these issues will be resolved, and, ultimately, how that resolution will influence policy recommendations at the international level going forward.

Conclusion

The drive to improve agricultural production and efficiency has been met by industrial agriculture over the past century. But as outlined in this chapter, it has come with a significant environmental cost not just at a local level, but also on a global scale. The dominant food system is organized and shaped by industrial agriculture and international economic forces such as the international trade in food and global financial activities. These complicate and add to the already significant tensions between food, fuel and land. Importantly, these activities can intensify environmental impacts.

To date, these complex and interrelated issues have not been dealt with ineffectively at the global scale. Political choices about how societies collectively organize agricultural systems are of overriding importance because these choices shape individual food choices in many ways. But efforts to forge a global cooperative strategy to promote more sustainable food and agricultural systems have been stalled due to a lack of consensus on what constitutes "sustainable agriculture" (on sustainability generally, see Chapter 16). The two leading interpretations of how a more environmentally friendly international organization of agricultural and food systems could be implemented are widely divergent. Some argue that agroecological methods and more locally oriented food systems will reduce the environmental damage caused by agriculture. But others are skeptical that such methods will be sufficient to feed the world's growing population.

The lack of agreement on a vision for a sustainable future for food and agricultural systems has meant that current agriculture, food and diet practices – many of which have negative environmental impacts – persist. This situation cannot last indefinitely, so long as the condition of the global environment worsens, and the demand for food from a growing global population increases. International institutions can promote more sustainable food systems in practice through the development of norms, rules and other incentives that shape collective societal choices on the organization of food and agriculture (see Chapter 8). But, as noted, debate has been polarized over different models of sustainable agriculture. This debate will be on full display at the 2021 UN Food System Summit but given concerns by civil society around this event thus far, it is unlikely to resolve differences between competing perspectives. The issue nonetheless remains urgent, since agricultural issues intersect with key issues such as climate change, pollution and energy policy, and directly impact the wellbeing and livelihood of the globe's poorest people.

References

Altieri, M.A. (2018). *Agroecology: the science of sustainable agriculture*. Boca Raton, USA: CRC Press.

Baldos, U. and Hertel, T. (2015). The role of international trade in managing food security risks from climate change. *Food Security*, 7(2): 275–290.

Barrett, C.B. and Maxwell, D.G. (2005). *Food Aid after Fifty Years: Recasting Its Role*, New York: Routledge.

Bioversity International. (2017). *Mainstreaming Agrobiodiversity in Sustainable Food Systems*. Rome: Bioversity International. Available at <https://cgspace.cgiar.org/handle/10568/89049>

Bonny, Sylvie. (2016). Genetically modified herbicide-tolerant crops, weeds, and herbicides: overview and impact. *Environmental Management*, 57(1): 31–48.

Buckley, K.J., Newton, P. Gibbs, H., McConnel, I. and Ehrmann, J. (2019). Pursuing sustainability through multi-stakeholder collaboration: A description of the governance, actions, and perceived impacts of the Roundtables for Sustainable Beef. *World Development*, 121: 203–217.

Campbell, B., Beare, D., Hall-Spencer, J.M., Ingram, J., Jaramillo, F., Ortiz, R., Ramankutty, N., Sayer, J.A. and Shindell, D. (2017). Agriculture production as a major driver of the Earth system exceeding planetary boundaries. *Ecology and Society*, 22(4): 8.

Carlson, K., Heilmayr, R., Gibbs, H., Noojipady, P., Burns, D., Morton, D., Walker, N., Paoli, G. and Kremen, C. (2018). Effect of oil palm Sustainability certification on deforestation and fire in Indonesia. *Proceedings of the National Academy of Sciences*, 115(1): 121–126.

Chemnitz, C., Schmidt-Landenberger, E., Cornely, B., Heinrich-Böll-Stiftung, and Friends of the Earth Europe. (2014). *Meat Atlas: Facts and Figures about the Animals We Eat*. Berlin: Heinrich-Böll-Stiftung.

Chen, K., Wang, Y., Zhang, R., Zhang, H. and Gao, C. (2019). CRISPR/Cas genome editing and precision plant breeding in agriculture. *Annual Review of Plant Biology*, 70(1): 667–697.

Civil Society Mechanism. (2016). *Connecting Smallholders to Markets: Analytical Guide.* Civil Society Mechanism (CSM). Available at <http://www.csm4cfs.org/wp-content/uploads/2016/10/ENG-ConnectingSmallholdersToMarketsweb.pdf>.

Clapp, J. (2020). *Food*, Cambridge: Polity.

Clapp, J. (2017). The trade-ification of the food sustainability agenda. *The Journal of Peasant Studies*, 44(2): 335–353.

Clapp, J. (2014). Financialization, distance and global food politics. *The Journal of Peasant Studies*, 41(5): 797–814.

Clapp, J. and Isakson, S.R. (2018). *Speculative Harvests: Financialization, Food and Agriculture*, Halifax: Fernwood.

Clapp, J. and Ruder, S. (2020). Precision technologies for agriculture: Digital farming, gene-edited crops, and the politics of sustainability. *Global Environmental Politics*, 20(3): 49–69.

Clapp, J. and Scott, C. (2018). The global environmental politics of food. *Global Environmental Politics*, 18(2): 1–11.

Commission on Genetic Resources for Food and Agriculture. (2010). *The Second Report on the State of the World's Plant Genetic Resources for Food and Agriculture*, Rome: Food and Agriculture Organization. Available at at//www.fao.org/agriculture/seed/sow2/

Cronon, W. (1992). *Nature's Metropolis: Chicago and the Great West*, New York: W.W. Norton.

Dalin, C. and Rodríguez-Iturbe, I. (2016). Environmental impacts of food trade via resource use and greenhouse gas emissions. *Environmental Research Letters*, 11(3): 035012.

Dauvergne, P. (2018). The global politics of the business of 'sustainable' palm oil. *Global Environmental Politics*, 18(2): 34–52.

D'Odorico, P., Carr, J.A., Laio, F., Ridolfi, L. and Vandoni, S. (2014). Feeding humanity through global food trade. *Earth's Future*, 2(9): 458–469.

Drabenstott, M. (1998). This little piggy went to market: Will the new pork industry call the heartland home? *Economic Review – Federal Reserve Bank of Kansas City*, 83: 79–97.

Fairbairn, M. (2020). *Fields of Gold: Financing the Global Land Rush*, Ithaca, NY: Cornell University Press.

FAO. (2020). *How the World's Food Security Depends on Biodiversity.* Rome: FAO. Available at<at//www.fao.org/3/cb0416en/CB0416EN.pdf>

FAO. (2019). *Meat Market Review*, March 2019. Rome: FAO. Available at <at//www.fao.org/3/ca3880en/ca3880en.pdf>

FAO. (2016). *The State of Food and Agriculture 2016: Climate Change, Agriculture and Food Security.* Rome. Available at <at//www.fao.org/3/a-i6030e.pdf>

FAO. (2015). *The Second Report on the State of the World's Animal Genetic Resources for Food and Agriculture.* Rome: FAO. Available at <at//www.fao.org/3/a-i4787e.pdf>

Foley, J.A., Ramankutty, N., Brauman, K.A., Cassidy, E.S., Gerber, J.S., Johnston, M., et al. (2011). Solutions for a cultivated planet. *Nature*, 478(7369): 337–342.

Douglas, L. (2020). Mapping Covid-19 Outbreaks in the Food System. Food and Environment Reporting Collective. Available at <https://thefern.org/2020/04/mapping-covid-19-in-meat-and-food-processing-plants/>

Garnett, T. (2013). Food sustainability: Problems, perspectives and solutions. *The Proceedings of the Nutrition Society*, 72(1): 29–39.

Gilligan, J. (2019). Modelling diet choices. *Nature Sustainability*, 2(8): 661–662.

Grossi, G., Goglio, P., Vitali, A. and Williams, A. (2019). Livestock and climate change: Impact of livestock on climate and mitigation strategies. *Animal Frontiers*, 9(1): 69–76.

HLPE. (2016). *Sustainable agricultural development for food security and nutrition: what roles for livestock?* A report by the High Level Panel of Experts on Food Security and Nutrition of the Committee on World Food Security. HLPE report 10. Rome. 140 pp. Available at <http://www.fao.org/3/a-i5795e.pdf>.

Henry, R. (2020). Innovations in plant genetics adapting agriculture to climate change. *Current Opinion in Plant Biology*, 56: 168–173.

Herrero, M., Thornton, P.K., Gerber, P. and Reid, R.S. (2009). Livestock, livelihoods and the environment: Understanding the trade-offs. *Current Opinion in Environmental Sustainability*, 1(2): 111–120.

Herrero, M., Thornton, P.K., Notenbaert, A.M., Wood, S., Msangi, S., Freeman, H.A., et al. (2010). Smart investments in sustainable food production: revisiting mixed crop-livestock systems. *Science*, 327(5967): 822–825.

IPCC (Intergovernmental Panel on Climate Change). (2019). Summary for Policymakers. In: *Climate Change and Land: an IPCC Special Report on Climate Change, Desertification, Land Degradation, Sustainable Land Management, Food Security, and Greenhouse Gas Fluxes in Terrestrial Ecosystems*. IPCC. Available at at <https://www.ipcc.ch/site/assets/uploads/sites/4/2020/02/SPM_Updated-Jan20.pdf>

Intergovernmental Science-Policy Platform on Biodiversity and Ecosystem Services (IPBES). (2019). *Summary for Policymakers of the Global Assessment Report on Biodiversity and Ecosystem Services of the Intergovernmental Science-Policy Platform on Biodiversity and Ecosystem Services*. Bonn, Germany: IPBES Secretariat. Available at <https://ipbes.net/global-assessment-report-biodiversity-ecosystem-services>

ISAAA. (2019). Biotech crops drive socio-economic development and sustainable environment in the new frontier: Executive summary. *ISAAA Brief* 55-2019. Available at https://www.isaaa.org/resources/publications/briefs/55/executivesummary/pdf/B55-ExecSum-English.pdf

Katz-Rosene, R. (2020). How do livestock impact the climate? In R. Katz Rosene and S.J. Martin (eds) *Green Meat? Sustaining Eaters, Animals and the Planet*, Montreal: McGill Queen's University Press: 43–63.

Katz-Rosene, R. and Martin, S.J. (2020). Which way(s) forward? In R. Katz Rosene and S.J. Martin (eds) *Green Meat? Sustaining Eaters, Animals and the Planet*, Montreal: McGill Queen's University Press.

Keating, B.A., Carberry, P.S., Bindraban, P., Asseng, S., Meinke, H. and Dixon, J. (2010). Eco-efficient agriculture: Concepts, challenges, and opportunities. *Crop Science*, 50(Supplement 1): S109–S119.

Kirchhelle, C. (2018). Pharming animals: a global history of antibiotics in food production (1935–2017). *Palgrave Communications*, 4(1): 1–13.

Mason, P. and Lang, T. (2017). *Sustainable Diets: How Ecological Nutrition can Transform Consumption and the Food System*, London: Routledge.

McMichael, P. (2010). Agrofuels in the food regime. *The Journal of Peasant Studies*, 37(4): 609–629.

McNeill, J.R. (2000). *Something New under the Sun: An Environmental History of the Twentieth-Century World*, New York: W.W. Norton.

Novek, J. (2003). Intensive livestock operations, disembedding, and community polarization in Manitoba. *Society and Natural Resources*, 16(7): 567–581.

Pellegrini, P. and Fernández, R. (2018). Crop intensification, land use, and on-farm energy-use efficiency during the worldwide spread of the green revolution. *Proceedings of the National Academy of Sciences*, 115(10): 2335–2340.

Pew Commission. (2009). *Putting Meat on the Table: Industrial Farm Animal Production in America*. A Report of the Pew Commission on Industrial Farm Animal Production. Executive Summary. A Project of the Pew Charitable Trusts and Johns Hopkins Bloomberg School of Public Health.

Pimentel, D., McLaughlin, L., Zepp, A., Lakitan, B., Kraus, T., Kleinman, P., et al. (1993). Environmental and economic impacts of reducing US agricultural pesticide use. In D. Pimentel and H. Lehman (eds) *The Pesticide Question*, New York: Springer US: 223–278.

Pimentel, D. and Burgess, M. (2014). Environment and economic costs of the application of pesticides primarily in the United States. In D. Pimentel and R. Reshin (eds) *Integrated Pest Management: Pesticide Problems. Volume 3*. New York: Springer: 47–71.

Rhodes, V.J. (1995). The industrialization of hog production. *Review of Agricultural Economics*, 17(2): 107–118.

Rotz, S., Duncan, E., Small, M., Botschner, J., Dara, R., Mosby, I., Reed, M. and Fraser, E. (2019). The politics of digital agricultural technologies: A preliminary review. *Sociologia Ruralis*, 59(2): 203–229.

Schmitz, C., Biewald, A., Lotze-Campen, H., Popp, A., Dietrich, J., Bodirsky, B., Krause, M. and Weindl, I. (2011). Trading more food: Implications for land use, greenhouse gas emissions, and the food system. *Global Environmental Change*, 22: 189–209.

Scott, C. (2020). Does meat belong in a sustainable diet? In R. Katz Rosene and S.J. Martin (eds) *Green Meat? Sustaining Eaters, Animals and the Planet*, Montreal: McGill Queen's University Press.

Sharma, A., Shukla, A., Attri, K., Kumar, M., Kumar, P., Suttee, A., Singh, G., Barnwal, R.P. and Singla, N. (2020). Global trends in pesticides: A looming threat and viable alternatives. *Ecotoxicology and Environmental Safety* 201: 110812.

Springmann, M., Clark, M., Mason-D'Croz, D., Wiebe, K., Bodirsky, B.L., Lassaletta, L., de Vries, W. et al. (2018). Options for keeping the food system within environmental limits. *Nature*, 562(7728): 519–525.

Stull, D.D. and Broadway, M.J. (2004). *Slaughterhouse Blues: The Meat and Poultry Industry in North America*, Belmont, CA: Thomson/Wadsworth.

Swinburn, B. Kraak VI, Allender S, Atkins, V.J., Baker, P.I., Bogard, J.R., Brinsden, H. et al. (2019). The global syndemic of obesity, undernutrition and climate change: The *Lancet* Commission report. *The Lancet*, 393(10173): 791–846.

Thu, K.M. and Durrenberger, E.P. (1998). *Pigs, Profits, and Rural Communities*, Albany, NY: SUNY Press.

UNEP (UN Environment Programme). (2019). *Global Environment Outlook – GEO-6: Healthy Planet, Healthy People*, 1st ed. Cambridge: Cambridge University Press.

USDA (US Department of Agriculture). (2020). Feedgrains Sector at a Glance. Available at <https://www.ers.usda.gov/topics/crops/corn-and-other-feedgrains/feedgrains-sector-at-a-glance/>

Van der Ploeg, J.D., Jingzhong, Y. and Schneider, S. (2012). Rural development through the construction of new, nested, markets: Comparative perspectives from China, Brazil and the European Union. *The Journal of Peasant Studies*, 39(1): 133–173.

Weindl, I., Popp, A, Bodirsky, B., Rolinski, S. Lotze-Campen, H., Biewald, A., Humpenöder, F., Dietrich, J. and Stevanović, M. (2017). Livestock and Human Use of Land: Productivity Trends and Dietary Choices as Drivers of Future Land and Carbon Dynamics. *Global and Planetary Change* 159: 1–10.

Weersink, A., Fraser, E., Pannell, D., Duncan, E. and Rotz, S. (2018). Opportunities and challenges for big data in agricultural and environmental analysis. *Annual Review of Resource Economics*, 10(1): 19–37.

Weis, A.J. (2020). Confronting meatification. In R. Katz Rosene and S.J. Martin (eds) *Green Meat? Sustaining Eaters, Animals and the Planet*, Montreal: McGill Queen's University Press.

Weis, A.J. (2013). *The Ecological Hoofprint: The Global Burden of Industrial Livestock*, London: Zed.

Wezel, A., Herren B.G., Bezner Kerr, R., Barrios, E., Rodrigues Gonçalves, A.L., and Sinclair, F. (2020). Agroecological Principles and Elements and Their Implications for Transitioning to Sustainable Food Systems. A Review. *Agronomy for Sustainable Development*, 40 (6). https://doi.org/10.1007/s13593-020-00646-z.

Willett, W., Rockström, J., Loken, B., Springmann, M., Lang, T., Vermeulen, S., Garnett, T. et al. (2019). Food in the anthropocene: The eat–lancet commission on healthy diets from sustainable food systems. *The Lancet*, 393(10170): 447–492.

Wittman, H., Desmarais, A. and Wiebe, N., (2010). The origins and potential of food sovereignty. In H. Wittman, A. Desmarais and N. Weibe (eds) *Food Sovereignty: Reconnecting Food, Nature and Community*, Halifax: Fernwood: 1–14.

World Bank. (2014). *The Practice of Responsible Investment Principles in Larger-Scale Agricultural Investments*, Washington, DC: World Bank. Available at <https://unctad.org/system/files/official-document/wb_unctad_2014_en.pdf>.

Part VI
Conclusion

45

The promises of global environmental politics

Prospects for study and practice

Paul G. Harris

In the eight years that have passed since the first edition of the *Routledge Handbook of Global Environmental Politics* was published, much has happened in the real world and in scholarly research. In this final chapter, my objective is to briefly reflect on both the practice and study of global environmental politics. I do this by following the organization of this book, partly to highlight developments since the first edition and partly to consider prospects for the domain in the future, in the process highlighting some important issues and themes. One thing has not changed since the first edition appeared and is worth keeping in mind: as pollution increases and the environment changes, global environmental politics is a subject that becomes more important, and more apparent, with each passing year. The urgency of understanding the human aspects of global environmental changes is increasing rapidly, as is the need to do vastly more to change polluting practices and ecologically harmful behaviors.

A fundamental question for real-world practitioners of global environmental politics is whether politics, including the making and implementation of environmental (and related) policies at international, national and subnational levels, will ever be able to keep pace with rapid environmental changes often caused by politics itself. While the environment changes rapidly, politics and policy tend to evolve slowly. A key question for scholars is whether it is enough to describe and explain global environmental politics. Given the enormous significance of many environmental changes – they can result in truly profound and widespread harm to humans, other species and entire ecosystems – it may be that the time has come for many more scholars of global environmental politics to be less impartial in their work, to move away from merely explaining events and to join forces with activists and others who are trying to proactively shape a more environmentally sustainable future. Put another way, while practitioners of global environmental politics can and should learn from scholars in the field, it may be time for more of the scholars to become practitioners themselves.

Explaining and understanding global environmental politics

The chapters in Part II have looked at ways of explaining and understanding global environmental politics. The purposes of those chapters, along with the introduction in Part I, are to frame global environmental politics as a scholarly discipline and unique field of practice, to

DOI: 10.4324/9781003008873-51

highlight what prominent theories – both traditional and "alternative" – can tell us, and to describe some of the ways in which scholars in the field go about their work. When thinking about how to explain and understand global environmental politics, one is likely to encounter problematic questions (which one finds in many other scholarly fields as well): Should scholars and researchers focus on empirical analyses, specifically undertaking studies of past events as a means to explain global environmental politics and possibly to predict its likely future? To maintain scientific rigor, and to claim impartiality, the answer that most scholars give to this question is "yes." However, should scholars instead be stepping into the more contested area of normative research, to draw on reality and theory in ways that help us to imagine, and even to bring about, an alternative – in this case, less-polluted and more environmentally sustainable – future? The answer that many scholars give to this question, and indeed the answer that politicians and most citizens may expect them to give, is something along the lines of, "No, that's not our job. That's the job of ordinary citizens, policymakers, industries, et cetera." This "problem" of what the field of global environmental politics is about, and how those who do global environmental politics research should behave, at least needs to be debated.

Most scholars of global environmental politics are teachers of some sort, whether as university professors or as researchers who are increasingly pushed by employers and funding agencies to make their findings digestible by the public, which in practice often means the media. Many scholars do both of these things simultaneously or at different times in their careers. This raises questions of how to portray global environmental issues and related politics and policies. Should scholars be "realistic" in how they teach about these topics, meaning probably sounding rather pessimistic and emphasizing how things are getting worse, at least with respect to most environmental issues? Or should they be "positive," highlighting the successes in addressing problems of pollution and environmental change? To do the former may put students and publics off the subject; to do the latter may be to mislead them into thinking that environmental problems are easier to solve than is actually the case, which may be what many industries and their political supporters want people to believe. One is tempted to say that scholars should strike a balance, but finding a balance that does justice to ongoing environmental changes and the world's responses to them is not simple or easy. Indeed, when the prospects are catastrophic, as they are with climate change, striving for balance may be downright immoral.

There is no denying that things are getting worse in most respects, especially if we think in global terms and notably if we focus on the countless places where pollution is now horrendous (some developing-world cities come to mind) and those where the environment and species are in rapid decline (as in many forests and coastal seas, which suffer from severe contamination and over-exploitation). Having said that, there are cases of success at all levels: local, such as cleaner surface water in many developed states; regional, for example substantial declines in pollution causing acid rain in North America and Europe; and global, most prominently, perhaps, signs that international agreements to protect Earth's stratospheric ozone layer seem to be working.

Actors and institutions in global environmental politics

Part III's chapters examine a number of important actors in global environmental politics, notably states, international organizations, corporations and nongovernmental organizations, as well as institutions, broadly defined, that are important in this issue area, such as environmental regimes, international law, domestic institutions and foreign policy bureaucracies. Among the many ways of considering the roles of actors and institutions is to emphasize how they have changed in significance with time. For much of its history, global environmental politics has really been about *international* (more accurately, inter-*state*) politics: international environmental negotiations, environmental diplomacy, environmental treaty making and the like. Even today, states, and more specifically their governments, are at the center of global environmental politics, and there are all manner of ongoing international environmental negotiations underway at any given time, with newly agreed and long-standing environmental agreements, conventions and treaties being implemented, more or less, around the world. This sort of activity is especially evident in the context of climate change, which is perhaps the most tortuous example of international diplomacy in human history, not least because it involves almost every state and more than a few major international organizations (such as the European Union and the United Nations). States have not gone away and will not do so anytime soon. Consequently, the institutions of states – their organizations and behavioral practices, manifested in bureaucratic agencies that have roles in domestic and foreign policymaking, as well as international regimes and international law – remain extremely important to global environmental politics. But states are by no means the only actors that are important today, and their importance may decline in the future insofar as they are sometimes unable to address problems effectively. One might argue that the importance of states is already declining as that of subnational and non-state actors has grown.

Local and domestic environmental politics have for more than half a century been dominated by nongovernmental actors: industries created pollution and citizens, frequently via groups and organizations, have acted to limit or even stop them from doing so, often through protests, political campaigns, targeted voting and direct action. Put another way, the genesis of global environmental politics is located among people, not among governments. Arguably, the future of global environmental politics will be largely characterized by the growing salience of nongovernmental actors, and indeed of individuals. Climate change is a case in point. Three decades of negotiations among diplomats to produce an effective international response to the problem have failed to stem the global increase in greenhouse gas emissions, particularly carbon dioxide. This is not to say that national governments have done nothing – some of them have started to implement policies that mitigate and in an increasing number of cases reduce emissions – but globally the pollutants causing climate change are still on the rise, as are the temperatures of the atmosphere and oceans. This failure by states to do enough to combat climate change has stimulated, and re-stimulated year on year, the creation and activity of nongovernmental organizations and growing responses by subnational governments. Environmental groups actively lobby governments to respond to the problem. They educate publics and encourage them to both reduce individual environmental impacts (which up to now have been of questionable widespread efficacy) and vote for politicians with pro-environment platforms (as evident by the influence of Green parties in Western Europe). In turn, what some might call "anti-environment" organizations push in the opposite direction, discouraging environmentally sustainable economic development, lobbying against energy regulation, and in many cases doing their best to confuse publics (and politicians) about environmental science (the most blatant recent examples being presidencies of Donald Trump in the United States and Jair Bolsonaro in Brazil). These and other activities of nongovernmental actors will only increase and may in large measure supplant the role of states in issue areas where governments find it too challenging to respond quickly enough to protect the environment, or where they actively side with polluters to destroy it.

Paul G. Harris

Industries are often rightly viewed as harmful to the environment, particularly those that make things or that unearth natural resources. Thus, often for the wrong reasons, they are absolutely central to global environmental politics. They may extract materials from the environment in ways that have extremely harmful consequences. For example, drilling and mining fossil fuels, especially coal, and their consumption in factories, power plants and (in the case of petroleum) automobiles, are absolutely devastating to the environment, not to mention the direct harmful impacts that their use has on human health. To continue with the example of climate change, coal and petroleum companies have been among those industries that have fought the hardest to prevent successful international action and national legislation. Likewise, industries that depend on living natural resources, whether these be trees, fish or other species, also cause environmental harm and routinely use their influence to prevent governments from imposing regulations for environmental protection (because sustainable use of natural resources may require new business models, possibly ones that are less profitable, at least in the short term). This is the general trend in how industries behave, and one assumes that it will continue as long as governments allow it. Indeed, around the world, many of the most damaging practices of industries – not least, again, the extraction of fossil fuels – are heavily subsidized by governments, often the same governments that claim to want to reduce or eliminate the environmentally adverse impacts of those industries.

That said, some businesses are doing the opposite: encouraging environmental sustainability and pushing for widespread regulation of their practices (often to internationalize domestic practices so that their international competitors are required to adhere to similar regulations). Much of this is "greenwashing" by otherwise polluting companies: making harmful practices look more environmentally sustainable than they really are, or encouraging people to consume more "green" products rather than consume less "stuff" altogether. Examples include the airline industry's attempts to paint itself as being "sustainable" and the auto industry's marketing of "environmentally friendly" cars and trucks, which, amazingly, governments have sometimes endorsed by lowering taxes on such vehicles, thereby encouraging more use of them when using them far less is what is required to eliminate pollution. Even when these industries are genuine in their desire to protect the environment, they almost never voluntarily do what would be best for the environment: close down. (It is hard to think of an essential reason for maintaining the global airline industry, for example. It is good for business and leisure travel, but hardly a necessity, as demonstrated during the global coronavirus pandemic.) However, alongside cynical attempts to jump on the environmental sustainability bandwagon are industries that are profiting by promoting genuine sustainability. A good example of this phenomenon can be found in some businesses working to expand alternative-energy capacity. While they may be in the minority, their impact is growing very rapidly. Furthermore, some well-established industries are now recognizing that global environmental change poses a threat to their business models. For example, the global reinsurance industry is concerned about the future impacts of climate change and therefore supports action by governments and adaptation by the businesses that they insure (even as they may profit from selling more insurance in uncertain times).

What is key here is that governments are now being pushed and pulled in different directions, often by conservative industries to continue allowing polluting practices, but increasingly by new ones, alongside environmental-protection organizations, to support sustainable alternatives. This is a trend that is going in the right direction – the balance may be shifting toward environmental sustainability – albeit certainly not at a pace that keeps up with the scale and speed of many adverse ecological changes.

Ideas and themes in global environmental politics

Part IV of the handbook highlights 15 ideas and themes that permeate global environmental politics, including the Anthropocene, sustainability, consumption, expertise, uncertainty, security, governance, diplomacy, North–South relations, globalization, justice (both international and environmental), ethics, public participation and citizenship. While the real-world significance of each of these concepts is explained in the chapters, it is important to bear in mind that their effects can overlap considerably. For example, on one hand, the spread of material consumption through processes of globalization is having an enormous adverse impact on the natural environment. On the other hand, it may be possible for the notion of environmental sustainability to spread via the same processes, potentially counteracting or at least mitigating the environmental impacts of global economic integration. Expertise can reduce uncertainty, which, in turn, can impact public opinion and the degree to which people are willing to think and act as environmental citizens. Realistically, politics always intervenes, with dubious "experts" routinely deployed to support business-as-usual and the status quo. North–South relations in global environmental politics are inseparable from international justice: developing states blame developed ones for global environmental decline and expect to be compensated for their resulting suffering. They also expect developed states to stop polluting the global environment before they are required to do likewise. A consequence is continued disagreement about how to cooperate internationally to address climate change and other pressing environmental problems. Environmental diplomacy, and specifically international environmental conferences among diplomats, is therefore permeated by discussions and debates about what is fair and just. To put it mildly, this makes reaching productive agreement on some issues all but impossible.

While all of the ideas and themes addressed in Part IV are vitally important, arguably none is more so than consumption. It is direct and indirect consumption of environmental resources, particularly by industries and individuals, which is driving most global environmental change. In past decades it has been common to blame population growth for the environmental harm done by humanity. To be sure, the number of people on Earth does matter greatly. All things being equal, more people mean more consumption, more pollution and further decline in environmental commons (shared ecological spaces and resources). There is no question that a growing global population is bad for the environment and those individuals – human and nonhuman – that depend upon it for survival and well-being. But things are more complicated than this; all things are not equal: some people consume more than others. Thus, the biggest problem is found in the highly consumptive lifestyles that took hold in the West in the twentieth century, most obviously the prototypical American way of life and its dependence on vast quantities of resources, fossil fuels and consumer products, characterized most perversely by the private automobile. If this lifestyle were restricted to a relatively few people – say, some millions worldwide – the global environment would be able to cope. But the consumer society is not restricted at all; it has expanded rapidly around the world from its already gargantuan base in North America, Europe and the rich states of other regions. As states of the developing world escape poverty, their populations – often not the majority, but tens or even hundreds of millions of people (as is especially the case in China but is also evident in India, for example) – are adopting lifestyles akin to those that still prevail in most of the Western world. Even as some states in the West have belatedly started to shift toward less materially intensive and less-polluting ways of life, material consumption and resulting pollution has exploded in parts of the developing world, especially Asia.

It may not be hyperbole to say that the future of global environmental politics will be a function of the extent to which this spread of material consumption continues. To be sure, ways of reducing and mitigating the impacts of consumption are being devised all the time. Alas, they are happening on a scale and at a pace that are at least an order of magnitude less than what is required. Consequently, even as the importance of environmental sustainability has become accepted and the means for realizing it have matured, pollution and ecological destruction have continued to grow, and apparently will do so for some decades at least.

Key issues and policies in global environmental politics

As the chapters in Part V reveal, global environmental politics encompasses a great many issues and related policies, including (but not limited to) air pollution, acid raid, stratospheric ozone depletion, climate change, energy, transport, persistent organic pollutants, hazardous wastes, plastics, water pollution, river systems and wetlands, seas and oceans, fisheries, marine mammals, biodiversity, migratory species, forests, desertification, food and agriculture. Again, these issues overlap, as do (or should) associated policies. For example, energy use and resulting pollution is a function of transport policy (or the lack of it); acid rain is a function of the type of energy used, notably the burning of coal; pollution of rivers eventually leads to the pollution of seas, which, in turn, can adversely affect fisheries, meaning that an effective policy to protect fisheries cannot be divorced in reality from an effective policy to prevent pollution of rivers, which is often a result of agricultural policy because farms produce much of the pollution that finds its way into rivers; biodiversity cannot be adequately protected without policies to protect the forests where millions of species reside, but protecting forests cannot be divorced from agricultural policies because forests are often sacrificed to make way for agricultural land. These and countless other interactions among environmental issues, and, in turn, the complexities that arise for effective policymaking, not least the many overlapping political actors with interests in policy outcomes, make global environmental politics extraordinarily complex for practitioners, and more than a little difficult for scholars to understand and explain. These interactions also mean that fostering an alternative, more environmentally sustainable future is incredibly challenging for activists.

Part V has only one chapter dedicated to climate change. This should not be interpreted as suggesting that it is of equal significance to other issues addressed in this part of the book. It can be argued that it is the most important issue facing practitioners and scholars of global environmental politics. The argument can go even further: it is the most important issue ever to face humanity (not to mention the nonhuman world, at least in the current epoch). This is because nearly all other environmental and natural resource problems are connected to climate change in some way. More often than not, this means that they are made more significant and more difficult by climate change. Energy policies can never again be divorced from climate change; the types of energy used, and how much, will be the largest determinant of global warming and other manifestations of climate change in the future. Depletion of Earth's stratospheric ozone layer is tightly connected to climate change. This is because classes of ozone-destroying chemicals that replaced earlier compounds are extremely powerful greenhouse gases, meaning that reducing the use of these new compounds is a very important step in addressing climate change. This example also highlights another complication: by solving one environmental problem, in this case by regulating ozone-destroying chemicals, another problem – climate change in this example – may be made much worse.

Similarly, air pollution and acid rain are tightly connected to climate change: they are very often caused by the same activities and pollutants, especially the burning of coal for the production of electricity. Likewise, transport – the vehicles used to move people and things, as well as the infrastructure enabling that movement – is tightly connected to climate change; the burning of fossil fuels for transport is a major contributor to global greenhouse gas emissions, and despite new transport technologies, those emissions are still on the rise globally. The health of the oceans is being degraded by climate change. Carbon dioxide emissions not only contribute to warming of Earth's atmosphere; they also contribute to both warming and acidification of the oceans, thereby threatening the marine food chain. This has obvious significance for fisheries: climate change may result in major declines in global catches, and decimation of them in some places, with potential implications for the welfare of billions of people (not to mention creatures that rely on the same fish). More generally, food production will suffer as climate change becomes more chronic in coming decades. What is more, as people eat more meat, climate change will worsen because meat production and consumption are major sources of greenhouse gas pollution, possibly exceeding that of all forms of transport combined. Climate change also means increasing losses of habitats and extinction of more species, reducing global biodiversity. Biodiversity loss and climate change are both exacerbated by deforestation: the loss of forests means loss of habitat for many species, and it means the loss of forests' function as enormous "sinks" for carbon, not to mention increased carbon pollution that results when felled forests are burned or decay. The links between climate change, especially global and regional warming, have clear implications for desertification: it is likely to spread. The list of connections between climate change and other environmental issues is almost endless.

Difficult questions for global environmental politics

The problem of climate change, its causes and the manner in which it is being addressed around the world, as well as the way in which scholars study it, demonstrates the enduring importance of global environmental politics for everyone. A key question is whether global environmental politics as a practice will be able to catch up with the pace of ecological changes. In some past instances it has done so – policymakers have successfully devised and implemented ways of protecting local air and water, and of limiting regional acid rain, and they have created a process for protecting Earth's stratospheric ozone layer. Enormous effort is going into devising effective means for mitigating global climate change, and it is possible that mechanisms for adapting to climate change will be widely implemented in the future. However, as things stand now and certainly in the near future, in broad terms the pace of ecological changes continues to be much faster than practitioners' responses to them. Policymakers and diplomats have just as many failures as successes. In many places, pollution is increasing and the environment is suffering from worsening exploitation. Despite more than three decades of international negotiations aimed at cutting global greenhouse gas pollution, that goal has remained elusive. Successes on one region – say, reducing local air pollution in North America – are often more than matched by failures in others – grotesque levels of air pollution being endured in many parts of eastern and southern Asia, for example. Perhaps the best that one could accurately say about the success of *doing* global environmental politics is that "the glass is only half empty."

A similar question can be asked of global environmental politics as a scholarly discipline: is it up to the task of not only explaining environmental problems but also providing workable solutions to them, solutions that are not just possible to envision but which have a good chance of being implemented relatively soon in a world of competing priorities and politically powerful interests resistant to change? What more can scholars of global environmental politics do to fill the glass from half empty to half full and more? For starters, knowledge of and concern about global environmental politics need to expand. A primary purpose of this volume is to serve as a solid foundation for such knowledge. But knowledge is not enough; there may be an obligation – call it a moral one – for scholars to do more than help us understand global environmental politics. There may be an obligation for them to become proactive advocates for, and participants in, efforts to protect the environment, even if that means diverting attention from traditional styles of research and teaching.

Is it any longer acceptable for scholars to say, in so many words (or behave as though), "It's my job to explain the politics of environmental change, and the job of others to do something about it"? At what point will environmental changes be so bad, or be forecast to be so bad, that everyone, including scholars of global environmental politics, must become environmentalists, even environmental activists? How many millions of hectares of forest will have to be lost to logging and ranching? How rare will clean water from rivers and aquifers have to be? How much plastic will have to clog the oceans and how acidic will they have to become due to carbon dioxide pollution? How many hundreds of thousands or millions of people will have to suffer and die from environmental changes and impacts resulting from climate change? How slow will governments have to be relative to the pace and scale of environmental changes before scholars shift from being expert observers to expert activists? Or is it the job of scholars to simply chronicle environmental decline, to leave an academic legacy so that their grandchildren will be absolutely certain that humanity knew exactly why it was destroying the global environment?

Is it the job of both practitioners and scholars to, in effect, save us from ourselves – to "save the environment" from the human, social and indeed political causes of its accelerating decline? The field of activity and scholarship that we call global environmental politics generates the information and ideas that force us to ask such a question. It can also generate answers to it, in the process identifying new pathways by which humanity might one day make the question much less necessary.

Prospects for global environmental politics

What are the prospects for global environmental politics in the future? That is, how likely are scholars and students to garner enough insight into global environmental problems and the world's responses to them to be able (in the case of scholars) to advise policymakers and businesses to be better environmental stewards and (in the case of students) to become sufficiently aware of the environmental crises facing the world to become genuine environmental citizens, and to devote their energies to solving environmental problems, or at the very least to greatly limit their own personal contributions to environmental problems? What are the prospects for the practice of global environmental politics? Will governments and other actors learn from past mistakes and choose to give the natural environment a much higher priority? Will those actors that have failed to do so (meaning most of them) soon realize that the wellbeing of whole societies is intimately linked to environmental sustainability – locally, nationally and globally?

We cannot predict the future with certainty, but in trying to parse these questions we are very likely to arrive at a mixed bag of answers, at best. The world has seen some real progress in addressing environmental problems. In this respect, we might say that global environmental politics has worked. But there is no escaping the stark reality that, broadly and globally speaking, and in countless individual locations across the planet, environmental problems continue to grow worse. Climate change is a case in point: despite decades of very serious international negotiations, many resulting treaties, some credible efforts to limit greenhouse gas emissions in some places, and even some limited progress toward helping those people who are and will be affected by climate change, greenhouse gas pollution continues to increase globally (when it needs to fall very rapidly), the impacts have grown worse (as have the predictions regarding them) and the resulting suffering is on the rise. Developed states and people living within them have done too little to reduce their use of fossil fuels. Developing states and their citizens are in the process of becoming addicted to them just as happened in the West. Sometimes this has been necessary; the world's poor have needed inexpensive energy to escape poverty. But this is not the only path to development. The world's affluent people, including the many millions of new middle-class consumers (not to mention the super-wealthy) in developing states, need not make the same mistakes of people in the developed world. The path toward sustainability ought to be followed by all who are capable of doing so. Up to now, too few people around the world have followed this path.

Thus, it seems that the work of global environmental politics – the work of government officials, environmental activists and others involved in its practice, as well as the work of scholars who study what those actors do – will be more of the same for the time being. This will involve a growing array of successful efforts by governments around the world to cooperate to address environmental problems and resource scarcities. These agreements will seldom come easily, will require payoffs to vested interests with stakes in continued pollution and overuse of resources, and will no doubt sometimes, perhaps even often, meet with too-limited success. But they will be signs of progress in global environmental politics. Similarly, efforts to implement environmental sustainability will spread, thereby reducing the human impact on the environment compared to what it would be without such efforts. However, it is likely that coming decades will see increasing environmental pollution at all levels, from the local to the global, as well as the increasing overuse of natural resources and the unsustainable exploitation of environmental commons. In short, the tide of environmental pollution and decline will not be stemmed anytime soon.

Yet, there may still be some room for hope. Scholars and students will continue to observe and learn about what is happening. The tools for doing so will likely improve. Sometimes the work of scholars and analysts, and the understanding of students and future generations, will positively influence policy and the real-world behavior of industries and individuals. At other times they may understand what is happening but be helpless to do much about it. Insofar as that happens, the scholarship of global environmental politics will be a chronicle of global environmental decline. In the hope that this does not happen, one aim of this handbook is to give those who practice and study global environmental politics some of the knowledge that they will need to build the foundations of an environmentally sustainable future. It cannot come too soon.

Index

For Product Safety Concerns and Information please contact our EU
representative GPSR@taylorandfrancis.com
Taylor & Francis Verlag GmbH, Kaufingerstraße 24, 80331 München, Germany

www.ingramcontent.com/pod-product-compliance
Lightning Source LLC
Chambersburg PA
CBHW081211220326
41598CB00037B/6745